Analysis and Probability

Analysis and Probability

Aurel Spătaru
Romanian Academy
Bucharest
Romania

ELSEVIER

AMSTERDAM • BOSTON • HEIDELBERG • LONDON • NEW YORK • OXFORD
PARIS • SAN DIEGO • SAN FRANCISCO • SINGAPORE • SYDNEY • TOKYO

Elsevier
32, Jamestown Road, London NW1 7BY, UK
225 Wyman Street, Waltham, MA 02451, USA

First Edition 2013

British Library Cataloguing-in-Publication Data
A catalogue record for this book is available from the British Library

Library of Congress Cataloging-in-Publication Data
Application submitted

ISBN: 978-0-12-401665-1

For information on all Elsevier publications
visit our website at store.elsevier.com

This book has been manufactured using Print On Demand technology. Each copy is produced to order and is limited to black ink. The online version of this book will show color figures where appropriate.

Contents

Preface

First and foremost, this book is designed as a probability theory text for graduate students and as a reference work in this field. The reader will require a solid background in abstract analysis in order to tackle the development of probability theory from a measure-theoretic perspective, which is the main focus of the book. For this reason, Part One of the book – Chapters 1 to 5 – systematically presents concepts and results from analysis with which every reader should be familiar before embarking on the study of probability theory; this part should also be useful for those interested in topology, measure theory, real analysis and functional analysis. Part Two – Chapters 6 to 12 – then presents the concepts, methodology and fundamental results that a graduate-level student of probability theory would be expected to know. Probability theory is a rapidly expanding field and so the author has necessarily made some subjective judgments about which topics will be included and which ones omitted. For instance, the important notion of Markov dependence is not discussed here, though the subject is touched upon during the presentation of the concepts of the Markov triple and the Markov shift. Similarly, neither stochastic processes in continuous time nor multiparameter processes are considered in this book; indeed, each of these topics merits an entire monograph of its own. Nonetheless, it is worth noting that the book does contain material about subjects, such as f-expansions, Fuk-Nagaev inequalities and Markov triples, that typically do not appear in other graduate-level probability texts.

Extra care has been taken to ensure that the exposition is kept as self-contained as possible. Except for elementary analysis, there are no other prerequisites for reading the book. Consequently, the book should lend itself well to individual study. To facilitate understanding and to offer interesting extensions of the theory, every section ends with a large collection of exercises. Since some of the exercises will be needed for subsequent definitions and proofs in the main text, the reader is encouraged to work carefully through the exercises.

A brief description of each of the twelve chapters may be useful at this point. Chapter 1 contains a review of some essential concepts and results from set theory upon which the development of probability theory in later chapters will be based. Section 1 deals mainly with operations on sets, Section 2 presents several types of functions and the concept of a Cartesian product, and Section 3 introduces the concepts of equivalence relation and partial ordering. Special attention is paid to the classification of sets according to the number of elements they contain. Some equivalent forms of the axiom of choice are presented at the end of the chapter. It should be noted that the treatment of set-theoretic notions in Chapter 1 is not axiomatic; rather the presentation relies on intuition and elementary logic, and does so without introducing any sets that could lead to paradoxes.

Chapter 2 is motivated by the prominent role that topology plays in measure theory – and, hence, in probability theory which, to some extent, may be viewed as a branch of measure theory. In this chapter we introduce and discuss most of the topological concepts that will be used in the book. In particular, Section 4 deals with various methods of constructing topological spaces, while Section 5 and Section 6 present important properties of topological spaces that will be used in subsequent chapters. In particular, Section 6 includes some deep results, such as Urysohn's embedding theorem, the Stone-Weierstrass theorem and the Arzelà-Ascoli theorem.

Chapter 3 is devoted to the study of three fundamental notions: measurable space, measurable function and measure. While the definitions of measurable space and measurable function do not inherently depend on the existence of a measure, measurable spaces and measurable functions, in practice, are almost always associated with a measure; this justifies the term "measurable" in their names. In Section 7 we present a detailed study of algebras and σ-algebras and introduce a special family of sets called a Dynkin system. In most situations, a Dynkin system can replace the notion of a monotone class (which is defined in various books on probability and measure theory) and proves to be a highly efficient tool. In Section 8 and Section 9 we discuss measurable functions and measures and then, in Section 10, we develop the technique of extending σ-finite measures, culminating in Carathéodory's extension theorem and the construction of Lebesgue-Stieltjes measures.

The material in Chapter 3 enables us to introduce and study the important notion of integration in Chapter 4. In Section 11 we define the integral and examine its remarkable properties. Section 12 deals with the Radon-Nikodým theorem and with the Lebesgue decomposition, and Section 13 presents the function spaces $\mathfrak{L}^p$. In Section 14 we define and study three new modes of convergence for sequences of function: convergence almost everywhere, in measure and in $\mathfrak{L}^p$.

Chapter 5 is devoted to the construction of measures on product σ-algebras. In Section 15 we construct the product of a finite number of σ-finite measures. This is extended to infinite products of probability measures in Section 16, which also presents the Daniell-Kolmogorov extension theorem.

In Chapter 6 we introduce the first notions of probability theory. In Section 17 we present the primary probabilistic concepts of events, random variables, random vectors, random elements, moments and variances. Section 18 begins with a presentation of elementary properties of conditional probabilities, and proceeds to consider the concept of independence for σ-algebras and for random variables and elements, and to establish such basic results as the Kolmogorov zero-one law, the Borel-Cantelli lemma and the Hewitt-Savage zero-one law.

Chapter 7 is devoted to two important tools in the study of random variables, namely, the distribution function and the characteristic function. Section 19 deals with distribution functions and sequences of distribution functions for both random variables and random vectors. The convolution of two distribution functions, the concept of weak convergence of a sequence of distribution functions and the related concept of complete convergence are studied in this section, which also includes a discussion of various types of distribution functions and the presentation of the important Helly-Bray theorem. The main purpose of Section 20 is an exposition of characteristic functions and their properties and how they may be used to establish results in probability theory. One such result is the continuity theorem, due to P. Lévy, which replaces the convergence of a sequence of distribution functions by the convergence of the corresponding sequence of characteristic functions, the latter generally being much more tractable than the former.

Chapter 8 puts a focus on the theory of probability on abstract metric spaces. The connection between probability theory and topology is illustrated in the first part of Section 21 and culminates in Lusin's theorem on the approximation of measurable functions by continuous functions. This section also presents an examination of the relationship between nonnegative linear functionals and probabilities, including a version of the famous Riesz representation theorem. In Section 22 we topologize the space of all probabilities on the σ-algebra of Borel subsets of a metric space and then investigate convergence properties of sequences of probabilities. Further, we proceed to analyze the question of metrization of this topology and derive a well-known result of Yu. V. Prohorov characterizing the compact subsets of probabilities.

By making use of powerful tools and techniques, such as characteristic functions and symmetrization, we present in Chapter 9 the crowning achievement of classical probability theory: the central limit theorem. Given a triangular array of row-wise independent random variables, the central limit problem seeks conditions under which the distribution functions of the row sums converge weakly to some distribution function. It turns out that the totality of such limit distribution functions forms a special set called the class of infinitely divisible distribution functions. Section 23 is devoted to a detailed study of this class and introduces the notions of self-decomposable and stable distribution functions. In Section 24 we present necessary and sufficient conditions for the weak convergence of the row sums to a given infinitely divisible distribution function, and we apply this convergence criterion to three important limit distribution functions: the degenerate, the Poisson and the normal distribution function.

The behaviour of sums of independent random variables is the main concern of Chapter 10. Making use of results from the preceding chapter, we focus in Section 25 on the weak convergence of row sums to zero, which leads to the weak law of large numbers. In Section 26 we investigate the almost sure convergence of series of independent random variables, culminating in the Kolmogorov three-series theorem and the two-series theorem. Section 27 is devoted to obtaining various strong laws of large numbers, i.e theorems establishing conditions under which stabilized sums of independent random variables converge almost surely to zero. Further, in Section 28, we investigate the order of magnitude of the fluctuations of sums of independent random variables by establishing two famous results of classical probability theory, namely, the Kolmogorov law of the iterated logarithm and the Hartman-Wintner law of the iterated logarithm.

While Chapter 10 is devoted to the limiting behaviour of sums of independent random variables, Chapter 11 considers the more general setting in which a sequence of random variables is conditioned with respect to a sequence of σ-algebras. Such sequences are important in Markov theory, semimartingale theory and ergodic theory. The first part of Section 29 introduces the concepts of conditional expectation and conditional probability with respect to a given σ-algebra and develops their most important properties. This section concludes with discussion about the notions of conditional independence and Markov triple. In Section 30 we apply results from the preceding section to the theory of martingales and semimartingales in discrete time. After an exposition of the basic theory of these processes, starting with the concept of stopping time, we proceed to establish a number of powerful results, including Doob's optional sampling theorem, supremum inequalities, the submartingale convergence theorem, an extension of the Borel-Cantelli lemma to conditional probabilities, and Wald's equations.

Classical ergodic theorems for measure-preserving transformation and their role in the theory of stationary sequences are the focus of Chapter 12. Section 31 formulates and proves the fundamental ergodic theorems due to G. D. Birkhoff and J. von Neumann. To this end, we first introduce and study such basic concepts as measure-preserving transformations, invariant σ-algebras, invariant random variables and ergodic transformations. This section concludes with some results concerning the related notions of weak-mixing and strong-mixing. In Section 32 we present some useful results for general stationary sequences of random elements, which are sequences that encompass a dependence structure of great importance in probability theory.

It is a pleasure to take this opportunity to thank a number of individuals who were instrumental, in one way or another, in the writing of this book. First, my thanks go to my teachers, especially to Ciprian Foiaş, whose courses on abstract and functional analysis were essential to my education, and to Ioan Cuculescu, whose courses on probability theory and stochastic processes guided me rapidly to the point where I could understand and write research papers. I am very fortunate to have had the opportunity to work with and to have been mentored by Marius Iosifescu, who has offered moral support and encouragement throughout my career. I am also grateful to my many colleagues at the Institute of Mathematical Statistics and Applied Mathematics (IMSAM) of the Romanian Academy who have been a source of support and inspiration for me over the years, and especially to my friends Gheorghiţă Zbaganu and Gheorghe Răuţu for many fruitful discussions and helpful suggestions. I thank the IMSAM for providing such a friendly and stimulating environment and for giving me the freedom and opportunity to complete this work.

Jim Tomkins kindly improved the presentation of an essential part of the text, and I warmly thank him for his help in this matter. I do wish to thank the following reviewers: George Roussas, University of California at Davis; Mark Pinsky, Northwestern University; and Peter Spreij, University of Amsterdam. Finally, I am grateful to the editors for accepting this project in the Elsevier Insights, and to the Elsevier production team for their kind and efficient cooperation in publishing.

Romanian Academy, Bucharest, Romania
March 2012

Aurel Spătaru

PART ONE: ANALYSIS

1 Elements of Set Theory

This chapter contains a review of some essential concepts and results of set theory on which lean the following chapters. Section 1 deals mainly with operations on sets. In Section 2 we define several types of functions, and discuss the notion of a Cartesian product. In Section 3 we study the concepts of equivalent relation and partial ordering. A special attention is paid to the classification of sets according to the number of their elements. At the end of this section we state some equivalent forms of the axiom of choice.

The treatment of the set-theoretic notions is not axiomatic, but relies on intuition and elementary logic. However, this will not introduce sets which lead to paradoxes.

1 Sets and Operations on Sets

The notion of a set and some notations 1.1. We assume that the reader is familiar with the primary concepts of a set and of an element of a set. Generally, elements of sets will be denoted by small letters $a, b, c, \ldots$, and sets will be denoted by capital letters $A, B, C, \ldots$ A set will be identified either by writing down its elements between braces, or by specifying the common property its elements possess. Thus we will write $A = \{a : a \text{ has property } P\}$ to define the set of all elements a which possess property P. Appealing to intuition, in several places sets will be called spaces, and elements of sets will be called points. We will frequently encounter families of sets, that is sets whose elements are themselves sets. Families of sets will be denoted by script letters $\mathcal{A}, \mathcal{B}, \mathcal{C}, \ldots$

We will write $a \in A$ if a is an element of A, and we will write $a \notin A$ if a is not an element of A. A special set is the set which has no elements. This set is the empty set which will be denoted by $\emptyset$. Thus, for example, $\emptyset = \{a : a \text{ is a natural number and } a < a\}$.

The following sets will be often encountered in this book: the set N of all positive integers, the set Z of all integers, the set Q of all rational numbers, the set R of all real numbers, and the set C of all complex numbers. Throughout this text, $K = R$ or $K = C$. We assume a knowledge on the part of the reader of the ordering structure and the algebraic structure of these sets. The reader is also assumed to know elementary analysis concerning these sets. Another important set will be now introduced.

Definition 1.2. Let $-\infty$ and ∞ be two distinct objects such that $-\infty \notin R$ and $\infty \notin R$. The set $\overline{R} = \{a : a \in R \text{ or } a \in \{-\infty, \infty\}\}$ is called the set of *extended*

Analysis and Probability. http://dx.doi.org/10.1016/B978-0-12-401665-1.00001-1

real numbers. We assume that $-\infty < \infty$, and $-\infty < a < \infty$ for each $a \in R$. Sums and products in $\overline{R}$ are defined by the following rules. If $a \in R$ and $b \in R$, then $a+b$ and ab are defined as usual. For any $a \in R$, we define $-\infty + a = a + (-\infty) = -\infty$ and $\infty + a = a + \infty = \infty$. We put $-\infty + (-\infty) = -\infty$, $\infty + \infty = \infty$, but we do not define $-\infty + \infty$ and $\infty + (-\infty)$. If $a \in \overline{R}$ and $a < 0$, we define $(-\infty) \cdot a = a \cdot (-\infty) = \infty$ and $\infty \cdot a = a \cdot \infty = -\infty$. (Particularly, $-(-\infty) = \infty$ and $-(\infty) = -\infty$.) We set $(-\infty) \cdot 0 = 0 \cdot (-\infty) = \infty \cdot 0 = 0 \cdot \infty = 0$. If $a \in \overline{R}$ and $a > 0$, we define $(-\infty) \cdot a = a \cdot (-\infty) = -\infty$ and $\infty \cdot a = a \cdot \infty = \infty$. We put $1/-\infty = 1/\infty = 0$. The absolute value $|a|$ of an element $a \in \overline{R}$ is defined by $|a| = \begin{cases} -a \text{ if } a < 0 \\ a \quad \text{if } a \geqslant 0 \end{cases}$. For $a \in R$, we write $[a]$ for the largest integer $\leqslant a$.

Definitions 1.3. If A and B are sets such that $x \in A$ implies $x \in B$, then we say that A is a *subset of* B, or that A is included in B; we then write $A \subset B$ or $B \supset A$. Since $\emptyset$ has no elements, the inclusion $\emptyset \subset A$ holds for any set A. If $A \subset B$ and $B \subset A$, then we say that A and B are *equal sets*, and we write $A = B$. The inclusion $A \subset B$ is said to be strict whenever $A \neq B$.

If $A \subset X$, and the elements of A possess property P, then we will sometimes write $A = \{x \in X : x \text{ has property } P\}$ to emphasize that A is a subset of X.

For any set X, the family of all subsets of X denoted by $\mathcal{P}(X)$ is called the *power set of* X.

Definitions 1.4. Let A and B be sets. The *union of* A *and* B denoted by $A \cup B$ is the set $\{x : x \in A \text{ or } x \in B\}$

The *intersection of* A *and* B denoted by $A \cap B$, or simply AB, is the set $\{x : x \in A \text{ and } x \in B\}$. If $A \cap B = \emptyset$, then A and B are said to be *disjoint*. The *difference* $A - B$ is the set $\{x : x \in A \text{ and } x \notin B\}$. The *symmetric difference of* A *and* B denoted by $A \Delta B$ is defined by $A \Delta B = (A - B) \cup (B - A)$.

Definition 1.5. If $A \subset X$, then the set $A^c = X - A$ is called the *complement of* A relative to X. In most situations, when we will meet complements of sets, we will assume that they are subsets of some given "universal" space X.

Theorem 1.6. *Let* A, B, C *be sets. Then we have:*

(i) $A \cup B = B \cup A$, $A \cap B = B \cap A$;

(ii) $A \cup (B \cup C) = (A \cup B) \cup C$, $A \cap (B \cap C) = (A \cap B) \cap C$;

(iii) $A \cap (B \cup C) = (A \cap B) \cup (A \cap C)$, $A \cup (B \cap C) = (A \cup B) \cap (A \cup C)$.

Proof. These identities, like other similar identities, are proved by elementary logical operations, taking into account the fact that two sets X and Y are equal if and only if $X \subset Y$ and $Y \subset X$. □

Exercise 1.7. If X has n elements, show that $\mathcal{P}(X)$ contains 2^n sets.

Exercise 1.8. For any sets A, B, C prove:

(a) the inclusion $A \subset B$ is equivalent to each of the equalities $A \cup B = B$ and $A \cap B = A$;

(b) $A - B = A \cap B^c$;

(c) $A \Delta B = B \Delta A$;

(d) $A\Delta B = \emptyset$ if and only if $A = B$;
(e) $A\Delta B = (A \cup B) - (A \cap B)$;
(f) $A\Delta(B\Delta C) = (A\Delta B)\Delta C$;
(g) $A \cup (B\Delta C) \supset (A \cup B)\Delta(A \cup C)$;
(h) $(A\Delta B) \cap (A\Delta C) \subset A\Delta(B \cup C) \subset (A\Delta B) \cup (A\Delta C)$;
(i) $A \cap (B\Delta C) = (A \cap B)\Delta(A \cap C)$;
(j) $(A\Delta B) \cap (A\Delta C) \subset A\Delta(B \cap C) \subset (A\Delta B) \cup (A\Delta C)$;
(k) $X = (A \cap B) \cup (A^c \cap B^c) \cup (A\Delta B)$;
(l) $(A^c)^c = A$;
(m) $A \subset B$ if and only if $A^c \supset B^c$.

2 Functions and Cartesian Products

In the first part of this section we introduce the concept of a function, and we define several types of functions. Many other classes of functions will be studied subsequently. By making use of the axiom of choice, in the last part of the present section we discuss the notion of a Cartesian product.

Definitions 2.1. Let X and Y be sets. A *function* [*mapping*, *transformation*] f from X into Y denoted by $f : X \to Y$ is a rule by which we assign to every element $x \in X$ a unique element $f(x) \in Y$. The element $f(x)$ is called the *image of x under f* [*value of f at x*], the set X is called the *domain of f*, and the set Y is called the *range of f*.

We note that a function is completely specified by giving its domain and the value of the function at each point of the domain. Thus we say that f and g are *equal functions*, and we write $f = g$, if the domain of f coincides with the domain of g, and $f(x) = g(x)$ for any x in this domain.

Definitions 2.2. Let $f : X \to K$ and $g : X \to K$ be functions. The function $f + g : X \to K$ is defined by $(f + g)(x) = f(x) + g(x)$ for any $x \in X$. The function $fg : X \to K$ is defined by $(fg)(x) = f(x)g(x)$ for any $x \in X$. Whatever $\alpha \in K$, the function $\alpha f : X \to K$ is defined by $(\alpha f)(x) = \alpha f(x)$ for any $x \in X$. The function $|f| : X \to R$ is defined by $|f|(x) = |f(x)|$ for any $x \in X$. If $f(x) \neq 0$ whenever $x \in X$, then the function $\frac{1}{f} : X \to K$ is defined by $(\frac{1}{f})(x) = \frac{1}{f(x)}$ for any $x \in X$. For $f : X \to C$, the function $\overline{f} : X \to C$ is defined by $\overline{f}(x) = \overline{f(x)}$ for any $x \in X$, where $\overline{z} = x - iy$ stands for the complex conjugate of $z = x + iy$, and the functions $\operatorname{Re} f : X \to R$ and $\operatorname{Im} f : X \to R$ are defined by $\operatorname{Re} f = \frac{1}{2}(f + \overline{f})$ and $\operatorname{Im} f = \frac{1}{2i}(f - \overline{f})$. We note that $f = \operatorname{Re} f + i\operatorname{Im} f$.

Let $f : X \to \overline{R}$ and $g : X \to \overline{R}$ be functions, and let $\alpha \in \overline{R}$. If $\{x : f(x) = -\infty\} \cap \{x : g(x) = \infty\} = \{x : f(x) = \infty\} \cap \{x : g(x) = -\infty\} = \emptyset$, then the function $f + g : X \to \overline{R}$ is defined by $(f + g)(x) = f(x) + g(x)$ for any $x \in X$. The functions $fg : X \to \overline{R}, \alpha f : X \to \overline{R}$, and $|f| : X \to \overline{R}$ are defined as in the preceding paragraph. If $f(x) \neq 0$ whenever $x \in X$, then the function $\frac{1}{f} : X \to R$ is defined by $(\frac{1}{f})(x) = \frac{1}{f(x)}$ for any $x \in X$. If $f :\to R$, and $f(x) \geqslant 0$ for every $x \in X$, then, whatever $\beta > 0$, the function $f^\beta : X \to R$ is defined by $f^\beta(x) = f(x)^\beta$ for any $x \in X$.

Whatever $\alpha \in Y$, the function $h : X \to Y$ defined by $h(x) = \alpha$ for any $x \in X$ is called the *constant function with value* α. This function will be written simply as α, since we consider the reader will be able to distinguish it from the element α wherever necessary.

Definition 2.3. Let $f : X \to Y$ and $g : X' \to Y$ be functions. If $X \subset X'$, and $f(x) = g(x)$ for any $x \in X$, then f is called the *restriction of* g (to X), and g is called an *extension of* f (to X'). Sometimes we shall write $g|_X$ for the restriction of g to X. If $A \subset X \cap X'$ and $f(a) = g(a)$ for any $a \in A$, then we say that $f = g$ on A.

Definition 2.4. Let $f : X \to Y$ and $g : Y \to Z$ be functions. The function $g \circ f : X \to Z$ defined by $(g \circ f)(x) = g(f(x))$ for any $x \in X$ is called the *composition of f and g.*

Definitions 2.5. Let $f : X \to Y$ be a function, and let $A \subset X$, $B \subset Y$ be sets. The set $f(A) = \{y \in Y : y = f(x) \text{ for some } x \in A\}$ is called the *image of A under f*, and the set $f^{-1}(B) = \{x \in X : f(x) \in B\}$ is called the *inverse image of B under f*. Thus $f(A)$ consists of the images of all points in A under f, and $f^{-1}(B)$ is the set of all elements in the domain of f whose images under f belong to B.

Let $\mathcal{B} \subset \mathcal{P}(Y)$ be a family of sets. The *inverse image of $\mathcal{B}$ under f* denoted by $f^{-1}(\mathcal{B})$ is the family $\{A \in \mathcal{P}(X) : A = f^{-1}(B) \text{ for some } B \in \mathcal{B}\}$. Thus $f^{-1}(\mathcal{B})$ consists of the inverse images of all sets in $\mathcal{B}$ under f.

Theorem 2.6. *Let $f : X \to Y$ and $g : Y \to Z$ be functions, and let $\mathcal{C} \subset \mathcal{P}(Z)$. Then $(g \circ f)^{-1}(\mathcal{C}) = f^{-1}(g^{-1}(\mathcal{C}))$.*

Proof. The equality of the two families is immediate, since $(g \circ f)^{-1}(C) = f^{-1}(g^{-1}(C))$ for any $C \subset Z$. □

Definitions 2.7. Let $f : X \to Y$ be a function. We say that f is *surjective* [from X *onto* Y], if $f(X) = Y$. If $f(x_1) \neq f(x_2)$ whenever $x_1 \neq x_2$, then f is said to be *injective.* If f is both surjective and injective, then it is said to be *bijective.* If $f : X \to Y$ is bijective, then the function $f^{-1} : Y \to X$ defined by $f^{-1}(y) = x$, where $f(x) = y$, is called the *inverse of f*.

Example 2.8. Let A be a subset of an arbitrary set X. The function $i_A : A \to X$ given by $i_A(x) = x$ for any $x \in A$ is called the *inclusion function* (from A into X). Plainly, i_A is injective. i_A is surjective, and so bijective, if and only if $A = X$. In this case i_A is called the *identity function.*

Now let $\mathcal{B}$ be a family of subsets of X. Then $i_A^{-1}(\mathcal{B}) = \{A \cap B : B \in \mathcal{B}\}$ is a family of subsets of A which is called the *family on A induced by $\mathcal{B}$*. For the sake of convenience, we will often write $A \cap \mathcal{B}$ instead of $i_A^{-1}(\mathcal{B})$. In several occasions we will encounter induced families in this text (see (4.16) and (7.12)).

Definition 2.9. Let A be a subset of X. The function $1_A : X \to \{0, 1\}$ given by

$$1_A(x) = \begin{cases} 1 \text{ if } x \in A \\ 0 \text{ if } x \in A^c \end{cases}$$

is called the *indicator function of A*. We note that the mapping that assigns to every subset A the indicator function 1_A is a bijective mapping from $\mathcal{P}(X)$ onto the set of all functions from X into $\{0, 1\}$. This fact allows replacing operations on subsets by equivalent operations on indicator functions, and so makes the indicator function an extremely useful tool in measure theory.

The concept of an indexed set we introduce now permits, among other things, to extend the notions of a union and of an intersection of two sets defined in (1.4).

Definition 2.10. Let I be a set, called a set of indices, let X be an arbitrary set, and let $f : I \to X$ be a function. For every $i \in I$, denote $x_i = f(i)$. The image of I under f may be written as $\{x_i : i \in I\}$ even if the points x_i in X may not be distinct. $\{x_i : i \in I\}$ is called an *indexed set* (by $i \in I$). Although the indexed set $\{x_i : i \in I\}$ is considered as a subset of X with possibly coinciding elements, it should be identified with the function $f : I \to X$.

Example 2.11. If $I = N$, then the indexed set $\{x_n : n \in N\}$ is called a *sequence.* If $I = \{n \in Z : n \geqslant m\}$ for some $m \in Z$, then the indexed set $\{x_i : i \in I\}$ will be usually denoted as $\{x_n : n \geqslant m\}$ and will be also called a sequence. Since there is a bijective mapping from N onto $\{n \in Z : n \geqslant m\}$, this name is justified. A sequence $\{x_n : n \in N\}$ will be sometimes denoted as $\{x(n) : n \in N\}$.

Now let $f : N \to X$ and $g : N \to N$ be sequences such that $i < j$ implies $g(i) < g(j)$. For any $n \in N$, denote $x_n = f(n)$, and for any $k \in N$, denote $n_k = g(k)$. Then the sequence $f \circ g : N \to X$, written $\{x_{n_k} : k \in N\}$, is called a *subsequence* of the sequence $\{x_n : n \in N\}$.

Notation 2.12. Let $\{x_i : i \in I\}$ be an indexed set and let $J \subset I$. A statement P that holds for each x_j in this set, where $j \in J$, will be usually denoted by "$P(x_j)$, $j \in J$."

Examples 2.13. (a) Let $\{x_n : n \geqslant 1\}$ be a sequence, and let A be a set. We will write "$x_n \in A, n \geqslant 1$" to state that $x_n \in A$ for any $n \geqslant 1$. (Here $I = J = N$.)

(b) Let $\{A_n : n \geqslant 1\}$ and $\{B_n : n \geqslant 1\}$ be sequences of sets. To state that $A_n = B_n$ for any odd positive integer n, we will write "$A_n = B_n, n = 1, 3, \ldots$". (Here $I = N, J = \{1, 3, \ldots\}$, and the indexed set to which (2.12) refers is obtained by assigning the ordered pair (A_n, B_n) to every $n \in N$.)

(c) Whatever $m \in N$ and $n \in N$, let $f_{mn} : R \to R$ and $g_{mn} : R \to R$ be functions. Let A and B be subsets of N. To state that $f_{mn}(x) < g_{mn}(x)$ for any x in the open interval $]0, 1[$, $m \in A$ and $n \in B$, we will write "$f_{mn}(x) < g_{mn}(x), x \in]0, 1[, m \in A, n \in B$". (Here I is the set of all ordered triples (x, m, n), where $x \in R, m \in N, n \in N$, J is the set of all ordered triples (x, m, n), where $x \in]0, 1[, m \in A, n \in B$, and the indexed set to which (2.12) refers is obtained by assigning the ordered pair $(f_{mn}(x), g_{mn}(x))$ to each $(x, m, n) \in I$.)

Definitions 2.14. Let $\{A_i : i \in I\}$ be an indexed family of sets. The set $\cup_{i \in I} A_i = \{x : x \in A_i \text{ for some } i \in I\}$ is called the *union of all sets of* $\{A_i : i \in I\}$, and the set $\cap_{i \in I} A_i = \{x : x \in A_i \text{ for all } i \in I\}$ is called the *intersection of all sets of* $\{A_i : i \in I\}$. If $I = \{n \in Z : n \geqslant m\}$ for some $m \in Z$, then $\cup_{i \in I} A_i$ and $\cap_{i \in I} A_i$ will be denoted as $\cup_{n \geqslant m} A_n$ and $\cap_{n \geqslant m} A_n$, respectively. If $I = \{1, \ldots, n\}$ for some $n \in N$, then $\cup_{i \in I} A_i$ and $\cap_{i \in I} A_i$ will be written as $\cup_{i=1}^n A_i$ and $\cap_{i=1}^n A_i$, respectively.

Now let $\mathcal{A}$ be an arbitrary family of sets. The set $\cup\mathcal{A} = \{x : x \in A \text{ for some } A \in \mathcal{A}\}$ is called the *union of all sets of* $\mathcal{A}$, and the set $\cap\mathcal{A} = \{x : x \in A \text{ for all } A \in \mathcal{A}\}$ is called the *intersection of all sets of* $\mathcal{A}$.

Definition 2.15. A *partition of* X is any family $\mathcal{A} \subset \mathcal{P}(X)$ such that:
(i) $\cup\mathcal{A} = X$;
(ii) $A, B \in \mathcal{A}$ and $A \neq B$ imply $A \cap B = \emptyset$.[1]

Definitions 2.16. Let $\{A_n : n \geqslant 1\}$ be a sequence of sets. The set $\liminf_n A_n = \cup_{n\geqslant 1} \cap_{k\geqslant n} A_k$ is called the *inferior limit of* $\{A_n : n \geqslant 1\}$, and the set $\limsup_n A_n = \cap_{n\geqslant 1} \cup_{k\geqslant n} A_k$ is called the *superior limit of* $\{A_n : n \geqslant 1\}$. We note that $x \in \liminf_n A_n$ if and only if x belongs to all but a finite number of the sets A_n; also $x \in \limsup_n A_n$ if and only if x belongs to infinitely many A_n.

De Morgan's laws 2.17. If $\{A_i : i \in I\}$ is a family of subsets of X, then:
(i) $(\cup_{i\in I} A_i)^c = \cap_{i\in I} A_i^c$;
(ii) $(\cap_{i\in I} A_i)^c = \cup_{i\in I} A_i^c$.

The proof of these identities is easy and is left to the reader.

Remark 2.18. Let X be a set, and let $\emptyset \subset \mathcal{P}(X)$ be the empty family of subsets of X. Then it is obvious that $\cup\emptyset = \emptyset$. To be in accordance with (2.17), we make the convention that $\cap\emptyset = X$.

Theorem 2.19. *Let $f : X \to Y$ be a function, let $\{B_i : i \in I\} \subset \mathcal{P}(Y)$ and $B \subset Y$. Then we have:*
(i) $f^{-1}(\cup_{i\in I} B_i) = \cup_{i\in I} f^{-1}(B_i)$;
(ii) $f^{-1}(\cap_{i\in I} B_i) = \cap_{i\in I} f^{-1}(B_i)$;
(iii) $(f^{-1}(B))^c = f^{-1}(B^c)$.

The proof of this theorem is easy and is left to the reader.

In what follows we want to introduce the notion of a Cartesian product of a nonempty family of sets. To do this, we will first enunciate an important axiom of set theory which has many applications in mathematics.

Axiom of choice 2.20. Let $\mathcal{A}$ be a nonempty family of nonempty sets. Then there is a function $f : \mathcal{A} \to \cup\mathcal{A}$, called a *choice function,* such that $f(A) \in A$ for any $A \in \mathcal{A}$. Thus this axiom asserts that one can choose simultaneously an element from each nonempty set of a nonempty family of sets. The axiom of choice is consistent with the other axioms of set theory. In this form, or in other equivalent forms (see (3.29)–(3.33), we will use it throughout whenever convenient.

Definitions 2.21. Let I be a nonempty set, and let $\{X_i : i \in I\}$ be an indexed family of sets. The *Cartesian product of the sets of* $\{X_i : i \in I\}$, written $\prod_{i\in I} X_i$, is the set of all functions $f : I \to \cup_{i\in I} X_i$ such that $f(i) \in X_i$ for any $i \in I$. If $I = \{n \in Z : n \geqslant m\}$ for some $m \in Z$, then $\prod_{i\in I} X_i$ will be usually denoted as $\prod_{n\geqslant m} X_n$; also, if $I = \{1, \ldots, n\}$, then we will write $\prod_{i=1}^n X_i$ or $X_1 \times \cdots \times X_n$

[1] If $\mathcal{A}$ is a partition of X, then $\mathcal{A} - \{\emptyset\}$ and $\mathcal{A} \cup \{\emptyset\}$ are different partitions of X that will be considered to be identical throughout this text.

instead of $\prod_{i\in I} X_i$. By virtue of the axiom of choice, we see that $\prod_{i\in I} X_i \neq \emptyset$ if and only if $X_i \neq \emptyset, i \in I$.

For each $f \in \prod_{i\in I} X_i$ and $i \in I$, the image $x_i = f(i)$ is called the *i-th coordinate of f*. Thus an element of the Cartesian product $\prod_{i\in I} X_i$ may be written as $(x_i)_{i\in I}$, where $x_i \in X_i$ for each $i \in I$. By specifying its coordinates, an element of the Cartesian product $\prod_{n\geqslant m} X_n$ will be conventionally denoted $(x_m, x_{m+1}, \ldots)$, where $x_n \in X_n, n \geqslant m$; also, an element of $\prod_{i=1}^n X_i$ will be written as an ordered n-tuple $(x_1, \ldots, x_n)$, where $x_i \in X_i$ for $i = 1, \ldots, n$.

Example 2.22. If there is a set X such that $X_i = X, i \in I$, then the Cartesian product $\prod_{i\in I} X_i$, written X^I, coincides with the set of all functions $f : I \to X$. If $I = N$, then we will write X^∞ instead of X^I; also, if $I = \{1, \ldots, n\}$, then X^I will be denoted X^n. Thus $R^\infty = \{(x_1, x_2, \ldots) : x_n \in R \text{ for } n \geqslant 1\}$, and $R^n = \{(x_1, \ldots, x_n) : x_i \in R \text{ for } i = 1, \ldots, n\}$.

Definition 2.23. Let I and J be nonempty sets such that $J \subset I$, and let $\{X_i : i \in I\}$ be an indexed family of sets. The function $\pi_J^I : \prod_{i\in I} X_i \to \prod_{j\in J} X_j$ defined by $\pi_J^I((x_i)_{i\in I}) = (x_j)_{j\in J}$ for any $(x_i)_{i\in I} \in \prod_{i\in I} X_i$ is called the *projection from* $\prod_{i\in I} X_i$ *onto* $\prod_{j\in J} X_j$. For each $i \in I$, the projection $\pi_{\{i\}}^I$ from $\prod_{i\in I} X_i$ onto X_i will be denoted π_i^I. If there is no ambiguity as to which set is I, we will write π_J and π_i instead of π_J^I and π_i^I.

Theorem 2.24. *Let $\prod_{i\in I} X_i$ be a Cartesian product of sets. Then $\emptyset \neq K \subset J \subset I$ implies $\pi_K^J \circ \pi_J^I = \pi_K^I$.*

Proof. Whatever $(x_i)_{i\in I} \in \prod_{i\in I} X_i$, we have

$$(\pi_K^J \circ \pi_J^I)((x_i)_{i\in I}) = \pi_K^J((x_j)_{j\in J}) = (x_k)_{k\in K} = \pi_K^I((x_i)_{i\in I}). \qquad \square$$

Definitions 2.25. Let I be nonempty set, and let $f_i : X \to Y_i, i \in I$, be functions. The *Cartesian product of the functions of* $\{f_i : i \in I\}$ is the function $(f_i)_{i\in I} : X \to \prod_{i\in I} Y_i$ given by $(f_i)_{i\in I}(x) = (f_i(x))_{i\in I}$ for any $x \in X$. If $I = \{n \in Z : n \geqslant m\}$ for some $m \in N$, then we will write $(f_n)_{n\geqslant m}$ instead of $(f_i)_{i\in I}$; also, if $I = \{1, \ldots, n\}$, then we will write $(f_1, \ldots, f_n)$ instead of $(f_i)_{i\in I}$. The functions $f_i, i \in I$, are called the *coordinate of* $(f_i)_{i\in I}$. We note that $\pi_j \circ (f_i)_{i\in I} = f_j$ for any $j \in I$.

Example 2.26. Let $\prod_{i\in I} X_i$ be a Cartesian product of sets. Then, for $\emptyset \neq J \subset I$, the coordinates of π_J^I are π_j^I, $j \in J$. Actually, we have $\pi_J^I = (\pi_j^I)_{j\in J}$.

Theorem 2.27. *Let $\prod_{i\in I} X_i$ be a Cartesian product of sets, and let $\{J_a : a \in A\}$ be a partition of I such that $J_a \neq \emptyset, a \in A$. Then the function $(\pi_{J_a})_{a\in A} : \prod_{i\in I} X_i \to \prod_{a\in A} \prod_{j\in J_a} X_j$ is bijective.*

Proof. For any $(x_i)_{i\in I} \in \prod_{i\in I} X_i$, we have $(\pi_{J_a})_{a\in A}((x_i)_{i\in I}) = ((x_j)_{j\in J_a})_{a\in A}$, whence we see easily that $(\pi_{J_a})_{a\in A}$ is surjective and injective. $\square$

Remarks 2.28. (a) (2.27) shows that a one to one correspondence from the collection of all families of subsets of $\prod_{i\in I} X_i$ onto the collection of all families of subsets of $\prod_{a\in A} \prod_{j\in J_a} X_j$ can be established by means of the mapping $(\pi_{J_a})_{a\in A}$. Also, whatever

$\mathcal{C} \subset \mathcal{P}(\prod_{i \in I} X_i)$, a one to one correspondence from the set of all functions whose domain is $\mathcal{C}$ onto the set of all functions whose domain is $\{(\pi_{J_a})_{a \in A}(C) : C \in \mathcal{C}\}$ can be established by means of $(\pi_{J_a})_{a \in A}$. These correspondences are essentially involved in the study of products of measure spaces, and explain why some authors identify the Cartesian products $\prod_{i \in I} X_i$ and $\prod_{a \in A} \prod_{j \in J_a} X_j$. The existence of the bijective mapping $(\pi_{J_a})_{a \in A}$may be viewed as an associativity property of the Cartesian product $\prod_{i \in I} X_i$. Comments on the associativity of Cartesian products of sets can be found in *Measure Theory* (Halmos, 1974).

(b) Notation is as in (2.25). If $\{J_a : a \in A\}$ is a partition of I such that, for any $a \in A$, $J_a \neq \emptyset$ and π_{J_a} stands for the projection from $\prod_{i \in I} Y_i$ onto $\prod_{j \in J_a} Y_j$, then $((f_j)_{j \in J_a})_{a \in A} = (\pi_{J_a})_{a \in A} \circ (f_i)_{i \in I}$. This equality expresses the associativity property of the Cartesian product $(f_i)_{i \in I}$.

Definitions 2.29. Let $\emptyset \neq J \subset I$ be such that $I - J \neq \emptyset$, let $\{X_i : i \in I\}$ be a family of sets, and let $(x_j)_{j \in J} \in \prod_{j \in J} X_j$. For $A \subset \prod_{i \in I} X_i$, the set $A_{(x_j)_{j \in J}} = \{(x_i)_{i \in I-J} \in \prod_{i \in I-J} X_i : (x_i)_{i \in I} \in A\}$ is called the *section of A at $(x_j)_{j \in J}$*. Thus $A_{(x_j)_{j \in J}} \subset \prod_{i \in I-J} X_i$.

Let $f : \prod_{i \in I} X_i \to Y$ be a function. The *section of f at $(x_j)_{j \in J}$* is the function $f_{(x_j)_{j \in J}} : \prod_{i \in I-J} X_i \to Y$ defined by $f_{(x_j)_{j \in J}}((x_i)_{i \in I-J}) = f((x_i)_{i \in I})$ for any $(x_i)_{i \in I-J} \in \prod_{i \in I-J} X_i$.

Examples 2.30. (a) Let $I = \{1, \ldots, 7\}$, $J = \{2, 3, 5\}$, and $(x_2, x_3, x_5) \in X_2 \times X_3 \times X_5$. For $A \subset \prod_{i=1}^{7} X_i$, we have $A_{(x_2,x_3,x_5)} = \{(x_1, x_4, x_6, x_7) \in X_1 \times X_4 \times X_6 \times X_7 : (x_1, \ldots, x_7) \in A\}$. Whatever $f : \prod_{i=1}^{7} X_i \to Y$, we have $f_{(x_2,x_3,x_5)}(x_1, x_4, x_6, x_7) = f(x_1, \ldots, x_7)$ for any $(x_1, x_4, x_6, x_7) \in X_1 \times X_4 \times X_6 \times X_7$.

(b) Let I, J, $\{X_i : i \in I\}$ and $(x_j)_{j \in J}$ be as in (2.29). For $A_i \subset X_i, i \in I$, we have

$$\left(\prod_{i \in I} A_i\right)_{(x_j)_{j \in J}} = \begin{cases} \prod_{i \in I-J} A_i & \text{if } (x_j)_{j \in J} \in \prod_{i \in J} A_j \\ \emptyset & \text{if } (x_j)_{j \in J} \notin \prod_{i \in J} A_j \end{cases}.$$

Exercise 2.31. Let $X \subset X'$, and let $f : X \to Y, g : X' \to Y$ be functions. Show that $f = g \circ i_X$, where i_X is the inclusion function from X into X', if and only if f is the restriction of g to X.

Exercise 2.32. Let $f : A \to B, g : B \to C, h : C \to D$ be functions. Prove that $h \circ (g \circ f) = (h \circ g) \circ f$.

Exercise 2.33. Let $f : X \to Y$ be a function, and let $A \subset X$, $B \subset Y$ be sets. Prove:

(a) $f(A \cap f^{-1}(B)) = f(A) \cap B$;

(b) $f(f^{-1}(A)) \subset A \subset f^{-1}(f(A))$;

(c) $f(A) \subset B$ if and only if $A \subset f^{-1}(B)$.

Exercise 2.34. Let $f : X \to Y$ be a function and let $B \subset Y$. Show that $1_B \circ f = 1_{f^{-1}(B)}$.

Exercise 2.35. Prove:

(a) $1_A + 1_{A^C} = 1$;

(b) $1_{A \cap B} = 1_A 1_B$;

(c) $1_{A-B} = 1_A(1 - 1_B)$;
(d) $1_{A\cup B} = 1_A + 1_B - 1_A 1_B$;
(e) $1_{A\Delta B} = |1_A - 1_B| = 1_A + 1_B - 2 \cdot 1_A 1_B$;
(f) if $A_1 \subset X_1$ and $A_2 \subset X_2$, then $1_{A_1 \times A_2}(x_1, x_2) = 1_{A_1}(x_1) 1_{A_2}(x_2)$ for any $(x_1, x_2) \in X_1 \times X_2$.

Exercise 2.36. If $J \subset I$, then $\cup_{j\in J} A_j \subset \cup_{i\in I} A_i$ and $\cap_{j\in J} A_j \supset \cap_{i\in I} A_i$.

Exercise 2.37. Let $\{A_i : i \in I\}$ be a family of sets, and let $\{J_a : a \in A\}$ be a family of subsets of I such that $\cup_{a\in A} J_a = I$. Show that $\cup_{i\in I} A_i = \cup_{a\in A} \cup_{j\in J_a} A_j$ and $\cap_{i\in I} A_i = \cap_{a\in A} \cap_{j\in J_a} A_j$.

Exercise 2.38. Let $f : X \to Y$ be a function, and let $\mathcal{B}_i \subset \mathcal{P}(Y)$, $i \in I$. Prove that $f^{-1}(\cup_{i\in I} \mathcal{B}_i) = \cup_{i\in I} f^{-1}(\mathcal{B}_i)$, but $f^{-1}(\cap_{i\in I} \mathcal{B}_i)$ may be different from $\cap_{i\in I} f^{-1}(\mathcal{B}_i)$.

Exercise 2.39. Whatever $k \in N$, set $B_n = A_{n+k}$, $n \geqslant 1$. Show that $\liminf_n A_n = \liminf_n B_n$ and $\limsup_n A_n = \limsup_n B_n$.

Exercise 2.40. Prove:
(a) $\liminf_n A_n \subset \limsup_n A_n$;
(b) $(\liminf_n A_n)^c = \limsup_n A_n^c$, $(\limsup_n A_n)^c = \liminf_n A_n^c$.

Exercise 2.41. Let A be a set, and let $\{B_i : i \in I\}$ be a family of sets. Prove:
(a) $A \cup (\cup_{i\in I} B_i) = \cup_{i\in I}(A \cup B_i)$;
(b) $A \cap (\cap_{i\in I} B_i) = \cap_{i\in I}(A \cap B_i)$;
(c) $A \cap (\cup_{i\in I} B_i) = \cup_{i\in I}(A \cap B_i)$;
(d) $A \cup (\cap_{i\in I} B_i) = \cap_{i\in I}(A \cup B_i)$;
(e) $A - (\cup_{i\in I} B_i) = \cap_{i\in I}(A - B_i)$;
(f) $A - (\cap_{i\in I} B_i) = \cup_{i\in I}(A - B_i)$;
(g) $\cap_{i\in I}(A \Delta B_i) \subset A \Delta (\cup_{i\in I} B_i) \subset \cup_{i\in I}(A \Delta B_i)$;
(h) $\cap_{i\in I}(A \Delta B_i) \subset A \Delta (\cap_{i\in I} B_i) \subset \cup_{i\in I}(A \Delta B_i)$.

Exercise 2.42. Let $f : X \to Y$ be a function, and let B, C be subsets of Y. Show that $f^{-1}(B \Delta C) = f^{-1}(B) \Delta f^{-1}(C)$.

Exercise 2.43. Let $f : X \to Y$ be a function. Prove that f is surjective if and only if $\mathcal{B}_1, \mathcal{B}_2 \subset \mathcal{P}(Y)$ and $f^{-1}(\mathcal{B}_1) = f^{-1}(\mathcal{B}_2)$ imply $\mathcal{B}_1 = \mathcal{B}_2$.

Exercise 2.44. Let I and J be nonempty sets, and let X_{ij}, $i \in I$, $j \in J$, be sets. Show that $\cap_{j\in J} \prod_{i\in I} X_{ij} = \prod_{i\in I} \cap_{j\in J} X_{ij}$.

Exercise 2.45. Let I, J, $\{X_i : i \in I\}$ and $(x_j)_{j\in J}$ be as in (2.29). Let $\{A_h : h \in H\}$ be a family of subsets of $\prod_{i\in I} X_i$, and let A, B be subsets of $\prod_{i\in I} X_i$. Prove:
(a) $(\cup_{h\in H} A_h)_{(x_j)_{j\in J}} = \cup_{h\in H} (A_h)_{(x_j)_{j\in J}}$;
(b) $(\cap_{h\in H} A_h)_{(x_j)_{j\in J}} = \cap_{h\in H} (A_h)_{(x_j)_{j\in J}}$;
(c) $(A - B)_{(x_j)_{j\in J}} = A_{(x_j)_{j\in J}} - B_{(x_j)_{j\in J}}$;
(d) $(A^c)_{(x_j)_{j\in J}} = (A_{(x_j)_{j\in J}})^c$;
(e) if $\emptyset \neq K \subset J$ and $J - K \neq \emptyset$, then $A_{(x_j)_{j\in J}} = (A_{(x_k)_{k\in K}})_{(x_j)_{j\in J-K}}$;
(f) if $f : \prod_{i\in I} X_i \to Y$ is a function, and C is a subset of Y, then $(f^{-1}(C))_{(x_j)_{j\in J}} = f^{-1}_{(x_j)_{j\in J}}(C)$.

Exercise 2.46. Find a function $f : R \to R$ such that $(f \circ f)(x) + f(x) + x = 0$ for any $x \in R$.

Exercise 2.47. Let $X = \{a, b\}$, $Y = \{0\} \cup N$, and define $f : X \to Y$ by $f(a) = 0$ and $f(b) = 1$. For $n \geqslant 1$, let $\mathcal{A}_n$ be the partition of N formed from all sets $\{2^n k + 1, 2^n k + 2, \ldots, 2^n k + 2^n\}$ with $k = 0, 1, \ldots$, and let $\mathcal{B}_n \subset \mathcal{P}(Y)$ consist of $\{0\}$ and unions of sets in $\mathcal{A}_n$. Prove the following.

(a) $\mathcal{B}_n \supset \mathcal{B}_{n+1}, n \geqslant 1$.
(b) $\cap_{n \geqslant 1} \mathcal{B}_n = \{\{0\}\}$.
(c) $f^{-1}(\cap_{n \geqslant 1} \mathcal{B}_n) \neq \cap_{n \geqslant 1} f^{-1}(\mathcal{B}_n)$.

Exercise 2.48. Let $f : X \to Y$ be a function and let $\{A_i : i \in I\} \subset \mathcal{P}(X)$.

(a) Show that $f(\cup_{i \in I} A_i) = \cup_{i \in I} f(A_i)$.
(b) Prove that $f(\cap_{i \in I} A_i) \subset \cap_{i \in I} f(A_i)$, and that this inclusion may be strict.

3 Equivalent Relations and Partial Orderings

If X_1 and X_2 are sets, then in agreement with (2.21) an element of the Cartesian product $X_1 \times X_2$ is written as an ordered pair (x_1, x_2), where $x_1 \in X_1$ and $x_2 \in X_2$. Suppressing indices, if X and Y are sets, then the Cartesian product of X and Y is the set $X \times Y = \{(x, y) : x \in X \text{ and } y \in Y\}$. In this section we discuss two special classes of subsets of $X \times X$ that will be needed in the sequel, namely the class of equivalent relations on X and the class of partial orderings on X.

Definitions 3.1. Any subset $\mathbb{R}$ of a Cartesian product $X \times Y$ is called a *relation*. Following conventional notation, we will write $x\mathbb{R}y$ instead of $(x, y) \in \mathbb{R}$. The subset $\mathbb{R}^* = \{(y, x) : x\mathbb{R}y\}$ of the Cartesian product $Y \times X$ is called the *inverse of* $\mathbb{R}$.

Example 3.2. Let $f : X \to Y$ be a function. The relation $\mathbb{R} = \{(x, f(x)) : x \in X\} \subset X \times Y$ is called the *graph of* f. Many authors identify a function with its graph.

Definitions 3.3. Let X be a set, and let $\mathbb{R} \subset X \times X$ be a relation. $\mathbb{R}$ is said to be *reflexive* if $x\mathbb{R}y$ for any $x \in X$. If $x\mathbb{R}y$ implies $y\mathbb{R}x$, then $\mathbb{R}$ is said to be *symmetric*. $\mathbb{R}$ is said to be *antisymmetric* if $x\mathbb{R}y$ and $y\mathbb{R}x$ imply $x = y$. If $x\mathbb{R}y$ and $y\mathbb{R}z$ imply $x\mathbb{R}z$, then $\mathbb{R}$ is said to be *transitive.*

Examples 3.4. (a) Let X be a nonempty set. Then the relation $\mathbb{R} = X \times X$ is reflexive, symmetric and transitive. $\mathbb{R}$ is antisymmetric if and only if X has a single element.

(b) Let X be a set, and let $\mathbb{D} = \{(x, x) : x \in X\}$ be the *diagonal* of the Cartesian product $X \times X$. The relation $\mathbb{D}$ is reflexive, symmetric, antisymmetric and transitive.

(c) The relation $\mathbb{R} = \{(x, y) \in R^2 : x \leqslant y\} \subset R^2$ is reflexive, antisymmetric and transitive. This relation is the usual ordering on R (see (3.19.a)). Notice that $\mathbb{R}$ is not symmetric.

(d) The relation $\mathbb{R} = \{(x, y) \in R^2 : x < y\} \cup \{(0, 0)\} \subset R^2$ is antisymmetric and transitive. $\mathbb{R}$ is neither reflexive nor symmetric.

Remark 3.5. Let X be a set, and let $\mathbb{R} \subset X \times X$ be a relation. We note that $\mathbb{R}$ is reflexive if and only if $\mathbb{R}^*$ is reflexive, $\mathbb{R}$ is symmetric if and only if $\mathbb{R}^*$ is symmetric,

$\mathbb{R}$ is antisymmetric if and only if $\mathbb{R}^*$ is antisymmetric, and $\mathbb{R}$ is transitive if and only if $\mathbb{R}^*$ is transitive.

Definitions 3.6. A relation $\sim \subset X \times X$ is called an *equivalent relation on X* if $\sim$ is reflexive, symmetric and transitive. For each $x \in X$, the subset $\widetilde{x} = \{y \in X : x \sim y\}$ is called the *equivalence class of x*. The family of all equivalent classes induced by $\sim$ on X will be denoted by $X/\sim$.

Example 3.7. Let $\mathcal{A}$ be a family of sets. We define a relation $\sim \subset \mathcal{A} \times \mathcal{A}$ by saying that $A \sim B$ if there exists a bijective function $f : A \to B$. For any $A \in \mathcal{A}$, the identity function $i_A : A \to A$ is bijective, and so $A \sim A$. Thus $\sim$ is reflexive. If $A \sim B$, and $f : A \to B$ is bijective, then the inverse function $f^{-1} : B \to A$ is bijective, and so $B \sim A$. Consequently, $\sim$ is symmetric. If $A \sim B$, $B \sim C$, and $f : A \to B$, $g : B \to C$ are bijective, then the composition function $g \circ f : A \to C$ is bijective, and so $A \sim C$. Hence $\sim$ is transitive. Therefore, $\sim$ is an equivalence relation on $\mathcal{A}$.

Theorem 3.8. *Let $\sim$ be an equivalent relation on X. Then the family $X/\sim$ is a partition of X.*

Proof. Whatever $x \in X$, we have $x \sim x$, and so $x \in \widetilde{x}$. Therefore, $\cup X/\sim = X$. Now let $\widetilde{x}, \widetilde{y} \in X/\sim$ be such that $\widetilde{x} \neq \widetilde{y}$. If z were in $\widetilde{x} \cap \widetilde{y}$, then we would have $x \sim z$ and $y \sim z$, and so we would get $y \sim x$ by symmetry and transitivity. Further, for any $x' \in \widetilde{x}$, we have $x \sim x'$, and so it would follow $y \sim x'$ by transitivity, whence $x' \in \widetilde{y}$. Consequently, we would have $\widetilde{x} \subset \widetilde{y}$ and, similarly, $\widetilde{y} \subset \widetilde{x}$, and so $\widetilde{x} = \widetilde{y}$. Thus $\widetilde{x} \cap \widetilde{y} = \emptyset$. □

Definitions 3.9. A set X is said to be *finite* if either $X = \emptyset$ or there exists a bijective function $f : \{1, \ldots, n\} \to X$ for some $n \in N$. Any set that is not finite is said to be *infinite.* A set X is said to be *denumerable* if there exists a bijective function $f : N \to X$. Any set that is either finite or denumerable is said to be *countable.* A set X is said to have the *continuum power* if there exists a bijective function $f : R \to X$.

Remarks 3.10. (a) Obviously, any finite set neither is denumerable nor has the continuum power. G. Cantor proved that any denumerable set has not the continuum power. We will present this result in (3.12). Also, in (3.11) we will show that any infinite set has a denumerable subset.

(b) Assume that the family $\mathcal{A}$ in Example (3.7) is sufficiently large; for instance, suppose that $\mathcal{A}$ consists of all sets encountered in this book excepting $\mathcal{A}$ itself. Then, for each $n \in N$, we may consider the family of all sets in $\mathcal{A}$ that are equivalent to $\{1, \ldots, n\}$; also we may consider the family of all sets in $\mathcal{A}$ that are equivalent to N, and we may consider the family of all sets in $\mathcal{A}$ that are equivalent to R. The equivalence classes in $\mathcal{A}/\sim$ are called *cardinal numbers* (relative to $\mathcal{A}$). Notice that the notion of a cardinal number we have just introduced is restrictive, since it depends on the chosen family $\mathcal{A}$. There is no way of eliminating this restriction by letting $\mathcal{A}$ be the family of all sets, since such a family does not exist. The interested reader can find an axiomatic definition of the notion of a cardinal number in *Naive Set Theory* (Halmos, 1974).

Theorem 3.11. *Any infinite set has a denumerable subset.*

Proof. Let X be an infinite set. We prove by induction that X has a subset consisting of n elements for each $n \in N$. Since $X \neq \emptyset$, there is $x_1 \in X$, and so $\{x_1\} \subset X$. If $\{x_1, \ldots, x_n\} \subset X$, then, since X is infinite, there exists $x_{n+1} \in X - \{x_1, \ldots, x_n\}$, and so $\{x_1, \ldots, x_n, x_{n+1}\} \subset X$. Hence $\{x_1, x_2, \ldots\} \subset X$. Since the function $f : N \to \{x_1, x_2, \ldots\}$ defined by $f(n) = x_n$ for any $n \in N$ is bijective, the set $\{x_1, x_2, \ldots\}$ is denumerable. □

Theorem 3.12 (Cantor). *Every denumerable set has not the continuum power.*

Proof. To get a contradiction, assume that there exists a denumerable set which has the continuum power. Then, in accordance with (3.9), there is a bijective function from N onto R. Since the function f defined by $f(x) = \tan(\pi x - \pi/2)$ is a bijective mapping from the open interval $]0, 1[$ onto R, there exists a bijective function $g : N \to]0, 1[$. Now each $x \in]0, 1[$ has a unique expansion in the form $x = \sum_{k \geqslant 1} a_k/10^k$, where $a_k \in \{0, 1, \ldots, 9\}$, $k \geqslant 1$, and, whatever $l \geqslant 1$, there is $m \geqslant l$ such that $a_m \neq 9$. For every $n \in N$, let $g(n) = \sum_{k \geqslant 1} a_{nk}/10^k$ denote this expansion of $g(n)$. We may choose a sequence $\{b_k : k \geqslant 1\}$ such that $b_k \in \{0, 1, \ldots, 9\}$, $k \geqslant 1$, $b_1 \neq 0$, $b_k \neq a_{kk}$, $k \geqslant 1$, and, whatever $l \geqslant 1$, there exists $m \geqslant l$ for which $b_m \neq 9$. Let $y = \sum_{k \geqslant 1} b_k/10^k$. Then $y \in]0, 1[$, and $y \neq g(n)$, $n \in N$. This contradicts the assertion that g is surjective. □

Remark 3.13. Theorem (3.12) emphasizes existence of infinite sets which are not denumerable.

Our immediate aim is to prove that the union of any countable family of countable sets is a countable set. The exact statement of this result is Theorem (3.17). To do this, we need the following lemmas.

Lemma 3.14. *If A is a countable set and $B \subset A$, then B is countable.*

Proof. If B is finite, there is nothing to prove. If B is infinite, then A is denumerable, since every subset of a finite set is finite. In this case there is a bijective function $f : N \to A$. Since f is surjective, (2.33.a) shows that the subset $f^{-1}(B)$ of N is infinite. Since N is well ordered (see (3.27) and (3.28)), there exists a bijective function $g : N \to f^{-1}(B)$. Actually, let $g(1)$ be the smallest element of $f^{-1}(B)$. If $g(1), \ldots, g(n)$ have been defined, then, since $f^{-1}(B) - \{g(1), \ldots, g(n)\} \neq \emptyset$, let $g(n+1)$ be the smallest element of $f^{-1}(B) - \{g(1), \ldots, g(n)\}$. Plainly, the function g defined in this manner is bijective. Then the function $h : N \to B$ given by $h(n) = f(g(n))$ for any $n \in N$ is bijective, and so B is denumerable. □

Lemma 3.15. *Let I be a nonempty countable set, and let $\{A_i : i \in I\}$ be a family of sets. Then there exists a family $\{B_i : i \in I\}$ such that $B_i \subset A_i$, $i \in I$, $B_i \cap B_j = \emptyset$ for $i \neq j$ and $\cup_{i \in I} A_i = \cup_{i \in I} B_i$.*

Proof. We prove this lemma in case I is denumerable. The case when I is finite needs only minor changes and is left to the reader. Since there is a bijective mapping from N onto I, we may assume without loss of generality that $I = N$. We define the family of sets $\{B_n : n \in N\}$ as follows. Set $B_1 = A_1$ and $B_{n+1} = A_{n+1} - \cup_{i=1}^{n} A_i$, $n \geqslant 1$.

Evidently, $B_n \subset A_n, n \in N$, and $B_i \cap B_j = \emptyset$ for $i \neq j$. If $x \in \cup_{n \in N} A_n$, then there exists $m \in N$ such that $x \in A_m$ and $x \notin A_i$ for $i < m$, and so $x \in B_m$. Consequently, $\cup_{n \in N} A_n \subset \cup_{n \in N} B_n$. Since the inclusion $\cup_{n \in N} A_n \supset \cup_{n \in N} B_n$ is immediate, it follows that $\cup_{n \in N} A_n = \cup_{n \in N} B_n$. □

Lemma 3.16. *Let I be a nonempty countable set, and let $\{B_i : i \in I\}$ be a family of sets. If B_i is countable for any $i \in I$, and $B_i \cap B_j = \emptyset$ for $i \neq j$, then $\cup_{i \in I} B_i$ is countable.*

Proof. Obviously, we may admit that $B_i \neq \emptyset$ for each $i \in I$. As in (3.15), the proof in case I is finite is left to the reader; also, without any loss of generality, we may assume that $I = N$.

First we consider the case when B_k is finite for any $k \in N$. Then, for every $k \in N$, there exist $n_k \in N$ and a bijective function $f_k : \{1, \ldots, n_k\} \to B_k$. We define a function $f : N \to \cup_{k \in N} B_k$ as follows. For each $n \in N$, there is an integer $k \geqslant 0$ such that $\sum_{l=0}^{k} n_l < n \leqslant \sum_{l=1}^{k+1} n_l$, where $n_0 = 0$. We set $f(n) = f_{k+1}(n - \sum_{l=0}^{k} n_l), n \in N$. Since f is bijective, $\cup_{k \in N} B_k$ is denumerable.

Now we consider the general case. For every $k \in N$, there exists a bijective function $f_k : X_k \to B_k$, where either $X_k = N$ or $X_k = \{1, \ldots, n_k\}$ for some $n_k \in N$. Whatever $m \in N$, define $C_m = \{f_k(n) : k + n = m + 1\}$. Then $C_m \neq \emptyset, m \in N, C_i \cap C_j = \emptyset$ for $i \neq j$, and $\cup_{k \in N} B_k = \cup_{m \in N} C_m$. For any $m \in N, C_m$ has at most m elements, and so, on account of the preceding paragraph, $\cup_{k \in N} B_k$ is denumerable. □

Theorem 3.17. *Let I be a nonempty countable set, and let $\{A_i : i \in I\}$ be a family of sets. If A_i is countable for every $i \in I$, then $\cup_{i \in I} A_i$ is countable.*

Proof. This result follows at once from Lemmas (3.15), (3.14) and (3.16). □

The ordering we discuss in what follows is a very important kind of relation.

Definitions 3.18. A relation $\leqslant \subset X \times X$ is called a *partial ordering on*X if $\leqslant$ is reflexive, antisymmetric and transitive. If, in addition, $x, y \in X$ implies $x \leqslant y$ or $y \leqslant x$, then $\leqslant$ is called a *total* [*linear*] ordering on X. If $x \leqslant y$ and $x \neq y$, then we will write $x < y$. An ordered pair $(X, \leqslant)$,where X is a set and $\leqslant$ is a partial ordering on X, is called a *partially ordered set.* If $\leqslant$ is a total ordering on X, then $(X, \leqslant)$ is called a *totally ordered set.*

Examples 3.19. (a) Let $\overline{R}$ be the set of extended real numbers. Considering the usual ordering $\leqslant$ on R, and defining $-\infty < \infty$, $-\infty < x < \infty$ for any $x \in R$, $(\overline{R}, \leqslant)$ is a totally ordered set.

Whatever $a, b \in \overline{R}$ such that $a < b$, the sets $]a, b[= \{x \in \overline{R} : a < x < b\}$, $[a, b[= \{x \in \overline{R} : a \leqslant x < b\}$, $]a, b] = \{x \in \overline{R} : a < x \leqslant b\}$ and $[a, b] = \{x \in \overline{R} : a \leqslant x \leqslant b\}$ are called *intervals.*

(b) Let X be a set, and let $\overline{R}^X$ be the set of all functions from X into $\overline{R}$. For $f, g \in \overline{R}^X$, we write $f \leqslant g$ if $f(x) \leqslant g(x)$ for any $x \in X$. Then $\leqslant$ is a partial ordering on $\overline{R}^X$. If X has at least two elements, then it is easily seen that $(\overline{R}^X, \leqslant)$ is not a totally ordered set. If $X = \{1, \ldots, n\}$, then $\leqslant$ is a partial ordering on $\overline{R}^n$.

(c) Let $\mathcal{A}$ be a family of sets. The set inclusion $\subset$ is a partial ordering on $\mathcal{A}$. In general $(\mathcal{A}, \subset)$ is not a totally ordered set. For example, if $\mathcal{A}$ is a partition that has at least two nonempty sets, then $\subset$ is not a total ordering on $\mathcal{A}$.

Definitions 3.20. Let $(X, \leqslant)$ be a partially ordered set, and let $A \subset X$. An element $x \in X$ is called an *upper bound of* A if $a \leqslant x$ for any $a \in A$. If y is an upper bound for A such that $y \leqslant x$ whenever x is an upper bound for A, then y is called a *supremum of* A, and we write $y = \sup A$. An element $x \in X$ is called a *lower bound of* A if $x \leqslant a$ for all $a \in A$. If y is a lower bound for A such that $x \leqslant y$ whenever x is a lower bound for A, then y is called an *infimum of* A, and we write $y = \inf A$. An element $z \in X$ is called a *maximal element of* X if $x \in X$ and $z \leqslant x$ imply $x = z$. An element $z \in X$ is called a *minimal element of* X if $x \in X$ and $x \leqslant z$ imply $x = z$.

Remarks 3.21. (a) By virtue of (3.5), if $\leqslant$ is a partial ordering on X, then the inverse relation $\geqslant$ is a partial ordering on X. Thus to each statement concerning the partially ordered set $(X, \leqslant)$ corresponds a statement concerning the partially ordered set $(X, \geqslant)$. For example, $(X, \leqslant)$ is a totally ordered set if and only if $(X, \geqslant)$ is a totally ordered set; also, y is a supremum of A relative to $\leqslant$ if and only if y is an infimum of A relative to $\geqslant$, and z is a maximal element of X relative to $\leqslant$ if and only if z is a minimal element of X relative to $\geqslant$.

(b) In (3.22) we will show that if $\leqslant$ is a partial ordering on X, and $A \subset X$, then A may fail to have an upper bound, and so A may fail to have a supremum. Furthermore, even if A has upper bounds, A may fail to have a supremum. Since $\leqslant$ is antisymmetric, if A has a supremum, then it is unique. Analogous statements concerning lower bounds and infimum of A hold.

(c) Let $\leqslant$ be the usual ordering on $\overline{R}$ defined in (3.19.a). Since each element of $\overline{R}$ is both an upper bound and a lower bound of $\emptyset \subset \overline{R}$, it follows that $\sup \emptyset = -\infty$ and $\inf \emptyset = \infty$.

(d) Let $(Y, \leqslant)$ be a partially ordered set and $\{y_i : i \in I\} \subset Y$. If the indexed set $\{y_i : i \in I\}$ has a supremum, then we write $\sup_{i \in I} y_i = \sup\{y_i : i \in I\}$; also if $\{y_i : i \in I\}$ has an infimum, we write $\inf_{i \in I} y_i = \inf\{y_i : i \in I\}$. If $I = \{n \in Z : n \geqslant m\}$ for some $m \in Z$, then we sometimes write $\sup_{n \geqslant m} y_n$ or $\vee_{n \geqslant m} y_n$ instead of $\sup_{i \in I} y_i$, and $\inf_{n \geqslant m} y_n$ or $\wedge_{n \geqslant m} y_n$ instead of $\inf_{i \in I} y_i$; also, if $I = \{1, \ldots, n\}$, then we will write $\vee_{i=1}^n y_i$ or $y_1 \vee \cdots \vee y_n$ for $\sup_{i \in I} y_i$, and $\wedge_{i=1}^n y_i$ or $y_1 \wedge \cdots \wedge y_n$ for $\inf_{i \in I} y_i$. If $f : X \to Y$ is a function and $A \subset X$, then, since $f(A) \subset Y$ may be considered as an indexed set (by $x \in A$), we write $\sup_{x \in A} f(x) = \sup f(A)$ and $\inf_{x \in A} f(x) = \inf f(A)$ whenever $\sup f(A)$ and $\inf f(A)$ exist.

Examples 3.22. (a) Let $(X, \leqslant)$ be a partially ordered set. Then $\sup\{x\} = \inf\{x\} = x$ for any $x \in X$.

(b) Let $\leqslant$ be the usual ordering on R. Then R has neither an upper bound nor a lower bound; also, there is neither a maximal nor a minimal element of R.

(c) Let $[a, b[$ and $]c, d]$ be intervals included in R such that $b < c$. Let $\leqslant$ be the usual ordering on $X = [a, b[\cup]c, d]$. Then any $x \in]c, d]$ is an upper bound for $[a, b[$, and any $x \in [a, b[$ is a lower bound for $]c, d]$. Nevertheless, $[a, b[$ has no supremum, and $]c, d]$ has no infimum. At the same time, $a = \inf[a, b[$ and $d = \sup]c, d]$.

(d) Let $(\mathcal{A}, \subset)$ be a partially ordered set, where $\mathcal{A}$ is a partition consisting of nonempty sets. Then any $A \in \mathcal{A}$ is both a maximal and a minimal element of $\mathcal{A}$.

Definitions 3.23. Let $(X, \leqslant)$ and $(Y, \preceq)$ be partially ordered sets, and let $f : X \to Y$ be a function. We say that f is *nondecreasing* if $x_1 \leqslant x_2$ implies $f(x_1) \preceq f(x_2)$. We say that f is *strictly increasing* if $x_1 < x_2$ implies $f(x_1) \prec f(x_2)$. If $x_1 \leqslant x_2$ implies $f(x_1) \succeq f(x_2)$, then f is said to be *nonincreasing.* If $x_1 < x_2$ implies $f(x_1) \succ f(x_2)$, then f is said to be *strictly decreasing.* We say that f is *monotone* if it is either nondecreasing or nonincreasing.

Examples 3.24. (a) Let $A \subset R$ and $f : A \to \overline{R}$. The terms nondecreasing, strictly increasing, etc. attached to f will be used with respect to the usual orderings on A and $\overline{R}$. For instance, we say that a sequence $\{x_n : n \in N\} \subset \overline{R}$ is nondecreasing if $m \leqslant n$ implies $x_m \leqslant x_n$.

(b) A sequence $\{f_n : n \in N\} \subset \overline{R}^X$ is said to be nondecreasing if $m \leqslant n$ implies $f_m \leqslant f_n$. A sequence $\{f_n : n \in N\} \subset \overline{R}^X$ is said to be nonincreasing if $m \leqslant n$ implies $f_m \geqslant f_n$.

(c) Let $(\mathcal{A}, \subset)$ be a partially ordered set and $\{A_n : n \in N\} \subset \mathcal{A}$. We say that $\{A_n : n \in N\}$ is nondecreasing if $m \leqslant n$ implies $A_m \subset A_n$. We say that $\{A_n : n \in N\}$ is nonincreasing if $m \leqslant n$ implies $A_m \supset A_n$.

In what follows we list a number of equivalent forms of the axiom of choice (2.20). To do this, we need the following definitions.

Definitions 3.25. Let $(X, \leqslant)$ be a partially ordered set. A set $L \subset X$ is called a *chain in* X if $x, y \in L$ implies $x \leqslant y$ or $y \leqslant x$. Let $\mathcal{L}$ be the family of all chains in X. A maximal element of $\mathcal{L}$ relative to the set inclusion $\subset$ is called a *maximal chain in* X.

Definition 3.26. We say that a family of sets $\mathcal{A}$ is of *finite character* if, for each set A, we have $A \in \mathcal{A}$ if and only if every finite subset of A belongs to $\mathcal{A}$. We observe that $\mathcal{L}$ in (3.25) is a family of finite character.

Definition 3.27. Let $(X, \leqslant)$ be a partially ordered set. We say that X is *well ordered* relative to $\leqslant$, if for each nonempty subset A of X, there is an element $x \in A$ (called the *smallest element of* A) such that $x \leqslant a$ for any $a \in A$.

Example 3.28. The set N is well ordered relative to the usual ordering on it. We used this in the proof of Lemmas (3.14) and (3.15).

The assertions (3.29)–(3.33) below are equivalent to the axiom of choice, and so they are all equivalent. We do not expose here the proof of these equivalences. This proof is rather complicated, and it is remote from the main purpose of the present book. For a detailed discussion of the assertions (2.20) and (3.29)–(3.33), the reader can see *General Topology* (Kelley, 1975) and *Real and Abstract Analysis* (Hewitt and Stromberg, 1975).

Hausdorff's maximal principle 3.29. *Every partially ordered set contains a maximal chain.*

Kuratowski's lemma 3.30. *Let* $(X, \leqslant)$ *be a partially ordered set. Each chain in* X *is included in a maximal chain in* X.

Zorn's lemma 3.31. *Let $(X, \leqslant)$ be a partially ordered set. If each chain in X has an upper bound, then there is a maximal element of X.*

Tukey's lemma 3.32. *Every nonempty family of finite character has a maximal element relative to the set inclusion.*

Well ordering principle 3.33 (Zermelo). *If X is a set, then there is a partial ordering on X relative to which X is well ordered.*

Exercise 3.34. Prove that $X \times Y = \cup_{x \in X}(\{x\} \times Y) = \cup_{y \in Y}(X \times \{y\})$.

Exercise 3.35. Let $\mathbb{R}$ be the graph of a function $f : X \to X$, and let $A = \{x : f(x) = x\}$. Prove the following.

(a) $\mathbb{R}$ is reflexive if and only if $f = i_X$.
(b) $\mathbb{R}$ is symmetric if and only if f is bijective and $f = f^{-1}$.
(c) $\mathbb{R}$ is antisymmetric if and only if $(f \circ f)(x) \neq x$ for any $x \in A^c$.
(d) $\mathbb{R}$ is transitive if and only if $f = f \circ f$.

Exercise 3.36. Let $\mathbb{R} \subset X \times X$ be a relation. Show that $\mathbb{R}$ is symmetric if and only if $\mathbb{R} = \mathbb{R}^*$.

Exercise 3.37. A *permutation* of a set X is any bijective mapping from X onto X. Let $\mathfrak{P}$ be the set of all permutations of X. For $f, g \in \mathfrak{P}$, we write $f \sim g$ if there is $h \in \mathfrak{P}$ such that $f = h \circ g \circ h^{-1}$.

(a) Prove that $\sim$ is an equivalent relation on $\mathfrak{P}$.
(b) For $f \in \mathfrak{P}$, describe the equivalence class $\widetilde{f}$.

Exercise 3.38. Let $f : X \to Y$ be a function and let $A \subset X$. Use the axiom of choice and Lemma (3.14) to show that $f(A)$ is countable whenever A is countable.

Exercise 3.39. Notation is as in (3.17). Show that $\cup_{i \in I} A_i$ is denumerable if and only if one of the following conditions hold:

(i) there is $j \in I$ such that A_j is denumerable;
(ii) there is a denumerable subset J of I such that $j, k \in J$ and $j \neq k$ imply $A_j \neq A_k$.

Exercise 3.40. Let X be a set and let $\mathcal{A} = \{A \in \mathcal{P}(X) : A \text{ is finite}\}$. Prove that $\mathcal{A}$ is countable whenever X is countable.

Exercise 3.41. Show that the sets Z and Q are denumerable.

Exercise 3.42. If X is a denumerable set, and Y is a nonempty countable set, then $X \times Y$ is denumerable.

Exercise 3.43. Let I be a nonempty set, let $f_i : X \to R, i \in I$, be functions, and let A be a nonempty subset of R. Let $\mathcal{J}$ denote the family of all nonempty finite subsets of I. For $J \in \mathcal{J}$ and any integer $n \geqslant 0$, let $S_n(J)$ be the set of all elements $(n_j)_{j \in J}$ whose coordinates $n_j, j \in J$, are nonnegative integers such that $\sum_{j \in J} n_j \leqslant n$. If $J \in \mathcal{J}$, then a polynomial of degree $\leqslant n$ in functions from $\{f_j : j \in J\}$ with coefficients in A is any function of the form[2]

[2]For any function $f : X \to K$, we assume that f^0 is the constant function with value 1.

(i)
$$\sum_{(n_j)_{j\in J}\in S_n(J)} \alpha((n_j)_{j\in J}) \prod_{j\in J} f_j^{n_j},$$

where $\alpha((n_j)_{j\in J}) \in A$ for each $(n_j)_{j\in J} \in S_n(j)$. Let $\mathfrak{P}_n(J)$ denote the set of all functions of the form (i).

(a) For $J \in \mathcal{J}$ and $n \geqslant 0$, let $\overline{\overline{S_n(J)}}$ be the number of elements of the set $S_n(J)$. Prove that there is a bijective mapping form $\overline{\overline{S_n(J)}}$ onto $\mathfrak{P}_n(J)$.

(b) If $J \in \mathcal{J}$, then a polynomial in functions from $\{f_j : j \in J\}$ with coefficients in A is any element of the set $\mathfrak{P}(J) = \cup_{n\geqslant 0}\mathfrak{P}_n(J)$. A *polynomial in functions from* $\{f_i : i \in I\}$ *with coefficients in* A is any element of the set $\mathfrak{P}(I) = \cup_{J\in\mathcal{J}}\mathfrak{P}(J)$. Use (3.42), (a), (3.17) and (3.40) to prove that $\mathfrak{P}(I)$ is countable whenever A and I are countable.

Exercise 3.44. Let X be an infinite set, and let A be a countable set. Use (3.14), (3.11), and (3.17) to show that there is a bijective mapping form $X \cup A$ onto X.

Exercise 3.45. (a) Prove that any interval $]a, b[\subset R$ has the continuum power.

(b) Use (a) and (3.44) to show that any interval $I \subset \overline{R}$ has the continuum power.

Exercise 3.46. Let $k \in N$ and $X = \{0, 1, \ldots, k\}^\infty$.

(a) Prove that the set $A = \{(a_n)_{n\in N} \in X :$ there is $m \in N$ such that $a_n = k$ for $n \geqslant m\}$ is denumerable.

(b) Use (3.45.b), (a) and (3.44) to show that X has the continuum power. [Hint. The function $f : A^c \to [0, 1[$ defined by $f((a_n)_{n\in N}) = \sum_{n\geqslant 1} a_n/(k+1)^n$ for any $(a_n)_{n\in N} \in A^c$ is bijective.]

Exercise 3.47. Let X be a denumerable set. Use (2.9) and (3.46.b) to prove that $\mathcal{P}(X)$ has the continuum power.

Exercise 3.48. Let $\mathcal{D}$ be the collection of all partitions of a set X. For $\mathcal{A}, \mathcal{B} \in \mathcal{D}$, we write $\mathcal{A} \leqslant \mathcal{B}$ if each element of $\mathcal{B}$ is a subset of an element of $\mathcal{A}$.

(a) Prove that $\leqslant$ is a partial ordering on $\mathcal{D}$.

(b) Assume that X is a well ordered set. For $\emptyset \neq A \subset X$, let x_A denote the smallest element of A. Show that the function $f : \mathcal{D} \to \mathcal{P}(\mathcal{X})$ defined by $f(\mathcal{A}) = \{x_A : A \in \mathcal{A}\}$ for any $\mathcal{A} \in \mathcal{D}$ is strictly increasing, where $\mathcal{P}(X)$ is endowed with the partial ordering $\subset$.

Exercise 3.49. Let $(\mathcal{P}(X), \subset)$ be a partially ordered set and let $\mathcal{A} \subset \mathcal{P}(\mathcal{X})$. Show that $\sup \mathcal{A} = \cup\mathcal{A}$ and $\inf \mathcal{A} = \cap\mathcal{A}$.

Exercise 3.50. Let $(X, \leqslant)$ be a partially ordered set and let $B \subset A \subset X$. Prove the following.

(a) If A and B have suprema, then $\sup B \leqslant \sup A$.

(b) If A and B have infima, then $\inf A \leqslant \inf B$.

Exercise 3.51. (a) If $\{A_n : n \geqslant 1\}$ is a nondecreasing sequence of sets, then $\liminf_n A_n = \limsup_n A_n = \cup_{n\geqslant 1} A_n$.

(b) If $\{A_n : n \geqslant 1\}$ is a nonincreasing sequence of sets, then $\liminf_n A_n = \limsup_n A_n = \cap_{n\geqslant 1} A_n$.

Exercise 3.52. Let $(X, \leqslant)$ and $(Y, \preceq)$ be partially ordered sets, and let $f : X \to Y$ be a strictly increasing function. If $y \in f(X)$ is a maximal element of Y, then the set $f^{-1}(\{y\})$ consists of maximal elements of X.

Exercise 3.53. If X is well ordered relative to $\leqslant$, then $(X, \leqslant)$ is a totally ordered set.

Exercise 3.54. Let $(\mathcal{P}(X), \subset)$ be a partially ordered set, and let $\mathcal{A}$ be a chain in $\mathcal{P}(X)$. For $x, y \in X$, define $x \preceq y$ if either $x = y$ or there is $A \in \mathcal{A}$ such that $x \in A$ and $y \notin A$. Prove the following.

(a) $\preceq$ is a partial ordering on X.

(b) If $\mathcal{A}$ is a maximal chain, then $(X, \preceq)$ is a totally ordered set.

Exercise 3.55. Let $(\mathcal{P}(X), \subset)$ be a partially ordered set, where X is denumerable. Find a chain in $\mathcal{P}(X)$ that has the continuum power.

Exercise 3.56. Let $(X, \leqslant)$ be a partially ordered set. If each chain in X has a lower bound, then there is a minimal element of X. [Hint. Use (3.21.a) and Zorn's lemma.]

References

Halmos, P. (1974). *Measure Theory*. New York, Heidelberg, Berlin: Springer.

Hewitt, E., & Stromberg, K. (1975). *Real and Abstract Analysis*. New York, Heidelberg, Berlin: Springer.

Kelley, J. (1975). *General Topology*. New York, Heidelberg, Berlin: Springer.

2 Topological Preliminaries

This chapter is motivated by the prominent role the topology plays in measure theory, and so in probability theory which may be viewed as a branch with specific problems and applications of measure theory. At the same time, comparing Chapter 2 with Chapter 3 shows the existence of certain similarities, that cannot be ignored, between some topological notions and some measure-theoretic notions, such as between topological space and measurable space, and between continuous function and measurable function. In this chapter we introduce and discuss most topological concepts we use in this book. Section 4 deals with methods of constructing topological spaces. In Section 5 and Section 6 we examine important properties of topological spaces that will be used throughout the following chapters. Section 5 is concerned with such properties that do not involve the notion of a metric, while Section 6 is devoted to metric spaces. The content of Section 6 includes some deep results such as Urysohn's embedding theorem, the Stone-Weierstrass theorem and the Arzelà-Ascoli theorem.

4 Construction of Some Topological Spaces

In this section we introduce the notion of a topological space, and we present three important methods of constructing topological spaces. A usual method of constructing a topological space X involves assigning to each $x \in X$ a family of subsets, satisfying certain properties, that will be the system of overneighborhoods of X. Another method consists in describing the topology generated by an arbitrary family of subsets. A third method we present here shows how one can build new topological spaces from others originally given. This method enables us to define the concepts of a relative topology and of a product topology.

Definitions 4.1. A *topology for* X is any family $\mathcal{T} \subset \mathcal{P}(X)$ satisfying:
(i) $\emptyset \in \mathcal{T}$ and $X \in \mathcal{T}$;
(ii) $U, V \in \mathcal{T}$ implies $U \cap V \in \mathcal{T}$;
(iii) $\mathcal{U} \subset T$ implies $\cup\mathcal{U} \in T$.

A pair $(X, \mathcal{T})$, where X is a set and $\mathcal{T}$ is a topology for X, is called a *topological space*. When no confusion seems possible, we will call X itself a topological space. The elements of $\mathcal{T}$ are called *open sets* (relative to $\mathcal{T}$), and the complements of the open sets are called *closed sets* (relative to $\mathcal{T}$).

Examples 4.2. (a) Let X be a set. The family $\mathcal{T} = \{\emptyset, X\}$ is a topology for X called the *trivial topology for* X.

Analysis and Probability. http://dx.doi.org/10.1016/B978-0-12-401665-1.00002-3

(b) The family $\mathcal{P}(X)$ is a topology for X called the *discrete topology for X*.

(c) Let $\mathcal{T}$ be the family of all sets $U \subset R$ with the property that for each $x \in U$ there is $\varepsilon > 0$ such that $]x-\varepsilon, x+\varepsilon[\subset U$. Then $\mathcal{T}$ is a topology for R called the *usual topology for R*. For $a, b \in \overline{R}$ such that $a < b$, the interval $]a, b[$ is an open set, and for $a, b \in R$ such that $a < b$, the interval $[a, b]$ is a closed set.

(d) Let $\mathcal{T}$ be the family of all sets $U \subset C$ with the property that for each $x \in U$ there is $\varepsilon > 0$ such that $\{y \in C : |x-y| < \varepsilon\} \subset U$. Then $\mathcal{T}$ is a topology for C called the *usual topology for C*. Obviously, the sets $\{y \in C : |x-y| < \varepsilon\}$, $x \in C$, $\varepsilon > 0$, are open relative to this topology. We will always assume that C is endowed with its usual topology unless the contrary is specifically stated.

(e) The family $\mathcal{U} = \{\emptyset,]-\infty, \infty]\} \cup \{]a, \infty] : a \in R\}$ is a topology for $]-\infty, \infty]$.

Remarks 4.3. (a) Let X be a topological space. According to De Morgan's laws (2.17), it follows immediately that any union of a finite number of closed subsets of X is a closed set, and any intersection of closed subsets of X is a closed set; also, X and $\emptyset$ are closed sets.

(b) Any intersection of topologies for X is also a topology for X.

Before introducing and examining other concepts relative to the basic notion of an open set, we will present several methods of constructing a topological space. To do this, we need the next definitions.

Definitions 4.4. Let X be a topological space and $x \in X$. A *neighborhood of x* is any open set U such that $x \in U$. An *overneighborhood of x* is any subset of X that contains a neighborhood of x. The family of all overneighborhoods of x is called the *system of overneighborhoods of x*.

Theorem 4.5. *Let X be a topological space, and let $\mathcal{V}(x)$ be the system of overneighborhoods of $x \in X$. Then:*

(i) *$x \in V$ for each $V \in \mathcal{V}(x)$;*

(ii) *$V_1, V_2 \in \mathcal{V}(x)$ implies $V_1 \cap V_2 \in \mathcal{V}(x)$;*

(iii) *$V \in \mathcal{V}(x)$ and $V \subset W$ imply $W \in \mathcal{V}(x)$;*

(iv) *if $V \in \mathcal{V}(x)$, then there is $U \in \mathcal{V}(x)$ such that $U \subset V$ and $U \in \mathcal{V}(y)$ for each $y \in U$.*

Proof. Properties (i) and (iii) are obvious. If $V_1, V_2 \in \mathcal{V}(x)$, then there are two open sets U_1 and U_2 such that $x \in U_1 \subset V_1$ and $x \in U_2 \subset V_2$. Since $U_1 \cap U_2$ is open and $x \in U_1 \cap U_2 \subset V_1 \cap V_2$, it follows that $V_1 \cap V_2 \in \mathcal{V}(x)$. This proves (ii). Since any open set is an overneighborhood of each of its points, assertion (iv) follows at once from the definition of an overneighborhood of x. □

Theorem (4.5) indicates what properties must satisfy a family of subsets associated to each point x of an arbitrary set X in order to form the system of overneighborhoods of x relative to some topology for X. As (4.7) will show, these properties are sufficient to construct a topology for X. First we will prove the next technical result.

Theorem 4.6. *Let $(X, \mathcal{T}_1)$ and $(X, \mathcal{T}_2)$ be topological spaces, and let $\mathcal{V}_i(x)$ be the system of overneighborhoods of $x \in X$ relative to $\mathcal{T}_i$, $i = 1, 2$. If $\mathcal{V}_1(x) = \mathcal{V}_2(x)$ for any $x \in X$, then $\mathcal{T}_1 = \mathcal{T}_2$.*

Proof. It suffices to prove that $\mathcal{V}_1(x) \subset \mathcal{V}_2(x)$, $x \in X$, implies $\mathcal{T}_1 \subset \mathcal{T}_2$. Let $U \in \mathcal{T}_1$ and $x \in U$. Then $U \in \mathcal{V}_1(x)$ and so $U \in \mathcal{V}_2(x)$. Therefore, there is $V_x \in \mathcal{T}_2$ such that $x \in V_x \subset U$. Since $U = \cup_{x \in U} V_x$, it follows that $U \in \mathcal{T}_2$. □

Theorem 4.7. *Let X be a set, and let $\mathcal{V}$ be a mapping that assigns to each element $x \in X$ a nonempty family $\mathcal{V}(x) \subset \mathcal{P}(X)$ satisfying properties* (i)–(iv) *of* (4.5). Then there exists a unique topology for X such that the system of overneighborhoods of x coincides with $\mathcal{V}(x)$ for any $x \in X$.

Proof. Let $\mathcal{T} = \{U \in \mathcal{P}(X) : U \in \mathcal{V}(x) \text{ for each } x \in U\}$. Let us show that $\mathcal{T}$ is a topology for X. It is trivial that $\emptyset \in \mathcal{T}$. For any $x \in X$, (iii) implies that $X \in \mathcal{V}(x)$, and so $X \in \mathcal{T}$. If $U, V \in \mathcal{T}$ and $x \in U \cap V$, then (ii) implies that $U \cap V \in \mathcal{V}(x)$, and so $U \cap V \in \mathcal{T}$. If $\mathcal{U} \subset \mathcal{T}$ and $x \in \cup\mathcal{U}$, then there is $U \in \mathcal{U}$ such that $U \in \mathcal{V}(x)$. Hence (iii) implies that $\cup\mathcal{U} \in \mathcal{V}(x)$, and so $\cup\mathcal{U} \in \mathcal{T}$. Consequently, $\mathcal{T}$ is a topology for X. Let $\mathcal{O}(x)$ be the system of overneighborhoods of $x \in X$ relative to $\mathcal{T}$. We show now that $\mathcal{O}(x) = \mathcal{V}(x)$ for any $x \in X$. If $V \in \mathcal{O}(x)$, then there is $U \in \mathcal{T}$ such that $x \in U \subset V$, and so (iii) implies $V \in \mathcal{V}(x)$. Conversely, if $V \in \mathcal{V}(x)$, then, by (iv), there is $U \in \mathcal{T}$ such that $x \in U \subset V$, and so $V \in \mathcal{O}(x)$. By virtue of Theorem (4.6), $\mathcal{T}$ is the unique topology for X such that $\mathcal{O}(x) = \mathcal{V}(x)$, $x \in X$. □

Example 4.8. For $\mathbf{x} = (x_1, \ldots, x_n) \in R^n$ and $\varepsilon > 0$, define $U_\varepsilon(\mathbf{x}) = \{(y_1, \ldots, y_n) \in R^n : |x_i - y_i| < \varepsilon \text{ for } i = 1, \ldots, n\}$. Let $\mathcal{V}(\mathbf{x}) = \{V \in \mathcal{P}(R^n) : \text{there is } \varepsilon > 0 \text{ such that } U_\varepsilon(\mathbf{x}) \subset V\}$, $\mathbf{x} \in R^n$. It is easily verified that $\mathcal{V}$ satisfies properties (4.5.i)–(4.5.iv), and so $\mathcal{V}$ defines a topology for R^n called the *usual topology for* R^n. From the proof of Theorem (4.7) it follows that a set $U \subset R^n$ is open relative to this topology if and only if for each $\mathbf{x} \in U$ there is $\varepsilon > 0$ such that $U_\varepsilon(\mathbf{x}) \subset U$. Particularly, the sets $U_\varepsilon(\mathbf{x})$, $\mathbf{x} \in R^n$, $\varepsilon > 0$, are open. We will always suppose that R^n is equipped with its usual topology unless the contrary is specifically stated.

Another class of topological spaces is the one in which the topologies are generated by families of subsets of given sets. The exact definition follows.

Definition 4.9. For an arbitrary set X and a family $\mathcal{S} \subset P(X)$, let $\tau(\mathcal{S})$ enote the intersection of all topologies for X that contain $\mathcal{S}$. Obviously, $\tau(\mathcal{S})$ is a topology for X called the *topology generated by* $\mathcal{S}$. $\tau(\mathcal{S})$ is the smallest topology for X containing $\mathcal{S}$, that is $\tau(\mathcal{S})$ is included in each topology for X that contains $\mathcal{S}$.

The next theorem describes the topology generated by a family of subsets.

Theorem 4.10. *For $\mathcal{S} \subset P(X)$ define $\mathcal{B} = \{\cap\mathcal{A} : \mathcal{A} \subset S \text{ and } \mathcal{A}\ isfinite\}$. Then* $\tau(\mathcal{S}) = \{\cup\mathcal{C} : \mathcal{C} \subset \mathcal{B}\}$.

Proof. Denote $\mathcal{T} = \{\cup\mathcal{C} : \mathcal{C} \subset \mathcal{B}\}$. Since any topology for X containing $\mathcal{S}$ contains the family $\mathcal{T}$, we have $\tau(\mathcal{S}) \supset \mathcal{T}$. Therefore, to complete the proof, it will suffice to show that $\mathcal{T}$ is a topology for X such that $\mathcal{S} \subset \mathcal{T}$. Since $\emptyset \subset \mathcal{B}$, we get $\emptyset = \cup\emptyset \in \mathcal{T}$. Since $\emptyset \subset \mathcal{S}$ and $\emptyset$ is finite, by the convention of (2.18), we have $X = \cap\emptyset \in \mathcal{B}$, and so $X \in \mathcal{T}$. If $\mathcal{C}_i \subset \mathcal{B}$, $i = 1, 2$, then $(\cup\mathcal{C}_1) \cap (\cup\mathcal{C}_2) = \cup\mathcal{C}$, where $\mathcal{C} = \{C_1 \cap C_2 : C_1 \in \mathcal{C}_1 \text{ and } C_2 \in \mathcal{C}_2\} \subset \mathcal{B}$. Hence $\mathcal{T}$ satisfies (4.1.ii). Since the union of the sets of any subfamily of $\mathcal{T}$ is a union of sets belonging to $\mathcal{B}$, property (4.1.iii) is verified. Consequently, $\mathcal{T}$ is a topology for X. Finally, it is clear that $\mathcal{S} \subset \mathcal{B} \subset \mathcal{T}$. □

Example 4.11. Let $\mathcal{S}$ be the family of all intervals of the form $[-\infty, a[$ or $]b, \infty]$, where $a, b \in R$. Then $\tau(\mathcal{S})$ is a topology for $\overline{R}$ called the *usual topology for* $\overline{R}$. By virtue of Theorem (4.10), a subset of $\overline{R}$ is open relative to this topology if and only if it is a union of intervals of the form $[-\infty, a[$, $]a, b[$, or $]b, \infty]$, where $a, b \in R$. If not otherwise stated, we will assume throughout that $\overline{R}$ is endowed with its usual topology.

Definitions 4.12. Let $(X, \mathcal{T})$ be a topological space. A family $\mathcal{B} \subset \mathcal{T}$ is called a *base for* $\mathcal{T}$ if for any $U \in \mathcal{T}$ there is $\mathcal{C} \subset B$ such that $U = \cup \mathcal{C}$. We say that X has a *countable base* if there is a base for $\mathcal{T}$ which is a countable family. A family $\mathcal{S} \subset \mathcal{T}$ is called a *subbase for* $\mathcal{T}$ if the family of all finite intersections of sets in $\mathcal{S}$ is a base for $\mathcal{T}$.

Examples 4.13. (a) Notation is as in (4.10). Then $\mathcal{B}$ is a base for $\tau(\mathcal{S})$, and $\mathcal{S}$ is a subbase for $\tau(\mathcal{S})$.

(b) The family of all open intervals $]a, b[$, where $a, b \in R$, is a base for the *usual topology for* R, and the family of all intervals of the form $]-\infty, a[$ or $]b, \infty[$, where $a, b \in R$, is a subbase for the *usual topology for* R.

Remarks 4.14. (a) Let $(X, \mathcal{T})$ be a topological space, and let $\mathcal{B}$ be a base for $\mathcal{T}$. Every family $\mathcal{A}$ such that $\mathcal{B} \subset \mathcal{A} \subset \mathcal{T}$ is a base for $\mathcal{T}$ too. Hence it follows at once that each base for $\mathcal{T}$ is a subbase for $\mathcal{T}$. Therefore, there are several different bases and subbases for $\mathcal{T}$. (Notice that $\mathcal{B} - \{\emptyset\}$ and $\mathcal{B} \cup \{\emptyset\}$ are distinct bases for $\mathcal{T}$.)

(b) As Example (4.13.b) shows, not every subbase for a topology is a base for that topology. Also, it is easily seen that not every subfamily of a topology is a subbase for that topology, and so not every subfamily of a topology is a base for that topology. Furthermore, not every family of subsets of X can be a base for a topology for X.

In contrast with Theorems (4.7) and (4.10), in what follows we construct new topological spaces from old ones.

Theorem 4.15. *Let $(Y, \mathcal{T})$ be a topological space, and let $f : X \to Y$ be a function. Then the family $f^{-1}(\mathcal{T})$ is a topology for X.*

Proof. Applying (2.19.i) and (2.19.ii), this statement is immediate. □

Example 4.16. Let $(X, \mathcal{T})$ be a topological space and $A \subset X$. Then $A \cap \mathcal{T} = i_A^{-1}(\mathcal{T})$ is a topology for A called the *relative topology on A induced by $\mathcal{T}$*. Thus a set $U \subset A$ is open relative to this topology if and only if $U = A \cap V$ for some $V \in \mathcal{T}$. We will always assume that a subset of a topological space is equipped with the relative topology on it.

Theorem 4.17. *Let $f : X \to Y$ be a function and $\mathcal{S} \subset P(Y)$. Then $f^{-1}(\tau(\mathcal{S})) = \tau(f^{-1}(\mathcal{S}))$.*

Proof. By virtue of Theorem (4.15), $f^{-1}(\tau(\mathcal{S}))$ is a topology for X that contains $f^{-1}(\mathcal{S})$, and so $f^{-1}(\tau(\mathcal{S})) \supset \tau(f^{-1}(\mathcal{S}))$. Now let $\mathcal{U} = \{U \in \tau(\mathcal{S}) : f^{-1}(U) \in \tau(f^{-1}(\mathcal{S}))\}$. Clearly,

$$\mathcal{S} \subset \mathcal{U} \subset \tau(\mathcal{S}). \tag{1}$$

Since $\tau(\mathcal{S})$ and $\tau(f^{-1}(\mathcal{S}))$ are topologies, by making use of (2.19.i) and (2.19.ii), it is easily verified that $\mathcal{U}$ is a topology for Y. From (1) we get $\tau(\mathcal{S}) = \mathcal{U}$, and so $f^{-1}(\tau(\mathcal{S})) \subset \tau(f^{-1}(\mathcal{S}))$. □

Remark 4.18. It is visible that another proof of Theorem (4.17) may be given on account of Theorems (4.10) and (2.19). We have preferred the above proof, since it emphasizes a type of reasoning which is applicable in other contexts too (see (7.13)).

Definition 4.19. Let $\{(Y_i, \mathcal{T}_i) : i \in I\}$ be an indexed family of topological spaces, and let $f_i : X \to Y_i, i \in I$, be functions. The topology $\tau(\cup_{i\in I} f_i^{-1}(\mathcal{T}_i))$ is a topology for X called the *topology generated by* $\{f_i : i \in I\}$.

Example 4.20. Let I be a nonempty set, and let $\{(X_i, \mathcal{T}_i) : i \in I\}$ be an indexed family of topological spaces. The topology generated by $\{\pi_i : i \in I\}$ (see (2.23)) is a topology for $\prod_{i\in I} X_i$ called the *product topology on* $\prod_{i\in I} X_i$. By virtue of Theorem (4.10), a subset of $\prod_{i\in I} X_i$ is open relative to this topology if and only if it is a union of sets of the form $\prod_{i\in I} U_i$, where $U_i \in \mathcal{T}_i, i \in I$, and $\{i \in I : U_i \neq X_i\}$ is finite. We will always suppose that $\prod_{i\in I} X_i$ is endowed with the product topology unless the contrary is specifically stated.

The following theorem has important applications.

Theorem 4.21. *Let I be a nonempty set, let $\{(Y_i, \mathcal{T}_i) : i \in I\}$ be a family of topological spaces, and let $f : X \to \prod_{i\in I} Y_i$ be a function. Let $\mathcal{T}$ denote the product topology on $\prod_{i\in I} Y_i$. Then $f^{-1}(\mathcal{T})$ coincides with the topology generated by $\{\pi_i \circ f : i \in I\}$, where π_i stands for the projection from $\prod_{i\in I} Y_i$ onto Y_i.*

Proof. On account of (4.17), (2.38), and (2.6), we have

$$f^{-1}(\mathcal{T}) = \tau\left(f^{-1}\left(\bigcup_{i\in I} \pi_i^{-1}(\mathcal{T}_i)\right)\right) = \tau\left(\bigcup_{i\in I} f^{-1}(\pi_i^{-1}(\mathcal{T}_i))\right)$$

$$= \tau\left(\bigcup_{i\in I} (\pi_i \circ f)^{-1}(\mathcal{T}_i)\right). \quad \square$$

Corollary 4.22. *Let $\{(Y_i, \mathcal{T}_i) : i \in I\}$ be a nonempty family of topological spaces, and let $f_i : X \to Y_i, i \in I$, be functions. Let $\mathcal{T}$ denote the product topology on $\prod_{i\in I} Y_i$. Then $(f_i)_{i\in I}^{-1}(\mathcal{T})$ coincides with the topology generated by $\{f_i : i \in I\}$.*

Proof. As noted in (2.25), we have $\pi_j \circ (f_i)_{i\in I} = f_j$ for each $j \in I$, and so the corollary follows from (4.21). □

Lemma 4.23. *Let X be a set, and let $\mathcal{S}_a \subset \mathcal{P}(X)$, $a \in A$. Then $\tau(\cup_{a\in A}\tau(\mathcal{S}_a)) = \tau(\cup_{a\in A}\mathcal{S}_a)$.*

Proof. $\cup_{a\in A}\tau(\mathcal{S}_a) \supset \cup_{a\in A}\mathcal{S}_a$, and so $\tau(\cup_{a\in A}\tau(\mathcal{S}_a)) \supset \tau(\cup_{a\in A}\mathcal{S}_a)$. Conversely, $\cup_{a\in A}\tau(\mathcal{S}_a) \subset \tau(\cup_{a\in A}\mathcal{S}_a)$, and so $\tau(\cup_{a\in A}\tau(\mathcal{S}_a)) \subset \tau(\cup_{a\in A}\mathcal{S}_a)$. □

The following theorem may be viewed as an associativity property of the product topology.

Theorem 4.24. *Let $\{(X_i, \mathcal{T}_i) : i \in I\}$ be a family of topological spaces, and let $\{J_a : a \in A\}$ be a partition of I such that $J_a \neq \emptyset, a \in A$. For each $a \in A$, let $\mathcal{U}_a$ denote the product topology on $\prod_{j \in J_a} X_j$. Let $\mathcal{W}$ be the product topology on $\prod_{a \in A}(\prod_{j \in J_a} X_j)$, and let $\mathcal{T}$ be the product topology on $\prod_{i \in I} X_i$. Then $(\pi_{J_a})_{a \in A}^{-1}(\mathcal{W}) = \mathcal{T}$.*

Proof. Corollary (4.22) shows that $(\pi_{J_a})_{a \in A}^{-1}(\mathcal{W}) = \tau(\cup_{a \in A}\pi_{J_a}^{-1}(\mathcal{U}_a))$. For any $a \in A$, by virtue of (2.26) and (4.22), we get $\pi_{J_a}^{-1}(\mathcal{U}_a) = \tau(\cup_{j \in J_a}\pi_j^{-1}(\mathcal{T}_j))$. Therefore, applying (4.23), we have

$$(\pi_{J_a})_{a \in A}^{-1}(\mathcal{W}) = \tau\left(\bigcup_{a \in A} \tau\left(\bigcup_{j \in J_a} \pi_j^{-1}(\mathcal{T}_j)\right)\right) = \tau\left(\bigcup_{a \in A}\left(\bigcup_{j \in J_a} \pi_j^{-1}(\mathcal{T}_j)\right)\right)$$

$$= \tau\left(\bigcup_{i \in I} \pi_i^{-1}(\mathcal{T}_i)\right) = \mathcal{T}. \qquad \square$$

Exercise 4.25. Let X be a set. Show that the family $\mathcal{T} = \{\emptyset\} \cup \{A \in \mathcal{P}(X) : A^c \text{ is finite}\}$ is a topology for X.

Exercise 4.26. Let $\mathcal{S}$ be the family of all closed intervals $[a, b]$, where $a, b \in R$. Show that the topology for R generated by $\mathcal{S}$ is the *discrete topology for* R.

Exercise 4.27. Prove that there is a countable base for a topology if and only if there is a countable subbase for that topology.

Exercise 4.28. Let $(X, \mathcal{T})$ be a topological space and $\mathcal{S} \subset \mathcal{P}(X)$. Show that $\mathcal{S}$ is a subbase for $\mathcal{T}$ if and only if $\mathcal{T} = \tau(\mathcal{S})$.

Exercise 4.29. Let $X = \{0, 1, 2\}$, and let $\mathcal{A} = \{\{0, 1\}, \{1, 2\}\}$. Show that $\mathcal{A}$ cannot be a base for a topology for X.

Exercise 4.30. Notation is as in (4.16). A set $B \subset A$ is closed with respect to the relative topology on A if and only if $B = A \cap F$, where F is closed with respect to $\mathcal{T}$.

Exercise 4.31. Let $(X, \mathcal{T})$ be a topological space and $B \subset A \subset X$. Prove that the relative topology on B induced by $\mathcal{T}$ coincides with the relative topology on B induced by $A \cap \mathcal{T}$.

Exercise 4.32. Prove that the relative topology on R induced by the *usual topology for* C is the *usual topology for* R.

Exercise 4.33. Show that the relative topology on R induced by the *usual topology for* $\overline{R}$ is the *usual topology for* R.

Exercise 4.34. Let $\alpha, \beta \in R$ be such that $\alpha < \beta$. Use (4.28) and (4.17) to prove that the family of all intervals of the form $[\alpha, a[$ or $]b, \beta]$, where $\alpha < a \leqslant \beta$ and $\alpha \leqslant b < \beta$, is a subbase for the relative topology on $[\alpha, \beta]$ induced by the *usual topology for* R.

Exercise 4.35. Notation is as in (4.20). Let $I = \{1, \dots, n\}$ and let $(X_i, \mathcal{T}_i) = (R, \mathcal{T}), 1 \leqslant i \leqslant n$, where $\mathcal{T}$ is the *usual topology for* R. Show that the product topology on R^n coincides with the *usual topology for* R^n.

Exercise 4.36. Notation is as in (4.20).

(a) Use (2.45.a) and (2.30.b) to prove that every section of an open subset of $\prod_{i \in I} X_i$ is open.

(b) Find a set $A \subset \prod_{i \in I} X_i$ such that every section of A is open, but A is not open. [Hint. Let $(X, \mathcal{T})$ be the topological space in (4.25), where X is infinite, and let $(X_i, \mathcal{T}_i) = (X, \mathcal{T}), i = 1, 2$. Then every section of $\mathbb{D}$ is closed, but the diagonal $\mathbb{D}$ is not closed.]

Exercise 4.37. Let $\{(X_i, \mathcal{T}_i) : i \in I\}$ be a nonempty family of topological spaces, and let $A_i \subset X_i, i \in I$. For each $i \in I$, assume that A_i is equipped with the relative topology on A_i induced by $\mathcal{T}_i$. Use (4.21) to prove that the relative topology on $\prod_{i \in I} A_i$ coincides with the product topology on $\prod_{i \in I} A_i$.

Exercise 4.38. Notation is as in (4.19). Assume that I is countable, and there is a countable base for $\mathcal{T}_i, i \in I$.

(a) Use (4.27), (4.28), (4.17) and (4.23) to show that there is a countable base for the topology generated by $\{f_i : i \in I\}$.

(b) Apply (a) to prove that $\prod_{i \in I} Y_i$ has a countable base.

Exercise 4.39. Let X be a set and $x \in X$. Show that the family $\mathcal{T} = \{A \in \mathcal{P}(X) : A^c \text{ is finite or } x \in A^c\}$ is a topology for X.

5 General Properties of Topological Spaces

This section contains properties of topological spaces that will be needed in the sequel. Only those useful properties which do not involve the notion of a metrizable topological space are included here. The special class of topological spaces in which the topology is derived from a metric will be discussed in the next section.

We begin with the notion of a limit of a sequence in a topological space.

Definitions 5.1. Let X be a topological space, let $\{x_n : n \in N\} \subset X$ and $x \in X$. We say that $\{x_n : n \in N\}$ *converges to* x [*has limit* x] if for each neighborhood U of x there is $n(U) \in N$ such that $n \geqslant n(U)$ implies $x_n \in U$. We will write $\lim_n x_n = x$ [$x_n \to x$] if $\{x_n : n \in N\}$ converges to x; also, we will write $x_n \nrightarrow x$ if $\{x_n : n \in N\}$ does not converge to x.

Let A be a set, and let $f_n : A \to X, n \in N, f : A \to X$ be functions. We say that the sequence $\{f_n : n \in N\}$ *converges to* f [*has limit* f] on $B \subset A$ if $f_n(b) \to f(b)$ for any $b \in B$. We will write $\lim_n f_n = f$ [$f_n \to f$] if $\{f_n : n \in N\}$ converges to f on A.

Let $]a, b[\subset R, t_0 \in [a, b]$, and let $f_t : A \to X, t \in]a, b[, f : A \to X$ be functions. We say that $\{f_t : t \in]a, b[\}$ *converges to* f as $t \to t_0$, and we write $\lim_{t \to t_0} f_t = f$ [$f_t \to f$ as $t \to t_0$], if $f_{t_n} \to f$ whenever $\{t_n : n \in N\} \subset]a, b[- \{t_0\}$ is such that $t_n \to t_0$.

Remarks 5.2. (a) A sequence $\{x_n : n \in N\}$ converges to x if and only if any subsequence of $\{x_n : n \in N\}$ converges to x.

(b) Let $\{\mathbf{x}(n) : n \in N\} = \{(x_1(n), \ldots, x_m(n)) : n \in N\} \subset R^m$ and $\mathbf{x} = (x_1, \ldots, x_m) \in R^m$. From (4.8) it follows that $\mathbf{x}(n) \to \mathbf{x}$ if and only if for each $\varepsilon > 0$ there is $n(\varepsilon) \in N$ such that $n \geqslant n(\varepsilon)$ implies $|x_i(n) - x_i| < \varepsilon, i = 1, \ldots, m$.

(c) Let $(X, \mathcal{T})$ be a topological space, where $\mathcal{T}$ is the *trivial topology for* X, and let $\{x_n : n \in N\} \subset X$. Then $x_n \to x$ for any $x \in X$. This shows that a sequence in a topological space may converge to several different points.

Now we introduce a class of topological spaces in which the limit of a sequence is unique whenever it exists.

Definition 5.3. A topological space is called a *Hausdorff space* if whenever x and y are distinct points of X there exist a neighborhood U of x and a neighborhood V of y such that $U \cap V = \emptyset$.

Example 5.4. The space R^n with its usual topology is a Hausdorff space.

Remark 5.5. Each sequence in a Hausdorff space converges to at most one point.

Theorem 5.6. *Let $(X, \mathcal{T})$ be a topological space, and let $\mathcal{S}$ be a subbase for $\mathcal{T}$. For $\{x_n : n \in N\} \subset X$ and $x \in X$, the following assertions are equivalent:*

(i) $x_n \to x$;

(ii) *whatever $V \in \mathcal{S}$ with $x \in V$ there is $n(V) \in N$ such that $n \geqslant n(V)$ implies $x_n \in V$.*

Proof. It is obvious that (i) implies (ii). Suppose that (ii) holds, and let U be a neighborhood of x. Then there is a finite family $\mathcal{A} \subset \mathcal{S}$ such that $x \in \cap\mathcal{A} \subset U$. For any $V \in \mathcal{A}$, let $n(V) \in N$ be such that $n \geqslant n(V)$ implies $x_n \in V$. Since $\mathcal{A}$ is finite, there is $n(U) \in N$ $(n(U) \geqslant n(V), V \in \mathcal{A})$ such that $n \geqslant n(U)$ implies $x_n \in \cap\mathcal{A} \subset U$. Thus (ii) implies (i). □

Corollary 5.7. *Let $\{(X_i, \mathcal{T}_i) : i \in I\}$ be a nonempty family of topological spaces, and assume that $\prod_{i\in I} X_i$ is equipped with the product topology. For $\{x_i(n)_{i\in I} : n \in N\} \subset \prod_{i\in I} X_i$ and $(x_i)_{i\in I} \in \prod_{i\in I} X_i$, the following assertions are equivalent:*

(i) $(x_i(n))_{i\in I} \to (x_i)_{i\in I}$;

(ii) $x_i(n) \to x_i$ *for each* $i \in I$.

Proof. Assume that (i) holds, and consider $U_j \in \mathcal{T}_j$ such that $x_j \in U_j$. Since $\pi_j^{-1}(U_j)$ is a neighborhood of $(x_i)_{i\in I}$, there is $n(U_j) \in N$ such that $n \geqslant n(U_j)$ implies $(x_i(n))_{i\in I} \in \pi_j^{-1}(U_j)$. Therefore, $n \geqslant n(U_j)$ implies $x_j(n) \in U_j$. Thus (ii) follows from (i). Suppose now that (ii) holds. For any $j \in I$, and any $U_j \in \mathcal{T}_j$ such that $(x_i)_{i\in I} \in \pi_j^{-1}(U_j)$, we have $x_j \in U_j$. Consequently, there exists $n(U_j) \in N$ such that $n \geqslant n(U_j)$ implies $x_j(n) \in U_j$, and so $(x_i(n))_{i\in I} \in \pi_j^{-1}(U_j)$. Since $\cup_{i\in I}\pi_i^{-1}(\mathcal{T})$ is a subbase for the product topology on $\prod_{i\in I} X_i$, applying (5.6), we see that (ii) implies (i). □

In what follows we describe various important results which are related to the notion of a limit of a sequence in $\overline{R}$. The reader is assumed to know that every bounded[1] monotone sequence in R has a limit in R. From this it follows at once that every monotone sequence in $\overline{R}$ is convergent.

Theorem 5.8. *If $\leqslant$ is the usual ordering on $\overline{R}$, then there exist* $\sup A$ *and* $\inf A$ *for any $A \subset \overline{R}$.*

Proof. If $\sup A$ exists for any $A \subset \overline{R}$, then it is easily verified that $-\sup(-A) = \inf A$, where $-A = \{-x : x \in A\}$. Therefore, it suffices to prove that $\sup A$ exists for any $A \subset \overline{R}$.

If $A = \emptyset$ or $A = \{-\infty\}$, then each element of $\overline{R}$ is an upper bound for A, and so $-\infty = \sup A$. If $\emptyset \neq A \subset]-\infty, \infty]$, then ∞ is an upper bound for A. In case ∞ is the only upper bound for A, then clearly $\infty = \sup A$. Thus the case when $\emptyset \neq A \subset R$ and there is an upper bound $b \in R$ for A remains to be analyzed. In this case choose $a \in R$ such that $a < b$ and $[a, b] \cap A \neq \emptyset$. Denote $l = b - a$. We define inductively a sequence of intervals $\{[a_n, b_n] : n \geqslant 1\}$ such that, whatever $n \geqslant 1$,

$$a_n \leqslant a_{n+1} < b_{n+1} \leqslant b_n, \quad b_n - a_n = l/2^{n-1}, \tag{1}$$

where b_n is an upper bound for A, and $[a_n, b_n] \cap A \neq \emptyset$. Put $[a_1, b_1] = [a, b]$. Let $[a_2, b_2]$ be one of the intervals $[a, (a+b)/2]$ and $[(a+b)/2, b]$ such that b_2 is an upper bound for A and $[a_2, b_2] \cap A \neq \emptyset$. Evidently, $a_1 \leqslant a_2 < b_2 \leqslant b_1$ and $b_2 - a_2 = l/2$. If the intervals $[a_1, b_1], \ldots, [a_n, b_n]$ have been specified, then let $[a_{n+1}, b_{n+1}]$ be one of the intervals $[a_n, (a_n + b_n)/2]$ and $[(a_n + b_n)/2, b_n]$ such that b_{n+1} is an upper bound for A and $[a_{n+1}, b_{n+1}] \cap A \neq \emptyset$. It is clear that $a_n \leqslant a_{n+1} < b_{n+1} \leqslant b_n$ and $b_{n+1} - a_{n+1} = l/2^n$. By virtue of (1), there is $\alpha \in R$ such that $a_n \to \alpha$ and $b_n \to \alpha$. For each $x \in A$, we have $x \leqslant b_n, n \geqslant 1$, whence $x \leqslant \lim_n b_n = \alpha$. Hence α is an upper bound for A. Now let β be an arbitrary upper bound for A. Since $[a_n, b_n] \cap A \neq \emptyset, n \geqslant 1$, it follows $a_n \leqslant \beta, n \geqslant 1$, and so $\alpha = \lim_n a_n \leqslant \beta$. Consequently, $\alpha = \sup A$. □

Corollary 5.9. *Let X be a set, and let $\leqslant$ be the partial ordering on $\overline{R}^X$ introduced in* (3.19.b). *Then there exist* $\sup \mathfrak{F}$ *and* $\inf \mathfrak{F}$ *for any $\mathfrak{F} \subset \overline{R}^X$.*

Proof. It is easily verified that the function $\sup \mathfrak{F} : X \to \overline{R}$ defined by $(\sup \mathfrak{F})(x) = \sup\{f(x) : f \in \mathfrak{F}\}$ for any $x \in X$ is the supremum of $\mathfrak{F}$. Similarly, the function $\inf \mathfrak{F} : X \to \overline{R}$ defined by $(\inf \mathfrak{F})(x) = \inf\{f(x) : f \in \mathfrak{F}\}$ for any $x \in X$ is the infimum of $\mathfrak{F}$. □

Definitions 5.10. Let $\{x_n : n \in N\} \subset \overline{R}$. The number $\limsup_n x_n = \inf_{n \geqslant 1} (\sup_{k \geqslant n} x_k)$ is called the *limit superior of* $\{x_n : n \in N\}$, and the number $\liminf_n x_n = \sup_{n \geqslant 1}(\inf_{k \geqslant n} x_k)$ is called the *limit inferior of* $\{x_n : n \in N\}$.

Let X be a set, and let $\{f_n : n \in N\} \subset \overline{R}^X$. The function $\limsup_n f_n = \inf_{n \geqslant 1} (\sup_{k \geqslant n} f_k)$ is called the *limit superior of* $\{f_n : n \in N\}$, and the function

[1] A set $A \subset R^n$ is said to be *bounded* if there is $k \in N$ such that $A \subset U_k(\mathbf{0})$, where $U_k(\mathbf{0})$ was defined in (4.8).

$\liminf_n f_n = \sup_{n\geqslant 1}(\inf_{k\geqslant n} f_k)$ is called the *limit inferior of* $\{f_n : n \in N\}$. In accordance with (5.9), we have $(\limsup_n f_n)(x) = \limsup_n f_n(x)$ and $(\liminf_n f_n)(x) = \liminf_n f_n(x)$ for any $x \in X$.

Theorem 5.11. *Let* $\{x_n : n \in N\} \subset \overline{R}$, *and let* $L = \{x \in \overline{R} : \lim_k x_{n_k} = x$ *for some subsequence* $\{x_{n_k} : k \in N\}$ *of* $\{x_n : n \in N\}\}$. *Then we have* $\limsup_n x_n \in L$, $\limsup_n x_n = \sup L$, *and* $\liminf_n x_n \in L$, $\liminf_n x_n = \inf L$.

Proof. We prove only the former assertion, the proof of the latter one being similar. Denote $x = \limsup_n x_n$, and set $y_n = \sup_{k\geqslant n} x_k, n \geqslant 1$, so that $x = \inf_{n\geqslant 1} y_n$. Therefore, for $a < x$, the set $\{n \in N : a < x_n\}$ is infinite, and for $x < b$, the set $\{n \in N : b \leqslant x_n\}$ is finite. If $x = \infty$, then there exists a subsequence $\{x_{n_k} : k \in N\}$ of $\{x_n : n \in N\}$ such that $x_{n_k} > k, k \in N$, and so $x_{n_k} \to \infty$. Actually, $\{n \in N : 1 < x_n\}$ is infinite, and so there is $n_1 \in N$ such that $x_{n_1} > 1$. If $n_1, \ldots, n_k$ have been chosen, then, since $\{n \in N : k+1 < x_n\}$ is infinite, there is $n_{k+1} \in N$; such that $n_{k+1} > n_k$ and $x_{n_{k+1}} > k+1$. Consequently, $x = \infty \in L$, and so $x = \sup L$. If $x \in R$, then $\{n \in N : x - 1/k < x_n < x + 1/k\}$ is infinite for any $k \in N$. As above, we may choose inductively a subsequence $\{x_{n_k} : k \in N\}$ of $\{x_n : n \in N\}$ such that $|x_{n_k} - x| < 1/k, k \in N$, and so $x_{n_k} \to x$. Thus $x \in L$. Since $\{n \in N : b \leqslant x_n\}$ is finite for any $b > x$, it follows immediately that $x = \sup L$. If $x = -\infty$, then, whatever $b \in R$, there is $n(b) \in N$ such that $n \geqslant n(b)$ implies $x_n < b$. Therefore, $x_n \to -\infty$, and so $\{-\infty\} = L$ (see (5.2.a)). □

The set L in (5.11) is called the set of *limit points of* $\{x_n : n \in N\}$.

Corollary 5.12. *A sequence* $\{x_n : n \in N\} \subset \overline{R}$ *has limit* x *if and only if* $x = \limsup_n x_n = \liminf_n x_n$.

Proof. Exercise. □

A useful property of topological spaces that have countable bases is presented in what follows. We begin with some definitions.

Definitions 5.13. Let A be a set. A family of sets $\mathcal{A}$ is called a *cover of* A if $A \subset \cup\mathcal{A}$. If $\mathcal{A}$ and $\mathcal{B}$ are covers of A such that $\mathcal{B} \subset \mathcal{A}$, then $\mathcal{B}$ is called a *subcover of* A.

Let $(X, \mathcal{T})$ be a topological space and $A \subset X$. A cover $\mathcal{A}$ of A is said to be open if $\mathcal{A} \subset \mathcal{T}$.

Theorem 5.14 (Lindelöf). *Let* $(X, \mathcal{T})$ *be a topological space that has a countable base, and let* $A \subset X$. *Then any open cover of* A *contains a countable subcover of* A.

Proof. Let $\mathcal{A}$ be a open cover of A, and let $\mathcal{B}$ be a countable base for $\mathcal{T}$. Since any set in $\mathcal{A}$ is a union of sets in $\mathcal{B}$, there is a cover $\mathcal{C} \subset \mathcal{B}$ of A such that each set in $\mathcal{C}$ is included in some set in $\mathcal{A}$. For every $C \in \mathcal{C}$, we choose $D_C \in \mathcal{A}$ such that $C \subset D_C$. Denote $\mathcal{D} = \{D_C : C \in \mathcal{C}\}$. By virtue of (3.14) and (3.38), the family $\mathcal{D}$ is countable. Since $\mathcal{C} \subset \mathcal{D} \subset \mathcal{A}$, $\mathcal{D}$ is a subcover of A. □

Another interesting property of topological spaces that have countable bases will be proved in (5.20). For this we proceed to new definitions.

Definitions 5.15. Let X be a topological space and $A \subset X$. Let A^- denote the intersection of all closed subsets of X containing A. In accordance with (4.3), A^- is a closed set containing A which is called the *adherence* [*closure*] *of* A. Obviously, A^- is the smallest closed set containing A, and A is closed if and only if $A = A^-$. The set $\partial(A) = A^- \cap (A^c)^-$ is called the *boundary of* A. A point $x \in X$ is said to be *adherent to* A if $A \cap U \neq \emptyset$ for any neighborhood U of x.

Examples 5.16. (a) Let $(X, \mathcal{T})$ be a topological space, where $\mathcal{T}$ is the *discrete topology for* X. Then $A^- = A$, $\partial(A) = \emptyset$, and $\{x : x \text{ is adherent to } A\} = A$ for any $A \subset X$.

(b) Let $I \subset \overline{R}$ be an interval of the form $]a, b[$, $[a, b[$, or $]a, b]$. Then $I^- = [a, b]$. Particularly, $R^- = \overline{R}$.

(c) Relative to the *usual topology for* R, $\partial(]-\infty, a[) = \partial(]-\infty, a]) = \{a\}$ for any $a \in R$.

(d) We assume the reader knows that for each $x \in R$ there exists $\{x_n : n \in N\} \subset Q$ such that $x_n \to x$. From this it follows at once that $\{x \in R : x \text{ is adherent to } Q\} = R$.

Theorem 5.17. *Let X be a topological space and $A \subset X$. Then $A^- = \{x : x$ is adherent to $A\}$*

Proof. Denote $B = \{x : x \text{ is adherent to } A\}$. If $x \in (A^-)^c$, then, since $(A^-)^c$ is a neighborhood of x and $(A^-)^c \cap A = \emptyset$, we have $x \in B^c$. Thus $A^- \supset B$. Conversely, for each $x \in B^c$ we choose a neighborhood U_x of x such that $U_x \subset A^c$. Since $B^c = \cup_{x \in B^c} U_x \subset A^c$, B is a closed set containing A, and so $B \supset A^-$. □

Definitions 5.18. Let X be a topological space. A set $A \subset X$ is said to be *dense in* X if $A^- = X$. We say that X is *separable* if there is a countable subset dense in X.

Examples 5.19. (a) Let $(X, \mathcal{T})$ be a topological space, where X is an uncountable set and $\mathcal{T}$ is the *discrete topology for* X. In accordance with (5.16.a), for each countable set $A \subset X$, we have $A^- = A \neq X$, and so X is not separable.

(b) The space R with its usual topology is separable. Actually, according to (3.41), Q is a denumerable set, and from (5.17) and (5.16.d) it follows that $Q^- = R$. Analogously, using (5.17) and (5.7), it follows that $(Q^n)^- = R^n$. By virtue of (3.42), Q^n is denumerable, and so R^n is separable.

(c) By making use of (3.11), it is easily seen that the topological space $(X, \mathcal{T})$ in (4.25) is separable.

Theorem 5.20. *Any topological space that has a countable base is separable.*

Proof. Let $\mathcal{B}$ be a countable base of the topological space X. Then, in accordance with (4.14.a), the family $\mathcal{B}' = \mathcal{B} - \{\emptyset\}$ is a countable base of X formed from nonempty sets. For each $B \in \mathcal{B}'$ select $x_B \in B$. By virtue of (3.38), the set $A = \{x_B : B \in \mathcal{B}'\}$ is countable. Moreover, $A \cap U \neq \emptyset$ for any open nonempty set U, and so (5.17) shows that A is dense in X. □

Remark 5.21. There exist separable spaces that have no countable bases. For example, let $(X, \mathcal{T})$ be the topological space in (4.25), where X is an uncountable set. As stated in (5.19.c), X is separable. Assume that there is a countable base $\mathcal{B}$ for $\mathcal{T}$. For a fixed point $x \in X$, let $\mathcal{U}_x$ denote the family of all neighborhoods of x, and put $\mathcal{B}_x = \{B \in \mathcal{B} : x \in B\}$. Then it is clear that $\cap\mathcal{B}_x = \cap\mathcal{U}_x = \{x\}$. From this it follows that $(\cap\mathcal{B}_x)^c$ is an uncountable set. On the other hand, in view of (2.17) and (3.17), the set $(\cap\mathcal{B}_x)^c$ is countable. This contradiction shows that X has no countable base.

Now we introduce a very important class of topological spaces.

Definitions 5.22. A topological space $(X, \mathcal{T})$ is said to be *compact* if each open cover of X contains a finite subcover of X. A set $A \subset X$ is said to be *compact* if $(A, A \cap \mathcal{T})$ is compact. By virtue of (4.16), A is compact if and only if every cover of A by sets in $\mathcal{T}$ has a finite subcover of A.

Before examining some useful properties of topological spaces, in (5.24) and (5.25) we present two characterizations of compactness.

Definition 5.23. A family of sets $\mathcal{A}$ has the *finite intersection property* if $\cap\mathcal{B} \neq \emptyset$ for each finite family $\mathcal{B} \subset \mathcal{A}$.

Theorem 5.24. *A topological space X is compact if and only if $\cap\mathcal{F} \neq \emptyset$ for any family $\mathcal{F}$ of closed subsets of X which has the finite intersection property.*

Proof. According to De Morgan's laws (2.17), $\mathcal{A}$ is an open cover of X if and only if $\mathcal{B} = \{A^c : A \in \mathcal{A}\}$ is a family of closed sets such that $\cap\mathcal{B} = \emptyset$. Therefore, X is compact if and only if each family $\mathcal{B}$ of closed sets having the property $\cap\mathcal{B} = \emptyset$ contains a finite subfamily $\mathcal{C}$ such that $\cap\mathcal{C} = \emptyset$, and so if and only if each family $\mathcal{F}$ of closed sets having the finite intersection property is such that $\cap\mathcal{F} \neq \emptyset$. □

Alexander's theorem 5.25. *Let $(X, \mathcal{T})$ be a topological space, and let $\mathcal{S}$ be a subbase for $\mathcal{T}$. Then the following assertions are equivalent:*

(i) *X is compact;*

(ii) *each cover of X by sets in $\mathcal{S}$ contains a finite subcover of X.*

Proof. It is clear that (i) implies (ii). Assume now that (ii) holds and X is not compact. Then the collection A of all open covers of X that do not admit finite subcovers is a nonempty collection which is partially ordered by inclusion. The union of the elements of any chain in A is obviously an element of A. Applying (3.49), Zorn's lemma (3.31) shows that A has a maximal element $\mathcal{A}$. Set $\mathcal{B} = \mathcal{A} \cap \mathcal{S}$. Since $\mathcal{A}$ does not contain finite subcovers of X, it follows that $\mathcal{B}$ does not admit finite subcovers of X. Thus, in view of (ii), $\cup\mathcal{B} \neq X$. For a point $x \in X - (\cup\mathcal{B})$, choose a set A in the cover $\mathcal{A}$ such that $x \in A$. Since $\mathcal{S}$ is a subbase, there are $n \in N$ and $S_1, \ldots, S_n \in \mathcal{S}$ such that $x \in \cap_{i=1}^n S_i \subset A$. Since $x \notin \cup\mathcal{B}$, we have $S_i \notin \mathcal{A}$, $i = 1, \ldots, n$. Hence, $\mathcal{A}$ being a maximal cover, the cover $\mathcal{A} \cup \{S_i\}$ admits a finite subcover of X for each i. Therefore, for any $i = 1, \ldots, n$, there exists a set A_i, which is a union of a finite number of sets in $\mathcal{A}$, such that $A_i \cup S_i = X$. Since $(\cup_{i=1}^n A_i) \cup A \supset (\cup_{i=1}^n A_i) \cup (\cap_{i=1}^n S_i) = X$, it follows that X is covered by a finite number of sets in $\mathcal{A}$. This contradicts the assertion that $\mathcal{A} \in$A. □

We now examine several useful properties of compact spaces.

Theorem 5.26. *Any closed subset of a compact space is compact.*

Proof. Let X be a compact space, let $A \subset X$ be a closed set, and let $\mathcal{A}$ be an open cover of A. Then $\mathcal{A} \cup \{A^c\}$ is an open cover of X. Since X is compact, $\mathcal{A} \cup \{A^c\}$ contains a finite subfamily $\mathcal{B}$ which covers X. Then $\mathcal{B} - \{A^c\}$ is a finite subfamily of $\mathcal{A}$ that covers A, and so A is compact. □

Theorem 5.27. *Any compact subset of a Hausdorff space is closed.*

Proof. Let X be a Hausdorff space, let $A \subset X$ be a compact set, and let $y \in A^c$. Since X is Hausdorff, for each $x \in A$, there exist a neighborhood U_x of x and a neighborhood V_x of y such that

$$U_x \cap V_x = \emptyset. \tag{1}$$

The family $\{U_x : x \in A\}$ is plainly an open cover of A. Since A is compact, there are $n \in N$ and $x_1, \ldots, x_n \in A$ such that $A \subset \cup_{i=1}^n U_{x_i}$. Denote $V(y) = \cap_{i=1}^n V_{x_i}$. Using (1), we get $A \cap V(y) = \emptyset$, and so $V(y)$ is an open set such that $V(y) \subset A^c$. Hence $A^c = \cup_{y \in A^c} V(y)$, and so A is closed. □

Tihonov's theorem 5.28. *If $\{(X_i, \mathcal{T}_i) : i \in I\}$ is a nonempty family of compact topological spaces, then $\prod_{i \in I} X_i$ is a compact topological space.*

Proof. Let $\mathcal{A}$ be a cover of $\prod_{i \in I} X_i$ such that $\mathcal{A} \subset \cup_{i \in I} \pi_i^{-1}(\mathcal{T}_i)$. Since $\cup_{i \in I} \pi_i^{-1}(\mathcal{T}_i)$ is a subbase for the product topology on $\prod_{i \in I} X_i$, applying Alexander's theorem (5.25), it suffices to show that $\mathcal{A}$ contains a finite subcover of $\prod_{i \in I} X_i$. Let $\mathcal{A}_i = \{A \in \mathcal{T}_i : \pi_i^{-1}(A) \in \mathcal{A}\}, i \in I$. Suppose that $\cup \mathcal{A}_i \neq X_i$ for any $i \in I$ and select $x_i \in X_i - (\cup \mathcal{A}_i)$. Then $(x_i)_{i \in I} \notin \cup \mathcal{A}$, that is $\mathcal{A}$ is not a cover of $\prod_{i \in I} X_i$. This contradiction shows that there is $j \in I$ such that $\cup \mathcal{A}_j = X_j$. Since X_j is compact, there are $A_1, \ldots, A_n \in \mathcal{A}_j$ such that $\cup_{k=1}^n A_k = X_j$. Consequently, $\{\pi_j^{-1}(A_1), \ldots, \pi_j^{-1}(A_n)\}$ is a finite subfamily of $\mathcal{A}$ which covers $\prod_{i \in I} X_i$. □

We will use Tihonov's theorem to characterize the compact subsets of R^n. We first prove the next lemma.

Lemma 5.29. *Any closed interval $[a, b]$ in R is compact.*

Proof. Let $\mathcal{A}$ be an open cover of $[a, b]$, and let $c = \sup\{x \in [a, b] : [a, x]$ is covered by a finite subfamily of $\mathcal{A}\}$. Since any open set containing a contains an interval $[a, a + \varepsilon]$ too, it follows that $a < c \leqslant b$. Choose $U \in \mathcal{A}$ such that $c \in U$, and choose $d \in]a, c[$ such that $[d, c] \subset U$. By the definition of supremum, there is a finite family $\mathcal{B} \subset \mathcal{A}$ such that $[a, d] \subset \cup \mathcal{B}$. Therefore, $\mathcal{B} \cup \{U\}$ is a finite subfamily of $\mathcal{A}$ which covers $[a, c]$. Since there is an interval $[c, c + \varepsilon] \subset U$, it follows that $c = b$. Thus $\mathcal{B} \cup \{U\}$ covers $[a, b]$. □

Theorem 5.30 (Heine-Borel-Lebesgue). *Let $A \subset R^n$. Then A is compact if and only if A is closed and bounded.*

Proof. Assume that A is compact. Since R^n is a Hausdorff space, applying (5.27), it follows that A is closed. Obviously, $A \subset \cup_{j \in N} U_j(\mathbf{0}) = R^n$. Since A is compact, there is $k \in N$ such that $A \subset U_k(\mathbf{0})$, that is A is bounded.

Conversely, assume that A is closed and bounded. Since A is bounded, there is $k \in N$ such that $A \subset U_k(\mathbf{0}) \subset [-k, k]^n$. Applying (5.29), using Tihonov's theorem (5.28), and taking into account (4.37), it follows that $[-k, k]^n$ is a compact subset of R^n.

In accordance with (4.30), A is closed with respect to the relative topology on $[-k, k]^n$. Thus, by virtue of (5.26), A is a compact subset of $[-k, k]^n$. In view of (4.31), A is a compact subset of R^n. □

A useful version of compactness will be introduced in what follows. We need the next definitions.

Definitions 5.31. Let X be a topological space and $A \subset X$. A point $x \in X$ is a *limit point of* A if $A \cap (U - \{x\}) \neq \emptyset$ for each neighborhood U of x. Let A^d denote the set of all *limit points of* A. A^d is sometimes called the *derived set of* A. Obviously, each limit point of A is adherent to A, but a point adherent to a set may fail to be a limit point of that set.

Examples 5.32. (a) Let $(X, \mathcal{T})$ be a topological space, where $\mathcal{T}$ is the *discrete topology for* X. Then $A^d = \emptyset$ for any $A \subset X$.

(b) Let X be a Hausdorff space. Then $A^d = \emptyset$ for any finite set $A \subset X$.

(c) Relative to the *usual topology for* R, we have $Z^d = \emptyset$, $\{1/n : n \in N\}^d = \{0\}$, and $I^d = I^-$ for any interval $I \subset R$.

Theorem 5.33. *Let X be a topological space and $A \subset X$. Then A is closed if and only if $A^d \subset A$.*

Proof. If A is closed, then, using (5.17), we have $A^d \subset A^- = A$.

Conversely, assume that $A^d \subset A$. Since $A^- - A \subset A^d \subset A$, it follows that $A^- - A = \emptyset$. Therefore, $A = A^-$, and so A is closed. □

Definitions 5.34. A topological space $(X, \mathcal{T})$ is said to be *Fréchet compact* if $A^d \neq \emptyset$ for any infinite set $A \subset X$. A set $B \subset X$ is said to be *Fréchet compact* if $(B, B \cap \mathcal{T})$ is Fréchet compact. Obviously, B is Fréchet compact if and only if every infinite subset of B has a limit point in B relative to $\mathcal{T}$.

Theorem 5.35. *Every compact topological space is Fréchet compact.*

Proof. Let X be a compact space. Assume that there is an infinite set $A \subset X$ such that $A^d = \emptyset$. Then, according to (5.33), A is closed. Moreover, for each $a \in A$, there exists a neighborhood U_a of a such that $A \cap U_a = \{a\}$. Therefore, $\{U_a : a \in A\} \cup \{A^c\}$ is an open cover of X which contains no finite subcover of X. The contradiction we get shows that X is Fréchet compact. □

The concept of continuity we will now discuss is extremely important.

Definitions 5.36. Let X and Y be topological spaces, and let $f : X \to Y$ be a function. We say that f is *continuous at* $x \in X$ if for each neighborhood V of $f(x)$ there is a neighborhood U of x such that $f(U) \subset V$. We say that f is *continuous on* X if f is continuous at every $x \in X$.

Theorem 5.37. *Let X, Y and f be as in* (5.36), *let $x \in X$, and let $\{x_n : n \in N\} \subset X$ be such that $x_n \to x$. If f is continuous at x, then $f(x_n) \to f(x)$.*

Proof. Consider a neighborhood V of $f(x)$, and choose a neighborhood U of x such that $f(U) \subset V$. Since $x_n \to x$, there is $n(U) \in N$ such that $n \geqslant n(U)$ implies $x_n \in U$. Therefore, $n \geqslant n(U)$ implies $f(x_n) \in f(U) \subset V$. Thus $f(x_n) \to f(x)$. □

Theorem 5.38. *Let X, Y and Z be topological spaces, let $f : X \to Y$ and $g : Y \to Z$ be functions, and let $x \in X$. If f is continuous at x, and g is continuous at $f(x)$, then $g \circ f$ is continuous at x.*

Proof. Let W be a neighborhood of $(g \circ f)(x) = g(f(x))$. Since g is continuous at $f(x)$, there is a neighborhood V of $f(x)$ such that $g(V) \subset W$. Since f is continuous at x, there is a neighborhood U of x such that $f(U) \subset V$. Hence $(g \circ f)(U) = g(f(U)) \subset g(V) \subset W$, and so $g \circ f$ is continuous at x. □

Corollary 5.39. *Let X, Y, Z, f and g be as in* (5.38). *If f is continuous on X, and g is continuous on Y, then $g \circ f$ is continuous on X.*

Theorem 5.40. *Let $(X, \mathcal{T})$ and $(Y, \mathcal{U})$ be topological spaces, and let $f : X \to Y$ be a function. Then f is continuous on X if and only if $f^{-1}(\mathcal{U}) \subset \mathcal{T}$.*

Proof. Assume that f is continuous on X. Let $V \in \mathcal{U}$ and $x \in f^{-1}(V)$. Since f is continuous at x, there is a neighborhood U_x of x such that $f(U_x) \subset V$. Hence, by (2.33.c), $U_x \subset f^{-1}(V)$. Therefore, $f^{-1}(V) = \cup_{x \in f^{-1}(V)} U_x \in \mathcal{T}$. Thus $f^{-1}(\mathcal{U}) \subset \mathcal{T}$.

Conversely, suppose that $f^{-1}(\mathcal{U}) \subset \mathcal{T}$. Then, for any $x \in X$ and any neighborhood V of $f(x)$, $f^{-1}(V)$ is a neighborhood of x. According to (2.33.b), $f(f^{-1}(V)) \subset V$, and so f is continuous at x. □

Corollary 5.41. *Notation is as in* (4.19). *Then the topology generated by $\{f_i : i \in I\}$ is the smallest topology for X relative to which f_i, $i \in I$, are continuous on X.*

Corollary 5.42. *Let $(X, \mathcal{T})$, $(Y, \mathcal{U})$ and f be as in* (5.40), *and let $\mathcal{S}$ be a subbase for $\mathcal{U}$. Then f is continuous on X if and only if $f^{-1}(\mathcal{S}) \subset \mathcal{T}$.*

Proof. If f is continuous on X, then $f^{-1}(\mathcal{S}) \subset f^{-1}(\mathcal{U}) \subset \mathcal{T}$.

Conversely, if $f^{-1}(\mathcal{S}) \subset \mathcal{T}$, then, using (4.28) and (4.17), we have $f^{-1}(\mathcal{U}) = f^{-1}(\tau(\mathcal{S})) = \tau(f^{-1}(\mathcal{S})) \subset \mathcal{T}$, and so f is continuous on X. □

Remarks 5.43. (a) Let $(X, \mathcal{T})$, $(Y, \mathcal{U})$ and f be as in (5.40), and let $A \subset X$. Assume that f is continuous on X. Then the restriction of f to A is continuous on A, where A is endowed with the relative topology $A \cap \mathcal{T}$. Actually, using (2.31), (2.6) and (5.40), we have $(f \circ i_A)^{-1}(\mathcal{U}) = i_A^{-1}(f^{-1}(\mathcal{U})) \subset i_A^{-1}(\mathcal{T})$.

(b) Let $(X, \mathcal{T})$, $(Y, \mathcal{U})$ and f be as in (5.40), and let $f(X) \subset B \subset Y$. Suppose that f is continuous on X. Then the function $g : X \to B$ defined by $g(x) = f(x)$ for any $x \in X$ is continuous on X, where B is equipped with the relative topology $B \cap \mathcal{U}$. Actually, $i_B \circ g = f$, and so $g^{-1}(i_B^{-1}(\mathcal{U})) = f^{-1}(\mathcal{U}) \subset \mathcal{T}$.

Theorem 5.44. *Let $\{(Y_i, \mathcal{U}_i) : i \in I\}$ be a nonempty family of topological spaces, and let $\mathcal{U}$ be the product topology on $\prod_{i \in I} Y_i$. Let $(X, \mathcal{T})$ be a topological space, and let $f : X \to \prod_{i \in I} Y_i$ be a function. Then f is continuous on X if and only if $\pi_i \circ f$, $i \in I$, are continuous on X.*

Proof. According to (5.41), $\mathcal{U}$ is the smallest topology for $\prod_{i\in I} Y_i$ relative to which the projections π_i, $i \in I$, are continuous on $\prod_{i\in I} Y_i$. Consequently, if f is continuous on X, then, applying (5.39), the functions $\pi_i \circ f$, $i \in I$ are continuous on X.

Conversely, assume that $\mathcal{T}$ is a topology for X relative to which $\pi_i \circ f$, $i \in I$, are continuous on X. Then, using (4.21) and (5.41), we have $f^{-1}(\mathcal{U}) \subset \mathcal{T}$, and so f is continuous on X. □

Theorem 5.45. *Let X, Y and f be as in* (5.36). If X is compact, and f is continuous on X, then $f(X)$ is compact.

Proof. Let $\mathcal{B}$ be an open cover of $f(X)$. Then $f^{-1}(\mathcal{B})$ is an open cover of X. Since X is compact, there is a finite family $\mathcal{C} \subset B$ such that $X = \cup f^{-1}(\mathcal{C}) = f^{-1}(\cup\mathcal{C})$. Therefore, $f(X) \subset \cup\mathcal{C}$, and so $f(X)$ is compact. □

Definitions 5.46. A set $A \subset K$ is said to be *bounded* if there exists $M > 0$ such that $|a| \leqslant M$ for any $a \in A$. A function $f : X \to K$ is said to be *bounded* if $f(X)$ is bounded.

Theorem 5.47. *Let X be a compact space, and let $f : X \to R$ be a function which is continuous on X. Then f is bounded, and there exist $a, b \in X$ such that $f(a) = \sup_{x\in X} f(x)$ and $f(b) = \inf_{x\in X} f(x)$.*

Proof. By virtue of (5.45), $f(X)$ is a compact subset of R. Thus, according to (5.30), $f(X)$ is bounded and closed. Therefore, f is bounded. Denote $\alpha = \sup f(X)$ and $\beta = \inf f(X)$. Since $f(X)$ is closed, we have $\alpha, \beta \in f(X)$. Consequently, there are $a, b \in R$ such that $f(a) = \alpha$ and $f(b) = \beta$. □

The following notions play an important role in studying topological spaces.

Definitions 5.48. Let X and Y be topological spaces. A function $f : X \to Y$ is called a *homeomorphism* from X onto Y if f is bijective, f is continuous on X, and the inverse function f^{-1} is continuous on Y. We say that X and Y are *homeomorphic* if there exists a homeomorphism from X onto Y.

Example 5.49. Let $a, b \in R$ be such that $a < b$. Then $]a, b[$ and R are homeomorphic. Actually, it is easily verified that the function $f :]a, b[\to R$ defined by $f(x) = \frac{1}{x-a} + \frac{1}{x-b}$ for any $x \in]a, b[$ is a homeomorphism from $]a, b[$ onto R.

Remarks 5.50. (a) Let X, Y and Z be topological spaces. Then the identity function i_X is a homeomorphism from X onto X. If f is a homeomorphism from X onto Y, then f^{-1} is a homeomorphism from Y onto X. If f is a homeomorphism from X onto Y, and g is a homeomorphism from Y onto Z, then $g \circ f$ is a homeomorphism from X onto Z.

(b) If X and Y are homeomorphic topological spaces, then to each statement concerning X corresponds a statement concerning Y. For example, X has a countable base if and only if Y has a countable base; also, X is separable if and only if Y is separable, and X is compact if and only if Y is compact.

Definition 5.51. Let X and Y be topological spaces. A function $f : X \to Y$ is called a *homeomorphism* from X *into* Y if the function $g : X \to f(X)$ defined by $g(x) = f(x)$ for any $x \in X$ is a homeomorphism from X onto $f(X)$, where $f(X)$ is endowed with the relative topology. According to (5.48) and (5.43.b), f is a homeomorphism from X into Y if f is injective, f is continuous on X, and the inverse of g is continuous on $f(X)$.

Exercise 5.52. Let $(X, \mathcal{T})$ be a topological space, let $\{x_n : n \in N\} \subset A$ and $x \in A$. Prove that $x_n \to x$ relative to $\mathcal{T}$ if and only if $x_n \to x$ relative to $A \cap \mathcal{T}$.

Exercise 5.53. (a) Let $\{x_n : n \in N\}$ be a sequence, and let $\{N_1, \ldots, N_m\}$ be a partition of N such that N_i, $1 \leqslant i \leqslant m$, are denumerable. Show that $\{x_n : n \in N\}$ converges to x if and only if each subsequence $\{x_n : n \in N_i\}$, $1 \leqslant i \leqslant m$, has limit x.

(b) Construct a sequence $\{x_n : n \in N\}$ and a denumerable partition $\{N_i : i \geqslant 1\}$ of N such that N_i, $i \geqslant 1$, are denumerable, each subsequence $\{x_n : n \in N_i\}$ has limit x, but $\{x_n : n \in N\}$ fails to converge to x.

Exercise 5.54. A sequence $\{x_n : n \in N\}$ converges to x if and only if every subsequence $\{x_{n_k} : k \in N\}$ contains a further subsequence which converges to x.

Exercise 5.55. Show that any finite subset of a Hausdorff space is closed.

Exercise 5.56. Let X be a topological space, and let $\mathbb{D}$ be the diagonal of the Cartesian product $X \times X$. Prove that X is a Hausdorff space if and only if $\mathbb{D}$ is closed.

Exercise 5.57. Let $\{X_i : i \in I\}$ be a nonempty family of topological spaces. Show that $\prod_{i \in I} X_i$ is a Hausdorff space if and only if each X_i, $i \in I$, is a Hausdorff space.

Exercise 5.58. Use (3.50) to prove the following assertions.

(a) If $A = \cup_{i \in I} A_i \subset \overline{R}$, then $\sup A = \sup_{i \in I}(\sup A_i)$ and $\inf A = \inf_{i \in I}(\inf A_i)$.

(b) If $\mathfrak{F} = \cup_{i \in I} \mathfrak{F}_i \subset \overline{R}^X$, then $\sup \mathfrak{F} = \sup_{i \in I}(\sup \mathfrak{F}_i)$ and $\inf \mathfrak{F} = \inf_{i \in I}(\inf \mathfrak{F}_i)$.

Exercise 5.59. Let $f_i : X \to \overline{R}, i \in I$, and $h : \Omega \to X$ be functions. Show that $(\sup_{i \in I} f_i) \circ h = \sup_{i \in I}(f_i \circ h)$ and $(\inf_{i \in I} f_i) \circ h = \inf_{i \in I}(f_i \circ h)$.

Exercise 5.60. Let $\{x_n : n \in N\}$, $\{y_n : n \in N\}$ and $\{z_n : n \in N\}$ be sequences in $\overline{R}$ such that $x_n \leqslant y_n, n \in N$, and $\lim_n z_n = 1$. Prove the following.

(a) $\sup_{n \geqslant 1} x_n \leqslant \sup_{n \geqslant 1} y_n$, $\inf_{n \geqslant 1} x_n \leqslant \inf_{n \geqslant 1} y_n$.

(b) $\limsup_n x_n \leqslant \limsup_n y_n$, $\liminf_n x_n \leqslant \liminf_n y_n$.

(c) $\inf_{n \geqslant 1} x_n \leqslant \liminf_n x_n \leqslant \limsup_n x_n \leqslant \sup_{n \geqslant 1} x_n$.

(d) $\limsup_n x_n z_n = \limsup_n x_n$, $\liminf_n x_n z_n = \liminf_n x_n$.

Exercise 5.61. Let $\{x_n : n \in N\} \subset \overline{R}$.

(a) If $\{x_n : n \in N\}$ is nondecreasing, then $\lim_n x_n = \sup_{n \geqslant 1} x_n$.

(b) If $\{x_n : n \in N\}$ is nonincreasing, then $\lim_n x_n = \inf_{n \geqslant 1} x_n$.

(c) $\limsup_n (-x_n) = -\liminf_n x_n$, $\liminf_n (-x_n) = -\limsup_n x_n$.

Exercise 5.62. Let $a_{mn} \in R, m, n \in N$, be such that, for each $m \in N$, the sequence $\{a_{mn} : n \in N\}$ is nondecreasing, and for each $n \in N$, the sequence $\{a_{mn} : m \in N\}$ is nondecreasing. Prove that $\lim_m \lim_n a_{mn} = \lim_n \lim_m a_{mn}$.

Exercise 5.63. Let X be a set, let $\{f_n : n \in N\} \subset \overline{R}^X$, and let $a \in R$. Prove the following.

(a) $\{x \in X : \{f_n(x) : n \in N\} \text{ has limit}\} = \{x \in X : (\limsup_n f_n)(x) = (\liminf_n f_n)(x)\}$.

(b) $\{\limsup_n f_n > a\} \subset \limsup_n\{f_n > a\} \subset \limsup_n\{f_n \geqslant a\} \subset \{\limsup_n f_n \geqslant a\}$.

(c) $\{\liminf_n f_n < a\} \subset \limsup_n\{f_n < a\} \subset \limsup_n\{f_n \leqslant a\} \subset \{\liminf_n f_n \leqslant a\}$. [Use (b) and (5.61.c).]

Exercise 5.64. (a) Let $\mathcal{A}$ be a family of sets. Show that $1_{\cup\mathcal{A}} = \sup_{A\in\mathcal{A}} 1_A$ and $1_{\cap\mathcal{A}} = \inf_{A\in\mathcal{A}} 1_A$.

(b) Let $\{A_n : n \geqslant 1\}$ be a sequence of sets. Prove that $1_{\liminf_n A_n} = \liminf_n 1_{A_n}$ and $1_{\limsup_n A_n} = \limsup_n 1_{A_n}$.

Exercise 5.65. Let $(X, \mathcal{T})$ be a topological space and $B \subset A \subset X$. Use (4.30) to show that the adherence of B relative to $A \cap \mathcal{T}$ is the intersection of A and the adherence of B relative to $\mathcal{T}$.

Exercise 5.66. For subsets A and B of a topological space, prove:

(a) $(A^-)^- = A^-$;
(b) if $A \subset B$, then $A^- \subset B^-$;
(c) $(A \cup B)^- = A^- \cup B^-$;
(d) $(A \cap B)^- \subset A^- \cap B^-$;
(e) $\partial(\partial(A)) = \emptyset$;
(f) $\partial(A \cup B) \subset \partial(A) \cup \partial(B)$;
(g) $\partial(A \cap B) \subset \partial(A) \cup \partial(B)$.

Exercise: Interior of a set 5.67. Let X be a topological space and $A \subset X$. The union of all open subsets of A, written A°, is called the *interior of* A. For $A, B \subset X$, prove:

(a) A° is the largest open subset of A, and A is open if and only if $A = A^\circ$;
(b) $(A \cup B)^\circ \supset A^\circ \cup B^\circ$;
(c) $(A \cap B)^\circ = A^\circ \cap B^\circ$;
(d) $(A^\circ)^c = (A^c)^-$;
(e) $\partial(A) = A^- - A^\circ$.

Exercise 5.68. Let X and Y be topological spaces, and let $A \subset X$, $B \subset Y$. Prove the following.

(a) $(A \times B)^- = A^- \times B^-$.
(b) $\partial(A \times B) = (\partial(A) \times B^-) \cup (A^- \times \partial(B))$.
(c) $(A \times B)^\circ = A^\circ \times B^\circ$.

Exercise 5.69. Use (3.14) and (3.12) to show that a subset A of R is dense in R whenever A^c is countable.

Exercise 5.70. Let $U \neq \emptyset$ be an open subset of R. Prove that there exists a unique partition of U formed from open intervals, and show that this partition is countable. The following steps may be helpful.

(a) For $x \in U$, put $a_x = \inf\{a :]a, x] \subset U\}$ and $b_x = \sup\{b : [x, b[\subset U\}$. Then $]a_x, b_x[\subset U$.

(b) For $x \in U$, $a_x \notin U$ and $b_x \notin U$.

(c) Let $\mathcal{I} = \{]a_x, b_x[: x \in U\}$. Then any two intervals in $\mathcal{I}$ are either identical or disjoint. [Use (b).]

(d) $\cup\mathcal{I} = U$.

(e) If $\mathcal{J}$ is a partition of U formed from open intervals, then $\mathcal{I} = \mathcal{J}$.

(f) $\mathcal{I}$ is countable. [Use (5.16.d) to assign to each $I \in \mathcal{J}$ a point $x_I \in Q \cap I$. Show that the function $f : \mathcal{I} \to Q$ defined by $f(I) = x_I$, $I \in \mathcal{I}$, is injective.]

Exercise 5.71. Let I be a nonempty set, and let $\{X_i : i \in I\}$ be a family of topological spaces. Prove the following statements.

(a) If $\prod_{i \in I} X_i$ is separable, then each X_i, $i \in I$, is separable.

(b) If I is countable, and X_i is separable for every $i \in I$, then $\prod_{i \in I} X_i$ is separable. [Hints. For $i \in I$, let A_i be a countable subset dense in X_i. Fix a point $(y_i)_{i \in I} \in \prod_{i \in I} X_i$. Then $\{(x_i)_{i \in I}$: there is a finite set $J \subset I$ such that $x_j \in A_j$, $j \in J$, and $x_i = y_i$, $i \notin J\}$ is a countable set ((3.42), (3.40) and (3.17)) dense in $\prod_{i \in I} X_i$.]

Exercise 5.72. If A and B are compact subsets of a topological space, then $A \cup B$ is compact.

Exercise 5.73. Use (5.24) to show that R is not a compact space.

Exercise 5.74. Use (4.11) and (5.29) to prove that the space $\overline{R}$ with its usual topology is compact.

Exercise 5.75. Let X be a topological space, and let $f : X \to R$ be a continuous function on X. Prove the following.

(a) The function αf is continuous on X for any $\alpha \in R$.

(b) The function $|f|$ is continuous on X.

(c) If $f(x) \neq 0$ for each $x \in X$, then the function $\frac{1}{f}$ is continuous on X.

(d) For any $a \in R$, the sets $\{x : f(x) < a\}$ and $\{x : f(x) > a\}$ are open.

Exercise 5.76. Let X be a topological space, and let $f : X \to R$ and $g : X \to R$ be continuous functions on X.

(a) Show that the functions $f + g$ and fg are continuous on X.

(b) Prove that $f \vee g = (f + g + |f - g|)/2$ and $f \wedge g = (f + g - |f - g|)/2$, and infer that the functions $f \vee g$ and $f \wedge g$ are continuous on X.

(c) Show that the set $\{x : f(x) < g(x)\}$ is open.

Exercise 5.77. Prove the following.

(a) The function $f : R^2 \to R$ defined by $f(x, y) = x + y$ for any $x, y \in R$ is continuous on R^2.

(b) The function $g : R^2 \to R$ defined by $g(x, y) = x - y$ for any $x, y \in R$ is continuous on R^2.

(c) The function $h : R^2 \to R$ defined by $h(x, y) = xy$ for any $x, y \in R$ is continuous on R^2.

Exercise 5.78. Notation is as in (5.40). Then f is continuous on X if and only if $f^{-1}(B)$ is closed whenever B is a closed subset of Y.

Exercise 5.79. Let $\{X_i : i \in I\}$ be a nonempty family of topological spaces, let Y be a topological space, and let $f : \prod_{i \in I} X_i \to Y$ be a function.

(a) If f is continuous on $\prod_{i \in I} X_i$, then every section of f is continuous. [Use (5.40), (2.45.f) and (3.46.a).]

(b) Show that the converse of the statement (a) is not in general true.

Exercise 5.80. Let X be a topological space, let $f : X \to Y$ be a function, let $\{Z_i : i \in I\}$ be a family of topological spaces, and let $g_i : Y \to Z_i$, $i \in I$, be functions. Assume Y is endowed with the topology generated by the family $\{g_i : i \in I\}$. Prove that f is continuous on X if and only if $g_i \circ f, i \in I$, are continuous on X.

Exercise 5.81. Let X and Y be topological spaces, and let $f : X \to Y$ and $g : X \to Y$ be continuous functions on X. Use (5.56), (5.80) and (5.78) to show that $\{x : f(x) = g(x)\}$ is closed whenever Y is a Hausdorff space.

Exercise 5.82. Let X and Y be topological spaces, let $f : X \to Y$ be a continuous function on X, let A be a subset dense in X, and let B be a closed subset of Y. If $f(A) \subset B$, then $f(X) \subset B$.

Exercise 5.83. For $f \in K^X$, the number $\|f\|_u = \sup_{x \in X} |f(x)|$ is called the *uniform norm of f*. Let $\alpha \in K$ and $f, g \in K^X$. Prove the following.

(a) $\|0\|_u = 0$ and $\|f\|_u > 0$ if $f \neq 0$.
(b) $\|\alpha f\|_u = |\alpha| \, \|f\|_u$.
(c) $\|f + g\|_u \leqslant \|f\|_u + \|g\|_u$.
(d) $\|fg\|_u \leqslant \|f\|_u \, \|g\|_u$.
(e) f is bounded if and only if $\|f\|_u < \infty$.
(f) For $f \in C^X$, f is bounded if and only if Re f and Im f are bounded.

Exercise 5.84. Let $x \in C$ and $r > 0$. Use (5.29) and (5.45) to show that the *circle* $\{y \in C : |x - y| = r\}$ is a compact subset of C.

Exercise 5.85. (a) Show that R^2 and C are homeomorphic topological spaces.
(b) Let X be a topological space, and let $f : X \to C$ be a function. Prove that f is continuous on X if and only if Re f and Im f are continuous on X.

Exercise 5.86. Let $(X, \mathcal{T})$ and $(Y, \mathcal{U})$ be topological spaces, and let $f : X \to Y$ be a homeomorphism from X into Y. Show that $f^{-1}(\mathcal{U}) = \mathcal{T}$.

Exercise 5.87. Let Y and Z be topological spaces, let $g : Y \to Z$ be a homeomorphism from Y into Z, let X be a set, and let $f_n : X \to Y, n \in N$, $f : X \to Y$ be functions. Prove that $f_n \to f$ if and only if $g \circ f_n \to g \circ f$.

Exercise 5.88. Let X be a set, let $\mathcal{T} = \{\emptyset\} \cup \{A \in \mathcal{P}(X) : A^c \text{ is countable }\}$, and let $\{x_n : n \in N\} \subset X$ and $x \in X$. Prove the following.

(a) $\mathcal{T}$ is a topology for X.
(b) $x_n \to x$ relative to $\mathcal{T}$ if and only if $x_n \to x$ relative to $\mathcal{P}(X)$.

Exercise 5.89. Let $(X, \mathcal{T})$ be a topological space, and let $\mathcal{T}^s$ denote the family of all subsets U of X with the property that $x \in U$ and $x_n \to x$ relative to $\mathcal{T}$ imply that there is $n(U) \in N$ such that $x_n \in U$ whenever $n \geqslant n(U)$. Show that $\mathcal{T}^s$ is a topology for X.

Exercise 5.90. Let X be a topological space and $A \subset X$. Show that $\{x : 1_A \text{ is not continuous at } x\} = \partial(A)$.

Exercise 5.91. Let $x \in]0, 1[\cap Q$. Find a sequence $\{x_n : n \in N\} \subset]0, 1[$ such that $\lim_n \frac{1}{n} \sum_{i=1}^n x_i = \lim_n \frac{1}{n} \sum_{i=1}^n x_i^i = x$.

Exercise 5.92. Define recursively the sequence of functions $\{P_n : n \geqslant 1\} \subset R^R$ by $P_1(x) = x$ for any $x \in R$, and $P_{2n}(x) = (x+1)P_{2n-1}(x) + x$, $P_{2n+1}(x) = (x+1)P_{2n}(x)$, $n \geqslant 1$, for any $x \in R$. For $n \geqslant 1$, let x_{2n-1} be the least real root of the equation $P_{2n-1}(x) + 1 = 0$, and let x_{2n} be the least real root of the equation $P_{2n}(x) = 0$. Show that the sequence $\{x_n : n \geqslant 1\}$ is convergent. [Hint. Prove that $\{x_n : n \geqslant 1\}$ is bounded and monotone.]

Exercise 5.93. Let $\{x_n : n \in N\} \subset [0, \infty]$ and $\{y_n : n \in N\} \subset [0, \infty]$. Prove the following.

(a) $\liminf_n x_n + \liminf_n y_n \leqslant \liminf_n (x_n + y_n) \leqslant \limsup_n (x_n + y_n) \leqslant \limsup_n x_n + \limsup_n y_n$.

(b) $\lim_n (x_n + y_n) \leqslant \liminf_n x_n + \limsup_n y_n$ whenever $\{x_n + y_n : n \in N\}$ is convergent.

Exercise: One-point compactification 5.94. Let $(X, \mathcal{T})$ be a topological space, and let $X^* = X \cup \{\delta\}$, where $\delta \notin X$. Define $\mathcal{T}^* = \{A \subset X^* : A \in \mathcal{T}$ or $X^* - A \in \mathcal{T}$ and $X^* - A$ is compact$\}$. Prove the following.

(a) $\mathcal{T}^*$ is a topology for X^*.

(b) $\mathcal{T}$ is the relative topology on X induced by $\mathcal{T}^*$.

(c) $(X^*, \mathcal{T}^*)$ is compact.

Exercise: Lower semicontinuous function 5.95. Let $(X, \mathcal{T})$ be a topological space, and let $f : X \to]-\infty, \infty]$ be a function. We say that f is *lower semicontinuous at* $x \in X$ if for every $a < f(x)$ there is a neighborhood U of x such that $f(U) \subset]a, \infty]$. f is called *lower semicontinuous on* X if it is lower semicontinuous at any $x \in X$. Prove the following.

(a) If f is continuous at x, then f is lower semicontinuous at x.

(b) If $f(X) \subset R$, and both f and $-f$ are lower semicontinuous at x, then f is continuous at x.

(c) If f is lower semicontinuous at x, then αf is lower semicontinuous at x for any $\alpha > 0$.

(d) For $U \subset X$, 1_U is lower semicontinuous on X if and only if $U \in \mathcal{T}$.

(e) f is lower semicontinuous on X if and only if $f^{-1}(\mathcal{U}) \subset \mathcal{T}$, where $\mathcal{U}$ is as in (4.2.e).

(f) If $f : X \to]-\infty, \infty]$ and $g : X \to]-\infty, \infty]$ are lower semicontinuous on X, then $f + g$ and $f \wedge g$ are lower semicontinuous on X.

(g) If $\{f_i : i \in I\} \subset]-\infty, \infty]^X$ is a family of lower semicontinuous functions on X, then $\sup_{i \in I} f_i$ is lower semicontinuous on X.

(h) If $f : X \to]a, b[$ is lower semicontinuous on X, and $h :]a, b[\to]c, d[$ is strictly increasing and surjective, then $h \circ f$ is lower semicontinuous on X.

Exercise 5.96. For $i \in I \neq \emptyset$, let X_i and Y_i be topological spaces, and let $h_i : X_i \to Y_i$ be a homeomorphism from X_i into Y_i. Define $h : \prod_{i \in I} X_i \to \prod_{i \in I} Y_i$ by $h((x_i)_{i \in I}) = (h_i(x_i))_{i \in I}$. Use (5.80) and (4.37) to show that h is a homeomorphism from $\prod_{i \in I} X_i$ into $\prod_{i \in I} Y_i$.

Exercise 5.97. Let $\{b_n : n \in N\} \subset]0, \infty[$ be such that $b_n \to \infty$and $b_{n+1}/b_n \to 1$. Prove the following.

(a) If $c \in]0, 1[$, then there exists a subsequence $\{m_n : n \in N\} \subset N$ such that $m_n \to \infty$, $n - m_n \to \infty$ and $b_{m_n}/b_n \to c$. [Hint. Let $n_c \in N$ be such that $b_{n-1}/b_n \geqslant c$, $n \geqslant n_c$. For $n \geqslant n_c$, define $m_n = \inf\{m < n : b_m/b_n \geqslant c\}$.]

(b) If $c \in]1, \infty[$, then there exists a subsequence $\{m_n : n \in N\} \subset N$ such that $m_n - n \to \infty$ and $b_{m_n}/b_n \to c$. [Define $m_n = \inf\{m < n : b_m/b_n \geqslant c\}, n \in N$.]

Exercise 5.98 (Dini). Let $\{d_n : n \geqslant 1\} \subset]0, \infty[$ be such that $\sum_{n\geqslant 1} d_n = \infty$, set $D_n = \sum_{k=1}^{n} d_k$, $n \geqslant 1$, and let $\delta > 0$. Prove the following.

(a) $\sum_{n\geqslant 1} \frac{d_n}{D_n} = \infty$.

(b) $\sum_{n\geqslant 1} \frac{d_n}{D_n^{1+\delta}} < \infty$. [Show that $\frac{d_n}{D_n^{1+\delta}} \leqslant \frac{1}{\delta}(\frac{1}{D_{n-1}^{\delta}} - \frac{1}{D_n^{\delta}})$, $n > 1$.]

(c) If $D_{n_0} > 1$, then $\sum_{n\geqslant n_0} \frac{d_n}{D_n(\log D_n)^{1+\delta}} < \infty$. [Show that $\frac{d_n}{D_n(\log D_n)^{1+\delta}} \leqslant \frac{1}{\delta}(\frac{1}{(\log D_{n-1})^{\delta}} - \frac{1}{(\log D_n)^{\delta}})$, $n > n_0$.]

6 Metric Spaces

In Section 5 we have already exhibited several properties of topological spaces. In the present section we extend the study of topological notions by introducing a special class of topological spaces in which the topology is determined by means of a metric. We examine here basic properties of metric spaces which we will encounter in various guises throughout the rest of the book, and we prove some deep results, such as Urysohn's embedding theorem, the Stone-Weierstrass theorem and the Arzelà-Ascoli theorem.

Definitions 6.1. A *metric* [*distance-function*] for X is any function $d : X \times X \to R$ such that for $x, y, z \in X$:

(i) $d(x, y) \geqslant 0$;

(ii) $d(x, y) = 0$ if and only if $x = y$;

(iii) $d(x, y) = d(y, x)$;

(iv) $d(x, z) \leqslant d(x, y) + d(y, z)$ (the *triangle inequality*).

A pair (X, d), where X is a set and d is a metric for X, is called a *metric space*. When confusion appears impossible, we will call X itself a metric space.

Examples 6.2. (a) Let X be a set. The function $d : X \times X \to R$ defined by

$$d(x, y) = \begin{cases} 0 \text{ if } x = y \\ 1 \text{ if } x \neq y \end{cases}$$

is a metric for X called the *discrete metric for* X.

(b) For $x, y \in C$, define $d(x, y) = |x - y|$. Then d is a metric for C called the *usual metric for* C.

(c) For $\mathbf{x} = (x_1, \ldots, x_n) \in R^n$ and $\mathbf{y} = (y_1, \ldots, y_n) \in R^n$, define $d(\mathbf{x}, \mathbf{y}) = (\sum_{i=1}^{n}(x_i - y_i)^2)^{1/2}$. Then d is a metric for R^n called the *Euclidean metric for* R^n. (The triangle inequality (6.1.iv) is a special case of Minkowski's inequality (13.7).)

(d) For $\mathbf{x}, \mathbf{y} \in R^n$, define $d(\mathbf{x}, \mathbf{y}) = \sup_{1\leqslant i\leqslant n} |x_i - y_i|$. Then (R^n, d) is a metric space.

(e) Let X be a set, and let $\mathfrak{B}(X)$ denote the set of all bounded functions from X into K. For $f, g \in \mathfrak{B}(X)$, define $d(f, g) = \|f - g\|_u$. Then d is a metric for $\mathfrak{B}(X)$ called the *uniform metric for* $\mathfrak{B}(X)$. Unless otherwise stated, we will always assume that $\mathfrak{B}(X)$ is equipped with the uniform metric.

Definitions 6.3. Let (X, d) be a metric space. For $x \in X$ and $r > 0$, the set $S(x, r) = \{y : d(x, y) < r\}$ is called the *open ball* [*sphere*] *with center* x *and radius* r, and the set $\overline{S}(x, r) = \{y : d(x, y) \leqslant r\}$ is called the *closed ball* [*sphere*] *with center* x *and radius* r. We will write $S_d(x, r)$ and $\overline{S}_d(x, r)$ wherever we will want to emphasize that $S(x, r)$ and $\overline{S}(x, r)$ depend on d.

Examples 6.4. (a) Notation is as in (6.2.a). Then $S(x, 1) = \{x\}$, and $\overline{S}(x, 1) = X$ for any $x \in X$.

(b) Notation is as in (6.2.c). For $n = 1$, whatever $x \in R$ and $r > 0$, $S(x, r) =]x - r, x + r[$, and $\overline{S}(x, r) = [x - r, x + r]$.

(c) Notation is as in (6.2.d). For $\mathbf{x} = (x_1, \ldots, x_n) \in R^n$ and $r > 0$, $S(\mathbf{x}, r) = U_r(\mathbf{x})$ and $\overline{S}(\mathbf{x}, r)$ is the *cube* $[x_1 - r, x_1 + r] \times \cdots \times [x_n - r, x_n + r]$.

Definition 6.5. Let (X, d) be a metric space, and let $\mathcal{S}_d = \{S(x, r) : x \in X, r > 0\}$. The topology $\mathcal{T}_d = \tau(\mathcal{S}_d)$ is called the *topology generated by* d. We will always suppose that X is endowed with the topology $\mathcal{T}_d$, unless we make some explicit statement to the contrary.

Remarks 6.6. (a) Each metric space is a Hausdorff space.

(b) Notation is as in (6.5). Then the following are equivalent:

(i) $U \in \mathcal{T}_d$;

(ii) For each $x \in U$ there is $\varepsilon > 0$ such that $S(x, \varepsilon) \subset U$.

Actually, it is clear that (ii) implies (i). Conversely, let $U \in \mathcal{T}_d$ and $x \in U$. In view of (4.10), there are $x_1, \ldots, x_n \in X$ and $r_1 > 0, \ldots, r_n > 0$ such that $x \in \bigcap_{i=1}^n S(x_i, r_i) \subset U$. Thus $d(x_i, x) < r_i$, $1 \leqslant i \leqslant n$. Choose $\varepsilon = \inf_{1 \leqslant i \leqslant n}(r_i - d(x_i, x))$. For $y \in S(x, \varepsilon)$ and every $i = 1, \ldots, n$, we have $d(x_i, y) \leqslant d(x_i, x) + d(x, y) < d(x_i, x) + \varepsilon \leqslant r_i$, so that $y \in S(x_i, r_i)$. Consequently, $S(x, \varepsilon) \subset \bigcap_{i=1}^n S(x_i, r_i) \subset U$.

(c) Let (X, d_1) and (X, d_2) be metric spaces. If for any $x \in X$ and $\varepsilon > 0$ there is $\delta > 0$ such that $S_{d_1}(x, \varepsilon) \supset S_{d_2}(x, \delta)$, then, applying (b), it follows that $\mathcal{T}_{d_1} \subset \mathcal{T}_{d_2}$. Therefore $\mathcal{T}_{d_1} = \mathcal{T}_{d_2}$ if and only if for each $x \in X$ and $\varepsilon > 0$ there is $\delta > 0$ such that $S_{d_1}(x, \varepsilon) \supset S_{d_2}(x, \delta)$ and $S_{d_2}(x, \varepsilon) \supset S_{d_1}(x, \delta)$. This happens whenever $cd_1 \leqslant d_2 \leqslant Cd_1$ for some $c, C \in]0, \infty[$.

Examples 6.7. (a) Notation is as in (6.2.a). Then $\mathcal{T}_d$ is the *discrete topology for* X.

(b) The usual metric for C generates the *usual topology for* C.

(c) Let d be the Euclidean metric for R^n. Then, for any $\mathbf{x} \in R^n$ and $r > 0$, we have $U_{r/\sqrt{n}}(\mathbf{x}) \subset S_d(\mathbf{x}, r) \subset U_r(\mathbf{x})$. Thus, using (6.6.c), (6.6.b), (6.4.c) and (4.8), it follows that both metrics introduced in (6.2.c) and (6.2.d) generate the *usual topology for* R^n.

Definition 6.8. A topological space $(X, \mathcal{T})$ is said to be metrizable if there exists a metric d for X such that $\mathcal{T} = \mathcal{T}_d$.

As (6.6.a) indicates, not every topological space is metrizable. Theorems (6.9) and (6.11) below show two simple examples of metrizable topological spaces.

Theorem 6.9. *Let (X, d) be a metric space and $A \subset X$. Then $(A, A \cap \mathcal{T}_d)$ is metrizable.*

Proof. Let d_A be the restriction of d to $A \times A$. Then d_A is a metric for A. Moreover, for any $x \in A$ and $r > 0$,

$$S_{d_A}(x, r) = A \cap S_d(x, r). \tag{1}$$

Using (4.16), (6.6.b) and (1), it follows that the relative topology on A induced by $\mathcal{T}_d$ coincides with the topology generated by d_A. □

Notation being as in (6.9), we will always assume that A is endowed with the metric d_A.

Lemma 6.10. *Let (X, d) be a metric space and let $d' = 1 \wedge d$. Then d' is a metric for X such that $\mathcal{T}_d = \mathcal{T}_{d'}$.*

Proof. Properties (6.1.i)–(6.1.iii) are obvious for d'. Further, if a, b, c are nonnegative numbers such that $a \leqslant b + c$, then $1 \wedge a \leqslant (1 \wedge b) + (1 \wedge c)$. From this it follows immediately that d' satisfies the triangle inequality. Thus d' is a metric for X. Now, for each $x \in X$, we have

$$S_{d'}(X, r) = \begin{cases} S_d(x, r) & \text{if } r \leqslant 1 \\ X & \text{if } r > 1 \end{cases}.$$

Therefore, by virtue of (6.6.c), $\mathcal{T}_d = \mathcal{T}_{d'}$. □

Theorem 6.11. *Let I be a nonempty countable set, and let $\{(X_i, d_i) : i \in I\}$ be a family of metric spaces. Then $\prod_{i \in I} X_i$ with the product topology on it is metrizable.*

Proof. We prove this theorem in case I is denumerable. The case when I is finite is simpler and is left to the reader. Without any loss of generality, assume that $I = N$. Set $d'_n = 1 \wedge d_n$, $n \in N$. For $\mathbf{x} = (x_n)_{n \in N} \in \prod_{n \in N} X_n$ and $\mathbf{y} = (y_n)_{n \in N} \in \prod_{n \in N} X_n$, define $d(\mathbf{x}, \mathbf{y}) = \sup_{n \in N} d'_n(x_n, y_n)/n$. Since d'_n is a metric for X_n, $n \in N$, it is easily verified that d is a metric for $\prod_{n \in N} X_n$. For any $r > 0$, and $n > 1/r$, we have $d'_n(x_n, y_n)/n \leqslant 1/n < r$. Thus, for each $\mathbf{x} \in \prod_{n \in N} X_n$ and $r > 0$, $\{\mathbf{y} : d(\mathbf{x}, \mathbf{y}) < r\} = \{y : d'_n(x_n, y_n) < rn \text{ for } n \leqslant 1/r\}$. Therefore

$$S_d(\mathbf{x}, r) = \bigcap_{n \leqslant 1/r} \pi_n^{-1}(S_{d'_n}(x_n, rn)). \tag{1}$$

Applying (6.10), from (1) it follows that $S_d(\mathbf{x}, r)$ is open relative to the product topology on $\prod_{n \in N} X_n$. Consequently, the product topology on $\prod_{n \in N} X_n$ includes the topology generated by d. To prove that the two topologies coincide, it will suffice to show that, for any $m \in N$ and $V \in \mathcal{T}_{d_m}$, $\pi_{m^{-1}}(V) \in \mathcal{T}_d$. Let $\mathbf{x} \in \pi_{m^{-1}}(V)$ and $x_m = \pi_m(\mathbf{x})$. Then $x_m \in V$. By (6.6.b) and (6.10), there is $\varepsilon \in]0, 1]$ such that

$$S_{d'_m}(x_m, \varepsilon) = S_{d_m}(x_m, \varepsilon) \subset V. \tag{2}$$

From (1) and (2) we get

$$S_d(\mathbf{x}, \varepsilon/m) = \bigcap_{n \leqslant m/\varepsilon} \pi_n^{-1}(S_{d'_n}(x_n, \varepsilon n/m)) \subset \pi_m^{-1}(S_{d'_m}(x_m, \varepsilon)) \subset \pi_m^{-1}(V).$$

By virtue of (6.6.b), this shows that $\pi_m^{-1}(V) \in \mathcal{T}_d$. □

Notation being as in (6.11), we will always suppose that $\prod_{i \in I} X_i$ is equipped with the metric d.

In (5.20) we proved that each topological space which has a countable base is separable, and in (5.21) we showed that the converse of this statement is not generally true. However, for metric spaces the following theorem holds.

Theorem 6.12. *Every separable metric space has a countable base.*

Proof. Let (X, d) be a metric space, and let A be a countable subset dense in X. Then, according to (3.41), (3.42) and (3.38), the family $\mathcal{B} = \{S(x, r) : x \in A, r \in Q \cap]0, \infty[\}$ is countable. We will show that $\mathcal{B}$ is a base for $\mathcal{T}_d$. Let $U \in \mathcal{T}_d$ and $x \in U$. By (6.6.b), there is $\varepsilon > 0$ such that $S(x, \varepsilon) \subset U$. Since A is dense in X, (5.17) shows that there is $y \in A \cap S(x, \varepsilon/3)$. Now select $r \in Q$ such that $\varepsilon/3 < r < 2\varepsilon/3$. Then, for any $z \in S(y, r)$, we have $d(x, z) \leqslant d(x, y) + d(y, z) < (\varepsilon/3) + r < \varepsilon$, and so $z \in S(x, \varepsilon)$. At the same time, $d(y, x) < \varepsilon/3 < r$. Thus $x \in S(y, r) \subset S(x, \varepsilon) \subset U$. Therefore, U is a union of balls in $\mathcal{B}$. □

For a metric space (X, d), the convergence of sequences in X can describe completely the topology $\mathcal{T}_d$. Corollary (6.15) illustrates this statement.

Remark 6.13. Let (X, d) be a metric space, $\{x_n : n \in N\} \subset X$ and $x \in X$. Then, according to (6.6.b), $x_n \to x$ if and only if $d(x, x_n) \to 0$.

Theorem 6.14. *Let X be a metric space and $A \subset X$. Then the following are equivalent:*

(i) $x \in A^-$;

(ii) *there is a sequence* $\{x_n : n \in N\} \subset A$ *such that* $x_n \to x$.

Proof. If $x \in A^-$, then, using (5.17), choose $x_n \in A \cap S(x, 1/n)$ for any $n \in N$. Applying (6.13), we have $x_n \to x$, and so (i) implies (ii). Plainly, (ii) implies (i). □

Corollary 6.15. *Let X be a metric space and $A \subset X$. Then the following are equivalent:*

(i) *A is closed;*

(ii) *for any sequence* $\{x_n : n \in N\} \subset A$ *such that* $x_n \to x \in X$, *we have* $x \in A$.

Proof. Assume that A is closed, and consider a sequence $\{x_n : n \in N\} \subset A$ such that $x_n \to x \in X$. Then, by (6.14), $x \in A^- = A$. Thus (i) implies (ii). Suppose now that (ii) holds, and let $x \in A^-$. According to (6.14), there is a sequence $\{x_n : n \in N\} \subset A$ such that $x_n \to x$, and so $x \in A$. Therefore, $A = A^-$, and so A is closed. □

The above result does not generalize to arbitrary topological spaces. To provide examples confirming this statement, objects more general than sequences are required.

In what follows we discuss the concept of compactness for metric spaces.

Definitions 6.16. A topological space $(X, \mathcal{T})$ is said to be *sequentially compact* if each sequence in X contains a convergent subsequence. A set $A \subset X$ is said to be *sequentially compact* if $(A, A \cap \mathcal{T})$ is sequentially compact. By virtue of (5.52), A is sequentially compact if and only if every sequence in A contains a subsequence converging to some point of A.

Lemma 6.17. *Let X be a metric space and $A \subset X$. Then x is a limit point of A if and only if $A \cap U$ is infinite for any neighborhood U of x.*

Proof. Exercise. □

Lemma 6.18. *Any sequentially compact metric space is separable.*

Proof. Let (X, d) be a sequentially compact metric space. For $n \in N$, consider $\mathcal{A}_n = \{A \subset X : d(x, y) \geqslant 1/n \text{ for } x, y \in A, x \neq y\}$. Then $\mathcal{A}_n$ is a family of finite character, and so, by Tukey's lemma (3.32), $\mathcal{A}_n$ has a maximal element A_n. The set A_n is finite, since otherwise, applying (3.11), there would exist a sequence $\{x_k : k \geqslant 1\} \subset A_n$ such that $d(x_k, x_l) \geqslant 1/n$ for $k \neq l$. Therefore, according to (6.13), $\{x_k : k \geqslant 1\}$ would contain no convergent subsequence. Hence the set $A = \cup_{n \in N} A_n$ is countable. We will show that A is dense in X. If $x \in (A^-)^c$, then, since $(A^-)^c$ is open, there would exist $\varepsilon > 0$ such that $S(x, \varepsilon) \subset (A^-)^c$. Select $n \in N$ such that $1/n < \varepsilon$. Then $d(x, y) \geqslant 1/n$ for any $y \in A_n$, and so $A_n \cup \{x\} \in \mathcal{A}_n$. This would contradict the assertion that A_n is maximal. Consequently, $A^- = X$. □

Theorem 6.19. *Let X be a metric space. Then the following assertions are equivalent:*
(i) *X is compact;*
(ii) *X is Fréchet compact;*
(iii) *X is sequentially compact.*

Proof. We showed in (5.35) that (i) implies (ii). Assume that (ii) holds, and consider a sequence $\{x_n : n \in N\} \subset X$. If $\{x_n : n \in N\}$ has a finite number of distinct points, then obviously there exists a strictly increasing sequence $\{n_k : k \in N\}$ such that $x_{n_k} = x_{n_1}, k \in N$. In this case the subsequence $\{x_{n_k} : k \in N\}$ converges to x_{n_1}. If the set $\{x_n : n \in N\}$ is infinite, then it has a limit point x. We will construct inductively a subsequence $\{x_{n_k} : k \in N\}$ of $\{x_n : n \in N\}$ such that $\lim_k x_{n_k} = x$. Put $x_{n_1} = x_1$. If $x_{n_1}, \dots, x_{n_l}$ have been defined, then, according to (6.17), we may choose $n_{l+1} > n_l$ such that $x_{n_{l+1}} \in S(x, 1/(l+1))$. It is clear that $x_{n_k} \to x$. Therefore, (ii) implies (iii). Suppose now that (iii) holds, and let $\mathcal{A}$ be an open cover of X. Then, from (6.18) and (6.12), it follows that X has a countable base. Hence, according to Lindelöf's theorem (5.14), $\mathcal{A}$ contains a countable subfamily $\mathcal{B}$ that covers X. If $\mathcal{B}$ is finite, then the proof is complete. If $\mathcal{B}$ is denumerable, then set $\mathcal{B} = \{B_i : i \in N\}$. Let $C_n = \cup_{i=1}^n B_i$, $n \in N$. To get a contradiction, assume that $C_n^c \neq \emptyset, n \in N$, and choose $x_n \in C_n^c$ for each $n \in N$. Then there are $x \in X$ and a subsequence $\{x_{n_k} : k \in N\}$ such that $x_{n_k} \to x$. Since $\mathcal{B}$ is a cover of X, there is $m \in N$ such that $x \in B_m \subset C_m$. Since $\{C_n^c : n \in N\}$ is a nonincreasing sequence of sets, C_m is a neighborhood of x such that $x_n \notin C_m$ for $n \geqslant m$. This contradicts the assertion that $x_{n_k} \to x$. Therefore, there exists $n \in N$ such that $X = C_n$, and so X is covered by a finite number of sets in $\mathcal{A}$. Thus (iii) implies (i). □

In what follows we introduce the notion of a complete metric space.

Definition 6.20. Let (X, d) be a metric space. A sequence $\{x_n : n \in N\} \subset X$ is said to be a *Cauchy [fundamental] sequence* if for any $\varepsilon > 0$ there is $n_\varepsilon \in N$ such that $m, n \geqslant n_\varepsilon$ implies $d(x_m, x_n) < \varepsilon$.

Examples 6.21. (a) Each convergent sequence in a metric space is a Cauchy sequence.

(b) Let (X, d) be a metric space, where d is the discrete metric for X, and let $\{x_n : n \in N\} \subset X$. Then $\{x_n : n \in N\}$ is a Cauchy sequence if and only if there is $n_0 \in N$; such that $x_n = x_{n_0}$ for $n \geqslant n_0$.

Remarks 6.22. (a) Let (X, d) be a metric space and $\{x_n : n \in N\} \subset X$. Then the following are equivalent:

(i) $\{x_n : n \in N\}$ is a Cauchy sequence;
(ii) for any $\varepsilon > 0$ there is $n_\varepsilon \in N$ such that $n \geqslant n_\varepsilon$ implies $d(x_{n_\varepsilon}, x_n) < \varepsilon$;
(iii) $\sup_{m>n} d(x_m, x_n) \to 0$.

(b) If (X, d_1) and (X, d_2) are metric spaces such that $\mathcal{T}_{d_1} = \mathcal{T}_{d_2}$, then a sequence $\{x_n : n \in N\} \subset X$ may be a Cauchy sequence relative to d_1, but $\{x_n : n \in N\}$ may fail to be a Cauchy sequence relative to d_2. For example, let $X = \{1/n : n \in N\}$, let d_1 be the usual metric for X, and let d_2 be the discrete metric for X. It is easily seen that both d_1 and d_2 generate the *discrete topology for* X. Let $x_n = 1/n, n \in N$. Then $\{x_n : n \in N\}$ is a Cauchy sequence relative to d_1. However, in view of (6.21.b), $\{x_n : n \in N\}$ is not a Cauchy sequence relative to d_2.

Definition 6.23. A metric space X is said to be complete if each Cauchy sequence in X is convergent.

Example 6.24. According to (6.21.b), any set endowed with the discrete metric is a complete metric space.

Remarks 6.25. (a) If d_1 and d_2 are metrics for X that generate the same topology for X, then X may be complete relative to d_1, but X may fail to be complete relative to d_2. For example, let $X = \{1/n : n \in N\}$, let d_1 be the discrete metric for X, and let d_2 be the usual metric for X. As stated in (6.24), (X, d_1) is complete. Since $\{1/n : n \in N\}$ is a Cauchy sequence relative to d_2 and $0 \notin X$, it follows that (X, d_2) is not complete.

(b) Let (X, d) be a complete metric space and $A \subset X$. Using (6.15) and (6.21.a), it is easily seen that (A, d_A) is complete if and only if A is closed.

(c) A metric space X is complete if each Cauchy sequence in X contains a convergent subsequence.

Definition 6.26. Let (X, d) be a metric space, and let A be a nonempty subset of X. The number $\operatorname{diam}(A) = \sup\{d(x, y) : x, y \in A\}$ is called the diameter of A. We define $\operatorname{diam}(\emptyset) = 0$.

Theorem (6.28) below is an important characterization of the notion of completeness. We need the next lemma.

Lemma 6.27. *Let (X, d) be a metric space, and let A be a nonempty subset of X. Then* $\operatorname{diam}(A) = \operatorname{diam}(A^-)$.

Proof. Clearly,

$$\operatorname{diam}(A) \leqslant \operatorname{diam}(A^-). \tag{1}$$

For $x, y \in A^-$ and each $\varepsilon > 0$, there are $x' \in A \cap S(x, \varepsilon/2)$ and $y' \in S(y, \varepsilon/2)$. Thus, using the triangle inequality, we have $d(x, y) \leqslant d(x', y') + \varepsilon \leqslant \operatorname{diam}(A) + \varepsilon$, and so $\operatorname{diam}(A^-) \leqslant \operatorname{diam}(A) + \varepsilon$. Since ε is arbitrary, it follows that

$$\operatorname{diam}(A^-) \leqslant \operatorname{diam}(A). \tag{2}$$

From (1) and (2) we get $\operatorname{diam}(A) = \operatorname{diam}(A^-)$. □

Theorem 6.28 (Cantor). *Let (X, d) be a metric space. Then the following are equivalent:*

(i) *X is complete;*

(ii) *for every nonincreasing sequence $\{A_n : n \in N\}$ of nonempty closed subsets of X such that* $\operatorname{diam}(A_n) \to 0$, *there exists $x \in X$ such that $\bigcap_{n\in N} A_n = \{x\}$.*

Proof. Assume that X is complete, and consider a nonincreasing sequence $\{A_n : n \in N\}$ of nonempty closed subsets of X such that $\operatorname{diam}(A_n) \to 0$. Choose $x_n \in A_n$ for each $n \in N$. If $\varepsilon > 0$, then there is $n_\varepsilon \in N$ such that $\operatorname{diam}(A_{n_\varepsilon}) < \varepsilon$. Hence, for $m, n \geqslant n_\varepsilon$, we have $d(x_m, x_n) \leqslant \operatorname{diam}(A_{n_\varepsilon}) < \varepsilon$. Therefore, $\{x_n : n \in N\}$ is a Cauchy sequence, and so there is $x \in X$ such that $x_n \to x$. Whatever $m \in N$, we have $x_n \in A_m$ for $n \geqslant m$, and so, in view of (6.15), $x \in A_m$. Consequently, $x \in \bigcap_{n\in N} A_n$. If $y \in \bigcap_{n\in N} A_n$, then $d(x, y) \leqslant \operatorname{diam}(A_n)$, $n \in N$. Since $\operatorname{diam}(A_n) \to 0$, it follows that $d(x, y) = 0$, and so $x = y$. Therefore, $\bigcap_{n\in N} A_n = \{x\}$. Thus (i) implies (ii). Suppose now that (ii) holds, and let $\{x_n : n \in N\} \subset X$ be a Cauchy sequence. Whatever $n \in N$, denote $A_n = \{x_m : m \geqslant n\}^-$. Obviously, $\{A_n : n \in N\}$ is a nonincreasing sequence of nonempty closed sets. For $\varepsilon > 0$, there is $n_\varepsilon \in N$ such that $m, n \geqslant n_\varepsilon$ implies $d(x_m, x_n) < \varepsilon$. Then, by virtue of (6.27), it follows that

$$\operatorname{diam}(A_{n_\varepsilon}) \leqslant \varepsilon. \tag{1}$$

Therefore, $\{A_n : n \in N\}$ being nonincreasing, it follows that $\operatorname{diam}(A_n) \to 0$. Let $\{x\} = \bigcap_{n\in N} A_n$. If $\varepsilon > 0$, then, using (1), we have $d(x, x_n) \leqslant \operatorname{diam}(A_{n_\varepsilon}) \leqslant \varepsilon$ for any $n \geqslant n_\varepsilon$. Thus $x_n \to x$, and so (ii) implies (i). □

Corollary 6.29. *Any compact metric space is complete.*

Proof. Let X be a compact metric space, and let $\{A_n : n \in N\}$ be a nonincreasing sequence of nonempty closed subsets of X such that $\operatorname{diam}(A_n) \to 0$. Since $\{A_n : n \in N\}$ has the finite intersection property, applying (5.24), it follows that $\bigcap_{n\in N} A_n \neq \emptyset$. Since $\operatorname{diam}(A_n) \to 0$, $\bigcap_{n\in N} A_n$ contains a single point. Thus, by (6.28), X is complete. □

Theorem (6.31) below exhibits the relation between compact metric spaces and complete metric spaces.

Definitions 6.30. A metric space (X, d) is said to be *totally bounded* if, for each $r > 0$, X is covered by a finite number of balls with radius r. A set $A \subset X$ is said to be *totally bounded* if (A, d_A) is totally bounded. Obviously, A is totally bounded if and only if, for each $r > 0$, A can be covered by a finite number of balls in X with radius r.

Theorem 6.31 (Hausdorff). *Let X be a metric space. Then the following are equivalent:*

(i) *X is compact;*
(ii) *X is totally bounded and complete.*

Proof. Suppose that X is compact. Then X is complete (6.29). Further, for each $r > 0$, the family $\{S(x, r) : x \in X\}$ is an open cover of X which contains a finite subcover, and so X is *totally bounded*. Assume now that (ii) holds. In view of (6.19), it will suffice to prove that X is Fréchet compact. Let A be an infinite subset of X. Since X is *totally bounded*, there is a ball S_1 with radius 1 such that $A_1 = A \cap S_1$ is infinite. If the infinite sets $A_1, \ldots, A_n$ have been chosen, then, since X is *totally bounded*, there is a ball S_{n+1} with radius $1/(n+1)$ such that $A_{n+1} = A_n \cap S_{n+1}$ is infinite. The sequence $\{A_n^- : n \in N\}$ is nonincreasing, and $\operatorname{diam}(A_n^-) = \operatorname{diam}(A_n) \leqslant 2/n \to 0$. Thus, by virtue of Cantor's theorem (6.28), there is $x \in X$ such that $\bigcap_{n \in N} A_n^- = \{x\}$. For $\varepsilon > 0$, select $n \in N$ such that $2/n < \varepsilon$. Then, for any $y \in A_n$, we have $d(x, y) \leqslant \operatorname{diam}(A_n) \leqslant 2/n < \varepsilon$, and so $A_n \subset S(x, \varepsilon)$. Since $A \cap S(x, \varepsilon) \supset A_n \cap S(x, \varepsilon) = A_n$, and A_n is infinite, it follows that $x \in A^d$ (6.17). Therefore, X is Fréchet compact. □

Theorem 6.32. *The space R^n is complete relative to the Euclidean metric.*

Proof. Let $\{\mathbf{x}_k : k \in N\} \subset R^n$ be a Cauchy sequence. Choose $k_0 > 1$ such that $d(\mathbf{x}_{k_0}, \mathbf{x}_k) < 1$ for any $k \geqslant k_0$. Let $\rho = 1 \vee d(\mathbf{x}_{k_0}, \mathbf{x}_1) \vee \cdots \vee d(\mathbf{x}_{k_0}, \mathbf{x}_{k_0 - 1})$ and $r = d(\mathbf{0}, \mathbf{x}_{k_0}) + \rho$. Then $d(\mathbf{x}_{k_0}, \mathbf{x}_k) \leqslant \rho$, $k \in N$, and so $\{\mathbf{x}_k : k \in N\} \subset \overline{S}(\mathbf{0}, r) \subset [-r, r]^n$. From (5.30) and (6.9) it follows that $[-r, r]^n$ is a compact metric space, and so $[-r, r]^n$ is complete relative to the Euclidean metric (6.29). Since $\{\mathbf{x}_k : k \in N\}$ is a Cauchy sequence in $[-r, r]^n$, there is $\mathbf{x} \in [-r, r]^n$ such that $d(\mathbf{x}, \mathbf{x}_k) \to 0$. Consequently, R^n is complete. □

Corollary 6.33. *Let X be a set, and let $\mathfrak{B}_r(X)$ be the set of all bounded functions from X into R. Then $\mathfrak{B}_r(X)$ is a complete space relative to the uniform metric.*

Proof. Let $\{f_n : n \in N\} \subset \mathfrak{B}_r(X)$ be a Cauchy sequence. Then, whatever $x \in X$, $\{f_n(x) : n \in N\}$ is a Cauchy sequence in R, and so $\{f_n(x) : n \in N\}$ is convergent (6.32). Let $f(x) = \lim_n f_n(x)$, $x \in X$. We will show that $f \in \mathfrak{B}_r(X)$ and that $f_n \to f$. For $\varepsilon > 0$, choose $n_\varepsilon \in N$ such that $m, n \geqslant n_\varepsilon$ implies $\|f_m - f_n\|_u < \varepsilon/2$. For fixed $x \in X$, select $m \geqslant n_\varepsilon$ such that $|f(x) - f_m(x)| < \varepsilon/2$. Then we have $|f(x)| \leqslant |f(x) - f_m(x)| + |f_m(x) - f_{n_\varepsilon}(x)| + |f_{n_\varepsilon}(x)| < (\varepsilon/2) + \|f_m - f_{n_\varepsilon}\|_u + \|f_{n_\varepsilon}\|_u < \varepsilon + \|f_{n_\varepsilon}\|_u$. Thus $\|f\|_u \leqslant \varepsilon + \|f_{n_\varepsilon}\|_u < \infty$, and so $f \in \mathfrak{B}_r(X)$ (5.83.e). If $n \geqslant n_\varepsilon$, then we get $|f(x) - f_n(x)| \leqslant |f(x) - f_m(x)| + |f_m(x) - f_n(x)| < (\varepsilon/2) + \|f_m - f_n\|_u < \varepsilon$. Therefore, $\|f - f_n\|_u \leqslant \varepsilon$ for any $n \geqslant n_\varepsilon$. Thus $f_n \to f$. □

The concept of continuity is next discussed within the framework of metric spaces. We begin with some remarks.

Remarks 6.34. Notation is as in (5.36).

(a) Assume that the topology for X is generated by d. Then f is continuous at x if and only if for any neighborhood V of $f(x)$ there is $\delta > 0$ such that $d(x, y) < \delta$ implies $f(x) \in V$.

(b) Suppose that the topology for Y is generated by ρ. Then f is continuous at x if and only if for each $\varepsilon > 0$ there is a neighborhood U of x such that $y \in U$ implies $\rho(f(x), f(y)) < \varepsilon$.

(c) Assume that the topology for X is generated by d, and the topology for Y is generated by ρ. Then f is continuous at x if and only if for each $\varepsilon > 0$ there is $\delta > 0$ such that $d(x, y) < \delta$ implies $\rho(f(x), f(y)) < \varepsilon$.

Theorem 6.35. *Let X, Y and f be as in* (5.36). *Assume that X is a metric space, and let $x \in X$. Then the following are equivalent:*

(i) *f is continuous at x;*

(ii) *for any sequence $\{x_n : n \in N\} \subset X$ such that $x_n \to x$, we have $f(x_n) \to f(x)$.*

Proof. We showed in (5.37) that (i) implies (ii). Suppose now that (i) does not hold. Then, according to (6.34.a), there exists a neighborhood V of $f(x)$ such that, for every $n \in N$, there is $x_n \in X$ with $d(x, x_n) < 1/n$ and $f(x_n) \notin V$. Thus $x_n \to x$, but $f(x_n) \not\to f(x)$. Therefore, (ii) does not hold. Consequently, (ii) implies (i). □

Definition 6.36. Let (X, d) be a metric space, let A be a nonempty subset of X, and let $x \in X$. The number $d(x, A) = \inf\{d(x, y) : y \in A\}$ is called the *distance from x to A*.

Theorem 6.37. *Let (X, d) be a metric space.*

(i) *If $\emptyset \neq A \subset X$, then $A^- = \{x : d(x, A) = 0\}$.*

(ii) *If $\emptyset \neq A \subset X$, then the function $f(x) = d(x, A)$, $x \in X$, is continuous on X.*

(iii) *If A and B are disjoint closed subsets of X, then there exists a continuous function $g : X \to [0, 1]$ such that $g(A) = \{0\}$ and $g(B) = \{1\}$.*

Proof. If $x \in A^-$, then $A \cap S(x, 1/n) \neq \emptyset$ for any $n \in N$. Let $x_n \in A \cap S(x, 1/n)$, $n \in N$. Then $d(x, A) \leqslant d(x, x_n) < 1/n$, $n \in N$, and so $d(x, A) = 0$. Hence $A^- \subset \{x : d(x, A) = 0\}$. Conversely, if $x \notin A^-$, then there is $\varepsilon > 0$ such that $A \cap S(x, \varepsilon) = \emptyset$, and so $d(x, A) \geqslant \varepsilon$. Therefore, $\{x : d(x, A) = 0\} \subset A^-$. This proves (i). Let $f(x) = d(x, A)$, $x \in X$. For any $x, y \in X$ and $z \in A$, we have $f(x) \leqslant d(x, z) \leqslant d(x, y) + d(y, z)$, whence

$$f(x) - d(x, y) \leqslant f(y). \tag{1}$$

Upon interchanging x and y, and using (6.1.iii), (1) shows that $|f(x) - f(y)| \leqslant d(x, y)$. Thus, according to (6.34.c), f is continuous on X. This proves (ii). If A and B are disjoint closed subsets of X, then, by (i), $d(x, A) + d(x, B) > 0$ for any $x \in X$. Hence, using (ii), (5.76.a) and (5.76.c), the function $g : X \to [0, 1]$ defined by

$$g(x) = \frac{d(x, A)}{d(x, A) + d(x, B)}, \quad x \in X,$$

is continuous on X. Moreover, $g(A) = \{0\}$ and $g(B) = \{1\}$. This proves (iii). □

The following theorem is a special case of Urysohn's embedding theorem.

Theorem 6.38 (Urysohn). *Let (X, d) be a separable metric space. Then there exist a compact metric space Y and a homeomorphism f from X into Y.*

Proof. Let A be a countable subset dense in X. Then there exist a countable set I and a bijective function from I onto A. For each $i \in I$, let a_i denote the image of i under this function. Put $Y = [0, 1]^I$. Since $[0, 1]$ is compact relative to the Euclidean metric (5.29), Tihonov's theorem (5.28) and (6.11) show that Y is a compact metric space. Let h be a homeomorphism from R onto $]0, 1[$. Whatever $i \in I$, consider the function $f_i : X \to [0, 1]$ defined by $f_i(x) = h(d(x, a_i))$ for any $x \in X$. By virtue of (6.37.ii) and (5.39), f_i is continuous on X for each $i \in I$. Set $f = (f_i)_{i \in I}$. Then, according to (5.80), f is continuous on X. Now let $x \in X$, and let $\{x_n : n \in N\} \subset X$ be a sequence such that $x_n \not\to x$. Then there are $\varepsilon > 0$ and a subsequence $\{x_{n_k} : k \in N\}$ such that $d(x, x_{n_k}) \geqslant \varepsilon$ for any $k \in N$. Since A is dense in X, there is $a_j \in A$ such that $d(x, a_j) < \varepsilon/3$. Therefore, by the triangle inequality, $d(x_{n_k}, a_j) > 2\varepsilon/3$ for any $k \in N$. Consequently, $d(x_{n_k}, a_j) > d(x, a_j) + \varepsilon/3, k \in N$. This means that $d(x_n, a_j) \not\to d(x, a_j)$. Since h^{-1} is continuous at $f_j(x)$, (5.37) shows that $f_j(x_n) \not\to f_j(x)$. Then, according to (5.7), it follows that $f(x_n) \not\to f(x)$. Further, let $z \in X$ such that $z \neq x$, and put $x_n = z, n \in N$. Then, since $x_n \not\to x$, we have $f(z) = f(x_n) \not\to f(x)$, and so $f(z) \neq f(x)$. Thus f is injective. Finally, let $y \in f(X)$, and let $\{y_n : n \in N\} \subset f(X)$ be such that $y_n \to y$. Let $x \in X$ be such that $y = f(x)$. For each $n \in N$, let $x_n \in X$ be such that $y_n = f(x_n)$. Since $y_n \to y$ implies $x_n \to x$, (5.50) and (6.35) show that the proof is complete. □

We next introduce the notion of uniform convergence.

Definition 6.39. Let X be a set, and let $f_n : X \to K, n \in N, f : X \to K$ be functions. We say that the sequence $\{f_n : n \in N\}$ *converges uniformly to* f if $\|f - f_n\|_u \to 0$. We will write $f_n \xrightarrow{u} f$ if $\{f_n : n \in N\}$ converges uniformly to f.

Remarks 6.40. (a) For $f_n \in \mathfrak{B}(X), n \in N$, and $f \in \mathfrak{B}(X)$, $f_n \xrightarrow{u} f$ if and only if $f_n \to f$ relative to the uniform metric.

(b) Let $f_n, n \in N$, and f be as in (6.39). If $f_n \xrightarrow{u} f$, then $f_n(x) \to f(x)$ for all $x \in X$. The converse of this statement does not generally hold. For example, let $X = [0, 1]$, let $f_n(x) = x^n$ for any $x \in X, n \in N$, and let $f = 1_{\{1\}}$. It is clear that $f_n(x) \to f(x)$ for each $x \in X$. However, it can be directly verified, and it follows from (6.41), that $\{f_n : n \in N\}$ does not converge uniformly to f.

Theorem 6.41. *Let X be a topological space, and let $f_n \in K^X, n \in N, f \in K^X$ be functions such that $f_n \xrightarrow{u} f$. If f_n is continuous at $x \in X$ for any $n \in N$, then f is continuous at x.*

Proof. For $\varepsilon > 0$, choose $n \in N$ such that $\|f - f_n\|_u < \varepsilon/3$. Since f_n is continuous at x, (6.34.b) shows that there is a neighborhood U of x such that $y \in U$ implies $|f_n(x) - f_n(y)| < \varepsilon/3$. Therefore, for $y \in U$, we have $|f(x) - f(y)| \leqslant |f(x) - f_n(x)| + |f_n(x) - f_n(y)| + |f_n(y) - f(y)| \leqslant 2\,\|f - f_n\|_u + |f_n(x) - f_n(y)| < \varepsilon$. Thus f is continuous at x. □

Corollary 6.42. *Let $f_n, n \in N$, and f be as in* (6.41). *If f_n is continuous on X for any $n \in N$, then f is continuous on X.*

Theorem 6.43. *Let X be a topological space, and let $\mathfrak{C}_r(X)$ be the set of all bounded continuous functions from X into R. Then $\mathfrak{C}_r(X)$ is a complete space relative to the uniform metric.*

Proof. As $\mathfrak{C}_r(X) \subset \mathfrak{B}_r(X)$, and $\mathfrak{B}_r(X)$ is complete relative to the uniform metric (6.33), it suffices to show that $\mathfrak{C}_r(X)$ is a closed subset of $\mathfrak{B}_r(X)$ (6.25.b). Let $f_n \in \mathfrak{C}_r(X)$, $n \in N$, and let $f \in \mathfrak{B}_r(X)$ be such that $f_n \to f$ relative to the uniform metric. Then, by (6.40.a), $f_n \overset{u}{\to} f$. Therefore, in view of (6.42), $f \in \mathfrak{C}_r(X)$. Thus $\mathfrak{C}_r(X)$ is closed. □

Theorem 6.44 (Dini). *Let X be a compact space, let $\{f_n : n \in N\} \subset R^X$ be a monotone sequence, and let $f \in R^X$ be such that $f_n(x) \to f(x)$ for any $x \in X$. If f_n is continuous on X for any $n \in N$, and f is continuous on X, then $f_n \overset{u}{\to} f$.*

Proof. Set $g_n = |f - f_n|$, $n \in N$. Then the sequence $\{g_n : n \in N\}$ is nonincreasing, and $g_n(x) \to 0$ for each $x \in X$. Therefore, for $\varepsilon > 0$, we have $X = \cup_{n \in N}\{x : g_n(x) < \varepsilon\}$. Since g_n is continuous on X for any $n \in N$, (5.75.d) shows that $\{\{x : g_n(x) < \varepsilon\} : n \in N\}$ is an open cover of X. Since X is compact, and $\{g_n : n \in N\}$ is nonincreasing, there is $n_\varepsilon \in N$ such that $X = \cup_{k=1}^{n_\varepsilon}\{x : g_k(x) < \varepsilon\} = \{x : g_{n_\varepsilon}(x) < \varepsilon\}$. Consequently, if $n \geqslant n_\varepsilon$, then, for each $x \in X$, we have $|f(x) - f_n(x)| \leqslant g_{n_\varepsilon}(x) < \varepsilon$. Thus $\|f - f_n\|_u \leqslant \varepsilon$ for any $n \geqslant n_\varepsilon$, and so $f_n \overset{u}{\to} f$. □

One of the essential results of abstract analysis is the Stone-Weierstrass theorem which we will prove in (6.57). This theorem illustrates the strong interplay between algebra and contemporary analysis. We begin by defining a few algebraic structures.

Definition 6.45. Any function $\top : X \times X \to X$ is called a *binary operation on X*. Following the usual custom, we will write $x \top y$ instead of $\top(x, y)$.

Definition 6.46. A *group* is a pair $(G, +)$, where G is a set, and $+$ is a binary operation on G satisfying:

(i) $x + (y + z) = (x + y) + z$ for any $x, y, z \in G$;
(ii) there is an element $0 \in G$ such that $0 + x = x + 0 = x$ for any $x \in G$;
(iii) for each $x \in G$ there is $-x \in G$ such that $(-x) + x = x + (-x) = 0$.

If, moreover,

(iv) $x + y = y + x$ for any $x, y \in G$, then $(G, +)$ is called a *commutative* [*Abelian*] *group*. For $x, y \in G$, we will write $x - y$ instead of $x + (-y)$.

Definitions 6.47. A *linear* [*vector*] *space over K* is a triple $(X, +, *)$, where $(X, +)$ is a commutative group, and $*$ is a function from $K \times X$ into X, whose value at (α, x) is denoted αx, such that for $\alpha, \beta \in K$ and $x, y \in X$:

(i) $\alpha(x + y) = \alpha x + \alpha y$;
(ii) $(\alpha + \beta)x = \alpha x + \beta x$;
(iii) $\alpha(\beta x) = (\alpha\beta)x$;
(iv) $1x = x$.

We note that $\alpha 0 = 0x = 0$, $\alpha \in K$, $x \in X$, and that $(-1)x = -x$, $x \in X$.

A set $Y \subset X$ is called a *linear* [*vector*] *subspace of* X if $x, y \in Y$ implies $x+y \in Y$, and $(\alpha, x) \in K \times Y$ implies $\alpha x \in Y$. It is easily seen that Y is a linear subspace of X if and only if Y is a linear space over K relative to the restriction of $+$ to $Y \times Y$ and the restriction of $*$ to $K \times Y$.

Examples 6.48. (a) For $\mathbf{x} = (x_1, \ldots, x_n) \in K^n$ and $\mathbf{y} = (y_1, \ldots, y_n) \in K^n$, define $\mathbf{x}+\mathbf{y} = (x_1+y_1, \ldots, x_n+y_n)$. For $\alpha \in K$ and $\mathbf{x} \in K^n$, define $\alpha\mathbf{x} = (\alpha x_1, \ldots, \alpha x_n)$. Then K^n is a linear space over K.

(b) For $\alpha \in R$ and $\mathbf{x}, \mathbf{y} \in C^n$, define $\mathbf{x}+\mathbf{y}$ as in (a), and put $\alpha\mathbf{x} = (\alpha x_1, \ldots, \alpha x_n)$. Then C^n is a linear space over R. Moreover, R^n is a linear subspace of C^n.

(c) Let X be a set. For $\alpha \in K$ and $f, g \in K^X$, define $f+g$ and αf as in (2.2). Then K^X is a linear space over K, and $\mathfrak{B}(X)$ is a linear subspace of K^X. If X is a topological space, then $\mathfrak{C}_r(X)$ is a linear subspace of $\mathfrak{B}_r(X)$.

(d) Notation is as in (3.43). If $A = R$, then $\mathfrak{P}(I)$ is a linear subspace of R^X; also, for every $J \in \mathcal{J}$, $\mathfrak{P}(J)$ is a linear subspace of $\mathfrak{P}(I)$, and $\mathfrak{P}_n(J)$ is a linear subspace of $\mathfrak{P}(J)$.

Definition 6.49. A *ring* is a triple $(X, +, \cdot)$, where $(X, +)$ is a commutative group, and $\cdot$ is a binary operation on X such that for $x, y, z \in X$:

(i) $x \cdot (y \cdot z) = (x \cdot y) \cdot z$;

(ii) $x \cdot (y+z) = (x \cdot y) + (x \cdot z)$;

(iii) $(x+y) \cdot z = (x \cdot z) + (y \cdot z)$.

Definitions 6.50. An *algebra over* K is a quadruple $(X, +, \cdot, *)$, where $(X, +, \cdot)$ is a ring and $(X, +, *)$ is a linear space over K such that for $\alpha \in K$ and $x, y \in X$:

(i) $\alpha(x \cdot y) = (\alpha x) \cdot y = x \cdot (\alpha y)$.

A set $Y \subset X$ is called a *subalgebra of* X, if $x, y \in Y$ implies $x \cdot y \in Y$, and Y is a linear subspace of X. Plainly, Y is a subalgebra of X if and only if Y is an algebra over K relative to the restrictions of $+$ and $\cdot$ to $Y \times Y$ and the restriction of $*$ to $K \times Y$.

Examples 6.51. (a) For $\alpha \in K$ and $\mathbf{x}, \mathbf{y} \in K^n$, define $\mathbf{x}+\mathbf{y}$ and $\alpha\mathbf{x}$ as in (6.48.a), and set $\mathbf{x} \cdot \mathbf{y} = (x_1 y_1, \ldots, x_n y_n)$. Then K^n is an algebra over K.

(b) Let X be a set. For $\alpha \in K$ and $f, g \in K^X$, define $f+g$, αf and $f \cdot g$ as in (2.2). Then K^X is an algebra over K, and $\mathfrak{B}(X)$ is a subalgebra of K^X. If X is a topological space, then $\mathfrak{C}_r(X)$ is a subalgebra of $\mathfrak{B}_r(X)$.

(c) Let X be a topological space, and let $\mathfrak{C}_c(X)$ be the set of all bounded continuous functions from X into C. Then $\mathfrak{C}_c(X)$ is a subalgebra of C^X.

(d) Notation is as in (3.43). Suppose that $A = R$. Then $\mathfrak{P}(I)$ is a subalgebra of R^X. For $J \in \mathcal{J}$, $\mathfrak{P}(J)$ is a subalgebra of $\mathfrak{P}(I)$, but in general $\mathfrak{P}_n(J)$, $n \geqslant 1$, is not an algebra over R. If X is a topological space, and $f_i \in \mathfrak{C}_r(X)$ for each $i \in I$, then $\mathfrak{P}(I)$ is a subalgebra of $\mathfrak{C}_r(X)$.

(e) Let $X \subset C - \{0\}$, and let $\mathfrak{A}$ be the set of all functions from X into C of the form $f(x) = \sum_{k=-n}^{n} \alpha_k x^k$, $x \in X$, where $\alpha_k \in C$ for each $k = -n, \ldots, 0, \ldots, n$, and $n \in N$. Then $\mathfrak{A}$ is an algebra over C. If there are $a, b > 0$ such that $a \leqslant |x| \leqslant b$ for any $x \in X$, then $\mathfrak{A}$ is a subalgebra of $\mathfrak{C}_c(X)$.

To prove the Stone-Weierstrass theorem, we need the next four lemmas.

Lemma 6.52. *For $M > 0$, consider the function $\varphi : [-M, M] \to R$ defined by $\varphi(x) = |x|, x \in [-M, M]$. Then there exists a sequence $\{p_n : n \in N\} \subset R^{[-M,M]}$ of polynomials with real coefficients such that $p_n(0) = 0, n \in N$, and $p_n \xrightarrow{u} \varphi$.*

Proof. Set $p_0(x) = 0, x \in [-M, M]$, and define

$$p_{n+1}(x) = p_n(x) + (x^2 - p_n^2(x))/2M, \quad x \in [-M, M], \ n \geqslant 0. \tag{1}$$

Then, by induction, p_n is a polynomial with real coefficients such that $p_n(0) = 0, n \in N$, and

$$0 \leqslant p_n(x) \leqslant p_{n+1}(x) \leqslant |x|, \quad x \in [-M, M], \ n \geqslant 0. \tag{2}$$

From (2) it follows that the sequence $\{p_n(x) : n \in N\}$ is convergent for each $x \in [-M, M]$. Hence, letting $n \to \infty$ in (1), we get $p_n^2(x) \to x^2$, $x \in [-M, M]$. Therefore, $p_n \to \varphi$. Since $[-M, M]$ is compact, and $\{p_n : n \in N\}$ is nondecreasing, Dini's theorem (6.44) shows that $p_n \xrightarrow{u} \varphi$. □

Throughout (6.53)–(6.58), X is a topological space, and $\mathfrak{C}_r(X)$ is the algebra over R of all bounded continuous functions from X into R.Moreover, $\mathfrak{C}_r(X)$ is endowed with the uniform metric.

Lemma 6.53. *If $\mathfrak{A}$ is a subalgebra of $\mathfrak{C}_r(X)$, then $\mathfrak{A}^-$ is a subalgebra of $\mathfrak{C}_r(X)$.*

Proof. According to (6.50) and (6.47), we must show that $\alpha \in R$ and $f, g \in \mathfrak{A}^-$ imply $\alpha f, f+g, fg \in \mathfrak{A}^-$. Applying (6.14), there are $\{f_n : n \in N\} \subset \mathfrak{A}$ and $\{g_n : n \in N\} \subset \mathfrak{A}$ such that $f_n \xrightarrow{u} f$ and $g_n \xrightarrow{u} g$. Then, by (5.83.b), $\alpha f_n \xrightarrow{u} \alpha f$, and so $\alpha f \in \mathfrak{A}^-$. Further, by (5.83.c), $f_n+g_n \xrightarrow{u} f+g$, and so $f+g \in \mathfrak{A}^-$. Using (5.83.b)–(5.83.d), we have $\|fg - f_n g_n\|_u \leqslant \|f - f_n\|_u \|g\|_u + \|f - f_n\|_u \|g - g_n\|_u + \|f\|_u \|g - g_n\|_u$, $n \in N$, and so $f_n g_n \xrightarrow{u} fg$. Since $f_n g_n \in \mathfrak{A}$, $n \in N$, (6.14) shows that $fg \in \mathfrak{A}^-$. Thus $\mathfrak{A}^-$ is a subalgebra of $\mathfrak{C}_r(X)$. □

Lemma 6.54. *Let $\mathfrak{A}$ be a subalgebra of $\mathfrak{C}_r(X)$ such that $\mathfrak{A}$ is a closed subset of $\mathfrak{C}_r(X)$. Then, for $f, g \in \mathfrak{A}$, we have:*
(i) $|f| \in \mathfrak{A}$;
(ii) $f \vee g \in \mathfrak{A}, \quad f \wedge g \in \mathfrak{A}$.

Proof. Since f is bounded, there is $M > 0$ such that $f(X) \subset [-M, M]$. Let $p_n, n \in N$, and φ be as in (6.52). Then, since $p_n(0) = 0$, $n \in N$, it follows that $p_n \circ f \in \mathfrak{A}, n \in N$. Since $p_n \xrightarrow{u} \varphi$, we have $p_n \circ f \xrightarrow{u} \varphi \circ f = |f|$. Therefore, according to (6.14), $|f| \in \mathfrak{A}$. Thus (i) is proved. The assertion (ii) follows at once from (5.76.b) and (i). □

Lemma 6.55. *Let X be a compact space, and let $\mathfrak{A}$ be a subset of $\mathfrak{C}_r(X)$ such that $f, g \in \mathfrak{A}$ implies $f \vee g \in \mathfrak{A}$ and $f \wedge g \in \mathfrak{A}$. If $f \in \mathfrak{C}_r(X)$ is such that for any $x, y \in X$ there is $f_{x,y} \in \mathfrak{A}$ satisfying $f_{x,y}(x) = f(x)$ and $f_{x,y}(y) = f(y)$, then $f \in \mathfrak{A}^-$.*

Proof. For $x, y \in X$ and $\varepsilon > 0$, let $U_{x,y} = \{z \in X : f(z) - \varepsilon < f_{x,y}(z)\}$ and $V_{x,y} = \{z \in X : f_{x,y}(z) < f(z) + \varepsilon\}$. Since $x \in U_{x,y}$, (5.76.c) shows that the family

$\{U_{x,y} : x \in X\}$ is an open cover of X for each $y \in X$. Since X is compact, there are $x_1, \ldots, x_n \in X$ such that $X = \cup_{i=1}^n U_{x_i,y}$. For $y \in X$, denote $V_y = \cap_{i=1}^n V_{x_i,y}$ and $f_y = \sup_{1 \leqslant i \leqslant n} f_{x_i,y}$. If $z \in X$, then $z \in U_{x_i,y}$ for some i, and so

$$f(z) - \varepsilon < f_{x_i,y}(z) \leqslant f_y(z). \tag{1}$$

If $z \in V_y$, then $f_{x_i,y}(z) < f(z) + \varepsilon$, $1 \leqslant i \leqslant n$, and so

$$f_y(z) < f(z) + \varepsilon. \tag{2}$$

Since $y \in V_y$, (5.76.c) shows that the family $\{V_y : y \in X\}$ is an open cover of X. Since X is compact, there are $y_1, \ldots, y_m \in X$ such that $X = \cup_{j=1}^m V_{y_j}$. Define $f_\varepsilon = \inf_{1 \leqslant j \leqslant m} f_{y_j}$. If $z \in X$, then, applying (1), $f(z) - \varepsilon < f_{y_j}(z)$, $1 \leqslant j \leqslant m$, and so

$$f(z) - \varepsilon < f_\varepsilon(z); \tag{3}$$

also, if $z \in X$, then $z \in V_{y_j}$ for some j, and so, by (2),

$$f_\varepsilon(z) \leqslant f_{y_j}(z) < f(z) + \varepsilon. \tag{4}$$

From (3) and (4) we get

$$\|f - f_\varepsilon\|_u \leqslant \varepsilon. \tag{5}$$

Further, using (5.58.b), we have $f_y \in \mathfrak{A}$, $y \in X$, and $f_\varepsilon \in \mathfrak{A}$. Since ε is arbitrary, (6.6.b) and (5) show that $f \in \mathfrak{A}^-$. □

Definition 6.56. Let X be a set and $\mathfrak{A} \subset K^X$. We say that $\mathfrak{A}$ *separates points of* X if whenever x and y are distinct points of X there is $f \in \mathfrak{A}$ such that $f(x) \neq f(y)$.

Stone-Weierstrass theorem 6.57. *Let X be a compact space, and let $\mathfrak{A}$ be a subalgebra of $\mathfrak{C}_r(X)$ such that:*

(i) $1 \in \mathfrak{A}$;

(ii) $\mathfrak{A}$ *separates points of* X.

Then $\mathfrak{A}^- = \mathfrak{C}_r(X)$.

Proof. Since $\mathfrak{A}$ is a subalgebra of $\mathfrak{C}_r(X)$, (6.53) shows $\mathfrak{A}^-$ is a subalgebra of $\mathfrak{C}_r(X)$. Therefore, $f, g \in \mathfrak{A}^-$ implies $f \vee g \in \mathfrak{A}^-$ and $f \wedge g \in \mathfrak{A}^-$ (6.54). Let $f \in \mathfrak{C}_r(X)$ and $x, y \in X$. If $x = y$, then the constant function $f_{x,y} = f(x)$ satisfies $f_{x,y}(x) = f(x)$ and $f_{x,y}(y) = f(y)$. Moreover, since $1 \in \mathfrak{A}$, $f_{x,y} \in \mathfrak{A} \subset \mathfrak{A}^-$. If $x \neq y$, then, since $\mathfrak{A}$ separates points of X, there is $g \in \mathfrak{A}$ such that $g(x) \neq g(y)$. Consequently, there are $a, b \in R$ (depending on x and y) such that $ag(x) + b = f(x)$ and $ag(y) + b = f(y)$. In this case, set $f_{x,y} = ag + b$. Evidently, $f_{x,y}(x) = f(x)$ and $f_{x,y}(y) = f(y)$. Since $1 \in \mathfrak{A}$, we have $f_{x,y} \in \mathfrak{A} \subset \mathfrak{A}^-$. Thus, by virtue of Lemma (6.55), $f \in (\mathfrak{A}^-)^- = \mathfrak{A}^-$. Consequently, $\mathfrak{A}^- = \mathfrak{C}_r(X)$. □

Theorem 6.58. *Let (X, d) be a compact metric space. Then $\mathfrak{C}_r(X)$ is a separable metric space.*

Proof. By virtue of (6.19) and (6.18), X is separable, and so there exists a countable set A such that $A^- = X$. For $z \in A$, consider the function $d_z : X \to R$ defined by $d_z(x) = d(x, z)$ for any $x \in X$. Then, by (6.37.ii) and (5.47), $d_z \in \mathfrak{C}_r(X)$ for each $z \in A$. Moreover, the set $\{d_z : z \in A\}$ separates points of X. Actually, if $x \neq y$, then $0 = d_x(x) \neq d_x(y) = d(y, x)$. Let now $\mathfrak{A}$ be the set of all polynomials in functions from $\{d_z : z \in A\}$ with coefficients in R. Then, according to (6.51.d), $\mathfrak{A}$ is a subalgebra of $\mathfrak{C}_r(X)$. Obviously, $1 \in \mathfrak{A}$. Moreover, since $\{d_z : z \in A\} \subset \mathfrak{A}$, the set $\mathfrak{A}$ separates points of X. Hence, by the Stone-Weierstrass theorem (6.57), $\mathfrak{A}^- = \mathfrak{C}_r(X)$. Further, let $\mathfrak{B}$ be the set of all polynomials in functions from $\{d_z : z \in A\}$ with coefficients in Q. Since $Q^- = R$, (6.14) and (5.82) show that for any $f \in \mathfrak{A}$, there is $\{f_n : n \in N\} \subset \mathfrak{B}$ such that $f_n \overset{u}{\to} f$. Consequently, $\mathfrak{A} \subset \mathfrak{B}^-$. Since $\mathfrak{B} \subset \mathfrak{A}$, (5.66.a) and (5.66.b) show that $\mathfrak{B}^- = \mathfrak{A}^- = \mathfrak{C}_r(X)$. Since Q is denumerable, from (3.43) it follows that $\mathfrak{B}$ is denumerable, and so $\mathfrak{C}_r(X)$ is separable. □

We next prove the complex version of the Stone-Weierstrass theorem.

Theorem 6.59. *Let X be a compact space, and let $\mathfrak{A}$ be a subalgebra of $\mathfrak{C}_c(X)$ such that:*

(i) $1 \in \mathfrak{A}$;

(ii) $\mathfrak{A}$ *separates points of X;*

(iii) $f \in \mathfrak{A}$ *implies* $\overline{f} \in \mathfrak{A}$.

Then $\mathfrak{A}^- = \mathfrak{C}_c(X)$, where $\mathfrak{C}_c(X)$ is endowed with the uniform metric.

Proof. Let $\mathfrak{B} = \{f \in \mathfrak{A} : f(X) \subset R\}$. Then, using (4.32) and (5.43.b), it follows that $\mathfrak{B}$ is a subalgebra of $\mathfrak{C}_r(X)$. Plainly, $1 \in \mathfrak{B}$. Moreover, $\mathfrak{B}$ separates points of X. Actually, if x and y are distinct points of X, then, by (ii), there is $f \in \mathfrak{A}$ such that $f(x) \neq f(y)$. Consequently, either $(\mathrm{Re}\, f)(x) \neq (\mathrm{Re}\, f)(y)$ or $(\mathrm{Im}\, f)(x) \neq (\mathrm{Im}\, f)(y)$. Since $\mathrm{Re}\, f = \frac{1}{2}(f + \overline{f})$ and $\mathrm{Im}\, f = \frac{1}{2i}(f - \overline{f})$, (iii) shows that $\mathrm{Re}\, f \in \mathfrak{B}$ and $\mathrm{Im}\, f \in \mathfrak{B}$. Thus $\mathfrak{B}$ separates points of X. Let now $g \in \mathfrak{C}_c(X)$. Then, by (5.83.f) and (5.85.b), $\mathrm{Re}\, g \in \mathfrak{C}_r(X)$ and $\mathrm{Im}\, g \in \mathfrak{C}_r(X)$. For any $\varepsilon > 0$, by virtue of (6.57), there are $f_1, f_2 \in \mathfrak{B}$ such that $\|\mathrm{Re}\, g - f_1\|_u < \varepsilon/2$ and $\|\mathrm{Im}\, g - f_2\|_u < \varepsilon/2$. Therefore, we have $\|g - (f_1 + if_2)\|_u \leqslant \|\mathrm{Re}\, g - f_1\|_u + \|\mathrm{Im}\, g - f_2\|_u < \varepsilon$. Since $f_1 + if_2 \in \mathfrak{A}$, it follows that $g \in \mathfrak{A}^-$, and so $\mathfrak{A}^- = \mathfrak{C}_c(X)$. □

The notion of uniform continuity we introduce now is closely related to the concept of continuity.

Definition 6.60. Let (X, d) and (Y, ρ) be metric spaces, and let $f : X \to Y$ be a function. We say that f is *uniformly continuous* if for any $\varepsilon > 0$ there is $\delta > 0$ such that $d(x, y) < \delta$ implies $\rho(f(x), f(y)) < \varepsilon$.

Remark 6.61. From (6.34.c) it follows at once that each uniformly continuous function is continuous. The converse of this assertion fails in general. For example, the function $f : R \to R$ defined by $f(x) = x^2$, $x \in R$, is continuous on R but not uniformly continuous. The next theorem presents a special case in which the converse of this statement holds.

Theorem 6.62. *Let (X, d), (Y, ρ) and f be as in* (6.60). *If X is compact, and f is continuous on X, then f is uniformly continuous.*

Proof. Let $\varepsilon > 0$. Then, according to (6.34.c), for each $x \in X$ there is $\delta_x > 0$ such that

$$f(S_d(x, \delta_x)) \subset S_\rho(f(x), \varepsilon/2). \tag{1}$$

Since X is compact, and $\{S_d(x, \delta_x/2) : x \in X\}$ is an open cover of X, there are $x_1, \ldots, x_n \in X$ such that $X = \cup_{i=1}^n S_d(x_i, \delta_{x_i}/2)$. Put $\delta = \inf_{1 \leqslant i \leqslant n} \delta_{x_i}/2$. Now let $x, y \in X$ be such that $d(x, y) < \delta$. Then $x \in S_d(x_i, \delta_{x_i}/2) \subset S_d(x_i, \delta_{x_i})$ for some i. Consequently, we have $d(x_i, y) \leqslant d(x_i, x) + d(x, y) < (\delta_{x_i}/2) + \delta \leqslant \delta_{x_i}$, and so $y \in S_d(x_i, \delta_{x_i})$. Therefore, applying (1), we get $\rho(f(x), f(y)) \leqslant \rho(f(x), f(x_i)) + \rho(f(x_i), f(y)) < (\varepsilon/2) + (\varepsilon/2) = \varepsilon$. Thus f is uniformly continuous. □

If X is a compact metric space, then the Arzelà-Ascoli theorem we will prove in (6.67) characterizes the compact subsets of $\mathfrak{C}_r(X)$. We begin with some definitions.

Definition 6.63. A set $\mathfrak{F} \subset R^X$ is said to be *equibounded* if there is $M > 0$ such that $\|f\|_u \leqslant M$ for all $f \in \mathfrak{F}$.

Definition 6.64. Let (X, d) and (Y, ρ) be metric spaces, and let $\mathfrak{F} \subset Y^X$. We say that $\mathfrak{F}$ is *uniformly equicontinuous* if for each $\varepsilon > 0$ there is $\delta > 0$ such that $d(x, y) < \delta$ implies $\rho(f(x), f(y)) < \varepsilon$ for all $f \in \mathfrak{F}$.

Example 6.65. Let X and Y be metric spaces. Then any finite set of uniformly continuous functions from X into Y is uniformly equicontinuous.

To prove the Arzelà-Ascoli theorem, we need the next lemma.

Lemma 6.66. *Let (X, d) be a metric space and let $A \subset X$. Then the following are equivalent:*

(i) *A^- is compact;*

(ii) *each sequence in A contains a convergent subsequence.*

Proof. If A^- is compact, then A^- is sequentially compact (6.19). Hence (i) implies (ii). Assume now that (ii) holds, and consider a sequence $\{x_n : n \in N\} \subset A^-$. For every $n \in N$, according to (6.14), choose $x'_n \in A^-$ such that $d(x_n, x'_n) < 1/n$. Then there are $x \in A^-$ and a subsequence $\{x'_{n_k} : k \in N\}$ such that $x'_{n_k} \to x$. Since $d(x, x_{n_k}) \leqslant d(x, x'_{n_k}) + 1/n_k$, it follows that $x_{n_k} \to x$. Therefore, A^- is sequentially compact. Thus, by (6.19), A^- is compact, and so (ii) implies (i). □

Arzelà-Ascoli theorem 6.67. *Let (X, d) be a compact metric space and let $\mathfrak{F} \subset \mathfrak{C}_r(X)$, where $\mathfrak{C}_r(X)$ is endowed with the uniform metric. Then $\mathfrak{F}^-$ is compact if and only if $\mathfrak{F}$ is equibounded and uniformly equicontinuous.*

Proof. Assume that $\mathfrak{F}^-$ is compact, and let $g \in \mathfrak{F}^-$. Then, by (6.37.ii), the function $d_g : \mathfrak{F}^- \to R$ defined by $d_g(f) = \|f - g\|_u$, $f \in \mathfrak{F}^-$, is continuous on $\mathfrak{F}^-$. By virtue of (5.47), there is $h \in \mathfrak{F}^-$ such that $\|f - g\|_u \leqslant \|h - g\|_u$ for any $f \in \mathfrak{F}^-$. Therefore, for each $f \in \mathfrak{F}$, we have $\|f\|_u \leqslant \|f - g\|_u + \|g\|_u \leqslant \|h - g\|_u + \|g\|_u$. This shows that $\mathfrak{F}$ is equibounded. Further, since $\mathfrak{F}^-$ is compact, (6.31) shows that $\mathfrak{F}^-$ is *totally bounded*. Consequently, for $\varepsilon > 0$, there are $f_1, \ldots, f_n \in \mathfrak{C}_r(X)$

such that

$$\mathfrak{F}^{-} \subset \bigcup_{i=1}^{n} \{f \in \mathfrak{C}_r(X) : \|f_i - f\|_u < \varepsilon/3\}. \tag{1}$$

Since $f_1, \ldots, f_n$ are continuous on X, (6.62) and (6.65) show that $\{f_1, \ldots, f_n\}$ is uniformly equicontinuous. Therefore, there is $\delta > 0$ such that $d(x, y) < \delta$ implies

$$|f_i(x) - f_i(y)| < \varepsilon/3, \quad i = 1, \ldots, n. \tag{2}$$

If $f \in \mathfrak{F}$, then, by (1), $\|f_i - f\|_u < \varepsilon/3$ for some i. Hence, if $d(x, y) < \delta$, then, using (2), we have $|f(x) - f(y)| \leqslant |f(x) - f_i(x)| + |f_i(x) - f_i(y)| + |f_i(y) - f(y)| \leqslant 2\|f_i - f\|_u + |f_i(x) - f_i(y)| < \varepsilon$. This shows that $\mathfrak{F}$ is uniformly equicontinuous.

Conversely, suppose that $\mathfrak{F}$ is equibounded and uniformly equicontinuous. Let $\{f_n : n \in N\} \subset \mathfrak{F}$. Then, according to (6.66), it will suffice to show that $\{f_n : n \in N\}$ contains a convergent subsequence. Since $\mathfrak{F}$ is equibounded, there is $M > 0$ such that $|f_n(x)| \leqslant M$, $x \in X$, $n \in N$. Since X is compact, (6.19) and (6.18) show that X is separable, and so there exists a countable set A dense in X. The case in which A is finite is simpler and is left to the reader. If A is denumerable, then put $A = \{x_i : i \in N\}$. We choose inductively a sequence

$$\begin{aligned}
&(f_{n_{11}}, f_{n_{12}}, \ldots, f_{n_{1k}}, f_{n_{1,k+1}}, \ldots)\\
&(f_{n_{21}}, f_{n_{22}}, \ldots, f_{n_{2k}}, f_{n_{2,k+1}}, \ldots)\\
&\vdots\\
&(f_{n_{k1}}, f_{n_{k2}}, \ldots, f_{n_{kk}}, f_{n_{k,k+1}}, \ldots)\\
&(f_{n_{k+1,1}}, f_{n_{k+1,2}}, \ldots, f_{n_{k+1,k}}, f_{n_{k+1,k+1}}, \ldots)\\
&\vdots
\end{aligned} \tag{3}$$

of subsequences of $\{f_n : n \in N\}$ as follows. The first row in (3) is a subsequence of $\{f_n : n \in N\}$ so chosen that the sequence $\{f_{n_{1m}}(x_1) : m \in N\}$ is convergent. Since $[-M, M]$ is compact, and $f_n(x_1) \in [-M, M], n \in N$, (6.19) shows that such a subsequence must exist. If kth row in (3) has been defined, then $(k+1)$st row in (3) is a subsequence of $\{f_{n_{km}} : m \in N\}$ so chosen that the sequence $\{f_{n_{k+1,m}}(x_{k+1}) : m \in N\}$ is convergent. Such a subsequence must exist, since $f_{n_{km}}(x_{k+1}) \in [-M, M], m \in N$, and $[-M, M]$ is compact. Set $n_k = n_{kk}, k \in N$. Then $\{f_{n_k} : k \in N\}$ is a subsequence of $\{f_n : n \in N\}$. Moreover, for each $i \in N, \{f_{n_k} : k \geqslant i\}$ is a subsequence of $\{f_{n_{im}} : m \in N\}$, and so the sequence $\{f_{n_k}(x_i) : k \in N\}$ is convergent. Let now $\varepsilon > 0$. Since $\mathfrak{F}$ is uniformly equicontinuous, there is $\delta > 0$ such that $d(x, y) < \delta$ implies

$$|f(x) - f(y)| < \varepsilon/3, \quad f \in \mathfrak{F}. \tag{4}$$

Further, since X is compact, X is *totally bounded* (6.31), and so there exists a finite family $\{S_1, \ldots, S_l\}$ of open balls with radius $\delta/2$ such that $X = \cup_{j=1}^{l} S_j$. Since A is dense in X, we can select $a_j \in A \cap S_j$ for each $j = 1, \ldots, l$. Since the sequences $\{f_{n_k}(a_j) : k \in N\}$, $1 \leqslant j \leqslant l$, are convergent, (6.21.a) shows that there is $k_\varepsilon \in N$ such that $p, q \geqslant k_\varepsilon$ implies

$$|f_{n_p}(a_j) - f_{n_q}(a_j)| < \varepsilon/3, \quad j = 1, \ldots, l. \tag{5}$$

If $x \in X$, then $x \in S_j$ for some j, and so $d(x, a_j) < \delta$. Therefore, for $p, q \geqslant k_\varepsilon$, using (4) and (5), we have $|f_{n_p}(x) - f_{n_q}(x)| \leqslant |f_{n_p}(x) - f_{n_p}(a_j)| + |f_{n_p}(a_j) - f_{n_q}(a_j)| + |f_{n_q}(a_j) - f_{n_q}(x)| < \frac{\varepsilon}{3} + \frac{\varepsilon}{3} + \frac{\varepsilon}{3} = \varepsilon$. Consequently, $\|f_{n_p} - f_{n_q}\|_u \leqslant \varepsilon$ for $p, q \geqslant k_\varepsilon$. This means that $\{f_{n_k} : k \in N\}$ is a Cauchy sequence in $\mathfrak{C}_r(X)$. Since $\mathfrak{C}_r(X)$ is complete relative to the uniform metric (6.43), the subsequence $\{f_{n_k} : k \in N\}$ is convergent. □

The process of obtaining the subsequence $\{f_{n_k} : k \in N\}$ in (6.67) is called the *Cantor diagonal method.*

Exercise 6.68. Notation is as in (6.3).

(a) Prove that $\overline{S}(x, r)$ is a closed set.

(b) Give an example to show that adherence of $S(x, r)$ may be different from $\overline{S}(x, r)$.

Exercise 6.69. Prove that $\overline{R}$ with its usual topology is metrizable.

Exercise 6.70. Let (X, d) be a separable metric space and let $A \subset X$. Use (6.12) and (5.20) to show that (A, d_A) is a separable metric space.

Exercise 6.71. Use (6.12) and (5.14) to prove that each open subset of a separable metric space is a union of countably many closed balls.

Exercise 6.72. Let X be a metric space and let $\{x_n : n \in N\} \subset X$. If $\{x_n : n \in N\}$ contains no convergent subsequence, then any subset of $\{x_n : n \in N\}$ is closed.

Exercise 6.73. Let X be a metric space and let $A \subset X$. Show that A is *totally bounded* if and only if A^- is *totally bounded.*

Exercise 6.74. If (X, d) is a metric space, then d is continuous on $X \times X$.

Exercise 6.75. If X is a compact metric space, then $\operatorname{diam}(X) < \infty$.

Exercise 6.76. Let (X, d) be a metric space, and let F be a closed subset of X. Construct a nonincreasing sequence $\{f_n : n \in N\} \subset [0, 1]^X$ of uniformly continuous functions such that $f_n \to 1_F$. [For $F \neq \emptyset$, set $f_n(x) = e^{-nd(x,F)}$, $x \in X$, $n \in N$.]

Exercise 6.77. Let (X, d) be a metric space. Use (5.41) and (5.40) to prove that the topology generated by d coincides with the topology generated by $\{d_x : x \in X\}$, where d_x stands for the section of d at x.

Exercise 6.78. Notation is as in (3.37). Show that, with the composition operation $\circ$, $\mathfrak{P}$ is a group.

Exercise 6.79. Let X be a topological space, and let $\mathfrak{A} \subset K^X$ be a set of continuous functions that separates points of X. Prove that X is a Hausdorff space.

Exercise 6.80. Let (X, d) be a metric space, and define the function $\omega : \mathfrak{B}(X) \times]0, \infty[\to R$ by $\omega(f, \delta) = \sup\{|f(x) - f(y)| : x, y \in X,\ d(x, y) < \delta\}$ for any $f \in$

$\mathfrak{B}(X)$ and $\delta \in]0, \infty[$. For $f \in \mathfrak{B}(X)$ and $\delta \in]0, \infty[$, the number $\omega(f, \delta)$ is called the *δ-oscillation of f*. Prove the following.

(a) For each $f \in \mathfrak{B}(X)$, the section of ω at f is nondecreasing.

(b) Whatever $\delta \in]0, \infty[$, let ω_δ be the section of ω at δ. For $f, g \in \mathfrak{B}(X)$, $|\omega_\delta(f) - \omega_\delta(g)| \leqslant 2\|f - g\|_u$, and so ω_δ is uniformly continuous.

Exercise 6.81. Let X be a topological space, let Y be a metric space, and let $f : X \to Y$ be a function. For $x \in X$, the number $\omega(x) = \inf\{\text{diam} f(U) : U \text{ is a neighborhood of } x\}$ is called the *oscillation of f at x*. Prove the following.

(a) f is continuous at x if and only if $\omega(x) = 0$.

(b) For any $a > 0$, the set $\{x : \omega(x) < a\}$ is open.

(c) The set $\{x : f \text{ is continuous at } x\}$ is a countable intersection of open sets.

Exercise 6.82. Let X be a set, let $\{f_n : n \in N\} \subset \mathfrak{B}_r(X)$, and let $f \in \mathfrak{B}_r(X)$. Show that $\{f_n : n \in N\}$ is equibounded whenever $f_n \xrightarrow{u} f$.

Exercise 6.83. Let (X, d) and (Y, ρ) be metric spaces, and let $\mathfrak{F} \subset Y^X$. We say that $\mathfrak{F}$ is *equicontinuous* if for each $x \in X$ and $\varepsilon > 0$ there is $\delta > 0$ (depending on x and ε) such that $d(x, y) < \delta$ implies $\rho(f(x), f(y)) < \varepsilon$ for all $f \in \mathfrak{F}$. Mimic the proof of (6.62) to show that if X is compact, then any equicontinuous subset of Y^X is uniformly equicontinuous.

Exercise 6.84. Let $[a, b] \subset R$ be a closed interval, let $\{f_n : n \in N\} \subset R^{[a,b]}$ be a sequence of nondecreasing functions, and let $f \in R^{[a,b]}$ be such that $f_n(x) \to f(x)$ for any $x \in [a, b]$. Use (5.29) and (6.62) to prove that $f_n \xrightarrow{u} f$ whenever f is continuous on $[a, b]$.

Exercise 6.85. Use (6.37) and (5.75.d) to prove that each closed subset of a metric space is an intersection of countably many open sets.

Exercise 6.86. Let (X, d) and (Y, ρ) be metric spaces, and let $f : X \to Y$ be a function. For $\varepsilon, \delta > 0$, put $A_{\varepsilon,\delta} = \{x \in X : \text{there exist } y, z \in S_d(x, \delta) \text{ such that } \rho(f(y), f(z)) \geqslant \varepsilon\}$. Prove the following.

(a) The set $A_{\varepsilon,\delta}$ is open.

(b) $\{x : f \text{ is continuous at } x\} = \bigcap_{m \geqslant 1} \bigcup_{n \geqslant 1} A^c_{1/m,1/n}$.

Exercise 6.87. Let $[a, b] \subset R$, and let $\mathfrak{A}$ be the set of all polynomials of the form $p(x) = \sum_{k=0}^{n} \alpha_k x^k$, $x \in [a, b]$, where $\alpha_k \in R$ for $k = 0, \ldots, n$, and $n \in N$. Verify that $\mathfrak{A}$ is dense in $\mathfrak{C}_r([a, b])$.

Exercise 6.88. Let (X, d) be a metric space, and set $d'' = d/(1 + d)$.

(a) Show that d'' is a metric for X.

(b) Use (6.6.c) and (6.10) to prove that $\mathcal{T}_d = \mathcal{T}_{d''}$. [Hint. $d'/2 \leqslant d'' \leqslant d'$, where d' is as in (6.10).]

Exercise 6.89. Let $(X^*, \mathcal{T}^*)$ be as in (5.94), where $X = R$. Show that $(X^*, \mathcal{T}^*)$ is metrizable. [Hint. Verify that $R \cup \{\delta\}$ and the circle $\{(x_1, x_2) \in R^2 : x_1^2 + x_2^2 = 1\}$ (equipped with the Euclidean metric) are homeomorphic topological spaces.]

Exercise 6.90. If A and B are disjoint closed subsets of a metric space, then there are disjoint open sets U and V with $A \subset U$ and $B \subset V$.

Exercise 6.91. Let (X, d) be a metric space, let $A \subset X$ be closed, let $n \in N$, and set $B_n = \{x \in X : d(x, A) \geqslant 1/n\}$. Let $\varphi(t) = \begin{cases} t \text{ if } 0 \leqslant t \leqslant 1 \\ 1 \text{ if } t > 1 \end{cases}$, and define $f_n(x) = \varphi(nd(x, A)), x \in X$. Prove the following.

(a) B_n is closed and $\cup_{n \in N} B_n = A^c$.

(b) $f_n(A) = \{0\}$ and $f_n(B) = \{1\}$.

(c) f_n is uniformly continuous. [Show that $|f_n(x) - f_n(y)| \leqslant nd(x, y), x, y \in X$.]

Exercise 6.92. Let X be a metric space, and let $f : X \to R$ be a continuous function. For $a > 0$, put $A = \{x : |f(x)| \leqslant a\}$, and define $g = f 1_A + a(f/|f|) 1_{A^c}$. Show that g is continuous on X.

Exercise 6.93. Let (X, d) be a metric space. For $\emptyset \neq A \subset X$ and $\varepsilon > 0$, write $A_\varepsilon = \cup_{x \in A} \overline{S}(x, \varepsilon)$. Let $\mathcal{F}$ denote the family of all closed subsets of X. For $A, B \in \mathcal{F} - \{\emptyset\}$, define $d_H(A, B) = \inf\{\varepsilon > 0 : A_\varepsilon \subset B$ and $B_\varepsilon \subset A\}$. Prove that d_H is a metric for $\mathcal{F} - \{\emptyset\}$. d_H is called the *Hausdorff distance*.

Exercise 6.94. Let X be a metric space, let $\{x_n : n \in N\} \subset X$, and let $x \in X$. Use (6.37.iii) to show that $x_n \to x$ whenever $f(x_n) \to f(x)$ for all $f \in \mathfrak{C}_r(X)$.

Exercise 6.95. Let (X, d) be a metric space. For $\emptyset \neq A \subset X$ and $\varepsilon > 0$, write $A^\varepsilon = \{x \in X : d(x, A) < \varepsilon\}$.

(a) Prove that $\cap_{\varepsilon > 0} A^\varepsilon = A^-$.

(b) For $\varepsilon, \varepsilon' > 0$, show that $(A^\varepsilon)^{\varepsilon'} \subset A^{\varepsilon + \varepsilon'}$, and that this inclusion may be strict.

Exercise 6.96. Let (X, d) be a metric space, and let $f : X \to R$ be a function. Prove the following.

(a) f is lower semicontinuous at $x \in X$ if and only if $\liminf_n f(x_n) \geqslant f(x)$ whenever $\{x_n : n \in N\} \subset X$ and $x_n \to x$.

(b) If f is lower semicontinuous on X, then there is a nondecreasing sequence $\{f_n : n \in N\} \subset R^X$ of continuous functions such that $f_n \to f$ and $|f_n| \leqslant \|f\|_u, n \in N$. [Hints. If $f(X) \subset [0, \infty[$, take $f_n(x) = \inf_{y \in X}(f(y) + nd(x, y)), x \in X$. If $|f| \leqslant M < \infty$, take $f_n = g_n - M$, where $\{g_n : n \in N\} \subset [0, \infty[^X$ is a nondecreasing sequence of continuous functions such that $g_n \to f + M$. In general, apply (5.95.h) with $h(z) = \arctan z + \pi/2, z \in R$.]

Exercise 6.97. Let X be a topological space, let $A \subset X$ be compact, and let $\{f_n : n \in N\} \subset]-\infty, 0]^X$ be a nondecreasing sequence of lower semicontinuous functions on X such that $f_n \to 0$. Prove that $f_n|_A \xrightarrow{u} 0$. [Mimic the proof of Dini's theorem (6.44).]

Exercise 6.98. Let $(X_i, d_i), 1 \leqslant i \leqslant n$, be metric spaces, and define $d(\mathbf{x}, \mathbf{y}) = \sup_{1 \leqslant i \leqslant n} d_i(x_i, y_i)$ for $\mathbf{x} = (x_1, \ldots, x_n)$ and $\mathbf{y} = (y_1, \ldots, y_n)$.

(a) Show that $\mathcal{T}_d$ coincides with the product topology on $\prod_{i=1}^n X_i$.

(b) If (X_i, d_i) is complete for1 $\leqslant i \leqslant n$, prove that $(\prod_{i=1}^n X_i, d)$ is complete.

Exercise 6.99. Let $\{Y_i : i \in I\}$ be a countable family of sets, put $Y = \prod_{i \in I} Y_i$, and let $\boldsymbol{\alpha} = (\alpha_i)_{i \in I} \in]0, \infty[^I$ be such that $\sum_{i \in I} \alpha_i < \infty$. For $\mathbf{y} = (y_i)_{i \in I} \in Y$ and

$\mathbf{y}' = (y_i')_{i \in I} \in Y$, define $d_{\boldsymbol{\alpha},H}(\mathbf{y}, \mathbf{y}')$ to be the sum of the values α_i over those indices for which $y_i \neq y_i'$. Prove the following.

(a) $d_{\boldsymbol{\alpha},H}$ is a metric for Y. $d_{\boldsymbol{\alpha},H}$ is called the $\boldsymbol{\alpha}$-*Hamming distance*.

(b) For $\mathbf{y}_n = (y_{ni})_{i \in I} \in Y, n \in N$, and $\mathbf{y} = (y_i)_{i \in I} \in Y$, $d_{\boldsymbol{\alpha},H}(\mathbf{y}, \mathbf{y}_n) \to 0$ if and only if, for each $i \in I$, there is $n(i)$ such that $y_{ni} = y_i$, $n \geqslant n(i)$.

(c) For $i \in I$, consider the topological space $(Y_i, \mathcal{T}_{d_i})$, where d_i is the discrete metric for Y_i. Then $\mathcal{T}_{d_{\boldsymbol{\alpha},H}}$ coincides with the product topology on Y. [Use (b) and (6.77).]

(d) $(Y, d_{\boldsymbol{\alpha},H})$ is separable if and only if Y_i is finite for any $i \in I$.

3 Measure Spaces

This chapter is devoted to the study of three fundamental notions: measurable space, measurable function and measure. The concepts of a measurable space and of a measurable function depend in no way upon any measure. However, in practice measurable spaces and measurable functions are almost always used in the presence of measure. This justifies the adoption of the term "measurable" for theirs names.

7 Measurable Spaces

This section is devoted to a detailed study of algebras and σ-algebras of sets, which are families of sets indispensable for introducing the notions of measurability and measure. As we stated in the beginning of Chapter 2, there is a certain analogy between the concepts of a topological space and of a measurable space that can be easily detected by a parallel reading of Sections 4 and 7. At the same time, a straight relation between the two concepts is revealed by introducing the notion of a Borel set. In the final part of the present section we introduce a special family of sets which is called a Dynkin system. The Dynkin system replaces mostly the notion of a monotone class encountered in various books dealing with measure theory, and proves to be a highly efficient tool.

Definitions 7.1. Let X be a set, and let $\mathcal{A}$ be a nonempty family of subsets of X such that:

(i) $A \in \mathcal{A}$ implies $A^c \in \mathcal{A}$;

(ii) $A, B \in \mathcal{A}$ implies $A \cup B \in \mathcal{A}$.

Then $\mathcal{A}$ is called an *algebra of sets*. If (ii) is replaced by

(iii) $\{A_n : n \in N\} \subset \mathcal{A}$ implies $\bigcup_{n \in N} A_n \in \mathcal{A}$, then $\mathcal{A}$ is called a *σ-algebra of sets*.

A pair $(X, \mathcal{A})$, where X is a set and $\mathcal{A}$ is a σ-algebra of subsets of X, is called a *measurable space*. The sets in $\mathcal{A}$ are called *$\mathcal{A}$-measurable sets*.

Remarks 7.2. (a) Each σ-algebra of sets is an algebra of sets, and each finite algebra of sets is a σ-algebra of sets.

(b) Let $\mathcal{A}$ be an algebra of subsets of X. Then, since $\mathcal{A}$ is nonempty, $X \in \mathcal{A}$ and $\emptyset \in \mathcal{A}$. If $A, B \in \mathcal{A}$, then, applying De Morgan's laws (2.17), it follows readily that $A \cap B \in \mathcal{A}$, $A - B \in \mathcal{A}$, and $A \Delta B \in \mathcal{A}$; also, if $\mathcal{A}$ is a σ-algebra, and $\{A_n : n \in N\} \subset \mathcal{A}$, then $\bigcap_{n \in N} A_n \in \mathcal{A}$.

(c) Any intersection of algebras of subsets of X is an algebra of subsets of X, and any intersection of σ-algebras of subsets of X is a σ-algebra of subsets of X.

Analysis and Probability. http://dx.doi.org/10.1016/B978-0-12-401665-1.00003-5

Examples 7.3. (a) Let X be a set. The family $\{\emptyset, X\}$ is a σ-algebra called the *trivial σ-algebra*, and the family $\mathcal{P}(X)$ is a σ-algebra called the *discrete σ-algebra*.

(b) Let X be a set and $A \subset X$. Then $\mathcal{A} = \{A, A^c, \emptyset, X\}$ is a σ-algebra. Obviously, $\mathcal{A}$ is the smallest σ-algebra of subsets of X containing A.

(c) Let X be a set. The family $\mathcal{A} = \{A \in \mathcal{P}(X) : A \text{ or } A^c \text{ is countable}\}$ is a σ-algebra.

(d) Let $\mathcal{B}$ be the family of all subsets B or R of the form $B = \cup_{i=1}^{n} I_i$, where I_i is an interval of the form $]a, b]$ or $]a, \infty[$ for $i = 1, \ldots, n$, and $I_i \cap I_j = \emptyset$ for $i \neq j$. Then the family $\mathcal{A} = \{\emptyset\} \cup \mathcal{B}$ is an algebra of sets. Since $]0, 1 - 1/n] \in \mathcal{A}$ for each $n \geqslant 2$, but $\cup_{n \geqslant 2}]0, 1 - 1/n] =]0, 1[\notin \mathcal{A}$, it follows that $\mathcal{A}$ is not a σ-algebra.

In most cases it is not possible to describe explicitly the sets of a σ-algebra. Only a subfamily of a σ-algebra is sufficiently well known in general, and this subfamily generates the whole σ-algebra as follows.

Definition 7.4. Let X be a set and $\mathcal{M} \subset \mathcal{P}(X)$. The intersection of all σ-algebras of subsets of X that contain $\mathcal{M}$, written $\sigma(\mathcal{M})$, is a σ-algebra called the *σ-algebra generated by $\mathcal{M}$*. Plainly, $\sigma(\mathcal{M})$ is the smallest σ-algebra of subsets of X containing $\mathcal{M}$.

Remarks 7.5. (a) Notation is as in (7.4). Put $\mathcal{M}_c = \{A^c : A \in \mathcal{M}\}$. Then $\sigma(\mathcal{M}) = \sigma(\mathcal{M}_c)$. This shows that different families may generate the same σ-algebra.

(b) Let $\mathcal{M}_1, \mathcal{M}_2 \subset \mathcal{P}(X)$. Then $\sigma(\mathcal{M}_1) \subset \sigma(\mathcal{M}_2)$ if and only if $\mathcal{M}_1 \subset \sigma(\mathcal{M}_2)$. Therefore, $\sigma(\mathcal{M}_1) = \sigma(\mathcal{M}_2)$ if and only if $\mathcal{M}_1 \subset \sigma(\mathcal{M}_2)$ and $\mathcal{M}_2 \subset \sigma(\mathcal{M}_1)$; also, $\sigma(\mathcal{M}_1) \subset \sigma(\mathcal{M}_2)$ whenever $\mathcal{M}_1 \subset \mathcal{M}_2$.

(c) For $\mathcal{M} \subset \mathcal{P}(X)$, we have $\sigma(\mathcal{M}) = \sigma(\mathcal{L})$, where $\mathcal{L}$ is the family of all finite intersections of sets in $\mathcal{M}$. Observe that $L_1, L_2 \in \mathcal{L}$ implies $L_1 \cap L_2 \in \mathcal{L}$, and that $\mathcal{L}$ is countable whenever so is $\mathcal{M}$. Therefore, when a σ-algebra is generated by a countable family we may always assume it is generated by a countable family closed under intersection.

Examples 7.6. (a) Let X be a set and let $\{A_i : i \in I\}$ be a partition of X, i.e. $\cup_{i \in I} A_i = X$ and $A_i \cap A_j = \emptyset$ for $i \neq j$. If I is countable, then $\sigma(\{A_i : i \in I\}) = \{\cup_{j \in J} A_j : J \subset I\}$.

(b) Let $(X, \mathcal{T})$ be a topological space. When confusion appears impossible, the σ-algebra generated by $\mathcal{T}$ will be denoted $\mathcal{B}(X)$. The elements of $\mathcal{B}(X)$ will be called *Borel sets of X*. Unless otherwise stated, we will always assume that X is endowed with the σ-algebra $\mathcal{B}(X)$.

In (7.9) and (7.10) we will present two important results concerning the examples of (7.6).

Definition 7.7. Let $\mathcal{A}$ be an algebra of sets. A nonempty set $A \in \mathcal{A}$ is called an *atom of $\mathcal{A}$* if $B \in A$ and $\emptyset \neq B \subset A$ imply $B = A$.

Lemma 7.8. *Let $\mathcal{A}$ be an algebra of sets. For $B \in \mathcal{A}$, let $\mathcal{A}_B$ be the family of all atoms A such that $A \subset B$. If $\mathcal{A}$ is finite, then $\mathcal{A}_B \neq \emptyset$ whenever $B \neq \emptyset$.*

Proof. Let $B \in \mathcal{A}$ be such that $B \neq \emptyset$. To get a contradiction, assume that $\mathcal{A}_B = \emptyset$. Then, since B is not an atom, there is $B_1 \in \mathcal{A}$ such that $\emptyset \neq B_1 \subset B$ and $B_1 \neq B$. If the sets $B_1, \ldots, B_n$ have been chosen, then, since B_n is not an atom, there is $B_{n+1} \in \mathcal{A}$ such that $\emptyset \neq B_{n+1} \subset B_n$ and $B_{n+1} \neq B_n$. Therefore, we may construct inductively a strictly decreasing sequence $\{B_n : n \in N\} \subset \mathcal{A}$. Since $\mathcal{A}$ is finite, this is a contradiction. Thus $\mathcal{A}_B \neq \emptyset$. □

Theorem 7.9. *Let $\mathcal{A}$ be a finite algebra of subsets of X. Then $\mathcal{A}_X$ is a partition of X, and $\mathcal{A} = \sigma(\mathcal{A}_X)$.*

Proof. As the case $X = \emptyset$ is trivial, we will suppose $X \neq \emptyset$. Let $B \in \mathcal{A}$ be such that $B \neq \emptyset$. Then

$$\bigcup \mathcal{A}_B = B. \tag{1}$$

Actually, $\cup\mathcal{A}_B \subset B$. Since $\mathcal{A}_B$ is finite, $B - (\cup\mathcal{A}_B) \in \mathcal{A}$. If $B - (\cup\mathcal{A}_B) \neq \emptyset$, then, according to (7.8), there would exist an atom $A \subset B - (\cup\mathcal{A}_B)$. Since $A \in \mathcal{A}_B$, the very inclusion would be impossible. Thus (1) is established. In particular, we have

$$\bigcup \mathcal{A}_X = X. \tag{2}$$

Now let $A, B \in \mathcal{A}_X$ be such that $A \neq B$. Assume that $A \cap B \neq \emptyset$. Then, since $A \cap B \subset A$, and A is an atom of $\mathcal{A}$, it follows that $A \cap B = A$. Therefore, we have $A \subset B$, and so $A = B$. The contradiction we got shows that

$$A \cap B = \emptyset. \tag{3}$$

From (2) and (3) it follows that $\mathcal{A}_X$ is a partition of X. Evidently, $\mathcal{A}_X$ is finite. Thus, using (1) and (7.6.a), we obtain $\mathcal{A} = \sigma(\mathcal{A}_X)$. □

Theorem 7.10. *The following families of subsets of R generate $\mathcal{B}(R)$:*
(i) *the family of all closed subsets;*
(ii) *the family of all open intervals;*
(iii) *the family of all closed intervals;*
(iv) *the family of all intervals of the form $]a, b]$, where $a \in R$;*
(v) *the family of all intervals of the form $[a, b[$, where $b \in R$;*
(vi) *the family of all intervals of the form $]a, \infty[$;*
(vii) *the family of all intervals of the form $[a, \infty[$;*
(viii) *the family of all intervals of the form $]-\infty, b[$;*
(ix) *the family of all intervals of the form $]-\infty, b]$;*
(x) *any family in* (iv)–(ix) *with $a \in Q$ or $b \in Q$;*
(xi) *the family of all intervals;*
(xii) *the algebra of sets $\mathcal{A}$ in* (7.3.d).

Proof. This theorem is proved by making use of (7.5.b). From (7.5.a) it follows at once that the family (i) generates $\mathcal{B}(R)$. In view of (5.70), the family (ii) generates $\mathcal{B}(R)$. Let $a, b \in R$ be such that $a < b$. Since $[a, \infty[= \cap_{n\in N}]a - 1/n, \infty[$, $]-\infty, b] = \cap_{n\in N}]-\infty, b+1/n[$, $[a, b] = [a, \infty[\cap]-\infty, b]$, and $R = \cup_{n\in N}]-n, n[$, it follows that the σ-algebra generated by the family (ii) contains the σ-algebra generated by the family (iii). Further, since $]a, \infty[= \cup_{n\in N}[a+1/n, \infty[$, $]-\infty, b[= \cup_{n\in N}]-\infty, b-1/n]$, $]a, b[=]a, \infty[\cap]-\infty, b[$, and $R = \cup_{n\in N}[-n, n]$, it follows that the σ-algebra generated by the family (ii) is included in the σ-algebra generated by the family (iii). Hence both families (ii) and (iii) generate the σ-algebra $\mathcal{B}(R)$. In a similar manner one can easily show that each of the families (iv)–(xi) generated $\mathcal{B}(R)$. □

In what follows we construct new algebras and σ-algebras of sets from other algebras and σ-algebras of sets originally given.

Theorem 7.11. *Let $\mathcal{A}$ be an algebra of subsets of Y, and let $f : X \to Y$ be a function. Then $f^{-1}(\mathcal{A})$ is an algebra of subsets of X. If $\mathcal{A}$ is a σ-algebra, then $f^{-1}(\mathcal{A})$ is a σ-algebra.*

Proof. This is an immediate consequence of (2.19.i) and (2.19.iii). □

Example 7.12. Let $\mathcal{A}$ be an algebra of subsets of X, and let $A \subset X$. Then $A \cap \mathcal{A} = i_A^{-1}(\mathcal{A})$ is an algebra of subsets of A called the *relative algebra on A induced by $\mathcal{A}$*. If $\mathcal{A}$ is a σ-algebra, then $A \cap \mathcal{A}$ is a σ-algebra of subsets of A called the *relative σ-algebra on A induced by $\mathcal{A}$*. Obviously, $C \in A \cap \mathcal{A}$ if and only if $C = A \cap B$ for some $B \in \mathcal{A}$. We will always suppose that A is endowed with the algebra of sets $A \cap \mathcal{A}$.

Theorem 7.13. *Let $f : X \to Y$ be a function and $\mathcal{M} \subset \mathcal{P}(Y)$. Then $f^{-1}(\sigma(\mathcal{M})) = \sigma(f^{-1}(\mathcal{M}))$.*

Proof. In view of Theorem (7.11), $f^{-1}(\sigma(\mathcal{M}))$ is a σ-algebra of subsets of X which contains $f^{-1}(\mathcal{M})$, and so $f^{-1}(\sigma(\mathcal{M})) \supset \sigma(f^{-1}(\mathcal{M}))$. Now let $\mathcal{B} = \{B \in \sigma(\mathcal{M}) : f^{-1}(B) \in \sigma(f^{-1}(\mathcal{M}))\}$. Evidently, we have $\mathcal{B} \neq \emptyset$, and

$$\mathcal{M} \subset \mathcal{B} \subset \sigma(\mathcal{M}). \tag{1}$$

Since $\sigma(\mathcal{M})$ and $\sigma(f^{-1}(\mathcal{M}))$ are σ-algebras, using (2.19.i) and (2.19.iii), it follows at once that $\mathcal{B}$ is a σ-algebra of subsets of Y. From (1) we get $\sigma(\mathcal{M}) = \mathcal{B}$, and so $f^{-1}(\sigma(\mathcal{M})) \subset \sigma(f^{-1}(\mathcal{M}))$. □

Corollary 7.14. *Let X be a set, let $\mathcal{M} \subset \mathcal{P}(X)$ and $A \subset X$. Then the relative σ-algebra on A induced by the σ-algebra generated by $\mathcal{M}$ coincides with the σ-algebra generated by the family on A induced by $\mathcal{M}$.*

Proof. By virtue of Theorem (7.13), we have $i_A^{-1}(\sigma(\mathcal{M})) = \sigma(i_A^{-1}(\mathcal{M}))$. □

Definition 7.15. Let $\{(Y_i, \mathcal{A}_i) : i \in I\}$ be an indexed family of measurable spaces, and let $f_i : X \to Y_i, i \in I$, be functions. Then $\sigma(\cup_{i\in I} f_i^{-1}(\mathcal{A}_i))$ is a σ-algebra of subsets of X called the *σ-algebra generated by* $\{f_i : i \in I\}$. Sometimes we will write $\sigma(f_i, i \in I)$ for $\sigma(\cup_{i\in I} f_i^{-1}(\mathcal{A}_i))$; if $I = \{n \in Z : n \geqslant m\}$ for some $m \in Z$, then we will write $\sigma(f_n, n \geqslant m)$ instead of $\sigma(\cup_{n\geqslant m} f_n^{-1}(\mathcal{A}_n))$; also, if $I = \{1, \ldots, n\}$, then we will write $\sigma(f_1, \ldots, f_n)$ instead of $\sigma(\cup_{i=1}^n f_i^{-1}(\mathcal{A}_i))$.

Example 7.16. Let I be a nonempty set, and let $\{(X_i, \mathcal{A}_i) : i \in I\}$ be a family of measurable spaces. For each $i \in I$, let π_i denote the projection from $\prod_{i\in I} X_i$ onto X_i. The σ-algebra generated by $\{\pi_i : i \in I\}$, written $\otimes_{i\in I}\mathcal{A}_i$, is a σ-algebra of subsets of $\prod_{i\in I} X_i$ called the *product σ-algebra*. The pair $(\prod_{i\in I} X_i, \otimes_{i\in I}\mathcal{A}_i)$ is called the *product measurable space*. If $I = \{n \in Z : n \geqslant m\}$ for some $m \in Z$, then we will write $\otimes_{n\geqslant m}\mathcal{A}_n$ instead of $\otimes_{i\in I}\mathcal{A}_i$; also, if $I = \{1, \ldots, n\}$, then we will write $\otimes_{i=1}^n\mathcal{A}_i$ or $\mathcal{A}_1 \otimes \cdots \otimes \mathcal{A}_n$ instead of $\otimes_{i\in I}\mathcal{A}_i$. Unless otherwise stated, we will always suppose that $\prod_{i\in I} X_i$ is endowed with the σ-algebra $\otimes_{i\in I}\mathcal{A}_i$.

Remarks 7.17. Notation is as in (7.16).

(a) Set $\mathcal{A} = \{\prod_{i\in I} A_i : A_i \in \mathcal{A}_i, i \in I,$ and $\{i \in I : A_i \neq X_i\}$ is finite$\}$. Since $\pi_i^{-1}(A) = \prod_{j\in I} A_j$, where $A_j = \begin{cases} A & \text{if } j = i \\ X_j & \text{if } j \neq i \end{cases}$, it follows readily that $\otimes_{i\in I}\mathcal{A}_i = \sigma(\mathcal{A})$. The sets in $\mathcal{A}$ are called *product cylinders.* More generally, if $J \subset I$ is nonempty and finite, and $B \in \otimes_{j\in J}\mathcal{A}_j$, then the set $\pi_J^{-1}(B) = B \times \prod_{i\in I-J} X_i$ is called a *cylinder* with *base* B.

(b) Let $\mathcal{A}_0$ denote the family of all finite unions of pairwise disjoint product cylinders. Then $\mathcal{A}_0$ is an algebra of subsets of $\prod_{i\in I} X_i$, and $\otimes_{i\in I}\mathcal{A}_i = \sigma(\mathcal{A}_0)$.

A knowledge of the structure of product σ-algebras is vital in order to construct product measure spaces. This is why we embark now on a detailed study of product σ-algebras.

Theorem 7.18. *Let I be a nonempty set, let $\{(Y_i, \mathcal{A}_i) : i \in I\}$ be a family of measurable spaces, and let $f : X \to \prod_{i\in I} Y_i$ be a function. Then $f^{-1}(\otimes_{i\in I}\mathcal{A}_i)$ coincides with $\sigma(\pi_i \circ f, i \in I)$, where π_i stands for the projection from $\prod_{i\in I} Y_i$ onto Y_i.*

Proof. By making use of (7.13), (2.38), and (2.6), we have

$$f^{-1}\left(\bigotimes_{i\in I}\mathcal{A}_i\right) = \sigma\left(f^{-1}\left(\bigcup_{i\in I}\pi_i^{-1}(\mathcal{A}_i)\right)\right) = \sigma\left(\bigcup_{i\in I}(\pi_i \circ f)^{-1}(\mathcal{A}_i)\right). \qquad \square$$

Corollary 7.19. *Let $\{(Y_i, \mathcal{A}_i) : i \in I\}$ be a nonempty family of measurable spaces, and let $f_i : X \to Y_i, i \in I$, be functions, Then $(f_i)_{i\in I}^{-1}(\otimes_{i\in I}\mathcal{A}_i)$ coincides with $\sigma(f_i, i \in I)$.*

Proof. Since $\pi_j \circ (f_i)_{i\in I} = f_j, j \in I$, the corollary follows immediately from (7.18). $\square$

Theorem 7.20. *Notation is as in* (7.16). *For each nonempty set $J \subset I$, put $\mathcal{A}_J = \pi_J^{-1}(\otimes_{j\in J}\mathcal{A}_j)$. Then $\emptyset \neq J_1 \subset J_2 \subset I$ implies $\mathcal{A}_{J_1} \subset \mathcal{A}_{J_2} \subset \otimes_{i\in I}\mathcal{A}_i$.*

Proof. For any $i \in I$, let π_i be the projection from $\prod_{i\in I} X_i$ onto X_i. Let J be a nonempty subset of I. Then, according to (2.26), we have $\pi_J = (\pi_j)_{j\in J}$. Therefore, in view of (7.19), $\mathcal{A}_J$ coincides with the σ-algebra generated by $\{\pi_j : j \in J\}$. Hence $\emptyset \neq J_1 \subset J_2 \subset I$ implies $\mathcal{A}_{J_1} \subset \mathcal{A}_{J_2} \subset \otimes_{i\in I}\mathcal{A}_i$. $\square$

Theorem 7.21. *Notation is as in* (7.16). *Let $\mathcal{J}$ be the family of all nonempty finite subsets of I. Set $\mathcal{A} = \cup_{J\in\mathcal{J}}\mathcal{A}_J$, where $\mathcal{A}_J$ is as in* (7.20). *Then $\mathcal{A}$ is an algebra of subsets of $\prod_{i\in I} X_i$, and $\sigma(\mathcal{A}) = \otimes_{i\in I}\mathcal{A}_i$.*

Proof. If $A \in \mathcal{A}$, then $A \in \mathcal{A}_J$ for some $J \in \mathcal{J}$. Since $\mathcal{A}_J$ is a σ-algebra $A^c \in \mathcal{A}_J \subset \mathcal{A}$. If $A_1, A_2 \in \mathcal{A}$, then there are $J_1, J_2 \in \mathcal{J}$ such that $A_1 \in \mathcal{A}_{J_1}$ and $A_2 \in \mathcal{A}_{J_2}$.

Consequently, $A_1, A_2 \in \mathcal{A}_{J_1 \cup J_2}$ (7.20). Since $\mathcal{A}_{J_1 \cup J_2}$ is a σ-algebra, $A_1 \cup A_2 \in \mathcal{A}_{J_1 \cup J_2} \subset \mathcal{A}$. Thus $\mathcal{A}$ is an algebra of sets. By virtue of (7.20), $\mathcal{A} \subset \otimes_{i \in I} \mathcal{A}_i$, and so $\sigma(\mathcal{A}) \subset \otimes_{i \in I} \mathcal{A}_i$. Since $\mathcal{A} \supset \cup_{i \in I} \mathcal{A}_{\{i\}} = \cup_{i \in I} \pi_i^{-1}(\mathcal{A}_i)$, it follows that $\sigma(\mathcal{A}) \supset \otimes_{i \in I} \mathcal{A}_i$. Hence $\sigma(\mathcal{A}) = \otimes_{i \in I} \mathcal{A}_i$. □

Theorem 7.22. *Notation is as in* (7.16). *Let $\mathcal{J}^*$ be the family of all nonempty countable subsets of I. Then $\otimes_{i \in I} \mathcal{A}_i = \cup_{J \in \mathcal{J}^*} \mathcal{A}_J$, where $\mathcal{A}_J$ is as in* (7.20).

Proof. Put $\mathcal{A} = \cup_{J \in \mathcal{J}^*} \mathcal{A}_J$. As in the proof of (7.21), we have $\otimes_{i \in I} \mathcal{A}_i \supset \mathcal{A} \supset \cup_{i \in I} \pi_i^{-1}(\mathcal{A}_i)$. Therefore, to complete the proof, it will be sufficient to show that $\mathcal{A}$ is a σ-algebra. If $A \in \mathcal{A}$, then $A \in \mathcal{A}_J$ for some $J \in \mathcal{J}^*$, and so $A^c \in \mathcal{A}_J \subset \mathcal{A}$. Now let $\{A_n : n \in N\} \subset \mathcal{A}$. Then, for each $n \in N$, there is $J_n \in \mathcal{J}^*$ such that $A_n \in \mathcal{A}_{J_n} \subset \mathcal{A}_{\cup_{n \in N} J_n}$. Since $\mathcal{A}_{\cup_{n \in N} J_n}$ is a σ-algebra, and $\cup_{n \in N} J_n \in \mathcal{J}^*$ (3.17), we have $\cup_{n \in N} A_n \in \mathcal{A}_{\cup_{n \in N} J_n} \subset \mathcal{A}$. Thus $\mathcal{A}$ is a σ-algebra. □

Lemma 7.23. *Let X be a set, and let $\mathcal{M}_a \subset \mathcal{P}(X)$, $a \in A$. Then $\sigma(\cup_{a \in A} \sigma(\mathcal{M}_a)) = \sigma(\cup_{a \in A} \mathcal{M}_a)$.*

The proof of this lemma is similar to the proof of (4.23).

Corollary 7.24. *Notation is as in* (7.15). *If $\mathcal{A}_i = \sigma(\mathcal{M}_i)$, $i \in I$, then the σ-algebra generated by $\{f_i : i \in I\}$ coincides with $\sigma(\cup_{i \in I} f_i^{-1}(\mathcal{M}_i))$.*

Proof. By virtue of (7.13) and (7.23), we have

$$\sigma\left(\bigcup_{i \in I} f_i^{-1}(\sigma(\mathcal{M}_i))\right) = \sigma\left(\bigcup_{i \in I} \sigma(f_i^{-1}(\mathcal{M}_i))\right) = \sigma\left(\bigcup_{i \in I} f_i^{-1}(\mathcal{M}_i)\right). \qquad \square$$

The next theorem expresses the associativity property of the product σ-algebra.

Theorem 7.25. *Let $\{(X_i, \mathcal{A}_i) : i \in I\}$ be a family of measurable spaces, and let $\{J_a : a \in A\}$ be a partition of I such that $J_a \neq \emptyset$, $a \in A$. Then $(\pi_{J_a})_{a \in A}^{-1}(\otimes_{a \in A}(\otimes_{j \in J_a} \mathcal{A}_j)) = \otimes_{i \in I} \mathcal{A}_i$.*

Proof. Corollary (7.19) shows that

$$(\pi_{J_a})_{a \in A}^{-1}\left(\bigotimes_{a \in A}\left(\bigotimes_{j \in J_a} \mathcal{A}_j\right)\right) = \sigma\left(\bigcup_{a \in A} \pi_{J_a}^{-1}\left(\bigotimes_{j \in J_a} \mathcal{A}_j\right)\right).$$

As shown in the proof of Theorem (7.20), we have

$$\pi_{J_a}^{-1}\left(\bigotimes_{j \in J_a} \mathcal{A}_j\right) = \sigma\left(\bigcup_{j \in J_a} \pi_j^{-1}(\mathcal{A}_j)\right), \quad a \in A.$$

Therefore, according to (7.23), we get

$$(\pi_{J_a})_{a\in A}^{-1}\left(\bigotimes_{a\in A}\left(\bigotimes_{j\in J_a}\mathcal{A}_j\right)\right)=\sigma\left(\bigcup_{a\in A}\left(\bigcup_{j\in J_a}\pi_j^{-1}(\mathcal{A}_j)\right)\right)$$
$$=\sigma\left(\bigcup_{i\in I}\pi_i^{-1}(\mathcal{A}_i)\right)=\bigotimes_{i\in I}\mathcal{A}_i. \qquad \square$$

In what follows we concern ourselves with Borel sets of a Cartesian product. We begin with the next lemmas.

Lemma 7.26. *Let $(X,\mathcal{T})$ be a topological space that has a countable base $\mathcal{B}$. Then $\mathcal{B}(X)=\sigma(\mathcal{B})$.*

Proof. Since $\mathcal{T}\subset\sigma(\mathcal{B})\subset\sigma(\mathcal{T})$ we have $\mathcal{B}(X)=\sigma(\mathcal{B})$. □

Lemma 7.27. *Let $\mathcal{M}$ be a countable family of subsets of X. Then $\sigma(\mathcal{M})\supset\tau(\mathcal{M})$.*

Proof. Exercise. □

Theorem 7.28. *Let I be a nonempty countable set, and let $\{(X_i,\mathcal{T}_i): i\in I\}$ be a family of topological spaces. If there is a countable base for $\mathcal{T}_i, i\in I$, then $\otimes_{i\in I}\mathcal{B}(X_i)=\mathcal{B}(\prod_{i\in I}X_i)$.*

Proof. Let $\mathcal{T}$ be the product topology on $\prod_{i\in I}X_i$. Since the projections $\pi_i, i\in I$, are continuous on $\prod_{i\in I}X_i$, we have $\cup_{i\in I}\pi_i^{-1}(\mathcal{T}_i)\subset\mathcal{T}$. Hence, according to (7.24), we get

$$\bigotimes_{i\in I}\mathcal{B}(X_i)=\sigma\left(\bigcup_{i\in I}\pi_i^{-1}(\mathcal{T}_i)\right)\subset\sigma(\mathcal{T})=\mathcal{B}\left(\prod_{i\in I}X_i\right).$$

Now let $\mathcal{B}_i$ be a countable base for $\mathcal{T}_i, i\in I$. Then, using (7.24) and (7.26), we have

$$\bigotimes_{i\in I}\mathcal{B}(X_i)=\sigma\left(\bigcup_{i\in I}\pi_i^{-1}(\mathcal{B}_i)\right). \tag{1}$$

Further, applying (4.14.a), (4.28), (4.17) and (4.23), we may write

$$\mathcal{T}=\tau\left(\bigcup_{i\in I}\pi_i^{-1}(\mathcal{T}_i)\right)=\tau\left(\bigcup_{i\in I}\pi_i^{-1}(\tau(\mathcal{B}_i))\right)$$
$$=\tau\left(\bigcup_{i\in I}\tau(\pi_i^{-1}(\mathcal{B}_i))\right)=\tau\left(\bigcup_{i\in I}\pi_i^{-1}(\mathcal{B}_i)\right). \tag{2}$$

Since $\cup_{i\in I}\pi_i^{-1}(\mathcal{B}_i)$ is a countable family, using (1), (2) and (7.27), we get $\otimes_{i\in I}\mathcal{B}(X_i)\supset\mathcal{T}$, and so $\otimes_{i\in I}\mathcal{B}(X_i)\supset\mathcal{B}(\prod_{i\in I}X_i)$. Thus $\otimes_{i\in I}\mathcal{B}(X_i)=\mathcal{B}(\prod_{i\in I}X_i)$. □

Example 7.29. For any $i \in N$, set $R_i = R$. By virtue of (5.19.b) and (6.12), R has a countable base. Consequently, $\otimes_{i \in N} \mathcal{B}(R_i) = \mathcal{B}(R^\infty)$, and $\otimes_{i=1}^n \mathcal{B}(R_i) = \mathcal{B}(R^n)$ for each $n \in N$. Observe also that the family $\{]-\infty, r_1] \times \cdots \times]-\infty, r_n] : (r_1, \ldots, r_n) \in Q^n\}$ is denumerable and generates $\mathcal{B}(R^n)$, and the family $\cup_{n \in N}\{]-\infty, r_1] \times \cdots \times]-\infty, r_n] \times R \times R \times \cdots : (r_1, \ldots, r_n) \in Q^n\}$ is denumerable and generates $\mathcal{B}(R^\infty)$.

Remark 7.30. Let $\{(X_i, \mathcal{T}_i) : i \in I\}$ be a nonempty family of topological spaces. Then, as the proof of Theorem (7.28) shows, we always have $\otimes_{i \in I} \mathcal{B}(X_i) \subset \mathcal{B}\left(\prod_{i \in I} X_i\right)$. If either I is an uncountable set or there are spaces X_i that have no countable bases, then the σ-algebra $\otimes_{i \in I} \mathcal{B}(X_i)$ may be different from $\mathcal{B}(\prod_{i \in I} X_i)$. For example, let $\{(X_i) : i \in I\}$ be a family of Hausdorff spaces, where I is an uncountable set. For each $i \in I$, assume that X_i contains at least two points. Let $\mathbf{x} = (x_i)_{i \in I}$ be an arbitrary point of $\prod_{i \in I} X_i$. Then, according to (5.57) and (5.55), the set $\{\mathbf{x}\}$ is closed, and so $\{\mathbf{x}\} \in \mathcal{B}(\prod_{i \in I} X_i)$. To get a contradiction, suppose that $\{\mathbf{x}\} \in \otimes_{i \in I} \mathcal{B}(X_i)$. Then, in view of (7.22), there are a nonempty countable set $J \subset I$ and a set $B \in \otimes_{j \in J} \mathcal{B}(X_j)$ such that $\{\mathbf{x}\} = \pi_J^{-1}(B)$. Since I is uncountable, and X_i has at least two elements, $i \in I$, it follows that $\pi_J^{-1}(B)$ is an infinite set. This contradiction shows that $\{\mathbf{x}\} \notin \otimes_{i \in I} \mathcal{B}(X_i)$. Therefore, in this case the σ-algebra $\otimes_{i \in I} \mathcal{B}(X_i)$ is strictly included in $\mathcal{B}(\prod_{i \in I} X_i)$ even if X_i has a countable base for each $i \in I$.

We next introduce a class of families of sets which turns out to be very important in measure theory.

Definition 7.31. Let X be a set, and let $\mathcal{D} \subset \mathcal{P}(X)$ be such that:

(i) $X \in \mathcal{D}$;

(ii) $A, B \in \mathcal{D}$ and $A \supset B$ imply $A - B \in \mathcal{D}$;

(iii) $\{A_n : n \in N\} \subset \mathcal{D}$ and $A_m \cap A_n = \emptyset$ for $m \neq n$ imply $\cup_{n \in N} A_n \in \mathcal{D}$.

Then $\mathcal{D}$ is called a *Dynkin system* (of subsets of X).

Example 7.32. Any σ-algebra is a Dynkin system.

Theorem 7.33. *Let $\mathcal{D}$ be a Dynkin system such that $A, B \in \mathcal{D}$ implies $A \cap B \in \mathcal{D}$. Then $\mathcal{D}$ is a σ-algebra.*

Proof. If $A \in \mathcal{D}$, then by (7.31.i) and (7.31.ii), it follows that $A^c \in \mathcal{D}$. Let $\{A_n : n \in N\} \subset \mathcal{D}$. As in the proof of Lemma (3.15), set $B_1 = A_1$ and $B_{n+1} = A_{n+1} \cap (\cap_{i=1}^n A_i^c)$, $n \geqslant 1$. Then $B_n \in \mathcal{D}$ for any $n \in N$, and $B_m \cap B_n = \emptyset$ for $m \neq n$. Therefore, $\cup_{n \in N} A_n = \cup_{n \in N} B_n \in \mathcal{D}$. Thus $\mathcal{D}$ is a σ-algebra. □

Definition 7.34. Let X be a set and $\mathcal{M} \subset \mathcal{P}(X)$. The intersection of all Dynkin systems of subsets of X that contain $\mathcal{M}$, written $d(\mathcal{M})$, is a Dynkin system called the *Dynkin system generated by* $\mathcal{M}$. $d(\mathcal{M})$ is the smallest Dynkin system of subsets of X containing $\mathcal{M}$.

Lemma 7.35. *Let $\mathcal{D}$ be a Dynkin system of subsets of X and $\mathcal{M} \subset \mathcal{D}$. Then the family $\mathcal{D}' = \{A \in \mathcal{D} : A \cap C \in \mathcal{D}$ for each $C \in \mathcal{M}\}$ is a Dynkin system.*

Proof. Since $\mathcal{M} \subset \mathcal{D}$, we have $X \in \mathcal{D}'$. Let C be an arbitrary set of $\mathcal{M}$. Let $A, B \in \mathcal{D}'$ be such that $A \supset B$. Since $(A - B) \cap C = (A \cap C) - (B \cap C)$, it follows that $A - B \in \mathcal{D}'$. Further, consider $\{A_n : n \in N\} \subset \mathcal{D}'$ such that $A_m \cap A_n = \emptyset$ for $m \neq n$. Since $(\cup_{n\in N} A_n) \cap C = \cup_{n\in N}(A_n \cap C)$, we have $\cup_{n\in N} A_n \in \mathcal{D}'$. Thus $\mathcal{D}'$ is a Dynkin system. □

Theorem 7.36. *Let $\mathcal{M} \subset \mathcal{P}(X)$ be such that $A, B \in \mathcal{M}$ implies $A \cap B \in \mathcal{M}$. Then $d(\mathcal{M}) = \sigma(\mathcal{M})$.*

Proof. (7.32) shows that $d(\mathcal{M}) \subset \sigma(\mathcal{M})$. Hence, according to (7.33), it will suffice to show that $A, B \in d(\mathcal{M})$ implies $A \cap B \in d(\mathcal{M})$. Put $\mathcal{D} = \{A \in d(\mathcal{M}) : A \cap C \in d(\mathcal{M})$ for each $C \in \mathcal{M}\}$. Then $\mathcal{D}$ is a Dynkin system (7.35) that contains $\mathcal{M}$. Therefore,

$$d(\mathcal{M}) = \mathcal{D}. \tag{1}$$

Let $\mathcal{D}' = \{A \in d(\mathcal{M}) : A \cap C \in d(\mathcal{M})$ for each $C \in d(\mathcal{M})\}$. From (1) we get $\mathcal{M} \subset \mathcal{D}'$. Consequently, using (7.35), we have $d(\mathcal{M}) = \mathcal{D}'$. Thus $A, B \in d(\mathcal{M})$ implies $A \cap B \in d(\mathcal{M})$. □

Exercise 7.37. Let $\mathcal{A}$ be a σ-algebra, and let $\{A_n : n \in N\} \subset \mathcal{A}$. Show that $\liminf_n A_n \in \mathcal{A}$ and $\limsup_n A_n \in \mathcal{A}$.

Exercise 7.38. Let X be a set, and let $\mathcal{A} = \{A \in \mathcal{P}(X) : A$ or A^c is finite$\}$. Prove the following.

(a) $\mathcal{A}$ is an algebra of sets.

(b) If X is infinite, then $\mathcal{A}$ is not a σ-algebra.

(c) In X is denumerable, then $\mathcal{A}$ is denumerable. Compare this with (7.2.a).

Exercise 7.39. Notation is as in (7.4). Let M be the collection of all countable subfamilies of $\mathcal{M}$. Show that $\sigma(\mathcal{M}) = \cup_{\mathcal{M}'\in \mathrm{M}}\sigma(\mathcal{M}')$.

Exercise 7.40. If $\mathcal{M}$ has n elements, then $\sigma(\mathcal{M})$ contains at most 2^{2^n} sets.

Exercise 7.41. (a) Use (7.40) and (3.47) to prove that any σ-algebra generated by a countable partition is not denumerable.

(b) Show that each σ-algebra is not denumerable. [Hints. Assume that there exists a denumerable σ-algebra $\mathcal{A} = \{A_1, A_2, \ldots\}$. Put $\varphi = (1_{A_n})_{n\in N}$. For $\mathbf{i} = (i_1, i_2, \ldots) \in \{0, 1\}^\infty$, we have $\varphi^{-1}(\{\mathbf{i}\}) = \cap_{n\in N} B_n \in \mathcal{A}$, where $B_n = \begin{cases} A_n & \text{if } i_n = 1 \\ A_n^c & \text{if } i_n = 0 \end{cases}$. Then the family $\mathcal{M} = \{\varphi^{-1}(\{\mathbf{i}\}) : \mathbf{i} \in \{0, 1\}^\infty\}$ is a partition, and $\mathcal{A} = \sigma(\mathcal{M})$.]

Exercise 7.42. Let $\mathcal{M}$ be a nonempty family of subsets of X. Set $\varphi = (1_A)_{A\in\mathcal{M}}$, and consider the family $\mathcal{L} = \{\varphi^{-1}(\{\mathbf{i}\}) : \mathbf{i} \in \{0, 1\}^{\mathcal{M}}\}$. Prove the following.

(a) $\mathcal{L}$ is a partition of X.

(b) If $\mathcal{M}$ is countable, then $\sigma(\mathcal{M}) \supset \sigma(\mathcal{L})$. Give an example to show that $\sigma(\mathcal{M})$ may be different from $\sigma(\mathcal{L})$.

(c) If $\mathcal{M}$ is countable, and each set in $\mathcal{M}$ is countable, then $\mathcal{L}$ is countable and $\sigma(\mathcal{M}) = \sigma(\mathcal{L})$.

Exercise 7.43. Notation is as in (7.6.a). Let $\mathcal{A} \subset \sigma(\{A_i : i \in I\})$ be a σ-algebra of subsets of X. Use (3.48), Tukey's Lemmas (3.32) and (3.52) to prove that there exists a countable partition of X which generates $\mathcal{A}$. [Hint. Let D be the collection of all partition $\mathcal{D}$ of X such that $\mathcal{D} \subset \mathcal{A}$. Show that D has a maximal element $\mathcal{D}'$ such that $\sigma(\mathcal{D}') = \mathcal{A}$.]

Exercise 7.44. Let $(X, \mathcal{T})$ be the topological space in (2.1.25). Prove the following.

(a) Both $\mathcal{B}(X)$ and $\sigma(\{\{x\} : x \in X\})$ coincide with the σ-algebra in (7.3.c).

(b) If X is an uncountable set, then $\mathbb{D} \notin \mathcal{B}(X) \otimes \mathcal{B}(X)$. [Hints. From (a) and (7.24) it follows that $\mathcal{B}(X) \otimes \mathcal{B}(X) = \sigma(\{\{x\} \times X : x \in X\} \cup \{X \times \{x\} : x \in X\})$. Then use (7.39) and (7.6.a).]

Exercise 7.45. Find an algebra of sets that has no atoms.

Exercise 7.46. Let $\mathcal{A}$ be an algebra of sets and $\emptyset \neq A \in \mathcal{A}$. Show that A is an atom of $\mathcal{A}$ if and only if the relative algebra on A induced by $\mathcal{A}$ coincides with $\{\emptyset, A\}$.

Exercise 7.47. (a) Find a denumerable family of sets that generates $\mathcal{B}(R)$.

(b) Use (7.42.c) and (7.6.a) to show that there is no countable family of sets which generates $\sigma(\{\{x\} : x \in R\})$. Contrast this with (7.43).

Exercise 7.48. Let $\mathcal{A}$ be an algebra of subsets of X, and let $A \subset X$. Show that $A \cap \mathcal{A} \subset \mathcal{A}$ if and only if $A \in \mathcal{A}$.

Exercise 7.49. Let $\{A_i : i \in I\}$ be a countable cover of X, and let $\mathcal{M} \subset \mathcal{P}(X)$. For $i \in I$, let $\mathcal{A}_i$ denote the σ-algebra generated by the family on A_i induced by $\mathcal{M}$. Use (7.14) and (7.48) to prove that $\sigma(\cup_{i \in I} \mathcal{A}_i) = \sigma(\mathcal{M})$ if and only if $A_i \in \sigma(\mathcal{M})$ for each $i \in I$.

Exercise 7.50. (a) Prove that a subset of $\overline{R}$ is $\mathcal{B}(\overline{R})$-measurable if and only if it is a union of sets of the form $\{-\infty\}$, A or $\{\infty\}$, where $A \in \mathcal{B}(R)$.

(b) Use (7.49) to show that the family of all intervals of the form $[-\infty, a]$, where $a \in R$, generates $\mathcal{B}(\overline{R})$.

Exercise: Sum measurable space 7.51. Let $\{(X_i, \mathcal{A}_i) : i \in I\}$ be a family of measurable spaces such that $X_i \cap X_j = \emptyset$ for $i \neq j$. Show that the family $\oplus_{i \in I} \mathcal{A}_i = \{\cup_{i \in I} A_i : A_i \in \mathcal{A}_i \text{ for each } i \in I\}$ is a σ-algebra of subsets of $\cup_{i \in I} X_i$. $\oplus_{i \in I} \mathcal{A}_i$ is called a *sum σ-algebra*, and the pair $(\cup_{i \in I} X_i, \oplus_{i \in I} \mathcal{A}_i)$ is called a *sum measurable space*.

Exercise 7.52. Let $\{X_i, : i \in I\}$ be a countable family of sets such that $X_i \cap X_j = \emptyset$ for $i \neq j$, and let $\mathcal{M}_i \subset \mathcal{P}(X_i)$, $i \in I$. Prove that $\oplus_{i \in I} \sigma(\mathcal{M}_i) = \sigma(\cup_{i \in I} \mathcal{M}_i)$ if and only if $X_i \in \sigma(\cup_{i \in I} \mathcal{M}_i)$ for each $i \in I$.

Exercise 7.53. Let $(X, \mathcal{A})$ be a measurable space, and let $\{A_i : i \in I\}$ be a countable partition of X. Show that $\oplus_{i \in I}(A_i \cap \mathcal{A}) = \mathcal{A}$ if and only if $A_i \in \mathcal{A}$ for each $i \in I$.

Exercise 7.54. Let $\{(X_i, \mathcal{A}_i) : i \in I\}$ be a nonempty family of measurable spaces, and let $A_i \subset X_i, i \in I$. Use (7.18) to show that the relative σ-algebra on $\prod_{i \in I} A_i$ induced by $\otimes_{i \in I} \mathcal{A}_i$ coincides with $\otimes_{i \in I}(A_i \cap \mathcal{A}_i)$.

Exercise 7.55. Let $\{(X_i, \mathcal{A}_i) : i \in I\}$ be a nonempty family of measurable spaces, and let A_i be an atom of $\mathcal{A}_i, i \in I$. Use (7.46) and (7.54) to prove that $\prod_{i \in I} A_i$ is an atom of $\otimes_{i \in I} \mathcal{A}_i$.

Exercise 7.56. Let $\mathcal{B}$ be the family of all subsets of R^n of the form $\prod_{i=1}^{n} I_i$, where I_i is an interval of the form $]a, b]$ or $]a, \infty[$, and let $\mathcal{A}$ be the family of all finite unions of disjoint sets in $\mathcal{B}$. Prove the following.

(a) $\mathcal{A}$ is an algebra of sets.

(b) $\sigma(\mathcal{A}) = \mathcal{B}(R^n)$.

Exercise 7.57. Find a σ-algebra that has no atoms. [Do not forget (7.30).]

Exercise 7.58. Let X be a nonempty set, and put $R_x = R$, $x \in X$. Prove the following.
(a) If X is countable, then $\mathfrak{B}_r(X) \in \otimes_{x\in X}\mathcal{B}(R_x)$.
(b) If X is uncountable, then $\mathfrak{B}_r(X) \notin \otimes_{x\in X}\mathcal{B}(R_x)$, but $\mathfrak{B}_r(X) \in \mathcal{B}(R^X)$.
(c) If $X = [a, b]$, where $a, b \in R$, then $\mathfrak{C}_r(X) \notin \otimes_{x\in X}\mathcal{B}(R_x)$, but $\mathfrak{C}_r(X) \in \mathcal{B}(R^X)$.

Exercise 7.59. Find a Dynkin system which is not a σ-algebra.

Exercise 7.60. For $x \in R$ and $A \subset R$, denote $x + A = A + x = \{x + y : y \in A\}$ and $xA = Ax = \{xy : y \in A\}$. Use (7.36) and (7.10) to show that $x + A$ and xA are $\mathcal{B}(R)$-measurable whenever $A \in \mathcal{B}(R)$. [Hint. The family $\{A \in \mathcal{P}(R) : x+A \in \mathcal{B}(R)\}$ is a Dynkin system that contains $\{]a, \infty[: a \in R\}$.]

Exercise 7.61. Let $\mathcal{A}$ and $\mathcal{B}$ be σ-algebras of subsets of X, and let $\{A_n : n \in N\}$ be a nondecreasing sequence of subsets of X such that $\cup_{n\in N} A_n = X$. Prove that $\mathcal{A} = \mathcal{B}$ whenever $A_n \cap \mathcal{A} = A_n \cap \mathcal{B}$ for each $n \in N$.

Exercise 7.62. Let $x, y \in X$ and $\mathcal{M} \subset \mathcal{P}(X)$. Show that $1_A(x) = 1_A(y)$ for any $A \in \mathcal{M}$ if and only if $1_A(x) = 1_A(y)$ for any $A \in \sigma(\mathcal{M})$.

Exercise 7.63. Let $x \in X$ and $\mathcal{M} \subset \mathcal{P}(X)$. Use (7.62) to prove that $\cap\{A \in \sigma(\mathcal{M}) : x \in A\} \in \sigma(\mathcal{M})$ whenever $\mathcal{M}$ is countable.

Exercise 7.64. Let $\{(Y_i, \mathcal{A}_i) : i \in I\}$ be a family of measurable spaces such that $Y_i \cap Y_j = \emptyset$ for $i \neq j$, and let $f : X \to \cup_{i\in I} Y_i$ be a function. Show that $f^{-1}(\oplus_{i\in I}\mathcal{A}_i) = \oplus_{i\in I} f^{-1}(\mathcal{A}_i)$.

Exercise 7.65. Let $\{(X_i, \mathcal{A}_i) : i \in I\}$ be as in (7.51). Prove the following.
(a) If I is countable, then $\oplus_{i\in I}\mathcal{A}_i = \sigma(\cup_{i\in I}\mathcal{A}_i)$.
(b) If I is an uncountable set, then $\oplus_{i\in I}\mathcal{A}_i$ may be different from $\sigma(\cup_{i\in I}\mathcal{A}_i)$.

Exercise 7.66. Let $J_1, \ldots, J_n$ be countable sets, and put $J = J_1 \times \cdots \times J_n$. For each $i = 1, \ldots, n$, let $\{(X_{ij}, \mathcal{A}_{ij}) : j \in J_i\}$ be a family of measurable spaces such that $X_{ij} \cap X_{ik} = \emptyset$ for $j \neq k$. Assume that $\cup_{j\in J_1} X_{1j} = \cdots = \cup_{j\in J_n} X_{nj}$. For $\mathbf{j} = (j_1, \ldots, j_n) \in J$, set $X_\mathbf{j} = X_{1j_1} \cap \cdots \cap X_{nj_n}$. Show that $\sigma(\cup_{i=1}^n(\oplus_{j\in J_i}\mathcal{A}_{ij})) = \sigma(\cup_{\mathbf{j}\in J}(\cup_{i=1}^n(X_\mathbf{j} \cap \mathcal{A}_{ij_i})))$.

Exercise 7.67. Let $J_1, \ldots, J_n$, J and $\{(X_{ij}, \mathcal{A}_{ij}) : j \in J_i\}$, $1 \leqslant i \leqslant n$, be as in (7.66). We do not assume that $\cup_{j\in J_1} X_{1j} = \cdots = \cup_{j\in J_n} X_{nj}$. Use (7.64), (7.66) and (7.52) to prove that $\otimes_{i=1}^n(\oplus_{j\in J_i}\mathcal{A}_{ij}) = \oplus_{\mathbf{j}\in J}(\otimes_{i\in I}\mathcal{A}_{ij_i})$.

Exercise 7.68. (a) Use (7.29), (7.10) and (7.24) to show that $\mathcal{B}(R^n) = \sigma(\{\prod_{i=1}^n] -\infty, b_i] : (b_1, \ldots, b_n) \in R^n\}) = \sigma(\{\prod_{i=1}^n]a_i, b_i] : (a_1, \ldots, a_n), (b_1, \ldots, b_n) \in R^n\})$.
(b) On account of (7.10), find other Cartesian products generating $\mathcal{B}(R^n)$.

Exercise: Monotone class 7.69. Let X be a set, and let $\emptyset \neq \mathcal{C} \subset \mathcal{P}(X)$ be such that:
(i) $A_n \in \mathcal{C}$ and $A_n \subset A_{n+1}$, $n \in N$, imply $\cup_{n\in N} A_n \in \mathcal{C}$;
(ii) $A_n \in \mathcal{C}$ and $A_n \supset A_{n+1}$, $n \in N$, imply $\cap_{n\in N} A_n \in \mathcal{C}$.

Then $\mathcal{C}$ is called a *monotone class.* The intersection of all monotone classes containing $\mathcal{M} \subset \mathcal{P}(X)$, written $c(\mathcal{M})$, is a monotone class called the *monotone class generated* by $\mathcal{M}$. Let $\mathcal{A} \subset \mathcal{P}(X)$ be an algebra of sets. Prove that $c(\mathcal{A}) = \sigma(\mathcal{A})$. The following steps may be helpful.

(a) $c(\mathcal{A}) \subset \sigma(\mathcal{A})$.

(b) If $\mathcal{A}$ is a monotone class, then $\mathcal{A}$ is a σ-algebra.

(c) For $A \subset X$, put $\mathcal{C}_A = \{B \subset X : A - B, B - A, A \cup B \in c(\mathcal{A})\}$. Then $\mathcal{C}_A$ is a monotone class.

(d) For $A, B \in \mathcal{P}(X)$, $A \in \mathcal{C}_B$ if and only if $B \in \mathcal{C}_A$.

(e) $c(\mathcal{A}) \subset \mathcal{C}_A$, $A \in \mathcal{A}$. [$\mathcal{A} \subset \mathcal{C}_A$, $A \in \mathcal{A}$.]

(f) $c(\mathcal{A}) \subset \mathcal{C}_A$, $A \in c(\mathcal{A})$. [Hint. $A \in \mathcal{A}$ and $B \in c(\mathcal{A})$ imply $B \in \mathcal{C}_A$ by (e), and so $A \in \mathcal{C}_B$ by (d). Thus $B \in c(\mathcal{A})$ implies $\mathcal{A} \subset \mathcal{C}_B$.]

(g) $c(\mathcal{A})$ is a σ-algebra. [$c(\mathcal{A})$ is an algebra by (f). Then apply (b).]

Exercise 7.70. Let $\mathcal{B}$ be the family of all subsets B or R of the form $B = \cup_{i=1}^{n} I_i$, where I_i is an interval of the form $]a, b]$ or $]a, \infty[$ with $a \in Q \cup \{-\infty\}$ and $b \in Q$ for $i = 1, \ldots, n$, and $I_i \cap I_j = \emptyset$ for $i \neq j$. Put $\mathcal{A} = \{\emptyset\} \cup \mathcal{B}$. Prove that $\mathcal{A}$ is a denumerable algebra of sets and $\sigma(\mathcal{A}) = \mathcal{B}(R)$.

Exercise 7.71. Let $f : X \to Y$ be a function, and let $\{\mathcal{A}_n : n \in N\}$ be a nonincreasing sequence of σ-algebras of subsets of Y. Prove that $f^{-1}(\cap_{n \in N} \mathcal{A}_n) = \cap_{n \in N} f^{-1}(\mathcal{A}_n)$. Contrast this with (2.47).

Exercise: Tail [remote] σ-algebra and exchangeable [permutable] σ-algebra 7.72. Let $(Y, \mathcal{B})$ be a measurable space. For $n \geqslant 1$, let $\mathfrak{P}_n$ stand for the set of all permutations $p : N \to N$ such that $p(m) = m, m \geqslant n$. For $p \in \cup_{n \geqslant 1} \mathfrak{P}_n$, define $\widetilde{p}(y_1, y_2, \ldots) = (y_{p(1)}, y_{p(2)}, \ldots)$, $(y_1, y_2, \ldots) \in Y^{\infty}$. For $n \geqslant 1$, let $\mathcal{T}_n$ be the σ-algebra of subsets of Y^{∞} generated by the projections $\pi_m, m \geqslant n$, and put $\mathcal{E}_n = \{B \in \mathcal{T}_1 : B = \widetilde{p}(B)$ for any $p \in \mathfrak{P}_n\}$.

(a) Show that $\widetilde{p}$ is a bijective mapping from Y^{∞} onto Y^{∞}.

(b) Prove that $\{\mathcal{E}_n : n \geqslant 1\}$ is a nonincreasing sequence of σ-algebras.

(c) Let $f_n : X \to Y, n \geqslant 1$, be functions, and set $\mathbf{f} = (f_1, f_2, \ldots)$. Show that $\mathbf{f}^{-1}(\mathcal{T}) \subset \mathbf{f}^{-1}(\mathcal{E})$, where $\mathcal{T} = \cap_{n \geqslant 1} \mathcal{T}_n$ and $\mathcal{E} = \cap_{n \geqslant 1} \mathcal{E}_n$. [Hint. $\mathcal{T}_n \subset \mathcal{E}_n, n \geqslant 1$.] $\mathbf{f}^{-1}(\mathcal{T})$ is called the *tail* [*remote*] *σ-algebra* of $\{f_n : n \geqslant 1\}$, and $\mathbf{f}^{-1}(\mathcal{E})$ is called the *exchangeable* [*permutable*] *σ-algebra* of $\{f_n : n \geqslant 1\}$.

(d) For $n \geqslant 1$, let $\mathcal{F}_n$ be the σ-algebra generated by $f_m, m \geqslant n$. Use (7.71) and (7.13) to prove that $\mathbf{f}^{-1}(\mathcal{T}) = \cap_{n \geqslant 1} \mathcal{F}_n$.

(e) If $Y = R$ and $\{b_n : n \in N\} \subset]0, \infty[$ is such that $b_n \to \infty$, show that $\liminf_n f_n$, $\limsup_n f_n$, $\liminf_n S_n/b_n$ and $\limsup_n S_n/b_n$ are $\mathbf{f}^{-1}(\mathcal{T})$-measurable.

Exercise 7.73. Notation is as in (7.72) with $Y = \{-1, 1\}$. Let $E = \{(y_1, y_2, \ldots) \in Y^{\infty} : y_1 + \cdots + y_{2n} = 0$ for infinitely many $n\}$. Show that $E \in \mathcal{E}$, but $E \notin \mathcal{T}$.

8 Measurable Functions

The notion of a measurable function is closely related to the concept of a measurable space and plays a central role in the theory of integration. The descriptive definition of a measurable function resembles the definition of the notion of a continuous function.

Since most measurable functions encountered in the present book are mappings from an arbitrary set X, endowed with a σ-algebra $\mathcal{A}$, into the set of extended real numbers $\overline{R}$, equipped with the σ-algebra $\mathcal{B}(\overline{R})$, in this section special attention is paid to the class of measurable functions from X into $\overline{R}$. We include here the constructive definition of measurability, and we explain the proof technique called the indicator function method.

Definition 8.1. Let $(X, \mathcal{A})$ and $(Y, \mathcal{B})$ be measurable spaces. A function $f : X \to Y$ is said to be $(\mathcal{A}, \mathcal{B})$-*measurable* if $f^{-1}(\mathcal{B}) \subset \mathcal{A}$. When there is no doubt about the σ-algebra $\mathcal{B}$ under discussion, we will say that f is $\mathcal{A}$-*measurable*; if there is no confusion possible regarding both σ-algebras $\mathcal{A}$ and $\mathcal{B}$, we will say simply that f is *measurable*.

Examples 8.2. (a) Let $(X, \mathcal{A})$ and $(Y, \mathcal{B})$ be measurable spaces. If either $\mathcal{A}$ is the discrete σ-algebra or $\mathcal{B}$ is the trivial σ-algebra, then any function from X into Y is $(\mathcal{A}, \mathcal{B})$-measurable.

(b) Let $(X, \mathcal{A})$ be a measurable space and $A \subset X$. Then 1_A is $\mathcal{A}$-measurable if and only if $A \in \mathcal{A}$.

(c) Let $(X, \mathcal{T})$ and $(Y, \mathcal{U})$ be topological spaces, and let $f : X \to Y$ be a function which is continuous on X. Then f is measurable. Actually, using (7.13) and (5.40), we have $f^{-1}(\mathcal{B}(Y)) = \sigma(f^{-1}(\mathcal{U})) \subset \sigma(T) = B(X)$.

Remarks 8.3. (a) Let $(X, \mathcal{A})$, $(Y, \mathcal{B})$ and $(Z, \mathcal{C})$ be measurable spaces, and let $f : X \to Y$ and $g : Y \to Z$ be functions. If f is $(\mathcal{A}, \mathcal{B})$-measurable, and g is $(\mathcal{B}, \mathcal{C})$-measurable, then $g \circ f$ is $(\mathcal{A}, \mathcal{C})$-measurable.

(b) Let $(X, \mathcal{A})$ and $(Y, \mathcal{B})$ be measurable spaces, let $f : X \to Y$ be a function, let $A \subset X$, and let $f(X) \subset B \subset Y$. Then the restriction of f to A is $(A \cap \mathcal{A}, \mathcal{B})$-measurable, and the function $g : X \to B$ given by $g(x) = f(x)$ for any $x \in X$ is $(\mathcal{A}, B \cap \mathcal{B})$-measurable.

(c) Let $(X, \mathcal{A})$ and $(Y, \sigma(\mathcal{M}))$ be measurable spaces, and let $f : X \to Y$ be a function. Then, according to (7.13), f is $(\mathcal{A}, \sigma(\mathcal{M}))$-measurable if and only if $f^{-1}(\mathcal{M}) \subset \mathcal{A}$.

(d) Notation is as in (7.15). Then the σ-algebra generated by $\{f_i : i \in I\}$ is the smallest σ-algebra of subsets of X relative to which $f_i, i \in I$, are measurable.

The next theorem has important applications.

Theorem 8.4. *Let $(X, \mathcal{A})$ be a measurable space, let $\{(Y_i, \mathcal{B}_i) : i \in I\}$ be a nonempty family of measurable spaces, and let $f : X \to \prod_{i \in I} Y_i$ be a function. Then f is $(\mathcal{A}, \otimes_{i \in I} \mathcal{B}_i)$- measurable if and only if $\pi_i \circ f$ is $(\mathcal{A}, \mathcal{B}_i)$-measurable for any $i \in I$.*

Proof. According to (8.3.d), $\otimes_{i \in I} \mathcal{B}_i$ is the smallest σ-algebra of subsets of $\prod_{i \in I} Y_i$ relative to which the projections $\pi_i, i \in I$, are measurable. Consequently, if f is $(\mathcal{A}, \otimes_{i \in I} \mathcal{B}_i)$-measurable, then, applying (8.3a), $\pi_i \circ f$ is $(\mathcal{A}, \mathcal{B}_i)$-measurable for any $i \in I$.

Conversely, assume that $(\pi_i \circ f)^{-1}(\mathcal{B}_i) \subset \mathcal{A}, i \in I$. Then, using (7.18) and (8.3.d), we get $f^{-1}(\otimes_{i \in I} \mathcal{B}_i) \subset \mathcal{A}$. □

Corollary 8.5. *Let $(X, \mathcal{A})$ and $\{(Y_i, \mathcal{B}_i) : i \in I\}$ be as in* (8.4), *and let $f_i : X \to Y_i, i \in I$, be functions. Then $(f_i)_{i \in I}$ is $(\mathcal{A}, \otimes_{i \in I} \mathcal{B}_i)$-measurable if and only if f_i is $(\mathcal{A}, \mathcal{B}_i)$- measurable for each $i \in I$.*

Discussion 8.6. Most measurable functions encountered in this book are mappings from an arbitrary set X, endowed with a σ-algebra $\mathcal{A}$, into the set of extended real numbers $\overline{R}$, endowed with the σ-algebra $\mathcal{B}(\overline{R})$. This is why we investigate the class of measurable functions from X into $\overline{R}$ in what follows. For $f \in \overline{R}^X$ and $B \subset \overline{R}$, we will write $\{f \in B\} = \{x : f(x) \in B\}$. Specially, for $a, b \in \overline{R}$, we will write $\{a < f\} = \{x : a < f(x)\}$, $\{a \leqslant f\} = \{x : a \leqslant f(x)\}$ and $\{a < f < b\} = \{x : a < f(x) < b\}$; other similar notations, such as $\{f < a\}$ or $\{a < f \leqslant b\}$, will have an obvious meaning. For $f, g \in \overline{R}^X$, we will write $\{f < g\} = \{x : f(x) < g(x)\}$; other similar notations, such as $\{f \leqslant a < g\}$ or $\{f = g\}$, will have an obvious meaning as well.

Throughout (8.7)–(8.15), $(X, \mathcal{A})$ is a fixed measurable space. We begin with the next remark.

Remark 8.7. Let $f : X \to \overline{R}$ be a function. Then, in view of (7.50.a), f is measurable if and only if $f^{-1}(\{-\infty\}) \in \mathcal{A}$, $f^{-1}(\{\infty\}) \in \mathcal{A}$ and $f^{-1}(B) \in \mathcal{A}$ for any $B \in \mathcal{B}(R)$.

Theorem 8.8. *Let $f : X \to \overline{R}$ be a function, and let $A \subset R$ be such that $A^- = R$. Then the following are equivalent:*

(i) *f is measurable;*

(ii) *$\{a < f\} \in \mathcal{A}$ for any $a \in A$;*

(iii) *$\{a \leqslant f\} \in \mathcal{A}$ for any $a \in A$;*

(iv) *$\{f < a\} \in \mathcal{A}$ for any $a \in A$;*

(v) *$\{f \leqslant a\} \in \mathcal{A}$ for any $a \in A$.*

Proof. Assume that f is measurable. Then, for each $a \in A$, we have $]a, \infty] \in \mathcal{B}(\overline{R})$, and so $\{a < f\} = f^{-1}(]a, \infty]) \in \mathcal{A}$. Thus (i) implies (ii). To show that (ii) implies (iii), let $a \in A$, and let $\{a_n : n \in N\} \subset A$ be such that $a_n < a, n \in N$, and $a_n \to a$. Then $\{a \leqslant f\} = \bigcap_{n \in N}\{a_n < f\}$. Since $\{f < a\} = \{a \leqslant f\}^c, a \in A$, (iii) implies (iv). To see that (iv) implies (v), let $a \in A$, and let $\{a_n : n \in N\} \subset A$ be such that $a < a_n, n \in N$, and $a_n \to a$. Then $\{f \leqslant a\} = \bigcap_{n \in N}\{f < a_n\}$. Suppose now that (v) holds, and let $a \in R$. Choose $\{a_n : n \in N\} \subset A$ such that $a \leqslant a_n, n \in N$, and $a_n \to a$. Then $\{f \leqslant a\} = \bigcap_{n \in N}\{f \leqslant a_n\} \in \mathcal{A}$. Hence, by virtue of (7.50.b) and (8.3.c), f is measurable. Thus (v) implies (i). □

Theorem 8.9. *Let $f : X \to \overline{R}$ be a measurable function. Then the following assertions hold.*

(i) *The function αf is measurable for any $\alpha \in \overline{R}$.*

(ii) *The function $|f|$ is measurable.*

(iii) *The function g defined by*

$$g(x) = \begin{cases} \frac{1}{f(x)} & \text{if } f(x) \neq 0 \\ a & \text{if } f(x) = 0 \end{cases},$$

where $a \in \overline{R}$, is measurable.

(iv) *If $f \geqslant 0$, then the function g defined by*

$$g(x) = \begin{cases} (f(x))^\beta & \text{if } f(x) < \infty \\ a & \text{if } f(x) = \infty \end{cases},$$

where $a \in \overline{R}$, is measurable for any $\beta > 0$.

Proof. To prove (i), consider the function $h : \overline{R} \to \overline{R}$ defined by $h(t) = \alpha t$ for any $t \in \overline{R}$. Using (3.2.2.c), it follows at once that h is measurable. Since $\alpha f = h \circ f$, (3.2.3.a) shows that the function αf is measurable. The assertions (ii)–(iv) are proved analogously, noticing that the functions thereof have the form $h \circ f$, where $h : \overline{R} \to \overline{R}$ is measurable. For (ii), define $h(t) = |t|$ for any $t \in \overline{R}$; for (iii), consider

$$h(t) = \begin{cases} 1/t & \text{if } t \neq 0 \\ a & \text{if } t = 0 \end{cases};$$

for (iv), let

$$h(t) = \begin{cases} -1 & \text{if } t < 0 \\ t^\beta & \text{if } 0 \leqslant t < \infty \\ a & \text{if } t = \infty \end{cases}. \qquad \square$$

Theorem 8.10. *Let $f : X \to \overline{R}$ and $g : X \to \overline{R}$ be measurable functions. Then the following assertions hold.*

(i) *The function h defined by*

$$h(x) = \begin{cases} a & \text{if } (f(x), g(x)) = \{-\infty, \infty\} \\ f(x) + g(x) & \text{if } (f(x), g(x)) \neq \{(-\infty, \infty), (\infty, -\infty)\} \\ b & \text{if } (f(x), g(x)) = \{\infty, -\infty\} \end{cases},$$

where $a, b \in \overline{R}$, is measurable.

(ii) *The functions fg, $f \vee g$ and $f \wedge g$ are measurable.*

Proof. According to (8.5), the function (f, g) is $(\mathcal{A}, \mathcal{B}(\overline{R}) \otimes \mathcal{B}(\overline{R}))$-measurable. To prove (i), consider the function $\varphi : \overline{R} \times \overline{R} \to \overline{R}$ defined by

$$\varphi(u, v) = \begin{cases} a & \text{if } (u, v) = \{-\infty, \infty\} \\ u + v & \text{if } (u, v) \neq \{(-\infty, \infty), (\infty, -\infty)\} \\ b & \text{if } (u, v) = \{\infty, -\infty\} \end{cases}.$$

Since $h = \varphi \circ (f, g)$, it will suffice to show that φ is $\mathcal{B}(\overline{R}) \otimes \mathcal{B}(\overline{R})$-measurable. Obviously, $\varphi^{-1}(\{-\infty\})$ and $\varphi^{-1}(\{\infty\})$ are sets of $\mathcal{B}(\overline{R}) \otimes \mathcal{B}(\overline{R})$. Further, applying (7.29), (7.46) and (7.54), we have

$$\mathcal{B}(R^2) = \mathcal{B}(R) \otimes \mathcal{B}(R) \subset \mathcal{B}(\overline{R}) \otimes \mathcal{B}(\overline{R}). \tag{1}$$

Using (5.77.a), (8.2.c), and (1), it follows that $\varphi^{-1}(B) \in \mathcal{B}(\overline{R}) \otimes \mathcal{B}(\overline{R})$ for any $B \in \mathcal{B}(R)$. Thus, in view of (8.7), φ is $\mathcal{B}(\overline{R}) \otimes \mathcal{B}(\overline{R})$-measurable. The assertion (ii) is proved in a similar manner, considering, respectively, the function $\varphi : \overline{R} \times \overline{R} \to \overline{R}$ defined by $\varphi(u, v) = uv$, $\varphi(u, v) = u \vee v$ and $\varphi(u, v) = u \wedge v$ for any $u, v \in \overline{R}$. $\square$

Theorem 8.11. *Let $\mathfrak{F} \subset \overline{R}^X$ be a countable set of measurable functions. Then* $\sup \mathfrak{F}$ *and* $\inf \mathfrak{F}$ *are measurable functions.*

Proof. For any $a \in R$, we have $\{a < \sup \mathfrak{F}\} = \cup_{f \in \mathfrak{F}} \{a < f\}$ and $\{\inf \mathfrak{F} < a\} = \cup_{f \in \mathfrak{F}} \{f < a\}$. Since $\mathfrak{F}$ is countable, (8.8) shows that $\sup \mathfrak{F}$ and $\inf \mathfrak{F}$ are measurable. □

Corollary 8.12. *Let $\{f_n : n \in N\} \subset \overline{R}^X$ be a sequence of measurable functions. Then the following assertions hold.*

(i) *The functions* $\limsup_n f_n$ *and* $\liminf_n f_n$ *are measurable.*

(ii) *If $f_n \to f$, then f is measurable.*

(iii) *If $\sum_{n \in N} f_n(x) = g(x)$ exists for any $x \in X$, then g is measurable.*

Proof. Since $\limsup_n f_n = \inf_{n \geqslant 1}(\sup_{k \geqslant n} f_k)$ and $\liminf_n f_n = \sup_{n \geqslant 1}(\inf_{k \geqslant n} f_k)$, the assertion (i) follows from (8.11). If $f_n \to f$, then, according to (5.12), $f = \limsup_n f_n = \liminf_n f_n$. Thus (ii) follows from (i). The assertion (iii) follows from (8.10.i) and (ii). □

We now introduce a class of measurable functions which plays an essential role in the theory of integration.

Definition 8.13. A function $f : X \to \overline{R}$ is said to be *simple* if it can be expressed as $f = \sum_{i=1}^{n} a_i 1_{A_i}$, where $a_i \in \overline{R}$ and $A_i \in \mathcal{A}$ for $i = 1, \ldots, n$. Evidently, if such an expression is possible, then $A_i \cap A_j = \emptyset$ whenever $a_i = -\infty$ and $a_j = \infty$.

Remarks 8.14. (a) Let $f = \sum_{i=1}^{n} a_i 1_{A_i}$ be a simple function. Then, according to (8.2.b), (8.9.i) and (8.10.i), f is measurable.

(b) Let $f = \sum_{i=1}^{n} a_i 1_{A_i}$ be a simple function. Then $f(X)$ is a finite set. If $f(X) = \{b_1, \ldots, b_m\}$, then $f = \sum_{j=1}^{m} b_j 1_{B_j}$, where $\{B_1, \ldots, B_m\}$ is a partition of X formed from $\mathcal{A}$-measurable sets. This shows that a simple function may have different representations.

Theorem 8.15. *Let $f : X \to [0, \infty]$ be a measurable function. Then there exists a nondecreasing sequence $\{f_n : n \in N\} \subset [0, \infty[^X$ of simple functions such that $f_n \to f$.*

Proof. For each $n \in N$, define

$$f_n = \sum_{k=1}^{n2^n} (k-1)2^{-n} 1_{\{(k-1)2^{-n} \leqslant f < k2^{-n}\}} + n 1_{\{f \geqslant n\}}.$$

Then, in view of (8.8), f_n is a simple function. Clearly $0 \leqslant f_n < f, n \in N$. If $(k-1)2^{-n} \leqslant f(x) < k2^{-n}$ for some $k \in \{1, 2, \ldots, n2^n\}$, then either $(2k-2)2^{-n-1} \leqslant f(x) < (2k-1)2^{-n-1}$ or $(2k-1)2^{-n-1} \leqslant f(x) < 2k \cdot 2^{-n-1}$, and so either $f_{n+1}(x) = f_n(x)$ or $f_{n+1}(x) = f_n(x) + 2^{-n-1}$. Therefore, in this case $f_n(x) \leqslant f_{n+1}(x)$. If $f(x) \geqslant n$, then it is also easily seen that $f_n(x) \leqslant f_{n+1}(x)$. Hence the sequence $\{f_n : n \in N\}$ is nondecreasing. If $f(x) < m$ for some $m \in N$, then $0 \leqslant f(x) - f_n(x) < 2^{-n}$ for $n \geqslant m$, and so $f_n(x) \to f(x)$. If $f(x) = \infty$, then $f_n(x) = n, n \in N$, and so also $f_n(x) \to f(x)$. Thus $f_n \to f$. □

Theorem (8.17) below has important applications. We need the next definition.

Definition 8.16. Let $f : X \to \overline{R}$ be a function. The function $f^+ = f \vee 0$ is called the *positive part of* f, and the function $f^- = -(f \wedge 0)$ is called the *negative part* *of* f. Notice that $f^+ = f1_{\{f\geqslant 0\}} \geqslant 0$, $f^- = -f1_{\{f\leqslant 0\}} \geqslant 0$, $f = f^+ - f^-$, and $|f| = f^+ + f^-$. If f is measurable, then, according to (8.11) and (8.9.i), f^+ and f^- are measurable.

Theorem 8.17 (Doob-Dynkin). *Let $(Y, \mathcal{B})$ be a measurable space, and let $f : X \to Y$ and $g : X \to \overline{R}$ be functions. If g is $f^{-1}(\mathcal{B})$-measurable, then there exists a $\mathcal{B}$-measurable function $h : Y \to \overline{R}$ such that $g = h \circ f$.*

Proof. (I) Assume first that $g = 1_A$, where $A \in f^{-1}(\mathcal{B})$. Then there is $B \in \mathcal{B}$ such that $A = f^{-1}(B)$, and so, by (2.34), $g = 1_A \circ f$.

(II) Consider now the case that $g = \sum_{i=1}^n a_i 1_{A_i}$, where $a_i \geqslant 0$ and $A_i \in f^{-1}(\mathcal{B})$ for $i = 1, \ldots, n$. Then there is $B_i \in \mathcal{B}$ such that $1_{A_i} = 1_{B_i} \circ f$ for each i, and so $g = (\sum_{i=1}^n a_i 1_{B_i}) \circ f$.

(III) Consider next the case that $g \geqslant 0$. We apply (8.15) to get a sequence $\{g_n : n \in N\} \subset [0, \infty[^X$ of $f^{-1}(\mathcal{B})$-measurable simple functions such that $g_n \to g$. Further, we apply (II) to get a sequence $\{h_n : n \in N\} \subset \overline{R}^Y$ of $\mathcal{B}$-measurable functions such that $g_n = h_n \circ f, n \in N$. Set $h = \limsup_n h_n$. Then, using (5.12), we have $g(x) = \lim_n g_n(x) = \lim_n h_n(f(x)) = h(f(x))$ for any $x \in X$. Thus $g = h \circ f$.

(IV) Finally, consider the general case. We apply (III) to get $\mathcal{B}$-measurable functions $h_1 : Y \to \overline{R}$ and $h_2 : Y \to \overline{R}$ such that $g^+ = h_1 \circ f$ and $g^- = h_2 \circ f$. Since $h_1 - h_2$ may be meaningless, some caution is needed in order to define the function h. We note that

$$\begin{aligned}\emptyset &= \{g^+ = -\infty\} \cup \{g^- = -\infty\} \cup (\{g^+ = \infty\} \cap \{g^- = \infty\}) \\ &= f^{-1}(\{h_1 = -\infty\} \cup \{h_2 = -\infty\} \cup (\{h_1 = \infty\} \cap \{h_2 = \infty\})). \qquad (1)\end{aligned}$$

Put $B = \{h_1 = -\infty\} \cup \{h_2 = -\infty\} \cup (\{h_1 = \infty\} \cap \{h_2 = \infty\})$, and define $h = h_1 1_{B^c} - h_2 1_{B^c}$. Obviously, h is $\mathcal{B}$-measurable. Using (1), we have $g(x) = g^+(x) - g^-(x) = h_1(f(x)) - h_2(f(x)) = (h_1 1_{B^c})(f(x)) - (h_2 1_{B^c})(f(x)) = (h \circ f)(x)$ for any $x \in X$. □

The type of reasoning used to prove (8.17) is important. Since many theorems in measure theory are proved in this way, it deserves to be displayed explicitly. Thus to show that a measurable function $f : X \to \overline{R}$ has property P, it suffices to work up the following steps: (I) if f is an indicator function, then f has property P; (II) if f is a nonnegative simple function, then f has property P; (III) if $f \geqslant 0$, then f has property P; (IV) f has property P. This technique will be called the *indicator function method.*

Exercise 8.18. Let $(X, \mathcal{A})$ and $(Y, \mathcal{B})$ be measurable spaces.

(a) Show that any constant function from X into Y is $(\mathcal{A}, \mathcal{B})$-measurable.

(b) Assume that Y has at least two points. If each $(\mathcal{A}, \mathcal{B})$-measurable function from X into Y is constant, then $\mathcal{A} = \{\emptyset, X\}$.

(c) Suppose that X has at least two points, and that $\mathcal{A} = \{\emptyset, X\}$. Prove that every $(\mathcal{A}, \mathcal{B})$-measurable function is constant if and only if whenever y_1 and y_2 are distinct points of Y there is $B \in \mathcal{B}$ such that $y_1 \in B$ and $y_2 \notin B$.

Exercise 8.19. Let $(X, \mathcal{A})$ be a measurable space, and let $f : X \to C$ be a function. Use (5.85.a) to show that f is measurable if and only if Re f and Im f are measurable.

Exercise 8.20. Let $\{(X_i, \mathcal{A}_i) : i \in I\}$ be a countable family of measurable spaces such that $X_i \cap X_j = \emptyset$ for $i \neq j$. Let $(Y, \mathcal{B})$ be a measurable space, and let $f_i : X_i \to Y, i \in I$, be functions. Consider the function $f : \cup_{i\in I} X_i \to Y$ defined by $f(x) = f_i(x)$ if $x \in X_i$. Prove that f is $(\oplus_{i\in I}\mathcal{A}_i, \mathcal{B})$-measurable if and only if f_i is $(\mathcal{A}_i, \mathcal{B})$-measurable for each $i \in I$.

Exercise 8.21. Let $(X, \mathcal{A})$ and $(Y, \mathcal{B})$ be measurable spaces, let $\{A_i : i \in I\} \subset \mathcal{A}$ be a countable partition of X, and let $f_i : X \to Y, i \in I$, be functions. Consider the function $f : X \to Y$ defined by $f(x) = f_i(x)$ if $x \in A_i$. Show that f is $(\mathcal{A}, \mathcal{B})$-measurable if and only if the restriction of f_i to A_i is $(A_i \cap \mathcal{A}, \mathcal{B})$-measurable for each $i \in I$.

Exercise 8.22. Let $(X, \mathcal{A})$ be a measurable space, let $f : X \to Y$ be a function, let $\{(Z_i, \mathcal{C}_i) : i \in I\}$ be a family of measurable spaces, and let $g_i : Y \to Z_i, i \in I$, be functions. Let $\mathcal{B}$ denote the σ-algebra generated by $\{g_i : i \in I\}$. Prove that f is $(\mathcal{A}, \mathcal{B})$-measurable if and only if $g_i \circ f$ is $(\mathcal{A}, \mathcal{C}_i)$-measurable for any $i \in I$.

Exercise: *f*-expansion 8.23. Consider either $1 < b \in N$ or $b = \infty$, and let $f : [0, b[\to [0, 1[$ be a strictly increasing surjective function.

(a) Show that f is continuous on $[0, b[$.

(b) If $1 < b \in N$, set $f(b) = 1$. For $\{t_n : n \in N\} \subset [0, b[$, put $\underline{x}_1(t_1) = f(t_1), \overline{x}_1(t_1) = f(t_1 + 1)$, and define inductively $\underline{x}_{n+1}(t_1, \dots, t_{n+1}) = f(t_1 + \underline{x}_n(t_2, \dots, t_{n+1})), \overline{x}_{n+1}(t_1, \dots, t_{n+1}) = f(t_1 + \overline{x}_n(t_2, \dots, t_{n+1})), n \in N$. Prove that $\{\underline{x}_n(t_1, \dots, t_n) : n \in N\}$ and $\{\overline{x}_n(t_1, \dots, t_n) : n \in N\}$ are monotone sequences such that $\underline{x}_n(t_1, \dots, t_n) \leqslant \lim_n \underline{x}_n(t_1, \dots, t_n) \leqslant \lim_n \overline{x}_n(t_1, \dots, t_n) \leqslant \overline{x}_n(t_1, \dots, t_n)$, $n \in N$.

(c) Let I be the set of all nonnegative integers that are less than b. For $x \in [0, 1[$, set $r_0(x) = x$ and define recursively the sequences $\{a_n(x) : n \in N\} \subset I$ and $\{r_n(x) : n \in N\} \subset [0, 1[$ by $r_{n-1}(x) = f(a_n(x) + r_n(x)), n \in N$. The numbers $a_n(x), n \in N$, are called the *digits of* x. For $i_1, \dots, i_n \in I$, show that $\{x : a_1(x) = i_1, \dots, a_n(x) = i_n\} = [\underline{x}_n(i_1, \dots, i_n), \overline{x}_n(i_1, \dots, i_n)[$.

(d) Prove that the function $(a_n)_{n\in N}$ is injective if and only if

$$\text{(i)} \qquad x = \lim_{n\to\infty} \underline{x}_n(a_1(x), \dots, a_n(x))$$

for each $x \in [0, 1[$. The representation (i) is called the *f-expansion of x*. [Hints. For $x \in [0, 1[, \underline{x}_n(a_1(x), \dots, a_n(x)) \leqslant x < \overline{x}_n(a_1(x), \dots, a_n(x)), n \in N$, and so $\lim_n \underline{x}_n(a_1(x), \dots, a_n(x)) \leqslant x \leqslant \lim_n \overline{x}_n(a_1(x), \dots, a_n(x))$. If $\lim_n \underline{x}_n(a_1(x), \dots, a_n(x)) < \lim_n \overline{x}_n(a_1(x), \dots, a_n(x))$, then there is $x' \neq x$ such that $\underline{x}_n(a_1(x), \dots, a_n(x)) \leqslant x' < \overline{x}_n(a_1(x), \dots, a_n(x)), n \in N$.]

(e) Assume that there exists $\gamma < 1$ such that

$$\text{(ii)} \qquad f(t') - f(t) \leqslant \gamma(t' - t)$$

whenever $0 \leqslant t < t' \leqslant b$. Prove that (i) holds for any $x \in [0, 1[$. (If $1 < b \in N$, and $f(t) = t/b, t \in [0, b[$, then condition (ii) is verified. In this case the representation (i) is called the *b-adic expansion of* x, and it becomes $x = \sum_{n\in N} a_n(x)/b^n$.) [Hint. For $i_1, \dots, i_n \in I, \overline{x}_n(i_1, \dots, i_n) - \underline{x}_n(i_1, \dots, i_n) \leqslant \gamma^n$.]

(f) Use (c) to show that a_n is $(\mathcal{B}([0, 1[), \mathcal{P}(I))$-measurable for any $n \in N$.

(g) Put $I_n = I$, $n \in N$. Use (7.19) and (8.3.d) to prove that $(a_n)_{n\in N}^{-1}(\otimes_{n\in N}\mathcal{P}(I_n)) = \mathcal{B}([0,1[)$ whenever $(a_n)_{n\in N}$ is injective. [Hints. For $i_1, \ldots, i_n \in I$, the interval $[0, \underline{x}_n(i_1, \ldots, i_n)[$ belongs to $\sigma(a_1, \ldots, a_n)$. If $(a_n)_{n\in N}$ is injective, then $[0, x[= \cup_{n\in N}[0, \underline{x}_n(a_1(x), \ldots, a_n(x))[$ for any $x \in]0,1[$.]

Exercise: Baire set 8.24. Let X be a topological space, and let $\mathcal{B}_0(X)$ be the σ-algebra generated by the set of all continuous functions from X into R. In other words, $\mathcal{B}_0(X)$ is the smallest σ-algebra of subsets of X that contains all sets of the form $f^{-1}(B)$, where $B \in \mathcal{B}(R)$ and f is a continuous function from X into R. The elements of $\mathcal{B}_0(X)$ are called the *Baire sets of* X. Prove the following.

(a) $\mathcal{B}_0(X)$ coincides with the σ-algebra generated by $\mathfrak{C}_r(X)$. [Hints. For $n \in N$, let $\varphi_n : R \to R$ be defined by

$$\varphi_n(x) = \begin{cases} -n & \text{if } x < n, \\ x & \text{if } -n \leqslant x \leqslant n. \\ n & \text{if } x > n. \end{cases}$$

If f is a continuous function from X into R, then $\{\varphi_n \circ f : n \in N\} \subset \mathfrak{C}_r(X)$ and $\varphi_n \circ f \to f$. Then use (8.12.ii) and (8.3.d).]

(b) $\mathcal{B}_0(X) \subset \mathcal{B}(X)$.

(c) If X is a metric space, then $\mathcal{B}_0(X) = \mathcal{B}(X)$. [Use (6.76).]

Exercise 8.25. Let (X, d) be a metric space, and let $\mathcal{A}$ be the σ-algebra generated by $\{d_x : x \in X\}$, where $d_x(y) = d(x, y)$, $x, y \in X$. Prove the following.

(a) $\mathcal{A} \subset \mathcal{B}(X)$.

(b) If X is separable, then $\mathcal{A} = \mathcal{B}(X)$.

(c) If X is not separable, then $\mathcal{A}$ may be different from $\mathcal{B}(X)$.

Exercise 8.26. Let $(X, \mathcal{A})$ be a measurable space, and let $f : X \to \overline{R}$ and $g : X \to \overline{R}$ be measurable functions. Use (8.8) to show that the sets $\{f < g\}$, $\{f \leqslant g\}$ and $\{f = g\}$ are $\mathcal{A}$-measurable.

Exercise 8.27. Let $(X, \mathcal{A})$ be a measurable space, let Y be a topological space that has a countable base, and let $f : X \to Y$ and $g : X \to Y$ be measurable functions. Use (8.5), (7.28) and (5.56) to prove that $\{x : f(x) = g(x)\} \in \mathcal{A}$ whenever Y is a Hausdorff space.

Exercise 8.28. (a) Let $(X, \mathcal{A})$, Y and f be as in (8.27), and let $\mathbb{R}$ be the graph of f. Show that $\mathbb{R} \in \mathcal{A} \otimes \mathcal{B}(Y)$ whenever Y is a Hausdorff space. [Hint. The function $\varphi : X \times Y \to Y \times Y$ defined by $\varphi(x, y) = (f(x), y)$, $(x, y) \in X \times Y$, is $(\mathcal{A} \otimes \mathcal{B}(Y), \mathcal{B}(Y \times Y))$-measurable, and $\mathbb{R} = \varphi^{-1}(\mathbb{D})$, where $\mathbb{D}$ is the diagonal of $Y \times Y$.]

(b) Find two measurable spaces $(X, \mathcal{A})$ and $(Y, \mathcal{B})$, and an $(\mathcal{A}, \mathcal{B})$-measurable function $f : X \to Y$ such that the graph of f is not an $(\mathcal{A} \otimes \mathcal{B})$-measurable set.

Exercise 8.29. Let $(X, \mathcal{A})$ be a measurable space, let (Y, d) be a metric space, and let $\{f_n : n \in N\} \subset Y^X$ be a sequence of measurable functions. If $f_n \to f$, then f is measurable. [Hint. If $B \subset Y$ is open, then $f^{-1}(B) = \cup_{m\geqslant 1} \cup_{n\geqslant 1} \cap_{k\geqslant n}\{x : d(f_k(x), B^c) \geqslant 1/n\}$.]

Exercise 8.30. Let $(X, \mathcal{A})$, (Y, d) and $\{f_n : n \in N\}$ be as in (8.29), and assume that Y is complete and separable. Use (6.74) and (7.28) to show that $\{x : \{f_n(x) : n \in N\}$ has a limit $\} \in \mathcal{A}$. [Hint. $\{x : \{f_n(x) : n \in N$ has a limit$\} = \bigcap_{m\geqslant 1} \bigcup_{n\geqslant 1} \bigcap_{k\geqslant n} \bigcap_{l\geqslant n} \{x : d(f_k(x), f_l(x)) < 1/m\}$.]

Exercise 8.31. Let $(X, \mathcal{A})$, Y and $\{f_n : n \in N\}$ be as in (8.30), let $A = \{x : \{f_n(x) : n \in N\}$ has a limit$\}$, and let $y \in Y$. Use (8.30), (8.29) and (8.20) to prove that the function $f : X \to Y$ defined by

$$f(x) = \begin{cases} \lim_n f_n(x) & \text{if } x \in A \\ y & \text{if } x \in A^c \end{cases}$$

is measurable.

Exercise: Elementary function 8.32. Let $(X, \mathcal{A})$ be a measurable space, let Y be a set, and let $f : X \to Y$ be a function. If $f(X)$ is countable, and $f^{-1}(\{y\}) \in \mathcal{A}$ for any $y \in f(X)$, then f is said to be elementary. If Y is a metric space, prove the following.

(a) If f is elementary, then f is measurable.

(b) If f is elementary, then there exists a sequence $\{f_n : n \in N\} \subset Y^X$ of measurable functions such that $f_n(X)$ is finite, $n \in N$, and $f_n \to f$.

(c) Assume that Y is separable. If f is measurable, then there exists a sequence $\{f_n : n \in N\} \subset Y^X$ of elementary functions such that $f_n \to f$. [Hints. For $n \in N$, let $\{B_i : i \in I\} \subset \mathcal{B}(Y)$ be a countable partition of Y such that $\operatorname{diam}(B_i) < 1/n, i \in I$. Choose $x_i \in f^{-1}(B_i)$ and define $f_n : X \to Y$ by $f_n(x) = f(x_i)$ if $x \in f^{-1}(B_i)$.]

(d) Suppose that Y is separable. If f is measurable, then there exists a sequence $\{f_n : n \in N\} \subset Y^X$ of measurable functions such that $f_n(X)$ is finite, $n \in N$, and $f_n \to f$. [Hints. If there is a denumerable subset dense in Y, consider g a homeomorphism from Y into $[0, 1]^\infty$ (6.38). For $n \in N$, the family $\mathcal{I}_n = \{[(k-1)2^{-n}, k2^{-n}[: 1 \leqslant k < 2^n\} \cup \{[1 - 2^{-n}, 1]\}$ is a partition of $[0, 1]$, and the family $\mathcal{C}_n$ of all sets of the form $\prod_{m\in N} I_m$, where $I_m \in \mathcal{I}_{n-m+1}$, $1 \leqslant m \leqslant n$, and $I_m = [0, 1]$, $m > n$, is a partition of $[0, 1]^\infty$, then $f^{-1}(g^{-1}(\mathcal{C}_n))$ is a finite partition of X made up from $\mathcal{A}$-measurable sets. For $B \in g^{-1}(\mathcal{C}_n)$, select $x_B \in f^{-1}(B)$. For $m \in N$, set $[0, 1]_m = [0, 1]$, and let π_m be the projection from $[0, 1]^\infty$ onto $[0, 1]_m$. Then $|\pi_m \circ g \circ f - \pi_m \circ g \circ f_n| \leqslant 2^{-n+m-1}$, $n \geqslant m$. Further, use (5.7) and (6.87).]

Exercise 8.33. Let $(Y, \mathcal{B})$ be a measurable space, let Z be a complete separable metric space, let $f : X \to Y$ be a function, and let $g : X \to Z$ be an $f^{-1}(\mathcal{B})$-measurable function. Use (8.32) and (8.31) to show that there exists a $\mathcal{B}$-measurable function $h : Y \to Z$ such that $g = h \circ f$.

Exercise 8.34. Let $f : R \to R$ be a monotone function. Prove that f is measurable.

Exercise 8.35. Let X be a compact metric space and put $R_x = R$, $x \in X$. Show that $\mathcal{B}(\mathfrak{C}_r(X)) = \mathfrak{C}_r(X) \cap (\otimes_{x\in X} \mathcal{B}(R_x))$. [Hint. Use (6.58) and (6.71) to prove that $\mathcal{B}(\mathfrak{C}_r(X)) \subset \mathfrak{C}_r(X) \cap (\otimes_{x\in X} \mathcal{B}(R_x))$, and use (8.2.c) to show that $\mathcal{B}(\mathfrak{C}_r(X)) \supset \mathfrak{C}_r(X) \cap (\otimes_{x\in X} \mathcal{B}(R_x))$.]

9 Definitions and Properties of the Measure

In this section we introduce and discuss the notions of a measure and of a signed measure. We present here some basic results that will be often referred to in the sequel.

Definitions 9.1. Let X be a set, and let $\mathcal{A}$ be an algebra of subsets of X. A *finitely additive measure* on $\mathcal{A}$ is any function $\mu : \mathcal{A} \to [0, \infty]$ satisfying:

(i) $\mu(\emptyset) = 0$;

(ii) $\mu(A \cup B) = \mu(A) + \mu(B)$ whenever $A, B \in \mathcal{A}$ and $A \cap B = \emptyset$.

If (ii) is replaced by

(iii) $\mu(\cup_{n \in N} A_n) = \sum_{n \in N} \mu(A_n)$ whenever $\{A_n : n \in N\} \subset \mathcal{A}$, $\cup_{n \in N} A_n \in \mathcal{A}$, and $A_m \cap A_n = \emptyset$ for $m \neq n$, then μ is called a *σ-additive measure* or simply a *measure* on $\mathcal{A}$.

A triple $(X, \mathcal{A}, \mu)$, where $(X, \mathcal{A})$ is a measurable space and μ is a measure on $\mathcal{A}$ is called a *measure space.*

Remarks 9.2. (a) Let $\mathcal{A}$ be an algebra of subsets of X, and let μ be a finitely additive measure on $\mathcal{A}$. If there is $A \in \mathcal{A}$ such that $\mu(A) < \infty$, then condition (9.1.i) is automatically satisfied. Actually, (9.1.ii) implies $\mu(A) = \mu(A) + \mu(\emptyset)$.

(b) Each σ-additive measure is a finitely additive measure.

(c) Let $\mathcal{A}$ and μ be as in (a), let $A \in \mathcal{A}$, and let $\mathcal{B}$ be an algebra of subsets of A such that $\mathcal{B} \subset A \cap \mathcal{A}$. Then the restriction of μ to $\mathcal{B}$, written $\mu_{\mathcal{B}}$, is a finitely additive measure on $\mathcal{B}$. If μ is σ-additive, then $\mu_{\mathcal{B}}$ is σ-additive. When confusion appears impossible, we will write simply μ instead of $\mu_{\mathcal{B}}$.

Examples 9.3. (a) Let $\mathcal{A}$ be an algebra of sets. The function $\mu : \mathcal{A} \to [0, \infty]$ defined by

$$\mu(A) = \begin{cases} 0 & \text{if } A = \emptyset \\ \infty & \text{if } A \neq \emptyset \end{cases}$$

is a measure on $\mathcal{A}$.

(b) Let $\mathcal{A}$ be an algebra of subsets of X. The function $\mu : \mathcal{A} \to [0, \infty]$ defined by

$$\mu(A) = \begin{cases} 0 & \text{if } A \text{ is finite} \\ \infty & \text{if } A \text{ is infinite} \end{cases}$$

is a finitely additive measure on $\mathcal{A}$. If X is infinite, then μ is not σ-additive.

(c) Let X be a set. For $A \subset X$, define

$$\mu(A) = \begin{cases} n & \text{if } A \text{ has } n \text{ elements} \\ \infty & \text{if } A \text{ is infinite} \end{cases}.$$

Then μ is a measure on $\mathcal{P}(X)$ called the *counting measure.*

(d) Let $\mathcal{A}$ be an algebra of subsets of X, let $\{x_i, : i \in I\}$ be a countable set of distinct points of X, and let $p : I \to [0, \infty]$ be a function whose value at i is denoted p_i. For $A \in \mathcal{A}$, define $\mu(A) = \sum_{i \in I} p_i 1_A(x_i)$. Then μ is a measure on $\mathcal{A}$ called the *discrete measure concentrated on* $\{x_i, : i \in I\}$. If $\{x_i, : i \in I\}$ contains a single point $x \in X$, and $p = 1$, then the corresponding measure μ, written ε_x, is called the *Dirac measure concentrated at* x. Thus $\varepsilon_x(A) = 1_A(x)$, $A \in \mathcal{A}$.

(e) Let $\mathcal{A}$ be as in (7.3.d), and let $F : R \to R$ be a nondecreasing function. Denote $F(-\infty) = \inf_{x\in R} F(x)$ and $F(\infty) = \sup_{x\in R} F(x)$. Put $\mu(\emptyset) = 0$. For $]a, b] \in \mathcal{A}$, define $\mu(]a, b]) = F(b) - F(a)$, and for $]a, \infty[\in \mathcal{A}$, define $\mu(]a, \infty[)$ $=F(\infty) - F(a)$. If $B = \cup_{i=1}^n I_i \in \mathcal{A}$, where I_i is an interval of the form $]a, b]$ or $]a, \infty[$, and $I_i \cap I_j = \emptyset$ for $i \neq j$, define $\mu(B) = \sum_{i=1}^n \mu(I_i)$. If B has a different representation $\cup_{k=1}^m I'_k$, where I'_k is an interval of the form $]a, b]$ or $]a, \infty[$, and $I'_k \cap I'_l = \emptyset$ for $k \neq l$, then it is plain that $\sum_{i=1}^n \mu(I_i) = \sum_{k=1}^m \mu(I'_k)$. Hence μ is well defined on $\mathcal{A}$. Obviously, μ is a finitely additive measure on $\mathcal{A}$.

(f) For $A \in \mathcal{P}(X)$, define

$$\mu(A) = \begin{cases} 0 & \text{if } A \text{ is countable,} \\ \infty & \text{if } A \text{ is uncountable} \end{cases}$$

Then μ is a measure on $\mathcal{P}(X)$.

Definitions 9.4. Let $\mathcal{A}$ be an algebra of subsets of X, and let μ be a measure on $\mathcal{A}$. We say that μ is *finite* if $\mu(X) < \infty$; if $\mu(X) = 1$, then μ is called a *probability* on $\mathcal{A}$ and $(X, \mathcal{A}, \mu)$ is called a *probability space.* We say that μ is σ*-finite* if there is a sequence $\{A_n : n \in N\} \subset \mathcal{A}$ such that $\cup_{n\in N} A_n = X$ and $\mu(A_n) < \infty, n \in N$. Evidently, each finite measure is σ-finite.

Examples 9.5. (a) Notation is as in (9.3.c). If X is countable, then μ is σ-finite. If X is uncountable, then, according to (3.17), μ is not σ-finite.

(b) Notation is as in (9.3.d). If $\sum_{i\in I} p_i = 1$, then μ is a probability on $\mathcal{A}$.

Theorem 9.6. *Let $\mathcal{A}$ be an algebra of sets, and let μ be a finitely additive measure on $\mathcal{A}$. Then the following assertions hold.*

(i) If $A, B \in \mathcal{A}$ and $A \subset B$, then $\mu(A) \leqslant \mu(B)$.

(ii) $\mu(A) + \mu(B) = \mu(A \cap B) + \mu(A \cup B)$ whenever $A, B \in \mathcal{A}$.

(iii) If $A, B \in \mathcal{A}$ and μ is finite, then $|\mu(A) - \mu(B)| \leqslant \mu(A\Delta B)$.

(iv) If $\{A_1, \ldots, A_n\} \subset \mathcal{A}$, then $\mu(\cup_{i=1}^n A_i) \leqslant \sum_{i=1}^n \mu(A_i)$.

(v) If μ is σ-additive, then $\mu(\cup_{n\in N} A_n) \leqslant \sum_{n\in N} \mu(A_n)$ whenever $\{A_n : n \in N\} \subset \mathcal{A}$ and $\cup_{n\in N} A_n \in \mathcal{A}$.

Proof. Let $A, B \in \mathcal{A}$. If $A \subset B$, then $B = A \cup (B - A)$, and so (9.1.ii) implies $\mu(B) = \mu(A) + \mu(B - A) \geqslant \mu(A)$. Thus (i) is proved. To prove (ii), we write $\mu(A) = \mu(A \cap B) + \mu(A - B)$ and $\mu(B) = \mu(A \cap B) + \mu(B - A)$, whence $\mu(A)+\mu(B) = \mu(A\cap B)+(\mu(A-B)+\mu(A\cap B)+\mu(B-A)) = \mu(A\cap B)+\mu(A\cup B)$. The proof of (iii) is immediate, since $\mu(A) - \mu(B) = \mu(A - B) - \mu(B - A)$. Now let $\{A_1, \ldots, A_n\} \subset \mathcal{A}$, and define $B_i, 1 \leqslant i \leqslant n$, as in the proof of (3.15). Then $\{B_1, \ldots, B_n\} \subset \mathcal{A}$ and $B_i \cap B_j = \emptyset$ for $i \neq j$. Consequently, using (9.1.ii) and (i), we have $\mu(\cup_{i=1}^n A_i) = \mu(\cup_{i=1}^n B_i) = \sum_{i=1}^n \mu(B_i) \leqslant \sum_{i=1}^n \mu(A_i)$. Hence (iv) is proved. The proof of assertion (v) is similar to the proof of (iv). □

With the help of the next two theorems we can verify, in given situations, whether certain finitely additive measures are σ-additive.

Theorem 9.7. *Let $\mathcal{A}$ and μ be as in* (9.6). *If $\mu(B_n) \to \mu(B)$ whenever $\{B_n : n \in N\}$ is a nondecreasing sequence of sets in $\mathcal{A}$ such that $\cup_{n\in N} B_n = B \in \mathcal{A}$, then μ is σ-additive.*

Proof. Let $\{A_n : n \in N\} \subset \mathcal{A}$ be such that $\cup_{n\in N} A_n \in \mathcal{A}$ and $A_m \cap A_n = \emptyset$ for $m \neq n$. Put $B_n = \cup_{i=1}^{n} A_i, n \in N$. Then $\{B_n : n \in N\}$ is a nondecreasing sequence of sets in $\mathcal{A}$ such that $\cup_{n\in N} B_n = \cup_{n\in N} A_n$. Consequently, we have $\sum_{i=1}^{n} \mu(A_i) = \mu(B_n) \to \mu(\cup_{n\in N} A_n)$. Thus μ is σ-additive. □

Theorem 9.8. *Let $\mathcal{A}$ and μ be as in* (9.6). *If $\mu(B_n) \to 0$ whenever $\{B_n : n \in N\}$ is a nonincreasing sequence of sets in $\mathcal{A}$ such that $\cap_{n\in N} B_n = \emptyset$, then μ is σ-additive.*

Proof. Let $\{A_n : n \in N\}$ be as in the proof of (9.7). Set $C_n = \cup_{i=1}^{n} A_i, n \in N$, and $B_n = (\cup_{n\in N} A_n) - C_n, n \in N$. Then $\{B_n : n \in N\}$ is a nonincreasing sequence of sets in $\mathcal{A}$ such that $\cap_{n\in N} B_n = \emptyset$. Since $\mu(\cup_{n\in N} A_n) = \mu(C_n) + \mu(B_n) = \sum_{i=1}^{n} \mu(A_i) + \mu(B_n)$, and $\mu(B_n) \to 0$, it follows that $\mu(\cup_{n\in N} A_n) = \sum_{i\geqslant 1} \mu(A_i)$. □

Theorem 9.9. *Let $\mathcal{A}$ be an algebra of sets, and let μ be a measure on $\mathcal{A}$. If $\{A_n : n \in N\}$ is a nondecreasing sequence of sets in $\mathcal{A}$ such that $\cup_{n\in N} A_n \in \mathcal{A}$, then $\mu(A_n) \to \mu(\cup_{n\in N} A_n)$.*

Proof. Put $A_0 = \emptyset$. Then $\cup_{n\in N} A_n = \cup_{n\in N}(A_n - A_{n-1})$. Using (9.1.iii), we have $\mu(\cup_{n\in N} A_n) = \sum_{n\in N} \mu(A_n - A_{n-1}) = \lim_m \sum_{n=1}^{m} \mu(A_n - A_{n-1}) = \lim_m \mu (\cup_{n=1}^{m}(A_n - A_{n-1})) = \lim_m \mu(A_m)$. □

Theorem 9.10. *Let $\mathcal{A}$ and μ be as in* (9.9). *If $\{A_n : n \in N\}$ is a nonincreasing sequence of sets in $\mathcal{A}$ such that $\cap_{n\in N} A_n \in \mathcal{A}$ and $\mu(A_m) < \infty$ for some $m \in N$, then $\mu(A_n) \to \mu(\cap_{n\in N} A_n)$.*

Proof. Obviously, $\{A_m - A_n : n \geqslant m\}$ is a nondecreasing sequence of sets in $\mathcal{A}$. On account of (2.41.f), we have $\cup_{n\geqslant m}(A_m - A_n) = A_m - (\cap_{n\geqslant m} A_n) = A_m - (\cap_{n\in N} A_n) \in \mathcal{A}$. Applying (9.9), we get $\lim_n \mu(A_m - A_n) = \mu(A_m - (\cap_{n\in N} A_n))$. Since $\mu(A_m) < \infty$, we may write $\lim_n(\mu(A_m) - \mu(A_n)) = \mu(A_m) - \mu(\cap_{n\in N} A_n)$, whence $\lim_n \mu(A_n) = \mu(\cap_{n\in N} A_n)$. □

Remark 9.11. The result of Theorem (9.10) cannot be proved in general without the assumption $\mu(A_m) < \infty$ for some $m \in N$. To see this, let μ be the counting measure on $\mathcal{P}(R)$, and consider $A_n = [n, \infty[, n \in N$. Then $\mu(A_n) = \infty, n \in N$, while $\mu(\cap_{n\in N} A_n) = \mu(\emptyset) = 0$.

We next introduce the notion of a *complete measure* space.

Definitions 9.12. Let $(X, \mathcal{A}, \mu)$ be a measure space. A set $A \in \mathcal{A}$ is said to be *μ-null* if $\mu(A) = 0$; if there is no confusion possible regarding μ, we will say simply that A is *null*. We say that μ is a *complete measure* and $(X, \mathcal{A}, \mu)$ is a *complete measure space* if $\mathcal{A}$ contains all subsets of μ-null sets.

Theorem 9.13. *Let $(X, \mathcal{A}, \mu)$ be a measure space, and let $\mathcal{N}$ denote the family of all subsets of μ-null sets. Define $\mathcal{A}^{\mu} = \{A \cup N : A \in \mathcal{A}, N \in \mathcal{N}\}$. For $A \cup N \in \mathcal{A}^{\mu}$, define $\overline{\mu}(A \cup N) = \mu(A)$. Then $(X, \mathcal{A}^{\mu}, \overline{\mu})$ is a complete measure space.*

Proof. Evidently, $\emptyset \neq \mathcal{A} \subset \mathcal{A}^\mu$. Let $A \in \mathcal{A}$ and $N \subset B$, where $B \in \mathcal{A}$ and $\mu(B) = 0$. Then $(A \cup N)^c = A^c \cap N^c = (A^c \cap B^c) \cup (A^c \cap N^c \cap B) \in \mathcal{A}^\mu$. Now let $\{A_n : n \in N\} \subset \mathcal{A}$ and $\{N_n : n \in N\} \subset \mathcal{N}$. For $n \in N$, choose $B_n \in \mathcal{A}$ such that $N_n \subset B_n$ and $\mu(B_n) = 0$. Using (9.6.v), we have $\mu(\cup_{n\in N} B_n) = 0$, and so $\cup_{n\in N}(A_n \cup N_n) = (\cup_{n\in N} A_n) \cup (\cup_{n\in N} N_n) \in \mathcal{A}^\mu$. Hence $\mathcal{A}^\mu$ is a σ-algebra. It remains to show that $\overline{\mu}$ is a complete measure on $\mathcal{A}^\mu$. Let $A, A' \in \mathcal{A}$ and $N, N' \in \mathcal{N}$ be such that $A \cup N = A' \cup N'$, and let $B, B' \in \mathcal{A}$ be such that $N \subset B$, $N' \subset B'$, and $\mu(B) = \mu(B') = 0$. Then $A - A' \subset N' \subset B'$ and $A' - A \subset N \subset B$. Therefore, we get $\mu(A) = \mu(A \cap A') + \mu(A - A') = \mu(A' \cap A) + \mu(A' - A) = \mu(A')$. This shows that $\overline{\mu}$ is well defined on $\mathcal{A}^\mu$. Obviously, $\overline{\mu}(\emptyset) = 0$. If $\{A_n : n \in N\} \subset \mathcal{A}$ and $\{N_n : n \in N\} \subset \mathcal{N}$ are such that $(A_m \cup N_m) \cap (A_n \cup N_n) = \emptyset$ for $m \neq n$, then $\overline{\mu}(\cup_{n\in N}(A_n \cup N_n)) = \overline{\mu}((\cup_{n\in N} A_n) \cup (\cup_{n\in N} N_n)) = \mu(\cup_{n\in N} A_n) = \sum_{n\in N} \mu(A_n) = \sum_{n\in N} \overline{\mu}(A_n \cup N_n)$. Consequently, $\overline{\mu}$ is a measure on $\mathcal{A}^\mu$. Now let $A \cup N \in \mathcal{A}^\mu$ be such that $\overline{\mu}(A \cup N) = \mu(A) = 0$, and let $M \subset A \cup N$. If $B \in \mathcal{A}$ is such that $N \subset B$ and $\mu(B) = 0$, then $M \subset A \cup B \in \mathcal{A}$ and $\mu(A \cup B) = 0$. Thus $M \in \mathcal{A}^\mu$, and so $\overline{\mu}$ is complete. □

Remarks 9.14. Notation is as in (9.13).

(a) $\mathcal{A}^\mu = \{A \Delta N : A \in \mathcal{A}, N \in \mathcal{N}\}$. Actually, if $A \in \mathcal{A}$ and $N \subset B$ with $B \in \mathcal{A}$ and $\mu(B) = 0$, then $A \Delta N = (A - B) \cup (B \cap (A \Delta N))$ and $A \cup N = (A - B)\Delta(B \cap (A \cup N))$.

(b) $\mathcal{A} = \mathcal{A}^\mu$ if and only if μ is complete.

(c) $\mathcal{A} \subset \mathcal{A}^\mu$ and $\overline{\mu}$ is an extension of μ to $\mathcal{A}^\mu$. It is easily seen that $\overline{\mu}$ is the unique measure on $\mathcal{A}^\mu$ such that $\overline{\mu}_{\mathcal{A}} = \mu$. $\overline{\mu}$ is called the *completion of* μ, and $(X, \mathcal{A}^\mu, \overline{\mu})$ is called the *completion of* $(X, \mathcal{A}, \mu)$.

(d) Clearly, $\mathcal{A}^\mu = \sigma(\mathcal{A} \cup \mathcal{N})$. If $\mathcal{B} \subset \mathcal{A}$ is a σ-algebra, we can also consider the σ-algebra $\sigma(\mathcal{B} \cup \mathcal{N})$, which is called the *$\mu$-completion of $\mathcal{B}$ in $\mathcal{A}$*. Let $\mathcal{C} = \{C \subset X : B \Delta C \in \mathcal{N} \text{ for some } B \in \mathcal{B}\}$. Then $\mathcal{C}$ is a σ-algebra and $\mathcal{C} = \sigma(\mathcal{B} \cup \mathcal{N})$. Indeed, as $B^c \Delta C^c = B \Delta C$, $\mathcal{C}$ is closed under complementation. Next, if $\{B_n : n \in N\} \subset \mathcal{B}$ and $\{C_n : n \in N\} \subset \mathcal{C}$ are such that $B_n \Delta C_n \in \mathcal{N}$ for each $n \in N$, then $(\cup_{n\in N} B_n)\Delta(\cup_{n\in N} C_n) \subset \cup_{n\in N}(B_n \Delta C_n) \in \mathcal{N}$. This shows that $\cup_{n\in N} C_n \in \mathcal{C}$. Hence $\mathcal{C}$ is a σ-algebra. Trivially, $\mathcal{C}$ contains $\mathcal{B}$ and $\mathcal{N}$, and so $\mathcal{C} \supset \sigma(\mathcal{B} \cup \mathcal{N})$. Conversely, if $C \in \mathcal{C}$ and $B \Delta C \in \mathcal{N}$ for some $B \in \mathcal{B}$, then $B - C, C - B \in \mathcal{N}$, whence $C = (C - B) \cup (B - (B - C)) \in \sigma(\mathcal{B} \cup \mathcal{N})$. Thus $\mathcal{C} \subset \sigma(\mathcal{B} \cup \mathcal{N})$. Observe that $\mathcal{B}^\mu \subset \sigma(\mathcal{B} \cup \mathcal{N}) \subset \mathcal{A}^\mu$.

In what follows we generalize the notion of a measure on a σ-algebra. Let R^* stand for either the interval $[-\infty, \infty[$ or the interval $]-\infty, \infty]$.

Definition 9.15. Let $(X, \mathcal{A})$ be a measurable space. A *signed measure* on $\mathcal{A}$ is any function $\mu : \mathcal{A} \to R^*$ satisfying:

(i) $\mu(\emptyset) = 0$;

(ii) $\mu(\cup_{n\in N} A_n) = \sum_{n\in N} \mu(A_n)$ whenever $\{A_n : n \in N\} \subset \mathcal{A}$ and $A_m \cap A_n = \emptyset$ for $m \neq n$. Obviously, any measure on $\mathcal{A}$ is a signed measure on $\mathcal{A}$.

Examples 9.16. (a) Let $(X, \mathcal{A})$ be a measurable space, and let μ_1 and μ_2 be measures on $\mathcal{A}$ such that either $\mu_1(\mathcal{A}) \subset [0, \infty[$ or $\mu_2(\mathcal{A}) \subset [0, \infty[$. The function $\mu : \mathcal{A} \to R^*$ defined by $\mu(A) = \mu_1(A) - \mu_2(A)$ for any $A \in \mathcal{A}$ is a signed measure on $\mathcal{A}$.

In (9.24) we will show that each signed measure can be expressed as the difference of two measures of which at least one is finite.

(b) Let I be a countable set, and let $\{p_i, : i \in I\} \subset R$ be such that $\sum_{i\in I} |p_i| < \infty$. For $J \subset I$, define $\mu(J) = \sum_{j\in J} p_j$. Then μ is a signed measure on $\mathcal{P}(I)$.

Remarks 9.17. (a) Let $\mathcal{A}$ be a σ-algebra, and let μ be a signed measure on $\mathcal{A}$. If $A, B \in \mathcal{A}$ and $A \subset B$, then $\mu(A) \in R$ whenever $\mu(B) \in R$. Actually, (9.15.ii) implies $\mu(B) = \mu(A) + \mu(B - A)$. Thus $\mu(B) \in R$ if and only if $\mu(A), \mu(B - A) \in R$.

(b) Let $\mathcal{A}$ be a σ-algebra, let $\{\mu_i, : i \in I\}$ be a finite set of signed measures on $\mathcal{A}$ such that $\mu_i(\mathcal{A}) \subset R, i \in I$, and let $\{a_i, : i \in I\} \subset R$. For $A \in \mathcal{A}$, define $(\sum_{i\in I} a_i\mu_i)(A) = \sum_{i\in I} a_i\mu_i(A)$. Then $\sum_{i\in I} a_i\mu_i$ is a signed measure on $\mathcal{A}$.

Definitions 9.18. Let $(X, \mathcal{A})$ be a measurable space, and let μ be a signed measure on $\mathcal{A}$. We say that $P \in \mathcal{A}$ is a *positive set for* μ if $\mu(A) \geqslant 0$ whenever $A \in \mathcal{A}$ and $A \subset P$. We say that $N \in \mathcal{A}$ is a *negative set for* μ if $\mu(A) \leqslant 0$ whenever $A \in \mathcal{A}$ and $A \subset N$. If P is a positive set for μ and P^c is a negative set for μ, then the pair (P, P^c) is called a *Hahn decomposition of* X (relative to μ).

Examples 9.19. (a) Let $(X, \mathcal{A})$ be a measurable space, and let μ be a measure on $\mathcal{A}$. Then X is a positive set for $\mu, \emptyset$ is a negative set for μ, and $(X, \emptyset)$ is a Hahn decomposition of X.

(b) Notation is as in (9.16.b). Put $A = \{i \in I : p_i \geqslant 0\}$ and $B = \{i \in I : p_i \leqslant 0\}$. Then A and B^c are positive sets for μ, A^c and B are negative sets for μ, and (A, A^c) and (B^c, B) are Hahn decompositions of X.

Lemma 9.20. *Let* $(X, \mathcal{A})$ *and* μ *be as in* (9.18). *If* $\{P_n : n \in N\}$ *is a sequence of positive sets for* μ, *then* $\cup_{n\in N} P_n$ *is a positive set for* μ.

Proof. If $P_m \cap P_n = \emptyset$ for $m \neq n$, then $\mu(A) = \sum_{n\in N} \mu(A \cap P_n) \geqslant 0$ whenever $A \in \mathcal{A}$ and $A \subset \cup_{n\in N} P_n$, and so $\cup_{n\in N} P_n$ is a positive set for μ. For $B \in \mathcal{A}$, it is clear that $P - B$ is a positive set for μ whenever P is a positive set for μ. Set $P_1' = P_1$ and $P_{n+1}' = P_{n+1} - \cup_{i=1}^n P_i, n \geqslant 1$. Then $\{P_n' : n \in N\}$ is a sequence of positive sets for μ, and $P_m' \cap P_n' = \emptyset$ for $m \neq n$. Hence $\cup_{n\in N} P_n = \cup_{n\in N} P_n'$ is a positive set for μ □

Theorem 9.21. *Let* $(X, \mathcal{A})$ *and* μ *be as in* (9.18). *Then there exists a Hahn decomposition of* X *relative to* μ.

Proof. We prove this theorem in case $R^* = [-\infty, \infty[$. The case when $R^* =]-\infty.\infty]$ requires obvious changes and is left to the reader. Let $\alpha = \sup\{\mu(P) : P$ is a positive set for $\mu\}$. Choose a sequence $\{P_n : n \in N\}$ of positive sets for μ such that $\mu(P_n) \to \alpha$. Put $P = \cup_{n\in N} P_n$. Then, in view of (9.20), P is a positive set for μ. Since $P \supset P_n, n \in N$, we have

$$\alpha \geqslant \mu(P) = \mu(P_n) + \mu(P - P_n) \geqslant \mu(P_n), \quad n \in N. \tag{1}$$

Upon letting $n \to \infty$ in (1), we obtain $\alpha = \mu(P) < \infty$. To get a contradiction, assume that P^c is not a negative set for μ, and choose $A \in \mathcal{A}$ such that $A \subset P^c$ and $\mu(A) > 0$. Since $\mu(P \cup A) = \alpha + \mu(A) > \alpha$, A is not a positive set for μ. Let n_1 be the smallest positive integer for which there exists $A_1 \in \mathcal{A}$ such that $A_1 \subset A$ and $\mu(A_1) \leqslant -1/n_1$. By (9.17.a), we have $\mu(A_1) \in R$, and so $\mu(A - A_1) = \mu(A) - \mu(A_1) > \mu(A) > 0$.

Since $\mu(P \cup (A - A_1)) > \alpha$, $A - A_1$ is not a positive set for μ. Let n_2 be the smallest positive integer for which there exists $A_2 \in \mathcal{A}$ such that $A_2 \subset A - A_1$ and $\mu(A_2) \leqslant -1/n_2$. Since $\mu(A-(A_1 \cup A_2)) = \mu(A)-\mu(A_1)-\mu(A_2) > 0$, $A-(A_1 \cup A_2)$ is not a positive set for μ. Continuing this process, we get a sequence $\{n_k : k \in N\} \subset N$ and a sequence $\{A_k : k \in N\} \subset \mathcal{A}$ such that, for each $k \in N$, $A_k \subset A - \cup_{i=1}^{k-1} A_i$, $\mu(A_k) \leqslant -1/n_k$, and $\mu(C) > -\frac{1}{n_k - 1}$ whenever $n_k > 1$, $C \in \mathcal{A}$ and $C \subset A - \cup_{i=1}^{k-1} A_i$. Since $\cup_{k \in N} A_k \subset A$, using (9.17), we have

$$\infty > \mu\left(A - \bigcup_{k \in N} A_k\right) = \mu(A) - \mu\left(\bigcup_{k \in N} A_k\right)$$
$$= \mu(A) - \sum_{k \in N} \mu(A_k) \geqslant \mu(A) + \sum_{k \in N} \frac{1}{n_k} > 0. \qquad (2)$$

Thus $\mu(P \cup (A - \cup_{k \in N} A_k)) = \alpha + \mu(A - \cup_{k \in N} A_k) > \alpha$, and so $A - \cup_{k \in N} A_k$ is not a positive set for μ. Consequently, there is $B \in \mathcal{A}$ such that $B \subset A - \cup_{k \in N} A_k$ and $\mu(B) < 0$. From (2) it follows that $\sum_{k \in N} \frac{1}{n_k} < \infty$, and so $n_k \to \infty$. Select $k \in N$ such that $n_k > 1$ and $\mu(B) \leqslant -1/n_k$. Since $A_k \cup B \subset A - \cup_{i=1}^{k-1} A_i$, we have $-\frac{1}{n_k - 1} < \mu(A_k \cup B) = \mu(A_k) + \mu(B) \leqslant -\frac{2}{n_k} \leqslant -\frac{1}{n_k - 1}$. This contradiction shows that P^c is a negative set for μ. Hence (P, P^c) is a Hahn decomposition of X. □

Remark 9.22. Let $(X, \mathcal{A})$ and μ be as in (9.18). Example (9.19.b) shows that there may exist several Hahn decompositions of X. However, if (P_1, P_1^c) and (P_2, P_2^c) are Hahn decompositions of X, then $\mu(A \cap P_1) = \mu(A \cap P_2)$ and $\mu(A \cap P_1^c) = \mu(A \cap P_2^c)$ for any $A \in \mathcal{A}$. Actually, for $A \in \mathcal{A}$, we have $A \cap (P_1 - P_2) \subset P_1$ and $A \cap (P_1 - P_2) \subset P_2^c$, whence $\mu(A \cap (P_1 - P_2)) = 0$. Analogously, we obtain $\mu(A \cap (P_2 - P_1)) = 0$. Therefore, $\mu(A \cap P_1) = \mu(A \cap P_1 \cap P_2) = \mu(A \cap P_2)$ and $\mu(A \cap P_1^c) = \mu(A \cap P_1^c \cap P_2^c) = \mu(A \cap P_2^c)$.

Definitions 9.23. Let $(X, \mathcal{A})$ and μ be as in (9.18), and let (P, P^c) be a Hahn decomposition of X relative to μ. The function $\mu^+ : \mathcal{A} \to [0, \infty]$ defined by $\mu^+(A) = \mu(A \cap P)$, $A \in \mathcal{A}$, is called the *positive variation of* μ. The function $\mu^- : \mathcal{A} \to [0, \infty]$ defined by $\mu^-(A) = -\mu(A \cap P^c)$, $A \in \mathcal{A}$, is called the *negative variation of* μ. The function $|\mu| = \mu^+ + \mu^-$ is called the *total variation of* μ. In view of (9.22), the functions μ^+, μ^- and $|\mu|$ are well defined on $\mathcal{A}$.

Theorem 9.24. *Notation is as in* (9.23). *Then the following assertions hold.*

(i) μ^+, μ^-, *and* $|\mu|$ *are measures on* $\mathcal{A}$.
(ii) *Either* μ^+ *or* μ^- *is finite.*
(iii) $\mu(A) = \mu^+(A) - \mu^-(A)$ *for any* $A \in \mathcal{A}$.

Proof. Plainly, $\mu^+(\emptyset) = 0$. If $\{A_n : n \in N\} \subset \mathcal{A}$ and $A_m \cap A_n = \emptyset$ for $m \neq n$, then $\mu^+(\cup_{n \in N} A_n) = \mu(\cup_{n \in N}(A_n \cap P)) = \sum_{n \in N} \mu(A_n \cap P) = \sum_{n \in N} \mu^+(A_n)$. Thus μ^+ is a measure on $\mathcal{A}$. The fact that μ^- is a measure on $\mathcal{A}$ is proved similarly. Therefore, it follows at once that $|\mu|$ is a measure on $\mathcal{A}$. Hence (i) is proved. To prove (ii), we write $\mu(X) = \mu(P) + \mu(P^c)$. Therefore, either $\mu(P) < \infty$ or $-\infty < \mu(P^c)$, and so either $\mu^+(X) < \infty$ or $\mu^-(X) < \infty$. For $A \in \mathcal{A}$, we have $\mu(A) = \mu(A \cap P) + \mu(A \cap P^c) = \mu^+(A) - \mu^-(A)$. Thus (iii) is proved. □

The expression $\mu = \mu^+ - \mu^-$ in (9.24.iii) is called the *Jordan decomposition of* μ.

Exercise 9.25. Let $\mathcal{A}$ be an algebra of subsets of X, and let $f : X \to [0, \infty]$ be a function. Show that the function $\mu : \mathcal{A} \to [0, \infty]$ defined by

$$\mu(A) = \begin{cases} 0 & \text{if } A = \emptyset \\ \sup\{f(x_1) + \cdots + f(x_n) : \{x_1, \ldots, x_n\} \subset A\} & \text{if } A \neq \emptyset \end{cases}$$

is a measure on $\mathcal{A}$.

Exercise 9.26. Let $(X, \mathcal{A}, \mu)$ be a measure space, let $(Y, \mathcal{B})$ be a measurable space, and let $f : X \to Y$ be an $(\mathcal{A}, \mathcal{B})$-measurable function. For $B \in \mathcal{B}$, define $(\mu f^{-1})(B) = \mu(f^{-1}(B))$. Prove that $(Y, \mathcal{B}, \mu f^{-1})$ is a measure space. The measure μf^{-1} is called the *image of* μ *under* f.

Exercise 9.27. Let $(X, \mathcal{A})$ be a measurable space, let $\{\mu_i, : i \in I\}$ be a countable set of measures on $\mathcal{A}$, and let $\{a_i, : i \in I\} \subset [0, \infty]$. For $A \in \mathcal{A}$, define $(\sum_{i \in I} a_i \mu_i)(A) = \sum_{i \in I} a_i \mu_i(A)$. Show that $\sum_{i \in I} a_i \mu_i$ is a measure on $\mathcal{A}$.

Exercise 9.28. Let $\{(X_i, \mathcal{A}_i, \mu_i) : i \in I\}$ be a countable family of measure spaces such that $X_i \cap X_j = \emptyset$ for $i \neq j$. For $\cup_{i \in I} A_i \in \oplus_{i \in I} \mathcal{A}_i$, define $(\oplus_{i \in I} \mu_i)(\cup_{i \in I} A_i) = \sum_{i \in I} \mu_i(A_i)$. Prove that $(\cup_{i \in I} X_i, \oplus_{i \in I} \mathcal{A}_i, \oplus_{i \in I} \mu_i)$ is a measure space. $\oplus_{i \in I} \mu_i$ is called a *sum measure,* and the triple $(\cup_{i \in I} X_i, \oplus_{i \in I} \mathcal{A}_i, \oplus_{i \in I} \mu_i)$ is called a *sum measure space.*

Exercise 9.29. Show that the sequence $\{A_n : n \in N\}$ in (9.4) may be chosen such that $A_m \cap A_n = \emptyset$ for $m \neq n$.

Exercise 9.30. Let $\mathcal{A}$ be an algebra of sets, and let μ be a finitely additive measure on $\mathcal{A}$. If $\mu(\cup_{n \in N} A_n) \leqslant \sum_{n \in N} \mu(A_n)$ whenever $\{A_n : n \in N\} \subset \mathcal{A}$, $\cup_{n \in N} A_n \in \mathcal{A}$, and $A_m \cap A_n = \emptyset$ for $m \neq n$, then μ is σ-additive.

Exercise 9.31. Let $\mathcal{A}$ be an algebra of subsets of X, let μ be a finitely additive measure on $\mathcal{A}$, and let $A \in \mathcal{A}$ be such that $\mu(A) = \mu(X)$. Prove the following.

(a) If $\mu(X) < \infty$, then $\mu(A \cap B) = \mu(B)$ for any $B \in \mathcal{A}$.

(b) If $\mu(X) = \infty$, then there may exist $B \in \mathcal{A}$ such that $\mu(A \cap B) \neq \mu(B)$.

Exercise 9.32. Let $(X, \mathcal{A}, \mu)$ be a measure space and $\{A_n : n \in N\} \subset \mathcal{A}$. Prove the following.

(a) $\mu(\liminf_n A_n) \leqslant \liminf_n \mu(A_n)$.

(b) If $\mu(\cup_{k \geqslant m} A_k) < \infty$ for some $m \in N$, then $\mu(\limsup_n A_n) \geqslant \limsup_n \mu(A_n)$.

Exercise 9.33. Notation is as in (9.13). Prove the following.

(a) $\mathcal{A}^\mu$ consists of all sets $A \subset X$ for which there are $B, C \in \mathcal{A}$ such that $B \subset A \subset C$ and $\mu(C - B) = 0$.

(b) $\mathcal{A}^\mu$ consists of all sets $A \subset X$ for which there are $B, C \in \mathcal{A}$ such that $A \Delta B \subset C$ and $\mu(C) = 0$. [Use (a).]

(c) A function $f : X \to \overline{R}$ is $\mathcal{A}^\mu$-measurable if and only if there exist $\mathcal{A}$-measurable functions $g : X \to \overline{R}$ and $h : X \to \overline{R}$ with $g \leqslant f \leqslant h$ and $\mu(\{g \neq h\}) = 0$. [Use (a) and the indicator function method.]

Exercise 9.34. Let $(X, \mathcal{A}, \mu)$ be a measure space, and let $\{\mathcal{A}_n : n \in N\}$ be a sequence of σ-algebras of subsets of X such that $\mathcal{A}_n \subset \mathcal{A}$, $n \in N$. Prove the following.

(a) If $\mathcal{A}_n \supset \mathcal{A}_{n+1}$, $n \in N$, then $\bigcap_{n\in N} \mathcal{A}_n^{\mu} = (\bigcap_{n\in N} \mathcal{A}_n)^{\mu}$. [Use (9.33.a).]

(b) If the sequence $\{\mathcal{A}_n : n \in N\}$ is not nonincreasing, then the σ-algebra $\bigcap_{n\in N} \mathcal{A}_n^{\mu}$ may be different from $(\bigcap_{n\in N} \mathcal{A}_n)^{\mu}$.

(c) $\sigma(\bigcup_{n\in N} \mathcal{A}_n^{\mu})$ may be different from $\sigma(\bigcup_{n\in N} \mathcal{A}_n)^{\mu}$ even if the sequence $\{\mathcal{A}_n : n \in N\}$ is nondecreasing. [Hints. Let X be an infinite set, and let $\{x_n : n \in N\}$ be a sequence of distinct points of X such that $X - \{x_1, x_2, \ldots\}$ contains at least two points. Set $\mathcal{A} = \mathcal{P}(X)$ and $\mu = \sum_{n\in N} \varepsilon_{x_n}$. For $n \in N$, let $\mathcal{A}_n = \sigma(\{\{x_1\}, \ldots, \{x_n\}\})$. Show that $\sigma(\bigcup_{n\in N} \mathcal{A}_n^{\mu}) = \sigma(\{\{x_n\} : n \in N\})$ and $\sigma(\bigcup_{n\in N} \mathcal{A}_n)^{\mu} = \mathcal{P}(X)$.]

Exercise 9.35. Let $\{(X_i, \mathcal{A}_i, \mu_i) : i \in I\}$ be as in (9.28). Show that $\oplus_{i\in I} \mathcal{A}_i^{\mu_i} = (\oplus_{i\in I} \mathcal{A}_i)^{\oplus_{i\in I}\mu_i}$ and $\oplus_{i\in I} \overline{\mu}_i = \overline{\oplus_{i\in I} \mu_i}$.

Exercise 9.36. Let $\mathcal{A}$ be an algebra of subsets of X, and let μ be a finitely additive measure on $\mathcal{A}$.

(a) For $A, B \in \mathcal{A}$, define $A \sqsubset B$ if $\mu(A - B) = 0$. Show that $\sqsubset$ is a reflexive and transitive relation.

(b) For $A, B \in \mathcal{A}$, we write $A \sim B$ if both $A \sqsubset B$ and $B \sqsubset A$. Prove that $\sim$ is an equivalence relation on $\mathcal{A}$.

(c) For $\widetilde{A}, \widetilde{B} \in \mathcal{A}/\sim$, define $\widetilde{A} \leqslant \widetilde{B}$ if $A \sqsubset B$. Show that $\leqslant$ is a well-defined partial ordering on $\mathcal{A}/\sim$.

(d) For $\widetilde{A}, \widetilde{B} \in \mathcal{A}/\sim$, put $d(\widetilde{A}, \widetilde{B}) = \mu(A \Delta B)$. Prove that d is a well-defined metric for $\mathcal{A}/\sim$.

(e) For $\widetilde{A} \in \mathcal{A}/\sim$, define $\widetilde{\mu}(\widetilde{A}) = \mu(A)$. Show that $\widetilde{\mu}$ is well defined on $\mathcal{A}/\sim$.

(f) If $\mu(X) < \infty$, then $\widetilde{\mu}$ is uniformly continuous.

Exercise: μ-atom 9.37. Notation is as in (9.36). An equivalence class $\widetilde{A} \neq \widetilde{\emptyset}$ is called a *μ-atom* if $\widetilde{\emptyset} \neq \widetilde{B} \leqslant \widetilde{A}$ implies $\widetilde{B} = \widetilde{A}$. Prove the following.

(a) $\widetilde{A}$ is a μ-atom if and only if $\mu(A) > 0$, and either $\mu(B) = 0$ or $\mu(A - B) = 0$ whenever $B \in \mathcal{A}$ and $B \subset A$.

(b) If $\widetilde{A_1}$ and $\widetilde{A_2}$ are distinct μ-atoms, then $\mu(A_1 \cap A_2) = 0$.

(c) If $\widetilde{A_1}, \ldots, \widetilde{A_n}$ are distinct μ-atoms, then $\mu(\bigcup_{i=1}^n A_i) = \sum_{i=1}^n \mu(A_i)$.

(d) If $\mu(X) < \infty$, then the set of all μ-atoms is countable. [Hint. For $n \in N$, the set $\{\widetilde{A} \in \mathcal{A}/\sim : \widetilde{A}$ is a μ-atom and $\mu(A) \geqslant \mu(X)/n\}$ has at most n elements.]

(e) Let $\{A_i : i \in I\} \subset \mathcal{A}$ be a countable partition of X, and let $\widetilde{A}$ be a μ-atom. If μ is σ-additive, then $A \sim A \cap A_i$ for some $i \in I$.

(f) If μ is a σ-finite measure on $\mathcal{A}$, then the set of all μ-atoms is countable. [Use (9.29), (a), (e) and (d).]

(g) If $\mathcal{A}$ is a σ-algebra, $\widetilde{A}$ is a μ-atom, and $f : X \to \overline{R}$ is an $\mathcal{A}$-measurable function, then there are $b \in \overline{R}$ and $B \in \mathcal{A}$ with $B \subset A$ and $\mu(B) = \mu(A)$ such that $f(B) = \{b\}$.

Exercise 9.38. Let $(X, \mathcal{A}, \mu)$ be a measure space, where μ is σ-finite, let $\{\widetilde{A_i} : i \in I\}$ be the set of all μ-atoms, and put $A = (\bigcup_{i\in I} A_i)^c$. We say that μ is *atomic* if $\mu(A) = 0$, and we say that μ is *nonatomic* if $I = \emptyset$.

(a) For $B \in \mathcal{A}$, define $\nu(B) = \mu(B - A)$. Show that ν is an atomic measure on $\mathcal{A}$.

(b) For $B \in \mathcal{A}$, define $\tau(B) = \mu(B \cap A)$. Prove that τ is a nonatomic measure on $\mathcal{A}$.

(c) Show that $\mu = \nu + \tau$.

(d) Let $C \subset A$ be such that $C \in \mathcal{A}$. Prove that the restriction of μ to $C \cap \mathcal{A}$ is nonatomic.

Exercise 9.39. Let μ be a measure on $\mathcal{B}(R^n)$. Prove the following.

(a) $\widetilde{A}$ is a μ-atom if and only if $A \sim \{x\}$ and $\mu(\{x\}) > 0$ for some $x \in A$.

(b) Assume that μ is σ-finite. μ is atomic if and only if μ is discrete.

(c) μ is nonatomic if and only if $\mu(\{x\}) = 0$ for any $x \in R^n$.

Exercise 9.40. Let $(X, \mathcal{A}, \mu)$ be a measure space, where μ is finite, and let $(\mathcal{A}/\sim, \leqslant)$ be as in (9.36.c). Prove that there exist $\sup \widetilde{\mathcal{A}}$ and $\inf \widetilde{\mathcal{A}}$ for any $\widetilde{\mathcal{A}} \subset \mathcal{A}/\sim$. [Hints. Let $\mathcal{B} = \{\cup_{i=1}^n A_i : \{\widetilde{A_1}, \ldots, \widetilde{A_n}\} \subset \widetilde{\mathcal{A}}\}$ and $\beta = \sup\{\mu(B) : B \in \mathcal{B}\}$. Choose a sequence $\{B_n : n \in N\} \subset \mathcal{B}$ such that $\mu(B_n) \to \beta$, and set $B = \cup_{n \in N} B_n$. For $\widetilde{A} \in \widetilde{\mathcal{A}}$, $\mu(A \cup B_n) \to \beta$, and so $\mu(A - B_n) \to 0$. Since $\mu(A - B) \leqslant \mu(A - B_n)$, $n \in N$, $\widetilde{B}$ is an upper bound for $\widetilde{\mathcal{A}}$. Further, show that $\widetilde{B} = \sup \widetilde{\mathcal{A}}$.]

Exercise 9.41. State and prove a result similar to (9.40) in case that μ is σ-finite.

Exercise 9.42. Let $(X, \mathcal{A}, \mu)$ be a finite measure space, let A be as in (9.38), and let $B \subset A$ be such that $B \in \mathcal{A}$ and $\mu(B) > 0$. Prove the following.

(a) There is a sequence $\{B_n : n \in N\} \subset \mathcal{A}$ such that $B_n \subset B$ and $0 < \mu(B_n) \leqslant \mu(B)/2^n$ for each $n \in N$.

(b) For any $c \in]\,0, \mu(B)[$, there is $C \subset B$ such that $C \in \mathcal{A}$ and $\mu(C) = c$. [Hints. Let $\leqslant$ and $\widetilde{\mu}$ be as in (9.36), put $\widetilde{\mathcal{C}} = \{\widetilde{C} \in \mathcal{A}/\sim : \widetilde{C} \leqslant \widetilde{B}$ and $\widetilde{\mu}(\widetilde{C}) \leqslant c\}$, and consider the partially ordered set $(\widetilde{\mathcal{C}}, \leqslant)$. Use (9.40) to show that each chain in $\widetilde{\mathcal{C}}$ has an upper bound, and let $\widetilde{D}$ be a maximal element of $\widetilde{\mathcal{C}}$.Apply (a) to prove that the inequality $\mu(B \cap D) < c$ contradicts the maximality of $\widetilde{D}$.]

(c) For each $\varepsilon > 0$, there exists a finite partition $\mathcal{E}$ of B such that $\mathcal{E} \subset \mathcal{A}$ and $\mu(E) \leqslant \varepsilon$, $E \in \mathcal{E}$.

Exercise 9.43. Let $(X, \mathcal{A})$ be a measurable space, let ν be an atomic measure on $\mathcal{A}$, and let τ be a nonatomic measure on $\mathcal{A}$. Prove that there is $A \in \mathcal{A}$ such that $\nu(A^c) = \tau(A) = 0$. [Hints. Let $\{\widetilde{A_i} : i \in I\}$ be the set of all ν-atoms. Applying (9.37.e), we may select $\{A_i : i \in I\}$ such that $\tau(A_i) < \infty$, $i \in I$. Using (9.38.d) and (9.42.c), choose inductively a nonincreasing sequence $\{A_{in} : n \in N\} \subset \mathcal{A}$ such that $A_{i1} \subset A_i$, $\nu(A_{in}) = \nu(A_i)$ and $\tau(A_{in}) \leqslant 1/n$ for any $n \in N$. Show that $A = \cup_{i \in I} \cap_{n \in N} A_{in}$ is the wanted set.]

Exercise 9.44. Let $(X, \mathcal{A})$ be a measurable space, let ν_1 and ν_2 be atomic measures on $\mathcal{A}$, and let τ_1 and τ_2 be nonatomic measures on $\mathcal{A}$ such that $\nu_1 + \tau_1 = \nu_2 + \tau_2$.

(a) For $B \in \mathcal{A}$, show that $\tau_1(B) = 0$ if and only if $\tau_2(B) = 0$.

(b) For $B \in \mathcal{A}$, show that $\nu_1(B) = 0$ if and only if $\nu_2(B) = 0$. [Use (9.43) and (a).]

(c) Prove that $\nu_1 = \nu_2$ and $\tau_1 = \tau_2$, and deduce that the decomposition of μ in (9.38.c) is unique. [Use (9.43), (a) and (b).]

Exercise 9.45. Let $(X, \mathcal{A})$ and μ be as in (9.18). If $\{N_i : i \in I\}$ is a countable family of negative sets for μ, then $\cup_{i \in I} N_i$ is a negative sets for μ.

Exercise 9.46. Notation is as in (9.23). If $\mu(\mathcal{A}) \subset R$, then μ^+, μ^- and $|\mu|$ are finite measures, and μ is a bounded function.

Exercise 9.47. Let $(X, \mathcal{A})$ and μ be as in (9.18). Show that $|\mu|(A) = \sup\{\sum_{i=1}^n |\mu(A_i)| : \{A_1, \dots, A_n\} \subset \mathcal{A}$ is a partition of $A\}$ for any $A \in \mathcal{A}$.

Exercise 9.48. Let $(X, \mathcal{A})$ and μ be as in (9.18). For $A \in \mathcal{A}$, prove that $\sup\{|\mu(B)| : B \in \mathcal{A},\ B \subset A\} \leqslant |\mu|(A) \leqslant 2\sup\{|\mu(B)| : B \in \mathcal{A},\ B \subset A\}$.

Exercise 9.49. Notation is as in (9.23). For $A \in \mathcal{A}$, show that $\mu^+(A) = \sup\{\mu(B) : B \in \mathcal{A},\ B \subset A\}$ and $\mu^-(A) = -\inf\{\mu(B) : B \in \mathcal{A},\ B \subset A\}$.

Exercise 9.50. Let $(X, \mathcal{A})$ be a measurable space, and let μ be a signed measure on $\mathcal{A}$ such that $\mu(X) = 0$. Prove the following.

(a) μ is a bounded function.

(b) $|\mu|(X) = 2\sup_{A\in\mathcal{A}} |\mu(A)|$.

Exercise 9.51. Let $(X, \mathcal{A})$ be a measurable space, and let μ_1 and μ_2 be measures on $\mathcal{A}$ such that $\mu_1(X) = \mu_2(X) < \infty$. Show that $\sup_{A\in\mathcal{A}} |\mu_1(A) - \mu_2(A)| = 2^{-1}|\mu_1 - \mu_2|(X)$.

Exercise 9.52. Let $\mathcal{A}$ be a σ-algebra, let μ, ν and τ be bounded signed measures on $\mathcal{A}$, and define $\mu \vee \nu = (\mu - \nu)^+ + \nu$. Prove the following.

(a) $\mu \vee \nu = \nu \vee \mu$.

(b) $\mu \vee (\nu \vee \tau) = (\mu \vee \nu) \vee \tau$.

(c) $(\mu \vee \nu)(A) \geqslant \mu(A)$ and $(\mu \vee \nu)(A) \geqslant \nu(A)$ for any $A \in \mathcal{A}$.

(d) If λ is a signed measure on $\mathcal{A}$ such that $\lambda(A) \geqslant \mu(A)$ and $\lambda(A) \geqslant \nu(A)$ for any $A \in \mathcal{A}$, then $\lambda(A) \geqslant (\mu \vee \nu)(A)$ for each $A \in \mathcal{A}$.

Exercise 9.53. Let $\mathcal{A}$, μ, ν and τ be as in (9.52), and define $\mu \wedge \nu = -(\mu - \nu)^- + \nu$. Prove the following.

(a) $\mu \wedge \nu = \nu \wedge \mu$.

(b) $\mu \wedge (\nu \wedge \tau) = (\mu \wedge \nu) \wedge \tau$.

(c) $(\mu \wedge \nu)(A) \leqslant \mu(A)$ and $(\mu \wedge \nu)(A) \leqslant \nu(A)$ for each $A \in \mathcal{A}$.

(d) If λ is a signed measure on $\mathcal{A}$ such that $\lambda(A) \leqslant \mu(A)$ and $\lambda(A) \leqslant \nu(A)$ for any $A \in \mathcal{A}$, then $\lambda(A) \leqslant (\mu \wedge \nu)(A)$ for each $A \in \mathcal{A}$.

Exercise 9.54. Let x' and $(X, \mathcal{T})$ be as in (2.1.39).

(a) Characterize the set of all continuous functions from X into R.

(b) Prove that $\mathcal{B}_0(X) = \sigma(\{\{x\} : x \in X - \{x'\}\})$.

(c) Show that $\mathcal{B}_0(X) = \mathcal{B}_0(X)^{\varepsilon_{x'}}$.

(d) Prove that $\mathcal{B}_0(X)$ may be different from $\mathcal{B}(X)$.

Exercise 9.55. Let $(X, \mathcal{A}, \mu)$ and $(Y, \mathcal{B})$ be as in (9.26), and let $f : X \to Y$, $g : X \to Y$ and $h : X \to Y$ be $(\mathcal{A}, \mathcal{B})$-measurable functions such that $\mu f^{-1} = \mu g^{-1}$. In case $(Y, \mathcal{B}) = (R, \mathcal{B}(R))$, does $\mu(fh)^{-1} = \mu(gh)^{-1}$ hold?

Exercise 9.56 (Poincaré). Let $(X, \mathcal{A}, \mu)$ be a finite measure space, let $A_1, \dots, A_n \in \mathcal{A}$, and set $m_k = \sum_{1\leqslant i_1<\dots<i_k\leqslant n} \mu(A_{i_1} \cap \dots \cap A_{i_k})$, $1 \leqslant k \leqslant n$. Show that $\mu(\cup_{k=1}^n A_k) = \sum_{k=1}^n (-1)^k m_k$.

Exercise 9.57. Let $\mathcal{A}$ be a σ-algebra, and let μ_1 and μ_2 be signed measures on $\mathcal{A}$ such that $\mu_i(\mathcal{A}) \subset R$, $i = 1, 2$. Use (9.47) to prove that $|\mu_1 + \mu_2| \leqslant |\mu_1| + |\mu_2|$.

10 Extending Certain Measures

In most cases a measure μ can be explicitly defined on an algebra of sets $\mathcal{A}$ which is not a σ-algebra. Since in practice the most useful measures are defined on σ-algebras, the problem of extending μ to a measure μ^* on $\sigma(\mathcal{A})$ is extremely important. As it will follow from (10.29), such an extension always exists, but it is not necessarily unique. However, if μ is σ-finite, then μ^* is unique. In the first part of the present section we develop the technique of extending σ-finite measures. This will culminate with Carathéodory's extension theorem. By making use of this theorem we will then construct a large class of measures on $\mathcal{B}(R^n)$, namely the so-called Lebesgue-Stieltjes measures. Other applications of Carathéodory's theorem will be given in Section 16.

Throughout (10.1)–(10.7), $\mathcal{A}$ is an algebra of subsets of X, and μ is a finite measure on $\mathcal{A}$.

Theorem 10.1. *Let $\{A_n : n \in N\} \subset \mathcal{A}$ and $\{A'_n : n \in N\} \subset \mathcal{A}$ be nondecreasing sequence such that $\cup_{n\in N} A_n \subset \cup_{n\in N} A'_n$. Then $\lim_n \mu(A_n) \leqslant \lim_n \mu(A'_n)$.*

Proof. For $m \in N$, $\{A_m \cap A'_n : n \in N\}$ is a nondecreasing sequence of sets in $\mathcal{A}$ such that $\cup_{n\in N}(A_m \cap A'_n) = A_m$. Consequently, applying (9.9) and (9.6i), we have

$$\mu(A_m) = \lim_n \mu(A_m \cap A'_n) \leqslant \lim_n \mu(A'_n). \tag{1}$$

On letting $m \to \infty$ in (1), we get the desired result. □

The first extension of μ is to the family of all countable unions of sets in $\mathcal{A}$.

Definition 10.2. Let $\mathcal{B}$ be the family of all countable unions of sets in $\mathcal{A}$. Obviously, $B \in \mathcal{B}$ if and only if there exists a nondecreasing sequence $\{A_n : n \in N\} \subset \mathcal{A}$ such that $B = \cup_{n\in N} A_n$. For $B \in \mathcal{B}$, define $\mu'(B) = \lim_n \mu(A_n)$, where $\{A_n : n \in N\} \subset \mathcal{A}$ is a nondecreasing sequence such that $B = \cup_{n\in N} A_n$. By virtue of (10.1), μ' is well defined on $\mathcal{B}$.

Theorem 10.3.

(i) *If $A \in \mathcal{A}$, then $\mu(A) = \mu'(A)$.*

(ii) *If $B_1, B_2 \in \mathcal{B}$ and $B_1 \subset B_2$, then $\mu'(B_1) \leqslant \mu'(B_2)$.*

(iii) *If $B_1, B_2 \in \mathcal{B}$, then $B_1 \cap B_2 \in \mathcal{B}$, $B_1 \cup B_2 \in \mathcal{B}$ and $\mu'(B_1) + \mu'(B_2) = \mu'(B_1 \cap B_2) + \mu'(B_1 \cup B_2)$.*

(iv) *If $\{B_n : n \in N\} \subset \mathcal{B}$ is a nondecreasing sequence, then $\cup_{n\in N} B_n \in \mathcal{B}$ and $\mu'(\cup_{n\in N} B_n) = \lim_n \mu'(B_n)$.*

Proof. Assertion (i) is obvious, and assertion (ii) follows from (10.1). To prove (iii), let $\{A_{n1} : n \in N\} \subset \mathcal{A}$ and $\{A_{n2} : n \in N\} \subset \mathcal{A}$ be nondecreasing sequences such that $B_1 = \cup_{n\in N} A_{n1}$ and $B_2 = \cup_{n\in N} A_{n2}$. Then $\{A_{n1} \cap A_{n2} : n \in N\}$ and

$\{A_{n1} \cup A_{n2} : n \in N\}$ are nondecreasing sequences of sets in $\mathcal{A}$ such that $B_1 \cap B_2 = \cup_{n \in N}(A_{n1} \cap A_{n2})$ and $B_1 \cup B_2 = \cup_{n \in N}(A_{n1} \cup A_{n2})$, and so $B_1 \cap B_2 \in \mathcal{B}$ and $B_1 \cup B_2 \in \mathcal{B}$. Further, using (9.6.ii), we have

$$\mu(A_{n1}) + \mu(A_{n2}) = \mu(A_{n1} \cap A_{n2}) + \mu(A_{n1} \cup A_{n2}), \quad n \in N. \tag{1}$$

To complete the proof of (iii), we let $n \to \infty$ in (1). To prove (iv), consider $\{A_{nm} : n \in N\} \subset \mathcal{A}$, $m \in N$, nondecreasing sequences such that $B_m = \cup_{n \in N} A_{nm}$, $m \in N$. For $n \in N$, define $A_n = \cup_{m=1}^{n} A_{nm}$. Evidently, $\{A_n : n \in N\}$ is a nondecreasing sequence of sets in $\mathcal{A}$ such that, for any $n \in N$,

$$A_{nm} \subset A_n \subset B_n,\ 1 \leqslant m \leqslant n. \tag{2}$$

Therefore, $B_m \subset \cup_{n \in N} A_n \subset \cup_{n \in N} B_n$, $m \in N$, and so $\cup_{n \in N} B_n = \cup_{n \in N} A_n$. Thus $\cup_{n \in N} B_n \in \mathcal{B}$. For each $n \in N$, by (2), (9.6.i), (i) and (ii), we have

$$\mu(A_{nm}) \leqslant \mu(A_n) \leqslant \mu'(B_n),\ 1 \leqslant m \leqslant n. \tag{3}$$

Letting $n \to \infty$ in (3), we get $\mu'(B_m) \leqslant \mu'(\cup_{n \in N} B_n) \leqslant \lim_n \mu'(B_n)$, $m \in N$, and so $\mu'(\cup_{n \in N} B_n) = \lim_n \mu'(B_n)$. □

Further, we extend μ to the family of all subsets of X.

Theorem 10.4. *For $A \subset X$, define $\mu^*(A) = \inf\{\mu'(B) : A \subset B \in \mathcal{B}\}$. Then the following assertions hold.*

(i) *If $B \in \mathcal{B}$, then $\mu'(B) = \mu^*(B)$.*
(ii) *If $A_1 \subset A_2$, then $\mu^*(A_1) \leqslant \mu^*(A_2)$.*
(iii) *$\mu^*(A_1) + \mu^*(A_2) \geqslant \mu^*(A_1 \cap A_2) + \mu^*(A_1 \cup A_2)$.*
(iv) *If $\{A_n : n \in N\}$ is a nondecreasing sequence, then $\mu^*(A_n) \to \mu^*(\cup_{n \in N} A_n)$.*

Proof. Assertion (i) follows from (10.3.ii), and assertion (ii) is a consequence of (3.50.b). To prove (iii), let $\varepsilon > 0$, and let $A_1 \subset B_1 \in \mathcal{B}$ and $A_2 \subset B_2 \in \mathcal{B}$ be such that $\mu^*(A_1) + \varepsilon/2 > \mu'(B_1)$ and $\mu^*(A_2) + \varepsilon/2 > \mu'(B_2)$. Then, applying (10.3.iii), we have

$$\begin{aligned}&\mu^*(A_1) + \mu^*(A_2) + \varepsilon > \mu'(B_1) + \mu'(B_2)\\ &= \mu'(B_1 \cap B_2) + \mu'(B_1 \cup B_2) \geqslant \mu^*(A_1 \cap A_2) + \mu^*(A_1 \cup A_2).\end{aligned} \tag{1}$$

Since ε is arbitrary, assertion (iii) follows from (1). To prove (iv) let $\varepsilon > 0$, and let $A_n \subset B_n \in \mathcal{B}$, $n \in N$, be such that

$$\mu'(B_n) < \mu^*(A_n) + \varepsilon 2^{-n}, \quad n \in N. \tag{2}$$

We will show by induction that

$$\mu'\left(\bigcup_{i=1}^{n} B_i\right) < \mu^*(A_n) + \sum_{i=1}^{n} \varepsilon 2^{-i}, \quad n \in N. \tag{3}$$

By (2), the inequality (3) is true for $n = 1$. For $n \in N$, in view of (10.3.iii), we may write

$$\mu'\left(\bigcup_{i=1}^{n+1} B_i\right) = \mu'\left(\bigcup_{i=1}^{n} B_i\right) + \mu'(B_{n+1}) - \mu'\left(\left(\bigcup_{i=1}^{n} B_i\right) \cap B_{n+1}\right).$$

Now, since $A_n = A_n \cap A_{n+1} \subset (\cup_{i=1}^n B_i) \cap B_{n+1} \in \mathcal{B}$, it follows that $\mu^*(A_n) \leqslant \mu'((\cup_{i=1}^n B_i) \cap B_{n+1})$. Therefore, using the induction hypothesis and (2), we obtain

$$\begin{aligned}\mu'\left(\bigcup_{i=1}^{n+1} B_i\right) &< \mu^*(A_n) + \sum_{i=1}^{n} \varepsilon 2^{-i} + \mu^*(A_{n+1}) + \varepsilon 2^{-n-1} - \mu^*(A_n) \\ &= \mu^*(A_{n+1}) + \sum_{i=1}^{n+1} \varepsilon 2^{-i}.\end{aligned}$$

Thus (3) is proved. Further, since $\cup_{n\in N} A_n \subset \cup_{n\in N} B_n \in \mathcal{B}$, using (10.3.iv) and (3), we have $\mu^*(\cup_{n\in N} A_n) \leqslant \mu'(\cup_{n\in N} B_n) = \lim_n \mu'(\cup_{i=1}^n B_i) \leqslant \lim_n \mu^*(A_n) + \varepsilon$. Since ε is arbitrary, it follows that $\mu^*(\cup_{n\in N} A_n) \leqslant \lim_n \mu^*(A_n)$. The reverse inequality $\mu^*(\cup_{n\in N} A_n) \geqslant \lim_n \mu^*(A_n)$ is a consequence of (ii), and so the proof of (iv) is complete. □

We next identify a σ-algebra of subsets of X on which μ^* is a measure.

Theorem 10.5. *Let $\mathcal{C} = \{A \subset X : \mu^*(A) + \mu^*(A^c) = \mu(X)\}$. Then the following assertions hold.*

(i) *$\mathcal{C} = \{A \subset X : \mu^*(A) + \mu^*(A^c) \leqslant \mu(X)\}$.*
(ii) *$\mathcal{A} \subset \mathcal{B} \subset \mathcal{C}$.*
(iii) *$\mathcal{C}$ is a σ-algebra, and μ^* is a measure on $\mathcal{C}$.*

Proof. For any $A \subset X$, using (10.4.iii), (10.4.i) and (10.3.i), we get

$$\mu^*(A) + \mu^*(A^c) \geqslant \mu(X). \tag{1}$$

Hence assertion (i) follows at once from (1). Assertion (ii) is obvious. To prove (iii), notice first that $X \in \mathcal{C}$, and that $A \in \mathcal{C}$ implies $A^c \in \mathcal{C}$. Now let $A_1, A_2 \in \mathcal{C}$. Then, according to (10.4.iii), we have

$$\mu^*(A_1) + \mu^*(A_2) \geqslant \mu^*(A_1 \cap A_2) + \mu^*(A_1 \cup A_2) \tag{2}$$

and

$$\mu^*(A_1^c) + \mu^*(A_2^c) \geqslant \mu^*((A_1 \cap A_2)^c) + \mu^*((A_1 \cup A_2)^c). \tag{3}$$

Taking into account (1), we get

$$\mu^*(A_1) + \mu^*(A_2) = \mu^*(A_1 \cap A_2) + \mu^*(A_1 \cup A_2) \tag{4}$$

and

$$\mu^*(A_1 \cup A_2) + \mu^*((A_1 \cup A_2)^c) = \mu(X), \tag{5}$$

since, otherwise, adding (2) and (3), it would follow $2\mu(X) < 2\mu(X)$. (5) shows that

$$A_1 \cup A_2 \in \mathcal{C}. \tag{6}$$

Further, consider $\{A_n : n \in N\} \subset \mathcal{C}$. Then, by virtue of (10.4.iv), $\mu^*(\cup_{i=1}^n A_n) \to \mu^*(\cup_{n\in N} A_n)$. Therefore, for $\varepsilon > 0$, there is $n_\varepsilon \in N$ such that

$$\mu^*\left(\bigcup_{n\in N} A_n\right) < \mu^*\left(\bigcup_{i=1}^n A_n\right) + \varepsilon, \quad n \geqslant n_\varepsilon. \tag{7}$$

On the other hand, applying (10.4.ii), we have

$$\mu^*\left(\left(\bigcup_{n\in N} A_n\right)^c\right) \leqslant \mu^*\left(\left(\bigcup_{i=1}^n A_n\right)^c\right), \quad n \geqslant n_\varepsilon. \tag{8}$$

Adding (7) and (8), and using (6), we get $\mu^*(\cup_{n\in N} A_n) + \mu^*((\cup_{n\in N} A_n)^c) < \mu(X) + \varepsilon$. Since ε is arbitrary, according to (i), it follows that $\cup_{n\in N} A_n \in \mathcal{C}$, and so $\mathcal{C}$ is a σ-algebra. Finally, from (4) it follows that μ^* is a finitely additive measure on $\mathcal{C}$. Hence, applying (9.7) and (10.4.iv), μ^* is a σ-additive measure on $\mathcal{C}$. □

Corollary 10.6. *μ^* is a measure on $\sigma(\mathcal{A})$ such that $\mu^* = \mu$ on $\mathcal{A}$.*

Proof. This follows at once from (10.5.iii), (10.5.ii), (10.4.i) and (10.3.i). □

The next theorem shows that there is very little difference between the σ-algebras $\mathcal{C}$ and $\sigma(\mathcal{A})$.

Theorem 10.7. *$(X, \mathcal{C}, \mu^*)$ is the completion of $(X, \sigma(\mathcal{A}), \mu^*_{\sigma(\mathcal{A})})$.*

Proof. For notational convenience, set $\nu = \mu^*_{\sigma(\mathcal{A})}$. First notice that μ^* is a complete measure on $\mathcal{C}$. Thus, applying (9.14.b), we have $\sigma(\mathcal{A})^\nu \subset \mathcal{C}^{\mu^*} \subset \mathcal{C}$. It remains to show that $\mathcal{C} \subset \sigma(\mathcal{A})^\nu$. Let $A \in \mathcal{C}$. For $n \in N$, choose $B_n \in \mathcal{B}$ such that $A \subset B_n$ and $\mu^*(A) \leqslant \mu^*(B_n) < \mu^*(A) + 1/n$. Therefore $\mu^*(A) \leqslant \mu^*(\cap_{n\in N} B_n) < \mu^*(A) + 1/n$, $n \in N$, and so

$$\mu^*(A) = \mu^*\left(\bigcap_{n\in N} B_n\right). \tag{1}$$

Analogously, select $\{C_n : n \in N\} \subset \mathcal{B}$ such that $A^c \subset C_n$, $n \in N$, and

$$\mu^*(A^c) = \mu^*\left(\bigcap_{n\in N} C_n\right). \tag{2}$$

Evidently, $\cap_{n\in N} B_n \in \sigma(\mathcal{A})$, $\cup_{n\in N} C_n^c \in \sigma(\mathcal{A})$, and $\cup_{n\in N} C_n^c \subset A \subset \cap_{n\in N} B_n$. Using (1) and (2), we get

$$\begin{aligned}\mu^*\left(\left(\bigcap_{n\in N} B_n\right) - \left(\bigcup_{n\in N} C_n^c\right)\right) &= \mu^*\left(\bigcap_{n\in N} B_n\right) - \mu^*\left(\bigcup_{n\in N} C_n^c\right)\\ &= \mu^*(A) - \mu(X) + \mu^*\left(\bigcap_{n\in N} C_n\right) = 0.\end{aligned}$$

Hence, according to (9.33.a), it follows that $A \in \sigma(\mathcal{A})^\nu$. Therefore, $\mathcal{C} \subset \sigma(\mathcal{A})^\nu$. □

To prove Carathéodory's extension theorem, we need the next lemma.

Lemma 10.8. *Let $\mathcal{M} \subset \mathcal{P}(X)$ be such that $A, B \in \mathcal{M}$ implies $A \cap B \in \mathcal{M}$, and let λ and μ be measures on $\sigma(\mathcal{M})$ such that $\lambda(X) = \mu(X)$ and $\lambda(A) = \mu(A)$ for any $A \in \mathcal{M}$. Then $\lambda = \mu$.*

Proof. Let $\mathcal{D} = \{A \in \sigma(\mathcal{M}) : \lambda(A) = \mu(A)\}$. Then $\mathcal{D}$ is a Dynkin system such that $\mathcal{M} \subset \mathcal{D} \subset \sigma(\mathcal{M})$. Thus, in view of (7.36), $\mathcal{D} = \sigma(\mathcal{M})$, and so $\lambda = \mu$. □

Carathéodory's extension theorem 10.9. *Let $\mathcal{A}$ be an algebra of subsets of X, and let μ be a σ-finite measure on $\mathcal{A}$. Then there exists a unique measure μ^* on $\sigma(\mathcal{A})$ such that $\mu^* = \mu$ on $\mathcal{A}$.*

Proof. According to (9.29), there is a countable partition $\{A_i : i \in I\} \subset \mathcal{A}$ of X such that $\mu(A_i) < \infty$, $i \in I$. For $i \in I$, denote $\mathcal{A}_i = A_i \cap \mathcal{A}$ and $\mu_i = \mu_{\mathcal{A}_i}$. Then, applying (10.6), there exists a measure μ_i^* on $\sigma(\mathcal{A}_i)$ such that $\mu_i^* = \mu_i$ on $\mathcal{A}_i$. Put $\mu^* = \oplus_{i\in I}\mu_i^*$. By virtue of (9.28), μ^* is a measure on $\oplus_{i\in I}\sigma(\mathcal{A}_i)$. Since, by (7.14) and (7.53), $\oplus_{i\in I}\sigma(\mathcal{A}_i) = \sigma(\mathcal{A})$, it follows that μ^* is a measure on $\sigma(\mathcal{A})$. For any $A \in \mathcal{A}$, we have

$$\mu^*(A) = \sum_{i\in I}\mu_i^*(A\cap A_i) = \sum_{i\in I}\mu_i(A\cap A_i) = \sum_{i\in I}\mu(A\cap A_i) = \mu(A),$$

and so $\mu^* = \mu$ on $\mathcal{A}$. Now let λ be a measure on $\sigma(\mathcal{A})$ such that $\lambda = \mu$ on $\mathcal{A}$. For $i \in I$, let λ_i be the restriction of λ to $\sigma(\mathcal{A}_i)$. Then, for each $B \in \mathcal{A}_i$, we have $\lambda_i(B) = \mu(B) = \mu_i^*(B) < \infty$. Thus, according to (10.8), we get $\lambda_i = \mu_i^*$, $i \in I$, and so $\lambda = \mu$. Therefore, μ^* is the unique measure on $\sigma(\mathcal{A})$ such that $\mu^* = \mu$ on $\mathcal{A}$. □

Remarks 10.10. (a) If μ in (10.9) is not σ-finite, then there may exist several extensions of μ to $\sigma(\mathcal{A})$. For example, let $X = Q$, let $\mathcal{A}$ be as in (7.3.d), and let $\mathcal{E} = Q \cap \mathcal{A}$. For $E \in \mathcal{E}$, define

$$\mu(E) = \begin{cases} 0 & \text{if } E = \emptyset \\ \infty & \text{if } E \neq \emptyset \end{cases}.$$

Then μ is a measure on $\mathcal{E}$ that is not σ-finite. Using (7.14) and (7.10), it follows immediately that $\sigma(\mathcal{E}) = \mathcal{P}(Q)$. Let μ_1 be the counting measure on $\mathcal{P}(Q)$, and set $\mu_2 = 2\mu_1$. Then $\mu_1 = \mu$ on $\mathcal{E}$, $\mu_2 = \mu$ on $\mathcal{E}$, but $\mu_1 \neq \mu_2$.

(b) Notation is as in the proof of (10.9). For each $i \in I$, starting from $\mathcal{A}_i$ and μ_i, we can construct a σ-algebra $\mathcal{C}_i$ of subsets of A_i as in (10.5). Then, according to (10.7) and (9.35), we have $\oplus_{i\in I}\mathcal{C}_i = \oplus_{i\in I}\sigma(\mathcal{A}_i)^{\mu_i^*} = \sigma(\mathcal{A})^{\mu^*}$.

By making use of Carathéodory's extension theorem we will construct in what follows a large class of measures on $\mathcal{B}(R^n)$, namely the so-called Lebesgue-Stieltjes measures. To do this, some definitions are in order.

Definition 10.11. Let X and Y be topological spaces, let $A \subset X$, $B \subset Y$, and let $f : A \to B$ be a function. Let $a \in A^d$ and $b \in Y$. We say that the *limit of f at a* is b, and we write $\lim_{x\to a} f(x) = b$ [$f(x) \to b$ as $x \to a$], if for each neighborhood V of b there is a neighborhood U of a such that $x \in (A - \{a\}) \cap U$ implies $f(x) \in V$.

Remarks 10.12. Notation is as in (10.11).

(a) Assume that X is a metric space. Then $\lim_{x\to a} f(x) = b$ if and only if $f(a_n) \to b$ whenever $\{a_n : n \in N\} \subset A - \{a\}$ is a sequence such that $a_n \to a$.

(b) Suppose that $a \in A$. Then f is continuous at a if and only if $\lim_{x\to a} f(x) = f(a)$.

Definitions 10.13. Let X, Y, A, B and f be as in (10.11). Assume that $X = \overline{R}^n$, let $\leqslant$ be the partial ordering on $\overline{R}^n$ introduced in (3.19.b), and let $\mathbf{a} \in A^d$. If $\mathbf{a}$ is a limit point of the set $A \cap \{\mathbf{x} \in \overline{R}^n : \mathbf{a} \leqslant \mathbf{x}\}$, and the restriction of f on $A \cap \{\mathbf{x} \in \overline{R}^n : \mathbf{a} \leqslant \mathbf{x}\}$ has limit $f(\mathbf{a}+)$ at $\mathbf{a}$, then we say that the *right-hand limit of f at* $\mathbf{a}$ is $f(\mathbf{a}+)$, and we write $\lim_{\mathbf{x}\searrow\mathbf{a}} f(\mathbf{x}) = f(\mathbf{a}+)$ $[f(\mathbf{x}) \to f(\mathbf{a}+)$ as $\mathbf{x} \searrow \mathbf{a}]$. If $\mathbf{a}$ is a limit point of the set $A \cap \{\mathbf{x} \in \overline{R}^n : \mathbf{x} \leqslant \mathbf{a}\}$, and the restriction of f to $A \cap \{\mathbf{x} \in \overline{R}^n : \mathbf{x} \leqslant \mathbf{a}\}$ has limit $f(\mathbf{a}-)$ at $\mathbf{a}$, then we say that the *left-hand limit of f at* $\mathbf{a}$ is $f(\mathbf{a}-)$, and we write $\lim_{\mathbf{x}\nearrow\mathbf{a}} f(\mathbf{x}) = f(\mathbf{a}-)$ $[f(\mathbf{x}) \to f(\mathbf{a}-)$ as $\mathbf{x} \nearrow \mathbf{a}]$.

Further, assume that $\mathbf{a} \in A$. We say that f is *right continuous at* $\mathbf{a}$ if $\lim_{\mathbf{x}\searrow\mathbf{a}} f(\mathbf{x}) = f(\mathbf{a})$. We say that f is *left continuous at* $\mathbf{a}$ if $\lim_{\mathbf{x}\nearrow\mathbf{a}} f(\mathbf{x}) = f(\mathbf{a})$.

Let $A \subset R$ be an open interval, let $f : A \to R$ be a function, and let $a \in A$. We say that f has a *jump at a* [a is a *point of jump* of f] if $f(a+)$ and $f(a-)$ both exist but are unequal. The number $f(a+) - f(a-)$ is called the *size of the jump.*

Remarks 10.14. Notation is as in (10.13).

(a) Suppose that $\mathbf{a} \in (A \cap \{\mathbf{x} \in \overline{R}^n : \mathbf{a} \leqslant \mathbf{x}\})^d$. Then $\lim_{\mathbf{x}\searrow\mathbf{a}} f(\mathbf{x}) = f(\mathbf{a}+)$ if and only if $f(\mathbf{a}_n) \to f(\mathbf{a}+)$ whenever $\{\mathbf{a}_n : n \in N\} \subset A$ is a sequence such that $\mathbf{a} < \mathbf{a}_n, n \in N$, and $\mathbf{a}_n \to \mathbf{a}$.

(b) Assume that $\mathbf{a} \in (A \cap \{\mathbf{x} \in \overline{R}^n : \mathbf{x} \leqslant \mathbf{a}\})^d$. Then $\lim_{\mathbf{x}\nearrow\mathbf{a}} f(\mathbf{x}) = f(\mathbf{a}-)$ if and only if $f(\mathbf{a}_n) \to f(\mathbf{a}-)$ whenever $\{\mathbf{a}_n : n \in N\} \subset A$ is a sequence such that $\mathbf{a}_n < \mathbf{a}, n \in N$, and $\mathbf{a}_n \to \mathbf{a}$.

(c) Suppose that $\mathbf{a} \in (A \cap \{\mathbf{x} \in \overline{R}^n : \mathbf{a} \leqslant \mathbf{x}\})^d \cap (A \cap \{\mathbf{x} \in \overline{R}^n : \mathbf{x} \leqslant \mathbf{a}\})^d$, and let $b \in Y$. For $n = 1$, $\lim_{\mathbf{x}\to\mathbf{a}} f(\mathbf{x}) = b$ if and only if $\lim_{\mathbf{x}\searrow\mathbf{a}} f(\mathbf{x}) = \lim_{\mathbf{x}\nearrow\mathbf{a}} f(\mathbf{x}) = b$. For $n > 1$, $\lim_{\mathbf{x}\to\mathbf{a}} f(\mathbf{x}) = b$ implies $\lim_{\mathbf{x}\searrow\mathbf{a}} f(\mathbf{x}) = \lim_{\mathbf{x}\nearrow\mathbf{a}} f(\mathbf{x}) = b$, but the converse of this assertion is not generally true. For example, consider the function $f : R^2 \to R$ defined by $f = 1_{[0,\infty[^2 \cup]-\infty,0]^2}$. Then $\lim_{\mathbf{x}\searrow\mathbf{0}} f(\mathbf{x}) = \lim_{\mathbf{x}\nearrow\mathbf{0}} f(\mathbf{x}) = 1$, but f has no limit at $\mathbf{0}$.

(d) Assume that $\mathbf{a} \in A \cap (A \cap \{\mathbf{x} \in \overline{R}^n : \mathbf{a} \leqslant \mathbf{x}\})^d \cap (A \cap \{\mathbf{x} \in \overline{R}^n : \mathbf{x} \leqslant \mathbf{a}\})^d$. For $n = 1$, f is continuous at $\mathbf{a}$ if and only if f is both right continuous and left continuous at $\mathbf{a}$. For $n > 1$, f is both right continuous and left continuous at $\mathbf{a}$ whenever f is continuous at $\mathbf{a}$, but the converse of this statement is not generally true. For example, the function f in (c) is right continuous and left continuous at $\mathbf{0}$, but f is not continuous at $\mathbf{0}$.

We first consider the class of Lebesgue-Stieltjes measures on $\mathcal{B}(R)$.

Theorem 10.15. *Let $F : R \to R$ be a nondecreasing function which is right continuous at any $x \in R$. Then there exists a unique measure μ_F on $\mathcal{B}(R)$ such that, for each bounded interval $]a, b] \subset R$,*

(i) $\mu_F(]a, b]) = F(b) - F(a)$.

Proof. Let $\mathcal{A}$ be as in (7.3.d), and let μ be as in (9.3.e). As stated in (9.3.e), μ is a finitely additive measure on $\mathcal{A}$. We next show that μ is σ-additive.

(I) First assume that $-\infty < F(-\infty) \leqslant F(\infty) < \infty$. Let $\varepsilon > 0$. For $]a, b] \in \mathcal{A}$, since F is right continuous at a in case $a \in R$, and $-\infty < F(-\infty)$ in case $a = -\infty$, there is $a_\varepsilon < b$ such that $[a_\varepsilon, b] \subset]a, b]$ and $\mu(]a, b]-]a_\varepsilon, b]) < \varepsilon$. Analogously, for $]a, \infty[\in \mathcal{A}$ there are $a_\varepsilon < b_\varepsilon$ such that $[a_\varepsilon, b_\varepsilon] \subset]a, \infty[$ and $\mu(]a, \infty[-]a_\varepsilon, b_\varepsilon]) < \varepsilon$. Therefore, for $B \in \mathcal{A}$, there is $C \in \mathcal{A}$ such that $C^- \subset B$, C^- is a compact subset of R, and $\mu(B - C) < \varepsilon$. Now let $\{B_n : n \in N\}$ be a nonincreasing sequence of sets in $\mathcal{A}$ such that $\cap_{n \in N} B_n = \emptyset$. For $n \in N$, choose $C_n \in \mathcal{A}$ such that $C_n^- \subset B_n$, C_n^- is a compact subset of R, and $\mu(B_n - C_n) < \varepsilon/2^n$. Since $\{C_1^- \cap C_n^- : n \in N\}$ is a sequence of closed subsets of C_1^-, and $\cap_{n \in N}(C_1^- \cap C_n^-) \subset \cap_{n\in N} B_n = \emptyset$, (5.24) shows that there is $n_\varepsilon \in N$ such that $\cap_{i=1}^{n_\varepsilon} C_i^- = \cap_{i=1}^{n_\varepsilon}(C_1^- \cap C_i^-) = \emptyset$. Consequently, for $n \geqslant n_\varepsilon$, using (9.6.i) and (9.6.iv), we have $\mu(B_n) = \mu(B_n - (\cup_{i=1}^n C_i)) = \mu(\cup_{i=1}^n (B_n - C_i)) \leqslant \sum_{i=1}^n \mu(B_i - C_i) < \varepsilon$. This means that $\mu(B_n) \to 0$. Hence, by (9.8), μ is σ-additive.

(II) Now consider the general case. For $n \in N$, let φ_n be as in (8.4.a) and put $F_n = \varphi_n \circ F$. Then F_n is a nondecreasing function which is right continuous at any $x \in R$. Moreover, F_n is bounded, and $F_n \to F$. For each $n \in N$, starting from F_n, we define μ_n as in (9.3.e). Applying (I), it follows that μ_n is a measure on $\mathcal{A}$. Obviously, for any $B \in \mathcal{A}$, we have $\mu_n(B) \leqslant \mu(B), n \in N$, and $\mu_n(B) \to \mu(B)$. Let $\{B_i : i \in N\} \subset \mathcal{A}$ be such that $\cup_{i \in I} B_i \in \mathcal{A}$ and $B_i \cap B_j = \emptyset$ for $i \neq j$.
Then, since μ_n is σ-additive, we get

$$\mu_n\left(\bigcup_{i\in N} B_i\right) = \sum_{i\in N} \mu_n(B_i) \leqslant \sum_{i \in N} \mu(B_i), \quad n \in N. \tag{1}$$

On letting $n \to \infty$ in (1), we have $\mu(\cup_{i\in N} B_i) \leqslant \sum_{i \in N} \mu(B_i)$. Thus, according to (9.30), μ is σ-additive.

Further, since $\cup_{n\in N}]-n, n] = R$, and $\mu(]-n, n]) = F(n) - F(-n) < \infty, n \in N$, it follows that μ is σ-finite. Therefore, using (7.10) and Carathéodory's extension theorem (10.9), there exists a unique measure μ_F on $\mathcal{B}(R)$ such that $\mu_F = \mu$ on $\mathcal{A}$. Consequently, applying (9.9), μ_F is the only measure on $\mathcal{B}(R)$ satisfying (i). □

Definitions 10.16. Notation is as in (10.15). The measure μ_F is called the *Lebesgue-Stieltjes measure on $\mathcal{B}(R)$ induced by F*, and the elements of the σ-algebra $\mathcal{B}(R)^{\mu_F}$ are called the *Lebesgue-Stieltjes sets of R induced by F*. If $F(x) = x, x \in R$, then μ_F is called the *Lebesgue measure on $\mathcal{B}(R)$*, and the elements of the σ-algebra $\mathcal{B}(R)^{\mu_F}$ are called the *Lebesgue sets of R*.

Remarks 10.17. (a) Let M be the set of all Lebesgue-Stieltjes measures on $\mathcal{B}(R)$, and let M' be the set of all measures μ on $\mathcal{B}(R)$ such that $\mu(I) < \infty$ for each bounded interval $I \subset R$. Evidently, if $\mu \in$M, then $\mu \in$ M'. Conversely, if $\mu \in$ M', define

$$F(x) = \begin{cases} \mu(]0, x]) & \text{if } x > 0 \\ 0 & \text{if } x = 0 \\ -\mu(]0, x]) & \text{if } x < 0 \end{cases}.$$

Then F is a nondecreasing function. Moreover, by (10.14.a), (9.9) and (9.10), F is right continuous at any $x \in R$. Let μ_F be as in (10.15). Then $\mu(]a, b]) = \mu_F(]a, b]) = F(b) - F(a)$ for any bounded interval $]a, b] \subset R$, and so $\mu = \mu_F \in$ M. Thus M = M'.

(b) Let $\mathfrak{F}$ be the set of all nondecreasing functions from R into R which are right continuous at any $x \in R$, and let M be as in (a). For $F, G \in \mathfrak{F}$, we write $F \sim G$ if $F - G = c$ for some $c \in R$. Plainly, $\sim$ is an equivalence relation on $\mathfrak{F}$. For $F, G \in \mathfrak{F}$, it is easily seen that $F \sim G$ if and only if $\mu_F = \mu_G$. Therefore, the mapping that assigns to the equivalence class $\widetilde{F} \in \mathfrak{F}/\sim$ the measure $\mu_F \in$ M is a well-defined bijective mapping from $\mathfrak{F}/\sim$ onto M.

(c) Let λ be the Lebesgue measure on $\mathcal{B}(R)$. For any $x \in R$ and $A \in \mathcal{B}(R)$, we have $\lambda(x + A) = \lambda(A)$ (see (7.60)). Actually, for $x \in R$, consider the measure λ_x on $\mathcal{B}(R)$ defined by $\lambda_x(A) = \lambda(x + A)$, $A \in \mathcal{B}(R)$. Then, according to (10.15), $\lambda_x = \lambda$. From this, using (9.13) and (7.60), it follows at once that $x + A \in \mathcal{B}(R)^\lambda$ and $\overline{\lambda}(x+A) = \overline{\lambda}(A)$ whenever $A \in \mathcal{B}(R)^\lambda$.

(d) Evidently, the σ-algebra $\mathcal{B}(R)^{\mu_F}$ depends on F. For example, if μ_F is a discrete measure, then $\mathcal{B}(R)^{\mu_F} = \mathcal{P}(R)$. However, in general $\mathcal{B}(R)^{\mu_F}$ does not coincide with $\mathcal{P}(R)$, and so $\mathcal{B}(R) \neq \mathcal{P}(R)$. Actually, let λ be the Lebesgue measure on $\mathcal{B}(R)$. For $x, y \in R$, we write $x \sim y$ if $x - y \in Q$. Clearly, $\sim$ is an equivalence relation on R. Also,

$$\widetilde{x} \cap \,]0, 1] \neq \emptyset, \quad x \in R. \tag{1}$$

By the axiom of choice (2.20), select an element from each equivalent class in $R/\sim$ to form a set A. Using (1), choose A such that $A \subset \,]0, 1]$. If $r, s \in Q$ and $r \neq s$, then $(r+A) \cap (s+A) = \emptyset$. Assume that $A \in \mathcal{B}(R)^\lambda$. Then, since $\cup_{r \in Q \cap]0,1]}(r+A) \subset \,]0, 2]$, using (c), we have $\sum_{r \in Q \cap]0,1]} \overline{\lambda}(r + A) \leqslant \overline{\lambda}(]0, 2]) = 2$, and so $\overline{\lambda}(A) = 0$. Further, since $R = \cup_{r \in Q}(r + A)$, applying (c), we get $\infty = \lambda(R) = \sum_{r \in Q} \overline{\lambda}(r + A) = 0$. This contradiction shows that A is not a Lebesgue set.

We now consider the class of Lebesgue-Stieltjes measures on $\mathcal{B}(R^n)$. Although similar to the construction of Lebesgue-Stieltjes measures on $\mathcal{B}(R)$, the construction of these measures is more involved. We begin with a definition.

Definition 10.18. Let $\mathbf{a} = (a_1, \ldots, a_n) \in R^n$ and $\mathbf{b} = (b_1, \ldots, b_n) \in R^n$. For each function $F : R^n \to R$, define $\Delta_{\mathbf{ba}} F = \sum_{k=0}^n (-1)^k F_k$, where F_k is the sum of all $\binom{n}{k}$ terms of the form $F(c_1, \ldots, c_n)$ with $c_i = a_i$ for exactly k values of i in $\{1, \ldots, n\}$ and $c_i = b_i$ for the remaining $n - k$ values of i. $\Delta_{\mathbf{ba}}$ is a mapping from R^{R^n} into R called the *difference operator.*

Example 10.19. For $n = 3$, we have $\Delta_{\mathbf{ba}} F = F(b_1, b_2, b_3) - F(a_1, b_2, b_3) - F(b_1, a_2, b_3) - F(b_1, b_2, a_3) + F(a_1, a_2, b_3) + F(a_1, b_2, a_3) + F(b_1, a_2, a_3) - F(a_1, a_2, a_3)$.

Remarks 10.20. (a) Let μ be a measure on $\mathcal{B}(R^n)$ such that $\mu(\prod_{i=1}^n I_i) < \infty$ whenever $I_1, \ldots, I_n$ are bounded intervals in R. Define a function $F : R^n \to R$ as follows. Let $(x_1, \ldots, x_n) \in R^n$. Set $F(x_1, \ldots, x_n) = 0$ whenever $x_j = 0$ for some j. If $x_1 \neq 0, \ldots, x_n \neq 0$, then put $y_i = \begin{cases} x_i \text{ if } x_i < 0 \\ 0 \ \text{ if } x_i > 0 \end{cases}$ and $z_i = \begin{cases} 0 \ \text{ if } x_i < 0 \\ x_i \text{ if } x_i > 0 \end{cases}$

for $i = 1, \ldots, n$, and define $F(x_1, \ldots, x_n) = (-1)^k \mu(\prod_{i=1}^n]y_i, z_i])$, where k is the number of negative coordinates of $(x_1, \ldots, x_n)$. Then it is easy to check that $\mu(\prod_{i=1}^n]a_i, b_i]) = \Delta_{\mathbf{ba}} F$ whenever $]a_1, b_1], \ldots,]a_n, b_n]$ are bounded intervals in R, where $\mathbf{a} = (a_1, \ldots, a_n)$ and $\mathbf{b} = (b_1, \ldots, b_n)$.

(b) For $\alpha \in R$ and $F, G \in R^{R^n}$, we have $\Delta_{\mathbf{ba}} \alpha F = \alpha \Delta_{\mathbf{ba}} F$ and $\Delta_{\mathbf{ba}}(F + G) = \Delta_{\mathbf{ba}} F + \Delta_{\mathbf{ba}} G$.

(c) Let $F : R^n \to R$ be a function. Assume that there is $i \in \{1, \ldots, n\}$ such that $F(x_1, \ldots, x_{i-1}, x_i, x_{i+1}, \ldots, x_n) = F(x_1, \ldots, x_{i-1}, x_i', x_{i+1}, \ldots, x_n)$ for any $x_i, x_i' \in R$, that is F is constant in x_i. Then $\Delta_{\mathbf{ba}} F = 0$.

(d) If $a_i = b_i$ for some i, then $\Delta_{\mathbf{ba}} F = 0$.

(e) Let $\mathbf{a}, \mathbf{b}, \mathbf{c} \in R^n$. Then $\Delta_{\mathbf{ca}} F$ is the sum of 2^n terms of the form $\Delta_{(\beta_1, \ldots, \beta_n)(\alpha_1, \ldots, \alpha_n)} F$, where $\beta_i = b_i$ or c_i, and $\alpha_i = \begin{cases} a_i \text{ if } \beta_i = b_i \\ b_i \text{ if } \beta_i = c_i \end{cases}$ for $i = 1, \ldots, n$.

Lemma 10.21. *Let $F : R^n \to R$ be a function such that $\Delta_{\mathbf{ba}} F \geqslant 0$ whenever $\mathbf{a} \leqslant \mathbf{b}$.*

(i) *If $\mathbf{a} \leqslant \mathbf{b} \leqslant \mathbf{c}$, then $\Delta_{\mathbf{ca}} F \geqslant \Delta_{\mathbf{cb}} F + \Delta_{\mathbf{ba}} F$.*

(ii) *If $\mathbf{a} \leqslant \mathbf{a}' \leqslant \mathbf{b}' \leqslant \mathbf{b}$, then $\Delta_{\mathbf{ba}} F \geqslant \Delta_{\mathbf{b'a'}} F$.*

Proof. If $\mathbf{a} \leqslant \mathbf{b} \leqslant \mathbf{c}$, then $\Delta_{\mathbf{ca}} F$ is the sum of 2^n nonnegative terms of the form $\Delta_{(\beta_1, \ldots, \beta_n)(\alpha_1, \ldots, \alpha_n)} F$ as in (10.20.e). Consequently, $\Delta_{\mathbf{ca}} F \geqslant \Delta_{\mathbf{cb}} F + \Delta_{\mathbf{ba}} F$. If $\mathbf{a} \leqslant \mathbf{a}' \leqslant \mathbf{b}' \leqslant \mathbf{b}$, then, applying (i), we have $\Delta_{\mathbf{ba}} F \geqslant \Delta_{\mathbf{ba'}} F \geqslant \Delta_{\mathbf{b'a'}} F$. □

Definition 10.22. Let $\mathcal{B}$ and $\mathcal{A}$ be as in (7.56), and let F be as in (10.21). Set $\mu(\emptyset) = 0$. Define $\mu(\prod_{i=1}^n]a_i, b_i]) = \Delta_{\mathbf{ba}} F$ whenever $]a_1, b_1], \ldots,]a_n, b_n]$ are bounded intervals in R, where $\mathbf{a} = (a_1, \ldots, , a_n)$ and $\mathbf{b} = (b_1, \ldots, b_n)$. Now let $I \subset R$ be an interval of the form $]a, b]$ or $]a, \infty[$. For $m \in N$, define

$$I(m) = \begin{cases} I & \text{if } I =]a, b] \text{ and } a \in R \\ ]b - m, b] & \text{if } I =]-\infty, b] \\ ]a, a + m] & \text{if } I =]a, \infty[\text{ and } a \in R \\ ]-m, m] & \text{if } I =]-\infty, \infty[\end{cases}.$$

For $B = \prod_{i=1}^n I_i \in \mathcal{B}$, denote $B_m = \prod_{i=1}^n I_i(m)$, $m \in N$, and define $\mu(B) = \lim_m \mu(B_m)$. In view of (10.21), the sequence $\{\mu(B_m) : m \in N\}$ is nondecreasing, and so $\lim_m \mu(B_m)$ exists. If $A = \cup_{j=1}^p B_j \in \mathcal{A}$, where $B_j \in \mathcal{B}$ and $B_j \cap B_k = \emptyset$ for $j \neq k$, define $\mu(A) = \sum_{j=1}^p \mu(B_j)$. If A may be also written as $\cup_{r=1}^q B_r'$, where $B_r' \in \mathcal{B}$ and $B_r' \cap B_s' = \emptyset$ for $r \neq s$, then, applying (10.20.e), it follows that $\sum_{j=1}^p \mu(B_j) = \sum_{r=1}^q \mu(B_r')$. Hence μ is well defined on $\mathcal{A}$. Obviously μ is a finitely additive measure on $\mathcal{A}$.

Lemma 10.23. *Let $\mathbf{b} \in R^n$, and let $F : R^n \to R$ be a function which is right continuous at any $\mathbf{x} \in R^n$. Then the function $H(\mathbf{x}) = \Delta_{\mathbf{bx}} F$, $\mathbf{x} \in R^n$, is right continuous at each $\mathbf{x} \in R^n$.*

Proof. This is an immediate consequence of (10.14.a). □

Theorem 10.24. *Let F be as in (10.21). Assume that F is right continuous at any $\mathbf{x} \in R^n$. Then there exists a unique measure μ_F on $\mathcal{B}(R^n)$ such that $\mu_F(\prod_{i=1}^n]a_i, b_i]) =$*

$\Delta_{\mathbf{ba}}F$ *whenever* $]a_1, b_1], \ldots,]a_n, b_n]$ *are bounded intervals in* R, *where* $\mathbf{a} = (a_1, \ldots, a_n)$ *and* $\mathbf{b} = (b_1, \ldots, b_n)$.

Proof. Let $\mathcal{B}$ and $\mathcal{A}$ be as in (7.56), and let μ be the finitely additive measure on $\mathcal{A}$ defined in (10.22). We show that μ is σ-additive.

(I) Suppose first that $\mu(R^n) < \infty$. Let $\varepsilon > 0$, and let $]a_1, b_1], \ldots,]a_n, b_n]$ be bounded intervals in R. Then, applying (10.23), there are $a_i(\varepsilon) < b_i$, $1 \leqslant i \leqslant n$, such that $\prod_{i=1}^n [a_i(\varepsilon), b_i] \subset \prod_{i=1}^n]a_i, b_i]$ and $\mu(\prod_{i=1}^n]a_i, b_i] - \prod_{i=1}^n]a_i(\varepsilon), b_i]) < \varepsilon$. Therefore, for $B \in \mathcal{B}$, there is $C \in \mathcal{B}$ such that $C^- \subset B$, C^- is a compact subset of R^n, and $\mu(B - C) < \varepsilon$. Consequently, for $A \in \mathcal{A}$, there is $D \in \mathcal{A}$ such that $D^- \subset A$, D^- is a compact subset of R^n, and $\mu(A - D) < \varepsilon$. Further, the proof that μ is σ-additive proceeds as in part (I) of (10.15).

(II) Now consider the general case. For $k \in N$, let φ_k be as in (8.24.a), and define $F_k(x_1, \ldots, x_n) = F(\varphi_k(x_1), \ldots, \varphi_k(x_n))$, $(x_1, \ldots, x_n) \in R^n$. Evidently, F_k is right continuous at any $\mathbf{x} \in R^n$, and $F_k \to F$. If $\mathbf{a} = (a_1, \ldots, a_n) \leqslant (b_1, \ldots, b_n) = \mathbf{b}$, then

$$\Delta_{\mathbf{ba}}F_k = \Delta_{(\varphi_k(b_1),\ldots,\varphi_k(b_n))(\varphi_k(a_1),\ldots,\varphi_k(a_n))}F \geqslant 0, \quad k \in N. \tag{1}$$

For $k \in N$, starting from F_k, define μ_k as in (10.22). Let $]a_1, b_1], \ldots,]a_n, b_n]$ be bounded intervals in R, and put $\mathbf{a} = (a_1, \ldots, a_n)$ and $\mathbf{b} = (b_1, \ldots, b_n)$. Using (1), we have

$$\mu_k\left(\prod_{i=1}^n]a_i, b_i]\right) = \Delta_{\mathbf{ba}}F_k = \mu\left(\left(\prod_{i=1}^n]a_i, b_i]\right) \bigcap\,]-k, k]^n\right), \quad k \in N. \tag{2}$$

From this it follows at once that $\mu_k(R^n) = \mu(]-k, k]^n) < \infty$, $k \in N$. Thus, according to (I), μ_k is a measure on $\mathcal{A}$ for each $k \in N$. Now let $B \in \mathcal{B}$, and define B_m, $m \in N$, as in (10.22). Then, for every $m \in N$, in view of (2), the sequence $\{\mu_k(B_m) : k \in N\}$ is nondecreasing, and $\mu_k(B_m) \to \mu(B_m)$. Therefore, applying (5.62), we have $\lim_k \mu(B) = \lim_k \lim_m \mu_k(B_m) = \lim_m \lim_k \mu_k(B_m) = \lim_m \mu(B_m) = \mu(B)$. Hence $\mu_k(A) \to \mu(A)$ for any $A \in \mathcal{A}$. Further, for each $k \in N$, by (2), we have $\mu_k(B_m) \leqslant \mu(B_m)$, $m \in N$, and so $\mu_k(B) \leqslant \mu(B)$. Consequently, $\mu_k(A) \leqslant \mu(A)$, $A \in \mathcal{A}$. Next the proof of the theorem is similar to the proof of (10.15) and is left to the reader. □

Definition 10.25. Notation is as in (10.24). The measure μ_F is called the *Lebesgue-Stieltjes measure on* $\mathcal{B}(R^n)$ *induced by* F, and the elements of the σ-algebra $\mathcal{B}(R^n)^{\mu_F}$ are called the *Lebesgue-Stieltjes sets of* R^n *induced by* F.

Example 10.26. For $i = 1, \ldots, n$, let $F_i : R \to R$ be a nondecreasing function which is right continuous at any $x \in R$. Define $F(x_1, \ldots, x_n) = \prod_{i=1}^n F_i(x_i)$, $(x_1, \ldots, x_n) \in R^n$. Then F is right continuous at any $\mathbf{x} \in R^n$. Moreover, for $\mathbf{a} = (a_1, \ldots, a_n) \leqslant (b_1, \ldots, b_n) = \mathbf{b}$, we have $\Delta_{\mathbf{ba}}F = \prod_{i=1}^n (F_i(b_i) - F_i(a_i)) \geqslant 0$. Thus F induces a Lebesgue-Stieltjes measure μ_F on $\mathcal{B}(R^n)$. If $F_i(x) = x$, $x \in R$, for $i = 1, \ldots, n$, then μ_F is called the *Lebesgue measure on* $\mathcal{B}(R^n)$, and the elements of the σ-algebra $\mathcal{B}(R^n)^{\mu_F}$ are called the *Lebesgue sets of* R^n.

Exercise: Outer measure 10.27. Let X be a set. A function $\mu^* : \mathcal{P}(X) \to [0, \infty]$ is called an *outer measure* on $\mathcal{P}(X)$ if:

(i) $\mu^*(\emptyset) = 0$;

(ii) $\mu^*(A) \leqslant \mu^*(B)$ whenever $A \subset B \subset X$;

(iii) $\mu^*(\cup_{n\in N} A_n) \leqslant \sum_{n\in N} \mu^*(A_n)$ whenever $A_n \subset X, n \in N$.

A set $A \subset X$ is said to be μ^* *-measurable* if $\mu^*(E) = \mu^*(E \cap A) + \mu^*(E \cap A^c)$ for any $E \subset X$. Let $\mathcal{M}$ be the family of all μ^*-measurable sets. Prove the following.

(a) $A \in \mathcal{M}$ if and only if $\mu^*(E) \geqslant \mu^*(E \cap A) + \mu^*(E \cap A^c)$ for any $E \subset X$.

(b) $\mathcal{M}$ is an algebra of sets.

(c) If $\{A_1, \ldots, A_n\} \subset \mathcal{P}(X)$ and $A_i \cap A_j = \emptyset$ for $i \neq j$, then $\mu^*(E \cap (\cup_{i=1}^n A_i)) = \sum_{i=1}^n \mu^*(E \cap A_i)$ for any $E \subset X$.

(d) $\mathcal{M}$ is a σ-algebra. [Hint. Let $\{A_n : n \in N\} \subset \mathcal{M}$ be such that $A_m \cap A_n = \emptyset$ for $m \neq n$. Use (b) and (c) to show that $\mu^*(E) \geqslant \sum_{n\in N} \mu^*(E \cap A_n) + \mu^*(E \cap (\cup_{n\in N} A_n)^c)$, $E \in \mathcal{P}(X)$.]

(e) $\{A \in \mathcal{P}(X) : \mu^*(A) = 0\} \subset \mathcal{M}$.

(f) μ^* is a complete measure on $\mathcal{M}$. [Hint. Let $\{A_n : n \in N\}$ be as in (d). Show that $\mu^*(E) = \sum_{n\in N} \mu^*(E \cap A_n) + \mu^*(E \cap (\cup_{n\in N} A_n)^c)$ for any $E \subset X$.]

Exercise 10.28. Let μ^* be an outer measure on $\mathcal{P}(X)$, let $\mathcal{M}$ be as in (10.27), and let $\{A_i : i \in I\}$ be a countable partition of X.

(a) For $i \in I$, let μ_i^* be the restriction of μ^* to $\mathcal{P}(A_i)$. Prove that μ_i^* is an outer measure on $\mathcal{P}(A_i)$.

(b) For $i \in I$, let $\mathcal{M}_i$ be the family of all μ_i^*-measurable sets. Show that $\oplus_{i\in I} \mathcal{M}_i = \mathcal{M}$ if and only if $A_i \in \mathcal{M}$ for any $i \in I$.

(c) Prove that $\oplus_{i\in I} \mu_i^* = \mu^*$ on $\mathcal{M}$.

Exercise 10.29. Let $\mathcal{A}$ be an algebra of subsets of X, and let μ be a measure on $\mathcal{A}$. For $A \subset X$, define $\mu^*(A) = \inf\{\sum_{n\in N} \mu(A_n) : A \subset \cup_{n\in N} A_n$ and $\{A_n : n \in N\} \subset \mathcal{A}\}$.

(a) Prove that μ^* is an outer measure on $\mathcal{P}(X)$.

(b) Show that $\mu^* = \mu$ on $\mathcal{A}$.

(c) Let $\mathcal{M}$ be as in (10.27). Prove that $\mathcal{A} \subset \mathcal{M}$.

(d) If μ is σ-finite, then $(X, \mathcal{M}, \mu^*)$ is the completion of $(X, \sigma(\mathcal{A}), \mu^*_{\sigma(\mathcal{A})})$. [Hint. If μ is finite, mimic the proof of (10.7). Further, use (9.35) and (10.28).]

Exercise 10.30. Let $\mathcal{A}$ be as in (7.56), let F and μ_F be as in (10.24), let μ^* be as in (10.29), and let $\mathcal{M}$ be as in (10.27). Show that $\mu_F = \mu^*_{\mathcal{B}(R^n)}$ and $\mathcal{B}(R^n)^{\mu_F} = \mathcal{M}$.

Exercise 10.31. Let $\alpha > 0$, let (X, d) be a metric space, and set $\mathcal{T}_d^\delta = \{U \in \mathcal{T}_d : \operatorname{diam}(U) \leqslant \delta\}$ for any $\delta > 0$. For $A \subset X$, define $\mu^*_{\alpha,\delta}(A) = \inf\{\sum_{n\in N} \operatorname{diam}(U_n)^\alpha : A \subset \cup_{n\in N} U_n$ and $U_n \in \mathcal{T}_d^\delta, n \in N\}$.

(a) Prove that $\mu^*_{\alpha,\delta}$ is an outer measure on $\mathcal{P}(X)$ for each $\delta > 0$.

(b) Show that $\mu^*_{\alpha,\delta}(A) \geqslant \mu^*_{\alpha,\delta'}(A)$ whenever $0 < \delta \leqslant \delta'$.

(c) For $A \subset X$, define $\mu^*_\alpha(A) = \lim_{\delta \searrow 0} \mu^*_{\alpha,\delta}(A)$. Prove that μ^*_α is an outer measure on $\mathcal{P}(X)$. μ^*_α is called the *Hausdorff α-dimensional (outer) measure* on $\mathcal{P}(X)$.

(d) If $\mu^*_\alpha(A) < \infty$, then $\mu^*_{\alpha'}(A) = 0$ whenever $\alpha' > \alpha$.

(e) For $A \subset X$, show that there is a number $H(A) \geqslant 0$ such that $\mu_\alpha^*(A) = \infty$ whenever $\alpha < H(A)$ and $\mu_\alpha^*(A) = 0$ whenever $\alpha > H(A)$. The number $H(A)$ is called the *Hausdorff dimension of* A.

(f) If A is countable, then $H(A) = 0$.

(g) If $A \subset B$, then $H(A) \leqslant H(B)$.

(h) Let $A_n \subset X, n \in N$. Show that $H(\cup_{n\in N} A_n) = \sup_{n\in N} H(A_n)$.

Exercise 10.32. Let R be equipped with its usual metric, and let $\mu_{\alpha,\delta}^*$, μ_α^*, and H be as in (10.31).

(a) For $A \subset R$, prove that $\mu_\alpha^*(A) = 0$ whenever $\alpha > 1$.

(b) Let $\mathcal{A}$ be as in (7.3.d), let λ be the Lebesgue measure on $\mathcal{A}$, and let λ^* be the outer measure on $\mathcal{P}(R)$ built from λ as in (10.29). Use (5.70) to show that $\lambda^* = \mu_{1,\delta}^* = \mu_1^*$ for any $\delta > 0$.

(c) Let λ be the Lebesgue measure on $\mathcal{B}(R)$ and $A \in \mathcal{B}(R)^\lambda$. Prove that $H(A) = 1$ whenever $\overline{\lambda}(A) > 0$. [Use (10.30), (10.29.d), (b) and (a).]

Exercise 10.33. Let $\mathcal{B}$ be an algebra of subsets of X, let μ be a measure on $\sigma(\mathcal{B})$ such that the restriction of μ to $\mathcal{B}$ is σ-finite, and let $\varepsilon > 0$. If $A \in \sigma(\mathcal{B})$ and $\mu(A) < \infty$, then there is $B \in \mathcal{B}$ such that $\mu(A \Delta B) < \varepsilon$. [Hints. Assume first that μ is finite. If A is a countable union of sets in $\mathcal{B}$, then apply (9.9). Further, if A is arbitrary, use (10.4) and (10.9). Finally, consider the general case.]

Exercise 10.34. Let $F : R \to R$ be a nondecreasing function. Prove the following.

(a) F has finite right-hand and left-hand limits at any $a \in R$.

(b) $\{a \in R : F \text{ is not continuous at } a\} = \{a \in R : \lim_{x \nearrow a} F(x) < \lim_{x \searrow a} F(x)\}$.

(c) $\{a \in R : F \text{ is not continuous at } a\}$ is a countable set.

Exercise 10.35. Let $f :]a, b[\to R$ be a monotone function. Use (10.12.a) and (5.61) to show that $\lim_{x\to a} f(x)$ and $\lim_{x\to b} f(x)$ exist.

Exercise 10.36. Let F and μ_F be as in (10.15).

(a) Express in terms of F the measure μ_F of the following intervals in R: $]a, b[$, $[a, b]$ and $[a, b[$.

(b) For $a \in R$, show that $\mu_F(\{a\}) = F(a) - F(a-)$, and deduce that $\mu_F(\{a\}) = 0$ if and only if F is continuous at a.

(c) If $F(R)$ is countable, then there exists a countable set $A \subset R$ such that $(A^-)^\circ = \emptyset$ and $\mu(A^c) = 0$.

(d) Prove that the converse of (c) is not generally true. [We may use (10.38.a) and (10.38.c).]

Exercise 10.37. Construct a function $F : R \to R$ which is discontinuous at any $x \in Q$ and is continuous at any $x \notin Q$. [Consider a discrete measure concentrated on Q, and then use (10.17.a) and (10.36.b).]

Exercise: Cantor-like set 10.38. We construct a set $C \subset [0, 1]$ as follows. Remove an open interval I_{11} of length < 1 from the center of $[0, 1]$. There remain two disjoint closed intervals J_{11} and J_{12}, each having length $< 1/2$. This completes the first step of the construction. If the nth step of the construction has been carried out, there remaining 2^n closed intervals $J_{n1}, \ldots, J_{n2^n}$ (numbered from left to right), each of length $< 1/2^n$, we perform the $(n+1)$st step by removing an open interval $I_{n+1,k}$ from the center of J_{nk} such that the length of $I_{n+1,k} <$ the length of J_{nk}, $1 \leqslant k \leqslant 2^n$. There remain 2^{n+1} closed intervals $J_{n+1,1}, \ldots, J_{n+1,2^{n+1}}$, each of length $<1/2^{n+1}$. Denote $B_n = \cup_{k=1}^{2^{n-1}} I_{nk}$, $n \in N$, and $C = [0, 1] - (\cup_{n \in N} B_n)$. The set C is called the *Cantor-like set.*

(a) Prove that $C = C^d$ and $C^\circ = \emptyset$.

(b) Let $\alpha \leqslant 1/3$. If the length of I_{nk} is α^n for $k = 1, \ldots 2^{n-1}$, then the Lebesgue measure of C is $\frac{1-3\alpha}{1-2\alpha}$.

(c) Assume that the length of I_{nk} is $1/3^n$ for $k = 1, \ldots 2^{n-1}$. In this case C is called the *Cantor set.* Show that C has the continuum power. [Hints. For $\mathbf{a} = (a_1, a_2, \ldots) \in \{0, 2\}^\infty$, define $\varphi(\mathbf{a}) = \sum_{n \in N} a_n/3^n$. Use 3-adic expansion (8.23) to prove that φ is a bijective mapping from $\{0, 2\}^\infty$ onto C. Then apply (3.46.b).]

Exercise 10.39. Let λ be the Lebesgue measure on $\mathcal{B}(R)$ and $x \in R$.

(a) Show that $\lambda(xA) = |x|\,\lambda(A)$ for any $A \in \mathcal{B}(R)$.

(b) Prove that $xA \in \mathcal{B}(R)^\lambda$ and $\overline{\lambda}(xA) = |x|\,\overline{\lambda}(A)$ whenever $A \in \mathcal{B}(R)^\lambda$.

Exercise 10.40. Let μ be a Lebesgue-Stieltjes measure on $\mathcal{B}(R)$ such that $\mu(x+I) = \mu(I)$ for any $x \in R$ and any bounded interval $I \subset R$. Show that $\mu = \alpha\lambda$, where $\alpha \in [0, \infty[$ and λ is the Lebesgue measure on $\mathcal{B}(R)$. [Hints. Let $\alpha = \mu(]0, 1])$. For $n \in N, \alpha = n\mu(]0, 1/n])$, and so $\mu(]0, 1/n]) = \alpha\lambda(]0, 1/n])$. Further, prove that $\mu(]a, b]) = \alpha\lambda(]a, b])$ for each bounded interval $]a, b] \subset R$, and use the uniqueness assertion in (10.15).]

Exercise 10.41. Consider $F : R^2 \to R$ defined by $F(x_1, x_2) = \begin{cases} 1 \text{ if } x_1 + x_2 \geqslant 0 \\ 0 \text{ if } x_1 + x_2 < 0 \end{cases}$.

Prove the following.

(a) F is right continuous at any $x \in R^2$.

(b) Each section of F is nondecreasing.

(c) $\Delta_{(0,1)(-1,0)} F < 0$.

Exercise 10.42. Let F be as in (10.21). For $i = 1, \ldots, n$, let $F(x_1, \ldots, x_{i-1}, x_{i+1}, \ldots, x_n) : R \to R$ be the section of F at $(x_1, \ldots, x_{i-1}, x_{i+1}, \ldots, x_n)$. Use (10.20.b) and (10.20.c) to show that the functions $F(x_1, \ldots, x_{i-1}, x_{i+1}, \ldots, x_n)$ may not be nondecreasing.

Exercise 10.43. Let M be the set of all Lebesgue-Stieltjes measures on $\mathcal{B}(R^n)$. Use (10.20.a) to prove that M coincides with the set of all measures μ on $\mathcal{B}(R^n)$ such that $\mu(\prod_{i=1}^n I_i) < \infty$ whenever $I_1, \ldots, I_n$ are bounded intervals in R.

Exercise 10.44. Let $\mathfrak{F} = \{F \in R^{R^n} : F$ is right continuous at any $\mathbf{x} \in R^n$, and $\Delta_{\mathbf{ba}} F \geqslant 0$ whenever $\mathbf{a} \leqslant \mathbf{b}\}$, and let M be as in (10.43). For $F, G \in \mathfrak{F}$, we write $F \sim G$ if $\Delta_{\mathbf{ba}} F = \Delta_{\mathbf{ba}} G$ whenever $\mathbf{a} \leqslant \mathbf{b}$. Construct a bijective mapping from $\mathfrak{F}/\!\sim$ onto M.

Exercise 10.45. Let $f : R \to R$ be a function that has finite right-hand and left-hand limits at any $a \in R$. Show that $\{a \in R : f$ is not continuous at $a\}$ is a countable set.

Exercise 10.46. Let $f :]a, b[\to [c, d]$ be a nondecreasing function such that $\lim_{x \to a} f(x) = c$ and $\lim_{x \to b} f(x) = d$.

(a) Consider the function $g :]c, d[\to]a, b[$ defined by $g(y) = \inf\{x : f(x) > y\}$ for any $y \in]c, d[$. Prove that g is right continuous at each $y \in]c, d[$. g is called the *right continuous quasi-inverse of* f.

(b) Consider the function $h :]c, d[\to]a, b[$ defined by $h(y) = \inf\{x : f(x) \geqslant y\}$ for any $y \in]c, d[$. Prove that h is right continuous at each $y \in]c, d[$. h is called the *left continuous quasi-inverse of* f.

Exercise 10.47. Let $a, b \in R$ be such that $a < b$, let $\mathfrak{F}_{ab}$ denote the set of all nondecreasing functions $f : R \to [a, b]$ such that $\lim_{x\to-\infty} f(x) = a$ and $\lim_{x\to\infty} f(x) = b$, let $\{g_n : n \in N\} \subset \mathfrak{F}_{ab}$, and let $g \in \mathfrak{F}_{ab}$ be such that $g_n(x) \to g(x)$ for any $x \in R$. Show that $g_n \stackrel{u}{\to} g$ whenever g is continuous on R.

Exercise 10.48. Let $\mathcal{M}$ be as in (10.8), and let $P_n, n \in N$, and P be probabilities on $\sigma(\mathcal{M})$ such that

$$\sum_{n\in N} |P_n(A) - P(A)| < \infty, \quad A \in \mathcal{M}. \tag{1}$$

Does (1) hold for any $A \in \sigma(\mathcal{M})$?

Exercise 10.49. Let $F : R^n \to R$ be a nondecreasing function. Prove that F is continuous at $\mathbf{a} \in R^n$ whenever F is right continuous and left continuous at $\mathbf{a}$.

Exercise 10.50. Let Y be a topological space, let $y_0 \in Y$, and let $f : R \to Y$ be a continuous function on R such that $\lim_{x\to-\infty} f(x) = \lim_{x\to\infty} f(x) = y_0$. As in (5.94), consider the one-point compactification $R \cup \{\delta\}$ of R. Use (5.94) and (5.29) to show that the function $g : R \cup \{\delta\} \to Y$ given by $g(x) = \begin{cases} f(x) & \text{if } x \in R \\ y_0 & \text{if } x = \delta \end{cases}$ is continuous on $R \cup \{\delta\}$.

4 The Integral

The material of Chapter 3 enables us to introduce and study the important notion of integral in this chapter. In Section 11 we define the integral, and we examine remarkable properties of the integral. Section 12 deals with the Radon-Nikodým theorem and with the Lebesgue decomposition. In Section 13 we study the function spaces $\mathfrak{L}^p$, and in Section 14 we define and study three new kinds of convergence for sequences of functions: almost everywhere, in measure and in $\mathfrak{L}^p$.

Throughout this chapter, unless otherwise stated, (X, $\mathcal{A}$, μ) is an arbitrary measure space, and a function is said to be measurable whenever it is $\mathcal{A}$-measurable.

11 Definitions and Properties of the Integral

In this section we present a constructive definition of the notion of an integral, and we study remarkable properties of the integral that are frequently applied in measure theory and in probability theory.

We define the operation of integration successively for wider and wider classes of measurable functions. Throughout this section, $\mathfrak{S}$ will be the class of all simple functions f from X into $[0, \infty]$, written in the form $f = \sum_{i=1}^{n} a_i 1_{A_i}$, where $\{A_1, \ldots, A_n\}$ is a partition of X, and $\mathfrak{S}' = \{f \in \mathfrak{S} : f(X) \subset [0, \infty[\}$.

Definition 11.1. Let $f = \sum_{i=1}^{n} a_i 1_{A_i} \in \mathfrak{S}$. The number $\int_X f \, d\mu = \sum_{i=1}^{n} a_i \mu(A_i)$ is called the *integral of* f (with respect to μ). If f is written in some other form $\sum_{j=1}^{m} b_j 1_{B_j}$, then $a_i = b_j$ whenever $A_i \cap B_j \neq \emptyset$. Consequently, $\sum_{i=1}^{n} a_i \mu(A_i) = \sum_{i=1}^{n} \sum_{j=1}^{m} a_i \mu(A_i \cap B_j) = \sum_{j=1}^{m} \sum_{i=1}^{n} b_j \mu(A_i \cap B_j) = \sum_{j=1}^{m} b_j \mu(B_j)$. Thus $\int_X f \, d\mu$ is independent of the way in which f is written.

Theorem 11.2.

(i) *If* $f, g \in \mathfrak{S}$, *then* $\int_X (f + g) d\mu = \int_X f \, d\mu + \int_X g \, d\mu$.
(ii) *If* $\alpha \geqslant 0$ *and* $f \in \mathfrak{S}$, *then* $\int_X \alpha f \, d\mu = \alpha \int_X f \, d\mu$.
(iii) *If* $f, g \in \mathfrak{S}$ *and* $f \leqslant g$, *then* $0 \leqslant \int_X f \, d\mu \leqslant \int_X g \, d\mu$.

Proof. Assertion (ii) is obvious. Let $f = \sum_{i=1}^{n} a_i 1_{A_i} \in \mathfrak{S}$ and $g = \sum_{j=1}^{m} b_j 1_{B_j} \in \mathfrak{S}$. Then $f + g = \sum_{i=1}^{n} \sum_{j=1}^{m} (a_i + b_j) 1_{A_i \cap B_j}$. Therefore, $\int_X (f + g) d\mu = \sum_{i=1}^{n} \sum_{j=1}^{m} (a_i + b_j) \mu(A_i \cap B_j) = \sum_{i=1}^{n} \sum_{j=1}^{m} a_i \mu(A_i \cap B_j) + \sum_{j=1}^{m} \sum_{i=1}^{n} b_j \mu(A_i \cap B_j) = \sum_{i=1}^{n} a_i \mu(A_i) + \sum_{j=1}^{m} b_j \mu(B_j) = \int_X f \, d\mu + \int_X g \, d\mu$. This proves (i). Clearly, $0 \leqslant \int_X f \, d\mu$. If $f \leqslant g$, then consider the function $h = \sum_{i=1}^{n} \sum_{j=1}^{m} c_{ij} 1_{A_i \cap B_j} \in \mathfrak{S}$,

Analysis and Probability. http://dx.doi.org/10.1016/B978-0-12-401665-1.00004-7

where $c_{ij} = \begin{cases} 0 & \text{if } (a_i, b_j) = (\infty, \infty) \\ b_j - a_i & \text{if } (a_i, b_j) \neq (\infty, \infty) \end{cases}$. Then $f + h = g$, and so $\int_X f\,d\mu \leqslant \int_X f\,d\mu + \int_X h\,d\mu = \int_X (f + h) d\mu = \int_X g\,d\mu$. This proves (iii). □

To define the integral of a nonnegative measurable function, we need the next result.

Theorem 11.3. *Let* $\{f_n : n \in N\} \subset \mathfrak{S}'$ *be a nondecreasing sequence, and let* $g \in \mathfrak{S}'$ *be such that* $\lim_n f_n \geqslant g$. *Then* $\lim_n \int_X f_n\,d\mu \geqslant \int_X g\,d\mu$.

Proof. Since the case $g = 0$ is trivial, we assume that $g = \sum_{i=1}^m a_i 1_{A_i} \neq 0$. Let $a = \inf\{a_i : a_i > 0\}$, choose $\varepsilon \in]0, a[$, and put $B_n = \{f_n \geqslant g - \varepsilon\}$, $n \in N$. Then $\{B_n : n \in N\}$ is a nondecreasing sequences of sets in $\mathcal{A}$ such that

$$\bigcup_{n \in N} B_n = X. \tag{1}$$

(I) Assume first that $\int_X g\,d\mu < \infty$. Then $\mu(\{g > 0\}) < \infty$. For $n \in N$, applying (11.2.iii) and (11.2.i), we have

$$\begin{aligned}
\int_X f_n\,d\mu &\geqslant \int_X f_n 1_{B_n \cap \{g>0\}}\,d\mu \geqslant \int_X (g - \varepsilon) 1_{B_n \cap \{g>0\}}\,d\mu \\
&= \int_X g 1_{B_n \cap \{g>0\}}\,d\mu - \varepsilon \mu(B_n \cap \{g > 0\}) \\
&\geqslant \int_X g 1_{B_n}\,d\mu - \varepsilon \mu(\{g > 0\}).
\end{aligned} \tag{2}$$

Further, using (1) and (9.9), we get $\int_X g 1_{B_n}\,d\mu = \sum_{i=1}^m a_i \mu(A_i \cap B_n) \to \sum_{i=1}^m a_i \mu(A_i) = \int_X g\,d\mu$. Therefore, there is $n_\varepsilon \in N$ such that

$$\int_X g 1_{B_n}\,d\mu \geqslant \int_X g\,d\mu - \varepsilon, \quad n \geqslant n_\varepsilon. \tag{3}$$

Combining (2) and (3), we have

$$\int_X f_n\,d\mu \geqslant \int_X g\,d\mu - \varepsilon - \varepsilon \mu(\{g > 0\}), \quad n \geqslant n_\varepsilon. \tag{4}$$

Letting $n \to \infty$ and then $\varepsilon \to 0$ in (4), we get the desired inequality.

(II) Suppose now that $\int_X g\,d\mu = \infty$. Then there is $i \in \{1, \ldots, m\}$ such that $a_i > 0$ and $\mu(A_i) = \infty$. Hence, using (11.2.iii), (1), and (9.9), we get $\lim_n \int_X f_n\,d\mu \geqslant \lim_n \int_X f_n 1_{B_n \cap A_i}\,d\mu \geqslant \lim_n (a_i - \varepsilon)\mu(B_n \cap A_i) = (a_i - \varepsilon)\mu(A_i) = \infty$. □

Corollary 11.4. *Let* $\{f_n : n \in N\} \subset \mathfrak{S}'$ *and* $\{g_n : n \in N\} \subset \mathfrak{S}'$ *be nondecreasing sequences such that* $\lim_n f_n = \lim_n g_n$. *Then* $\lim_n \int_X f_n\,d\mu = \lim_n \int_X g_n\,d\mu$.

Proof. For $m \in N$, we have $\lim_n f_n \geqslant g_m$. Consequently, applying (11.3), we get

$$\lim_n \int_X f_n\,d\mu \geqslant \int_X g_m\,d\mu, \quad m \in N. \tag{1}$$

On letting $m \to \infty$ in (1), we have $\lim_n \int_X f_n \, d\mu \geqslant \lim_m \int_X g_m \, d\mu$. Since the reverse inequality is proved similarly, we get the desired result. □

Definition 11.5. Let $f : X \to [0, \infty]$ be a measurable function. We define $\int_X f \, d\mu = \lim_n \int_X f_n \, d\mu$, where $\{f_n : n \in N\} \subset \mathfrak{S}'$ is a nondecreasing sequence such that $f_n \to f$. By Theorem (8.15), such a sequence always exists. In view of (11.2.iii), the sequence $\{\int_X f_n \, d\mu : n \in N\}$ is nondecreasing, and so it has a limit in $[0, \infty]$. Finally, by (11.4), the number $\int_X f \, d\mu$ is well defined. $\int_X f \, d\mu$ is called the *integral of* f (with respect to μ).

Theorem 11.6. *Let $f : X \to [0, \infty]$ and $g : X \to [0, \infty]$ be measurable functions. Then the following assertions hold.*

(i) $\int_X (f + g) d\mu = \int_X f \, d\mu + \int_X g \, d\mu$.
(ii) *If* $0 \leqslant \alpha < \infty$, *then* $\int_X \alpha f \, d\mu = \alpha \int_X f \, d\mu$.
(iii) *If* $f \leqslant g$, *then* $\int_X f \, d\mu \leqslant \int_X g \, d\mu$.

Proof. Let $\{f_n : n \in N\} \subset \mathfrak{S}'$ and $\{g_n : n \in N\} \subset \mathfrak{S}'$ be nondecreasing sequences such that $f_n \to f$ and $g_n \to g$. Then, according to (11.2.i), we have

$$\int_X (f_n + g_n) d\mu = \int_X f_n \, d\mu + \int_X g_n \, d\mu, \quad n \in N. \tag{1}$$

Since $\{f_n + g_n : n \in N\} \subset \mathfrak{S}'$ is a nondecreasing sequence, and $f_n + g_n \to f + g$, upon letting $n \to \infty$ in (1), we get assertion (i). Applying (11.2.ii), assertion (ii) is immediate. If $f \leqslant g$, then $f_m \leqslant \lim_n g_n$, $m \in N$. Hence (iii) is a consequence of (11.3). □

Definitions 11.7. Let $f : X \to \overline{R}$ be a measurable function, and let f^+ and f^- be as in (8.16). If $(\int_X f^+ \, d\mu, \int_X f^- \, d\mu) \neq (\infty, \infty)$, then we define the *integral of* f (with respect to μ) by $\int_X f \, d\mu = \int_X f^+ \, d\mu - \int_X f^- \, d\mu$, and we say that the integral of f (with respect to μ) exists. If $\int_X f \, d\mu \in R$, then f is said to be *integrable* [μ-*integrable*]. Evidently, f is integrable if and only if $\int_X f^+ \, d\mu < \infty$ and $\int_X f^- \, d\mu < \infty$. The set of all integrable functions will be denoted by $\mathfrak{L}^1(X, \mathcal{A}, \mu)$. When confusion appears impossible, we will write $\mathfrak{L}^1$ instead of $\mathfrak{L}^1(X, \mathcal{A}, \mu)$.

Examples 11.8. (a) A function $f = \sum_{i=1}^n a_i 1_{A_i} \in \mathfrak{S}$ is integrable if and only if $\mu(A_i) < \infty$ whenever $a_i > 0$.

(b) Suppose that $X = N$, $\mathcal{A} = \mathcal{P}(N)$, and μ is the counting measure on $\mathcal{A}$. Let $f : N \to \overline{R}$ be a function. By (8.2.a), f is $\mathcal{A}$-measurable. Set $N_1 = \{n \in N : f(n) \geqslant 0\}$ and $N_2 = \{n \in N : f(n) < 0\}$. Then $\int_X f^+ d\mu = \sum_{n \in N_1} f(n)$ and $\int_X f^- \, d\mu = -\sum_{n \in N_2} f(n)$. If $(\sum_{n \in N_1} f(n), \sum_{n \in N_2} f(n)) \neq (\infty, -\infty)$, then $\sum_{n \in N} f(n)$ exists, and $\sum_{n \in N} f(n) = \sum_{n \in N_1} f(n) + \sum_{n \in N_2} f(n) = \int_X f \, d\mu$.

Remarks 11.9. (a) Let $\mathcal{B} \subset \mathcal{A}$ be a σ-algebra of subsets of X, let $\mu_{\mathcal{B}}$ be the restriction of μ to $\mathcal{B}$, and let $f : X \to \overline{R}$ be a $\mathcal{B}$-measurable function. Then $\int_X f \, d\mu$ exists if and only if $\int_X f \, d\mu_{\mathcal{B}}$ exists. In this case, $\int_X f \, d\mu = \int_X f \, d\mu_{\mathcal{B}}$.

(b) Let $f : X \to \overline{R}$ be a measurable function such that $\int_X f \, d\mu$ exists. Since $|f| = f^+ + f^-$, using (11.6.i), we see that $\left|\int_X f \, d\mu\right| \leqslant \int_X |f| \, d\mu$.

(c) Let $f : X \to \overline{R}$ be a measurable function. Applying (11.6.i) and (11.6.ii), it follows that $f \in \mathfrak{L}^1$ if and only if $|f| \in \mathfrak{L}^1$. If $g : X \to \overline{R}$ is measurable, and $|g| \leqslant f$, then $g \in \mathfrak{L}^1$ whenever $f \in \mathfrak{L}^1$.

Definition 11.10. Let f be as in (11.7) and $A \in \mathcal{A}$. If the integral of $f 1_A$ exists, then define $\int_A f \, d\mu = \int_X f 1_A \, d\mu$. The number $\int_A f \, d\mu$ is called the *integral of f on A*. Since $(f 1_A)^+ = f^+ 1_A$ and $(f 1_A)^- = f^- 1_A$, it follows that the integral of $f 1_A$ exists whenever the integral of f exists, but the converse of this assertion does not generally hold. Also, $\int_A f \, d\mu \in R$ whenever $f \in \mathfrak{L}^1$. Following conventional notation, we will sometimes write $\int_A f(x) \, d\mu(x)$ or $\int_A f(x)\mu(dx)$ instead of $\int_A f \, d\mu$. If I is an interval in R with endpoints a, b, where $a < b$, and μ is the Lebesgue measure on $\mathcal{B}(R)$, then we will use the notation $\int_a^b f(x)dx$ for $\int_I f(x) \, d\mu(x)$. If $A \in \mathcal{B}(R^n)$ and μ is the Lebesgue measure on $\mathcal{B}(R^n)$, we shall also write $\int_A f(x_1, \ldots, x_n)dx_1 \ldots dx_n$ for $\int_A f \, d\mu$.

Remark 11.11. Let f and A be as in (11.10), and let μ_A be the restriction of μ to $A \cap \mathcal{A}$. According to (8.3.b), it is easily seen that $\int_A f \, d\mu = \int_A (f \circ i_A) d\mu_A$, in the sense that if one of the integral exists, so does the other, and the two integrals are equal.

Theorem 11.12. *Let $f : X \to \overline{R}$ and $g : X \to \overline{R}$ be measurable functions such that $\int_X f \, d\mu$ and $\int_X g \, d\mu$ exist. Then the following assertions hold.*

(i) *If $\{f = -\infty\} \cap \{g = \infty\} = \{f = \infty\} \cap \{g = -\infty\} = \emptyset$, and $(\int_X f \, d\mu, \int_X g \, d\mu) \notin \{(-\infty, \infty), (\infty, -\infty)\}$, then $\int_X (f+g) d\mu$ exists, and $\int_X (f+g) d\mu = \int_X f \, d\mu + \int_X g \, d\mu$.*

(ii) *If A and B are disjoint sets in $\mathcal{A}$, then $\int_{A \cup B} f \, d\mu = \int_A f \, d\mu + \int_B f \, d\mu$.*

(iii) *If $\alpha \in R$, then $\int_X \alpha f \, d\mu$ exists, and $\int_X \alpha f \, d\mu = \alpha \int_X f \, d\mu$.*

(iv) *If $f \leqslant g$, then $\int_X f \, d\mu \leqslant \int_X g \, d\mu$.*

Proof. If $\int_X f \, d\mu$, $\int_X g \, d\mu$ and $\int_X f \, d\mu + \int_X g \, d\mu$ exist, then $(\int_X f^+ \, d\mu + \int_X g^+ \, d\mu, \int_X f^- \, d\mu + \int_X g^- \, d\mu) \neq (\infty, \infty)$. Thus, since $(f+g)^+ \leqslant f^+ + g^+$ and $(f+g)^- \leqslant f^- + g^-$, (11.6.iii) and (11.6.i) show that $(\int_X (f+g)^+ \, d\mu, \int_X (f+g)^- \, d\mu) \neq (\infty, \infty)$. Therefore, $\int_X (f+g) \, d\mu$ exists. Further, we have $f + g = (f+g)^+ - (f+g)^- = f^+ - f^- + g^+ - g^-$, whence $(f+g)^+ + f^- + g^- = (f+g)^- + f^+ + g^+$. Applying (11.6.i), we get $\int_X (f+g)^+ \, d\mu + \int_X f^- \, d\mu + \int_X g^- \, d\mu = \int_X (f+g)^- \, d\mu + \int_X f^+ \, d\mu + \int_X g^+ \, d\mu$. Consequently, we have $\int_X (f+g) \, d\mu = \int_X (f+g)^+ \, d\mu - \int_X (f+g)^- \, d\mu = \int_X f^+ \, d\mu - \int_X f^- \, d\mu + \int_X g^+ \, d\mu - \int_X g^- \, d\mu = \int_X f \, d\mu + \int_X g \, d\mu$. This proves (i). If A and B are disjoint sets, then $f 1_{A \cup B} = f 1_A + f 1_B$. Hence, (ii) follows from (i). Since $(\alpha f)^+ = \begin{cases} \alpha f^+ & \text{if } \alpha \geqslant 0 \\ -\alpha f^- & \text{if } \alpha < 0 \end{cases}$ and $(\alpha f)^- = \begin{cases} \alpha f^- & \text{if } \alpha \geqslant 0 \\ -\alpha f^+ & \text{if } \alpha < 0 \end{cases}$, assertion (iii) follows at once from (11.6.ii). If $f \leqslant g$, then $f^+ + g^- \leqslant f^- + g^+$. Thus, by (11.6.i) and (11.6.iii), we have $\int_X f^+ \, d\mu + \int_X g^- \, d\mu \leqslant \int_X f^- \, d\mu + \int_X g^+ \, d\mu$, whence we get (iv). □

Definition 11.13. We say that a property P holds *μ-almost everywhere* [*μ-a.e.*] if there exists a μ-null set A such that P holds for all $x \in A^c$.

Examples 11.14. (a) Let $f : X \to \overline{R}$ and $g : X \to \overline{R}$ be functions. We say that $f \leqslant g$ μ-a.e. if there is $A \in \mathcal{A}$ such that $\mu(A) = 0$ and $f(x) \leqslant g(x)$ for any $x \in A^c$.

(b) Let $f : X \to \overline{R}$ be a function. We say that f is *bounded μ-a.e.* if there exists a μ-null set A such that $f(A^c)$ is bounded.

(c) Let $\mathfrak{L}$ be a set of measurable functions from X into $\overline{R}$. For $f, g \in \mathfrak{L}$, we write $f = g$ μ-a.e. if there is a μ-null set A such that $f(x) = g(x)$, $x \in A^c$. For $f, g \in \mathfrak{L}$, define $f \sim g$ if $f = g$ μ-a.e. Then $\sim$ is an equivalence relation on $\mathfrak{L}$.

Theorem 11.15. *If $f \in \mathfrak{L}^1$, then $\mu(\{|f| = \infty\}) = 0$, i.e. f is finite μ-a.e.*

Proof. If $f \in \mathfrak{L}^1$, then $|f| \in \mathfrak{L}^1$ (11.9.c), and so $\infty \cdot \mu(\{|f| = \infty\}) = \int_{\{|f|=\infty\}} |f|\, d\mu \leqslant \int_X |f|\, d\mu < \infty$. Therefore, $\mu(\{|f| = \infty\}) = 0$. □

Theorem 11.16. *Let $f : X \to [0, \infty]$ be a measurable function. Then $\int_X f\, d\mu = 0$ if and only if $f = 0$ μ-a.e.*

Proof. Assume that $\int_X f\, d\mu = 0$. For any $n \in N$, by (11.6.iii), we have $0 = \int_X f\, d\mu \geqslant \int_{\{f \geqslant 1/n\}} f\, d\mu \geqslant \mu(\{f \geqslant 1/n\})/n$, whence $\mu(\{f \geqslant 1/n\}) = 0$. Since $\{f > 0\} = \cup_{n \in N}\{f \geqslant 1/n\}$, (9.6.v) shows that $\mu(\{f > 0\}) = 0$, i.e. $f = 0$ μ-a.e.

Conversely, assume that $\mu(\{f > 0\}) = 0$. Then $0 \leqslant \int_X f\, d\mu \leqslant \int_X \infty \cdot 1_{\{f>0\}}\, d\mu = \infty \cdot \mu(\{f > 0\}) = 0$. □

Theorem 11.17. *Let $f : X \to \overline{R}$ and $g : X \to \overline{R}$ be measurable functions such that $f = g$ μ-a.e. If $\int_X f\, d\mu$ exists, then so does $\int_X g\, d\mu$, and $\int_X f\, d\mu = \int_X g\, d\mu$.*

Proof. We have $\mu(\{f \neq g\}) = 0$.

(I) Suppose first that $f \geqslant 0$ and $g \geqslant 0$. Then, according to (11.12.ii) and (11.16), we get $\int_X f\, d\mu = \int_{\{f=g\}} f\, d\mu + \int_{\{f \neq g\}} f\, d\mu = \int_{\{f=g\}} f\, d\mu = \int_{\{f=g\}} g\, d\mu = \int_X g\, d\mu$.

(II) Consider now the general case. Since $\{f^+ \neq g^+\} \cup \{f^- \neq g^-\} \subset \{f \neq g\}$, it follows that $f^+ = g^+$ μ-a.e. and $f^- = g^-$ μ-a.e. Thus, applying (I), $\int_X f^+\, d\mu = \int_X g^+\, d\mu$ and $\int_X f^- d\mu = \int_X g^-\, d\mu$. Therefore, if $\int_X f\, d\mu$ exists, then $\int_X g\, d\mu$ exists, and $\int_X f\, d\mu = \int_X g\, d\mu$. □

Theorem (11.17) admits the following converse.

Theorem 11.18. *Let $f, g \in \mathfrak{L}^1$ be such that $\int_A f\, d\mu = \int_A g\, d\mu$ for each $A \in \mathcal{A}$. Then $f = g$ μ-a.e.*

Proof. Set $A_n = \{f + 1/n \leqslant g\}$, $n \in N$. Using (11.12.i) and (11.12.iv), we have $\int_{A_n} f\, d\mu = \int_{A_n} g\, d\mu \geqslant \int_{A_n} (f + 1/n)\, d\mu = \int_{A_n} f\, d\mu + \mu(A_n)/n$, $n \in N$. Since $\int_{A_n} f\, d\mu \in R$ for any $n \in N$, it follows that $\mu(A_n) = 0$, $n \in N$. Hence, since $\{f < g\} \subset \cup_{n \in N} A_n$, we get $\mu(\{f < g\}) = 0$. Analogously we obtain $\mu(\{f > g\}) = 0$, and so $\mu(\{f \neq g\}) = \mu(\{f < g\}) + \mu(\{f > g\}) = 0$, i.e. $f = g$ μ-a.e. □

Remarks 11.19. (a) If $f, g \in \mathfrak{L}^1$ and $\int_{A \cap B} f\, d\mu = \int_{A \cap B} g\, d\mu$, $A \in \mathcal{A}$, for some $B \in \mathcal{A}$, then $f = g$ μ-a.e. on B.

(b) The hypothesis that $f, g \in \mathfrak{L}^1$ cannot be dropped in (11.18). For example, let μ be as in (9.3.a), and consider $f = 1$ and $g = 2$. Then $\int_A f\, d\mu = \int_A g\, d\mu$, $A \in \mathcal{A}$, but f and g do not coincide μ-a.e.

Remark 11.20. Let $f : X \to \overline{R}$ be a measurable function, let $A \in \mathcal{A}$, and let $\alpha \in \overline{R}$. Consider the function $g : X \to \overline{R}$ defined by $g(x) = \begin{cases} \alpha & \text{if } x \in A \\ f(x) & \text{if } x \in A^c \end{cases}$. Then, according to (8.3.b) and (8.21), g is measurable. If $\mu(A) = 0$ and $\int_X f\,d\mu$ exists, then (11.17) shows that $\int_X g\,d\mu$ exists, and $\int_X f\,d\mu = \int_X g\,d\mu$. Therefore, if $f \in \mathfrak{L}^1$, $A = \{|f| = \infty\}$, and $\alpha \in R$, then (11.15) shows that f may be replaced by a measurable function $g : X \to R$ such that $f = g$ μ-a.e. and $\int_X f\,d\mu = \int_X g\,d\mu$. Thus, without any loss of generality, we will assume $f(X) \subset R$ whenever $f \in \mathfrak{L}^1$.

Since it is convenient at times to consider functions defined only μ-a.e., we next extend the definition of the integral.

Definition 11.21. Let $A \in \mathcal{A}$ be such that $\mu(A^c) = 0$, let $f : A \to \overline{R}$ be an $A \cap \mathcal{A}$-measurable function, and let $g : X \to \overline{R}$ be a measurable extension of f. If $\int_X g\,d\mu$ exists, then define $\int_X f\,d\mu = \int_X g\,d\mu$. By (11.17), the number $\int_X f\,d\mu$ is well defined.

On account of this definition, since a countable union of μ-null sets is a μ-null set, in each integration theorem we may replace a countable set of functions defined μ-a.e. by an appropriate set of functions defined on X.

In what follows we present convergence properties of the integral. We begin with the next theorem due to B. Levi.

Monotone convergence theorem 11.22. *Let $\{f_n : n \in N\} \subset [0, \infty]^X$ be a nondecreasing sequence of measurable functions. Then $\lim_n \int_X f_n\,d\mu = \int_X (\lim_n f_n)\,d\mu$.*

Proof. For each $n \in N$, consider $\{f_{nm} : m \in N\} \subset \mathfrak{S}'$ a nondecreasing sequence such that $f_n = \lim_m f_{nm}$. For $m \in N$, define $g_m = \sup_{1 \leqslant n \leqslant m} f_{nm}$. Then $\{g_m : m \in N\}$ is a nondecreasing sequence of simple functions from X into $[0, \infty[$ such that, for any $m \in N$,

$$f_{nm} \leqslant g_m \leqslant f_m, \quad 1 \leqslant n \leqslant m. \tag{1}$$

On letting $m \to \infty$ and then $n \to \infty$ in (1), we get

$$\lim_n f_n = \lim_m g_m. \tag{2}$$

For each $n \in N$, by (1) and (11.6.iii), we have

$$\int_X f_{nm}\,d\mu \leqslant \int_X g_m\,d\mu \leqslant \int_X f_m\,d\mu, \quad 1 \leqslant n \leqslant m. \tag{3}$$

On letting $m \to \infty$ in (3), and using (2), we get $\int_X f_n\,d\mu \leqslant \int_X (\lim_n f_n)\,d\mu \leqslant \lim_m \int_X f_m\,d\mu$, $n \in N$, and so $\lim_n \int_X f_n\,d\mu = \int_X (\lim_n f_n)\,d\mu$. □

Corollary 11.23. *Let $\{f_n : n \in N\} \subset [0, \infty]^X$ be a sequence of measurable functions. Then $\sum_{n \in N} \int_X f_n\,d\mu = \int_X (\sum_{n \in N} f_n)\,d\mu$.*

Proof. By (11.6.i) and (11.22), we have $\sum_{n \in N} \int_X f_n\,d\mu = \lim_m \sum_{n=1}^m \int_X f_n\,d\mu = \lim_m \int_X (\sum_{n=1}^m f_n)\,d\mu = \int_X (\sum_{n \in N} f_n)\,d\mu$. □

Fatou's lemma 11.24. *Let $\{f_n : n \in N\} \subset [0, \infty]^X$ be a sequence of measurable functions. Then $\int_X (\liminf_n f_n)\,d\mu \leqslant \liminf_n \int_X f_n\,d\mu$.*

Proof. For $n \in N$, set $g_n = \inf_{k \geqslant n} f_k$. Then $\{g_n : n \in N\}$ is a nondecreasing sequence of measurable functions in $[0, \infty]^X$ such that $\liminf_n f_n = \lim_n g_n$ and $g_n \leqslant f_n$, $n \in N$. Hence, applying (11.22) and (5.60.b), we have $\int_X (\liminf_n f_n)\,d\mu = \lim_n \int_X g_n\,d\mu \leqslant \liminf_n \int_X f_n\,d\mu$. □

The next result, due to Lebesgue, is extremely important.

Dominated convergence theorem 11.25. *Let $f_n : X \to \overline{R}$, $n \in N$, be measurable functions, and let $g \in \mathfrak{L}^1$ be such that $|f_n| \leqslant g$ for any $n \in N$. Then the following assertions hold.*

(i) $\int_X (\liminf_n f_n)\,d\mu \leqslant \liminf_n \int_X f_n\,d\mu \leqslant \limsup_n \int_X f_n\,d\mu \leqslant \int_X (\limsup_n f_n)\,d\mu$.

(ii) *If $f_n \to f$, then $\int_X f_n\,d\mu \to \int_X f\,d\mu$.*

Proof. Since $|f_n| \leqslant g$, $n \in N$, it follows that $|\liminf_n f_n| \leqslant g$ and $\left|\limsup_n f_n\right| \leqslant g$. Thus, according to (11.9.c), we have $f_n \in \mathfrak{L}^1$, $n \in N$, and $\liminf_n f_n$, $\limsup_n f_n \in \mathfrak{L}^1$. Let $\alpha \in R$, and define $f'_n(x) = \begin{cases} \alpha & \text{if } g(x) = \infty \\ f_n(x) & \text{if } g(x) < \infty \end{cases}$, $n \in N$. Then, applying (11.15), we get $f'_n = f_n$ μ-a.e., $n \in N$, $\liminf_n f'_n = \liminf_n f_n$ μ-a.e. and $\limsup_n f'_n = \limsup_n f_n$ μ-a.e. Since $\left|f'_n\right| \leqslant g$, $n \in N$, we have $\{g + f'_n : n \in N\} \subset [0, \infty]^X$. Thus, using (11.17) and Fatou's lemma (11.24), we have $\int_X g\,d\mu + \int_X (\liminf_n f_n)\,d\mu = \int_X g\,d\mu + \int_X (\liminf_n f'_n)\,d\mu = \int_X (\liminf_n (g + f'_n))\,d\mu \leqslant \liminf_n \int_X (g + f'_n)\,d\mu = \int_X g\,d\mu + \liminf_n \int_X f'_n\,d\mu = \int_X g\,d\mu + \liminf_n \int_X f_n\,d\mu$. Therefore, since $\int_X g\,d\mu \in R$, we have $\int_X (\liminf_n f_n)\,d\mu \leqslant \liminf_n \int_X f_n\,d\mu$. The inequality $\limsup_n \int_X f_n\,d\mu \leqslant \int_X (\limsup_n f_n)\,d\mu$ is proved similarly, considering the sequence $\{g - f'_n : n \in N\}$ and using the equality $-\limsup_n x_n = \liminf_n(-x_n)$. This completes the proof of (i). If $f_n \to f$, then $\liminf_n f_n = \limsup_n f_n = f$ (5.12). Hence (ii) follows immediately from (i) and (5.12). □

Corollary 11.26. *Let $f_n : X \to R$, $n \in N$, $f : X \to R$ be measurable functions such that $f_n \xrightarrow{u} f$. Assume that $f \in \mathfrak{L}^1$ and $\mu(X) < \infty$. Then $\int_X f_n\,d\mu \to \int_X f\,d\mu$.*

Proof. For $\varepsilon > 0$, choose $n_\varepsilon \in N$ such that $n \geqslant n_\varepsilon$ implies $\|f - f_n\|_u < \varepsilon$. Therefore, $|f_n| \leqslant |f - f_n| + |f| < \varepsilon + |f|$, $n \geqslant n_\varepsilon$. Since $\varepsilon + |f| \in \mathfrak{L}^1$, the result follows at once from (11.25.ii). □

Remarks 11.27. (a) The condition $|f_n| \leqslant g \in \mathfrak{L}^1$, $n \in N$, in (11.25) is essential. To see this, let $X = R$, let μ be the Lebesgue measure on $\mathcal{B}(R)$, and let $f_n = n 1_{]0,1/n]}$, $n \in N$. Then $f_n \to 0$, but $\lim_n \int_R f_n\,d\mu = 1 \neq 0 = \int_R 0\,d\mu$.

(b) It is not difficult to see that if either the hypotheses of (11.22)–(11.25) hold only μ-a.e. or each f_n is defined only μ-a.e., then the conclusions therein remain unchanged.

The result of (11.22)–(11.25) is very often applied in measure theory and in probability theory. We will frequently encounter such applications in the rest of the book. For example, a consequence of (11.23) is the next theorem which generalizes (11.12.ii).

Theorem 11.28. *Let $f : X \to \overline{R}$ be a measurable function such that $\int_X f\,d\mu$ exists. For $A \in \mathcal{A}$, define $\nu(A) = \int_A f\,d\mu$. Then ν is a signed measure on $\mathcal{A}$.*

Proof. Evidently, $\nu(\emptyset) = 0$. Let $\{A_n : n \in N\} \subset \mathcal{A}$ be such that $A_m \cap A_n = \emptyset$ for $m \neq n$. Since either $\int_X f^+\,d\mu < \infty$ or $\int_X f^-\,d\mu < \infty$, using (11.23), we have

$$\begin{aligned}
\nu\Big(\bigcup_{n\in N} A_n\Big) &= \int_X f 1_{\bigcup_{n\in N} A_n}\,d\mu = \int_X f^+ 1_{\bigcup_{n\in N} A_n}\,d\mu - \int_X f^- 1_{\bigcup_{n\in N} A_n}\,d\mu \\
&= \int_X \Big(\sum_{n\in N} f^+ 1_{A_n}\Big)\,d\mu - \int_X \Big(\sum_{n\in N} f^- 1_{A_n}\Big)\,d\mu \\
&= \sum_{n\in N} \int_X f^+ 1_{A_n}\,d\mu - \sum_{n\in N} \int_X f^- 1_{A_n}\,d\mu \\
&= \sum_{n\in N} \Big(\int_X f^+ 1_{A_n}\,d\mu - \int_X f^- 1_{A_n}\,d\mu\Big) \\
&= \sum_{n\in N} \int_X f 1_{A_n}\,d\mu = \sum_{n\in N} \nu(A_n).
\end{aligned}$$
□

The following theorem allows the change of variable in an integral.

Theorem 11.29 (Change of variable formula). *Let $(Y, \mathcal{B}, \mu f^{-1})$ be as in (9.26), let $g : Y \to \overline{R}$ be a $\mathcal{B}$-measurable function, and let $B \in \mathcal{B}$. Then*
(i) $\int_{f^{-1}(B)} (g \circ f)\,d\mu = \int_B g d(\mu f^{-1})$,
in the sense that if either integral exists so does the other and then both are equal.

Proof. To prove this theorem, we will use the indicator function method. If $g = 1_C$, where $C \in \mathcal{B}$, then, using (2.34), (i) becomes $\mu(f^{-1}(B) \cap f^{-1}(C)) = (\mu f^{-1})(B \cap C)$, which is true by definition of μf^{-1}. Thus, if $g = \sum_{i=1}^n c_i 1_{C_i}$ is a simple function, where $c_i \in [0, \infty[$ for $i = 1, \ldots, n$, then, applying (11.6.i) and (11.6.ii), we get $\int_{f^{-1}(B)} (g \circ f)d\mu = \sum_{i=1}^n c_i \int_{f^{-1}(B)} (1_{C_i} \circ f)d\mu = \sum_{i=1}^n c_i \int_B 1_{C_i}\,d(\mu f^{-1}) = \int_B g\,d(\mu f^{-1})$. Now if $g \geqslant 0$, apply (8.15) to get a nondecreasing sequence $\{g_n : n \in N\} \subset [0, \infty[^Y$ of simple functions such that $g_n \to g$. We have

$$\int_{f^{-1}(B)} (g_n \circ f)\,d\mu = \int_B g_n d(\mu f^{-1}), \quad n \in N. \tag{1}$$

Since $\{g_n \circ f : n \in N\} \subset \mathfrak{S}'$ is a nondecreasing sequence, and $g_n \circ f \to g \circ f$, letting $n \to \infty$ in (1), we get (i). Finally, if g is arbitrary, then $\int_{f^{-1}(B)} (g^+ \circ f)\,d\mu = \int_B g^+\,d(\mu f^{-1})$ and $\int_{f^{-1}(B)} (g^- \circ f)\,d\mu = \int_B g^-\,d(\mu f^{-1})$. Thus, since $(g \circ f)^+ = g^+ \circ f$ and $(g \circ f)^- = g^- \circ f$, it is easily seen that if one of the integrals in (i) exists, then so does the other, and the two integrals are equal. □

In probability theory is important the special case when $(X, \mathcal{A}) = (R, \mathcal{B}(R))$, and $\mu = \mu_F$ is a Lebesgue-Stieltjes measure on $\mathcal{B}(R)$. If $f : R \to \overline{R}$ is measurable, $A \in \mathcal{B}(R)$, and $\int_A f\,d\mu_F$ exists, then this integral will be also denoted by $\int_A f\,dF$ or $\int_A f(x)\,dF(x)$, and it will be called the *Lebesgue-Stieltjes integral of f on A*. When $A =]a, b]$ or $A =]a, \infty[$, the sign $\int_A$ will be sometimes replaced by $\int_a^b$ or $\int_a^\infty$, respectively. In certain circumstances a Lebesgue-Stieltjes integral on an interval becomes a Riemann-Stieltjes integral. We assume on the part of the reader a knowledge of the Riemann-Stieltjes integral.

We use the symbol $(RS)\int$ to denote the Riemann-Stieltjes integral in the next result. Following conventional notation, this symbol will be dropped later on.

Theorem 11.30. *Let $f : R \to R$ be a function which is continuous on R, let μ_F be as in* (10.15), *and let $a, b \in R$ be such that $a < b$. Then the following assertions hold.*

(i) $\int_{]a,b]} f\,dF = (RS)\int_a^b f\,dF$.

(ii) *If $((RS)\int_{-\infty}^\infty f^+\,dF, (RS)\int_{-\infty}^\infty f^-\,dF) \neq (\infty, \infty)$, then $\int_R f\,dF = (RS)\int_{-\infty}^\infty f\,dF, \int_{]-\infty,a]} f\,dF = (RS)\int_{-\infty}^a f\,dF$ and $\int_{]a,\infty[} f\,dF = (RS)\int_a^\infty f\,dF$.*

Proof. For each $n \in N$, consider $a = x_{n0} < x_{n1} < \cdots < x_{n2^n} = b$ be a division of $[a, b]$ such that $x_{nk} - x_{n,k-1} = (b - a)/2^n$, $1 \leqslant k \leqslant 2^n$, and define $f_n :]a, b] \to R$ by $f_n = \sum_{k=1}^{2^n} f(x_{nk}) 1_{]x_{n,k-1},x_{nk}]}$. Since f is uniformly continuous on $[a, b]$ (6.62), it follows that $f_n \xrightarrow{u} f|_{]a,b]}$. Since $f(]a, b])$ is bounded (5.47), and $\mu_F(]a, b]) < \infty$, we have $\int_{]a,b]} f\,dF \in R$. Thus, applying (11.11) and (11.26), we get

$$\int_{]a,b]} f\,dF = \lim_n \int_{]a,b]} f_n\,dF = \lim_n \sum_{k=1}^{2^n} f(x_{nk})\mu_F(]x_{n,k-1}, x_{nk}])$$
$$= \lim_n \sum_{k=1}^{2^n} f(x_{nk})(F(x_{nk}) - F(x_{n,k-1})). \tag{1}$$

Since $\sum_{k=1}^{2^n} f(x_{nk})(F(x_{nk}) - F(x_{n,k-1}))$ is a Riemann-Stieltjes sum, and $(RS)\int_a^b f\,dF$ exists, assertion (i) follows from (1). To prove (ii), choose $\{a_n : n \in N\} \subset R$ and $\{b_n : n \in N\} \subset R$ such that $a_n \to -\infty$ and $b_n \to \infty$. Using the monotone convergence theorem (11.22) and (i), we have $\int_R f^+\,dF = \int_R (\lim_n f^+ 1_{]a_n,b_n]})dF = \lim_n \int_{]a_n,b_n]} f^+\,dF = \lim_n (RS)\int_{a_n}^{b_n} f^+\,dF = (RS)\int_{-\infty}^\infty f^+\,dF$. Similarly, we have $\int_R f^-\,dF = (RS)\int_{-\infty}^\infty f^-\,dF$, and so $\int_R f\,dF = (RS)\int_{-\infty}^\infty f^+\,dF - (RS)\int_{-\infty}^\infty f^-\,dF = (RS)\int_{-\infty}^\infty f\,dF$. The last two equalities in (ii) are proved in like manner. □

Exercise 11.31. Let $f : X \to [0, \infty]$ be a measurable function. For $A \in \mathcal{A}$, prove that $\int_A f\,d\mu = \sup\{\int_A g\,d\mu : g \in \mathfrak{S}, 0 \leqslant g \leqslant f\}$.

Exercise 11.32. Let $f : X \to \overline{R}$ be a measurable function such that $\int_X f\,d\mu$ exists, and let $a \in [0, \infty[$. Show that $\int_X f\,d(a\mu)$ exists, and $\int_X f\,d(a\mu) = a\int_X f\,d\mu$.

Exercise 11.33. Let μ_1 and μ_2 be measures on $\mathcal{A}$, and let $f : X \to \overline{R}$ be a measurable function such that $\int_X f\,d\mu_1$ and $\int_X f\,d\mu_2$ exist. If $(\int_X f\,d\mu_1, \int_X f\,d\mu_2) \notin \{(-\infty, \infty), (\infty, -\infty)\}$, prove that $\int_X f\,d(\mu_1 + \mu_2) = \int_X f\,d\mu_1 + \int_X f\,d\mu_2$.

Exercise 11.34. Let $f : X \to \overline{R}$ be a measurable function, let μ be a signed measure on $\mathcal{A}$, and let μ^+, μ^- and $|\mu|$ be as in (9.23). If $\int_X f d\mu^+$ and $\int_X f d\mu^-$ exist, and $(\int_X f\,d\mu^+, \int_X f\,d\mu^-) \notin \{(-\infty, -\infty), (\infty, \infty)\}$, then we define the *integral of* f (with respect to μ) by $\int_X f\,d\mu = \int_X f\,d\mu^+ - \int_X f\,d\mu^-$, and we say that the integral of f (with respect to μ) exists. Prove the following.

(a) $f \in \mathfrak{L}^1(X, \mathcal{A}, |\mu|)$ if and only if $f \in \mathfrak{L}^1(X, \mathcal{A}, \mu^+)$ and $f \in \mathfrak{L}^1(X, \mathcal{A}, \mu^-)$.

(b) If $\alpha \in R$ and $\int_X f d\mu$ exists, then $\int_X \alpha f d\mu$ exists, and $\int_X \alpha f d\mu = \alpha \int_X f d\mu$.

(c) Let $g : X \to \overline{R}$ be a measurable function such that $\{f = -\infty\} \cap \{g = \infty\} = \{f = \infty\} \cap \{g = -\infty\} = \emptyset$. If $\int_X f d\mu$ and $\int_X g d\mu$ exist, and $(\int_X f d\mu, \int_X f g d\mu) \notin \{(-\infty, \infty), (\infty, -\infty)\}$, then $\int_X (f + g)\,d\mu$ exists, and $\int_X (f + g)\,d\mu = \int_X f\,d\mu + \int_X g\,d\mu$.

Exercise 11.35. Let $f : X \to C$ be a measurable function. Apply (11.9.c) to show that $|f| \in \mathfrak{L}^1$ if and only if $\mathrm{Re} f \in \mathfrak{L}^1$ and $\mathrm{Im} f \in \mathfrak{L}^1$.

Exercise 11.36. Let $f : X \to C$ be a measurable function such that $|f| \in \mathfrak{L}^1$. Then define $\int_X f d\mu = \int_X \mathrm{Re} f d\mu + i \int_X \mathrm{Im} f d\mu$. Prove the following.

(a) Let $g : X \to C$ be a measurable function. If $|g| \in \mathfrak{L}^1$, then $|f + g| \in \mathfrak{L}^1$, and $\int_X (f + g)\,d\mu = \int_X f d\mu + \int_X g\,d\mu$.

(b) If $\alpha \in C$, then $|\alpha f| \in \mathfrak{L}^1$, and $\int_X \alpha f d\mu = \alpha \int_X f d\mu$.

(c) $\left|\int_X f\,d\mu\right| \leqslant \int_X |f|\,d\mu$. [Use polar coordinates.]

Exercise 11.37. Let $f : X \to \overline{R}$ be a measurable function. Prove that $f \in \mathfrak{L}^1$ if and only if $\sum_{n \geqslant 0} \mu(\{|f| \geqslant n\}) < \infty$.

Exercise 11.38. Assume that $(X, \mathcal{A}, \mu)$ is a complete measure space, and let $f, g \in \overline{R}^X$ be such that $f = g$ μ-a.e. Show that f is measurable if and only if g is measurable.

Exercise 11.39. Let $\mathcal{M} \subset \mathcal{A}$ be such that $\sigma(\mathcal{M}) = \mathcal{A}$, $X \in \mathcal{M}$, and $A, B \in \mathcal{M}$ implies $A \cap B \in \mathcal{M}$. Let $f, g \in \mathfrak{L}^1$ be such that $\int_A f\,d\mu = \int_A g\,d\mu$, $A \in \mathcal{M}$. Prove that $f = g$ μ-a.e.

Exercise 11.40. Let $f : X \to \overline{R}$ be a measurable function such that $\int_X f\,d\mu$ exists. If $\int_A f\,d\mu \geqslant 0$, $A \in \mathcal{A}$, then $f \geqslant 0$ μ-a.e.

Exercise 11.41. Suppose that μ is σ-finite, and let $f : X \to \overline{R}$ and $g : X \to \overline{R}$ be measurable functionssuch that $\int_X f\,d\mu$ and $\int_X g\,d\mu$ exist. If $\int_A f\,d\mu \leqslant \int_A g\,d\mu$ for any $A \in \mathcal{A}$, then $f \leqslant g$ μ-a.e. [Hints. Consider first the case that μ is finite and $f, g \geqslant 0$. In this case, for $n \in N$, put $X_n = \{f \leqslant n\}$, and apply (11.11) and (11.40) to the measure space $(X_n, X_n \cap \mathcal{A}, \mu_{X_n})$. Then consider the general case.]

Exercise 11.42. Let μ, f and g be as in (11.41).

(a) If $\int_A f\,d\mu = \int_A g\,d\mu$, $A \in \mathcal{A}$, then $f = g$ μ-a.e.

(b) Contrast (a) with (11.18).

Exercise 11.43. Let $\{f_i : i \in I\}$ be an indexed set of measurable functions from X into $\overline{R}$.

(a) Show that there exists a measurable function $f : X \to \overline{R}$ such that the following assertions hold:

(i) $f_i \leqslant f$ μ-a.e., $i \in I$;

(ii) if $g : X \to \overline{R}$ is a measurable function such that $f_i \leqslant g$ μ-a.e., $i \in I$, then $f \leqslant g$ μ-a.e.

[Hints. Let $\varphi(x) = \arctan x$, $x \in \overline{R}$. Let $\mathcal{J}$ denote the family of all nonempty countable subsets of I. For $J \in \mathcal{J}$, define $f_J = \sup_{j \in J} f_j$. Set $\alpha = \sup\{\int_X (\varphi \circ f_J)\, d\mu : J \in \mathcal{J}\}$. Choose $J \in \mathcal{J}$ such that $\alpha = \int_X (\varphi \circ f_J)\, d\mu$, and put $f = f_J$.]

(b) If $f' : X \to \overline{R}$ is a measurable function that satisfies (i) and (ii), then $f = f'$ μ-a.e.

(c) Prove that there exists a measurable function $f : X \to \overline{R}$ such that the following assertions hold:

(iii) $f_i \geqslant f$ μ-a.e., $i \in I$;

(iv) if $g : X \to \overline{R}$ is a measurable function such that $f_i \geqslant g$ μ-a.e., $i \in I$, then $f \geqslant g$ μ-a.e.

(d) If $f' : X \to \overline{R}$ is a measurable function that satisfies (iii) and (iv), then $f = f'$ μ-a.e.

The essentially unique f in (a) is denoted by $\operatorname{esssup}_{i \in I} f_i$, and the essentially unique f in (c) is denoted by $\operatorname{essinf}_{i \in I} f_i$.

Exercise 11.44. Let $f \in \mathcal{L}^1$. Show that for each $\varepsilon > 0$ there is $\delta > 0$ such that $A \in \mathcal{A}$ and $\mu(A) < \delta$ imply $\int_A |f|\, d\mu < \varepsilon$. [Hints. For $n \in N$, let φ_n be as in (8.24.a), and set $f_n = \varphi_n \circ |f|$. Then $\{f_n : n \in N\}$ is a nondecreasing sequence of measurable functions such that $f_n \to |f|$. By (11.22), $\int_X f_n\, d\mu \to \int_X |f|\, d\mu$. Choose n such that $\int_X (|f| - f_n)\, d\mu < \varepsilon/2$, and set $\delta = \varepsilon/2n$.]

Exercise 11.45. Let $]a, b[\subset R$, let $\{f_t : t \in]a, b[\} \subset [0, \infty]^X$ be a set of measurable functions, and let $f : X \to [0, \infty]$ be a measurable function. Apply (10.14) and (11.22) to prove the following.

(a) If $\lim_{t \searrow a} f_t(x) = f(x), x \in X$, and $a < s < t < b$ imply $f_s \geqslant f_t$, then $\lim_{t \searrow a} \int_X f_t\, d\mu = \int_X f\, d\mu$.

(b) If $\lim_{t \nearrow b} f_t(x) = f(x), x \in X$, and $a < s < t < b$ imply $f_s \leqslant f_t$, then $\lim_{t \nearrow b} \int_X f_t\, d\mu = \int_X f\, d\mu$.

Exercise 11.46. Let $]a, b[\subset R$, let $\{f_t : t \in]a, b[\} \subset \overline{R}^X$ be a set of measurable functions, and let $g \in \mathfrak{L}^1$ be such that $|f_t| \leqslant g, t \in]a, b[$. Let $t_0 \in [a, b]$, and let $f : X \to \overline{R}$ be a measurable function such that $\lim_{t \to t_0} f_t(x) = f(x)$ for any $x \in X$. Show that $\lim_{t \to t_0} \int_X f_t\, d\mu = \int_X f\, d\mu$.

Exercise 11.47. Notation is as in (11.28). Prove the following.

(a) If $P = \{f \geqslant 0\}$, then (P, P^c) is a Hahn decomposition of X relative to ν.

(b) $\nu^+(A) = \int_A f^+\, d\mu, A \in \mathcal{A}$.

(c) $\nu^-(A) = \int_A f^-\, d\mu, A \in \mathcal{A}$.

(d) $|\nu|(A) = \int_A |f|\, d\mu, A \in \mathcal{A}$.

Exercise 11.48. Let $f(x) = \begin{cases} x & \text{if } x < 1 \\ 1/x^2 & \text{if } x \geqslant 1 \end{cases}$ and $F(x) = \begin{cases} -1 & \text{if } x < 0 \\ x^2 & \text{if } x \geqslant 0 \end{cases}$. Use (11.30) to compute $\int_R f\,dF$ and $\int_{-1}^{2} f\,dF$.

Exercise 11.49. Let I be a countable set, let $f_i : X \to [a, b], i \in I$, be injective functions, where $a, b \in R$, and let $\mathcal{T}$ denote the topology generated by $\{f_i : i \in I\}$.

(a) Show that $(X, \mathcal{T})$ is metrizable. [Hint. Define $d(x, y) = \sum_{i \in I} a_i \,|f_i(x) - f_i(y)|$ for $x, y \in X$, where $a_i > 0, i \in I$, and $\sum_{i \in I} a_i < \infty$.]

(b) For $\{x_n : n \in N\} \subset X$ and $x \in X$, prove that $\lim_n x_n = x$ relative to $\mathcal{T}$ if and only if $\lim_n f_i(x_n) = f_i(x)$ for any $i \in I$. [Use (a) and (11.25.ii).]

Exercise 11.50. (a) Let X be a metric space, and let μ and ν be finite measures on $\mathcal{B}(X)$ such that $\int_X f\,d\mu = \int_X f d\nu$ for any $f \in \mathfrak{C}_r(X)$. Use (6.76), (11.22) and (10.8) to show that $\mu = \nu$.

(b) Prove that the conclusion of (a) still holds if $\mathfrak{C}_r(X)$ is replaced by a subset $\mathfrak{A}$ with $\mathfrak{A}^- = \mathfrak{C}_r(X)$.

Exercise 11.51. Let $[a, b] \subset R$, and let λ and μ be measures on $\mathcal{B}([a, b])$ such that $\lambda([a, b]) = \mu([a, b]) < \infty$ and $\int_{[a,b]} x^n d\lambda(x) = \int_{[a,b]} x^n\,d\mu(x), n \in N$. Prove that $\lambda = \mu$. [Hint. $\int_{[a,b]} p(x) d\lambda(x) = \int_{[a,b]} p(x)\,d\mu(x)$ for each polynomial $p(x)$ as in (6.87). This entails that $\int_{[a,b]} f d\lambda = \int_{[a,b]} f d\mu$ for any $f \in \mathfrak{C}_r([a, b])$, by (6.87) and (11.26).]

Exercise 11.52. Notation is as in (9.27). Let $f : X \to \overline{R}$ be a measurable function such that $(\sum_{i \in I} a_i \int_X f^+\,d\mu_i, \sum_{i \in I} a_i \int_X f^-\,d\mu_i) \neq (\infty, \infty)$. Show that $\int_X fd(\sum_{i \in I} a_i \mu_i) = \sum_{i \in I} a_i \int_X f d\mu_i$.

Exercise 11.53. Let μ be a measure on $\mathcal{B}(R)$ such that $\mu(h + A) = \mu(A)$ for some $h \in R$ and all $A \in \mathcal{B}(R)$, and let $f : R \to R$ be a measurable function. Prove that $\int_R f(x - h)\,d\mu(x) = \int_R f(x)\,d\mu(x)$, in the sense that if either integral exists so does the other and then both are equal.

Exercise 11.54. Let X be a metric space, and let μ and ν be finite measures on $\mathcal{B}(X)$ such that $\int_X f d\mu = \int_X f d\nu$ for any bounded uniformly continuous function $f : X \to R$. Show that $\mu = \nu$. [Hint. Let A, B_n and f_n be as in (6.91). Then $\mu(A^c) \geqslant \int_{A^c} f_n\,d\mu = \int_X f_n\,d\mu = \int_X f_n d\nu \geqslant \int_{B_n} f_n d\nu = \nu(B_n), n \in N$, and so $\mu(A^c) \geqslant \nu(A^c)$.]

Exercise 11.55. Let μ_1 and μ_2 be signed measures on $\mathcal{A}$ such that $|\mu_1|\,(X), |\mu_2|\,(X) < \infty$, and let $f \in \mathfrak{L}^1(X, \mathcal{A}, |\mu_i|), i = 1, 2$. Prove that $f \in \mathfrak{L}^1(X, \mathcal{A}, |\mu_1 + \mu_2|)$, and $\int_X f\,d(\mu_1 + \mu_2) = \int_X f\,d\mu_1 + \int_X f\,d\mu_2$. [Hint. Use (9.57), and notice that $(\mu_1 + \mu_2)^+ + \mu_1^- + \mu_2^- = (\mu_1 + \mu_2)^- + \mu_1^+ + \mu_2^+$.]

Exercise 11.56. Let X be a metric space, and let μ be a signed measure on $\mathcal{B}(X)$ such that $|\mu|\,(X) < \infty$ and $\int_X f\,d\mu = 0, f \in \mathfrak{C}_r(X)$. Show that $\mu = 0$.

Exercise 11.57. (a) Let $(X, \mathcal{T})$ be a topological space, let (Y, ρ) be a metric space, let $h : X \to Y$ be a function such that $\mathcal{T} = h^{-1}(\mathcal{T}_\rho)$, and let μ and ν be finite measures on $\mathcal{B}(X)$ such that $\int_X (g \circ h) d\mu = \int_X (g \circ h) d\nu$ for any $g \in \mathfrak{C}_r(Y)$. Mimic the solution of (11.50.a) to prove that $\mu = \nu$.

(b) Show that the conclusion of (a) still holds if $\mathfrak{C}_r(Y)$ is replaced by a subset $\mathfrak{B}$ with $\mathfrak{B}^- = \mathfrak{C}_r(Y)$.

12 Radon-Nikodým Theorem and the Lebesgue Decomposition

This section is devoted to the study of the notions of absolute continuity and singularity for pairs of measures defined on a fixed measurable space $(X, \mathcal{A})$.

Definition 12.1. Let $(X, \mathcal{A})$ be a measurable space, and let μ and ν be measures on $\mathcal{A}$. We say that ν is *absolutely continuous with respect to* μ, and we write $\nu \ll \mu$, if $\nu(A) = 0$ whenever $A \in \mathcal{A}$ and $\mu(A) = 0$.

Examples 12.2. (a) Let $(X, \mathcal{A})$ be a measurable space, and let μ be a measure on $\mathcal{A}$ such that $\mu(A) > 0$ whenever $\emptyset \neq A \in \mathcal{A}$. Then any measure on $\mathcal{A}$ is absolutely continuous with respect to μ.

(b) Notation is as in (11.28). If $f \geqslant 0$, then, according to (11.16), $\nu \ll \mu$.

Remarks 12.3. (a) Obviously, $\ll$ is a reflexive and transitive relation on the set of all measures on $\mathcal{A}$. $\ll$ is neither symmetric nor antisymmetric. If μ and ν are measures on $\mathcal{A}$ such that $\nu \ll \mu$ and $\mu \ll \nu$, then we say that ν and μ are *equivalent measures,* and we write $\nu \sim \mu$.

(b) If μ and ν are measures on $\mathcal{A}$ such that $\nu \ll \mu$ and $\mathcal{B} \subset \mathcal{A}$ is a σ-algebra, then $\nu_{\mathcal{B}} \ll \mu_{\mathcal{B}}$.

The Radon-Nikodým theorem is a converse of (12.2.b). To prove it, we need the next lemma.

Lemma 12.4. *Let μ and ν be measures on $\mathcal{A}$ such that $\nu \ll \mu$, $\mu(X) < \infty$ and $\nu(X) > 0$. Then there are $n \in N$ and $P_n \in \mathcal{A}$ such that $\mu(P_n) > 0$ and P_n is a positive set for the signed measure $\nu - \frac{1}{n}\mu$.*

Proof. For $n \in N$, let (P_n, P_n^c) be a Hahn decomposition of X relative to $\nu - \frac{1}{n}\mu$. Since $\bigcap_{n\in N} P_n^c \subset P_n^c$, $n \in N$, we have

$$0 \leqslant \nu\left(\bigcap_{n\in N} P_n^c\right) \leqslant \frac{1}{n}\mu\left(\bigcap_{n\in N} P_n^c\right), \quad n \in N. \tag{1}$$

Since μ is finite, (1) shows that $\nu(\bigcap_{n\in N} P_n^c) = 0$, and so $\nu(\bigcup_{n\in N} P_n) > 0$. Consequently, there is $n \in N$ such that $\mu(P_n) > 0$. □

Radon-Nikodým theorem 12.5. *Let μ be a σ-finite measure on $\mathcal{A}$, and let ν be a measure on $\mathcal{A}$ such that $\nu \ll \mu$. Then there exists a measurable function $f : X \to [0, \infty]$ such that*

(i) $\nu(A) = \int_A f\, d\mu, \quad A \in \mathcal{A}$.

Proof. (I) Assume first that μ and ν are finite measures. Let $\mathfrak{G}$ be the set of all measurable functions $g : X \to [0, \infty]$ such that $\int_A g\, d\mu \leqslant \nu(A)$ for each $A \in \mathcal{A}$.

Set $\alpha = \sup\{\int_X g\,d\mu : g \in \mathfrak{G}\}$, and choose $\{g_n : n \in N\} \subset \mathfrak{G}$ such that $\int_X g_n\,d\mu \to \alpha$. For $n \in N$, consider $f_n = \sup_{1 \leqslant i \leqslant n} g_i$. If $A \in \mathcal{A}$, then there is a partition $\{A_1, \ldots, A_n\} \subset \mathcal{A}$ of A such that $f_n = g_i$ on A_i, $1 \leqslant i \leqslant n$. Consequently, we have $\int_A f_n\,d\mu = \sum_{i=1}^n \int_{A_i} f_n\,d\mu = \sum_{i=1}^n \int_{A_i} g_i\,d\mu \leqslant \sum_{i=1}^n \nu(A_i) = \nu(A)$, and so $f_n \in \mathfrak{G}$. Let $f = \lim_n f_n$. On account of the monotone convergence theorem (11.22), we get $f \in \mathfrak{G}$ and $\alpha = \int_X f\,d\mu$. Define the measure λ on $\mathcal{A}$ by $\lambda(A) = \nu(A) - \int_A f d\mu$, and suppose that $\lambda(X) > 0$. Since $\lambda \ll \mu$, applying (12.4), choose $n \in N$ and $P_n \in \mathcal{A}$ such that $\mu(P_n) > 0$, and $\mu(A)/n \leqslant \lambda(A) = \nu(A) - \int_A f d\mu$ whenever $A \in \mathcal{A}$ and $A \subset P_n$. Let $g = f + 1_{P_n}/n$. Then, for $A \in \mathcal{A}$, we have $\int_A g d\mu = \int_A f d\mu + \mu(A \cap P_n)/n \leqslant \int_A f d\mu + \nu(A \cap P_n) - \int_{A \cap P_n} f d\mu = \int_{A \cap P_n^c} f\,d\mu + \nu(A \cap P_n) \leqslant \nu(A)$, and so $g \in \mathfrak{G}$. Therefore, we find $\alpha < \int_X f\,d\mu + \mu(P_n)/n = \int_X g\,d\mu \leqslant \alpha$. The contradiction we got shows that $\lambda(X) = 0$, and so (i) is proved in this case.

(II) Suppose now that μ is finite and ν is σ-finite. Then there is a countable partition $\{A_i : i \in I\} \subset \mathcal{A}$ of X such that $\nu(A_i) < \infty, i \in I$. For $i \in I$, denote $\mathcal{A}_i = A_i \cap \mathcal{A}$ and $\mu_i = \mu_{\mathcal{A}_i}$, $\nu_i = \nu_{\mathcal{A}_i}$. Applying (I), for each $i \in I$, there exists an $\mathcal{A}_i$-measurable function $f_i : A_i \to [0, \infty]$ such that

$$\nu_i(A) = \int_A f_i\,d\mu_i, \quad A \in \mathcal{A}_i. \tag{1}$$

Consider the function $f : X \to [0, \infty]$ defined by $f(x) = f_i(x)$ if $x \in A_i$. Then, by (1) and (11.11), we have $\nu(A) = \sum_{i \in I} \nu_i(A \cap A_i) = \sum_{i \in I} \int_{A \cap A_i} f_i\,d\mu_i = \sum_{i \in I} \int_{A \cap A_i} f\,d\mu = \int_A f\,d\mu$ for any $A \in \mathcal{A}$.

(III) Consider next the case that μ is a finite measure and ν is an arbitrary measure. For $A \in \mathcal{A}$, let μ_A and ν_A denote the restrictions of μ and ν to $A \cap \mathcal{A}$, respectively. Let $\mathcal{B} = \{B \in \mathcal{A} : \nu_B \text{ is } \sigma\text{-finite}\}$. Set $\beta = \sup\{\mu(B) : B \in \mathcal{B}\}$, and choose $\{B_n : n \in N\} \subset \mathcal{B}$ such that $\mu(B_n) \to \beta$. Let $B = \cup_{n \in N} B_n$. Evidently, $B \in \mathcal{B}$. Since $\mu(B_n) \leqslant \mu(B) \leqslant \beta, n \in N$, we have $\beta = \mu(B)$. Applying (II), there exists a $B \cap \mathcal{A}$-measurable function $f_B : B \to [0, \infty]$ such that, for any $A \in \mathcal{A}$,

$$\nu(A \cap B) = \nu_B(A \cap B) = \int_{A \cap B} f_B\,d\mu_B. \tag{2}$$

Now let $A \in \mathcal{A}$. If $\mu(A \cap B^c) = 0$, then $\nu(A \cap B^c) = 0$, and so

$$\nu(A \cap B^c) = \int_{A \cap B^c} \infty\,d\mu. \tag{3}$$

If $\mu(A \cap B^c) > 0$, then $\nu(A \cap B^c) = \infty$, and so (3) holds in this case as well. Actually, if $\nu(A \cap B^c) < \infty$, then $B \cup (A \cap B^c) \in \mathcal{B}$, and so we would get $\beta = \mu(B) < \mu(B) + \mu(A \cap B^c) = \mu(B \cup (A \cap B^c)) \leqslant \beta$. Consider the function $f : X \to [0, \infty]$ defined by $f(x) = \begin{cases} f_B(x) & \text{if } x \in B \\ \infty & \text{if } x \in B^c \end{cases}$. Then, using (2), (3) and (11.11), we have $\nu(A) = \nu(A \cap B) + \nu(A \cap B^c) = \int_{A \cap B} f_B\,d\mu_B + \int_{A \cap B^c} \infty\,d\mu = \int_{A \cap B} f\,d\mu + \int_{A \cap B^c} f\,d\mu = \int_A f\,d\mu$ for each $A \in \mathcal{A}$.

(IV) Finally we consider the general case. There exists a countable partition $\{A_i : i \in I\} \subset \mathcal{A}$ of X such that $\mu(A_i) < \infty$, $i \in I$. Then the proof is similar to the proof of case (II). □

Remarks 12.6. (a) The function f in (12.5) is unique, in the sense that if g is a measurable function from X into $[0, \infty]$ such that $\nu(A) = \int_A f d\mu$, $A \in \mathcal{A}$, then $f = g$ μ-a.e. Actually, this follows at once from (11.42.a).

(b) Notation is as in (12.5). If $\nu(X) < \infty$, then f is μ-integrable, and so f is finite μ-a.e. Thus, without loss of generality, we may assume $f(X) \subset [0, \infty[$ whenever ν is σ-finite.

Definition 12.7. The function f in the Radon-Nikodým theorem (12.5) is called the *Radon-Nikodým derivative of* ν *with respect to* μ. To express the relationship between μ, ν and f, we use the notation $f = d\nu/d\mu$, $d\nu = f\, d\mu$ or $\nu = f\mu$. As remarked in (12.6.a), the derivative $d\nu/d\mu$ is essentially unique.

Theorem 12.8. *Let* μ, ν *and* f *be as in* (12.5), *and let* $g : X \to \overline{R}$ *be a measurable function. Then*

(i) $\int_X g d\nu = \int_X g f d\mu$,

in the sense that if one of the integral exists, so does the other, and the two integrals are equal.

Proof. To prove this theorem, we will use the indicator function method. If $g = 1_A$, where $A \in \mathcal{A}$, then (i) becomes $\nu(A) = \int_A f d\mu$, which is true. Hence, if $g = \sum_{i=1}^n a_i 1_{A_i}$ is a simple function, where $a_i \in [0, \infty[$ for $i = 1, \ldots, n$, then (i) holds. Now if $g \geqslant 0$, choose a nondecreasing sequence $\{g_n : n \in N\} \subset [0, \infty[^X$ of simple functions such that $g_n \to g$. Since $\int_X g_n d\nu = \int_X g_n f d\mu$, $n \in N$, (11.22) shows that (i) holds in this case too. Finally, if g is arbitrary, then $\int_X g^+ d\nu = \int_X g^+ f d\mu$ and $\int_X g^- d\nu = \int_X g^- f d\mu$, and so if either integral in (i) exists so does the other one and then both are equal. □

Corollary 12.9 (Chain rule). *Let* λ *and* μ *be* σ*-finite measures on* $\mathcal{A}$ *such that* $\mu \ll \lambda$, *and let* ν *be a measure on* $\mathcal{A}$ *such that* $\nu \ll \mu$. *Then* $\frac{d\nu}{d\lambda} = \frac{d\nu}{d\mu} \cdot \frac{d\mu}{d\lambda}$ λ*-a.e.*

Proof. Set $f = d\nu/d\mu$ and $g = d\mu/d\lambda$. For any $A \in \mathcal{A}$, applying (12.8), we have $\nu(A) = \int_A f d\mu = \int_A f g d\lambda$, and so $d\nu/d\lambda = fg$ λ-a.e. □

The next notion is in a sense opposite to absolute continuity.

Definition 12.10. Let $(X, \mathcal{A})$ be a measurable space, and let μ and ν be measures on $\mathcal{A}$. We say that ν is *singular with respect to* μ, and we write $\nu \perp \mu$, if there is $B \in \mathcal{A}$ such that $\nu(B) = \mu(B^c) = 0$. Clearly, $\perp$ is a symmetric relation on the set of all measures on $\mathcal{A}$.

Example 12.11. Let $(X, \mathcal{A})$ be a measurable space. According to (9.43), each atomic measure on $\mathcal{A}$ is singular with respect to any nonatomic measure on $\mathcal{A}$.

Like the Radon-Nikodým theorem, the next theorem has important applications.

Theorem 12.12 (Lebesgue decomposition). *Let* μ *and* ν *be* σ*-finite measures on* $\mathcal{A}$. *Then* $\nu = \nu_1 + \nu_2$, *where* $\nu_1 \ll \mu$ *and* $\nu_2 \perp \mu$.

Proof. Consider the σ-finite measure $\lambda = \mu + \nu$. Evidently, $\mu \ll \lambda$. Let $f = d\mu/d\lambda$, and put $B = \{f > 0\}$. For $A \in \mathcal{A}$, define $\nu_1(A) = \nu(A \cap B)$ and $\nu_2(A) = \nu(A \cap B^c)$. Then ν_1 and ν_2 are measures on $\mathcal{A}$ such that $\nu = \nu_1 + \nu_2$. Now let $A \in \mathcal{A}$ be such that $\mu(A) = 0$. Then $\int_A f d\lambda = 0$, and so, by (11.16), $f = 0$ λ-a.e. on A. Then, since $f(x) > 0$ for any $x \in A \cap B$, we have $\lambda(A \cap B) = 0$, and so $\nu_1(A) = \nu(A \cap B) = 0$. Therefore, $\nu_1 \ll \mu$. Further, notice that $\nu_2(B) = 0$ and $\mu(B^c) = \int_{B^c} f d\mu = 0$, and so $\nu_2 \perp \mu$. □

Remark 12.13. The Lebesgue decomposition of ν with respect to μ in (12.12) is unique. Actually, assume that $\nu = \nu_1 + \nu_2 = \nu_1' + \nu_2'$, where $\nu_1 \ll \mu$, $\nu_1' \ll \mu$, and $\nu_2 \perp \mu$, $\nu_2' \perp \mu$. Choose $B, B' \in \mathcal{A}$ such that $\nu_2(B) = \nu_2'(B) = 0$ and $\mu(B^c) = \mu(B'^c) = 0$. Obviously,

$$\nu_2(A) = \nu_2'(A) = 0 \tag{1}$$

whenever $A \in \mathcal{A}$ and $A \subset B \cap B'$. Now if $A \in \mathcal{A}$ and $A \subset B^c \cup B'^c$, then $\mu(A) = 0$, and so $\nu_1(A) = \nu_1'(A) = 0$. Hence, in this case,

$$\nu_2(A) = \nu_2'(A) = \nu(A). \tag{2}$$

(1) and (2) show that $\nu_2 = \nu_2'$. Since ν is σ-finite, there exists a countable partition $\{A_i : i \in I\} \subset \mathcal{A}$ of X such that $\nu(A_i) < \infty, i \in I$. Since $\nu_2 = \nu_2'$, for any $A \in \mathcal{A}$ and $i \in I$, we have $\nu_1(A \cap A_i) + \nu_2(A \cap A_i) = \nu_1'(A \cap A_i) + \nu_2(A \cap A_i)$, and so $\nu_1(A \cap A_i) = \nu_1'(A \cap A_i)$. Thus $\nu_1(A) = \sum_{i \in I} \nu_1(A \cap A_i) = \sum_{i \in I} \nu_1'(A \cap A_i) = \nu_1'(A)$, $A \in \mathcal{A}$. □

Exercise 12.14. Let μ and ν be measures on $\mathcal{A}$ such that $\nu(X) < \infty$. Show that $\nu \ll \mu$ if and only if for each $\varepsilon > 0$ there is $\delta > 0$ such that $A \in \mathcal{A}$ and $\mu(A) < \delta$ imply $\nu(A) < \varepsilon$.

Exercise 12.15. Let μ be a σ-finite measure on $\mathcal{A}$, and let ν_1 and ν_2 be measures on $\mathcal{A}$ such that $\nu_1 \ll \mu$ and $\nu_2 \ll \mu$. Prove the following.

(a) $\frac{d(\nu_1+\nu_2)}{d\mu} = \frac{d\nu_1}{d\mu} + \frac{d\nu_2}{d\mu}$ μ-a.e.

(b) $\frac{d(a\nu_1)}{d\mu} = a\frac{d\nu_1}{d\mu}$ μ-a.e. for any $a \geqslant 0$.

Exercise 12.16. Let X be an uncountable set, let $\mathcal{A}$ be as in (7.3.c), and let μ be the counting measure on $\mathcal{A}$. For $A \in \mathcal{A}$, define $\nu(A) = \begin{cases} 0 & \text{if } A \text{ is countable} \\ \infty & \text{if } A^c \text{ is countable} \end{cases}$. Prove the following.

(a) ν is a measure on $\mathcal{A}$ such that $\nu \ll \mu$.

(b) There is no Radon-Nikodým derivative of ν with respect to μ.

Exercise 12.17. Let μ be the Lebesgue measure on $\mathcal{B}(R)$, and let $f : R \to [0, \infty]$ be a measurable function such that $\int_R f d\mu < \infty$. For $x \in R$, define $F(x) = \int_{-\infty}^x f(t)dt$.

(a) Show that F is a nondecreasing function which is continuous on R.

(b) Let μ_F be as in (10.15). Prove that $\mu_F \ll \mu$ and $f = d\mu_F/d\mu$.

Exercise 12.18. Let μ and ν be σ-finite measures on $\mathcal{A}$ such that $\nu \sim \mu$. Show that $d\nu/d\mu > 0$ μ-a.e. and ν-a.e., $d\mu/d\nu > 0$ μ-a.e. and ν-a.e., and $d\mu/d\nu = 1/(d\nu/d\mu)$ μ-a.e. and ν-a.e.

Exercise 12.19. Let μ, ν_1, and ν_2 be measures on $\mathcal{A}$. Prove the following.

(a) If $\nu_1 \perp \mu$ and $\nu_2 \perp \mu$, then $\nu_1 + \nu_2 \perp \mu$.

(b) If $\nu_1 \ll \mu$ and $\nu_2 \perp \mu$, then $\nu_1 \perp \nu_2$.

(c) If $\nu_1 \ll \mu$ and $\nu_1 \perp \mu$, then $\nu_1 = 0$.

Exercise: Cantor function 12.20. For $n \in N$, consider the sets I_{nk}, $1 \leqslant k \leqslant 2^{n-1}$, and B_n appearing in the construction of the Cantor set as in (10.38). We construct inductively a sequence $\{F_n : n \in N\} \subset [0, 1]^{[0,1]}$ of piecewise linear nondecreasing continuous functions. Define $F_1(0) = 0$, $F_1(1) = 1$, $F_1 = 1/2$ on I_{11}, and complete the graph of F_1 by linear segments. If F_n has been constructed, then we construct F_{n+1} as follows. Define $F_{n+1}(0) = 0$, $F_{n+1}(1) = 1$, $F_{n+1} = F_n$ on $\cup_{i=1}^n B_i$, $F_{n+1} = k/2^{n+1}$ on $I_{n+1,k}$, $1 \leqslant k \leqslant 2^n$, and complete the graph of F_{n+1} by linear segments.

(a) for $n \in N$, show that $|F_{n+1} - F_n| < 1/2^{n+1}$.

(b) Prove that $\{F_n : n \in N\}$ is a Cauchy sequence in $\mathfrak{C}_r([0, 1])$.

(c) Let $F = \lim_n F_n$. The function F is called the *Cantor function*. Show that F is nondecreasing and continuous on $[0, 1]$.

(d) Consider the function $G : R \to [0, 1]$ defined by

$$G(x) = \begin{cases} 0 & \text{if } x < 0 \\ F(x) & \text{if } x \in [0, 1] \\ 1 & \text{if } x > 1 \end{cases}.$$

Prove that μ_G is singular with respect to the Lebesgue measure on $\mathcal{B}(R)$.

Exercise 12.21. Let μ and ν be σ-finite measures on $\mathcal{A}$, let $\nu = \nu_1 + \nu_2$, where $\nu_1 \ll \mu$ and $\nu_2 \perp \mu$, and let $f = d\nu_1/d\mu$. Prove the following.

(a) $\mu \ll \nu$ if and only if $f > 0$ μ-a.e.

(b) $\mu \perp \nu$ if and only if $f = 0$ μ-a.e.

(c) If $\nu(X) < \infty$, then $\mu \ll \nu$ if and only if $\lim_{\beta \searrow 0} \int_X f^\beta d\mu = 1$. [Use (11.46) and (a).]

Exercise 12.22. Let μ and ν be σ-finite measures on $\mathcal{A}$ such that $\nu \ll \mu$, let $f = d\nu/d\mu$, let $\mu = \mu_1 + \mu_2$ be the Lebesgue decomposition of μ with respect to ν, where $\mu_1 \ll \nu$ and $\mu_2 \perp \nu$, and let $g = d\mu_1/d\nu$. Show that $fg = 1$ ν-a.e.

Exercise 12.23. Let μ, ν and ν' be σ-finite measures on $\mathcal{A}$ such that $\nu \ll \mu$ and $\nu' \ll \mu$, and put $f = d\nu/d\mu$ and $f' = d\nu'/d\mu$. For $A \in \mathcal{A}$, define $\nu_1(A) = \int_A f 1_{\{f'>0\}}\, d\mu$ and $\nu_2(A) = \int_A f 1_{\{f'=0\}}\, d\mu$. Prove the following.

(a) $\nu_1 \ll \nu'$ and $\nu_2 \perp \nu'$.

(b) $\nu = \nu_1 + \nu_2$ is the Lebesgue decomposition of ν with respect to ν'.

(c) $\int_X (ff')^{1/2}\, d\mu$ depends solely on ν and ν'.

(d) $\int_X (ff')^{1/2}\, d\mu = 0$ if and only if $\nu \perp \nu'$. [Use (12.21.b).]

Exercise 12.24. Let $\mathcal{A}$ be a σ-algebra, let $\{\mathcal{A}_n : n \in N\}$ be a sequence of σ-algebras such that $\mathcal{A}_n \subset \mathcal{A}_{n+1} \subset \mathcal{A}$, $n \in N$, let μ and ν be measures on $\mathcal{A}$, and put $\mu_n = \mu_{\mathcal{A}_n}$ and $\nu_n = \nu_{\mathcal{A}_n}$, $n \in N$. For $n \in N$, assume that μ_n and ν_n are σ-finite, and let f_n be the Radon-Nikodým derivative of the absolutely continuous part of ν_n with respect to μ_n. Show that $\int_A f_n d\mu \geqslant \int_A f_{n+1}\, d\mu$ for any $A \in \mathcal{A}_n$.

Exercise 12.25. Let μ and ν be as in (12.10), and assume that for each $\varepsilon > 0$ there is $B \in \mathcal{A}$ such that $\nu(B) < \varepsilon$ and $\mu(B^c) < \varepsilon$. Prove that $\nu \perp \mu$.

13 The Spaces $\mathfrak{L}^p$

The function spaces $\mathfrak{L}^p$, which we study in what follows, have remarkable properties and are of great importance in measure theory.

Definition 13.1. Let $(X, \mathcal{A}, \mu)$ be a measure space. For $0 < p < \infty$, the set of all measurable functions $f : X \to R$ such that $\int_X |f|^p \, d\mu < \infty$ will be denoted by $\mathfrak{L}^p(X, \mathcal{A}, \mu)$. When no confusion seems possible, we will write $\mathfrak{L}^p$ instead of $\mathfrak{L}^p(X, \mathcal{A}, \mu)$. For $f \in \mathfrak{L}^p$, set $\|f\|_p = (\int_X |f|^p \, d\mu)^{1/p}$. For $1 \leqslant p < \infty$, the number $\|f\|_p$ is called the *$\mathfrak{L}^p$-norm of f*.

We begin the study of the spaces $\mathfrak{L}^p$ with several important inequalities.

Lemma 13.2 (Young's inequality). *Let $f : [0, \infty[\to [0, \infty[$ be a strictly increasing continuous function such that $f([0, \infty[) = [0, \infty[$, and let $g = f^{-1}$. For $x \in [0, \infty[$, put $F(x) = \int_0^x f(u)du$ and $G(x) = \int_0^x g(v)dv$. Then*

(i) $ab \leqslant F(a) + G(b)$

for any $a, b \in [0, \infty[$.

Proof. Interpreting $F(a)$ and $G(b)$ as areas, the inequality (i) is immediate. □

For $p \in]1, \infty[$, define $p' \in]1, \infty[$ by $\frac{1}{p} + \frac{1}{p'} = 1$.

Corollary 13.3. *For $1 < p < \infty$ and $a, b \in [0, \infty[$, we have*

(i) $ab \leqslant \frac{a^p}{p} + \frac{b^{p'}}{p'}$.

Proof. For $u \in [0, \infty[$, define $f(u) = u^{p-1}$. The function $g = f^{-1}$ is defined by $g(v) = v^{\frac{1}{p-1}}$, $v \in [0, \infty[$. Thus (i) is a special case of Young's inequality. □

Theorem 13.4 (Hölder's inequality). *For $1 < p < \infty$, $f \in \mathfrak{L}^p$ and $g \in \mathfrak{L}^{p'}$, we have $fg \in \mathfrak{L}^1$ and*

(i) $\|fg\|_1 \leqslant \|f\|_p \|g\|_{p'}$.

Proof. If $fg = 0$ μ-a.e., then (i) is immediate. Otherwise we have $\|f\|_p \|g\|_{p'} > 0$, and so, applying (13.3), we may write

$$\frac{|f|}{\|f\|_p} \cdot \frac{|g|}{\|g\|_{p'}} \leqslant \frac{1}{p} \cdot \frac{|f|^p}{\|f\|_p^p} + \frac{1}{p'} \cdot \frac{|g|^{p'}}{\|g\|_{p'}^{p'}}.$$

Hence $fg \in \mathfrak{L}^1$, and

$$\frac{1}{\|f\|_p \|g\|_{p'}} \int_X |fg| \, d\mu \leqslant \frac{1}{p \|f\|_p^p} \int_X |f|^p \, d\mu + \frac{1}{p' \|g\|_{p'}^{p'}} \int_X |g|^{p'} \, d\mu$$

$$= \frac{1}{p} + \frac{1}{p'} = 1.$$

Therefore, $\|fg\|_1 \leqslant \|f\|_p \|g\|_{p'}$. □

For $p = p' = 2$, the inequality $\left|\int_X fg\,d\mu\right| \leqslant \|f\|_2 \|g\|_2$ is called *Schwarz's inequality.*

Corollary 13.5. *If $\mu(X) < \infty$ and $0 < p < q < \infty$, then $\mathfrak{L}^q \subset \mathfrak{L}^p$, and $\|f\|_p \leqslant \|f\|_q (\mu(X))^{\frac{1}{p}-\frac{1}{q}}$ for any $f \in \mathfrak{L}^q$.*

Proof. Put $r = q/p > 1$. Then, for $f \in \mathfrak{L}^q$, applying (13.4), we have $\int_X |f|^p\,d\mu \leqslant (\int_X |f|^{pr}\,d\mu)^{1/r}(\int_X 1^{r'}\,d\mu)^{1/r'} = (\int_X |f|^q\,d\mu)^{p/q}(\mu(X))^{\frac{q-p}{q}}$. This means that $f \in \mathfrak{L}^p$, and $\|f\|_p \leqslant \|f\|_q (\mu(X))^{\frac{q-p}{pq}} = \|f\|_q (\mu(X))^{\frac{1}{p}-\frac{1}{q}}$. □

Remark 13.6. Condition $\mu(X) < \infty$ in (13.5) is essential. To see this, let $X = [1, \infty[$, let μ be the Lebesgue measure on $\mathcal{B}([1, \infty[)$, and let $f(x) = 1/x$, $x \in [1, \infty[$. Then $f \in \mathfrak{L}^2([1, \infty[, \mathcal{B}([1, \infty[), \mu)$ but $f \notin \mathfrak{L}^1([1, \infty[, \mathcal{B}([1, \infty[), \mu)$.

Theorem 13.7 (Minkowski's inequality). *For $1 \leqslant p < \infty$, $f \in \mathfrak{L}^p$ and $g \in \mathfrak{L}^p$, we have $f + g \in \mathfrak{L}^p$ and*

(i) $\|f + g\|_p \leqslant \|f\|_p + \|g\|_p$.

Proof. Since $\int_X |f + g|\,d\mu \leqslant \int_X |f|\,d\mu + \int_X |g|\,d\mu$, the theorem is true for $p = 1$. Assume now that $p > 1$. Since $|f + g|^p \leqslant 2^p(|f|^p + |g|^p)$, it follows that $f + g \in \mathfrak{L}^p$, and so $|f + g|^{p-1} \in \mathfrak{L}^{p'}$. Therefore, using (13.4), we have

$$\begin{aligned}\|f+g\|_p^p &= \int_X |f+g|^p\,d\mu \leqslant \int_X |f+g|^{p-1}|f|\,d\mu + \int_X |f+g|^{p-1}|g|\,d\mu \\ &\leqslant \|f\|_p \left(\int_X |f+g|^{(p-1)p'}\,d\mu\right)^{1/p'} \\ &\quad + \|g\|_p \left(\int_X |f+g|^{(p-1)p'}\,d\mu\right)^{1/p'} \\ &= (\|f\|_p + \|g\|_p)\,\|f+g\|_p^{p/p'}.\end{aligned}$$

Thus $\|f + g\|_p^{p-p/p'} \leqslant \|f\|_p + \|g\|_p$. Since $p - p/p' = 1$, we get (i). □

Theorem 13.8 (Markov's inequality). *Let $f : X \to R$ be a measurable function, let $0 < p < \infty$, and let $0 < a < \infty$. Then*

(i) $\mu(\{|f| \geqslant a\}) \leqslant \int_X |f|^p\,d\mu/a^p$.

Proof. We have $\int_X |f|^p\,d\mu \geqslant \int_{\{|f|\geqslant a\}} |f|^p\,d\mu \geqslant a^p\mu(\{|f| \geqslant a\})$. □

For $p = 2$, the inequality (i) is called *Chebyshev's inequality.*

To prove Jensen's inequality below we need some definitions.

Definitions 13.9. Let $f :]a, b[\to R$ be a function, and let $x \in]a, b[$. If $\lim_{h\searrow 0}(f(x + h) - f(x))/h$ exists, then we say that f has a *right derivative at* x, and we write $f'_+(x) = \lim_{h \searrow 0}(f(x+h) - f(x))/h$. If $\lim_{h \nearrow 0}(f(x+h) - f(x))/h$ exists, then we say that f has a *left derivative at* x, and we write $f'_-(x) = \lim_{h \nearrow 0}(f(x+h) - f(x))/h$. If $f'_+(x)$ and $f'_-(x)$ exist and are equal, then we say that f has a *derivative at* x, and we denote by $f'(x)$ the common value $f'_+(x) = f'_-(x)$.

Remark 13.10. Notation is as in (13.9). If $f'_+(x)$ exists and is finite, then f is right continuous at x, and if $f'_-(x)$ exists and is finite, then f is left continuous at x.

Definition 13.11. Let $f :]a, b[\to R$ be a function. We say that f is *convex* if $f(tx + (1-t)y) \leqslant tf(x) + (1-t)f(y)$ whenever $x, y \in]a, b[$ and $t \in [0, 1]$.

Theorem 13.12. *Let $f :]a, b[\to R$ be a convex function. Then the following assertions hold.*

(i) *$f'_+(x)$ and $f'_-(x)$ exist and are finite for any $x \in]a, b[$.*

(ii) *f is continuous on $]a, b[$.*

(iii) $f'_-(x) \leqslant f'_+(x)$, $x \in]a, b[$.

(iv) *For each $x \in]a, b[$, $f'_+(x) = \inf_{y \in]x,b[} \frac{f(y)-f(x)}{y-x}$ and $f'_-(x) = \sup_{y \in]a,x[} \frac{f(y)-f(x)}{y-x}$.*

(v) *$f'_+(x)$ and $f'_-(x)$ are nondecreasing functions on $]a, b[$.*

(vi) *There exist $\{a_n : n \in N\} \subset R$ and $\{b_n : n \in N\} \subset R$ such that $f(y) = \sup_{n \in N}(a_n y + b_n)$ for any $y \in]a, b[$.*

Proof. Let $x \in]a, b[$ and let $0 < h_1 \leqslant h_2$ be such that $x + h_2, x - h_2 \in]a, b[$. Since f is convex, we have $f(x + h_1) \leqslant \frac{h_2-h_1}{h_2} f(x) + \frac{h_1}{h_2} f(x + h_2)$ and $f(x - h_1) \leqslant \frac{h_2-h_1}{h_2} f(x) + \frac{h_1}{h_2} f(x - h_2)$. Therefore,

$$(f(x + h_1) - f(x))/h_1 \leqslant (f(x + h_2) - f(x))/h_2 \tag{1}$$

and

$$(f(x - h_1) - f(x))/ - h_1 \geqslant (f(x - h_2) - f(x))/ - h_2. \tag{2}$$

(1) and (10.35) show that $f'_+(x)$ exists, and

$$f'_+(x) = \inf_{h \in]0, b-x[} (f(x + h) - f(x))/h. \tag{3}$$

(2) and (10.35) show that $f'_-(x)$ exists, and

$$f'_-(x) = \sup_{h \in]0, x-a[} (f(x - h) - f(x))/ - h. \tag{4}$$

Now let $h, h' > 0$ be such that $x - h, x + h' \in]a, b[$. Since f is convex, we have $f(x) \leqslant \frac{h'}{h+h'} f(x - h) + \frac{h}{h+h'} f(x + h')$, and so

$$(f(x - h) - f(x))/ - h \leqslant (f(x + h') - f(x))/h'. \tag{5}$$

From (3) and (5) it follows that $f'_+(x)$ is finite and from (4) and (5) it follows that $f'_-(x)$ is finite. Thus (i) is proved. Assertion (ii) follows from (i), (13.10) and (10.14.d).

Assertion (iii) follows from (3), (4) and (5), and assertion (iv) follows at once from (3) and (4). Let $x_1, x_2 \in]a, b[$ be such that $x_1 < x_2$. Then, applying (iv), we have $f'_+(x_1) \leqslant \frac{f(x_2)-f(x_1)}{x_2-x_1} = \frac{f(x_1)-f(x_2)}{x_1-x_2} \leqslant f'_-(x_2)$, and so, using (iii), we get $f'_+(x_1) \leqslant f'_+(x_2)$ and $f'_-(x_1) \leqslant f'_-(x_2)$. Hence (v) is proved. Further, according to (iv), we have

$$f(y) \geqslant f(x) + (y-x)f'_+(x), \quad y \in [x, b[, \tag{6}$$

and

$$f(y) \geqslant f(x) + (y-x)f'_-(x), \quad y \in]a, x]. \tag{7}$$

Choose c_x such that $f'_-(x) \leqslant c_x \leqslant f'_+(x)$. Then, by (6) and (7), we have

$$f(y) \geqslant f(x) + (y-x)c_x, \quad y \in]a, b[. \tag{8}$$

Now fix $y \in]a, b[$. Then, using (8), we get

$$f(y) \geqslant \sup_{x \in\,]a,b[\, \cap\, Q} (f(x) + (y-x)c_x). \tag{9}$$

Let $\{y_n : n \in N\} \subset]a, b[\cap Q$ be such that $y_n \to y$. Since f'_+ and f'_- are nondecreasing functions, the sequences $\{f'_+(y_n) : n \in N\}$ and $\{f'_-(y_n) : n \in N\}$ are bounded, and so $\{c_{y_n} : n \in N\}$ is a bounded sequence. Thus, applying (ii), we have

$$f(y) = \lim_n (f(y_n) + (y - y_n)c_{y_n}). \tag{10}$$

From (9) and (10) we infer that $f(y) = \sup_{x \in]a,b[\cap Q}(f(x) + (y-x)c_x) = \sup_{x \in]a,b[\cap Q}(c_x y + f(x) - xc_x)$. Since $]a, b[\cap Q$ is denumerable, assertion (vi) is proved. □

Theorem 13.13 (Jensen's inequality). *Let $f : X \to]a, b[$ be a measurable function, and let $g :]a, b[\to R$ be a convex function. If $0 < \mu(X) < \infty$ and $f \in \mathfrak{L}^1$, then*

(i) $\frac{1}{\mu(X)} \int_X (g \circ f)\, d\mu \geqslant g\left(\frac{1}{\mu(X)} \int_X f\, d\mu\right)$.

Proof. Notice first that, by virtue of (11.16), we have $\frac{1}{\mu(X)} \int_X f d\mu \in]a, b[$. Applying (13.12.vi), let $\{a_n : n \in N\} \subset R$ and $\{b_n : n \in N\} \subset R$ be such that $g(y) = \sup_{n \in N}(a_n y + b_n)$ for any $y \in]a, b[$. Therefore,

$$g \circ f = \sup_{n \in N}(a_n f + b_n) \geqslant a_n f + b_n, \quad n \in N. \tag{1}$$

Since $a_n f + b_n \in \mathfrak{L}^1$, (1) shows that $(g \circ f)^- \in \mathfrak{L}^1$, and so $\int_X (g \circ f)\, d\mu$ exists. Using (1), we have $\frac{1}{\mu(X)} \int_X (g \circ f)\, d\mu \geqslant a_n \frac{1}{\mu(X)} \int_X f\, d\mu + b_n, n \in N$, whence, by (13.12.vi), we get (i). □

For $1 \leqslant p < \infty$, the spaces $\mathfrak{L}^p$ can be endowed with a topological structure as follows. Let $\mathfrak{L}^p$ be as in (13.1), and let $\sim$ be the equivalence relation on $\mathfrak{L}^p$ defined in (11.14.c). Set $L^p = \mathfrak{L}^p/\sim$, and define $d(\widetilde{f}, \widetilde{g}) = \|f - g\|_p$ for $\widetilde{f}, \widetilde{g} \in L^p$. Then

d is a well-defined metric for L^p. Actually, if $f, f', g, g' \in \mathfrak{L}^p$, and $f \sim f', g \sim g'$, then $\|f - g\|_p = \|f' - g'\|_p$ by (11.17). It is trivial that d satisfies (6.1.i) and (6.1.iii). Further, d satisfies (6.1.ii) by (11.16), and d satisfies (6.1.iv) by Minkowski's inequality (13.7). We will always assume that L^p is equipped with this metric. L^p is also a linear space over R. Indeed, for $\alpha \in R$ and $\widetilde{f}, \widetilde{g} \in L^p$, define $\widetilde{f} + \widetilde{g} = \widetilde{f+g}$ and $\widetilde{\alpha f} = \alpha \widetilde{f}$. These definitions are correct, since $f+g \sim f'+g'$ and $\alpha f \sim \alpha f'$ whenever $f, f', g, g' \in \mathfrak{L}^p$ with $f \sim f'$ and $g \sim g'$. Sometimes we shall write $L^p(X, \mathcal{A}, \mu)$ in place of L^p.

Lemma 13.14. *For $0 < p < \infty$, $f \in \mathfrak{L}^p$, and $\varepsilon > 0$, there exists a simple function $g \in \mathfrak{L}^p$ such that $\|f - g\|_p < \varepsilon$.*

Proof. Since $f = f^+ - f^-$, $f^+ \leqslant |f|$ and $f^- \leqslant |f|$, (8.15) shows that there is a sequence $\{f_n : n \in N\} \subset R^X$ of simple functions such that $|f_n| \leqslant |f|$, $n \in N$, and $f_n \to f$. Therefore, $f_n \in \mathfrak{L}^p$ and $|f - f_n|^p \leqslant 2^{p+1} |f|^p \in \mathfrak{L}^1$ for any $n \in N$. Hence, in view of (11.25.ii), we obtain $\lim_n \int_X |f - f_n|^p \, d\mu = \int_X (\lim_n |f - f_n|^p) \, d\mu = 0$. Consequently, there is $n \in N$ such that $\|f - f_n\|_p < \varepsilon$. □

Theorem 13.15. *Assume that there exists a countable algebra $\mathcal{B}$ of subsets of X such that $\mathcal{A} = \sigma(\mathcal{B})$, and that the restriction of μ to $\mathcal{B}$ is σ-finite. Then, for $1 \leqslant p < \infty$, L^p is a separable metric space.*

Proof. Let $f \in \mathfrak{L}^p$ and $\varepsilon > 0$. Using (13.14), we choose a simple function $g = \sum_{i=1}^n a_i 1_{A_i} \in \mathfrak{L}^p$ such that

$$\|f - g\|_p < \varepsilon/3. \tag{1}$$

Without any loss of generality, assume $a_i \neq 0$ for $i = 1, \ldots, n$, and $A_i \cap A_j = \emptyset$ for $i \neq j$. Since $|g|^p = \sum_{i=1}^n |a_i|^p 1_{A_i} \in \mathfrak{L}^1$, we have $\mu(A_i) < \infty$ for $i = 1, \ldots, n$. Hence, applying (10.33), there is $B_i \in \mathcal{B}$ such that $\mu(A_i \Delta B_i) < (\varepsilon/3n\,|a_i|)^p$ for $i = 1, \ldots, n$. Therefore, $\mu(B_i) < \infty$, $i = 1, \ldots, n$, and so $\sum_{i=1}^n a_i 1_{B_i} \in \mathfrak{L}^p$. Applying (13.7), we have

$$\left\| g - \sum_{i=1}^n a_i 1_{B_i} \right\|_p \leqslant \sum_{i=1}^n |a_i| \left\| 1_{A_i} - 1_{B_i} \right\|_p = \sum_{i=1}^n |a_i| \left(\mu(A_i \Delta B_i)\right)^{1/p} < \varepsilon/3. \tag{2}$$

For $i = 1, \ldots, n$, without loss of generality, suppose that $\mu(B_i) > 0$, and choose $b_i \in Q$ such that $|a_i - b_i| < \varepsilon/3n(\mu(B_i))^{1/p}$. Evidently, $\sum_{i=1}^n b_i 1_{B_i} \in \mathfrak{L}^p$. Using (13.7), we get

$$\begin{aligned} \left\| \sum_{i=1}^n a_i 1_{B_i} - \sum_{i=1}^n b_i 1_{B_i} \right\|_p &\leqslant \sum_{i=1}^n |a_i - b_i| \left\| 1_{B_i} \right\|_p \\ &= \sum_{i=1}^n |a_i - b_i| \left(\mu(B_i)\right)^{1/p} < \varepsilon/3. \end{aligned} \tag{3}$$

Combining (1), (2) and (3), we have

$$\left\| f - \sum_{i=1}^{n} b_i 1_{B_i} \right\|_p < \varepsilon. \tag{4}$$

Since the set of all functions of the form $\sum_{i=1}^{n} b_i 1_{B_i}$, where $b_i \in Q$ and $B_i \in \mathcal{B}$ for $i = 1, \ldots, n$, is denumerable, (4) shows that L^p is a separable metric space. □

The next theorem has important applications.

Theorem 13.16. *For $1 \leqslant p < \infty$, L^p is a complete metric space.*

Proof. Let $\{\widetilde{f}_n : n \in N\} \subset L^p$ be a Cauchy sequence. Choose a strictly increasing sequence $\{n_k : k \in N\} \subset N$ such that, for each $k \in N$, $\left\| f_m - f_{n_k} \right\|_p < 2^{-k}$, $m \geqslant n_k$. Particularly, we have $\left\| f_{n_{k+1}} - f_{n_k} \right\|_p < 2^{-k}$ for any $k \in N$, and so

$$\sum_{k \geqslant 1} \left\| f_{n_{k+1}} - f_{n_k} \right\|_p < 1. \tag{1}$$

For $k \in N$, define $g_k = \left| f_{n_1} \right| + \sum_{i=1}^{k} \left| f_{n_{i+1}} - f_{n_i} \right|$. Using (13.7) and (1), we have

$$\int_X g_k^p d\mu = \|g_k\|_p^p \leqslant \left(\left\| f_{n_1} \right\|_p + \sum_{i=1}^{k} \left\| f_{n_{i+1}} - f_{n_i} \right\|_p \right)^p < (\left\| f_{n_1} \right\|_p + 1)^p \tag{2}$$

for any $k \in N$. Let $g = \left| f_{n_1} \right| + \sum_{i \geqslant 1} \left| f_{n_{i+1}} - f_{n_i} \right|$. Applying the monotone convergence theorem (11.22) and (2), we get $\int_X g^p \, d\mu = \int_X (\lim_k g_k^p) \, d\mu = \lim_k \int_X g_k^p \, d\mu \leqslant (\left\| f_{n_1} \right\|_p + 1)^p < \infty$, and so $g^p \in \mathfrak{L}^1$. Hence, in view of (11.15), $\left| f_{n_1} \right| + \sum_{i \geqslant 1} \left| f_{n_{i+1}} - f_{n_i} \right| < \infty$ μ-a.e. Therefore, there is $A \in \mathcal{A}$ such that $\mu(A^c) = 0$, and the series $f_{n_1}(x) + \sum_{i \geqslant 1}(f_{n_{i+1}}(x) - f_{n_i}(x))$ is convergent for any $x \in A$. Consequently, for each $x \in A$, there is $f(x) \in R$ such that $f_{n_k}(x) \to f(x)$. Define $f(x) = 0$ for any $x \in A^c$. Obviously, f is an $\mathcal{A}$-measurable function. Now let $\varepsilon > 0$. Select j sufficiently large that $\|f_m - f_n\|_p < \varepsilon$ for $m, n \geqslant n_j$. Therefore, for $m \geqslant n_j$ and $k \geqslant j$, we get

$$\left\| f_m - f_{n_k} \right\|_p < \varepsilon. \tag{3}$$

Applying Fatou's lemma (11.24) and (3), we have

$$\int_X |f - f_m|^p \, d\mu = \int_X (\liminf_k \left| f_{n_k} - f_m \right|^p) \, d\mu$$

$$\leqslant \liminf_k \int_X \left| f_{n_k} - f_m \right|^p \, d\mu \leqslant \varepsilon^p. \tag{4}$$

Thus, for $m \geqslant n_j$, $f - f_m \in \mathfrak{L}^p$, and so $f = f - f_m + f_m \in \mathfrak{L}^p$. (4) shows that $\|f - f_m\|_p \leqslant \varepsilon$ for any $m \geqslant n_j$, and so $\|f - f_m\|_p \to 0$. Consequently, L^p is complete. □

Exercise 13.17. If $f \in R^X$ is a measurable function, define $\|f\|_\infty = \inf\{a > 0 : \mu(\{|f| > a\}) > 0\}$. Let $\alpha \in R$, and let $f, g \in R^X$ be measurable functions. Prove the following.

(a) $|f| \leqslant \|f\|_\infty$ μ-a.e.

(b) $\|\alpha f\|_\infty = |\alpha| \, \|f\|_\infty$.

(c) $\|f+g\|_\infty \leqslant \|f\|_\infty + \|g\|_\infty$.

(d) $\|fg\|_\infty \leqslant \|f\|_\infty \|g\|_\infty$.

Exercise 13.18. Let $\mu(X) < \infty$, and let $f \in \bigcap_{p\in]r,\infty[} \mathfrak{L}^p$. Show that $\lim_{p\to\infty} \|f\|_p = \|f\|_\infty$.

Exercise: $\mathfrak{L}^\infty$ space 13.19. The set of all measurable functions $f : X \to R$ such that $\|f\|_\infty < \infty$ will be denoted by $\mathfrak{L}^\infty$. Assume that $\mu(X) < \infty$. Prove the following.

(a) $\mathfrak{L}^\infty \subset \mathfrak{L}^p$ for $0 < p < \infty$.

(b) $\mathfrak{L}^\infty$ may be different from $\cap_{p\in]0,\infty[} \mathfrak{L}^p$.

Exercise 13.20. Let $\mathfrak{L}^\infty$ be as in (13.19), and let $\sim$ be the equivalence relation on $\mathfrak{L}^\infty$ defined as in (11.14.c). Denote $L^\infty = \mathfrak{L}^\infty/\sim$.

(a) For $\widetilde{f}, \widetilde{g} \in L^\infty$, put $d(\widetilde{f}, \widetilde{g}) = \|f - g\|_\infty$. Show that d is a well-defined metric for L^∞.

(b) Let $\{\widetilde{f_n} : n \in N\} \subset L^\infty$ and $\widetilde{f} \in L^\infty$. Prove that $d(\widetilde{f}, \widetilde{f_n}) \to 0$ if and only if there is $A \in \mathcal{A}$ such that $\mu(A^c) = 0$ and $f_n|_A \xrightarrow{u} f|_A$.

(c) Let $\{\widetilde{f_n} : n \in N\} \subset L^\infty$. Show that $\{\widetilde{f_n} : n \in N\}$ is a Cauchy sequence if and only if there is $A \in \mathcal{A}$ such that $\mu(A^c) = 0$ and $\{f_n|_A : n \in N\}$ is a Cauchy sequence in $\mathfrak{B}_r(A)$.

(d) Use (c) and (6.33) to prove that L^∞ is a complete metric space.

Exercise 13.21. Let $a_i, b_i \in R$ for $i = 1, \ldots, n$, and let $p, p' \in]1, \infty[$ with $\frac{1}{p} + \frac{1}{p'} = 1$.

(a) Show that $\sum_{i=1}^n |a_i b_i| \leqslant (\sum_{i=1}^n |a_i|^p)^{1/p} (\sum_{i=1}^n |b_i|^{p'})^{1/p'}$. [Apply Hölder's inequality (13.4).]

(b) Prove that $\left|\frac{1}{n}\sum_{i=1}^n a_i\right|^p \leqslant \frac{1}{n}\sum_{i=1}^n |a_i|^p$. [Use (a).]

Exercise 13.22. Let $p, p' \in]1, \infty[$ be such that $\frac{1}{p} + \frac{1}{p'} = 1$, let $f_n \in \mathfrak{L}^p$, $n \in N$, $f \in \mathfrak{L}^p$ be such that $\|f - f_n\|_p \to 0$, and let $g_n \in \mathfrak{L}^{p'}$, $n \in N$, $g \in \mathfrak{L}^{p'}$ be such that $\|g - g_n\|_{p'} \to 0$. Prove that $\|fg - f_n g_n\|_1 \to 0$.

Exercise 13.23. (a) If $0 < p < 1$, and $f, g \in \mathfrak{L}^p$, then $\int_X |f+g|^p \, d\mu \leqslant \int_X |f|^p \, d\mu + \int_X |g|^p \, d\mu$, and so $f + g \in \mathfrak{L}^p$.

(b) If $1 \leqslant p < \infty$, and $f, g \in \mathfrak{L}^p$, then we have $\int_X |f+g|^p \, d\mu \leqslant 2^{p-1}(\int_X |f|^p \, d\mu + \int_X |g|^p \, d\mu)$.

Exercise 13.24. Let $f : [a, b] \to R$ be a nondecreasing function. For $x \in]a, b[$, define $g(x) = \int_a^x f(u) du$. Show that g is a convex function.

Exercise 13.25. Let $f : X \to]a, b[$ and $g :]a, b[\to [0, \infty[$ be measurable functions, and let $\alpha \in]a, b[$. Prove the following.

(a) If g is nonincreasing, then $g(\alpha)\mu(\{f \leqslant \alpha\}) \leqslant \int_X (g \circ f) \, d\mu$.

(b) If g is nondecreasing, then $g(\alpha)\mu(\{f \geqslant \alpha\}) \leqslant \int_X (g \circ f)\, d\mu$.

(c) If $a = -b < 0, \alpha \in [0, b[$ and g is even and nondecreasing on $[0, b[$, then $g(\alpha)\mu(\{|f| \geqslant \alpha\}) \leqslant \int_X (g \circ f)\, d\mu$.

Exercise 13.26. Notation is as in (12.23). Assume that $\nu(X) = \nu'(X) = 1$ and put $\rho(\nu, \nu') = \int_X (ff')^{1/2}\, d\mu$. Prove the following.

(a) $\rho(\nu, \nu') \leqslant 1$.

(b) $2(1 - \rho(\nu, \nu')) \leqslant \left|\nu - \nu'\right|(X) \leqslant 2(1 - \rho(\nu, \nu')^2)^{1/2}$. [Use (11.47.d).]

(c) $\rho(\nu, \nu') = 1$ if and only if $\nu = \nu'$.

(d) $\rho(\nu, \nu') = \inf\{\sum_{i \in I} (\nu(A_i)\nu'(A_i))^{1/2} : \{A_i : i \in I\} \subset \mathcal{A}$ is a countable partition of $X\}$. [Hints. If $\{A_i : i \in I\} \subset \mathcal{A}$ is a countable partition of X, then $\rho(\nu, \nu') \leqslant \sum_{i \in I} (\nu(A_i)\nu'(A_i))^{1/2}$ by Schwarz's inequality. For the opposite inequality, if $t > 1$ and $\mu = \nu + \nu'$, show that $\rho(\nu, \nu') \geqslant t^{-2} \sum_{m,n \in N} (\nu(A_{mn})\nu'(A_{mn}))^{1/2}$, where $A_{mn} = \{t^{-2m} < f \leqslant t^{-2(m-1)}\} \cap \{t^{-2n} < f' \leqslant t^{-2(n-1)}\}$, $m, n \in N$.]

Exercise 13.27. Let $f : X \to \overline{R}$ be a measurable function.

(a) Assume that $\mu(X) < \infty$. Show that $f \in \mathfrak{L}^p$ if and only if $\sum_{n \geqslant 1} n^{p-1} \mu(\{|f| \geqslant n\}) < \infty$.

(b) Prove that one of the implications in (a) may fail if $\mu(X) = \infty$.

Exercise 13.28. Assume $0 \leqslant f \in \mathfrak{L}^1$, and let $\varepsilon > 0$. Show that there exists $A \in \mathcal{A}$ such that $\mu(A) < \infty$and $\int_{A^c} f\, d\mu < \varepsilon$. [Use (11.25) and(13.8).]

Exercise 13.29. Let $f : X \to [0, \infty[$ be a measurable function with $\mu(\{f > 0\}) \neq 0$, and put $g(x) = \log \int_X f^x\, d\mu$, $x > 0$. Apply Hölder's inequality (13.4) to show that g is convex on $]0, \infty[$.

Exercise 13.30. Let $(X, \mathcal{A})$ be a measurable space, and let P be the set of all probabilities on $\mathcal{A}$. For $P, Q \in$P, define $H(P, Q) = (\int_X (\sqrt{p} - \sqrt{q})^2\, d\mu)^{1/2}$, where $p = dP/d\mu$, $q = dQ/d\mu$, and μ is any σ-finite measure on $\mathcal{A}$ such that $P + Q \ll \mu$. Prove the following.

(a) H is a metric for P independent of the choice of μ. [Use (12.23.c).] H is called the *Hellinger distance*.

(b) If $\mathcal{C} \subset \mathcal{A}$ is a σ-algebra, then $H(P_{\mathcal{C}}, Q_{\mathcal{C}}) \leqslant H(P, Q)$. [Use (13.26.d).]

(c) If $(Y, \mathcal{B})$ is a measurable space, and $h : X \to Y$ is $(\mathcal{A}, \mathcal{B})$-measurable, then $H(Ph^{-1}, Qh^{-1}) \leqslant H(P, Q)$. [Use (b).]

Exercise 13.31. Let $f \in \mathfrak{L}^p(R, \mathcal{B}(R), \lambda)$ with $1 \leqslant p < \infty$, where λ is the Lebesgue measure on $\mathcal{B}(R)$. Prove that $\int_R |f(x + h) - f(x)|^p\, dx \to 0$ as $h \to 0$. [Hints. If $f = g = \sum_{i=1}^n b_i 1_{]c_i, d_i]}$, where $b_i \in R$ and $]c_i, d_i]$ are bounded and disjoint, then $\int_R |g(x + h) - g(x)|^p\, dx \leqslant h \sum_{i=1}^n (|\alpha_i| + |\beta_i|)$, where α_i and β_i are the sizes of the jumps of g at c_i and d_i, respectively. For the general case, let $\varepsilon > 0$ and use (7.70) and the proof of (13.15) to select a function g as above with $\int_R |f(x) - g(x)|^p\, dx < \varepsilon$. Choose $\delta > 0$ so that $\int_R |g(x + h) - g(x)|^p\, dx < \varepsilon$ for $|h| < \delta$. Then, by (10.17.c) and (11.53),

$$\int_R |f(x + h) - f(x)|^p\, dx \leqslant \int_R |f(x + h) - g(x + h)|^p\, dx$$

$$+\int_R |g(x+h)-g(x)|^p\,dx$$

$$+\int_R |f(x)-g(x)|^p\,dx = 2\int_R |f(x)-g(x)|^p\,dx$$

$$+\int_R |g(x+h)-g(x)|^p\,dx < 3\varepsilon$$

for $|h| < \delta$.]

Exercise 13.32. Let $f :]a,b[\to R$ be a convex function. Show that there exists a nonincreasing sequence $\{f_n : n \in N\} \subset R^{]a,b[}$ of piecewise linear convex functions such that $f_n \to f$. [Hint. Let the graph of f_n be a polygonal line formed from successive chords.]

Exercise 13.33. Let M be a closed linear subspace of L^2, and let $f_0 \in \mathfrak{L}^2$. Prove that f_0 has an a.e. unique decomposition $f_0 = g_0 + h_0$, where $\widetilde{g_0} \in M$ and $\int_X gh_0\,d\mu = 0$ whenever $\widetilde{g} \in M$. The next steps may be helpful.

(a) Set $d = \inf\{\|f_0 - g\|_2 : \widetilde{g} \in M\}$ and choose $\{\widetilde{g_n} : n \in N\} \subset M$ such that $\|f_0 - g_n\|_2 \to d$. Then $\{\widetilde{g_n} : n \in N\}$ is a Cauchy sequence. [Hint. For $m, n \in N$, $\|g_m - g_n\|_2^2 = 2\|f_0 - g_m\|_2^2 + 2\|f_0 - g_n\|_2^2 - 4\|f_0 - (g_m + g_n)/2\|_2^2 \leqslant 2\|f_0 - g_m\|_2^2 + 2\|f_0 - g_n\|_2^2 - 4d^2$.]

(b) There is $\widetilde{g_0} \in M$ with $\|g_0 - g_n\|_2 \to 0$, and $\|f_0 - g_0\|_2 = d$. [Use (a), (13.16), and (6.15).]

(c) Let $h_0 = f_0 - g_0$. Then $\int_X gh_0\,d\mu = 0, \widetilde{g} \in M$. [Hint. For $\alpha \in R$ and $\widetilde{g} \in M$, $\|f_0 - g_0\|_2^2 \leqslant \|f_0 - (g_0 - \alpha g)\|_2^2 = \|f_0 - g_0\|_2^2 + 2\alpha\int_X (f_0 - g_0)g\,d\mu + \alpha^2\|g\|_2^2$, whence $0 \leqslant 2\alpha\int_X gh_0\,d\mu + \alpha^2\|g\|_2^2$. Hence $2\int_X gh_0\,d\mu + \alpha\|g\|_2^2 \geqslant 0$ or $\leqslant 0$ according as $\alpha > 0$ or < 0.]

(d) If $f_0 = g_0' + h_0'$, where $\widetilde{g_0'} \in M$ and $\int_X gh_0'\,d\mu = 0, \widetilde{g} \in M$, then $g_0 = g_0'$ a.e. and $h_0 = h_0'$ a.e. [Hint. $g_0 - g_0' = h_0 - h_0'$, and so $\int_X (g_0 - g_0')^2\,d\mu = 0$.]

14 Convergence for Sequences of Measurable Functions

In (5.1) and (6.39) we defined two kinds of convergence for sequences of functions. In this section, involving the notion of a measure, we introduce three new types of convergence for sequences of measurable functions and examine the relationship among them.

Definitions 14.1. Let $f_n : X \to \overline{R}$, $n \in N$, and $f : X \to \overline{R}$ be functions. We say that the sequence $\{f_n : n \in N\}$ *converges μ-a.e.* to f if there exists $A \in \mathcal{A}$ such that $\mu(A) = 0$ and $\{f_n : n \in N\}$ converges to f on A^c. Where no confusion seems possible, we will write $f_n \overset{a.e.}{\to} f$ if $\{f_n : n \in N\}$ converges μ-a.e. to f. We say that the sequence $\{f_n : n \in N\}$ is *Cauchy μ-a.e.*if there is $A \in \mathcal{A}$ with $\mu(A) = 0$ such that $\{f_n(x) : n \in N\} \subset R$ and $\{f_n(x) : n \in N\}$ is Cauchy for each $x \in A^c$.

Remarks 14.2. Let $f_n, n \in N$, and f be as in (14.1).

(a) Let $f' : X \to \overline{R}$ be a function. If $f_n \overset{a.e.}{\to} f$ and $f_n \overset{a.e}{\to} f'$, then $f = f'$ μ-a.e.

(b) For $n \in N$, let $f_n' : X \to \overline{R}$ be such that $f_n = f_n'$ μ-a.e. If $f_n \overset{a.e.}{\to} f$, then $f_n' \overset{a.e}{\to} f$.

(c) For each $n \in N$, assume that f_n is measurable. If $f_n \overset{a.e.}{\to} f$, and μ is complete, then f is measurable. This follows at once from (8.12.ii) and (11.38).

(d) If $\{f_n : n \in N\}$ is Cauchy μ-a.e., then $\{f_n : n \in N\}$ converges μ-a.e. to some $f : X \to R$.

Theorem (14.4) presents a somewhat surprising result. We need the next lemma.

Lemma 14.3. *Suppose that $\mu(X) < \infty$, and let $f_n : X \to R, n \in N$, and $f : X \to R$ be measurable functions. Then the following are equivalent:*

(i) $f_n \overset{a.e.}{\to} f$;

(ii) $\mu(\limsup_n \{|f_n - f| \geqslant 1/m\}) = 0$ *for each $m \in N$;*

(iii) $\lim_n \mu(\cup_{k \geqslant n}\{|f_k - f| \geqslant 1/m\}) = 0$ *for each $m \in N$;*

(iv) $\lim_n \mu(\sup_{k \geqslant n} |f_k - f| \geqslant 1/m) = 0$ *for each $m \in N$.*

Proof. We have $\{x : f_n(x) \nrightarrow f(x)\} = \cup_{m \geqslant 1} \cap_{n \geqslant 1} \cup_{k \geqslant n}\{|f_k - f| \geqslant 1/m\}$. Thus $f_n \overset{a.e.}{\to} f$ if and only if $0 = \mu(\cap_{n \geqslant 1} \cup_{k \geqslant n}\{|f_k - f| \geqslant 1/m\}) = \lim_n \mu(\cup_{k \geqslant n}\{|f_k - f| \geqslant 1/m\}) = \lim_n \mu(\sup_{k \geqslant n} |f_k - f| \geqslant 1/m) = 0$ for any $m \in N$. □

Theorem 14.4 (Egorov). *Let μ, $f_n, n \in N$, and f be as in* (14.3). *If $f_n \overset{a.e.}{\to} f$, then, for each $\varepsilon > 0$, there is $A \in \mathcal{A}$ such that $\mu(A) < \varepsilon$ and $f_n|_{A^c} \overset{u}{\to} f|_{A^c}$.*

Proof. Let $\varepsilon > 0$. For each $m \in N$, in view of (14.3), there is $n_m \in N$ such that $\mu(\cup_{k \geqslant n_m}\{|f_k - f| \geqslant 1/m\}) < \varepsilon/2^m$. Set $A_m = \cup_{k \geqslant n_m}\{|f_k - f| \geqslant 1/m\}$, $m \in N$, and $A = \cup_{m \in N} A_m$. Obviously, $\mu(A) < \varepsilon$. Now let $\delta > 0$, and choose $m \in N$ such that $1/m \leqslant \delta$. Then, for any $x \in A^c$, we have $|f_k(x) - f(x)| < 1/m \leqslant \delta$ whenever $k \geqslant n_m$. Hence $f_n|_{A^c} \overset{u}{\to} f|_{A^c}$. □

Remark 14.5. The condition $\mu(X) < \infty$ in (14.4) is essential. To see this, let $X = N$, let μ be the counting measure on $\mathcal{P}(N)$, and let $f_n = 1_{\{1,\ldots,n\}}, n \in N$. Then $f_n \to 1$, but the conclusion of Egorov's theorem does not hold.

We now introduce a new kind of convergence for sequences of measurable functions.

Definitions 14.6. Let $f_n : X \to R$, $n \in N$, and $f : X \to \overline{R}$ be measurable functions. We say that the sequence $\{f_n : n \in N\}$ *converges in μ-measure to* f, and we write $f_n \overset{\mu}{\to} f$, if for every $\varepsilon > 0$ we have $\mu(\{|f - f_n| \geqslant \varepsilon\}) \to 0$. We say that $\{f_n : n \in N\}$ is *Cauchy in μ-measure* if for any $\varepsilon, \delta > 0$ there is $n_{\varepsilon,\delta} \in N$ such that $m, n \geqslant n_{\varepsilon,\delta}$ implies $\mu(\{|f_m - f_n| \geqslant \varepsilon\}) < \delta$.

Remarks 14.7. Let $f_n, n \in N$, and f be as in (14.6).

(a) If $f_n \overset{\mu}{\to} f$, then f is finite μ-a.e. Actually, for $\varepsilon > 0$, we have $\mu(\{|f| = \infty\}) = \mu(\{|f - f_n| = \infty\}) \leqslant \mu(\{|f - f_n| \geqslant \varepsilon\}) \to 0$.

(b) Let $f' : X \to \overline{R}$ be a measurable function. If $f_n \overset{\mu}{\to} f$ and $f_n \overset{\mu}{\to} f'$, then $f = f'$ μ-a.e. Actually, for any $\varepsilon > 0$, we have $\mu(\{|f - f'| \geqslant \varepsilon\}) \leqslant \mu(\{|f - f_n| \geqslant$

$\varepsilon/2\}) + \mu(\{|f' - f_n| \geqslant \varepsilon/2\}) \to 0$, and so $\mu(\{f \neq f'\}) = \mu(\cup_{m\in N}\{|f - f'| \geqslant 1/m\}) = 0$.

(c) For $n \in N$, let $f_n' : X \to R$ be a measurable function such that $f_n = f_n'$ μ-a.e. If $f_n \xrightarrow{\mu} f$, then $f_n' \xrightarrow{\mu} f$.

(d) If $f_n \xrightarrow{\mu} f$, then $f_n 1_A \xrightarrow{\mu} f 1_A$ for any $A \in \mathcal{A}$.

(e) $\{f_n : n \in N\}$ is Cauchy in μ-measure if and only if $\sup_{m>n} \mu(\{|f_m - f_n| \geqslant \varepsilon\}) \to 0$ for any $\varepsilon > 0$.

(f) Assume $\mu(X) < \infty$. Then $f_n \xrightarrow{\mu} 0$ if and only if $\int_X (|f_n| \wedge 1)\, d\mu \to 0$. Actually, if $f_n \xrightarrow{\mu} 0$ and $\varepsilon > 0$, choose $n_\varepsilon \in N$ such that $\mu(\{|f_n| \geqslant \varepsilon\}) < \varepsilon$ for $n \geqslant n_\varepsilon$. Then $\int_X (|f_n|\wedge 1)\, d\mu = \int_{\{|f_n|<\varepsilon\}} (|f_n|\wedge 1)\, d\mu + \int_{\{|f_n|\geqslant\varepsilon\}} (|f_n|\wedge 1)\, d\mu \leqslant \varepsilon\mu(X) + \mu(\{|f_n| \geqslant \varepsilon\}) < \varepsilon(\mu(X) + 1)$, $n \geqslant n_\varepsilon$. Conversely, if $\int_X (|f_n| \wedge 1)\, d\mu \to 0$ and $0 < \varepsilon \leqslant 1$, then $\mu(\{|f_n| \geqslant \varepsilon\}) = \mu(\{|f_n| \wedge 1 \geqslant \varepsilon\}) \leqslant \int_X (|f_n| \wedge 1)\, d\mu/\varepsilon \to 0$ by Markov's inequality (13.8).

Theorem 14.8 (Riesz). *Let $f_n : X \to R$, $n \in N$, and $f : X \to \overline{R}$ be measurable functions such that $f_n \xrightarrow{\mu} f$. Then there exists a subsequence $\{f_{n_k} : k \in N\}$ such that $f_{n_k} \xrightarrow{a.e.} f$.*

Proof. Choose $n_1 \in N$ such that $\mu(\{|f - f_{n_1}| \geqslant 1\}) < 2^{-1}$. If $n_1, \ldots, n_k$ have been chosen, then choose $n_{k+1} \in N$ such that $n_{k+1} > n_k$ and $\mu(\{|f - f_{n_{k+1}}| \geqslant 1/(k+1)\}) < 2^{-k-1}$. For $i \in N$, define $A_i = \cup_{k\geqslant i}\{|f - f_{n_k}| \geqslant 1/k\}$. Put $A = \cap_{i\in N} A_i$. Since $\{A_i : i \in N\}$ is a nonincreasing sequence, and $\mu(A_1) < \sum_{k\geqslant 1} 2^{-k} = 1$, (9.10) shows that $\mu(A) = \lim_i \mu(A_i) \leqslant \lim_i \sum_{k\geqslant i} 2^{-k} = 0$. Now let $x \in A^c = \cup_{i\in N} A_i^c$. Then there is i_x such that $x \in A_{i_x}^c = \cap_{k\geqslant i_x}\{|f - f_{n_k}| < 1/k\}$. Thus, for $k \geqslant i_x$, we have $|f(x) - f_{n_k}(x)| < 1/k$, and so $f_{n_k}(x) \to f(x)$. Therefore, $f_{n_k} \xrightarrow{a.e.} f$. □

Theorem 14.9 (Lebesgue). *Assume $\mu(X) < \infty$. Let $f_n : X \to R$, $n \in N$, and $f : X \to R$ be measurable functions such that $f_n \xrightarrow{a.e.} f$. Then $f_n \xrightarrow{\mu} f$.*

Proof. Let $\varepsilon > 0$ and $\delta > 0$. Applying (14.4), choose $A \in \mathcal{A}$ such that $\mu(A) < \varepsilon$and $f_n|_{A^c} \xrightarrow{u} f|_{A^c}$. Then there is $n(\varepsilon, \delta) \in N$ such that $A^c \subset \{|f - f_n| < \delta\}$ for any $n \geqslant n(\varepsilon, \delta)$. Therefore, for $n \geqslant n(\varepsilon, \delta)$, we have $\mu(\{|f - f_n| \geqslant \delta\}) \leqslant \mu(A) < \varepsilon$. Hence $f_n \xrightarrow{\mu} f$. □

Remarks 14.10. (a) The condition $\mu(X) < \infty$ in (14.9) is essential. To see this, let $X = R$, let μ be the Lebesgue measure on $\mathcal{B}(R)$, and let $f_n = 1_{[n,n+1]}$, $n \in N$. Then $f_n \to 0$, but $\mu(\{|0 - f_n| \geqslant 1\}) = \mu([n, n+1]) = 1 \nrightarrow 0$.

(b) There exist sequences of measurable functions that converge in μ-measure, but do not converge μ-a.e. For example, let $X = [0, 1]$, and let μ be the Lebesgue measure on $\mathcal{B}([0, 1])$. For $n \in N$, define $f_n = 1_{[i2^{-k},(i+1)2^{-k}]}$, where $n = 2^k + i$, $0 \leqslant i < 2^k$. For each $\varepsilon > 0$, we have $\{f_n \geqslant \varepsilon\} \subset [i2^{-k}, (i+1)2^{-k}]$, and so $\mu(\{f_n \geqslant \varepsilon\}) \leqslant 2^{-k} \to 0$. Thus $f_n \xrightarrow{\mu} 0$. On the other hand, it is easily seen that $\liminf_n f_n = 0$ and $\limsup_n f_n = 1$, and so the sequence $\{f_n(x) : n \in N\}$ is convergent for no $x \in [0, 1]$.

Theorem 14.11. *Let μ, $f_n, n \in N$, and f be as in* (14.3). *Then the following are equivalent:*

(i) $f_n \xrightarrow{\mu} f$;

(ii) *each subsequence* $\{f_{n_k} : k \in N\}$ *contains a subsubsequence* $\{f_{n_{k_l}} : l \in N\}$ *such that* $f_{n_{k_l}} \xrightarrow{a.e.} f$.

Proof. If $f_n \xrightarrow{\mu} f$, then $f_{n_k} \xrightarrow{\mu} f$ for every subsequence $\{f_{n_k} : k \in N\}$. Thus (i) implies (ii) by (14.8). Suppose now that (ii) holds, and let $\varepsilon > 0$. For $n \in N$, set $a_n = \mu(\{|f - f_n| \geqslant \varepsilon\})$. Then, applying (14.9), each subsequence $\{a_{n_k} : k \in N\}$ contains a subsubsequence $\{a_{n_{k_l}} : l \in N\}$ such that $a_{n_{k_l}} \to 0$. Therefore, according to (5.54), $a_n \to 0$. Thus (ii) implies (i). □

In what follows we introduce another kind of convergence.

Definition 14.12. For $0 < p < \infty$, let $\{f_n : n \in N\} \subset \mathfrak{L}^p$ and $f \in \mathfrak{L}^p$. We say that the sequence $\{f_n : n \in N\}$ *converges in* $\mathfrak{L}^p$ *to* f, and we write $f_n \xrightarrow{\mathfrak{L}^p} f$, if $\|f - f_n\|_p \to 0$.

Remarks 14.13. Let $\{f_n : n \in N\}$ and f be as in (14.12).

(a) Let $f' \in \mathfrak{L}^p$. If $f_n \xrightarrow{\mathfrak{L}^p} f$ and $f_n \xrightarrow{\mathfrak{L}^p} f'$, then $f = f'$ μ-a.e. Actually, for $0 < p < 1$, (13.23.a) shows that $\int_X |f - f'|^p \, d\mu \leqslant \int_X |f - f_n|^p \, d\mu + \int_X |f' - f_n|^p \, d\mu \to 0$. For $1 \leqslant p < \infty$, the assertion is a consequence of Minkowski's inequality (13.7).

(b) For $n \in N$, let $f'_n \in \mathfrak{L}^p$ be such that $f_n = f'_n$ μ-a.e. If $f_n \xrightarrow{\mathfrak{L}^p} f$, then $f'_n \xrightarrow{\mathfrak{L}^p} f$.

(c) If $f_n \xrightarrow{\mathfrak{L}^p} f$, then $\int_X |f_n|^p \, d\mu \to \int_X |f|^p \, d\mu$. Actually, for $0 < p < 1$, (13.23.a) shows that $\left|\int_X |f|^p \, d\mu - \int_X |f_n|^p \, d\mu\right| \leqslant \int_X |f - f_n|^p \, d\mu \to 0$. For $p \geqslant 1$, the assertion follows immediately from (13.7).

(d) If $\mu(X) < \infty$ and $0 < q < p$, then, according to (13.5), $f_n \xrightarrow{\mathfrak{L}^q} f$ whenever $f_n \xrightarrow{\mathfrak{L}^p} f$.

(e) If $p \geqslant 1$ and $f_n \xrightarrow{\mathfrak{L}^p} f$, then $\int_X f_n \, d\mu \to \int_X f \, d\mu$.

Theorem 14.14. *For* $0 < p < \infty$, *let* $\{f_n : n \in N\} \subset \mathfrak{L}^p$ *and* $f \in \mathfrak{L}^p$ *be such that* $f_n \xrightarrow{\mathfrak{L}^p} f$. *Then* $f_n \xrightarrow{\mu} f$.

Proof. For each $\varepsilon > 0$, in view of Markov's inequality (13.8), we have $\mu(\{|f - f_n| \geqslant \varepsilon\}) \leqslant \int_X |f - f_n|^p \, d\mu / \varepsilon^p \to 0$. □

Remarks 14.15. (a) There are sequences in $\mathfrak{L}^p$ which converge either μ-a.e. or in μ-measure, but do not converge in $\mathfrak{L}^p$. For example, let $X = [0, 1]$ and let μ be the Lebesgue measure on $\mathcal{B}([0, 1])$. For $n \in N$, define $f_n = e^n 1_{[0,1/n]}$. Obviously, $f_n \in \mathfrak{L}^p$ for $0 < p < \infty$. Then $f_n \xrightarrow{a.e.} 0$, and so $f_n \xrightarrow{\mu} 0$ by (14.9). On the other hand, $\|f_n\|^p = \int_{[0,1]} f_n^p \, d\mu = e^{np}/n \nrightarrow 0$.

(b) There exist sequences in $\mathfrak{L}^p$ that converge in $\mathfrak{L}^p$, but do not converge μ-a.e. For example, let $(X, \mathcal{A}, \mu)$ and $f_n, n \in N$, be as in (14.10.b). Then $\|f_n\|_p^p = 2^{-k} \to 0$, but $\{f_n : n \in N\}$ does not converge μ-a.e. to 0.

(c) Let $\{f_n : n \in N\}$ and f be as in (14.12), and let $f' : X \to \overline{R}$ be a function. If $f_n \xrightarrow{\mathfrak{L}^p} f$ and $f_n \xrightarrow{a.e} f'$, then $f = f'$ μ-a.e. Indeed, on account of (14.14) and Riesz's theorem (14.8), $f_{n_k} \xrightarrow{a.e} f$ for some subsequence $\{n_k : k \in N\} \subset N$, and (14.2) applies.

The next theorem is a partial converse of Theorem (14.14).

Theorem 14.16 (Vitali). *Suppose that $\mu(X) < \infty$ and $0 < p < \infty$. Let $\{f_n : n \in N\} \subset \mathfrak{L}^p$, and let $f : X \to R$ be a measurable function such that $f_n \xrightarrow{\mu} f$. Then the following are equivalent:*

(i) *$f \in \mathfrak{L}^p$ and $f_n \xrightarrow{\mathfrak{L}^p} f$;*

(ii) *for each $\varepsilon > 0$ there is $\delta > 0$ such that $A \in \mathcal{A}$ and $\mu(A) < \delta$ imply $\int_A |f_n|^p \, d\mu < \varepsilon$ for any $n \in N$.*

Proof. We prove this theorem in case that $0 < p < 1$. The case that $1 \leqslant p < \infty$ is treated in like manner using Minkowski's inequality (13.7) instead of (13.23.a). Assume that (i) holds, and let $\varepsilon > 0$.Choose $n_\varepsilon \in N$ such that

$$\int_X |f - f_n|^p \, d\mu < \varepsilon/2, \quad n > n_\varepsilon. \tag{1}$$

Applying (11.44), choose $\delta_1 > 0$ such that $A \in \mathcal{A}$ and $\mu(A) < \delta_1$ imply $\int_A |f|^p \, d\mu < \varepsilon/2$. Applying again (11.44), choose $\delta_2 > 0$ such that $A \in \mathcal{A}$ and $\mu(A) < \delta_2$ imply

$$\int_A |f_n|^p \, d\mu < \varepsilon, \quad 1 \leqslant n \leqslant n_\varepsilon. \tag{2}$$

Let $\delta = \delta_1 \wedge \delta_2$. Then, using (13.23.a) and (1), $A \in \mathcal{A}$ and $\mu(A) < \delta$ imply

$$\int_A |f_n|^p \, d\mu \leqslant \int_A |f|^p \, d\mu + \int_A |f - f_n|^p \, d\mu < \frac{\varepsilon}{2} + \frac{\varepsilon}{2} = \varepsilon, \quad n > n_\varepsilon. \tag{3}$$

Hence (2) and (3) show that (i) implies (ii). Assume now that (ii) holds. For $\varepsilon > 0$, let δ be as in (ii). Let $\{f_{n_k} : k \in N\}$ be a subsequence of the sequence $\{f_n : n \in N\}$. Using (14.11), choose a subsubsequence $\{f_{n_{k_l}} : l \in N\}$ such that $f_{n_{k_l}} \xrightarrow{a.e.} f$. Applying Egorov's theorem (14.4), choose $A \in \mathcal{A}$ such that $\mu(A) < \delta$ and $f_{n_{k_l}}|_{A^c} \xrightarrow{u} f|_{A^c}$. Using Fatou's lemma (11.24), we have

$$\int_A |f|^p \, d\mu = \int_A \liminf_l \left|f_{n_{k_l}}\right|^p d\mu \leqslant \liminf_l \int_A \left|f_{n_{k_l}}\right|^p d\mu \leqslant \varepsilon. \tag{4}$$

Let $l_\varepsilon \in N$ be such that $l \geqslant l_\varepsilon$ implies $\left|f_{n_{k_l}}(x) - f(x)\right| < \varepsilon$ for any $x \in A^c$. Then, by (13.23.a) and (4), we have

$$\int_X \left|f - f_{n_{k_l}}\right|^p d\mu = \int_A \left|f - f_{n_{k_l}}\right|^p d\mu + \int_{A^c} \left|f - f_{n_{k_l}}\right|^p d\mu$$

$$\leqslant \int_A |f|^p \, d\mu + \int_A \left| f_{n_{k_l}} \right|^p d\mu + \int_{A^c} \left| f - f_{n_{k_l}} \right|^p d\mu < \varepsilon + \varepsilon + \varepsilon\mu(A^c)$$
$$= \varepsilon(2 + \mu(A^c))$$

for any $l \geqslant l_\varepsilon$. Thus $f = (f - f_{n_{k_l}}) + f_{n_{k_l}} \in \mathfrak{L}^p$ and $\lim_l \int_X \left| f - f_{n_{k_l}} \right|^p d\mu = 0$. According to (5.54), it follows that $\lim_n \int_X |f - f_n|^p \, d\mu = 0$. Consequently, (ii) implies (i). □

Exercise 14.17. Assume that μ is σ-finite. Let $f_n : X \to R, n \in N$, and $f : X \to R$ be measurable functions such that $f_n \xrightarrow{a.e.} f$. Show that there exists a countable family $\{A_i : i \in I\} \subset \mathcal{A}$ such that $\mu(\cap_{i \in I} A_i^c) = 0$ and $f_n|_{A_i} \xrightarrow{u} f|_{A_i}$ for each $i \in I$.

Exercise 14.18. Let $f_n : X \to R,\ n \in N$, and $f : X \to R$ be measurable functions such that $f_n \xrightarrow{a.e.} f$. If there is $g \in \mathfrak{L}^1$ such that $|f_n| \leqslant g, n \in N$, then, for each $\varepsilon > 0$, there exists $A \in \mathcal{A}$ such that $\mu(A) < \varepsilon$ and $f_n|_{A^c} \xrightarrow{u} f|_{A^c}$. [Hint. Use (11.25) and (13.8) to prove that $\lim_n \mu(\cup_{k \geqslant n}\{|f_k - f| \geqslant 1/m\}) = 0$ for any $m \in N$.]

Exercise 14.19. Let $(X, \mathcal{A}, \mu)$ be a measure space, where X is countable and $\mathcal{A} = \mathcal{P}(X)$. Let $f_n : X \to R,\ n \in N$, and $f : X \to R$ be functions such that $f_n \xrightarrow{\mu} f$. Show that $f_n \xrightarrow{a.e.} f$.

Exercise 14.20. Let $(X, \mathcal{A}, \mu)$ be a measure space, where X is countable, $\mathcal{A} = \mathcal{P}(X)$ and μ is the counting measure on $\mathcal{P}(X)$. For $f_n \in R^X, n \in N$, and $f \in R^X$, prove that $f_n \xrightarrow{\mu} f$ if and only if $f_n \xrightarrow{u} f$.

Exercise 14.21. Let $f_n : X \to R,\ n \in N$, and $f : X \to R$ be measurable functions such that $f_n \xrightarrow{a.e.} f$. If there is $g \in \mathfrak{L}^1$ such that $|f_n| \leqslant g, n \in N$, then $f_n \xrightarrow{\mu} f$.

Exercise 14.22. Let $f_n : X \to R, n \in N$, and $f : X \to R$ be measurable functions such that $f_n \xrightarrow{\mu} f$. Prove the following.

(a) $\alpha f_n \xrightarrow{\mu} \alpha f$ for any $\alpha \in R$.

(b) $|f_n| \xrightarrow{\mu} |f|$.

(c) For $\beta > 0$, $\{|f_n|^\beta : n \in N\}$ may fail to converge in μ-measure to $|f|^\beta$.

(d) If $f_n(x) \neq 0, x \in X$, for any $n \in N$, and $f(x) \neq 0, x \in X$, then $\{1/f_n : n \in N\}$ may fail to converge in μ-measure to $1/f$.

(e) If $\mu(X) < \infty$, and $\varphi : R \to R$ is continuous, then $\varphi \circ f_n \xrightarrow{\mu} \varphi \circ f$.

(f) If $\mu(X) < \infty$, and $f_n + a_n \xrightarrow{\mu} f$ for some sequence $\{a_n : n \in N\} \subset R$, then $a_n \to 0$. [Use (5.54) and (14.11).]

Exercise 14.23. Let $f_n : X \to R,\ g_n : X \to R, n \in N$, and $f : X \to R,\ g : X \to R$ be measurable functions such that $f_n \xrightarrow{\mu} f$ and $g_n \xrightarrow{\mu} g$. Prove the following.

(a) $f_n + g_n \xrightarrow{\mu} f + g$.

(b) $\{f_n g_n : n \in N\}$ may fail to converge in μ-measure to fg.

(c) $f_n \vee g_n \xrightarrow{\mu} f \vee g$ and $f_n \wedge g_n \xrightarrow{\mu} f \wedge g$.

(d) If $\mu(X) < \infty$, and $\varphi : R^2 \to R$ is continuous, then $\varphi \circ (f_n, g_n) \xrightarrow{\mu} \varphi \circ (f, g)$.

Exercise 14.24. Let $f_n : X \to R$, $n \in N$, and $f : X \to R$ be functions, and assume there exists a strictly increasing sequence $\{n_k : k \in N\} \subset N$ such that $f_{n_k} \overset{a.e.}{\to} f$ and $\sup_{n_{k-1}<n\leqslant n_k} \left|f_n - f_{n_{k-1}}\right| \overset{a.e.}{\to} 0$. Show that $f_n \overset{a.e.}{\to} f$.

Exercise 14.25. Let $f_n : X \to R, n \in N$, and $f : X \to R$ be measurable functions such that $f_n \overset{\mu}{\to} f$. If there is $g \in \mathfrak{L}^p$ with $0 < p < \infty$ such that $|f_n| \leqslant g, n \in N$, then $f \in \mathfrak{L}^p$ and $f_n \overset{\mathfrak{L}^p}{\to} f$. [Use (13.28), then (14.16) and (14.7.d).]

Exercise 14.26. For $0 < p < \infty$, construct $\{f_n : n \in N\} \subset \mathfrak{L}^p$ and $f \in \mathfrak{L}^p$ such that $f_n \overset{a.e.}{\to} f$, but $\int_X |f_n|^p \, d\mu \nrightarrow \int_X |f|^p \, d\mu$.

Exercise: Uniform integrability 14.27. For $i \in I$, let $f_i : X \to R$ be a measurable function. The set $\{f_i : i \in I\}$ is said to be *uniformly integrable* if $\{f_i : i \in I\} \subset \mathfrak{L}^1$ and $\sup_{i\in I} \int_{\{|f_i|\geqslant a\}} |f_i| \, d\mu \to 0$ as $a \to \infty$. Prove the following.

(a) If $\{g_i : i \in I\} \subset \mathfrak{L}^1$ is uniformly integrable, and $|f_i| \leqslant |g_i|, i \in I$, then $\{f_i : i \in I\}$ is uniformly integrable.

(b) If there exists $g \in \mathfrak{L}^1$ such that $|f_i| \leqslant g$, $i \in I$, then $\{f_i : i \in I\}$ is uniformly integrable.

(c) If $\{f_i : i \in I\} \subset \mathfrak{L}^p$ for some $p > 1$, and $\sup_{i\in I} \|f_i\|_p < \infty$, then $\{f_i : i \in I\}$ is uniformly integrable. [Use (13.4).]

(d) If $\{f_i : i \in I\} \subset \mathfrak{L}^p$ for some $p \geqslant 1$, and $\{|f_i|^p : i \in I\}$ is uniformly integrable, then $\{|f_i|^q : i \in I\}$ is uniformly integrable whenever $1 \leqslant q \leqslant p$.

Exercise 14.28. Assume that $\mu(X) < \infty$, and let $\{f_i : i \in I\} \subset \mathfrak{L}^1$. Show that $\{f_i : i \in I\}$ is uniformly integrable if and only if the following conditions hold:

(i) for each $\varepsilon > 0$ there is $\delta > 0$ such that $A \in \mathcal{A}$ and $\mu(A) < \delta$ imply $\int_A |f_i| \, d\mu < \varepsilon$ for any $i \in I$;

(ii) $\sup_{i\in I} \|f_i\|_1 < \infty$.

Exercise 14.29. Assume that $\mu(X) < \infty$, and let $\{f_i : i \in I\} \subset \mathfrak{L}^1$ and $\{g_i : i \in I\} \subset \mathfrak{L}^1$ be uniformly integrable. Prove that $\{f_i + g_i : i \in I\}$ is uniformly integrable.

Exercise 14.30. Let μ, p, $\{f_n : n \in N\}$ and f be as in (14.16). Show that the following assertions are equivalent:

(i) $\{|f_n|^p : n \in N\}$ is uniformly integrable;

(ii) $f \in \mathfrak{L}^p$ and $f_n \overset{\mathfrak{L}^p}{\to} f$;

(iii) $f \in \mathfrak{L}^p$ and $\int_X |f_n|^p \, d\mu \to \int_X |f|^p \, d\mu$.

[Hints. If (i) is true, choose a subsequence $\{f_{n_k} : k \in N\}$ such that $f_{n_k} \overset{a.e.}{\to} f$ (14.8). Then $\int_X |f|^p \, d\mu \leqslant \liminf_k \int_X \left|f_{n_k}\right|^p d\mu < \infty$ by Fatou's lemma (11.24) and (14.28.ii). Since $\int_X |f - f_n|^p \, d\mu \leqslant 2^p(\int_X |f|^p \, d\mu + \int_X |f_n|^p \, d\mu)$, the sequence $\{|f - f_n|^p : n \in N\}$ is also uniformly integrable. For $\varepsilon > 0$, $\int_X |f - f_n|^p \, d\mu \leqslant \varepsilon^p \mu(X) + \int_{\{|f-f_n|^p\geqslant\varepsilon\}} |f - f_n|^p \, d\mu$. This and (14.28.i) lead to (ii), as $\mu(\{|f - f_n|^p \geqslant \varepsilon\}) \to 0$. If (iii) holds, consider $a > 0$ and let $\varphi_a \in \mathfrak{C}_r(R)$ be the function that equals $|x|^p$ for $|x|^p \leqslant a$, is linear in $]-a-1, -a[$ and $]a, a+1[$, and equals 0 on $]-\infty, -a-1] \cup [a+1, \infty[$. Since $|x|^p \, 1_{\{x:|x|^p<a\}} \leqslant \varphi_a(x) \leqslant |x|^p \, 1_{\{x:|x|^p<a+1\}}$, $x \in R$, use (14.22.e) and (14.25) to get $\int_{\{|f|^p<a\}} |f|^p \, d\mu \leqslant \int_X (\varphi_a \circ f) \, d\mu = \lim_n \int_X (\varphi_a \circ f_n) d\mu$

$\leqslant \liminf_n \int_{\{|f_n|^p < a+1\}} |f_n|^p \, d\mu$. This entails $\limsup_n \int_{\{|f_n|^p \geqslant a+1\}} |f_n|^p \, d\mu \leqslant \int_{\{|f|^p \geqslant a\}} |f|^p \, d\mu$, in view of (iii). For $\varepsilon > 0$, choose $a_\varepsilon > 0$ so that $\int_{\{|f|^p \geqslant a\}} |f|^p \, d\mu < \varepsilon$ whenever $a \geqslant a_\varepsilon$, then select $n_\varepsilon \in N$ such that $\sup_{n \geqslant n_\varepsilon} \int_{\{|f_n|^p \geqslant a+1\}} |f_n|^p \, d\mu < \varepsilon$, $a \geqslant a_\varepsilon$.]

Exercise 14.31. Assume that $\mu(X) < \infty$, let $\{f_n : n \in N\} \subset \mathfrak{L}^1$, and let $f : X \to R$ be a measurable function such that $f_n \overset{\mu}{\to} f$. The following are equivalent:

(i) $\{f_n : n \in N\}$ is uniformly integrable;

(ii) if $\{A_k : k \in N\} \subset \mathcal{A}$ is a nonincreasing sequence such that $\bigcap_{k \in N} A_k = \emptyset$, then $\lim_k (\sup_{n \in N} \int_{A_k} |f_n| \, d\mu) = 0$.

Exercise 14.32. Assume that $\mu(X) < \infty$, let $\{f_n : n \in N\} \subset \mathfrak{L}^1$ be uniformly integrable, and let $f : X \to R$ be a measurable function such that $f_n \overset{\mu}{\to} f$. Prove that $\int_A f_n \, d\mu \to \int_A f \, d\mu$, $A \in \mathcal{A}$.

Exercise 14.33 (Scheffé). Let μ be a measure on $\mathcal{A}$, and let $f : X \to [0, \infty[$ be a measurable function. For $n \in N$, let μ_n be a measure on $\mathcal{A}$, and let $f_n : X \to [0, \infty[$ be a measurable function. Assume that $f_n \overset{\mu}{\to} f$, and $\int_X f_n \, d\mu = \int_X f d\mu < \infty$. Show that $\int_X |f - f_n| \, d\mu \to 0$. [Apply (14.23.c) and (14.25), as $(f - f_n)^+ \leqslant f, n \in N$.]

Exercise 14.34. Let $f_n : X \to R$, $n \in N$, and $f : X \to R$ be measurable functions such that $f_n \overset{\mu}{\to} f$. If $\mu(X) < \infty$ and $|f_n| \leqslant M$ μ-a.e. for some $M < \infty$, then $f_n \overset{\mathfrak{L}^p}{\to} f$ for any $0 < p < \infty$.

5 Measures on Product σ-Algebras

This chapter is devoted to the construction of measures on product σ-algebras. In Section 15 we construct the product of a finite number of σ-finite measures. The first part of Section 16 is the infinite version of Section 15, and in the second part of Section 16 we prove the Daniell-Kolmogorov extension theorem.

15 The Product of a Finite Number of Measures

In this section we construct the product of a finite number of σ-finite measures and prove versions of Fubini's theorem.

We first construct the product of two σ-finite measures.

Theorem 15.1. *Let $(X_1, \mathcal{A}_1)$ and (X_2, A_2) be measurable spaces, and let $A \in \mathcal{A}_1 \otimes \mathcal{A}_2$. Then*

(i) $A_{x_1} \in \mathcal{A}_2$ *for any* $x_1 \in X_1$

and

(ii) $A_{x_2} \in \mathcal{A}_1$ *for any* $x_2 \in X_2$,

where A_{x_i} is the section of A at x_i, $i = 1, 2$.

Proof. To prove (i), let $\mathcal{A} = \{A_1 \times A_2 : A_1 \in \mathcal{A}_1, A_2 \in \mathcal{A}_2\}$ and $\mathcal{M} = \{A \in \mathcal{A}_1 \otimes \mathcal{A}_2 : A_{x_1} \in \mathcal{A}_2$ for any $x_1 \in X_1\}$. For $A_1 \times A_2 \in \mathcal{A}$ and $x_1 \in X_1$, in view of (1.2.30.b), we have

$$(A_1 \times A_2)_{x_1} = \begin{cases} A_2 & \text{if } x_1 \in A_1, \\ \emptyset & \text{if } x_1 \notin A_1. \end{cases}$$

Thus $\mathcal{A} \subset \mathcal{M}$. If $\{A_n : n \in N\} \subset \mathcal{M}$, then $(\cup_{n \in N} A_n)_{x_1} = \cup_{n \in N} (A_n)_{x_1} \in \mathcal{A}_2$ for any $x_1 \in X_1$, and so $\cup_{n \in N} A_n \in \mathcal{M}$. If $A \in \mathcal{M}$, then $(A^c)_{x_1} = (A_{x_1})^c \in \mathcal{A}_2$ for any $x_1 \in X_1$, and so $A^c \in \mathcal{M}$. Therefore, $\mathcal{M}$ is a σ-algebra. Since $\mathcal{A}_1 \otimes \mathcal{A}_2 = \sigma(\mathcal{A})$ (7.17.a), we have $\mathcal{A}_1 \otimes \mathcal{A}_2 = \mathcal{M}$. This proves (i). The proof of (ii) is similar. □

Corollary 15.2. *Let $(X_1, \mathcal{A}_1)$ and (X_2, A_2) be measurable spaces, and let $f : X_1 \times X_2 \to \overline{R}$ be an $\mathcal{A}_1 \otimes \mathcal{A}_2$-measurable function. Then*

(i) f_{x_1} *is* $\mathcal{A}_2 -$ *measurable for any* $x_1 \in X_1$

Analysis and Probability. http://dx.doi.org/10.1016/B978-0-12-401665-1.00005-9

and

(ii) f_{x_2} *is* $\mathcal{A}_1$*−measurable for any* $x_2 \in X_2$,

where f_{x_i} *is the section of* f *at* $x_i, i = 1, 2$.

Proof. If $f = 1_A$, where $A \in \mathcal{A}_1 \otimes \mathcal{A}_2$, then $f_{x_i} = 1_{A_{x_i}}, i = 1, 2$. Hence in this case (i) and (ii) follow at once from (15.1). The corollary is further proved by making use of the indicator function method. □

Theorem 15.3. *Let* $(X_1, \mathcal{A}_1, \mu_1)$ *and* $(X_2, \mathcal{A}_2, \mu_2)$ *be measure spaces, where* μ_1 *and* μ_2 *are* σ*-finite and let* $A \in \mathcal{A}_1 \otimes \mathcal{A}_2$*. Then the following assertions hold.*

(i) *The function* $x_1 \to \mu_2(A_{x_1})$ *from* X_1 *into* $\overline{R}$ *is* $\mathcal{A}_1$*-measurable.*

(ii) *The function* $x_2 \to \mu_1(A_{x_2})$ *from* X_2 *into* $\overline{R}$ *is* $\mathcal{A}_2$*-measurable.*

Proof. Since the proofs of (i) and (ii) are similar, we prove only (i).

(I) Assume first that $\mu_2(X_2) < \infty$. Let $\mathcal{A}$ be as in (15.1), and let $\mathcal{D} = \{A \in \mathcal{A}_1 \otimes \mathcal{A}_2 :$ assertion (i) holds$\}$. For $A_1 \times A_2 \in \mathcal{A}$, we have $\mu_2((A_1 \times A_2)_{x_1}) = \mu_2(A_2)1_{A_1}(x_1), x_1 \in X_1$. Thus $\mathcal{A} \subset \mathcal{D}$. Evidently $X_1 \times X_2 \in \mathcal{D}$. If $A, B \in \mathcal{D}$ and $A \supset B$, then $\mu_2((A - B)_{x_1}) = \mu_2(A_{x_1} - B_{x_1}) = \mu_2(A_{x_1}) - \mu_2(B_{x_1}), x_1 \in X_1$, and so $A - B \in \mathcal{D}$. If $\{A_n : n \in N\} \subset \mathcal{D}$ and $A_m \cap A_n = \emptyset$ for $m \neq n$, then $\mu_2((\cup_{n\in N} A_n)_{x_1}) = \mu_2(\cup_{n\in N}(A_n)_{x_1}) = \sum_{n\in N} \mu_2((A_n)_{x_1}), x_1 \in X_1$, and so $\cup_{n\in N} A_n \in \mathcal{D}$. Therefore, $\mathcal{D}$ is a Dynkin system. Since $A_1 \times A_2 \in \mathcal{A}$ and $A_1' \times A_2' \in \mathcal{A}$ imply $(A_1 \times A_2) \cap (A_1' \times A_2') = (A_1 \cap A_1') \times (A_2 \cap A_2') \in \mathcal{A}$, (7.36) shows that $d(\mathcal{A}) = \sigma(\mathcal{A}) = \mathcal{A}_1 \otimes \mathcal{A}_2$. Consequently, $\mathcal{D} = \mathcal{A}_1 \otimes \mathcal{A}_2$. This proves (i).

(II) Consider now the general case. Let $\{B_j : j \in J\} \subset \mathcal{A}_2$ be a countable partition of X_2 such that $\mu_2(B_j) < \infty, j \in J$. For $A \in \mathcal{A}_1 \otimes \mathcal{A}_2$, using (7.14),we have $A\cap(X_1 \times B_j) \in i^{-1}_{X_1\times B_j}(\mathcal{A}_1 \otimes \mathcal{A}_2) = \mathcal{A}_1 \otimes i^{-1}_{B_j}(\mathcal{A}_2), j \in J$. Thus, according to (I), the function $x_1 \to \mu_2((A \cap (X_1 \times B_j))_{x_1})$ is $\mathcal{A}_1$-measurable for any $j \in J$. Therefore, the function $x_1 \to \mu_2(A_{x_1}) = \mu_2(\cup_{j\in J}(A\cap(X_1 \times B_j))_{x_1}) = \sum_{j\in J} \mu_2((A\cap(X_1 \times B_j))_{x_1})$ is $\mathcal{A}_1$-measurable. □

Corollary 15.4. *Let* $(X_1, \mathcal{A}_1, \mu_1)$ *and* $(X_2, \mathcal{A}_2, \mu_2)$ *be as in* (15.3), *and let* $f : X_1 \times X_2 \to [0, \infty]$ *be an* $\mathcal{A}_1 \otimes \mathcal{A}_2$*-measurable function. Then the following assertions hold.*

(i) *The function* $x_1 \to \int_{X_2} f(x_1, x_2)d\mu_2(x_2)$ *from* X_1 *into* $[0, \infty]$ *is* $\mathcal{A}_1$*-measurable.*

(ii) *The function* $x_2 \to \int_{X_1} f(x_1, x_2)d\mu_1(x_1)$ *from* X_2 *into* $[0, \infty]$ *is* $\mathcal{A}_2$*-measurable.*

Proof. If $f = 1_A$, where $A \in \mathcal{A}_1 \otimes \mathcal{A}_2$, then $\int_{X_2} 1_A(x_1, x_2)d\mu_2(x_2) = \int_{X_2} 1_{A_{x_1}}(x_2)\, d\mu_2(x_2) = \mu_2(A_{x_1}), x_1 \in X_1$, and $\int_{X_1} 1_A(x_1, x_2)d\mu_1(x_1) = \int_{X_1} 1_{A_{x_2}}(x_1)d\mu_1(x_1) = \mu_1(A_{x_2}), x_2 \in X_2$. Hence in this case (i) and (ii) follow immediately from (15.3). Therefore, if *f* is a simple function, then, using (11.6), (i) and (ii) obtain. Finally, in the general case choose a nondecreasing sequence $\{f_n : n \in N\} \subset [0, \infty[^{X_1 \times X_2}$ of simple functions such that $f_n \to f$. Applying (11.22), we have $\int_{X_2} f(x_1, x_2)d\mu_2(x_2) = \lim_n \int_{X_2} f_n(x_1, x_2)d\mu_2(x_2), x_1 \in X_1$, and $\int_{X_1} f(x_1, x_2)d\mu_1(x_1) = \lim_n \int_{X_1} f_n(x_1, x_2)d\mu_1(x_1), x_2 \in X_2$. Hence, using (8.12.ii), assertions (i) and (ii) are proved. □

Theorem 15.5. *Let $(X_1, \mathcal{A}_1, \mu_1)$ and $(X_2, \mathcal{A}_2, \mu_2)$ be as in (15.3). For $A \in \mathcal{A}_1 \otimes \mathcal{A}_2$, define $\mu(A) = \int_{X_1} \mu_2(A_{x_1}) d\mu_1(x_1)$ and $\mu'(A) = \int_{X_2} \mu_1(A_{x_2}) d\mu_2(x_2)$. Then μ and μ' are equal measures on $\mathcal{A}_1 \otimes \mathcal{A}_2$.*

Proof. Obviously, $\mu(\emptyset) = 0$. Let $\{A_n : n \in N\} \subset \mathcal{A}_1 \otimes \mathcal{A}_2$ be such that $A_m \cap A_n = \emptyset$ for $m \neq n$. Then, applying (11.23), we have $\mu(\cup_{n\in N} A_n) = \int_{X_1} \mu_2(\cup_{n\in N}(A_n)_{x_1}) d\mu_1 (x_1) = \sum_{n\in N} \int_{X_1} \mu_2((A_n)_{x_1}) d\mu_1(x_1) = \sum_{n\in N} \mu(A_n)$. Hence μ is a measure on $\mathcal{A}_1 \otimes \mathcal{A}_2$. The fact that μ' is a measure on $\mathcal{A}_1 \otimes \mathcal{A}_2$ is proved in like manner. Let $\mathcal{A}$ be as in (15.1). For $A_1 \times A_2 \in \mathcal{A}$, we obtain $\mu(A_1 \times A_2) = \int_{X_1} \mu_2((A_1 \times A_2)_{x_1}) d\mu_1(x_1) = \int_{X_1} \mu_2(A_2) 1_{A_1} d\mu_1(x_1) = \mu_1(A_1)\mu_2(A_2) = \int_{X_2} \mu_1(A_1) 1_{A_2} d\mu_2(x_2) = \int_{X_2} \mu_1 ((A_1 \times A_2)_{x_2}) d\mu_2(x_2) = \mu'(A_1 \times A_2)$. Therefore, μ and μ' coincide on $\mathcal{A}$. Let $\mathcal{B}$ be the family of all finite unions of disjoint sets in $\mathcal{A}$. Then $\mathcal{B}$ is an algebra of sets such that $\sigma(\mathcal{B}) = \mathcal{A}_1 \otimes \mathcal{A}_2$, and $\mu(B) = \mu'(B)$ for any $B \in \mathcal{B}$. Now let $\{A_i) : i \in I\} \subset \mathcal{A}_1$ be a countable partition of X_1 such that $\mu_1(A_i) < \infty, i \in I$, and let $\{B_j) : j \in J\} \subset \mathcal{A}_2$ be a countable partition of X_2 such that $\mu_2(B_j) < \infty, j \in J$. For $i \in I$ and $j \in J$, we have $A_i \times B_j \in \mathcal{A} \subset \mathcal{B}$ and $\mu(A_i \times B_j) = \mu'(A_i \times B_j) = \mu_1(A_i)\mu_2(B_j) < \infty$. Consequently, μ and μ' are σ-finite measures on $\mathcal{B}$. Thus, using (10.9), it follows that $\mu = \mu'$. □

Remark 15.6. The measure $\mu = \mu'$ in (15.5) is denoted by $\mu_1 \otimes \mu_2$, and it is called the *product of* μ_1 *and* μ_2. $\mu_1 \otimes \mu_2$ is the unique measure on $\mathcal{A}_1 \otimes \mathcal{A}_2$ such that $(\mu_1 \otimes \mu_2)(A_1 \times A_2) = \mu_1(A_1)\mu_2(A_2)$ for any $A_1 \in \mathcal{A}_1$ and $A_2 \in \mathcal{A}_2$.

The next theorem generalizes (15.5).

Theorem 15.7. *Let $(X_1, \mathcal{A}_1)$ and $(X_2, \mathcal{A}_2)$ be measurable spaces, let μ_1 be a σ-finite measures on $\mathcal{A}_1$, and let $\mu_2 : X_1 \times \mathcal{A}_2 \to [0, \infty]$ be a function such that the section of μ_2 at any $x_1 \in X_1$ is a measure on $\mathcal{A}_2$ and the section of μ_2 at any $B \in \mathcal{A}_2$ is $\mathcal{A}_1$-measurable. Assume that there exists a countable partition $\{B_j : j \in J\} \subset \mathcal{A}_2$ of X_2 such that $\mu_2(x_1, B_j) \leqslant k_j < \infty, x_1 \in X_1$, for any $j \in J$. Then there exists a unique measure μ on $\mathcal{A}_1 \otimes \mathcal{A}_2$ such that $\mu(A_1 \times A_2) = \int_{A_1} \mu_2(x_1, A_2) d\mu_1(x_1)$ for any $A_1 \in \mathcal{A}_1$ and $A_2 \in \mathcal{A}_2$.*

Proof. Let $A \in \mathcal{A}_1 \otimes \mathcal{A}_2$. Like in (15.3) it is proved that the function $x_1 \to \mu_2(x_1, A_{x_1})$ from X_1 into $\overline{R}$ is $\mathcal{A}_1$-measurable. Define $\mu(A) = \int_{X_1} \mu_2(x_1, A_{x_1}) d\mu_1 (x_1)$. Like in (15.5) it is proved that μ is a measure on $\mathcal{A}_1 \otimes \mathcal{A}_2$. The details and the rest of the proof are left to the reader. □

In what follows we prove two versions of Fubini's theorem.

Theorem 15.8 (Fubini). *Let $(X_1, \mathcal{A}_1, \mu_1)$ and $(X_2, \mathcal{A}_2, \mu_2)$ be as in (15.3), and let $f : X_1 \times X_2 \to [0, \infty]$ be an $\mathcal{A}_1 \otimes \mathcal{A}_2$-measurable function. Then*

$$\text{(i)} \quad \int_{X_1 \times X_2} f \, d(\mu_1 \otimes \mu_2) = \int_{X_1} \left(\int_{X_2} f(x_1, x_2) d\mu_2(x_2) \right) d\mu_1(x_1)$$

$$= \int_{X_2} \left(\int_{X_1} f(x_1, x_2) d\mu_1(x_1) \right) d\mu_2(x_2).$$

Proof. If $f = 1_A$, where $A \in \mathcal{A}_1 \otimes \mathcal{A}_2$, then $\int_{X_2 \times X_2} 1_A \, d(\mu_1 \otimes \mu_2) = (\mu_1 \otimes \mu_2)(A) = \int_{X_1} \mu_2(A_{x_1}) d\mu_1(x_1) = \int_{X_1} (\int_{X_2} 1_A(x_1, x_2) d\mu_2(x_2)) d\mu_1(x_1)$. Using (15.5), in like manner we get $\int_{X_2 \times X_2} 1_A \, d(\mu_1 \otimes \mu_2) = \int_{X_2} (\int_{X_1} 1_A(x_1, x_2) d\mu_1(x_1)) \, d\mu_2(x_2)$. Thus (i) holds in this case. Therefore, if f is a simple function, then, using (11.6), (i) obtains. Finally, in the general case choose a nondecreasing sequence $\{f_n : n \in N\} \subset [0, \infty[^{X_1 \times X_2}$ of simple functions such that $f_n \to f$. Applying (11.22), we have

$$\begin{aligned}
\int_{X_1 \times X_2} f \, d(\mu_1 \otimes \mu_2) &= \lim_n \int_{X_1 \times X_2} f_n \, d(\mu_1 \otimes \mu_2) \\
&= \lim_n \int_{X_1} \left(\int_{X_2} f_n(x_1, x_2) d\mu_2(x_2) \right) d\mu_1(x_1) \\
&= \int_{X_1} \left(\lim_n \int_{X_2} f_n(x_1, x_2) d\mu_2(x_2) \right) d\mu_1(x_1) \\
&= \int_{X_1} \left(\int_{X_2} f(x_1, x_2) d\mu_2(x_2) \right) d\mu_1(x_1).
\end{aligned}$$

Similarly we get $\int_{X_2 \times X_2} f \, d(\mu_1 \otimes \mu_2) = \int_{X_2} (\int_{X_1} f(x_1, x_2) d\mu_1(x_1)) d\mu_2(x_2)$, and so (i) is proved. □

Remark 15.9. The assumption that μ_1 and μ_2 are σ-finite in (15.8) is essential. To see this, let $X_1 = X_2 = [0, 1]$, $\mathcal{A}_1 = \mathcal{B}([0, 1])$, $\mathcal{A}_2 = \mathcal{P}([0, 1])$, let μ_1 be the Lebesgue measure on $\mathcal{B}([0, 1])$, let μ_2 be the counting measure on $\mathcal{P}([0, 1])$, and let $f = 1_{\mathbb{D}}$, where $\mathbb{D}$ is the diagonal of $[0, 1]^2$. Then $1 = \int_{X_1} (\int_{X_2} 1_{\mathbb{D}}(x_1, x_2) d\mu_2(x_2)) d\mu_1(x_1) \neq \int_{X_2} (\int_{X_1} 1_{\mathbb{D}}(x_1, x_2) d\mu_1(x_1)) d\mu_2(x_2) = 0$.

Theorem 15.10 (Fubini). *Let $(X_1, \mathcal{A}_1, \mu_1)$ and $(X_2, \mathcal{A}_2, \mu_2)$ be as in (15.3), and let $f : X_1 \times X_2 \to \overline{R}$ be an $\mathcal{A}_1 \otimes \mathcal{A}_2$-measurable function such that one of the integrals $\int_{X_1 \times X_2} |f| \, d(\mu_1 \otimes \mu_2)$, $\int_{X_1} (\int_{X_2} |f(x_1, x_2)| \, d\mu_2(x_2)) d\mu_1(x_1)$ and $\int_{X_2} (\int_{X_1} |f(x_1, x_2)| d\mu_1(x_1)) d\mu_2(x_2)$ is finite. Then the following assertions hold.*

(i) *There is $A_1 \in \mathcal{A}_1$ such that $\mu_1(A_1^c) = 0$ and $f_{x_1} \in \mathfrak{L}^1(X_2, \mathcal{A}_2, \mu_2)$ for any $x_1 \in A_1$.*

(ii) *There is $A_2 \in \mathcal{A}_2$ such that $\mu_2(A_2^c) = 0$ and $f_{x_2} \in \mathfrak{L}^1(X_1, \mathcal{A}_1, \mu_1)$ for any $x_2 \in A_2$.*

(iii) *The function $\varphi_1(x_1) = 1_{A_1}(x_1) \int_{X_2} f(x_1, x_2) d\mu_2(x_2) \in \mathfrak{L}^1(X_1, \mathcal{A}_1, \mu_1)$.*

(iv) *The function $\varphi_2(x_2) = 1_{A_2}(x_2) \int_{X_1} f(x_1, x_2) d\mu_1(x_1) \in \mathfrak{L}^1(X_2, \mathcal{A}_2, \mu_2)$.*

(v) $\int_{X_1 \times X_2} f \, d(\mu_1 \otimes \mu_2) = \int_{X_1} (\int_{X_2} f(x_1, x_2) d\mu_2(x_2)) d\mu_1(x_1) = \int_{X_2} (\int_{X_1} f(x_1, x_2) d\mu_1(x_1)) d\mu_2(x_2)$.

Proof. In view of (15.8), we have

$$\int_{X_1 \times X_2} |f|\, d(\mu_1 \otimes \mu_2) = \int_{X_1} \left(\int_{X_2} |f(x_1, x_2)|\, d\mu_2(x_2) \right) d\mu_1(x_1)$$
$$= \int_{X_2} \left(\int_{X_1} |f(x_1, x_2)|\, d\mu_1(x_1) \right) d\mu_2(x_2) < \infty. \tag{1}$$

Thus, applying (11.15), there is $A_1 \in \mathcal{A}_1$ such that $\mu_1(A_1^c) = 0$ and $\int_{X_2} |f(x_1, x_2)|\, d\mu_2(x_2) < \infty$ for any $x_1 \in A_1$. Therefore, $f_{x_1} \in \mathfrak{L}^1(X_2, \mathcal{A}_2, \mu_2)$ for any $x_1 \in A_1$, and so (i) is proved. Assertion (ii) is proved in like manner. Using (15.4), the functions $x_1 \to \int_{X_2} f^+(x_1, x_2) d\mu_2(x_2)$ and $x_1 \to \int_{X_2} f^-(x_1, x_2) d\mu_2(x_2)$ from X_1 into $[0, \infty]$ are $\mathcal{A}_1$-measurable, and so the function $\varphi_1(x_1) = 1_{A_1}(x_1) \int_{X_2} f(x_1, x_2) d\mu_2(x_2)$ is $\mathcal{A}_1$-measurable. As $|\varphi_1(x_1)| \leqslant \int_{X_2} |f(x_1, x_2)|\, d\mu_2(x_2)$, $x_1 \in X_1$, (1) shows that $\varphi_1 \in \mathfrak{L}^1(X_1, \mathcal{A}_1, \mu_1)$. Similarly it is shown that $\varphi_2 \in \mathfrak{L}^1(X_2, \mathcal{A}_2, \mu_2)$. Hence (iii) and (iv) are proved. Further, applying (15.8), we have

$$\int_{X_1 \times X_2} f\, d(\mu_1 \otimes \mu_2) = \int_{X_1 \times X_2} f^+\, d(\mu_1 \otimes \mu_2) - \int_{X_1 \times X_2} f^-\, d(\mu_1 \otimes \mu_2)$$
$$= \int_{X_1} \left(\int_{X_2} f^+(x_1, x_2) d\mu_2(x_2) \right) d\mu_1(x_1)$$
$$- \int_{X_1} \left(\int_{X_2} f^-(x_1, x_2) d\mu_2(x_2) \right) d\mu_1(x_1)$$
$$= \int_{X_1} \left(\int_{X_2} f^+(x_1, x_2) d\mu_2(x_2) \right.$$
$$\left. - \int_{X_2} f^-(x_1, x_2) d\mu_2(x_2) \right) d\mu_1(x_1)$$
$$= \int_{X_1} \left(\int_{X_2} f(x_1, x_2) d\mu_2(x_2) \right) d\mu_1(x_1).$$

In like manner we get $\int_{X_1 \times X_2} f\, d(\mu_1 \otimes \mu_2) = \int_{X_2} (\int_{X_1} f(x_1, x_2) d\mu_1(x_1)) d\mu_2(x_2)$, and so (v) is proved. □

Theorem 15.11. *Let $(X_1, \mathcal{A}_1)$, $(X_2, \mathcal{A}_2)$, μ_1, μ_2, and μ be as in (15.7), and let $f : X_1 \times X_2 \to [0, \infty]$ be an $\mathcal{A}_1 \otimes \mathcal{A}_2$-measurable function. Then the following assertions hold.*

(i) *The function $x_1 \to \int_{X_2} f(x_1, x_2) \mu_2(x_1, dx_2)$ from X_1 into $[0, \infty]$ is $\mathcal{A}_1$-measurable.*

(ii) $\int_{X_1 \times X_2} f \, d\mu = \int_{X_1} (\int_{X_2} f(x_1, x_2) \mu_2(x_1, dx_2)) d\mu_1(x_1)$.

Proof. Exercise. □

Theorem 15.12. *Let $(X_1, \mathcal{A}_1)$, $(X_2, \mathcal{A}_2)$, μ_1, μ_2 and μ be as in (15.7), and let $f : X_1 \times X_2 \to \overline{R}$ be an $\mathcal{A}_1 \otimes \mathcal{A}_2$-measurable function such that either $\int_{X_1 \times X_2} |f| \, d\mu < \infty$ or $\int_{X_1} (\int_{X_2} |f(x_1, x_2)| \mu_2(x_1, dx_2)) d\mu_1(x_1) < \infty$. Then the following assertions hold.*

(i) *There is $A \in \mathcal{A}_1$ such that $\mu_1(A^c) = 0$ and $\int_{X_2} f(x_1, x_2) \mu_2(x_1, dx_2) \in R$ for any $x_1 \in A$.*

(ii) *The function $\varphi_1(x_1) = 1_A(x_1) \int_{X_2} f(x_1, x_2) \mu_2(x_1, dx_2) \in \mathfrak{L}^1(X_1, \mathcal{A}_1, \mu_1)$.*

(iii) $\int_{X_1 \times X_2} f \, d\mu = \int_{X_1} (\int_{X_2} f(x_1, x_2) \mu_2(x_1, dx_2)) d\mu_1(x_1)$.

Proof. Exercise. □

We now construct the product of a finite number of σ-finite measures.

Theorem 15.13. *For $n \geqslant 2$, let $(X_i, \mathcal{A}_i, \mu_i)$, $1 \leqslant i \leqslant n$, be σ-finite measure spaces. Then there exists a unique measure $\mu_1 \otimes \cdots \otimes \mu_n$ $[\otimes_{i=1}^n \mu_i]$ on $\mathcal{A}_1 \otimes \cdots \otimes \mathcal{A}_n$ such that*

(i) $(\mu_1 \otimes \cdots \otimes \mu_n)(A_1 \times \cdots \times A_n) = \mu_1(A_1) \cdots \mu_n(A_n), \quad A_1 \in \mathcal{A}_1, \ldots, A_n \in \mathcal{A}_n$.

Proof. For $n = 2$ the result is true by (15.5) and (15.6). If the result is true for n, then, applying (15.5), there exists the measure $(\mu_1 \otimes \cdots \otimes \mu_n) \otimes \mu_{n+1}$ on $(\mathcal{A}_1 \otimes \cdots \otimes \mathcal{A}_n) \otimes \mathcal{A}_{n+1}$ such that $((\mu_1 \otimes \cdots \otimes \mu_n) \otimes \mu_{n+1})((A_1 \times \cdots \times A_n) \times A_{n+1}) = (\mu_1 \otimes \cdots \otimes \mu_n)(A_1 \times \cdots \times A_n)\mu_{n+1}(A_{n+1}) = \mu_1(A_1) \cdots \mu_n(A_n)\mu_{n+1}(A_{n+1})$ for $A_i \in \mathcal{A}_i$, $1 \leqslant i \leqslant n+1$. In accordance with (1.2.28) we identify the Cartesian products $(X_1 \times \cdots \times X_n) \times X_{n+1}$ and $X_1 \times \cdots \times X_n \times X_{n+1}$, and in view of (7.25) we identify the σ-algebras $(\mathcal{A}_1 \otimes \cdots \otimes \mathcal{A}_n) \otimes \mathcal{A}_{n+1}$ and $\mathcal{A}_1 \otimes \cdots \otimes \mathcal{A}_n \otimes \mathcal{A}_{n+1}$. Thus the existence of the measure $\mu_1 \otimes \cdots \otimes \mu_n$ satisfying (i) is proved by induction. The uniqueness of this measure is proved as in (15.5), by making use of (10.9). □

The next theorem generalizes (15.13).

Theorem 15.14. *For $n \geqslant 2$, let $(X_i, \mathcal{A}_i, \mu_i)$, $1 \leqslant i \leqslant n$, be measurable spaces, and let μ_1 be a σ-finite measure on $\mathcal{A}_1$. For $2 \leqslant i \leqslant n$, let $\mu_i : X_1 \times \cdots \times X_{i-1} \times \mathcal{A}_i \to [0, \infty]$ be a function such that the section of μ_i at any $(x_1, \ldots, x_{i-1}) \in X_1 \times \cdots \times X_{i-1}$ is a measure on $\mathcal{A}_i$ and the section of μ_i at any $A \in \mathcal{A}_i$ is $\mathcal{A}_1 \otimes \cdots \otimes \mathcal{A}_{i-1}$-measurable. For $2 \leqslant i \leqslant n$, assume that there is a countable partition $\{B_{ij} : j \in J_i\} \subset \mathcal{A}_i$ of X_i such that $\mu_i(x_1, \ldots, x_{i-1}, B_{ij}) \leqslant k_{ij} < \infty$, $(x_1, \ldots, x_{i-1}) \in X_1 \times \cdots \times X_{i-1}$, for any $j \in J_i$. Then there exists a unique measure μ on $\mathcal{A}_1 \otimes \cdots \otimes \mathcal{A}_n$ such that $\mu(A_1 \times \cdots \times A_n) = \int_{A_1} \mu_1(dx_1) \int_{A_2} \mu_2(x_1, dx_2) \cdots \int_{A_{n-1}} \mu_{n-1}(x_1, \ldots, x_{n-2}, dx_{n-1}) \int_{A_n} \mu_n(x_1, \ldots, x_{n-1}, dx_n)$ for any $A_1 \in \mathcal{A}_1, \ldots, A_n \in \mathcal{A}_n$.*

Proof. Using (15.7) and (15.11), the theorem is proved by induction as in (15.13). The details are left to the reader. □

Exercise 15.15. (a) Find measurable spaces $(X_1, \mathcal{A}_1)$ and $(X_2, \mathcal{A}_2)$, and $A \subset X_1 \times X_2$ such that $A_{x_1} \in \mathcal{A}_2$ for any $x_1 \in X_1$, $A_{x_2} \in \mathcal{A}_1$ for any $x_2 \in X_2$, but $A \notin \mathcal{A}_1 \otimes \mathcal{A}_2$.

(b) Find measurable spaces $(X_1, \mathcal{A}_1)$ and (X_2, A_2), and $f : X_1 \times X_2 \to \overline{R}$ such that f_{x_1} *is* $\mathcal{A}_2$-measurable for any $x_1 \in X_1$, f_{x_2} is $\mathcal{A}_1$-measurable for any $x_2 \in X_2$, but f is not $\mathcal{A}_1 \otimes \mathcal{A}_2$-measurable.

Exercise 15.16. Let $(X_1, \mathcal{A}_1)$ and (X_2, A_2) be measurable spaces, and let $\emptyset \neq A_1 \subset X_1$ and $\emptyset \neq A_2 \subset X_2$. Show that $A_1 \times A_2 \in \mathcal{A}_1 \otimes \mathcal{A}_2$ if and only if $A_1 \in \mathcal{A}_1$ and $A_2 \in \mathcal{A}_2$.

Exercise 15.17. Use (1.2.45.f) to give another proof of (15.2).

Exercise 15.18. Let λ be the Lebesgue measure on $\mathcal{B}(R)$. Prove that $\overline{\lambda} \otimes \overline{\lambda}$ is not a complete measure on $\mathcal{B}(R)^{\lambda} \otimes \mathcal{B}(R)^{\lambda}$. [Use (10.17.d) and (15.16).]

Exercise 15.19. Let $(X, \mathcal{A}, \mu)$ be a σ-finite measure space, let $f : X \to [0, \infty]$ be an $\mathcal{A}$-measurable function, and let $V_* f = \{(x, y) \in X \times [0, \infty] : y < f(x)\}$ and $V^* f = \{(x, y) \in X \times [0, \infty] : y \leqslant f(x)\}$. Prove the following.

(a) $V_* f \in \mathcal{A} \otimes \mathcal{B}([0, \infty])$, $V^* f \in \mathcal{A} \otimes \mathcal{B}([0, \infty])$.

(b) If λ is the Lebesgue measure on $\mathcal{B}([0, \infty])$, then $(\mu \otimes \lambda)(V_* f) = (\mu \otimes \lambda)(V^* f) = \int_X f \, d\mu$.

Exercise 15.20. Let $X_1 = X_2 = N$, $\mathcal{A}_1 = \mathcal{A}_2 = \mathcal{P}(N)$, and $\mu_1 = \mu_2 =$ counting measure on $\mathcal{P}(N)$. Define $f : X_1 \times X_2 \to R$ by $f(i, i) = i$, $f(i, i+1) = -i$, and $f(i, j) = 0$ if $j \neq i$ or $j \neq i + 1$. Show that $\int_{X_1} (\int_{X_2} f(x_1, x_2) d\mu_2(x_2)) d\mu_1 (x_1) = 0$ and $\int_{X_2} (\int_{X_1} f(x_1, x_2) d\mu_1(x_1)) d\mu_2(x_2) = \infty$.

Exercise 15.21. Let $f : R \to R$ be a bijective function, let μ_1 be the Lebesgue measure on $\mathcal{B}(R)$, let $\mu_2 : R \times \mathcal{B}(R) \to [0, \infty]$ be defined by $\mu_2(x, A) = \varepsilon_{f(x)}(A)$ for any $x \in R$ and $A \in \mathcal{B}(R)$, and let μ be as in (15.14). For $(x_1, x_2) \in R^2$, define

$$F(x_1, x_2) = \begin{cases} x_1 \wedge f^{-1}(x_2) & \text{if } f \text{ is strictly increasing} \\ (x_1 - f^{-1}(x_2)) \vee 0 & \text{if } f \text{ is strictly decreasing} \end{cases}.$$

Prove the following.

(a) F is continuous on R^2.

(b) $\Delta_{\mathbf{ba}} F \geqslant 0$ whenever $\mathbf{a}, \mathbf{b} \in R^2$ are such that $\mathbf{a} \leqslant \mathbf{b}$.

(c) $\mu = \mu_F$, where μ_F is as in (3.4.24).

(d) $\mu(\mathbb{R}^c) = 0$, where $\mathbb{R}$ is the graph of f.

Exercise 15.22. For $n \geqslant 2$ and $i = 1, \ldots, n$, let $(X_i, \mathcal{A}_i)$ be a measurable space, let μ_i and ν_i be σ-finite measures on $\mathcal{A}_i$ such that $\nu_i \ll \mu_i$, and let $f_i = d\nu_i/d\mu_i$. Put $f(x_1, \ldots, x_n) = f_1(x_1) \cdots f_n(x_n)$, $(x_1, \ldots, x_n) \in X_1 \times \cdots \times X_n$.

(a) Show that $\nu_1 \otimes \cdots \otimes \nu_n \ll \mu_1 \otimes \cdots \otimes \mu_n$.

(b) Use (11.28), (15.8) and (10.8) to prove that $f = d\nu_1 \otimes \cdots \otimes \nu_n / d\mu_1 \otimes \cdots \otimes \mu_n$.

Exercise 15.23. Let $(X, \mathcal{A}, \mu)$ be a measure space, let $f : X \to [0, \infty]$ be an $\mathcal{A}$-measurable function, and let $a, p \in]0, \infty[$.

(a) Use (15.19.a) and Fubini's theorem (15.8) to show that $\int_X f \, d\mu = \int_0^\infty \mu(\{f > x\}) dx = \int_0^\infty \mu(\{f \geqslant x\}) dx$, and that $\int_X (f - a)^+ \, d\mu = \int_a^\infty \mu(\{f > x\}) dx = \int_a^\infty \mu(\{f \geqslant x\}) dx$.

(b) If $\mu(X) < \infty$, prove that $\int_{\{f<a\}} f^p \, d\mu + a^p \mu(\{f \geqslant a\}) = p \int_0^a x^{p-1} \mu(\{f > x\}) dx = p \int_0^a x^{p-1} \mu(\{f \geqslant x\}) dx$, and that $\int_{\{f \leqslant a\}} f^p \, d\mu + a^p \mu(\{f > a\}) = p \int_0^a x^{p-1} \mu(\{f > x\}) dx = p \int_0^a x^{p-1} \mu(\{f \geqslant x\}) dx$.

Exercise 15.24. If $f \in \mathfrak{L}^p$ for some $0 < p < \infty$, show that $\int_X |f|^p \, d\mu = p \int_0^\infty x^{p-1} \mu(\{|f| > x\}) dx = p \int_0^\infty x^{p-1} \mu(\{|f| \geqslant x\}) dx$, and infer that $v^p \mu(\{|f| \geqslant v\}) \to 0$ as $v \to \infty$. [Hint. Use (15.23.a), then write $\mu(\{|f| \geqslant v\}) \int_u^v x^{p-1} \, dx \leqslant \int_u^v x^{p-1} \mu(\{|f| \geqslant x\}) dx \leqslant \int_u^\infty x^{p-1} \mu(\{|f| \geqslant x\}) dx, 0 < u < v < \infty$.]

Exercise 15.25. Assume that $\mu(X) < \infty$. Let $\mathfrak{L}^0$ be the set of all measurable functions from X into R, and let $\sim$ be the equivalence relation on $\mathfrak{L}^0$ defined as in (11.14.c). Put $L^0 = \mathfrak{L}^0/\sim$, and define $d_1(\tilde{f}, \tilde{g}) = \int_X (1 \wedge |f - g|) d\mu$ and $d_2(\tilde{f}, \tilde{g}) = \int_X |f - g| / (1 + |f - g|) d\mu$ for $\tilde{f}, \tilde{g} \in L^0$. Prove the following.

(a) d_1 and d_2 are well-defined metrics for L^0 generating equal topologies. [Do not forget (6.88).]

(b) For $\{\widetilde{f_n} : n \in N\} \subset L^0$ and $\tilde{f} \in L^0$, $f_n \xrightarrow{\mu} f$ if and only if $d_i(\tilde{f}, \widetilde{f_n}) \to 0, i = 1, 2$.

(c) $\{\widetilde{f_n} : n \in N\}$ is a Cauchy sequence in $(L^0, d_i), i = 1, 2$, if and only if $\{f_n : n \in N\}$ is Cauchy in μ-measure.

(d) (L^0, d_1) and (L^0, d_2) are complete metric spaces. [Hints. If $\{\widetilde{f_n} : n \in N\}$ is Cauchy relative to d_1, choose a strictly increasing sequence $\{n_k : k \geqslant 1\} \subset N$ such that $\int_X (1 \wedge |f_{n_{k+1}} - f_{n_k}|) d\mu < 2^{-k}, k \geqslant 1$, and so $\sum_{k \geqslant 1} \int_X (1 \wedge |f_{n_{k+1}} - f_{n_k}|) d\mu < \infty$. Use Fubini's theorem (15.8) to show that $\sum_{k \geqslant 1} |f_{n_{k+1}} - f_{n_k}| < \infty$ μ-a.e., and deduce that $\{f_{n_k} : k \geqslant 1\}$ is Cauchy μ-a.e. and converges μ-a.e. to some $f \in \mathfrak{L}^0$. Show that $f_n \xrightarrow{\mu} f$.]

Exercise 15.26. Let $(X, \mathcal{A}, \mu)$ be a measure space, let $f : X \to R$ be an $\mathcal{A}$-measurable function, and let $t \in \,]0, \infty[$. Prove that $\int_X e^{tf} \, d\mu = t \int_R e^{tx} \mu(\{f > x\}) dx = t \int_R e^{tx} \mu(\{f \geqslant x\}) dx$. [Use (15.8).]

16 The Product of Infinitely Many Measures

In this section we construct the product of infinitely many probabilities, and we prove the Daniell-Kolmogorov extension theorem.

In (15.13) we constructed the product of a finite number of σ-finite measures. In what follows we construct the product of infinitely many probabilities. We begin with a lemma.

Lemma 16.1. *Let $\{(X_i, \mathcal{A}_i, P_i) : i \in I\}$ be a finite set of probability spaces, and let $\emptyset \neq J \subset I$. Let P_I be the product of the probabilities $P_i, i \in I$, and let P_J be the product of the probabilities $P_i, i \in J$. Then $P_I(\pi_J^I)^{-1} = P_J$.*

Proof. In view of (15.13), P_I is a probability on $\otimes_{i \in I} \mathcal{A}_i$, and P_J is a probability on $\otimes_{i \in J} \mathcal{A}_i$. Thus $P_I(\pi_J^I)^{-1}$ and P_J are probabilities on $\otimes_{i \in J} \mathcal{A}_i$. For $i \in J$, let $A_i \in \mathcal{A}_i$.

Then $(\pi^I_J)^{-1}(\prod_{i\in J} A_i) = \prod_{i\in I} A'_i$, where $A'_i = \begin{cases} A_i & \text{if } i \in J \\ X_i & \text{if } i \notin J \end{cases}$. Therefore,

$$(P_I(\pi^I_J)^{-1})\left(\prod_{i\in J} A_i\right) = P_I\left(\prod_{i\in I} A'_i\right) = \prod_{i\in I} P_i(A'_i) = \prod_{i\in J} P_i(A_i) = P_J\left(\prod_{i\in J} A_i\right). \tag{1}$$

Since the family $\{\prod_{i\in J} A_i : A_i \in \mathcal{A}_i \text{ for } i \in J\}$ generates $\otimes_{i\in J}\mathcal{A}_i$, (1) and (10.8) show that $P_I(\pi^I_J)^{-1} = P_J$. □

Definition 16.2. Let $\{(X_i, \mathcal{A}_i, P_i) : i \in I\}$ be a nonempty set of probability spaces, and let $\mathcal{J}$ and $\mathcal{A}$ be as in (7.21). For each $J \in \mathcal{J}$, let P_J denote the product of the probabilities P_j, $j \in J$. For $A \in \mathcal{A}$, choose $J \in \mathcal{J}$ such that $A \in \mathcal{A}_J$. Therefore, $A = \pi_J^{-1}(B)$ for some $B \in \otimes_{j\in J}\mathcal{A}_j$. Define $Q(A) = P_J(B)$.

Theorem 16.3. *Notation is as in* (16.2). *Then Q is a well-defined finitely additive measure on $\mathcal{A}$.*

Proof. Let us show that Q is well defined on $\mathcal{A}$. For $A \in \mathcal{A}$, let $J_1 \in \mathcal{J}$ and $J_2 \in \mathcal{J}$ be such that $A \in \mathcal{A}_{J_1}$ and $A \in \mathcal{A}_{J_2}$. Therefore,

$$A = \pi_{J_1}^{-1}(B_1) = \pi_{J_2}^{-1}(B_2) \tag{1}$$

for some $B_1 \in \otimes_{j\in J_1}\mathcal{A}_j$ and $B_2 \in \otimes_{j\in J_2}\mathcal{A}_j$. Set $J = J_1 \cup J_2$. In view of (1.2.24), we have $\pi_{J_1} = \pi^J_{J_1} \circ \pi_J$ and $\pi_{J_2} = \pi^J_{J_2} \circ \pi_J$. Hence, using (1), we get

$$\pi_J^{-1}\left(\left(\pi^J_{J_1}\right)^{-1}(B_1)\right) = \pi_J^{-1}\left(\left(\pi^J_{J_2}\right)^{-1}(B_2)\right). \tag{2}$$

Since the function π_J is surjective, (2) shows that

$$\left(\pi^J_{J_1}\right)^{-1}(B_1) = \left(\pi^J_{J_2}\right)^{-1}(B_2). \tag{3}$$

By (16.1) and (3), we have $P_{J_1}(B_1) = (P_J(\pi^J_{J_1})^{-1})(B_1) = P_J((\pi^J_{J_1})^{-1}(B_1)) = P_J((\pi^J_{J_2})^{-1}(B_2)) = P_{J_2}(B_2)$. Thus Q is well defined on $\mathcal{A}$. Evidently, $Q(\emptyset) = 0$. Now let $A_1, A_2 \in \mathcal{A}$ be such that $A_1 \cap A_2 = \emptyset$. Choose $J_1 \in \mathcal{J}$ and $J_2 \in \mathcal{J}$ such that $A_1 \in \mathcal{A}_{J_1}$ and $A_2 \in \mathcal{A}_{J_2}$. Put $J = J_1 \cup J_2$. Then, in view of (7.20), we have $A_1, A_2 \in \mathcal{A}_J$. Therefore, $A_1 = \pi_J^{-1}(B_1)$ and $A_2 = \pi_J^{-1}(B_2)$, where $B_1, B_2 \in \otimes_{j\in J}\mathcal{A}_j$. We have

$$\emptyset = A_1 \cap A_2 = \pi_J^{-1}(B_1 \cap B_2). \tag{4}$$

Since π_J is surjective, (4) shows that $B_1 \cap B_2 = \emptyset$. Since $A_1 \cup A_2 = \pi_J^{-1}(B_1 \cup B_2)$, we have $Q(A_1 \cup A_2) = P_J(B_1 \cup B_2) = P_{J_1}(B_1) + P_{J_2}(B_2) = Q(A_1) + Q(A_2)$. Thus Q is finitely additive. □

Notation 16.4. Let $\mathcal{A}$ be as in (7.21), and let Q be as in (16.2). To emphasize the dependence of $\mathcal{A}$ and Q on the index set I we will write $\mathcal{A}^I$ instead of $\mathcal{A}$ and Q^I instead of Q.

Theorem 16.5. *Let $\{(X_i, \mathcal{A}_i, P_i) : i \in I\}$ be a set of probability spaces, let $A \in \mathcal{A}^I$, and let J be a nonempty finite subset of I. Then the following assertions hold:*

(i) $A_{\mathbf{x}} \in \mathcal{A}^{I-J}$ *for any* $\mathbf{x} \in \prod_{i \in J} X_i$.

(ii) $Q^I(A) = \int_{\prod_{i \in J} X_i} Q^{I-J}(A_{\mathbf{x}}) dP_J(\mathbf{x})$, *where P_J stands for the product of the probabilities P_i, $i \in J$.*

Proof. Let $\mathcal{J}$ be as in (7.21). There is $K \in \mathcal{J}$ such that $A \in \mathcal{A}_K$. Set $L = J \cup K$. Then, by (7.20), we have $A \in \mathcal{A}_L$. Therefore, $A = (\pi_L^I)^{-1}(B) = B \times \prod_{i \in I-L} X_i$, where $B \in \otimes_{i \in L} \mathcal{A}_i$. Let $\mathbf{x} \in \prod_{i \in J} X_i$. Since $\prod_{i \in I} X_i = \prod_{i \in J} X_i \times \prod_{i \in L-J} X_i \times \prod_{i \in I-L} X_i$, we have

$$\begin{aligned} A_{\mathbf{x}} &= \left\{ (\mathbf{y}, \mathbf{z}) \in \prod_{i \in L-J} X_i \times \prod_{i \in I-L} X_i : (\mathbf{x}, \mathbf{y}, \mathbf{z}) \in A = B \times \prod_{i \in I-L} X_i \right\} \\ &= \left\{ (\mathbf{y}, \mathbf{z}) \in \prod_{i \in L-J} X_i \times \prod_{i \in I-L} X_i : (\mathbf{x}, \mathbf{y}) \in B \right\} \\ &= B_{\mathbf{x}} \times \prod_{i \in I-L} X_i = (\pi_{L-J}^{I-J})^{-1}(B_{\mathbf{x}}). \end{aligned} \tag{1}$$

According to (15.1), we have $B_{\mathbf{x}} \in \otimes_{i \in L-J} \mathcal{A}_i$. Consequently, (1) shows that $A_{\mathbf{x}} \in \mathcal{A}^{I-J}$. Thus (i) holds. Moreover, we have

$$Q^{I-J}(A_{\mathbf{x}}) = P_{L-J}(B_{\mathbf{x}}), \tag{2}$$

where P_{L-J} is the product of the probabilities P_i, $i \in L-J$. Let P_L denote the product of the probabilities P_i, $i \in L$. Then, using (2), we have $Q^I(A) = P_L(B) = (P_J \otimes P_{L-J})(B) = \int_{\prod_{i \in J} X_i} P_{L-J}(B_{\mathbf{x}}) dP_J(\mathbf{x}) = \int_{\prod_{i \in J} X_i} Q^{I-J}(A_{\mathbf{x}}) dP_J(\mathbf{x})$. Hence (ii) holds. □

Theorem 16.6. *Let $\{(X_i, \mathcal{A}_i, P_i) : i \in I\}$ be a denumerable set of probability spaces. Then there exists a unique probability P on $\otimes_{i \in I} \mathcal{A}_i$ such that*

(i) $P\pi_J^{-1} = P_J$

for any nonempty finite set $J \subset I$, where P_J stands for the product of the probabilities P_j, $j \in J$.

Proof. Without loss of generality, assume that $I = N$. For $k \in N$, denote $I_k = \{k+1, k+2, \ldots\}$. Let $\mathcal{A}$ be as in (7.21), and let Q be as in (16.2). In view of (16.3), Q is a finitely additive measure on $\mathcal{A}$. We next show that Q is a σ-additive measure on $\mathcal{A}$. To do this, let $\{A_n : n \in N\} \subset \mathcal{A}$ be a nonincreasing sequence such that $\lim_n Q(A_n) > 0$. Then, applying (9.8), it will suffice to show that $\cap_{n \in N} A_n \neq \emptyset$. Using (16.5), we have $Q(A_n) = Q^I(A_n) = \int_{X_1} Q^{I_1}((A_n)_{x_1}) dP_1(x_1)$. Thus the dominated convergence theorem (11.25.ii) shows that $0 < \lim_n Q(A_n) = \int_{X_1} \lim_n Q^{I_1}((A_n)_{x_1}) dP_1(x_1)$. Therefore, there is $x_1 \in X_1$ such that $\lim_n Q^{I_1}((A_n)_{x_1}) > 0$. If $x_1 \in X_1, \ldots, x_k \in X_k$ have been chosen so that $\lim_n Q^{I_k}((A_n)_{(x_1,\ldots,x_k)}) > 0$, then, using (16.5) and (11.25.ii), we have

$$0 < \lim_n Q^{I_k}\left((A_n)_{(x_1,\ldots,x_k)}\right) = \int_{X_{k+1}} \lim_n Q^{I_{k+1}}\left((A_n)_{(x_1,\ldots,x_k,x_{k+1})}\right) dP_{k+1}(x_{k+1}).$$

Consequently, there is $x_{k+1} \in X_{k+1}$ such that $\lim_n Q^{I_{k+1}}((A_n)_{(x_1,\ldots,x_k,x_{k+1})}) > 0$. Thus we have proved inductively that there is a point $\mathbf{x} = (x_k)_{k\in N} \in \prod_{k\in N} X_k$ such that

$$\lim_n Q^{I_k}((A_n)_{(x_1,\ldots,x_k)}) > 0, \quad k \in N. \tag{1}$$

Equation (1) shows that

$$(A_n)_{(x_1,\ldots,x_k)} \neq \emptyset, \quad k, n \in N. \tag{2}$$

For $n \in N$, select a nonempty finite set $J \subset I$ such that $A_n \in \mathcal{A}_J$. Let $k = \sup J$. Then $A_n \in \mathcal{A}_{\{1,\ldots,k\}}$. Therefore, $A_n = \overset{-1}{\pi}_{\{1,\ldots,k\}}(B) = B \times \prod_{i>k} X_i$, where $B \in \mathcal{A}_1 \otimes \cdots \otimes \mathcal{A}_k$. Since

$$(A_n)_{(x_1,\ldots,x_k)} = \begin{cases} \prod_{i>k} X_i & \text{if } (x_1,\ldots,x_k) \in B \\ \emptyset & \text{if } (x_1,\ldots,x_k) \notin B \end{cases},$$

(2) shows that $(x_1,\ldots,x_k) \in B$. Consequently, $\mathbf{x} \in A_n$. Thus $\mathbf{x} \in \cap_{n\in N} A_n \neq \emptyset$. Therefore, Q is a σ-additive measure on $\mathcal{A}$. Obviously, $Q(\prod_{i\in I} X_i) = 1$. Using (7.21), Carathéodory's extension theorem (10.9) shows that there exists a unique probability P on $\otimes_{i\in I}\mathcal{A}_i$ such that $P = Q$ on $\mathcal{A}$. From this it follows at once that P is the only probability on $\otimes_{i\in I}\mathcal{A}_i$ that satisfies (i). □

The probability P in (16.6) is called the *product of the probabilities P_i, $i \in I$.*

Theorem 16.7. *Let $\{(X_i, \mathcal{A}_i, P_i) : i \in I\}$ and P be as in* (16.6). *If $A_i \in \mathcal{A}_i$ for any $i \in I$, then $P(\prod_{i\in I} A_i) = \prod_{i\in I} P_i(A_i)$.*

Proof. Without any loss of generality, suppose that $I = N$. For $n \in N$, put $B_n = A_1 \times \cdots \times A_n \times \prod_{i>n} X_i$. Then $\{B_n : n \in N\}$ is a nonincreasing sequence such that $\prod_{n\in N} A_n = \cap_{n\in N} B_n$. Therefore, using (16.6), we have $P(\prod_{n\in N} A_n) = \lim_n P(B_n) = \lim_n (P\pi_{\{1,\ldots,n\}^{-1}})(A_1 \times \cdots \times A_n) = \lim_n (P_1 \otimes \cdots \otimes P_n)(A_1 \times \cdots \times A_n) = \lim_n P_1(A_1)\cdots P_n(A_n) = \prod_{n\in N} P_n(A_n)$. □

The next theorem generalizes (16.6).

Theorem 16.8 (Ionescu Tulcea). *Let $\{(X_n, \mathcal{A}_n) : n \in N\}$ be a sequence of measurable spaces, and let P_1 be a probability on $\mathcal{A}_1$. For $n \geqslant 2$, let $P_n : X_1 \times \cdots \times X_{n-1} \times \mathcal{A}_n \to [0, 1]$ be a function such that the section of P_n at any $(x_1, \ldots, x_{n-1}) \in X_1 \times \cdots \times X_{n-1}$ is a probability on $\mathcal{A}_n$ and the section of P_n at any $A \in \mathcal{A}_n$ is $\mathcal{A}_1 \otimes \cdots \otimes \mathcal{A}_{n-1}$-measurable. Then there exists a unique probability P on $\otimes_{n\in N}\mathcal{A}_n$ such that $P\overset{-1}{\pi}_{\{1,\ldots,n\}} = P^{(n)}$, $n \in N$, where $P^{(n)}$ is the probability on $\mathcal{A}_1 \otimes \cdots \otimes \mathcal{A}_n$ constructed as in* (15.14) *starting from $P_1, \ldots, P_n$.*

The proof of this theorem is similar to the proof of (16.6) and is left to the reader. In what follows we extend Theorem (16.6) to the case where I is an arbitrary infinite set.

Lemma 16.9. *Let $\{(X_i, \mathcal{A}_i, P_i) : i \in I\}$ be a denumerable set of probability spaces, and let $\emptyset \neq J \subset I$. Let P_I be the product of the probabilities P_i, $i \in I$, and let P_J be the product of the probabilities P_j, $j \in J$. Then*

(i) $P_I(\pi_J^I)^{-1} = P_J$.

Proof. $P_I(\pi_J^I)^{-1}$ and P_J are probabilities on $\otimes_{j\in J}\mathcal{A}_j$. Let K be a nonempty finite subset of J, and let P_K be the product of the probabilities P_k, $k \in K$. Then, according to (16.6), we have $P_I(\pi_J^I)^{-1}(\pi_K^J)^{-1} = P_I(\pi_K^I)^{-1} = P_K$ and $P_J(\pi_K^J)^{-1} = P_K$. Thus the uniqueness assertion in (16.6) shows that (i) holds. □

Definition 16.10. Let $\{(X_i, \mathcal{A}_i, P_i) : i \in I\}$ be a nonempty set of probability spaces, and let $\mathcal{J}^*$ be as in (7.22). For each $J \in \mathcal{J}^*$, let P_J denote the product of the probabilities P_j, $j \in J$. As proved in (7.22), $\otimes_{i\in I}\mathcal{A}_i = \cup_{J\in\mathcal{J}^*}\mathcal{A}_J$, where $\mathcal{A}_J$ is as in (7.20). For $A \in \otimes_{i\in I}\mathcal{A}_i$, choose $J \in \mathcal{J}^*$ such that $A \in \mathcal{A}_J$. Therefore, $A = \pi_J^{-1}(B)$ for some $B \in \otimes_{j\in J}\mathcal{A}_j$. Define $P(A) = P_J(B)$. Using (16.9), a reasoning similar to the one in the proof of (16.3) shows that P is well defined.

Theorem 16.11. *Notation is as in* (16.10). *Then P is the unique probability on $\otimes_{i\in I}\mathcal{A}_i$ such that*

(i) $P\pi_J^{-1} = P_J$

for any nonempty finite set $J \subset I$, where P_J stands for the product of the probabilities P_j, $j \in J$.

Proof. Obviously, $P(\emptyset) = 0$ and $P(\prod_{i\in I} X_i) = 1$. Let $\{A_n : n \in N\} \subset \otimes_{i\in I}\mathcal{A}_i$ be such that $A_m \cap A_n = \emptyset$ for $m \neq n$. For $n \in N$, choose $J_n \in \mathcal{J}^*$ such that $A_n \in \mathcal{A}_{J_n}$. Set $J = \cup_{n\in N} J_n$. Then $J \in \mathcal{J}^*$, and $A_n \in \mathcal{A}_J$ for any $n \in N$. Therefore, for $n \in N$, we have $A_n = \pi_J^{-1}(B_n)$ for some $B_n \in \otimes_{j\in J}\mathcal{A}_j$. For $m \neq n$, we have $\emptyset = A_m \cap A_n = \pi_J^{-1}(B_m \cap B_n)$, and so $B_m \cap B_n = \emptyset$. Since $\cup_{n\in N} A_n = \pi_J^{-1}(\cup_{n\in N} B_n)$, we get $P(\cup_{n\in N} A_n) = P_J(\cup_{n\in N} B_n) = \sum_{n\in N} P_J(B_n) = \sum_{n\in N} P(A_n)$. Thus P is a probability on $\otimes_{i\in I}\mathcal{A}_i$. Evidently, P satisfies (i). Using (7.21) and (10.9), it is easily seen that P is the only probability on $\otimes_{i\in I}\mathcal{A}_i$ that satisfies (i). □

The probability P in (16.11) is called the *product of the probabilities* P_i, $i \in I$, and is denoted by $\otimes_{i\in I} P_i$. If $I = \{n \in Z : n \geqslant m\}$ for some $m \in Z$, then we will write also $\otimes_{n\geqslant m} P_n$ instead of $\otimes_{i\in I} P_i$.

One of the essential results of this chapter is the Daniell-Kolmogorov extension theorem which we will prove in (16.20). We begin with a definition.

Definition 16.12. A family $\mathcal{B} \subset \mathcal{P}(X)$ is said to be *compact* if $\{B_n : n \geqslant 1\} \subset \mathcal{B}$ and $\cap_{i=1}^n B_i \neq \emptyset$ for any $n \geqslant 1$ imply $\cap_{i\geqslant 1} B_i \neq \emptyset$.

Example 16.13. (a) The family of all finite subsets of an arbitrary set X is compact.

(b) Let X be a Hausdorff space, and let $\mathcal{B}$ be the family of all compact subsets of X. Then, according to (5.24) and (5.27), $\mathcal{B}$ is compact.

Remark 16.14. (a) If $\mathcal{B} \subset \mathcal{P}(X)$ is compact, then $\mathcal{B} \cup \{X\}$ is compact.

(b) If $\mathcal{B}$ is compact and $\mathcal{C} \subset \mathcal{B}$, then $\mathcal{C}$ is compact.

(c) For $i \in I \neq \emptyset$, let X_i be a set, and let $\mathcal{B}_i \subset \mathcal{P}(X_i)$ be a compact family. Then, in view of (2.44), the family $\cup_{i\in I}\pi_i^{-1}(\mathcal{B}_i) \subset \mathcal{P}(\prod_{i\in I} X_i)$ is compact.

Theorem 16.15. *Let $\mathcal{B}$ be a compact family, and let $\widetilde{\mathcal{B}}$ be the family of all denumerable intersections of finite unions of sets in $\mathcal{B}$. Then the following assertions hold.*

(i) *$\widetilde{\mathcal{B}}$ is compact.*

(ii) *If $\{A_n : n \geqslant 1\} \subset \widetilde{\mathcal{B}}$, then $\cap_{n\geqslant 1} A_n \in \widetilde{\mathcal{B}}$.*

(iii) *If $A, B \in \widetilde{\mathcal{B}}$, then $A \cup B \in \widetilde{\mathcal{B}}$.*

Proof. To prove (i), for $k \geqslant 1$, let $B^k = \cap_{i\geqslant 1} C_i^k \in \widetilde{\mathcal{B}}$, where $C_i^k = A_{i1}^k \cup \cdots \cup A_{il_i(k)}^k$, $i \geqslant 1$, is such that $\cap_{k=1}^n B^k \neq \emptyset$ whatever $n \geqslant 1$. We arrange the sets $C_i^k = \cup_{j=1}^{l_i(k)} A_{ij}^k$, $i, k \geqslant 1$, in a sequence $\{D^m : m \geqslant 1\}$, where $D^m = A_1^m \cup \cdots \cup A_{r_m}^m$, $m \geqslant 1$. Since $\emptyset \neq \cap_{k=1}^n B^k = \cap_{k=1}^n \cap_{i\geqslant 1} C_i^k$, $n \geqslant 1$, it follows that $\cap_{m=1}^n D^m \neq \emptyset$ for any $n \geqslant 1$. Consequently, if $n \geqslant 2$, there is s_1 with $1 \leqslant s_1 \leqslant r_1$ such that $A_{s_1}^1 \cap D^2 \cap \cdots \cap D^n \neq \emptyset$. If $s_1, \ldots, s_l$ have been chosen so that $A_{s_1}^1 \cap \cdots \cap A_{s_l}^l \cap D^{l+1} \cap \cdots \cap D^n \neq \emptyset$ for any $n \geqslant l+1$, then there is s_{l+1} with $1 \leqslant s_{l+1} \leqslant r_{l+1}$ such that $A_{s_1}^1 \cap \cdots \cap A_{s_l}^l \cap A_{s_{l+1}}^{l+1} \cap D^{l+2} \cap \cdots \cap D^n \neq \emptyset$, $n \geqslant l+2$. Thus it follows inductively that $\cap_{i=1}^l A_{s_i}^i \neq \emptyset$ for any $l \geqslant 1$. As $A_{s_i}^i \in \mathcal{B}$, $i \geqslant 1$, we have $\emptyset \neq \cap_{i\geqslant 1} A_{s_i}^i \subset \cap_{m\geqslant 1} D^m = \cap_{k\geqslant 1} B^k$. This completes the proof of (i). Assertion (ii) is obvious, and assertion (iii) follows from (ii). □

Definition 16.16. Let $\mathcal{A}$ be an algebra of sets, let $\mathcal{B} \subset \mathcal{A}$ be a compact family, and let μ be a finitely additive measure on $\mathcal{A}$. We say that μ is *regular* [*$\mathcal{B}$-regular*] if $\mu(A) = \sup\{\mu(B) : A \supset B \in \mathcal{B}\}$ for any $A \in \mathcal{A}$.

Theorem 16.17. *Let $\mathcal{A}$ be an algebra of subsets of X, let $\mathcal{B} \subset \mathcal{A}$ be a compact family, and let μ be a finitely additive measure on $\mathcal{A}$ such that $\mu(X) < \infty$. If μ is $\mathcal{B}$ regular, then μ is σ-additive.*

Proof. Let $\{A_n : n \in N\} \subset \mathcal{A}$ be such that $\cup_{n\in N} A_n \in \mathcal{A}$ and $A_m \cap A_n = \emptyset$ for $m \neq n$. Evidently, $\mu(\cup_{n\in N} A_n) \geqslant \sum_{n\in N} \mu(A_n)$. To get a contradiction assume that $\mu(\cup_{n\in N} A_n) > \sum_{n\in N} \mu(A_n)$, and let $0 < \varepsilon \leqslant \mu(\cup_{n\in N} A_n) - \sum_{n\in N} \mu(A_n)$. For $n \in N$, set $B_n = \cup_{i>n} A_i$. Then $B_n \supset B_{n+1}$, $B_n = \cup_{i\in N} A_i - (\cup_{i=1}^n A_i) \in \mathcal{A}$ and

$$\mu(B_n) = \mu\left(\bigcup_{i\in N} A_i\right) - \sum_{i=1}^{n} \mu(A_i) \geqslant \mu\left(\bigcup_{i\in N} A_i\right) - \sum_{i\in N} \mu(A_i) \geqslant \varepsilon, \quad n \in N. \tag{1}$$

For $n \in N$, select $C_n \in \mathcal{B}$ such that $C_n \subset B_n$ and

$$\mu(B_n) - \mu(C_n) \leqslant \varepsilon/2^{n+1}. \tag{2}$$

Then, using (1) and (2), we have $\mu(\cap_{i=1}^n C_i) = \mu(\cap_{i=1}^n B_i) - \mu(\cap_{i=1}^n B_i - (\cap_{i=1}^n C_i)) \geqslant \mu(B_n) - \mu(\cup_{i=1}^n (B_i - C_i)) \geqslant \varepsilon - \sum_{i=1}^n \varepsilon/2^{i+1} > \varepsilon/2, n \in N$. Therefore, $\cap_{i=1}^n C_i \neq \emptyset$ for any $n \in N$. Since $\mathcal{B}$ is compact, we get $\cap_{i\in N} C_i \neq \emptyset$. On the other hand, we have $\cap_{n\in N} C_n \subset \cap_{n\in N} B_n = \emptyset$. This contradiction shows that μ *is* σ-additive. □

Theorem 16.18. *Let $(X, \mathcal{A}, \mu)$ be a finite measure space, and let $\mathcal{B} \subset \mathcal{A}$ be such that $\{A_n : n \in N\} \subset \mathcal{B}$ implies $\cap_{n\in N} A_n \in \mathcal{B}$, and $A, B \in \mathcal{B}$ implies $A \cup B \in \mathcal{B}$. Let $\mathcal{C} = \{A \in \mathcal{A} : \mu(A) = \sup\{\mu(B) : A \supset B \in \mathcal{B}\}\}$. Then the family $\{A \in \mathcal{A} : A \in \mathcal{C}$ and $A^c \in \mathcal{C}\}$ is a σ-algebra unless it is empty.*

Proof. Let $\{A_n : n \in N\} \subset \mathcal{C}$. It suffices to show that $\cup_{n\in N} A_n \in \mathcal{C}$ and $\cap_{n\in N} A_n \in \mathcal{C}$. Let $\alpha < \mu(\cup_{n\in N} A_n)$, and choose $m \in N$ such that $\alpha < \mu(\cup_{n=1}^m A_n)$. For $n = 1, \ldots, m$, select $B_n \in \mathcal{B}$ such that $B_n \subset A_n$ and $\mu(A_n) - \mu(B_n) \leqslant (\mu(\cup_{n=1}^m A_n) - \alpha)/2^n$. Then we have

$$\begin{aligned}\mu\left(\bigcup_{n=1}^m B_n\right) &= \mu\left(\bigcup_{n=1}^m A_n\right) - \mu\left(\bigcup_{n=1}^m A_n - \left(\bigcup_{n=1}^m B_n\right)\right)\\ &\geqslant \mu\left(\bigcup_{n=1}^m A_n\right) - \sum_{n=1}^m \mu(A_n - B_n)\\ &\geqslant \mu\left(\bigcup_{n=1}^m A_n\right) - \left(\mu\left(\bigcup_{n=1}^m A_n\right) - \alpha\right) = \alpha. \end{aligned} \tag{1}$$

Since $\cup_{n=1}^m B_n \subset \cup_{n=1}^m A_n$ and $\cup_{n=1}^m B_n \in \mathcal{B}$, (1) shows that $\cup_{n\in N} A_n \in \mathcal{C}$. Now let $\varepsilon > 0$. For $n \in N$, select $C_n \in \mathcal{B}$ such that $C_n \subset A_n$ and $\mu(A_n) - \mu(C_n) \leqslant \varepsilon/2^n$. Then we have

$$\begin{aligned}\mu\left(\bigcap_{n\in N} C_n\right) &= \mu\left(\bigcap_{n\in N} A_n\right) - \mu\left(\bigcap_{n\in N} A_n - \left(\bigcap_{n\in N} C_n\right)\right)\\ &\geqslant \mu\left(\bigcap_{n\in N} A_n\right) - \sum_{n\in N} \mu(A_n - C_n)\\ &\geqslant \mu\left(\bigcap_{n\in N} A_n\right) - \varepsilon. \end{aligned} \tag{2}$$

Since $\cap_{n\in N} C_n \subset \cap_{n\in N} A_n$ and $\cap_{n\in N} C_n \in \mathcal{B}$, (2) shows that $\cap_{n\in N} A_n \in \mathcal{C}$. □

Theorem 16.19. *Let $\{(X_i, \mathcal{A}_i) : i \in I\}$ be a nonempty family of measurable spaces, and let μ be a finite measure on $\otimes_{i\in I} \mathcal{A}_i$ such that $\mu\pi_i^{-1}$ is regular for any $i \in I$. Then μ is regular.*

Proof. For $i \in I$, let $\mathcal{B}_i \subset \mathcal{A}_i$ be a compact family such that $\mu\pi_i^{-1}$ is $\mathcal{B}_i$-regular. In view of (16.14.a), we may and do assume that $X_i \in \mathcal{B}_i$. Let $\mathcal{B} = \{\prod_{i\in I} B_i : B_i \in \mathcal{B}_i, i \in I$, and $\{i \in I : B_i \neq X_i\}$ is finite$\}$. Obviously, $\mathcal{B} \subset \otimes_{i\in I} \mathcal{A}_i$. Using (1.2.44), it follows at once that $\mathcal{B}$ is a compact family. Let $\widetilde{\mathcal{B}}$ be as in (16.15). Evidently, $\widetilde{\mathcal{B}} \subset \otimes_{i\in I} \mathcal{A}_i$. We will show that μ is $\widetilde{\mathcal{B}}$-regular. Since $\otimes_{i\in I} \mathcal{A}_i = \sigma(\cup_{i\in I} \pi_i^{-1}(\mathcal{A}_i))$, and $A \in \cup_{i\in I} \pi_i^{-1}(\mathcal{A}_i)$ implies $A^c \in \cup_{i\in I} \pi_i^{-1}(\mathcal{A}_i)$, (16.18) shows that it suffices to prove that $\mu(A) = \sup\{\mu(B) : A \supset B \in \widetilde{\mathcal{B}}\}$ for any $A \in \cup_{i\in I} \pi_i^{-1}(\mathcal{A}_i)$. Let $A \in \pi_i^{-1}(\mathcal{A}_i)$. Therefore, $A = \pi_i^{-1}(B)$ for some $B \in \mathcal{A}_i$.

For $\varepsilon > 0$, choose $C \in \mathcal{A}_i$ such that $C \subset B$ and $(\mu\pi_i^{-1})(C) > (\mu\pi_i^{-1})(B) - \varepsilon$. Consequently

$$(\mu\pi_i^{-1})(C) > (\mu\pi_i^{-1})(B) - \varepsilon = \mu(A) - \varepsilon \quad (1)$$

Since $\pi_i^{-1}(C) \in \mathcal{B} \subset \widetilde{\mathcal{B}}$, (1) shows that μ is $\widetilde{\mathcal{B}}$-regular. □

Daniell-Kolmogorov extension theorem 16.20. *Let $\{(X_i, \mathcal{A}_i) : i \in I\}$ be a nonempty family of measurable spaces, and let $\mathcal{J}$ be as in* (7.21). *For any $J \in \mathcal{J}$, let μ_J be a finite measure on $\otimes_{j\in J}\mathcal{A}_j$ such that $\mu_J(\pi_K^J)^{-1} = \mu_K$ whenever $\emptyset \neq K \subset J$. If $\mu_{\{i\}}$ is regular for any $i \in I$, then there exists a unique measure μ on $\otimes_{i\in I}\mathcal{A}_i$ such that*

(i) $\mu\pi_J^{-1} = \mu_J$

for each $J \in \mathcal{J}$.

Proof. Let $\mathcal{A}$ be as in (7.21). For $A \in \mathcal{A}$, choose $J \in \mathcal{J}$ such that $A \in \mathcal{A}_J$. Therefore, $A = \pi_J^{-1}(B)$ for some $B \in \otimes_{j\in J}\mathcal{A}_j$. Define $\mu(A) = \mu_J(B)$. As in (16.3), it is proved that μ is a well-defined finitely additive measure on $\mathcal{A}$. For $i \in I$, let $\mathcal{B}_i \subset \mathcal{A}_i$ be a compact family such that $\mu_{\{i\}}$ is $\mathcal{B}_i$-regular. According to (16.14.a), we may and do assume that $X_i \in \mathcal{B}_i$. Let $\mathcal{B} = \{\prod_{i\in I} B_i : B_i \in \mathcal{B}_i, i \in I,$ and $\{i \in I : B_i \neq X_i\}$ is finite$\}$, and let $\widetilde{\mathcal{B}}$ be as in (16.15). Let $A = \pi_J^{-1}(B)$, where $B \in \otimes_{j\in J}\mathcal{A}_j$. In view of (16.19), there exists a compact family $\widetilde{\mathcal{B}_J} \subset \otimes_{j\in J}\mathcal{A}_j$ such that μ_J is $\widetilde{\mathcal{B}_J}$-regular. For $\varepsilon > 0$, select $C \in \widetilde{\mathcal{B}_J}$ such that $C \subset B$ and $\mu_J(C) > \mu_J(B) - \varepsilon$. Therefore,

$$\mu\pi_J^{-1}(C) > \mu\pi_J^{-1}(B) = \mu(A) - \varepsilon. \quad (1)$$

Since $\pi_J^{-1}(C) \subset \pi_J^{-1}(B)$, (1) shows that μ is regular with respect to the family $\cup_{J\in\mathcal{J}}\pi_J^{-1}(\widetilde{\mathcal{B}_J}) \subset \mathcal{A}$. As $\cup_{J\in\mathcal{J}}\pi_J^{-1}(\widetilde{\mathcal{B}_J}) \subset \widetilde{\mathcal{B}}$, (16.14.b) shows that $\cup_{J\in\mathcal{J}}\pi_J^{-1}(\widetilde{\mathcal{B}_J})$ is compact. Using (16.17), it follows that μ is σ-additive. Using (7.21), Carathéodory's extension theorem (10.9) shows that μ has a unique extension to $\otimes_{i\in I}\mathcal{A}_i$. From this it follows at once that μ is the only measure on $\otimes_{i\in I}\mathcal{A}_i$ that satisfies (i). □

Theorem (16.23) below shows that the Daniell-Kolmogorov extension theorem applies whenever each X_i is a complete separable metric space and $\mathcal{A}_i = \mathcal{B}(X_i)$; in particular, it applies whenever each X_i is a countable set and $\mathcal{A}_i = \mathcal{P}(X_i)$. We need the following intermediate results.

Theorem 16.21. *Let X be a metric space, let $\mathcal{B}$ be the family of all closed subsets of X and let μ be a finite measure on $\mathcal{B}(X)$. Then $\mu(A) = \sup\{\mu(B) : A \supset B \in \mathcal{B}\}$ for any $A \in \mathcal{B}(X)$.*

Proof. Let $\mathcal{A} = \mathcal{B}(X)$. Notice that $\mathcal{B}$ fulfills the assumptions in (16.18). Further, let $\mathcal{C}$ be as in (16.18), and put $\mathcal{D} = \{A \in \mathcal{A} : A \in \mathcal{C}$ and $A^c \in \mathcal{C}\}$. Clearly, it suffices to prove that $\mathcal{C} \supset \mathcal{B}(X)$. Let $A \in \mathcal{B}$. Plainly, $A \in \mathcal{C}$. Moreover, according to (6.85), there is a sequence $\{B_n : n \in N\} \subset \mathcal{B}$ such that $A^c = \cup_{n\in N} B_n$. Since $\cup_{i=1}^n B_i \in \mathcal{B}$, $n \in N$, and $\mu(A^c) = \sup \mu(\cup_{i=1}^n B_i)$ (9.9), it follows that $A^c \in \mathcal{C}$. Thus $A \in \mathcal{D}$, and so $\mathcal{D} \supset \mathcal{B}$. Hence, in view of (16.18), we see that $\mathcal{D} \supset \mathcal{B}(X)$. As $\mathcal{C} \supset \mathcal{D}$, we get $\mathcal{C} \supset \mathcal{B}(X)$. □

Theorem 16.22 (Ulam). *Let (X, d) be a complete separable metric space, and let μ be a finite measure on $\mathcal{B}(X)$. Then for each $\varepsilon > 0$, there exists a compact set $C \subset X$ such that $\mu(X) - \mu(C) < \varepsilon$.*

Proof. Since X is separable, there is a countable set A dense in X. The case in which A is finite is simpler and it is left to the reader. If A is denumerable, then put $\{A = x_i : i \geqslant 1\}$. For $i, n \geqslant 1$, set $S_{in} = \overline{S}(x_i, 1/n)$. For every $n \geqslant 1$, as A is dense in X we get $X = \cup_{i \geqslant 1} S_{in}$, and so, by (9.9), there exists $i_n \geqslant 1$ such that

$$\mu\left(\bigcup_{i=1}^{i_n} S_{in}\right) > \mu(X) - \varepsilon/2^n. \tag{1}$$

Denote $C_n = \cup_{i=1}^{i_n} S_{in}$, $n \geqslant 1$, and $C = \cap_{n \geqslant 1} C_n$. Obviously, C is a totally bounded set. Since C is closed, and X is complete, (6.25.b) shows that (C, d_C) is a complete metric space. Hence, in view of (6.31), it follows that C is compact. Finally, on account of (1), we have $\mu(X) - \mu(C) = \mu(\cup_{n \geqslant 1} C_n^c) \leqslant \sum_{n \geqslant 1} \mu(C_n^c) < \sum_{n \geqslant 1} \varepsilon/2^n = \varepsilon$. □

Theorem 16.23. *Let (X, d) and μ be as in* (16.22). *Then μ is regular.*

Proof. Let $\mathcal{C}$ be the family of all compact subsets of X. Then, in view of (16.13.b), $\mathcal{C}$ is compact. We will show that μ is $\mathcal{C}$-regular. Let $A \in \mathcal{B}(X)$ and $\varepsilon > 0$. Applying (16.21), select a closed set $B \subset X$ such that

$$\mu(A) - \mu(B) < \varepsilon/2. \tag{1}$$

Then, by (6.25.b) and (6.70), (B, d_B) is a complete separable metric space. On the other hand, according to (7.14) and (7.48), we have $\mathcal{B}(B) = B \cap \mathcal{B}(X) \subset \mathcal{B}(X)$. Hence (16.22) applies, and we can select a compact subset C of B such that

$$\mu(B) - \mu(C) < \varepsilon/2. \tag{2}$$

By virtue of (4.31), $C \in \mathcal{C}$. Moreover, combining (1) with (2), we get $\mu(A) - \mu(C) < \varepsilon$. Thus μ is $\mathcal{C}$-regular. □

Exercise 16.24. For $1 < b \in N$, let $I = \{0, \dots, b-1\}$. For $n \in N$, set $I_n = I$ and define a probability P_n on $\mathcal{P}(I_n)$ by $P_n(\{i\}) = 1/b$, $i \in I_n$. Let λ be the Lebesgue measure on $\mathcal{B}([0, 1[)$, and let P be the product of the probabilities P_n, $n \in N$. Prove that $\lambda(a_n)_{n \in N}^{-1} = P$, where the function $(a_n)_{n \in N}$ is as in (8.23).

Exercise 16.25 (Kakutani). For $n \in N$, let $(X_n, \mathcal{A}_n)$ be a measurable space, let P_n and Q_n be probabilities on $\mathcal{A}_n$ such that $P_n \ll Q_n$, and let $f_n = dP_n/dQ_n$. Let P and Q denote the products of the probabilities P_n, $n \in N$, and Q_n, $n \in N$, respectively.

(a) Show that $P \ll Q$ whenever $\prod_{n \in N} \int_{X_n} f_n^{1/2}\, dQ_n > 0$. [Hints. For $\mathbf{x} = (x_1, x_2, \dots) \in \prod_{n \in N} X_n$, define $g_n(\mathbf{x}) = (f_1(x_1) \cdots f_n(x_n))^{1/2}$. Use (11.29) and (15.8) to prove that $\{\widetilde{g_n} : n \in N\}$ is a Cauchy sequence in $L^2(\prod_{n \in N} X_n, \otimes_{n \in N} \mathcal{A}_n, Q)$, and let $g \in \mathfrak{L}^2(\prod_{n \in N} X_n, \otimes_{n \in N} \mathcal{A}_n, Q)$ be such that $\|g - g_n\|_2 \to 0$ (13.16). For $A \in \pi_{\{1, \dots, m\}}^{-1}(\mathcal{A}_1 \otimes \cdots \otimes \mathcal{A}_m)$, use (15.22) and (11.29) to show that $P(A) = \lim_n \int_A g_n^2\, dQ = \int_A g^2\, dQ$. Then apply (10.8).]

(b) Prove that $P \perp Q$ whenever $\prod_{n \in N} \int_{X_n} f_n^{1/2}\, dQ_n = 0$. [Hints. For $\varepsilon > 0$, choose $m \in N$ such that $\prod_{n=1}^{m} \int_{X_n} f_n^{1/2}\, dQ_n < \varepsilon$. Put $B = \{\mathbf{x} \in \prod_{n \in N} X_n : g_m(\mathbf{x}) \geqslant 1\}$. Show that $Q(B) < \varepsilon$ and $P(B^c) < \varepsilon$, and apply (12.25).]

(c) Let ρ be as in (13.26). Use (13.5) and (12.23.d) to prove that $\rho(P, Q) = \prod_{n \in N} \rho(P_n, Q_n)$.

Exercise 16.26. For $i \in I \neq \emptyset$, let $\mathcal{A}_i$ be a σ-algebra, and let P_i and Q_i be probabilities on $\mathcal{A}_i$. Let P and Q stand for the products of the probabilities $P_i, i \in I$, and $Q_i, i \in I$, respectively. Show that $P \ll Q$ implies that $P_i \ll Q_i$ for any $i \in I$.

Exercise 16.27. If $\mathcal{B} \subset \mathcal{P}(X)$ is finite, then $\mathcal{B}$ is compact.

Exercise 16.28. Let $\mathcal{B}_i \subset \mathcal{P}(X), i = 1, 2$, be compact families, and define $\mathcal{B} = \{B_1 \cup B_2 : B_1 \in \mathcal{B}_1 \text{ and } B_2 \in \mathcal{B}_2\}$. Is $\mathcal{B}$ compact?

Exercise 16.29. Let $\mathcal{A}$ and $\mathcal{B}$ be as in (16.16), and let $\mu_1, \ldots, \mu_n$ be $\mathcal{B}$-regular finitely additive measures on $\mathcal{A}$. Show that $\mu_1 + \cdots + \mu_n$ is regular.

Exercise 16.30. Let $(X, \mathcal{A}, \mu)$, (Y, B) and f be as in (9.26), let $\mathcal{C} \subset \mathcal{B}$ be a compact family, and assume that f is surjective. Prove the following.

(a) $f^{-1}(\mathcal{C})$ is a compact family.

(b) $\mu_{f^{-1}(\mathcal{B})}$ is $f^{-1}(\mathcal{C})$-regular if and only if μf^{-1} is $\mathcal{C}$-regular.

Exercise 16.31. Let $\mathcal{A}$ be a σ-algebra, let μ be a σ-finite regular measure on $\mathcal{A}$, and let ν be a finite measure on $\mathcal{A}$ such that $\nu \ll \mu$. Prove that ν is regular.

Exercise 16.32. Let X be a complete separable metric space, let μ be a finite measure on $\mathcal{B}(X)$ such that $\mu(\{x\}) = 0$ for any $x \in X$, let $B \in \mathcal{B}(X)$, and consider $c \in [0, \mu(B)]$. Use (9.42.b) and (16.23) to prove that there exists a sequence $\{C_n : n \in N\}$ of compact subsets of X such that $C_n \subset B, n \in N$, and $\mu(\cup_{n \in N} C_n) = c$.

Exercise 16.33. Find a metric space X and a measure μ on $\mathcal{B}(X)$ such that μ is not $\mathcal{B}$-regular, where $\mathcal{B}$ stands for the family of all compact subsets of X. [Do not forget (9.55).]

Exercise 16.34. Let (X, d) be a metric space, and let μ be a finite measure on $\mathcal{B}(X)$. Show that $\mu(A) = \inf\{\mu(U) : A \subset U \in \mathcal{T}_d\}$ for any $A \in \mathcal{B}(X)$.

Exercise 16.35. Let $\{(X_i, \mathcal{A}_i, P_i) : i \in I\}$ be a nonempty set of probability spaces, and let $\{J_a : a \in A\}$ be a partition of I such that $J_a \neq \emptyset, a \in A$. Prove that $\otimes_{i \in I} P_i = \otimes_{a \in A}(\otimes_{j \in J_a} P_j)$.

Exercise 16.36. For $i \in I \neq \emptyset$, let $(X_i, \mathcal{A}_i, P_i)$ be a probability space, let $(Y_i, \mathcal{B}_i)$ be a measurable space, and let $h_i : X_i \to Y_i$ be a measurable function. Define $h : \prod_{i \in I} X_i \to \prod_{i \in I} Y_i$ by $h((x_i)_{i \in I}) = (h_i(x_i))_{i \in I}$. Prove the following.

(a) For $B_i \subset Y_i, i \in I$, $h^{-1}(\prod_{i \in I} B_i) = \prod_{i \in I} h_i^{-1}(B_i)$.

(b) h is $(\otimes_{i \in I} \mathcal{A}_i, \otimes_{i \in I} \mathcal{B}_i)$-measurable. [Use (8.4) and (a).]

(c) $(\otimes_{i \in I} P_i) h^{-1} = \otimes_{i \in I} P_i h_i^{-1}$. [Hints. For nonempty finite $J \subset I$, let π_J^X denote the projection from $\prod_{i \in I} X_i$ onto $\prod_{j \in J} X_j$, let π_J^Y denote the projection from $\prod_{i \in I} Y_i$ onto $\prod_{j \in J} Y_j$, and let $B_j \in \mathcal{B}_j, j \in J$. Then, by (a) and (16.11),

$$\begin{aligned}
&\left(\left(\left(\bigotimes_{i\in I} P_i\right)h^{-1}\right)(\pi_J^Y)^{-1}\right)\left(\prod_{j\in J} B_j\right)\\
&= \left(\left(\bigotimes_{i\in I} P_i\right)(\pi_J^X)^{-1}\right)\left(\prod_{j\in J} h_j^{-1}(B_j)\right)\\
&= \left(\bigotimes_{j\in J} P_j\right)\left(\prod_{j\in J} h_j^{-1}(B_j)\right)\\
&= \prod_{j\in J} P_j\left(h_j^{-1}(B_j)\right) = \left(\bigotimes_{j\in J} P_j h_j^{-1}\right)\left(\prod_{j\in J} B_j\right)\\
&= \left(\left(\bigotimes_{i\in I} P_i h_i^{-1}\right)(\pi_J^Y)^{-1}\right)\left(\prod_{j\in J} B_j\right).
\end{aligned}$$

This means that $((\otimes_{i\in I} P_i)h^{-1})(\pi_J^Y)^{-1} = (\otimes_{i\in I} P_i h_i^{-1})(\pi_J^Y)^{-1}$. Apply next (16.11).]

Exercise 16.37. Let $\{(X_j, d_j) : j \in N\}$ be a sequence of metric spaces. For $n \in N$, let $\pi_{\{1,\ldots,n\}}$ denote the projection from $\prod_{j\in N} X_j$ onto $\prod_{j=1}^n X_j$, let $\mathcal{C}_n$ be the family of all compact subsets of $\prod_{j=1}^n X_j$, let $\mathcal{B}_n = \pi_{\{1,\ldots,n\}}^{-1}(\mathcal{C}_n)$, and put $\mathcal{B} = \cup_{n\in N}\mathcal{B}_n$. Show that the following assertions are equivalent:

(i) $\mathcal{B}$ is compact;

(ii) if $\{n_k : k \in N\} \subset N$ is strictly increasing and $B_k \in \mathcal{B}_{n_k}, k \in N$, with $\cap_{i=1}^k B_i \neq \emptyset$ for any $k \in N$, then $\cap_{i\in N} B_i \neq \emptyset$;

(iii) if $B_n \in \mathcal{B}_n, n \in N$, with $\cap_{i=1}^n B_i \neq \emptyset$ for any $n \in N$, then $\cap_{i\in N} B_i \neq \emptyset$.

[Hints. Let $\{B_n : n \in N\} \subset \mathcal{B}$ be such that $\cap_{i=1}^n B_i \neq \emptyset, n \in N$. To prove that (ii) implies (i) in case $\sup\{n \in N : B_m \in \mathcal{B}_n$ for some $m \in N\} < \infty$, use (5.27) and (5.24). To show that (iii) implies (ii), assume that $B_k = C_k \times (\prod_{j>n_k} X_j)$, where $C_k \in \mathcal{C}_{nk}$, and that $\cap_{i=1}^k B_i \neq \emptyset$ for any $k \in N$. Define $B_n' = B_k$ if $n = n_k$ for some k, and $B_n' = \pi_{\{1,\ldots,n\}}^{\{1,\ldots,n_k\}}(C_k) \times (\prod_{j>n} X_j)$ if $n_{k-1} < n < n_k$ (with $n_0 = 0$) for some k. Then $B_n' \in \mathcal{B}_n, n \in N$, by (5.45), and $B_n' \supset B_n, n \in N$. Apply (iii) to $\{B_n' : n \in N\}$.]

Exercise 16.38. Notation is as in (16.37). Prove that $\mathcal{B}$ is compact. [Hints. Let $B_n = C_n \times (\prod_{j>n} X_j)$, where $C_n \in \mathcal{C}_n$, with $\cap_{i=1}^n B_i \neq \emptyset$ for $n \in N$. Let $\mathbf{x}^n = (x_1^n, x_2^n, \ldots) \in \cap_{i=1}^n B_i, n \in N$. Then $\mathbf{x}^{n+p} \in B_n$, i.e. $(x_1^{n+p}, \ldots, x_n^{n+p}) \in C_n$, for any $n \in N$ and $p \geqslant 0$. Since $x_1^n \in C_1, n \in N$, and C_1 is sequentially compact (6.19), choose $\{n_{1k} : k \in N\} \subset N$ so that $\lim_k x_1^{n_{1k}} = x_1 \in C_1$. Since $(x_1^{n_{1k}}, x_2^{n_{1k}}) \in C_2, k \in N$, and C_2 is sequentially compact, choose $\{n_{2k} : k \in N\}$ a subsequence of $\{n_{1k} : k \in N\}$ such that $\lim_k (x_1^{n_{1k}}, x_2^{n_{1k}}) = (x_1, x_2) \in C_2$. Continue this procedure and show that $\mathbf{x} = (x_1, x_2, \ldots) = \lim_k \mathbf{x}^{n_{kk}} \in \cap_{i\in N} B_i$.]

Exercise 16.39. Let $\{(X_j, d_j) : j \in N\}$ be a sequence of complete separable metric spaces. For $n \in N$, let μ_n be a finite measure on $\mathcal{B}(\prod_{j=1}^n X_j)$ such that

$\mu_{n+1}(\pi_{\{1,\ldots,n\}}^{\{1,\ldots,n+1\}})^{-1} = \mu_n$. Prove that there exists a unique measure μ on $\mathcal{B}(\prod_{j\in N} X_j)$ such that $\mu\pi_{\{1,\ldots,n\}}^{-1} = \mu_n$, $n \in N$. The following steps may be useful.

(a) The family $\mathcal{A} = \cup_{n\in N}\pi_{\{1,\ldots,n\}}^{-1}(\mathcal{B}(\prod_{j=1}^{n} X_j))$ is an algebra of sets.

(b) For $A \in \mathcal{A}$, choose $n \in N$ so that $A = \pi_{\{1,\ldots,n\}}^{-1}(B)$ for some $B \in \mathcal{B}(\prod_{j=1}^{n} X_j)$, and put $\mu(A) = \mu_n(B)$. Then μ is a well-defined finitely additive measure on $\mathcal{A}$.

(c) μ is σ-additive. [Hint. μ is $\mathcal{B}$-regular, where $\mathcal{B}$ is as in (16.37) by (5.71.b), (6.98) and (16.23) applied to μ_n. Then use (16.17).]

(d) μ has a unique extension to $\mathcal{B}(\prod_{j\in N} X_j)$. [Use (10.9).]

Exercise 16.40. Let $I = N$ or Z, and let X be a countable set with at least two elements. For each $l \geqslant 1$ and $x_1, \ldots, x_l \in X$, suppose a nonnegative number $p_l(x_1, \ldots, x_l)$ is given such that:

(i) $\sum_{x\in X} p_1(x) < \infty$;

(ii) $\sum_{x\in X} p_{l+1}(x_1, \ldots, x_l, x) = p_l(x_1, \ldots, x_l)$;

(iii) $\sum_{x\in X} p_{l+1}(x, x_1, \ldots, x_l) = p_l(x_1, \ldots, x_l)$.

Let $\mathcal{J}$ denote the family of all finite subsets of I consisting of consecutive integers. If $J \in \mathcal{J}$ has l elements, define a measure P_J on $\mathcal{P}(X^J)$ by $P_J(\{(x_1, \ldots, x_l)\}) = p_l(x_1, \ldots, x_l)$, $(x_1, \ldots, x_l) \in X^J$. Prove that there is a unique finite measure P on $\otimes_{i\in I}\mathcal{P}(X_i)$, where $X_i = X$ for any $i \in I$, such that $P\pi_J^{-1} = P_J$, $J \in \mathcal{J}$. The next steps may be helpful.

(a) The family $\mathcal{A}_0 = \cup_{J\in\mathcal{J}}\pi_J^{-1}(\mathcal{P}(X^J))$ is an algebra of sets that generates $\otimes_{i\in I}\mathcal{P}(X_i)$.

(b) For $A \in \mathcal{A}_0$, choose $J \in \mathcal{J}$ so that $A = \pi_J^{-1}(B)$ for some $B \subset X^J$, and set $P(A) = P_J(B)$. Then P is a well-defined finitely additive measure on $\mathcal{A}_0$. [Use (i) and (ii).]

(c) μ is σ-additive. [Hints. Let $\mathcal{B}_i$ be the family of all finite subsets of X_i, where $X_i = X$ for any $i \in I$, and put $\mathcal{B} = \cup_{i\in I}\pi_i^{-1}(\mathcal{B}_i)$. Then $\mathcal{B}$ is compact by (16.13.a) and (16.14.c). Show that P is $\mathcal{B}$-regular and use (16.17).]

(d) P has a unique extension to $\otimes_{i\in I}\mathcal{P}(X_i)$. [Use (10.9).]

(e) $P(X^I) = \sum_{x\in X} p_1(x)$.

PART TWO: PROBABILITY

6 Elementary Notions in Probability Theory

In this chapter we introduce the first notions of the theory of probability, we begin the discussion about conditioning, and we make a detailed study of the concept of stochastic independence.

17 Events and Random Variables

In this section we introduce the primary notions of event, probability and random variable.

The σ-algebra of events 17.1. The starting point for probability theory is a set X whose elements represent the possible outcomes of an experiment. A basic notion in probability theory is that of *event*. The events are subsets of X. Thus the family $\mathcal{A}$ of events is included in $\mathcal{P}(X)$. We are not interested in the nature of an event $A \in \mathcal{A}$, but we are rather concerned with its occurrence or nonoccurrence. $\mathcal{A}$ contains two special events: the impossible event, denoted by $\emptyset$, and the sure event, denoted by X. If A and B are events such that $A \subset B$, then we say that A implies B. With the help of the logical operations expressed by the terms "not", "or" and "and" we can form new events. Thus to each event $A \in \mathcal{A}$ there corresponds the contrary event $A^c \in \mathcal{A}$. If $A, B \in \mathcal{A}$, then $A \cup B \in \mathcal{A}$ is the event A or B, and $A \cap B \in \mathcal{A}$ is the event A and B. If $A \cap B = \emptyset$, then we say that A and B are disjoint events. For practical and theoretical reasons we shall assume that the events family $\mathcal{A}$ is a σ-algebra.

Probabilities of events 17.2. Suppose that in n repetitions of an experiment the event $A \in \mathcal{A}$ occurs n_A times, so that the frequency of occurrence of A is n_A/n. If n_A/n approximate to a number $P(A)$ as $n \to \infty$, then $P(A)$ represents the probability of the event A. Obviously, $0 \leqslant P(A) \leqslant 1$. Also, we have $P(\emptyset) = 0$ and $P(X) = 1$. If A and B are disjoint events, then it is clear that $P(A \cup B) = P(A) + P(B)$. In order to get a rich mathematical theory we shall assume that P is σ-additive, and so $(X, \mathcal{A}, P)$ is a probability space. We shall say that a property holds P-*almost surely* [P-*a.s.*] if it holds P-almost everywhere. When confusion is impossible, we shall write simply a.s. instead of P-a.s. Also a.s. convergence will mean P-a.s. convergence. If $A, B \in \mathcal{A}$ with $1_A \leqslant 1_B$ a.s. or $1_A = 1_B$ a.s., we shall sometimes write $A \overset{a.s.}{\subset} B$ or $A \overset{a.s.}{=} B$, respectively. (With the notation of (9.36), observe that $A \overset{a.s.}{\subset} B$ if and only if $A \sqsubset B$, and $A \overset{a.s.}{=} B$ if and only if $A \sim B$.)

Analysis and Probability. http://dx.doi.org/10.1016/B978-0-12-401665-1.00006-0

The next notions play a central role in probability theory.

Definitions 17.3. Let $(X, \mathcal{A})$ be as in (17.1). If to each possible outcome $x \in X$ we assign a number $f(x) \in R$ such that the function $f : X \to R$ is $(\mathcal{A}, \mathcal{B}(R))$-measurable, then we say that f is a *random variable on* X. An $\overline{R}$- *valued random variable on* X is an $(\mathcal{A}, \mathcal{B}(\overline{R}))$-measurable function from X into $\overline{R}$. We drop "on X" when confusion is impossible.

Let $(Y, \mathcal{B})$ be a measurable space, and let $g : X \to Y$ be an $(\mathcal{A}, \mathcal{B})$-measurable function. In the context of probability theory g is called a *random element.* If $(Y, \mathcal{B}) = (\prod_{i\in I} Y_i, \otimes_{i\in I}\mathcal{B}_i)$ for some nonempty family $\{(Y_i, \mathcal{B}_i) : i \in I\}$ of measurable spaces, then g is called a *random vector.*

Definitions 17.4. Let $(X, \mathcal{A}, P)$ be a probability space, and let f be a random variable on X. If $\int_X f\,dP$ exists, then the number $Ef = \int_X f\,dP$ is called the *expectation of* f. (Sometimes we will write $E[f]$ instead of Ef.) If $g : X \to C$ is such that $|g| \in \mathfrak{L}^1$, the number $Eg = E[\mathrm{Re}\, g] + iE[\mathrm{Im}\, g]$ is called the *expectation of* g.

The following quantities are of special interest in probability theory.

Definitions 17.5. Let $(X, \mathcal{A}, P)$ and f be as in (17.4), and let $k > 0$. If f^k is defined and Ef^k exists, then Ef^k is called the *k-th moment of* f. The number $E\,|f|^k$ is called the *k-th absolute moment of* f. In accordance with (13.5), if $E\,|f|^k < \infty$, then $E\,|f|^j < \infty$ whenever $0 < j < k$. If $Ef \in R$, then the number $\mathrm{Var}\, f = E(f - Ef)^2 = Ef^2 - (Ef)^2$ is called the *variance of* f, and the number $\sigma[f] = (\mathrm{Var}\, f)^{1/2}$ is called the *standard deviation of* f. (Sometimes we will write $\mathrm{Var}[f]$ instead of $\mathrm{Var}\, f$.)

Definitions 17.6. Let $(X, \mathcal{A}, P)$ be a probability space, and let f and g be random variables on X such that $Ef \in R$, $Eg \in R$ and $E[fg] \in R$. The number $\mathrm{Cov}[f, g] = E[(f - Ef)(g - Eg)] = E[fg] - E[f]E[g]$ is called the *covariance of* f *and* g. If $\sigma[f]^2 = \mathrm{Var}\, f \in]0, \infty[$ and $\sigma[g]^2 = \mathrm{Var}\, g \in]0, \infty[$, the number $\rho[f, g] = \mathrm{Cov}[f, g]/\sigma[f]\sigma[g]$ is called the *correlation coefficient between* f *and* g. According to Schwarz's inequality, we have $-1 \leqslant \rho[f, g] \leqslant 1$.

Exercise 17.7. Let f be a random variable such that $Ef \in R$. $\mathrm{Var}\, f = 0$ if and only if $f = c$ a.s. for some $c \in R$.

Exercise 17.8. Let f be a random variable such that $m = Ef \in R$ and $\sigma^2 = \mathrm{Var}\, f \in]0, \infty[$. Show that $P(|f - m| \geqslant a\sigma) \leqslant a^{-2}$ whenever $0 < a < \infty$.

Exercise 17.9. For $i = 1, \ldots, n$, let f_i be a random variable such that $Ef_i^2 < \infty$, and let $a_i \in R$. Prove that $\mathrm{Var}[\sum_{i=1}^n a_i f_i] = \sum_{i=1}^n a_i^2 \mathrm{Var}\, f_i + 2\sum_{1\leqslant i<j\leqslant n1}^n a_i a_j \mathrm{Cov}[f_i, f_j] = \mathbf{a}C\mathbf{a}^T$, where $\mathbf{a} = (a_1, \ldots, a_n)$, $\mathbf{a}^T$ denote its transpose, and C is a matrix with entries $c_{ij} = \mathrm{Cov}[f_i, f_j]$, $1 \leqslant i, j \leqslant n$.

Exercise 17.10. Let $(X, \mathcal{A}, P)$ be a probability space, let f be a random variables on X, and let $\varphi : [0, \infty[\to [0, \infty[$ be a differentiable function such that $\varphi(0) = 0$ and the derivative $\varphi' \geqslant 0$. Assume that either φ' is nonincreasing, or φ' is nondecreasing and $\varphi'(x + 1) \leqslant C\varphi'(x)$, $x \geqslant x_0$, for some $x_0 > 0$ and $C > 0$. Show that $E[\varphi \circ |f|] < \infty$ if and only if $\sum_{n\geqslant 1} \varphi'(n)P(\{|f| \geqslant n\}) < \infty$; in particular, $E\,|f|^p < \infty$ for some $p > 0$ if and only if $\sum_{n\geqslant 1} n^{p-1}P(\{|f| \geqslant n\}) < \infty$. [Use (15.23.a).]

Exercise 17.11. Let f and g be random variables such that $\operatorname{Var} f = \operatorname{Var} g \in]0, \infty[$. Prove that $\operatorname{Cov}[f+g, f-g] = 0$.

Exercise 17.12 (Liapounov). Notation is as in (6.99). Set $\alpha_k = E\,|f|^k$, $k > 0$. If $0 < p \leqslant q \leqslant r$, then $\alpha_q^{r-p} \leqslant \alpha_p^{r-q}\alpha_r^{q-p}$.

Exercise 17.13. Let f and g be random variables, and assume that $P(\{f > x\} \cap \{g > y\}) \leqslant P(\{f > x\})P(\{g > y\})$ for any $x, y \in R$. Show that $P(\{-f > x\} \cap \{-g > y\}) \leqslant P(\{-f > x\})P(\{-g > y\})$.

Exercise 17.14. Let $(X, \mathcal{A}, P)$ be a probability space, let $\{A_n : n \in N\} \subset \mathcal{A}$ be such that $P(A_n) \to 1$, and let f be a P-integrable random variable on X with $Ef = 0$. Prove that $E[f 1_{A_n}] \to 0$.

Exercise 17.15. Let f be a nonnegative random variable such that $\operatorname{Var} f \in]0, \infty[$. Show that $Ef \leqslant (\operatorname{Var} f)/\operatorname{Var}\sqrt{f}$. [Use Liapounov's inequality (17.12).]

Exercise 17.16. Let f be a random variable, let $\psi(x) = x \log x$, $x \geqslant 1$, and let $\chi(x)$, $x \geqslant 0$, denote the inverse function of ψ. Prove the following.

(a) $\chi(x) \sim x/\log x$ as $x \to \infty$. (For the meaning of $\sim$, see (7.2.16).)

(b) $\log \chi(x) \sim \log x$ as $x \to \infty$.

(c) For $p \geqslant 0$, $E\,|f|^{p+1} < \infty$ if and only if $\sum_{n \geqslant 1} n^p (\log n)^{p+1} P(\{|f| \geqslant n \log n\}) < \infty$; in particular, $Ef^2 < \infty$ if and only if $\sum_{n\geqslant 1} \log n\, P(\{|f| \geqslant \sqrt{n \log n}\}) < \infty$. [Hint. Apply (17.10) with $\varphi(x) = x^{p+1}(\log x)^{p+1}$, $x \geqslant x_0$, to the random variable $\chi \circ |f|$, taking (a) and (b) into account.]

Exercise 17.17. Let $(X, \mathcal{A}, P)$ be a probability space, and let $\{A_n : n \in N\} \subset \mathcal{A}$ be such that $A_1 \overset{a.s.}{=} A_2 \overset{a.s.}{=} \cdots$ Show that $A_1 \overset{a.s.}{=} \bigcup_{n\in N} A_n \overset{a.s.}{=} \bigcap_{n \in N} A_n \overset{a.s.}{=} \liminf_n \overset{a.s.}{=} \limsup_n A_n$.

Exercise 17.18. Notation is as in (6.99), assume that Y_i is finite for any $i \in I$, and let $(X, \mathcal{A}, P)$ be a probability space. For $i \in I$, consider the measurable space $(Y_i, \mathcal{P}(Y_i))$, and let $f_i : X \to Y_i$ and $g_i : X \to Y_i$ be random elements.

(a) Prove that $d_{\alpha,H} \circ (\mathbf{f}, \mathbf{g})$ is measurable, where $\mathbf{f} = (f_i)_{i \in I}$ and $\mathbf{g} = (g_i)_{i\in I}$.

(b) Verify that $E[d_{\alpha,H} \circ (\mathbf{f}, \mathbf{g})] = \sum_{i \in I} \alpha_i P(f_i \neq g_i)$.

18 Conditioning and Independence

The concept of conditioning and independence is peculiar to probability theory. In this section we introduce the notion of conditional probability, and we study in detail the concept of independence.

From now on, unless otherwise stated, $(X, \mathcal{A}, P)$ is a fixed probability space, an event is an event in $\mathcal{A}$, and a random variable is a random variable on X.

Definition 18.1. Let $B \in \mathcal{A}$ be such that $P(B) > 0$. For $A \in \mathcal{A}$, define $P_B(A) = P(A \cap B)/P(B)$. Then P_B is a probability on $\mathcal{A}$ called the *conditional probability given B*. We shall also use the notation $P(A|B) = P_B(A)$.

Theorem 18.2. *The following assertions hold.*

(i) *If* $B_1, B_2 \in \mathcal{A}$ *and* $P(B_1 \cap B_2) > 0$, *then* $(P_{B_1})_{B_2} = P_{B_1 \cap B_2}$.

(ii) *If* $A, B \in \mathcal{A}$ *and* $P(B) > 0$, *then* $P(A|B) = P(A)/P(B)$ *whenever* $A \subset B$, *and* $P(A|B) = 1$ *whenever* $A \supset B$.

(iii) *If* $A, B \in \mathcal{A}$ *and* $P(A)P(B) > 0$, *then* $P(A)P(B|A) = P(B)P(A|B)$.

(iv) *If* $A_1, \ldots, A_n, A_{n+1} \in \mathcal{A}$ *and* $P(A_2 \cap \cdots \cap A_{n+1}) > 0$, *then* $P(A_1 \cap \cdots \cap A_n|A_{n+1}) = \prod_{i=1}^{n} P(A_i|A_{i+1} \cap \cdots \cap A_{n+1})$ *(product formula).*

(v) *If* $\{B_i, : i \in I\} \subset \mathcal{A}$ *is a countable partition of* X *such that* $P(B_i) > 0, i \in I$, *then* $P(A) = \sum_{i \in I} P(A|B_i)P(B_i)$ *for any* $A \in \mathcal{A}$ *(total probability formula).*

(vi) *If* $A \in \mathcal{A}$ *and* $P(A) > 0$, *and* $\{B_i, : i \in I\} \subset \mathcal{A}$ *is a countable partition of* X *such that* $P(B_i) > 0, i \in I$, *then* $P(B_j|A) = P(A|B_j)P(B_j)/\sum_{i \in I} P(A|B_i)P(B_i)$, $j \in I$ *(Bayes formula).*

(vii) *If* $B \in \mathcal{A}$ *and* $P(B) > 0$, *and* f *is a random variable on* X, *then*

$$\int_X f\,dP_B = \frac{1}{P(B)} \int_B f\,dP, \tag{1}$$

in the sense that if one of the integral exists, so does the other, and the two integrals are equal.

Proof. The proof of (i)–(vi) is easy and is left to the reader. If $f = 1_A$, where $A \in \mathcal{A}$, then (1) becomes $P_B(A) = P(A \cap B)/P(B)$, which is true. (vii) is further proved by making use of the indicator function method. □

Now we introduce the concept of stochastic independence.

Definition 18.3. Let $\{\mathcal{A}_i : i \in I\}$ be a set of σ-algebras such that $\mathcal{A}_i \subset \mathcal{A}$ for each $i \in I$. We say that $\mathcal{A}_i, i \in I$, are *independent* [*P-independent*] if, for every finite set $\{i_1, \ldots, i_n\} \subset I$ and $A_1 \in \mathcal{A}_{i_1}, \ldots, A_n \in \mathcal{A}_{i_n}$, we have $P(A_1 \cap \cdots \cap A_n) = P(A_1) \cdots P(A_n)$.

Remark 18.4. Let $\{\mathcal{A}_i : i \in I\}$ be as in (18.3).

(a) If $\mathcal{A}_i, i \in I$, are independent, then $\mathcal{A}_i, i \in J$, are independent whenever $J \subset I$.

(b) If $\mathcal{A}_i, i \in J$, are independent for any finite set $J \subset I$, then $\mathcal{A}_i, i \in I$, are independent.

(c) Let $\{\mathcal{B}_i : i \in I\}$ be a set of σ-algebras such that $\mathcal{B}_i \subset \mathcal{A}_i$ for any $i \in I$. If $\mathcal{A}_i, i \in I$, are independent, then $\mathcal{B}_i, i \in I$, are independent.

Definition 18.5. Let $\{f_i : i \in I\}$ be a set of random variables on X. We say that $f_i, i \in I$, are *independent* [*P-independent*] if $f_i^{-1}(\mathcal{B}(R)), i \in I$, are independent σ-algebras.

Remark 18.6. Let $\{f_i : i \in I\}$ be as in (18.5). For $i \in I$, let $g_i : R \to R$ be a measurable function. Then, in accordance with (18.4.c), $g_i \circ f_i, i \in I$, are independent.

Definition 18.7. Let $\{A_i : i \in I\} \subset \mathcal{A}$. We say that $A_i, i \in I$, are *independent* [*P-independent*] if $1_{A_i}, i \in I$, are independent random variables.

Remarks 18.8. (a) Let $\{A_i : i \in I\}$ be as in (18.7). Then $A_i, i \in I$, are independent if and only if $\sigma(\{A_i\}), i \in I$, are independent σ-algebras.

(b) Independence may be lost under conditioning. On the other hand, dependent objects may gain independence under conditioning. To see this, let $X = \{1, 2, 3, 4, 5\}$, $\mathcal{A} = \mathcal{P}(X)$, and set $P(\{1\}) = P(\{2\}) = P(\{3\}) = 1/4$ and $P(\{4\}) = P(\{5\}) = 1/8$. Put $A = \{1, 2\}$, $B = \{2, 3\}$, $C = \{2, 4\}$, and $D = \{4, 5\}$. Then $P(A \cap B) = P(A)P(B) = 1/4$, while $2/3 = P(A \cap B|C) \neq P(A|C)P(B|C) = 4/9$. On the other hand, $P(C \cap D) = 1/8 \neq 3/32 = P(C)P(D)$, while $P(C \cap D|C \cap D) = P(C|C \cap D)P(D|C \cap D) = 1$.

Lemma 18.9. *For $i = 1, \ldots, n$, let $\mathcal{M}_i \subset \mathcal{A}$ be such that $X \in \mathcal{M}_i$ and $A, B \in \mathcal{M}_i$ imply $A, B \in \mathcal{M}_i$. If $P(A_1 \cap \cdots \cap A_n) = P(A_1) \cdots P(A_n)$ for any $A_1 \in \mathcal{M}_1, \ldots, A_n \in \mathcal{M}_n$, then*

$$(i) \quad P(A_1 \cap \cdots \cap A_n) = P(A_1) \cdots P(A_n)$$

for any $A_1 \in \mathcal{M}_1, \ldots, A_{n-1} \in \mathcal{M}_{n-1}$, $A_n \in \sigma(\mathcal{M}_n)$.

Proof. Putting $A_n = X$ we get $P(A_1 \cap \cdots \cap A_{n-1}) = P(A_1) \cdots P(A_{n-1})$ for any $A_1 \in \mathcal{M}_1, \ldots, A_{n-1} \in \mathcal{M}_{n-1}$. If $P(A_1 \cap \cdots \cap A_{n-1}) = P(A_1) \cdots P(A_{n-1}) = 0$, then $P(A_1 \cap \cdots \cap A_{n-1} \cap A) = P(A_1) \cdots P(A_{n-1})P(A)$ for every $A \in \mathcal{A}$, and so (i) holds in this case. Assume now that $P(A_1 \cap \cdots \cap A_{n-1}) = P(A_1) \cdots P(A_{n-1}) \neq 0$, and consider the conditional probability $P_{A_1 \cap \cdots \cap A_{n-1}}$. For each $A_n \in \mathcal{M}_n$, we have $P_{A_1 \cap \cdots \cap A_{n-1}}(A) = P(A_1 \cap \cdots \cap A_{n-1} \cap A)/P(A_1 \cap \cdots \cap A_{n-1}) = P(A_1) \cdots P(A_{n-1}) P(A)/P(A_1) \cdots P(A_{n-1}) = P(A)$. Therefore, in view of (10.8), $P_{A_1 \cap \cdots \cap A_{n-1}}$ and P coincide on $\sigma(\mathcal{M}_n)$. Thus we have $P_{A_1 \cap \cdots \cap A_{n-1}}(A) = P(A)$ for any $A \in \sigma(\mathcal{M}_n)$, and so (i) obtains. □

Theorem 18.10. *For $i \in I$, let $\mathcal{M}_i \subset \mathcal{A}$ be such that $A, B \in \mathcal{M}_i$ implies $A, B \in \mathcal{M}_i$. If for any finite set $\{i_1, \ldots, i_n\} \subset I$ and $A_1 \in \mathcal{M}_{i_1}, \ldots, A_n \in \mathcal{M}_{i_n}$, we have $P(A_1 \cap \cdots \cap A_n) = P(A_1) \cdots P(A_n)$, then $\sigma(\mathcal{M}_i)$, $i \in I$, are independent.*

Proof. We may and do assume that $X \in \mathcal{M}_i$ for each $i \in I$, because replacing $\mathcal{M}_i$ by $\mathcal{M}_i \cup \{X\}$ leaves the hypothesis of the theorem unchanged. In accordance with (18.4.b), it suffices to consider only the case that I is finite. Hence, without loss of generality, we may take $I = \{1, \ldots, n\}$. Then applying (18.9) n times, it follows that $\sigma(\mathcal{M}_1), \ldots, \sigma(\mathcal{M}_n)$ are independent. □

Corollary 18.11. *Let $\{A_i : i \in I\}$ be as in* (18.7). Then A_i, $i \in I$, are independent if and only if, for every finite set $\{i_1, \ldots, i_n\} \subset I$, we have $P(A_{i_1} \cap \cdots \cap A_{i_n}) = P(A_{i_1}) \cdots P(A_{i_n})$.

Proof. The result is an immediate consequence of (18.8) and (18.10). □

Notation 18.12. For $i \in I$, let $\mathcal{A}_i$ be a σ-algebra of subsets of X. We will write $\vee_{i \in I} \mathcal{A}_i$ instead of $\sigma(\cup_{i \in I} \mathcal{A}_i)$ whenever convenient. If $I = \{n \in Z : n \geqslant m\}$ for some $m \in Z$, then we will write $\vee_{n \geqslant m} \mathcal{A}_n$ in place of $\vee_{i \in I} \mathcal{A}_i$; also, if $I = \{1, \ldots, n\}$, then we will write $\vee_{i=1}^n \mathcal{A}_i$ or $\mathcal{A}_1 \vee \cdots \vee \mathcal{A}_n$ instead of $\vee_{i \in I} \mathcal{A}_i$. If $\{I_a : a \in A\}$ is a partition of I such that $I_a \neq \emptyset$, $a \in A$, then, according to (7.23), $\vee_{a \in A}(\vee_{i \in I_a} \mathcal{A}_i) = \vee_{i \in I} \mathcal{A}_i$.

Theorem 18.13. *Let $(X, \mathcal{A})$ be a measurable space, let $\{\mathcal{A}_i : i \in I\}$ be a set of σ-algebras such that $\mathcal{A}_i \subset \mathcal{A}$, $i \in I$, and let P and Q be probabilities on $\mathcal{A}$ such that*

$P_{\mathcal{A}_i} = Q_{\mathcal{A}_i}$. If $\mathcal{A}_i, i \in I$, are both P-independent and Q-independent, then P and Q coincide on $\vee_{i\in I}\mathcal{A}_i$.

Proof. Set $\mathcal{M} = \{\cap_{i\in I} A_i : A_i \in \mathcal{A}_i, i \in I$, and $\{i \in I : A_i \neq X_i\}$ is finite$\}$. Evidently, $A, B \in \mathcal{M}$ implies $A \cap B \in \mathcal{M}$. As $\cup_{i\in I}\mathcal{A}_i \subset \mathcal{M} \subset \vee_{i\in I}\mathcal{A}_i$, and $P(A) = Q(A)$ for any $A \in \mathcal{M}$, the result follows at once from (10.8). □

The next theorem expresses the disassociativity property of the independence.

Theorem 18.14. *Let $\{\mathcal{B}_a : a \in A\}$ be a set of σ-algebras such that $\mathcal{B}_a \subset \mathcal{A}, a \in A$, and $\mathcal{B}_a, a \in A$, are independent. If for each $a \in A$ there exists a set of σ-algebras $\{\mathcal{A}_i : i \in I_a\}$ such that $\mathcal{A}_i \subset \mathcal{B}_a, i \in I_a$, and $\mathcal{A}_i, i \in I_a$, are independent, then $\mathcal{A}_i, i \in \cup_{a\in A} I_a$, are independent.*

Proof. Exercise. □

The following theorem expresses the associativity property of the independence.

Theorem 18.15. *Let $\{\mathcal{A}_i : i \in I\}$ be a set of σ-algebras such that $\mathcal{A}_i \subset \mathcal{A}, i \in I$, and $\mathcal{A}_i, i \in I$, are independent. If $\{I_a : a \in A\}$ is a partition of I such that $I_a \neq \emptyset, a \in A$, then $\vee_{i\in I_a}\mathcal{A}_i, a \in A$, are independent.*

Proof. For $a \in A$, put $\mathcal{M}_a = \{\cap_{i\in I_a} A_i : A_i \in \mathcal{A}_i, i \in I_a$, and $\{i \in I_a : A_i \neq X_i\}$ is finite$\}$. Obviously, $A, B \in \mathcal{M}_a$ implies $A \cap B \in \mathcal{M}_a$ and $\sigma(\mathcal{M}_a) = \vee_{i\in I_a}\mathcal{A}_i$. Since $\mathcal{A}_i, i \in I$, are independent, for any finite set $\{a_1, \ldots, a_n\} \subset A$ and $B_1 \in \mathcal{M}_{a_1}, \ldots, B_n \in \mathcal{M}_{a_n}$, we have $P(B_1\cap\cdots\cap B_n) = P(B_1)\cdots P(B_n)$. Hence, according to (18.10), the σ-algebras $\sigma(\mathcal{M}_a), a \in A$, are independent. □

Example 18.16. If f_1, f_2, f_3, f_4, f_5 are independent random variables, then $g_1 = f_1^2 + f_2^2 + f_3^2$ and $g_2 = f_4 f_5$ are independent. Actually, by (18.15), $\sigma(f_1, f_2, f_3)$ and $\sigma(f_4, f_5)$ are independent. In view of (8.10), we have $g_1^{-1}(\mathcal{B}(R)) \subset \sigma(f_1, f_2, f_3)$ and $g_2^{-1}(\mathcal{B}(R)) \subset \sigma(f_4, f_5)$. Thus (18.4.c) shows that g_1 and g_2 are independent.

Definition 18.17. Let $(X, \mathcal{A}, P)$ be a probability space. Define $\mathcal{O}_P = \{A \in \mathcal{A} : P(A) = 0$ or $1\}$. Clearly, $\mathcal{O}_P$ is a σ-algebra.

Theorem 18.18. *The following assertions hold.*

(i) *$\mathcal{A}$ and $\mathcal{O}_P$ are independent.*

(ii) *Let $\mathcal{B} \subset \mathcal{A}$ be a σ-algebra. $\mathcal{B} \subset \mathcal{O}_P$ if and only if $\mathcal{B}$ is independent of $\mathcal{B}$.*

(iii) *Let $\{\mathcal{A}_i : i \in I\}$ be a set of σ-algebras such that $\mathcal{A}_i \subset \mathcal{A}, i \in I$. If $\mathcal{A}_i, i \in I$, are independent, then $\mathcal{A}_i \vee \mathcal{O}_P, i \in I$, are independent.*

Proof. For $A \in \mathcal{A}$ and $B \in \mathcal{O}_P$, we have $P(A \cap B) = P(A)P(B)$. This proves (i). Let $\mathcal{B} \subset \mathcal{A}$ be a σ-algebra. If $\mathcal{B} \subset \mathcal{O}_P$, then $\mathcal{B}$ is independent of $\mathcal{B}$ by (i) and (18.4.c). Conversely, if $\mathcal{B}$ is independent of $\mathcal{B}$, then, for any $B \in \mathcal{B}$, we have $P(B) = P(B\cap B) = P(B)^2$, and so $P(B) = 0$ or 1. This shows that $\mathcal{B} \subset \mathcal{O}_P$. Thus (ii) is proved. To prove (iii), it suffices to consider only the case when I is finite. So, without loss of generality, we may take $I = \{1, \ldots, n\}$. Since $\mathcal{A}_i \subset \mathcal{A}, i \in I$, and $\mathcal{A}$ and $\mathcal{O}_P$ are independent, (18.14) shows that $\mathcal{A}_1, \ldots, \mathcal{A}_n, \mathcal{O}_P$ are independent σ-algebras. Therefore, by virtue of (18.15), it follows that $\mathcal{A}_1, \ldots, \mathcal{A}_{n-1}, \mathcal{A}_n \vee \mathcal{O}_P$ are independent. By repeating this reasoning $n - 1$ times we see that $\mathcal{A}_1 \vee \mathcal{O}_P, \ldots, \mathcal{A}_n \vee \mathcal{O}_P$ are independent. □

Theorem 18.19. *Let f be a random variable. Then f is $\mathcal{O}_P$-measurable if and only if $f = c$ a.s. for some $c \in R$.*

Proof. Assume first that f is $\mathcal{O}_P$-measurable, and consider the function $F(x) = P(\{f \leqslant x\})$, $x \in R$. Then $F(x) = 0$ or 1 for every $x \in R$. Let $c = \inf\{x : F(x) = 1\}$. Since $F(R) = \{0, 1\}$, we have $c \in R$. As F is nondecreasing, we get $F(x) = 0$ whenever $x < c$ and $F(x) = 1$ whenever $x > c$. Since $\{f = c\} = \bigcap_{n \in N}\{c - 1/n < f \leqslant c + 1/n\} = \bigcap_{n \in N}(\{f \leqslant c + 1/n\} - \{f \leqslant c - 1/n\})$, we have $P(\{f = c\}) = \lim_n(P(\{f \leqslant c+1/n\}) - P(\{f \leqslant c-1/n\}) = \lim_n(F(c+1/n) - F(c-1/n)) = 1$, and so $f = c$ a.s.

Assume now that there is $c \in R$ such that $f = c$ a.s., and let $B \in \mathcal{B}(R)$. If $c \in B$, we have $\{f = c\} \subset f^{-1}(B)$, and so $P(f^{-1}(B)) = 1$, that is $f^{-1}(B) \in \mathcal{O}_P$. Hence, if $c \in B^c$, then $f^{-1}(B^c) \in \mathcal{O}_P$, and so $f^{-1}(B) = (f^{-1}(B^c))^c \in \mathcal{O}_P$. Thus f is $\mathcal{O}_P$-measurable. □

The next result, due to A. N. Kolmogorov, has important applications.

Zero-one law 18.20. *Let $\{\mathcal{A}_n : n \in N\}$ be a sequence of σ-algebras such that $\mathcal{A}_n \subset \mathcal{A}$, $n \in N$. If $\mathcal{A}_n$, $n \in N$, are independent, then $\bigcap_{n \geqslant 1}(\vee_{k \geqslant n}\mathcal{A}_k) \subset \mathcal{O}_P$.*

Proof. Set $\mathcal{T} = \bigcap_{n \geqslant 1}(\vee_{k \geqslant n}\mathcal{A}_k)$. For $n \in N$, (18.15) shows that $\mathcal{A}_1, \ldots, \mathcal{A}_n, \vee_{k > n}\mathcal{A}_k$ are independent σ-algebras. Therefore, in view of (18.4.c), we see that $\mathcal{A}_1, \ldots, \mathcal{A}_n, \mathcal{T}$ are independent. Consequently, using (18.4.a) and (18.4.b), it follows that $\mathcal{A}_1, \ldots, \mathcal{A}_n, \ldots, \mathcal{T}$ are independent σ-algebras. Applying again (18.15), we see that $\vee_{k \geqslant n}\mathcal{A}_k$ and $\mathcal{T}$ are independent. Thus, by virtue of (18.4.c), $\mathcal{T}$ is independent of $\mathcal{T}$. Finally, (18.18.ii) shows that $\mathcal{T} \subset \mathcal{O}_P$. □

Example 18.21. Let f_n, $n \in N$, be independent random variables, and let $A = \{x \in X : \sum_{k \geqslant 1} f_k(x)$ is convergent$\}$. Then $P(A) = 0$ or 1. Actually, for each $n \in N$, $A = \{x \in X : \sum_{k \geqslant n} f_k(x)$ is convergent$\} \in \sigma(f_k, k \geqslant n)$ (8.30). Therefore, $A \in \bigcap_{n \geqslant 1}\sigma(f_k, k \geqslant n) \subset \mathcal{O}_P$. This may be rephrased as $A \in \mathcal{T} \subset \mathcal{O}_P$, where $\mathcal{T}$ is the tail σ-algebra of $\{f_n : n \in N\}$ (7.72).

Lemma 18.22. *Let f and g be nonnegative independent random variables. Then $E[fg] = E[f]E[g]$.*

Proof. If $f = \sum_{i=1}^n a_i 1_{A_i}$ and $g = \sum_{j=1}^m b_j 1_{B_j}$ are simple functions, where $\{f = a_i\} = A_i$, $1 \leqslant i \leqslant n$, and $\{g = b_j\} = B_j$, $1 \leqslant j \leqslant m$, then $E[fg] = E[\sum_{i=1}^n \sum_{j=1}^m a_i b_j 1_{A_i \cap B_j}] = \sum_{i=1}^n \sum_{j=1}^m a_i b_j P(A_i)P(B_j) = (\sum_{i=1}^n a_i P(A_i)) \sum_{j=1}^m b_j P(B_j) = E[f]E[g]$. If f and g are arbitrary, then, for every $n \in N$, define $f_n = \sum_{k=1}^{n2^n}(k-1)2^{-n}1_{\{(k-1)2^{-n} \leqslant f < k2^{-n}\}} + n1_{\{f \geqslant n\}}$ and $g_n = \sum_{k=1}^{n2^n}(k-1)2^{-n}1_{\{(k-1)2^{-n} \leqslant g < k2^{-n}\}} + n1_{\{g \geqslant n\}}$. Since f and g are independent, it follows that f_n and g_n are independent for any $n \in N$. Therefore,

$$E[f_n g_n] = E[f_n]E[g_n], \quad n \in N. \tag{1}$$

Letting $n \to \infty$ in (1), (8.15) and the monotone convergence theorem (11.22) show that $E[fg] = E[f]E[g]$. □

Theorem 18.23. *Let f and g be independent random variables. Then the following statements hold.*

(i) *If f and g are P-integrable, then fg is P-integrable, and $E[fg] = E[f]E[g]$.*

(ii) *If fg is P-integrable, and $P(\{f = 0\}) < 1$, then g is P-integrable.*

Proof. As f and g are independent, (18.6) shows that f^* and g^* are independent, where $f^* = f^+$ or f^- or $|f|$, and $g^* = g^+$ or g^- or $|g|$. If f and g are P-integrable, then, by (18.22), we have $E[|fg|] = E[|f|]E[|g|] < \infty$, and so fg *is* P-integrable. Moreover, according to (18.22), we get $E[fg] = E[(f^+ - f^-)(g^+ - g^-)] = E[f^+]E[g^+] - E[f^+]E[g^-] - E[f^-]E[g^+] + E[f^-]E[g^-] = E[f]E[g]$. This proves (i). If fg is P-integrable, then $E[|f|]E[|g|] = E[|fg|] < \infty$. Hence, if $P(\{f = 0\}) < 1$, then $E[|f|] > 0$, and so $E[|g|] < \infty$. This proves (ii). □

The following result is extremely important.

Borel-Cantelli lemma 18.24. *Let $\{A_n : n \in N\} \subset \mathcal{A}$. Then the following statements hold.*

(i) *If $\sum_{n \in N} P(A_n) < \infty$, then $P(\limsup_n A_n) = 0$.*

(ii) *If $A_n, n \in N$, are independent, and $\sum_{n \in N} P(A_n) = \infty$, then $P(\limsup_n A_n) = 1$.*

Proof. If $\sum_{n \in N} P(A_n) < \infty$, then $P(\limsup_n A_n) = P(\cap_{n \geqslant 1}(\cup_{k \geqslant n} A_k)) = \lim_n P(\cup_{k \geqslant n} A_k) \leqslant \lim_n \sum_{k \geqslant n} P(A_k) = 0$. This proves (i). If $A_n, n \in N$, are independent, and $\sum_{n \in N} P(A_n) = \infty$, then, for each $n \in N$, we have $P(\cap_{k \geqslant n} A_k^c) = \lim_m P(\cap_{k=n}^m A_k^c) = \lim_m \prod_{k=n}^m P(A_k^c) = \lim_m \prod_{k=n}^m (1 - P(A_k)) \leqslant \lim_m \prod_{k=n}^m \exp(-P(A_k)) = \lim_m \exp(-\sum_{k=n}^m P(A_k)) = \exp(-\sum_{k \geqslant n} P(A_k)) = 0$. Therefore, $P(\cup_{k \geqslant n} A_k) = 1, n \in N$, and so $P(\limsup_n A_n) = 1$. This proves (ii). □

Remark 18.25. The condition that $A_n, n \in N$, are independent in (18.24.ii) is important. To see this, let $X = [0, 1]$, let P be the Lebesgue measure on $\mathcal{B}([0, 1])$, and let $A_n = [0, 1/n]$. Then $\sum_{n \in N} P(A_n) = \infty$, but $P(\limsup_n A_n) = 0$. However, the independence condition can be relaxed to pairwise independence (see (18.46)).

The next definition generalizes (17.5).

Definition 18.26. Let $(X, \mathcal{A}, P)$ be a probability space, and let $\{(Y_i, \mathcal{B}_i) : i \in I\}$ be a nonempty family of measurable spaces. For $i \in I$, let $f_i : X \to Y_i$ be a random element. We say that $f_i, i \in I$, are *independent* [P-*independent*] if $f_i^{-1}(\mathcal{B}_i), i \in I$, are independent σ-algebras.

Example 18.27. Let $\{(X_i, \mathcal{A}_i, P_i) : i \in I\}$ be a nonempty set of probability spaces, and let P be the product of the probabilities $P_i, i \in I$. For $i \in I$, let π_i stand for the projection from $\prod_{i \in I} X_i$ onto X_i. Then $\pi_i, i \in I$, are P-independent. Indeed, for every finite set $J = \{i_1, \dots, i_n\} \subset I$ and $A_1 \in \mathcal{A}_{i_1}, \dots, A_n \in \mathcal{A}_{i_n}$, by virtue of (16.11), we have $P(\pi_{i_1}^{-1}(A_1) \cap \cdots \cap \pi_{i_n}^{-1}(A_n)) = P(\pi_J^{-1}(A_1 \times \cdots \times A_n)) = P_{i_1}(A_1) \cdots P_{i_n}(A_n) = P(\pi_{i_1}^{-1}(A_1)) \cdots P(\pi_{i_n}^{-1}(A_n))$.

Remark 18.28. Notation is as in (18.26). Let $\{I_a : a \in A\}$ be a partition of I such that $I_a \neq \emptyset, a \in A$.

(a) If $(f_i)_{i\in I_a}$, $a \in A$, are independent and, for each $a \in A$, $f_i, i \in I_a$, are independent, then, in accordance with (7.19) and (18.14), $f_i, i \in I$, are independent.

(b) If $f_i, i \in I$, are independent, then $(f_i)_{i\in I_a}$, $a \in A$, are independent by (7.19) and (18.15).

(c) For $i \in I$, let $(Z_i, \mathcal{C}_i)$ be a measurable space, and let $g_i : Y_i \to Z_i$ be a measurable function. If $f_i, i \in I$, are independent, then $g_i \circ f_i, i \in I$, are independent by (18.4.c).

Theorem 18.29. *Let $(X, \mathcal{A}, P)$, $\{(Y_i, \mathcal{B}_i) : i \in I\}$ and $\{f_i : i \in I\}$ be as in* (18.26), *and let $\mathbf{f} = (f_i)_{i\in I}$. Then $f_i, i \in I$, are independent if and only if $P\mathbf{f}^{-1}$ is the product of the probabilities $Pf_i^{-1}, i \in I$.*

Proof. Let Q denote the product of the probabilities $Pf_i^{-1}, i \in I$. Then $P\mathbf{f}^{-1}$ and Q are probabilities on $\otimes_{i\in I}\mathcal{B}_i$. Assume that $f_i, i \in I$, are independent, and put $\mathcal{B} = \{\prod_{i\in I} B_i : B_i \in \mathcal{B}_i, i \in I, \text{and } \{i \in I : B_i \neq Y_i\} \text{ is finite}\}$. Obviously, $A, B \in \mathcal{B}$ implies $A \cap B \in \mathcal{B}$. Let $\prod_{i\in I} B_i \in \mathcal{B}$, and set $J = \{i \in I : B_i \neq Y_i\}$. Then we have $(P\mathbf{f}^{-1})(\prod_{i\in I} B_i) = P(\cap_{i\in I} f_i^{-1}(B_i)) = P(\cap_{i\in J} f_i^{-1}(B_i)) = \prod_{i\in J} P(f_i^{-1}(B_i)) = (Q\pi_J^{-1})(\prod_{i\in J} B_i) = Q(\prod_{i\in I} B_i)$. Therefore, using (7.17.a) and (10.8), it follows that $P\mathbf{f}^{-1} = Q$.

Assume now that $P\mathbf{f}^{-1} = Q$. Then, for any finite set $J = \{i_1, \ldots, i_n\} \subset I$ and $B_1 \in \mathcal{B}_{i_1}, \ldots, B_n \in \mathcal{B}_{i_n}$, we have $P(f_{i_1}^{-1}(B_1) \cap \cdots \cap f_{i_n}^{-1}(B_n)) = P(\mathbf{f}^{-1}(\prod_{i\in I} B_i)) = Q(\prod_{i\in I} B_i)$, where $B_i = Y_i$ if $i \notin J$. Since $Q(\prod_{i\in I} B_i) = (Q\pi_J^{-1})(B_1 \times \cdots \times B_n) = P(f_{i_1}^{-1}(B_1)) \cdots P(f_{i_n}^{-1}(B_n))$, it follows that $f_i, i \in I$, are independent. □

Exercise 18.30. Let $(X, \mathcal{A}, P)$ be a complete probability space, and let $B \in \mathcal{A}$ be such that $P(B) > 0$. Show that $(B, B \cap \mathcal{A}, P_B)$ is a complete probability space.

Exercise 18.31. For $i = 1, \ldots, n$, let $\mathcal{A}_i \subset \mathcal{A}$ be a countable partition of X. Prove that $\sigma(\mathcal{A}_1), \ldots, \sigma(\mathcal{A}_n)$ are independent if and only if $P(A_1 \cap \cdots \cap A_n) = P(A_1) \cdots P(A_n)$ whenever $A_1 \in \mathcal{A}_1, \ldots, A_n \in \mathcal{A}_n$.

Exercise 18.32. Let $\{\mathcal{A}_i : i \in I\}$ be as in (18.15). Let $J \subset I$, and let $B \in \vee_{j\in J}\mathcal{A}_j$ be such that $P(B) > 0$. Prove the following.

(a) The restriction of P to $\mathcal{A}_i$ coincides with the restriction of P_B to $\mathcal{A}_i$ for any $i \in I - J$.

(b) $\mathcal{A}_i, i \in I - J$, are independent with respect to P_B.

Exercise 18.33. Let f be an $\overline{R}$-valued random variable on X. Show that f is $\mathcal{O}_P$-measurable if and only if f is constant a.s.

Exercise 18.34. Let $f_n, n \in N$, be independent random variables. Prove that $\liminf_n f_n$ and $\limsup_n f_n$ are constants a.s.

Exercise 18.35. Let $A_n, n \in N$, be independent events. Show that $P(\liminf_n A_n) \in \{0, 1\}$ and $P(\limsup_n A_n) \in \{0, 1\}$.

Exercise 18.36. Let $f_n, n \in N$, be as in (18.34), and let $A = \{x : \{f_n(x) : n \in N\} \text{ has a limit}\}$. Prove the following.

(a) $P(A) = 0$ or 1.

(b) If $P(A) = 1$, then there is $c \in \overline{R}$ such that $f_n \overset{a.s.}{\to} c$.

Exercise 18.37. Let $f_n, n \in N$, be as in (18.34), let $\{b_n : n \in N\} \subset]0, \infty[$ be such that $b_n \to \infty$, and put $g_n = (f_1 + \cdots + f_n)/b_n$. Let $A = \{x : \{g_n(x) : n \in N\}$ has a limit$\}$. Prove the following.

(a) $P(A) = 0$ or 1.

(b) If $P(A) = 1$, then there is $c \in \overline{R}$ such that $g_n \overset{a.s.}{\to} c$.

(c) $\liminf_n g_n$ and $\limsup_n g_n$ are constants a.s.

Exercise 18.38. Let $\{\mathcal{A}_n : n \in N\}$ be a sequence of σ-algebras such that $\mathcal{A}_{n+1} \subset \mathcal{A}_n \subset \mathcal{A}, n \in N$. Show that $\bigcap_{n\in N}(\mathcal{A}_n \vee \mathcal{O}_P) = (\bigcap_{n\in N}\mathcal{A}_n) \vee \mathcal{O}_P$.

Exercise 18.39. Let $A_i, 1 \leqslant i \leqslant n$, be independent events, and let $f_i, 1 \leqslant i \leqslant n$, be independent nonnegative random variables. Prove the following.

(a) $P(\bigcup_{i=1}^n A_i) \geqslant (\sum_{i=1}^n P(A_i))/(1+\sum_{i=1}^n P(A_i))$. [Make use of the inequalities $x/(1+x) \leqslant 1 - e^{-x} \leqslant x, x \geqslant 0$.]

(b) $P(\sup_{1\leqslant i\leqslant n} f_i \geqslant x) \geqslant (\sum_{i=1}^n P(f_i \geqslant x))/(1 + \sum_{i=1}^n P(f_i \geqslant x)), x \geqslant 0$.

(c) If $P(\sup_{1\leqslant i\leqslant n} f_i \geqslant x) \leqslant 1/2$, then $\sum_{i=1}^n P(f_i \geqslant x) \leqslant 2P(\sup_{1\leqslant i\leqslant n} f_i \geqslant x)$.

Exercise: Pairwise independence 18.40. For $i \in I$, let $\mathcal{A}_i \subset \mathcal{A}$ be a σ-algebra, let f_i be a random variable, and let $A_i \in \mathcal{A}$. If, for $i \neq j$ in I, $\mathcal{A}_i$ is independent of $\mathcal{A}_j$, f_i is independent of f_j, and A_i is independent of A_j, then we say that $\mathcal{A}_i, i \in I$, $f_i, i \in I$, and $A_i, i \in I$, are, respectively, *pairwise independent.*

(a) If $f_1, \ldots, f_n$ are pairwise independent and $Ef_i^2 < \infty, 1 \leqslant i \leqslant n$, show that $\operatorname{Var}[f_1 + \cdots + f_n] = \operatorname{Var} f_1 + \cdots + \operatorname{Var} f_n$.

(b) Prove that $A_i, i \in I$, are pairwise independent if and only if, for any $i \neq j$ in I, we have $P(A_i \cap A_j) = P(A_i)P(A_j)$.

Exercise 18.41 (Davis). Let $A_n, n \geqslant 1$, be pairwise independent events. Prove the following.

(a) $P(\bigcup_{n=1}^m A_n) \geqslant \sum_{n=1}^m P(A_n) - \sum_{n=1}^{m-1} P(A_n) \sum_{k=n+1}^m P(A_k), m \geqslant 2$.

(b) If $\sum_{n=1}^\infty P(A_n) < \infty$, then $P(\bigcup_{n=1}^\infty A_n) \geqslant \sum_{n=1}^\infty P(A_n) - \sum_{n=1}^\infty P(A_n) \sum_{k=n+1}^\infty P(A_k)$.

Exercise 18.42. Let $f : X \to C$ and $g : X \to C$ be measurable functions such that $|f|, |g| \in \mathfrak{L}^1(X, \mathcal{A}, P)$. Assume that f and g are independent. Prove the following.

(a) Any function in the set $\{\operatorname{Re} f, \operatorname{Im} f, |f|, \overline{f}\}$ is independent of any function in the set $\{\operatorname{Re} g, \operatorname{Im} g, |g|, \overline{g}\}$. [Use (8.2.c) and (18.28.c).]

(b) $|fg| \in \mathfrak{L}^1(X, \mathcal{A}, P)$ and $E[fg] = E[f]E[g]$. [Use (18.23) and (11.35).]

Exercise 18.43. Let a_n and $P_n, n \geqslant 1$, and λ be as in (16.24). Prove the following.

(a) $\lambda a_n^{-1} = P_n, n \geqslant 1$.

(b) $a_n, n \geqslant 1$, are independent with respect to λ.

Exercise 18.44. Let $f_n, n \in N$, be random variables such that $\sum_{n\in N} P(|f_n| \geqslant \varepsilon) < \infty$ for every $\varepsilon > 0$. Use (18.24.i) to show that $f_n \overset{a.s.}{\to} 0$.

Exercise 18.45. Let $f_n, n \in N$, be as in (18.34), and let f be a random variable such that $f_n \xrightarrow{P} f$. Prove that $f = c$ a.s. for some $c \in R$.

Exercise 18.46. Let $A_n, n \geqslant 1$, be pairwise independent events such that $\sum_{n\geqslant 1} P(A_n) = \infty$. Show that $P(\limsup_n A_n) = 1$. The following steps may be useful.

(a) For $n \geqslant 1$, set $p_n = P(A_n)$ and $I_n = \sum_{i=1}^n 1_{A_i}$. Then $\sigma[I_n]^2 = \sum_{i=1}^n p_i - \sum_{i=1}^n p_i^2$. [Use (18.40.a).]

(b) $E[I_n]/\sigma[I_n] \to \infty$. [Use (a).]

(c) For every $a > 0$, $P(I_n - E[I_n] > -a\sigma[I_n]) \geqslant 1 - a^{-2}, n \geqslant 1$. [$P(I_n - E[I_n] > -a\sigma[I_n]) \geqslant P(|I_n - E[I_n]| < a\sigma[I_n])$. Then use (17.8).]

(d) For $a > 0$, $P(I_n > E[I_n]/2) \geqslant 1 - a^{-2}$ whatever n large enough ($n \geqslant n(a)$). [Use (c) and (b).]

(e) $P(\limsup_n A_n) = P(\sum_{i\geqslant 1} 1_{A_i} = \infty) = P(\lim_n I_n = \infty) \geqslant 1 - a^{-2}, a > 0$. [Use (d).]

Exercise 18.47. Let $A, B \in \mathcal{A}$ be such that $0 < P(A), P(B) < 1$. Prove that A, B and $A \Delta B$ are pairwise independent if and only if $P(A) = P(B) = 2P(A \cap B) = 1/2$.

Exercise 18.48. Assume that P is nonatomic, and that $P(A) > 0$ for some $A \in \mathcal{A}$. Show that there are events $A_i \subset A, 1 \leqslant i \leqslant n$, such that $P(A_1 \cap \cdots \cap A_n) = P(A_1) \cdots P(A_n) \neq 0$. [Use (9.42.b).]

Exercise 18.49. Let P and Q be probabilities on $\mathcal{A}$ such that P is nonatomic, and assume that $P(A \cap B) = P(A)P(B)$ whenever $Q(A \cap B) = Q(A)Q(B)$ for $A, B \in \mathcal{A}$. Prove that $P \ll Q$. [Hint. If $Q(A) = 0$, then $P(A) = 0$ or 1. To rule out the latter, use (9.42.b).]

Exercise 18.50. Let P and Q be as in (18.49), and assume that $P(A \cap B) = P(A)P(B)$ is equivalent to $Q(A \cap B) = Q(A)Q(B)$ for $A, B \in \mathcal{A}$. Show that $P = Q$. [Hints. Let $f = dP/dQ$ (18.49). Use (18.48) and (18.49) to prove that $Q(\{1 + \delta < f \leqslant 1 + 2\delta\}) = 0$ for any $\delta > 0$. Then apply (12.26).]

Exercise 18.51. Let f and g be independent random variables such that $E|f + g|^p < \infty$ for some $p \in]0, \infty[$. Prove that $E|f|^p, E|g|^p < \infty$. [Hints. Choose $x_0 > 0$ so that $P(\{|g| < x\}) \geqslant 1/2$ for $x \geqslant x_0$. Then $P(\{|f| \geqslant 2x\}) \leqslant 2P(\{|f| \geqslant 2x\} \cap \{|g| < x\}) \leqslant 2P(\{|f + g| \geqslant x\}), x \geqslant x_0$. Next use (15.23.a).]

Exercise 18.52 (Hewitt-Savage). Notation is as in (7.72). Let Q be a probability on $\mathcal{B}$, and put $P = \otimes_{m\geqslant 1} Q_m$, where $Q_m = Q, m \geqslant 1$. Show that $\mathcal{E} \subset \mathcal{O}_P$. The following steps may be helpful.

(a) $P\widetilde{p} = P$ for any $p \in \cup_{n\geqslant 1}\mathfrak{P}_n$. [Hint. Let $p^{-1} : N \to N$ be the inverse function of p, and let $B_m \in \mathcal{B}, m \geqslant 1$. Then $(P\widetilde{p})(\prod_{m\geqslant 1} B_m) = P(\prod_{m\geqslant 1} B_{p^{-1}(m)}) = \prod_{m\geqslant 1} Q(B_{p^{-1}(m)}) = \prod_{m\geqslant 1} Q(B_m) = P(\prod_{m\geqslant 1} B_m)$. Apply next (10.8), taking into account that $\{\prod_{m\geqslant 1} B_m : B_m \in \mathcal{B}$ for $m \geqslant 1\}$ is closed under intersection and generates $\mathcal{T}_1$.]

(b) For $n \geqslant 1$, let $\mathcal{B}_n$ denote the σ-algebra of subsets of Y^∞ generated by $\pi_m, 1 \leqslant m \leqslant n$. For $\varepsilon > 0$ and $E \in \mathcal{E}$, apply (7.21) and (10.33) to choose $n_\varepsilon \geqslant 1$ and $E_\varepsilon \in \mathcal{B}_{n_\varepsilon}$ so that $P(E \Delta E_\varepsilon) < \varepsilon$. For $B_1, \ldots, B_n \in \mathcal{B}$, write $B^{(n)} = B_1 \times \cdots \times B_n \times Y \times Y \times \cdots$. Then

$$\left|P(E \cap B^{(n)}) - P(\widetilde{p}(E_\varepsilon) \cap B^{(n)})\right| < \varepsilon, \quad p \in \cup_{n \geqslant 1} \mathfrak{P}_n. \tag{1}$$

[Hint. $P(E \Delta \widetilde{p}(E_\varepsilon)) = P(\widetilde{p}(E) \Delta \widetilde{p}(E_\varepsilon)) = P(E \Delta E_\varepsilon) < \varepsilon$ by (a). Now remember (9.6.iii).]

(c) For $n \geqslant 1$, let $k \geqslant (n_\varepsilon \vee n)$, and define $p^* \in \mathfrak{P}_{2k+1}$ by

$$p^*(m) = \begin{cases} m + k \text{ if } 1 \leqslant m \leqslant k \\ m - k \text{ if } k + 1 \leqslant m \leqslant 2k \\ m \qquad \text{if } m \geqslant 2k + 1 \end{cases}.$$

Then

$$\left|P(E)P(B^{(n)}) - P(\widetilde{p^*}(E_\varepsilon) \cap B^{(n)})\right| < \varepsilon. \tag{2}$$

[Hint. $\widetilde{p^*}(E_\varepsilon) \in \mathcal{T}_{k+1}$, $B^{(n)} \in \mathcal{B}_k$, and $\mathcal{T}_{k+1}$ and $\mathcal{B}_k$ are independent.]

(d) $P(E \cap B) = P(E)P(B)$ for all $B \in \mathcal{T}_1$. [Hint. By (1) and (2), $P(E \cap B^{(n)}) = P(E)P(B^{(n)})$. Then use (10.8), as the family of all $B^{(n)}$ with $n \geqslant 1$ is closed under intersection and generates $\mathcal{T}_1$.]

(e) $P(E) = 0$ or 1.

Exercise 18.53. Let $(X, \mathcal{A}, P)$ be a probability space, let $(Y, \mathcal{B})$ be a measurable space, let $f_n : X \to Y, n \geqslant 1$, be random elements, and set $\mathbf{f} = (f_1, f_2, \ldots)$. Suppose that $f_n, n \geqslant 1$, are independent, and that $Pf_n^{-1} = Q$ for some probability Q on $\mathcal{B}$ and any $n \geqslant 1$. Use (18.29) and (18.52) to prove that $\mathbf{f}^{-1}(\mathcal{E}) \subset \mathcal{O}_P$.

Exercise 18.54 (Feller-Chung). For $\{A_n : n \geqslant 1\} \subset \mathcal{A}$ and $\{B_n : n \geqslant 1\} \subset \mathcal{A}$, prove the following.

(a) If B_n is independent of $A_n A_{n-1}^c \cdots A_0^c$ for all $n \geqslant 1$, where $A_0 = \emptyset$, then

$$P\left(\bigcup_{i=1}^n A_i B_i\right) \geqslant \left(\inf_{1 \leqslant i \leqslant n} P(B_i)\right) P\left(\bigcup_{i=1}^n A_i\right), \quad n \geqslant 1, \tag{1}$$

and

$$P\left(\bigcup_{i \geqslant 1} A_i B_i\right) \geqslant \left(\inf_{i \geqslant 1} P(B_i)\right) P\left(\bigcup_{i \geqslant 1} A_i\right). \tag{2}$$

(b) If B_n is independent of each event $A_n, A_n A_{n+1}^c, A_n A_{n+1}^c A_{n+2}^c, \ldots$ for any $n \geqslant 1$, then (1) and (2) also hold.

Exercise 18.55. Let $f_n, n \geqslant 1$, be pairwise independent random variables, and assume that $Pf_n^{-1} = Pf_1^{-1}, n \geqslant 1$. If $E|f_1|^p < \infty$ for some $p \in [1, 2[$ and $Ef_1 = 0$, then $E|S_n|/n^{1/p} \to 0$, where $S_n = f_1 + \cdots + f_n$. The following steps may be helpful.

(a) For $n \geqslant 1$, put $f_{n,i} = f_i 1_{\{|f_i| < n^{1/p}\}}$, $1 \leqslant i \leqslant n$, and $U_n = f_{n,1} + \cdots + f_{n,n}$. Then $EU_n/n^{1/p} \to 0$ and $E|S_n - U_n|/n^{1/p} \to 0$.

(b) $\mathrm{Var}[U_n/n^{1/p}] \to 0$. [Use the dominated convergence theorem (11.25.ii).]

(c) $E|U_n|/n^{1/p} \to 0$. [Write $E|U_n|/n^{1/p} \leqslant \sqrt{\mathrm{Var}[U_n/n^{1/p}] + (EU_n/n^{1/p})}$.]

Exercise 18.56. For $n \geqslant 1$, let f_n and S_n be as in (18.55). If $E[f_1^2/\log(2+|f_1|)] < \infty$ and $Ef_1 = 0$, then $E|S_n|/\sqrt{n\log n} \to 0$. [Mimic the solution to (18.55).]

Exercise 18.57. For $n \geqslant 2$, let $f_1, \ldots, f_n$ be random variables such that $\sigma(f_1, \ldots, f_{n-1})$ and $\sigma(f_n)$ are independent, $E|f_n|^p < \infty$ for some $p \geqslant 1$ and $Ef_n = 0$, and let $\varphi : R^{n-1} \to R$ be a measurable function. Show that $E[|\varphi \circ (f_1, \ldots, f_{n-1}) + f_n|^p 1_A] \geqslant E[|\varphi \circ (f_1, \ldots, f_{n-1})|^p 1_A]$ for $A \in \sigma(f_1, \ldots, f_{n-1})$; in particular, $E[|f_1 + \cdots + f_{n-1} + f_n|^p 1_A] \geqslant E[|f_1 + \cdots + f_{n-1})|^p 1_A]$, $A \in \sigma(f_1, \ldots, f_{n-1})$. [Hints. Let $A = (f_1, \ldots, f_{n-1})^{-1}(B)$ with $B \in \mathcal{B}(R^{n-1})$. By (11.29), (18.29) and (15.8), write

$$
\begin{aligned}
&E[|\varphi \circ (f_1, \ldots, f_{n-1}) + f_n|^p 1_A] \\
&\quad = \int_{B\times R} |\varphi(x_1, \ldots, x_{n-1}) + x_n|^p \, d(P(f_1, \ldots, f_n)^{-1})(x_1, \ldots, x_n) \\
&\quad = \int_B \left(\int_R |\varphi(x_1, \ldots, x_{n-1}) + x_n|^p \, d(Pf_n^{-1})(x_n)) \right. \\
&\qquad\qquad \left. \times\, d(P(f_1, \ldots, f_{n-1})^{-1} \right) (x_1, \ldots, x_{n-1}) \\
&\quad = \int_B E\,|\varphi(x_1, \ldots, x_{n-1}) + f_n|^p \, d(P(f_1, \ldots, f_{n-1})^{-1})(x_1, \ldots, x_{n-1}) \\
&\quad \geqslant \int_B |\varphi(x_1, \ldots, x_{n-1})|^p \, d(P(f_1, \ldots, f_{n-1})^{-1})(x_1, \ldots, x_{n-1}) \\
&\quad = E[|\varphi \circ (f_1, \ldots, f_{n-1})|^p 1_A],
\end{aligned}
$$

where the last inequality comes from $|x|^p = |E[x + f_n]|^p \leqslant E|x + f_n|^p$, $x \in R$.]

Exercise 18.58. Let $\{g_n : n \geqslant 1\}$ and $\{h_n : n \geqslant 1\}$ be sequences of random variables such that either $\sigma(h_n)$ is independent of $\sigma(g_1, \ldots, g_n)$ for all $n \geqslant 1$, or $\sigma(h_n)$ is independent of $\sigma(g_k, k \geqslant n)$ for any $n \geqslant 1$. Prove the following.

(a) For $\varepsilon_n, \delta_n \in R, n \geqslant 1$, $P(\cup_{k\geqslant n}\{g_k + h_k > \varepsilon_k\}) \geqslant (\inf_{k\geqslant n} P(h_k \geqslant -\delta_k)) \times P(\cup_{k\geqslant n}\{g_k > \varepsilon_k + \delta_k\})$, $n \geqslant 1$. [Hint. For $k \geqslant n \geqslant 1$, write $A_k = \{g_k > \varepsilon_k + \delta_k\}$ and $B_k = \{h_k \geqslant -\delta_k\}$. Then use (2) of (18.54).]

(b) For $\varepsilon, \delta \in R$, $P(\limsup_n (g_n + h_n) \geqslant \varepsilon) \geqslant (\liminf_n P(h_n \geqslant -\delta))P(\limsup_n g_n > \varepsilon + \delta)$. [Use (5.63.b) and (a).]

(c) For $\varepsilon, \delta \in R$, $P(\liminf_n (g_n + h_n) \leqslant -\varepsilon) \geqslant (\liminf_n P(h_n \leqslant \delta))P(\liminf_n g_n < -\varepsilon - \delta)$. [Use (b) and (5.61.c).]

Exercise 18.59 (Katona). Let f and g be independent random variables such that $Pf^{-1} = Pg^{-1}$. Prove that $P(|f + g| \geqslant x) \geqslant \frac{1}{2}P(|f| \geqslant x)^2$, $x \geqslant 0$.

Exercise 18.60. For $1 \leqslant i \leqslant m$, let $f_{in}, n \geqslant 1$, be independent random variables. Assume that $\sigma(f_{in}, n \geqslant 1)$, $1 \leqslant i \leqslant m$, are independent. For $n \geqslant 1$, let $\varphi_n : R^m \to R$

be a measurable function. Use (18.14), (18.28.b) and (18.28.c) to show that $\varphi_n \circ (f_{1n}, \ldots, f_{mn})$, $n \geqslant 1$, are independent.

Exercise 18.61. Let $(X, \mathcal{A}, P)$ be a probability space, let $f : X \to X$ be a measurable function such that $Pf^{-1} = P$, and let $\mathcal{B} \subset \mathcal{A}$ be a σ-algebra that is independent of $f^{-1}(\mathcal{A})$. Put $f_0 = i_X$ and $f_{n+1} = f_n \circ f$, $n \geqslant 0$. Prove the following.

(a) $f_n^{-1}(\mathcal{B})$ is independent of $f_{n+1}^{-1}(\mathcal{A})$ for any $n \geqslant 0$.

(b) For $n \geqslant 0$, $f_0^{-1}(\mathcal{B}), \ldots, f_n^{-1}(\mathcal{B}), f_{n+1}^{-1}(\mathcal{A})$ are independent. [Hint. If $f_0^{-1}(\mathcal{B}), \ldots, f_n^{-1}(\mathcal{B}), f_{n+1}^{-1}(\mathcal{A})$ are independent, then $f_1^{-1}(\mathcal{B}), \ldots, f_{n+1}^{-1}(\mathcal{B}), f_{n+2}^{-1}(\mathcal{A})$ are independent. Since $f_1^{-1}(\mathcal{B}) \vee \cdots \vee f_{n+1}^{-1}(\mathcal{B}) \vee f_{n+2}^{-1}(\mathcal{A}) \subset f^{-1}(\mathcal{A})$, use (18.14) to infer that $f_0^{-1}(\mathcal{B}), f_1^{-1}(\mathcal{B}), \ldots, f_{n+1}^{-1}(\mathcal{B}), f_{n+2}^{-1}(\mathcal{A})$ are independent.]

(c) $f_n^{-1}(\mathcal{B})$, $n \geqslant 0$, are independent.

7 Distribution Functions and Characteristic Functions

In this chapter we introduce and discuss two important tools in the study of random variables, namely the distribution function and the characteristic function. Distribution functions of random variables encompass the probabilities of all events involving them, and characteristic functions associated to distribution functions, while paralleling their properties, are easy to handle by methods and results in classical analysis.

19 Distribution Functions

Given a probability space $(X, \mathcal{A}, P)$, a random variable f is by definition an $(\mathcal{A}, \mathcal{B}(R))$-measurable mapping from X into R. By (9.26), it induces the probability Pf^{-1} on $\mathcal{B}(R)$. This measure provides the probabilities of all events involving f. In turn, Pf^{-1} determines a function F as in (10.15) such that the Lebesgue-Stieltjes measure μ_F coincides with Pf^{-1}. This section is mainly devoted to the study of such functions F, as it is easier to handle point functions, by means of classical analysis, rather than set functions.

Notation 19.1. To simplify the writing, if f is a random variable and $B \in \mathcal{B}(R)$, then we shall write $P(f \in B)$ instead of $P(\{f \in B\})$. Sometimes we shall write $P(A_1, \ldots, A_n)$ instead of $P(A_1 \cap \ldots \cap A_n)$ for $A_1, \ldots, A_n \in \mathcal{A}$. Thus, if $f_1, \ldots, f_n$ are random variables and $B_1, \ldots, B_n \in \mathcal{B}(R)$, sometimes we shall write $P(f_1 \in B_1, \ldots, f_n \in B_n)$ instead of $P(\{f_1 \in B_1\}, \ldots, \{f_n \in B_n\})$. If $f_n, n \geqslant 1$, are random variables and $B_n \in \mathcal{B}(R), n \geqslant 1$, then in place of $P(\limsup_n \{f_n \in B_n\})$ we shall sometimes write $P(f_n \in B_n \text{ i.o.})$, to be read "probability that $f_n \in B_n$ infinitely often". Similar conventions will apply to random vectors and random elements.

Definition 19.2. Let $(X, \mathcal{A}, P)$ be a probability space, and let f be a random variable on X. The probability Pf^{-1} on $\mathcal{B}(R)$ is called the *distribution of f*. This definition extends in the obvious way to random elements and random vectors (see (19.29)).

Examples 19.3. (a) If f is a random variable such that $f = a$ a.s. for some $a \in R$, then the distribution of f is the Dirac measure ε_a.

(b) If f is a random variable such that $P(f = -1) = P(f = 1) = 1/2$, then the distribution of f is $(\varepsilon_{-1} + \varepsilon_1)/2$. It is called the *symmetric Bernoulli [Rademacher] distribution.*

Analysis and Probability. http://dx.doi.org/10.1016/B978-0-12-401665-1.00007-2

(c) Let $f : X \to \{0, \ldots, n\}$ be a random variable such that $P(f = k) = \binom{n}{k} p^k(1-p)^{n-k}, k = 0, \ldots, n$, where $p \in]0, 1[$. Then the distribution of f is discrete and called the *binomial distribution*. Sometimes we shall write $b(k; n, p) = \binom{n}{k} p^k(1-p)^{n-k}$, to ease the notation.

(d) Let $f : X \to \{0, 1, \ldots\}$ be a random variable such that $P(f = a + bk) = e^{-\lambda}\lambda^k/k!, k = 0, 1, \ldots$, where $\lambda > 0, a \in R$, and $b \neq 0$. Then the distribution of f is discrete and called the *Poisson distribution*. The customary case is that when $a = 0$ and $b = 1$. We shall write also $p(k; \lambda) = e^{-\lambda}\lambda^k/k!$

(e) Let $f : X \to \{0, 1, \ldots\}$ be a random variable such that $P(f = k) = p(1-p)^k, k = 0, 1, \ldots$, where $p \in]0, 1[$. Then the distribution of f is discrete and called the *geometric distribution*.

(f) Let $m_1, m_2, n \in N$ with $n \leqslant m_1 + m_2$, and let $f : X \to \{0, \ldots, n\}$ be a random variable such that $P(f = k) = \binom{m_1}{k}\binom{m_2}{n-k}\Big/\binom{m_1+m_2}{n}, k = 0, \ldots, n$. Then the distribution of f is discrete and called the *hypergeometric distribution*.

Definition 19.4. Let f be a random variable. The function $F : R \to [0, 1]$ defined by $F(x) = P(f \leqslant x) = (Pf^{-1})(]-\infty, x])$ for any $x \in R$ is called the *distribution function of f*. Obviously, F is nondecreasing and right continuous on R, $\lim_{x\to-\infty} F(x) = 0$ and $\lim_{x\to\infty} F(x) = 1$.

Example 19.5. The distribution function of f in (19.3.a) is $F_a(x) = \begin{cases} 0 \text{ if } x < a \\ 1 \text{ if } x \geqslant a \end{cases}$. We say that F_a and f are *degenerate* (at a).

Remarks 19.6. (a) Notation is as in (19.4). For each bounded interval $]a, b] \subset R$, we have $(Pf^{-1})(]a, b]) = (Pf^{-1})(]-\infty, b]) - (Pf^{-1})(]-\infty, a]) = F(b) - F(a)$. Thus, according to (10.15), the distribution of f coincides with the Lebesgue-Stieltjes measure μ_F induced by F, and so $P(f \in B) = \mu_F(B), B \in \mathcal{B}(R)$.

(b) Distribution functions and random variables associated to celebrate distributions such as those in (19.3) will bear their names.

Theorem 19.7. *Let $F : R \to [0, 1]$ be a nondecreasing function which is right continuous on R and such that $\lim_{x\to-\infty} F(x) = 0$ and $\lim_{x\to\infty} F(x) = 1$. Then there exist a probability space $(X, \mathcal{A}, P)$ and a random variable f on X such that the distribution function of f is F.*

Proof. Consider $X =]0, 1[, \mathcal{A} = \mathcal{B}(]0, 1[)$, and let P be the Lebesgue measure on $\mathcal{B}(]0, 1[)$. Define $f :]0, 1[\to R$ by $f(y) = \sup\{x : F(x) < y\}$ for any $y \in]0, 1[$. Since $\lim_{x\to-\infty} F(x) = 0$, we have $\{x : F(x) < y\} \neq \emptyset, y \in]0, 1[$. As $\lim_{x\to\infty} F(x) = 1$, the set $\{x : F(x) < y\}$ has an upper bound in R for each $y \in]0, 1[$. Thus $f(y) \in R$ for any $y \in]0, 1[$. For $a \in R$, we have $\{y : f(y) \leqslant a\} = \{y : \sup\{x : F(x) < y\} \leqslant a\} = \{y : F(x) \geqslant y \text{ for any } x > a\} = \{y : F(a) \geqslant y\} =]0, F(a)]$. Hence $P(f \leqslant a) = F(a)$ for any $a \in R$. □

Remarks 19.8. (a) We may speak of any function F with the foregoing properties in (19.7) as being a distribution function (on R) without mentioning a specific random variable.

(b) Let μ be a probability on $\mathcal{B}(R)$. Then the function $F(x) = \mu(]-\infty, x])$, $x \in R$, is a distribution function called the *cumulative distribution function of* μ.

(c) An easier proof of (19.7) is as follows. Take $(X, \mathcal{A}, P) = (R, \mathcal{B}(R), \mu_F)$, and let $f = i_R$ be the identity function. Then $P(f \leqslant x) = \mu_F(\{i_R \leqslant x\}) = \mu_F(]-\infty, x]) = F(x)$ for any $x \in R$.

(d) Let $\{F_i : i \in I\}$ be a nonempty set of distribution functions. Then there exist a probability space $(X, \mathcal{A}, P)$ and a set $\{f_i : i \in I\}$ of independent random variables on X such that the distribution function of f_i is F_i for each $i \in I$. Actually, consider $(X, \mathcal{A}, P) = (\prod_{i\in I} X_i, \otimes_{i\in I}\mathcal{A}_i, \otimes_{i\in I}\mu_{F_i})$, where $(X_i, \mathcal{A}_i) = (R, \mathcal{B}(R))$, $i \in I$, and let $f_i = \pi_i$, $i \in I$. Then f_i, $i \in I$, are P-independent by (18.27). Moreover, in view of (16.11), $P(f_i \leqslant x) = P(\pi_i \leqslant x) = \mu_{F_i}(]-\infty, x]) = F_i(x)$ for any $i \in I$ and $x \in R$.

(e) If $F(x)$, $x \in R$, is a distribution function, then $F(ax+b)$, $x \in R$, is a distribution function for any $a > 0$ and $b \in R$.

(f) The function f in (19.7) has the following properties: (i) f is nondecreasing, (ii) f is left continuous on $]0, 1[$, (iii) $f(F(a)) \leqslant a$ for any $a \in R$, and (iv) $F(f(b)) \geqslant b$ for any $b \in]0, 1[$.

Definition 19.9. A measurable function $p : R \to [0, \infty[$ such that $\int_{-\infty}^{\infty} p(x)dx = 1$ is called a *probability density* (on R). Evidently, the function $F(x) = \int_{-\infty}^{x} p(t)dt$, $x \in R$, is a distribution function. In this case we say that F has density p. In fact, according to (12.17), p is the Radon-Nikodým derivative of μ_F with respect to the Lebesgue measure. A random variable f with distribution function F is also said to have density p and will bear the name of F. In this case the distribution Pf^{-1} will bear the name of F as well.

Examples 19.10. (a) Let $a, b \in R$ be such that $a < b$. Then the function $p : R \to [0, \infty[$ defined by $p(x) = \begin{cases} (b-a)^{-1} & \text{if } x \in [a, b] \\ 0 & \text{if } x \notin [a, b] \end{cases}$ is a probability density. The distribution function associated to p is called the *uniform distribution function* (on $[a, b]$).

(b) The function $p : R \to [0, \infty[$ defined by $p(x) = \begin{cases} \lambda e^{-\lambda x} & \text{if } x \geqslant 0 \\ 0 & \text{if } x < 0 \end{cases}$, where $\lambda > 0$, is a probability density. The distribution function corresponding to p is called the *exponential distribution function.*

(c) The function $p(x) = a/\pi(a^2+x^2)$, $x \in R$, where $a > 0$, is a probability density. The distribution function associated to p is called the *Cauchy distribution function.*

(d) For $m \in R$ and $\sigma > 0$, the function $p(x) = \sigma^{-1}(2\pi)^{-1/2}\exp\{-(x-m)^2/2\sigma^2\}$, $x \in R$, is a probability density. The distribution function corresponding to p is called the *normal* [*Gaussian*] *distribution function*. Sometimes we shall write $G_{m,\sigma}$ to designate this function. If $m = 0$ and $\sigma = 1$, then this distribution function is called the *standard normal* [*Gaussian*] *distribution function,* hereafter to be denoted by Φ.

(e) The function $p : R \to [0, \infty[$ defined by $p(x) = \begin{cases} b^a x^{a-1} e^{-bx} / \Gamma(a) & \text{if } x \geqslant 0 \\ 0 & \text{if } x < 0 \end{cases}$, where $a, b > 0$, is a probability density. The distribution function corresponding to p is called the *gamma distribution function.* If $a = 1$, we get an exponential distribution function. The special case $a = k/2$ and $b = 1/2\sigma^2$, where $k \in N$ and $\sigma > 0$, corresponds to the so-called χ^2 *distribution function* (with k degrees of freedom).

(f) The function $p : R \to [0, \infty[$ defined by $p(x) = \begin{cases} bc(x-a)^{b-1} e^{-c(x-a)^b} & \text{if } x \geqslant a \\ 0 & \text{if } x < a \end{cases}$, where $a \geqslant 0$ and $b, c > 0$, is a probability density. The distribution function associated to p is called the *Weibull distribution function.* If $a = 0$ and $b = 1$, we get an exponential distribution function.

(g) The function $p : R \to [0, \infty[$ defined by $p(x) = \begin{cases} (a - |x|)/a^2 & \text{if } x \in [-a, a] \\ 0 & \text{if } x \notin [-a, a] \end{cases}$, where $a > 0$, is a probability density. The distribution function corresponding to p is called the *triangular distribution function.*

(h) The function $p : R \to [0, \infty[$ defined by $p(x) = \begin{cases} (4 - x^2)^{1/2}/2\pi & \text{if } x \in [-2, 2] \\ 0 & \text{if } x \notin [-2, 2] \end{cases}$ is a probability density. The distribution function associated to p is called the *Wigner [semicircle] distribution function.*

Remarks 19.11. (a) Let μ_F be as in (10.15), and let λ be the Lebesgue measure on $\mathcal{B}(R)$. Then, for the distribution functions F corresponding to the random variables in (19.3), $\mu_F \perp \lambda$ by (12.17). These are special instances of the following general result. In fact, any measure μ_F induced by a distribution function F is uniquely expressible in the form

$$\mu_F = \mu_1 + \mu_2 + \mu_3, \tag{1}$$

where $\mu_1 \ll \lambda$, $\mu_2 \perp \lambda$ and $\mu_2(\{x\}) = 0$ for any $x \in R$, and μ_3 is discrete. Actually, by the Lebesgue decomposition (12.12) and (12.13), μ_F is uniquely expressible in the form

$$\mu_F = \mu_1 + \mu, \tag{2}$$

where $\mu_1 \ll \lambda$ and $\mu \perp \lambda$. Further, in view of (9.38) and (9.44.c), μ is uniquely decomposed in the form

$$\mu = \mu_2 + \mu_3, \tag{3}$$

where μ_2 is nonatomic and μ_3 is atomic. Finally, on account of (9.39), $\mu_2(\{x\}) = 0$, $x \in R$, and μ_3 is discrete. Now (1) follows from (2) and (3).

If we put $F_i(x) = \mu_i(]-\infty, x])$, $x \in R$, for $i = 1, 2, 3$, then (1) induces a unique decomposition of F in the form

$$F = F_1 + F_2 + F_3,$$

where F_1 is absolutely continuous, i.e. F_1 has a derivative F_1' almost everywhere with respect to λ such that $F_1' = d\mu_1/d\lambda$ λ-a.e., F_2 is continuous (see (10.17.a) and (10.36.b)) and singular, i.e. the derivative F_2' exists and equals 0 λ-a.e., and F_3 is a jump function. Basic facts about absolutely continuous functions and singular functions are to be found in *Theory of Functions of a Real Variable* (Natanson, 1955).

(b) Let f be a random variable on X with distribution function F, let $g : R \to R$ be a measurable function, and let $B \in \mathcal{B}(R)$. Then, by the change of variable formula (11.29) and (19.6.a), we have

$$\int_{f^{-1}(B)} (g \circ f) dP = \int_B g d(Pf^{-1}) = \int_B g d\mu_F = \int_B g dF, \tag{1}$$

in the sense that if any integral exists so do the others and then all are equal. (Sometimes we will write $\int_{\{x \in B\}} g(x) dF(x)$, in place of the last integral in (1).) Moreover, if there is a probability density p such that F is as in (19.9), and the last integral in (1) exists, then

$$\int_{f^{-1}(B)} (g \circ f) dP = \int_B g(x) p(x) d\lambda(x) \tag{2}$$

by (12.17.b) and (12.18), where λ is the Lebesgue measure on $\mathcal{B}(R)$. In particular, if the kth moment of f exists for some $k > 0$, then $Ef^k = \int_X f^k dP = \int_R x^k dF(x)$. Furthermore, if F has density p, then $Ef^k = \int_{-\infty}^{\infty} x^k p(x) dx$. (2) shows also that $P(f \in B) = \int_B p(x) dx$.

Throughout this part, $\mathfrak{F}$ will be the set of all distribution functions on R. Now we introduce a binary operation on $\mathfrak{F}$.

Definition 19.12. For $F_1, F_2 \in \mathfrak{F}$ define $F_1 * F_2 \in \mathfrak{F}$ by

$$(F_1 * F_2)(x) = \int_{-\infty}^{\infty} F_1(x - y) dF_2(y), \quad x \in R. \tag{1}$$

For each fixed $x \in R$, the function $F_1(x - y)$, $y \in R$, is measurable by (8.34), and so the integral in (1) makes sense. Moreover, in view of the dominated convergence theorem (11.25.ii), it follows that $F_1 * F_2$ is right continuous on R, $\lim_{x \to -\infty} (F_1 * F_2)(x) = 0$ and $\lim_{x \to \infty} (F_1 * F_2)(x) = 1$. Thus $F_1 * F_2$ is indeed a distribution function called the *convolution of* F_1 *and* F_2.

Remarks 19.13. Let $F_1, F_2 \in \mathfrak{F}$, and set $s(x_1, x_2) = x_1 + x_2$, $(x_1, x_2) \in R^2$.

(a) $F_1 * F_2$ coincides with the cumulative distribution function of $(\mu_{F_1} \otimes \mu_{F_2}) s^{-1}$. Actually, for $x \in R$, we have

$$\begin{aligned} ((\mu_{F_1} \otimes \mu_{F_2}) s^{-1})(]-\infty, x]) &= (\mu_{F_1} \otimes \mu_{F_2})(\{(x_1, x_2) \in R^2 : x_1 + x_2 \leqslant x\}) \\ &= \int_R \mu_{F_1}(]-\infty, x - x_2]) d\mu_{F_2}(x_2) \\ &= \int_{-\infty}^{\infty} F_1(x - x_2) dF_2(x_2) = (F_1 * F_2)(x). \end{aligned}$$

In view of (10.8), this also shows that $(\mu_{F_1} \otimes \mu_{F_2}) s^{-1} = \mu_{F_1 * F_2}$.

(b) If $\mu_{F_1} \ll \lambda$ and $d\mu_{F_1} = f d\lambda$, where λ is the Lebesgue measure on $\mathcal{B}(R)$, then $\mu_{F_1 * F_2} \ll \lambda$ and $d\mu_{F_1 * F_2} = g d\lambda$, where $g(x) = \int_R f(x - x_2) dF_2(x_2)$, $x \in R$. Actually, let $h : R \to [0, \infty[$ be a measurable function. Then, by (11.29), (15.8) and (12.8), we have

$$\int_R h d\mu_{F_1 * F_2} = \int_R hd((\mu_{F_1} \otimes \mu_{F_2}) s^{-1}) = \int_R (h \circ s) d(\mu_{F_1} \otimes \mu_{F_2})$$

$$= \int_R \left(\int_R h(x_1 + x_2) d\mu_{F_1}(x_1) \right) d\mu_{F_2}(x_2)$$

$$= \int_R \left(\int_R h(x_1 + x_2) f(x_1) d\lambda(x_1) \right) dF_2(x_2)$$

$$= \int_R \left(\int_R (h \circ s_{x_2})(x_1)(f \circ s_{x_2}^{-1} \circ s_{x_2})(x_1) d\lambda(x_1) \right) dF_2(x_2)$$

$$= \int_R \left(\int_R h(x)(f \circ s_{x_2}^{-1})(x) d(\lambda s_{x_2}^{-1})(x) \right) dF_2(x_2),$$

where s_{x_2} is the section of s at x_2. Since $\lambda s_{x_2}^{-1} = \lambda$ (10.17.c), applying again Fubini's theorem (15.8), we see that

$$\int_R h d\mu_{F_1 * F_2} = \int_R \left(\int_R h(x) f(x - x_2) d\lambda(x) \right) dF_2(x_2)$$

$$= \int_R h(x) \left(\int_R f(x - x_2) dF_2(x_2) \right) d\lambda(x).$$

The assertion now follows on putting $h = 1_A$, $A \in \mathcal{B}(R)$.

The primary interest of convolution for probability theory stems from the next result.

Theorem 19.14. *Let f_1 and f_2 be independent random variables with corresponding distribution functions F_1 and F_2. Then $f_1 + f_2$ has the distribution function $F_1 * F_2$.*

Proof. On account of (18.29) and (19.6.a), we have $P(f_1 + f_2)^{-1} = (P(f_1, f_2)^{-1})s^{-1} = ((P(f_1)^{-1} \otimes P(f_2)^{-1})s^{-1} = (\mu_{F_1} \otimes \mu_{F_2})s^{-1}$. Hence the result follows from (19.13.a). □

Corollary 19.15. *The binary operation $*$ is commutative and associative.*

Proof. This follows with the aid of (19.8.d), since the binary operation of addition of independent random variables has the two properties. □

The following remark helps to justify the definition of convergence of distribution functions.

Remark 19.16. Let $D \subset R$ be such that $D^- = R$, and let $F_1, F_2 \in \mathfrak{F}$ be such that $F_1(x) = F_2(x), x \in D$. Then $F_1 = F_2$. Actually, for $x \in R$, let $\{x_n : n \in N\} \subset D$ be such that $x < x_n, n \in N$, and $x_n \to x$. Such a sequence exists, as D is dense in R. Since F_1 and F_2 are right continuous on R, $F_1(x) = \lim_n F_1(x_n) = \lim_n F_2(x_n) = F_2(x)$.

Definitions 19.17. Let $\{F_n : n \in N\} \subset \mathfrak{F}$ and $F \in \mathfrak{F}$. We say that the sequence $\{F_n : n \in N\}$ *converges weakly to* F, and we write $F_n \xrightarrow{w} F$, if $F_n(x) \to F(x)$ for any $x \in C(F)$, where $C(F)$ denotes the set of continuity points of F, that is $C(F) = \{x \in R : F(x-) = F(x)\}$.

Let $]a, b[\subset R$, $t_0 \in [a, b]$, $\{F_t : t \in]a, b[\} \subset \mathfrak{F}$ and $F \in \mathfrak{F}$. We say that $\{F_t : t \in]a, b[\}$ *converges weakly to* F as $t \to t_0$, and we write $F_t \xrightarrow{w} F$, if $F_{t_n} \xrightarrow{w} F$ whenever $\{t_n : n \in N\} \subset]a, b[- \{t_0\}$ is such that $t_n \to t_0$.

For $n \in N$, let f_n be a random variable with distribution function F_n, and let f be a random variable with distribution function F. We say that $\{f_n : n \in N\}$ *converges in distribution to* f, and we write $f_n \xrightarrow{D} f$, if $\{F_n : n \in N\}$ converges weakly to F. Sometimes we will use the hybrid notation $f_n \xrightarrow{D} F$ instead of $f_n \xrightarrow{D} f$. The convergence in distribution depends only on the distribution functions F_n and F, and so the domains of f_n and f need not be the same. The same remark applies to convergence in distribution of sequences of random vectors and random elements (see (19.36.d) and (22.13)).

Remark 19.18. Let $\{F_n : n \in N\}$ and F be as in (19.17).

(a) Let $F' \in \mathfrak{F}$. If $F_n \xrightarrow{w} F$ and $F_n \xrightarrow{w} F'$, then $F = F'$. Actually, $C(F)^c \cup C(F')^c$ is countable by (10.34.c) and (3.17), and so $C(F) \cap C(F')$ is dense in R by (5.69). Therefore, as $F(x) = F'(x), x \in C(F) \cap C(F')$, $F = F'$ follows from (19.16).

(b) Let $\{A, B\} \subset \mathcal{A}$ be a partition of X, assume that $P(A) = P(B) = 1/2$, and set $f = 1_A$ and $g = 1_B$. Further, for $n \in N$, put $f_n = f$ and $g_n = g$. Then $f_n \xrightarrow{D} f$ and $f_n \xrightarrow{D} g$, but $f(x) \neq g(x)$ for any $x \in X$.

We have $f_n \xrightarrow{D} f$, $g_n \xrightarrow{D} f$, and $f_n + g_n \xrightarrow{D} 1$. Thus $\{f_n + g_n : n \in N\}$ does not converge in distribution to $2f$. In this connection see (19.14) and (20.12.b), noticing that f_n and g_n are not independent. Also $f_n g_n \xrightarrow{D} 0$, but $\{f_n g_n : n \in N\}$ does not converge in distribution to f^2.

(c) Let $f_n, n \in N$, and f be as in (19.17). If $f_n \xrightarrow{D} f$, then $f_n + a \xrightarrow{D} f + a$ and $af_n \xrightarrow{D} af$ for any $a \in R$ (see also (19.61)).

(d) If $C(F) = R$ and $F_n \xrightarrow{w} F$, then it follows from (6.84) that $F_n \xrightarrow{u} F$.

(e) $F_n \xrightarrow{w} F$ if and only if $F_n(x-) \to F(x)$ for any $x \in C(F)$.

(f) If $f_n, n \in N$, are random variables such that $P(|f_n| \geqslant a) \to 1$ for any $a > 0$, then there is no random variable f with $f_n \xrightarrow{D} f$.

In what follows we present necessary and sufficient conditions for weak convergence.

Theorem 19.19. *Let $\{F_n : n \in N\}$ and F be as in (19.17). Then $F_n \xrightarrow{w} F$ if and only if there is a set D dense in R such that $F_n(y) \to F(y)$, $y \in D$.*

Proof. If $F_n \xrightarrow{w} F$, then $F_n(x) \to F(x), x \in C(F)$, and $C(F)$ is dense in R.

Conversely, let D be dense in R, assume that $F_n(y) \to F(y)$ for any $y \in D$, and consider $x \in C(F)$. For $\varepsilon > 0$, choose $y, y' \in D$ such that $y < x < y'$ and $F(y') - F(y) < \varepsilon$. Then choose $n(\varepsilon) \in N$ such that $|F_n(y) - F(y)| < \varepsilon$ and $\left|F_n(y') - F(y')\right| < \varepsilon$ for $n \geqslant n(\varepsilon)$. Therefore, $|F_n(y) - F(x)| < 2\varepsilon$ and $\left|F_n(y') - F(x)\right| < 2\varepsilon$ for $n \geqslant n(\varepsilon)$. Hence $|F_n(x) - F(x)| < 2\varepsilon$ for $n \geqslant n(\varepsilon)$. □

The metric ρ for $\mathfrak{F}$ we introduce now is called the *Lévy metric.*

Theorem 19.20. *For $F, G \in \mathfrak{F}$, put $\rho(F, G) = \inf\{h > 0 : G(x-h)-h \leqslant F(x) \leqslant G(x+h)+h$ for all $x \in R\} = \inf A(F, G)$ (say). Then ρ is a metric for $\mathfrak{F}$, and $(\mathfrak{F}, \rho)$ is a separable metric space.*

Proof. It is rather obvious that $\rho(F, F) = 0$, $\rho(F, G) = \rho(G, F)$ and $\rho(F, G) = 0$ implies $F = G$. To prove the triangle inequality, let $H \in \mathfrak{F}$, and let $h \in A(F, G)$ and $h' \in A(G, H)$. Then $H(x - h - h') - h - h' \leqslant F(x) \leqslant H(x + h + h') + h + h'$ for all $x \in R$, and so $h + h' \in A(F, H)$, that is $h + h' \geqslant \rho(F, H)$. Thus $\rho(F, H) \leqslant \rho(F, G) + \rho(G, H)$. Hence ρ is a metric for $\mathfrak{F}$. Now let $\mathfrak{F}_1$ stand for the set of all distribution functions $F \in \mathfrak{F}$ such that $F(R)$ is a finite subset of Q, and let $\mathfrak{F}_0$ denote the set of those distribution functions in $\mathfrak{F}_1$ having jumps at points in Q. Further, let $F \in \mathfrak{F}$, let f be as in (19.7), and put $f(1) = \infty$. For $\varepsilon > 0$ and $1/\varepsilon < n \in N$, define $F_n(x) = 0$ when $x < f(1/n)$, and set $F_n(x) = i/n$ whenever $f(i/n) \leqslant x < f((i + 1)/n)$, $1 \leqslant i \leqslant n - 1$. Obviously, $F_n \in \mathfrak{F}_1$. If $x < f(1/n)$, then

$$F(x - \varepsilon) - \varepsilon < F(x) - \frac{1}{n} < 0 = F_n(x) < F(x + \varepsilon) + \varepsilon. \tag{1}$$

If $f(i/n) \leqslant x < f((i + 1)/n)$ for some i, then

$$F(x - \varepsilon) - \varepsilon < F(x) - \frac{1}{n} \leqslant \frac{i+1}{n} - \frac{1}{n} = F_n(x) < F(x + \varepsilon) + \varepsilon. \tag{2}$$

From (1) and (2) we infer that $\rho(F_n, F) \leqslant \varepsilon, n > 1/\varepsilon$, and so $\rho(F_n, F) \to 0$. This shows that $\mathfrak{F}_1^- = \mathfrak{F}$. Next, for $G = \sum_{j=1}^{m} a_j 1_{[x_j, x_{j+1}[} \in \mathfrak{F}_1$, where $-\infty < x_1 < \cdots < x_m < x_{m+1} = \infty$ and $0 < a_1 < \cdots < a_m = 1$, choose $x_1(n), \ldots, x_m(n) \in Q$ such that $x_j \leqslant x_j(n) < x_{j+1}$, $1 \leqslant j \leqslant m$, for each $n \in N$. Put $G_n = \sum_{j=1}^{m} a_j 1_{[x_j(n), x_{j+1}(n)[}$, where $x_{m+1}(n) = \infty$. Clearly, $G_n \in \mathfrak{F}_0$ for any $n \in N$. If $x_j(n) \to x_j$ for $1 \leqslant j \leqslant m$, then $G_n(x) \to G(x)$ for any $x \in R$. On account of (19.21) below, this entails that $\rho(G_n, G) \to 0$, and so $\mathfrak{F}_0^- = \mathfrak{F}_1^-$. Consequently, $\mathfrak{F}_0^- = \mathfrak{F}$. Since Q is denumerable, it follows that $\mathfrak{F}_0$ is denumerable by (3.40) and (3.42). Thus $(\mathfrak{F}, \rho)$ is separable. □

Theorem 19.21. *Let $\{F_n : n \in N\}$ and F be as in (19.17). Then $F_n \xrightarrow{w} F$ if and only if $\rho(F_n, F) \to 0$.*

Proof. Assume that $F_n \xrightarrow{w} F$ and let $\varepsilon > 0$. Choose $a, b \in C(F)$ such that $a < b$, $F(a) < \varepsilon/2$ and $1 - F(b) < \varepsilon/2$. Then consider $a = a_0 < a_1 < \cdots < a_m = b$ a division of $[a, b]$, where $a_j \in C(F), 0 \leqslant j \leqslant m$, and $a_j - a_{j-1} < \varepsilon, 1 \leqslant j \leqslant m$. For $0 \leqslant j \leqslant m$, determine $n_j(\varepsilon) \in N$ such that $n \geqslant n_j(\varepsilon)$ entails $|F_n(a_j) - F(a_j)| < \varepsilon/2$. Set $n(\varepsilon) = \vee_{j=0}^{m} n_j(\varepsilon)$ and let $n \geqslant n(\varepsilon)$. Then, if $x \leqslant a_0$, we have

$$F(x-\varepsilon)-\varepsilon < F(x)-\frac{\varepsilon}{2} < 0 \leqslant F_n(x) \leqslant F_n(a_0) < F(a_0)+\frac{\varepsilon}{2} < F(x+\varepsilon)+\varepsilon. \tag{1}$$

Analogously, if $x \geqslant a_m$, we get

$$F(x-\varepsilon)-\varepsilon < F(a_m)-\frac{\varepsilon}{2} < F_n(a_m) \leqslant F_n(x) \leqslant 1 < F(x)+\frac{\varepsilon}{2} < F(x+\varepsilon)+\varepsilon. \tag{2}$$

Finally, if $x \in [a_{j-1}, a_j]$ for some j, then

$$\begin{aligned} F(x-\varepsilon)-\varepsilon &< F(a_{j-1})-\frac{\varepsilon}{2} < F_n(a_{j-1}) \leqslant F_n(x) \leqslant F_n(a_j) \\ &< F(a_j)+\frac{\varepsilon}{2} < F(x+\varepsilon)+\varepsilon. \end{aligned} \tag{3}$$

(1), (2) and (3) show that $\rho(F_n, F) \leqslant \varepsilon, n \geqslant n(\varepsilon)$, and so $\rho(F_n, F) \to 0$.

Assume now that $\rho(F_n, F) \to 0$, and let $x_0 \in C(F)$ and $\varepsilon > 0$. Choose $\delta > 0$ such that $|x - x_0| \leqslant \delta$ entails $|F(x) - F(x_0)| < \varepsilon$, put $h = \varepsilon \vee \delta$, and select $n_h \in N$ such that $\rho(F_n, F) < h$ whenever $n \geqslant n_h$. Then, for $n \geqslant n_h$, we have $F_n(x_0) \leqslant F(x_0+h)+h \leqslant F(x_0+\delta)+\varepsilon < F(x_0)+2\varepsilon$ and $F_n(x_0) \geqslant F(x_0-h)-h \geqslant F(x_0-\delta)-\varepsilon > F(x_0)-2\varepsilon$. Thus $|F_n(x_0) - F(x_0)| < 2\varepsilon, n \geqslant n_h$. Hence $F_n \overset{w}{\to} F$. □

To prove the Helly-Bray theorem below, we need the next lemma.

Lemma 19.22. *Let $\{F_n : n \in N\}$ and F be as in* (19.17), *let $a, b \in C(F)$ be such that $a < b$, let $g : [a, b] \to R$ be a continuous function, and assume that $F_n \overset{w}{\to} F$. Then $\int_a^b g\,dF_n \to \int_a^b g\,dF$.*

Proof. Let $\varepsilon > 0$. Since g is uniformly continuous, choose $\delta > 0$ so that $|g(x) - g(y)| < \varepsilon$ whenever $x, y \in [a, b]$ and $|x - y| < \delta$. Then select $a = a_0 < a_1 < \cdots < a_m = b$ a division of $[a, b]$ such that $a_j \in C(F), 0 \leqslant j \leqslant m$, and $a_j - a_{j-1} < \delta, 1 \leqslant j \leqslant m$. For $0 \leqslant j \leqslant m$, consider $n_j(\varepsilon) \in N$ so that $n \geqslant n_j(\varepsilon)$ entails $|F_n(a_j) - F(a_j)| < \varepsilon/m$. Put $n(\varepsilon) = \vee_{j=0}^{m} n_j(\varepsilon)$ and let $n \geqslant n(\varepsilon)$. Then, for $1 \leqslant j \leqslant m$, we have

$$\begin{aligned} &\left| \int_{a_{j-1}}^{a_j} g\,dF_n - \int_{a_{j-1}}^{a_j} g\,dF \right| \\ &\leqslant \int_{a_{j-1}}^{a_j} |g(x) - g(a_j)|\,dF_n(x) + |g(a_j)|\,|F_n(a_j) \\ &\quad - F_n(a_{j-1}) - F(a_j) + F(a_{j-1})| + \int_{a_{j-1}}^{a_j} |g(a_j) - g(x)|\,dF(x) \end{aligned}$$

$$\leqslant \varepsilon(F_n(a_j) - F_n(a_{j-1}) + F(a_j) - F(a_{j-1}))$$
$$+ \|g\|_u \left(\left|F_n(a_j) - F(a_j)\right| + \left|F_n(a_{j-1}) - F(a_{j-1})\right|\right)$$
$$< \varepsilon(F_n(a_j) - F_n(a_{j-1}) + F(a_j) - F(a_{j-1})) + 2\,\|g\|_u\,\varepsilon/m.$$

Consequently, for $n \geqslant n(\varepsilon)$,

$$\left|\int_a^b g\,dF_n - \int_a^b g\,dF\right| \leqslant \sum_{j=1}^m \left|\int_{a_{j-1}}^{a_j} g\,dF_n - \int_{a_{j-1}}^{a_j} g\,dF\right| < 2\varepsilon + 2\,\|g\|_u\,\varepsilon. \qquad \square$$

Helly-Bray theorem 19.23. *Let $\{F_n : n \in N\}$ and F be as in* (19.17). *Then $F_n \overset{w}{\to} F$ if and only if $\int_{-\infty}^{\infty} g\,dF_n \to \int_{-\infty}^{\infty} g\,dF$ for any $g \in \mathfrak{C}_r(R)$.*

Proof. Assume that $F_n \overset{w}{\to} F$ and let $\varepsilon > 0$. Choose $a, b \in C(F)$ such that $a < b$, $F(a) < \varepsilon$, and $1 - F(b) < \varepsilon$. Since $F_n(a) \to F(a)$ and $F_n(b) \to F(b)$, determine $n_1 \in N$ such that $F_n(a) < \varepsilon$ and $1 - F_n(b) < \varepsilon$ for all $n \geqslant n_1$. Next, applying (19.22), select $n_2 \in N$ such that $n \geqslant n_2$ entails $\left|\int_a^b g\,dF_n - \int_a^b g\,dF\right| < \varepsilon$. Then, for $n \geqslant n_1, n_2$, we have

$$\left|\int_{-\infty}^{\infty} g\,dF_n - \int_{-\infty}^{\infty} g\,dF\right| \leqslant \int_{-\infty}^{a} |g|\,dF_n + \int_{-\infty}^{a} |g|\,dF + \int_{b}^{\infty} |g|\,dF_n + \int_{b}^{\infty} |g|\,dF$$
$$+ \left|\int_a^b g\,dF_n - \int_a^b g\,dF\right| < 4\,\|g\|_u\,\varepsilon + \varepsilon.$$

Conversely, assume that $\int_{-\infty}^{\infty} g\,dF_n \to \int_{-\infty}^{\infty} g\,dF$ for any $g \in \mathfrak{C}_r(R)$, and let $x \in C(F)$ and $\varepsilon > 0$. Select $\delta > 0$ so that $F(x+\delta) - F(x-\delta) < \varepsilon$. Then define $g_1 \in \mathfrak{C}_r(R)$ to be the function that equals 1 on $]-\infty, x-\delta]$, is linear in $]x-\delta, x[$, and equals 0 on $[x, \infty[$. Also define $g_2 \in \mathfrak{C}_r(R)$ to be the function that equals 1 on $]-\infty, x]$, is linear in $]x, x+\delta[$, and equals 0 on $[x+\delta, \infty[$. Clearly, $g_1 \leqslant 1_{]-\infty,x]} \leqslant g_2$, and so

$$\int_{-\infty}^{\infty} g_1\,dF_n \leqslant F_n(x) \leqslant \int_{-\infty}^{\infty} g_2\,dF_n, \quad n \in N.$$

This entails that

$$\int_{-\infty}^{\infty} g_1\,dF \leqslant \liminf_n F_n(x) \leqslant \limsup_n F_n(x) \leqslant \int_{-\infty}^{\infty} g_2\,dF. \tag{1}$$

At the same time,

$$\int_{-\infty}^{\infty} g_1\,dF \leqslant F(x) \leqslant \int_{-\infty}^{\infty} g_2\,dF. \tag{2}$$

Since

$$\int_{-\infty}^{\infty} g_2\,dF - \int_{-\infty}^{\infty} g_1\,dF = \int_{x-\delta}^{x+\delta} (g_2 - g_1)dF \leqslant F(x+\delta) - F(x-\delta) < \varepsilon,$$

and ε is at our disposal, (1) and (2) show that $\liminf_n F_n(x) = \limsup_n F_n(x) = F(x)$. □

Corollary 19.24. *Let $f_n, n \in N$, and f be as in* (19.17), *and assume that $f_n \xrightarrow{P} f$. Then $f_n \xrightarrow{D} f$.*

Proof. Let $g \in \mathfrak{C}_r(R)$. Then $g \circ f_n \xrightarrow{P} g \circ f$ by (14.22.e). Since $|g \circ f_n| \leqslant \|g\|_u$, $n \in N$, (14.25) shows that $\int_X (g \circ f_n)dP \to \int_X (g \circ f)dP$. In view of (19.11.b), this means that $\int_{-\infty}^{\infty} g\,dF_n \to \int_{-\infty}^{\infty} g\,dF$, where F_n is the distribution function of f_n, $n \in N$, and F is the distribution function of f, that is $f_n \xrightarrow{D} f$. □

Remarks 19.25. (a) Notation is as in (19.17). $F_n \xrightarrow{w} F$ if and only if $\int_{-\infty}^{\infty} g\,dF_n \to \int_{-\infty}^{\infty} g\,dF$ for any uniformly continuous $g \in \mathfrak{C}_r(R)$.

(b) If $f_n \xrightarrow{D} f$ and $h : R \to R$ is continuous, then $h \circ f_n \xrightarrow{D} h \circ f$ by (19.23) and (19.11.b).

(c) Convergence in distribution does not imply convergence in probability. To see this, let f and g be as in (19.18.b), and set $f_{2n-1} = f$, $f_{2n} = g$, $n \geqslant 1$. Then $f_n \xrightarrow{D} f$, but $P(|f_{2n} - f| > 1/2) = 1 \nrightarrow 0$. However, a partial converse to (19.24) is as follows.

Theorem 19.26. *Let $f_n, n \in N$, be random variables such that $f_n \xrightarrow{D} a$, where $a \in R$. Then $f_n \xrightarrow{P} a$.*

Proof. Let F_n be the distribution function of f_n, $n \in N$, let F_a be the degenerate distribution function, and let $\varepsilon > 0$. Then $P(|f_n - a| > \varepsilon) = P(f_n < a-\varepsilon) + P(f_n > a+\varepsilon) \leqslant F_n(a-\varepsilon) + 1 - F_n(a+\varepsilon) \to F_a(a-\varepsilon) + 1 - F_a(a+\varepsilon) = 0 + 1 - 1 = 0$. □

The following theorem is a special case of a more general result (see (22.19)).

Theorem 19.27 (Slutsky). *For $n \in N$, let f_n and g_n be random variables, let f be a random variable, and assume that $f_n \xrightarrow{D} f$ and $g_n \xrightarrow{P} 0$. Then $f_n + g_n \xrightarrow{D} f$.*

Proof. For $n \in N$, let F_n and G_n be the distribution functions of f_n and g_n respectively, let F be the distribution functions of f, and let $\varepsilon > 0$. Then select $n_1 \in N$ such that

$$\rho(F_n, F) < \varepsilon, \quad n \geqslant n_1. \tag{1}$$

Now, for any $n \in N$ and $x \in R$, we have $\{f_n \leqslant x - \varepsilon) \subset \{f_n + g_n \leqslant x) \cup \{g_n \geqslant \varepsilon)$ and $\{f_n + g_n \leqslant x) \subset \{f_n \leqslant x + \varepsilon) \cup \{g_n \leqslant -\varepsilon)$, and so

$$F_n(x-\varepsilon) - P(|g_n| \geqslant \varepsilon) \leqslant G_n(x) \leqslant F_n(x+\varepsilon) + P(|g_n| \geqslant \varepsilon). \tag{2}$$

Select $n_2 \in N$ so that $P(|g_n| \geqslant \varepsilon) < \varepsilon$ for $n \geqslant n_2$. Then (2) shows that

$$\rho(G_n, F_n) \leqslant \varepsilon, \quad n \geqslant n_2. \tag{3}$$

From (1) and (3) we infer that $\rho(G_n, F) \leqslant 2\varepsilon$ for $n \geqslant n_1, n_2$, i.e. $f_n + g_n \xrightarrow{D} f$. □

The next result should be compared with (14.30).

Theorem 19.28. *For $0 < p < \infty$, let $\{f_n : n \in N\} \subset \mathfrak{L}^p(X, \mathcal{A}, P)$ and $f \in \mathfrak{L}^p(X, \mathcal{A}, P)$ be such that $f_n \xrightarrow{D} f$. Then the following are equivalent:*
(i) $\{|f_n|^p : n \in N\}$ is uniformly integrable;
(ii) $E\,|f_n|^p \to E\,|f|^p$.

Proof. Assume that (i) holds, and let $\varepsilon > 0$. Then choose $a > 0$ so large that both

$$\sup_{n\in N} \int_{\{|x|\geqslant a\}} |x|^p \, dF_n(x) = \sup_{n\in N} \int_{\{|f_n|\geqslant a\}} |f_n|^p \, dP < \varepsilon \tag{1}$$

and

$$\int_{\{|x|\geqslant a\}} |x|^p \, dF(x) = \int_{\{|f|\geqslant a\}} |f|^p \, dP < \varepsilon, \tag{2}$$

where $F_n, n \in N$, and F are as in the proof of (19.24). Then define $g_a \in \mathfrak{C}_r(R)$ by $g_a(x) = \begin{cases} |x|^p & \text{if } |x| < a \\ a^p & \text{if } |x| \geqslant a \end{cases}$. Applying the Helly-Bray theorem (19.23), select $n_\varepsilon \in N$ such that

$$\left| \int_{-\infty}^{\infty} g_a dF_n - \int_{-\infty}^{\infty} g_a dF \right| < \varepsilon, \ n \geqslant n_\varepsilon. \tag{3}$$

On account of (1), (2) and (3), for $n \geqslant n_\varepsilon$, we have

$$\begin{aligned}
\big|E\,|f_n|^p - E\,|f|^p\big| &= \left| \int_{-\infty}^{\infty} |x|^p \, dF_n(x) - \int_{-\infty}^{\infty} |x|^p \, dF(x) \right| \\
&\leqslant \int_{-\infty}^{\infty} \big||x|^p - g_a(x)\big| \, dF_n(x) + \left| \int_{-\infty}^{\infty} g_a(x) dF_n(x) - \int_{-\infty}^{\infty} g_a(x) dF(x) \right| \\
&+ \int_{-\infty}^{\infty} \big|g_a(x) - |x|^p\big| \, dF(x) \leqslant 2 \int_{\{|x|\geqslant a\}} |x|^p \, dF_n(x) + \left| \int_{-\infty}^{\infty} g_a \, dF_n \right. \\
&\left. - \int_{-\infty}^{\infty} g_a \, dF \right| + 2 \int_{\{|x|\geqslant a\}} |x|^p \, dF(x) < 5\varepsilon.
\end{aligned}$$

Thus (i) implies (ii). Suppose now that (ii) holds, and define φ_a, $a > 0$, as in the hints for (14.30). Then $\int_X (\varphi_a \circ f_n) dP = \int_R \varphi_a \, dF_n \to \int_R \varphi_a \, dF = \int_X (\varphi_a \circ f) dP$, $a > 0$, by the Helly-Bray theorem (19.23). Further, the proof that (ii) implies (i) proceeds as in the hints for proving (iii) $\Longrightarrow$ (i) in (14.30) □

Now we deal briefly with the n-dimensional analog of the notion of distribution function.

Definitions 19.29. Let $(X, \mathcal{A}, P)$ be a probability space, and let $\mathbf{f} : X \to R^n$ be a random vector with coordinates $f_1, \ldots, f_n$. The probability $P\mathbf{f}^{-1}$ on $\mathcal{B}(R^n)$ is called the *distribution of* $\mathbf{f}$, and the probabilities Pf_i^{-1}, $1 \leqslant i \leqslant n$, are called *marginal distributions.*

Examples 19.30. (a) For $m \geqslant 2$, let $K = \{\mathbf{k} = (k_1, \ldots, k_n) : k_i \in \{0, \ldots, m\}, 1 \leqslant i \leqslant n, k_1 + \cdots + k_n = m\}$, and let $p_1, \ldots, p_n \in]0, 1[$ be such that $p_1 + \cdots + p_n = 1$. Let $\mathbf{f} : X \to K$ be a random vector such that $P(\mathbf{f} = \mathbf{k} = (k_1, \ldots, k_n)) = \frac{m!}{k_1! \ldots k_n!} p_1^{k_1} \ldots p_n^{k_n}$, $\mathbf{k} \in K$. Then the distribution of $\mathbf{f}$ is discrete and called the *multinomial distribution.* If $\mathbf{f} = (f_1, \ldots, f_n)$, then $P(f_i = k) = \binom{m}{k} p_i^k (1 - p_i)^{m-k}$, $k = 0, \ldots, m$, for $1 \leqslant i \leqslant n$, so that the marginal distributions corresponding to $\mathbf{f}$ are binomial distributions.

The model leading to the multinomial distribution is as follows. Suppose that we have $m \geqslant 2$ independent trials, each one of which resulting in any one of n outcomes, labelled $1, \ldots, n$, the probability of outcome i being p_i. For $1 \leqslant i \leqslant n$, consider the random variable f_i defined as the number of times for which the outcome i occurs. Then $P((f_1, \ldots, f_n) = (k_1, \ldots, k_n)) = \frac{m!}{k_1! \ldots k_n!} p_1^{k_1} \ldots p_n^{k_n}$ for any $(k_1, \ldots, k_n) \in K$.

(b) Let $\mathbf{f} : X \to R^2$ be a random vector with

$$P(\mathbf{f} = (k_1, k_2)) = e^{-(\lambda_1 + \lambda_2 + \lambda_{12})} \sum_{j=0}^{k_1 \wedge k_2} \frac{\lambda_1^{k_1 - j} \lambda_2^{k_2 - j} \lambda_{12}^j}{(k_1 - j)!(k_2 - j)! j!}, \quad k_1, k_2 = 0, 1, \ldots,$$

where $\lambda_1, \lambda_2 > 0$ and $\lambda_{12} \geqslant 0$. Then the distribution of $\mathbf{f}$ is discrete and called the *bivariate Poisson distribution.*

Definition 19.31. Let $\mathbf{f} : X \to R^n$ be a random vector. The function $F : R^n \to [0, 1]$ defined by $F(\mathbf{x}) = P(\mathbf{f} \leqslant \mathbf{x})$, $\mathbf{x} \in R^n$ is called the *distribution function of* $\mathbf{f}$.

Remarks 19.32. Let $\mathbf{f} = (f_1, \ldots, f_n)$ and $\mathbf{g} = (g_1, \ldots, g_n)$ be random vectors on X, let F be the distribution function of $\mathbf{f}$, and let F_i be the distribution function of f_i, $1 \leqslant i \leqslant n$.

(a) $f_1, \ldots, f_n$ are independent if and only if

$$F(x_1, \ldots, x_n) = F_1(x_1) \ldots F_n(x_n), \quad (x_1, \ldots, x_n) \in R^n. \tag{1}$$

Actually, if $f_1, \ldots, f_n$ are independent, then $F(x_1, \ldots, x_n) = P((f_1, \ldots, f_n) \leqslant (x_1, \ldots, x_n)) = P(f_1 \leqslant x_1, \ldots, f_n \leqslant x_n) = F_1(x_1) \ldots F_n(x_n)$ for any

$(x_1, \ldots, x_n) \in R^n$. Conversely, if (1) holds, then

$$(P\mathbf{f}^{-1})\left(\prod_{i=1}^{n}]-\infty, x_i]\right) = F(x_1, \ldots, x_n) = \prod_{i=1}^{n}(Pf_i^{-1})(]-\infty, x_i])$$

$$= \left(\bigotimes_{i=1}^{n} Pf_i^{-1}\right)\left(\prod_{i=1}^{n}]-\infty, x_i]\right), \quad (x_1, \ldots, x_n) \in R^n. \tag{2}$$

Since the family $\{\prod_{i=1}^{n}]-\infty, x_i] : (x_1, \ldots, x_n) \in R^n\}$ is closed under intersection and generates $\mathcal{B}(R^n)$ (see (7.68)), (2) and (10.8) show that $P\mathbf{f}^{-1} = \otimes_{i=1}^{n} Pf_i^{-1}$. The independence of $f_1, \ldots, f_n$ follows now from (18.29).

(b) If $f_1, \ldots, f_n$ are independent, $g_1, \ldots, g_n$ are independent, and $Pf_i^{-1} = Pg_i^{-1}$, $1 \leqslant i \leqslant n$, then $P\mathbf{f}^{-1} = P\mathbf{g}^{-1}$.

(c) F has the following properties: (i) F is right continuous at any $\mathbf{x} \in R^n$, (ii) $\lim_{x_i \to -\infty} F(x_1, \ldots, x_{i-1}x_i, x_{i+1}, \ldots, x_n) = 0, 1 \leqslant i \leqslant n$, (iii) $\lim_{\mathbf{x} \to \overline{\infty}} F(\mathbf{x}) = 1$, where $\overline{\infty} = (\infty, \ldots, \infty) \in \overline{R}^n$, and (iv) $\Delta_{\mathbf{ba}} F \geqslant 0$ for any $\mathbf{a}, \mathbf{b} \in R^n$ with $\mathbf{a} \leqslant \mathbf{b}$. This last property is a consequence of (10.20.d) and of the fact that $\Delta_{\mathbf{ba}} F = P(a_i < f_i \leqslant b_i, 1 \leqslant i \leqslant n) = (P\mathbf{f}^{-1})(\prod_{i=1}^{n}]a_i, b_i])$ whenever $]a_1, b_1], \ldots,]a_n, b_n]$ are bounded intervals in R, where $\mathbf{a} = (a_1, \ldots, a_n)$ and $\mathbf{b} = (b_1, \ldots, b_n)$. Hence, according to (10.24), the distribution of $\mathbf{f}$ coincides with the Lebesgue-Stieltjes measure μ_F on $\mathcal{B}(R^n)$ induced by F, and so $P(\mathbf{f} \in B) = \mu_F(B)$, $B \in \mathcal{B}(R^n)$.

Theorem 19.33. *Let $F : R^n \to [0, 1]$ be a function possessing the properties in* (19.32.c). *Then there exist a probability space $(X, \mathcal{A}, P)$ and a random vector $\mathbf{f}$ on X such that the distribution function of $\mathbf{f}$ is F.*

Proof. Take $(X, \mathcal{A}, P) = (R^n, \mathcal{B}(R^n), \mu_F)$, and let $\mathbf{f} = i_{R^n}$ be the identity function on R^n. For $\mathbf{x} = (x_1, \ldots, x_n) \in R^n$, we have

$$P(\mathbf{f} \leqslant \mathbf{x}) = \mu_F(i_{R^n} \leqslant \mathbf{x}) = \mu_F\left(\prod_{i=1}^{n}]-\infty, x_i]\right) = \lim_k \mu_F\left(\prod_{i=1}^{n}]-k, x_i]\right)$$
$$= \lim_k \Delta_{\mathbf{x}(-k,\ldots,-k)} F = F(x_1, \ldots, x_n) \tag{1}$$

since F satisfies (ii). Further, (1) and (iii) imply that P is actually a probability measure. □

On account of this theorem, we may speak of any function F having the properties (i)–(iv) in (19.32.c) as being a distribution function (on R^n) without mentioning a specific random vector. For instance, if μ is a probability measure on $\mathcal{B}(R^n)$ and $F(x_1, \ldots, x_n) = \mu(\prod_{i=1}^{n}]-\infty, x_i])$, $(x_1, \ldots, x_n) \in R^n$, then it is easily verified that F has the properties (i)-(iv), and so F is a distribution function. It is the *cumulative distribution function of* μ.

Definition 19.34. A measurable function $p : R^n \to [0, \infty[$ such that $\int_{R^n} p(\mathbf{x}) d\lambda(\mathbf{x}) = 1$, where λ is the Lebesgue measure on $\mathcal{B}(R^n)$, is called a probability density (on R^n). It is easily seen that the function $F(x_1, \ldots, x_n) = \int_{]-\infty,x_1]\times\ldots\times]-\infty,x_n]}$

$p(\mathbf{t})d\lambda(\mathbf{t})$, $(x_1, \ldots, x_n) \in R^n$, satisfies (i)–(iv) in (19.32.c), and so it is a distribution function. In this case we say that F has density p. A random vector $\mathbf{f}$ with distribution function F is also said to have density p. If $]a_1, b_1], \ldots,]a_n, b_n]$ and $\mathbf{a}, \mathbf{b}$ are as in (19.32.c), then $\mu_F(\prod_{i=1}^n]a_i, b_i]) = \Delta_{\mathbf{ba}} F = \int_{]a_1,b_1]\times\ldots\times]a_n,b_n]} p(\mathbf{t})d\lambda(\mathbf{t})$. This together with (11.28) and (10.8) entail that $\mu_F(A) = \int_A p(\mathbf{t})d\lambda(\mathbf{t})$, $A \in \mathcal{B}(R^n)$, since the family $\{\prod_{i=1}^n]a_i, b_i] : -\infty < a_i < b_i < \infty, 1 \leqslant i \leqslant n\}$ is closed under intersection and generates $\mathcal{B}(R^n)$. Hence, by (12.2.b) and (12.5), $\mu_F \ll \lambda$ and $p = d\mu_F/d\lambda$.

Examples 19.35. (a) The function $p : R^2 \to [0, \infty[$ defined by $p(\mathbf{x}) = \begin{cases} 1/4a_1a_2 & \text{if } \mathbf{x} \in [-a_1, a_1] \times [-a_2, a_2] \\ 0 & \text{if } \mathbf{x} \notin [-a_1, a_1] \times [-a_2, a_2] \end{cases}$, where $a_1, a_2 > 0$, is a probability density on R^2. The distribution function associated to p is called the *uniform distribution function* (on $[-a_1, a_1] \times [-a_2, a_2]$).

(b) Let $\mathbf{m} = (m_1, \ldots, m_n) \in R^n$, and let $V = (v_{ij})$ be an $(n \times n)$ symmetric and positive-definite matrix, i.e. $v_{ij} = v_{ji}$ and $\mathbf{a}V\mathbf{a}^T > 0$ whenever $\mathbf{0} \neq \mathbf{a} \in R^n$, where $\mathbf{a}^T$ is the transpose of $\mathbf{a}$. Then the function $p(\mathbf{x}) = (2\pi)^{-n/2}(\det V)^{-1/2} \exp \{-(\mathbf{x} - \mathbf{m})V^{-1}(\mathbf{x} - \mathbf{m})^T/2\}$, $\mathbf{x} \in R^n$, where $\det V$ is the determinant of V, is a probability density on R^n. The distribution function corresponding to p is called the *multinormal distribution function.*

(c) The function $p(x_1, x_2,) = 1/2\pi(a^2 + x_1^2 + x_2^2)^{3/2}$, $(x_1, x_2,) \in R^2$, where $a \in R$, is a probability density. The distribution function associated to p is called the *2-dimensional Cauchy distribution function.*

(d) The function $p(x_1, x_2, x_3) = 1/\pi^2(1 + x_1^2 + x_2^2 + x_3^2)^2$, $(x_1, x_2, x_3) \in R^3$, is a probability density. The distribution function associated to p is called the *symmetric 3-dimensional Cauchy distribution function.*

Remarks 19.36. (a) If the random vector $\mathbf{f} = (f_1, \ldots, f_n)$ has density p, then f_i has density $p_i(x_i) = \int_{R^{n-1}} p(x_1, \ldots, x_n)dx_1 \ldots dx_{i-1}dx_{i+1} \ldots dx_n$, $x_i \in R$, for $1 \leqslant i \leqslant n$. Moreover, if $p(x_1, \ldots, x_n) = p_1(x_1) \ldots p_n(x_n)$, $(x_1, \ldots, x_n) \in R^n$, then $f_1, \ldots, f_n$ are independent by (19.32.a).

(b) Assume that $f_1, \ldots, f_n$ are independent random variables and that f_i has density p_i for $1 \leqslant i \leqslant n$. Then (19.32.a) shows that $p(x_1, \ldots, x_n) = p_1(x_1) \ldots p_n(x_n)$, $(x_1, \ldots, x_n) \in R^n$, is a density of $(f_1, \ldots, f_n)$.

(c) Let F_1 and F_2 be distribution functions on R^n, put $s(\mathbf{x}_1, \mathbf{x}_2) = \mathbf{x}_1 + \mathbf{x}_2$, $\mathbf{x}_1, \mathbf{x}_2 \in R^n$, and consider the probability $(\mu_{F_1} \otimes \mu_{F_2})s^{-1}$ on $\mathcal{B}(R^n)$. By analogy with (19.13.a), the cumulative distribution function of $(\mu_{F_1} \otimes \mu_{F_2})s^{-1}$ is the *convolution of* F_1 *and* F_2 to be denoted $F_1 * F_2$. Also, if $\mathbf{f}_1$ and $\mathbf{f}_2$ are independent random vectors with corresponding distribution functions F_1 and F_2, then $\mathbf{f}_1 + \mathbf{f}_2$ has the distribution function $F_1 * F_2$.

(d) For $k \in N$, let $\mathbf{f}_k : X \to R^n$ be a random vector with distribution function F_k, and let $\mathbf{f} : X \to R^n$ be a random vector with distribution function F. As in (19.17), the notations $F_k \overset{w}{\to} F$, $\mathbf{f}_k \overset{D}{\to} \mathbf{f}$, or $\mathbf{f}_k \overset{D}{\to} F$ mean that $F_k(\mathbf{x}) \to F(\mathbf{x})$ at all continuity points $\mathbf{x}$ of F. In accordance with (10.49), F is continuous at $\mathbf{x}$ if and only if F is left continuous at $\mathbf{x}$. It can be shown that $F_k \overset{w}{\to} F$ if and only if $\mu_{F_k}(B) \to \mu_F(B)$ whenever $B \in \mathcal{B}(R^n)$ and $\mu(\partial(B)) = 0$; also $F_k \overset{w}{\to} F$ if and only if $\int_{R^n} g \, d\mu_{F_k} \to \int_{R^n} g \, d\mu_F$ for any $g \in \mathfrak{C}_r(R^n)$. These are special instances of general results concerning convergence of probability measures on metric spaces to be discussed in Section 22.

Exercise 19.37. Let f be as in (19.3.c). Compute Ef, $\operatorname{Var} f$, Ef^3 and Ef^4.

Exercise 19.38. Let $f_1, \ldots, f_n$ be independent random variables such that $p = P(f_i = 1) = 1 - P(f_i = 0) \in]0, 1[$, $1 \leqslant i \leqslant n$. Show that $f_1 + \cdots + f_n$ has a binomial distribution.

Exercise 19.39 (Bernoulli). Notation is as in (19.38). Put $S_n = f_1 + \cdots + f_n$ and let $\varepsilon > 0$. Prove that $P(|S_n/n - p| \geqslant \varepsilon) \to 0$. [Use (19.37) and (17.8).]

Exercise 19.40. Notation is as in (19.39). Show that $\sum_{n \in N} P(|S_n/n - p| \geqslant \varepsilon) < \infty$, and infer that $S_n/n \overset{a.s.}{\to} p$. [Hint. Use Markov's inequality (13.8) $P(|S_n - np| \geqslant \varepsilon n) \leqslant E(S_n - np)^4/\varepsilon^4 n^4$, and apply (19.37).]

Exercise 19.41. Let f be as in (19.3.d). Verify that $Ef = a + b\lambda$, $Ef^2 = (a + b\lambda)^2 + b^2\lambda$ and $Ef^3 = (a + b\lambda)^3 + 3b^2\lambda(a + b\lambda) + b^3\lambda$.

Exercise 19.42 (Poisson). If $\lim_n np_n = \lambda > 0$, then $\lim_n b(k; n, p_n) = e^{-\lambda}\lambda^k/k! = p(k; \lambda)$ for each $k \geqslant 0$.

Exercise 19.43. Notation is as in (19.42). Prove that $\sum_{k \geqslant 0} |b(k; n, p_n) - p(k; \lambda)| \to 0$ as $n \to \infty$. [Use (19.42) and (14.33).]

Exercise 19.44. Let f be as in (19.3.e). Show that $P((1 - p)f > x) \to e^{-x}$ as $p \to 1$ for any $x > 0$.

Exercise 19.45. Let f be a random variable with density p as in (19.10.b). Compute Ef and $\operatorname{Var} f$.

Exercise 19.46. Let f be a random variable with density p as in (19.10.c). Verify that $Ef = \infty$.

Exercise 19.47. Assume f has density p as in (19.10.d), and set $g = (f - m)/\sigma$. Prove that g has a standard normal distribution.

Exercise 19.48. Let f and g be as in (19.47). Show that $Eg = 0$ and $\operatorname{Var} g = 1$, and deduce that $Ef = m$ and $\operatorname{Var} f = \sigma^2$. [Hint. Use polar coordinates to prove that $\int_{R^2} e^{-(x_1^2 + x_2^2)/2} dx_1\, dx_2 = 2\pi$.]

Exercise 19.49. Let g be as in (19.47). For $n \geqslant 1$, show that $Eg^{2n-1} = 0$ and $Eg^{2n} = 1 \cdot 3 \cdots (2n - 1)$. [Hint. $Eg^{2n} = (2n - 1)Eg^{2n-2}$.]

Exercise 19.50. Let Φ be as in (19.10.d). Prove that $(x^{-1} - x^{-3})e^{-x^2/2} \leqslant \sqrt{2\pi}(1 - \Phi(x)) \leqslant x^{-1}e^{-x^2/2}$, $x > 0$.

Exercise 19.51. Let $F_1, F_2 \in \mathfrak{F}$, let $g \in \mathfrak{L}^1(X, \mathcal{A}, \mu_{F_1 * F_2})$, and let $B \in \mathcal{B}(R)$. Use (19.13.a), (11.29) and (15.10) to verify that

$$\int_B g\, d(F_1 * F_2) = \int_R \left(\int_{B - x_1} g(x_1 + x_2) dF_2(x_2) \right) dF_1(x_1)$$

$$= \int_R \Big(\int_{B-x_2} g(x_1+x_2)dF_1(x_1)\Big)dF_2(x_2).$$

Exercise 19.52. For $F \in \mathfrak{F}$, construct a sequence $\{F_n : n \in N\} \subset \mathfrak{F}$ such that $F_n(R)$ is finite, $n \in N$, and $F_n \xrightarrow{w} F$. [Hints. Let D be a denumerable set dense in R, and assume that $D = \cup_{n\geqslant 1} D_n$, where $D_n \subset D_{n+1}$. Suppose that $D_n = \{x_1^{(n)}, \ldots, x_n^{(n)}\}$ with $x_1^{(n)} < \cdots < x_n^{(n)}$. Define $F_n(x) = 0$ for $x < x_1^{(n)}$, $F_n(x) = F(x_k^{(n)})$ for $x_k^{(n)} \leqslant x < x_{k+1}^{(n)}$, $1 \leqslant k < n$, and $F_n(x) = 1$ for $x \geqslant x_n^{(n)}$.]

Exercise 19.53. Show that $(\mathfrak{F}, \rho)$ is a complete metric space, where ρ is the Lévy metric.

Exercise 19.54. Let $\{F_n : n \in N\}$, F and $C(F)$ be as in (19.17). Assume that $F_n \xrightarrow{w} F$ and $F_n(x-) \to F(x-)$, $x \in R - C(F)$. Prove that $F_n \xrightarrow{u} F$.

Exercise 19.55. Let $\mathbf{f} = (f_1, \ldots, f_n)$ be a random vector with density p as in (19.35.b). Prove the following.

(a) For $1 \leqslant i \leqslant n$, f_i has a normal distribution with $Ef_i = m_i$ and $\operatorname{Var} f_i = v_{ii}$.

(b) For $1 \leqslant i, j \leqslant n$, $\operatorname{Cov}[f_i, f_j] = v_{ij}$.

(c) If $n = 2$ and ρ is the correlation coefficient between f_1 and f_2, then $p(x_1, x_2,) = (2\pi)^{-1}(1-\rho)^{-1/2} \exp\left\{-\frac{1}{1-\rho^2}\left[\frac{(x_1-m_1)^2}{2\sigma[f_1]^2} - \rho\frac{(x_1-m_1)(x_2-m_2)}{\sigma[f_1]\sigma[f_2]} + \frac{(x_2-m_2)^2}{2\sigma[f_2]^2}\right]\right\}$, $(x_1, x_2,) \in R^2$.

Exercise 19.56. Let $\{F_n : n \in N\}$ and F be as in (19.17). Prove that $F_n \xrightarrow{w} F$ if and only if every subsequence $\{F_{n_k} : k \in N\}$ has itself a subsequence $\{F_{n_{k_l}} : l \in N\}$ such that $F_{n_{k_l}} \xrightarrow{w} F$. [Use (5.54) and (19.23).]

Exercise 19.57. Let $\{F_n : n \in N\} \subset \mathfrak{F}$, and assume there is a function $F : R \to [0, 1]$, which is nondecreasing and right continuous on R, such that $F_n(x) \to F(x)$ whenever x is a continuity point of F. Show that $\int_{-\infty}^{\infty} g\,dF_n \to \int_{-\infty}^{\infty} g\,dF$ for any $g \in \mathfrak{C}_r(R)$ with $\lim_{x\to-\infty} g(x) = \lim_{x\to\infty} g(x) = 0$.

Exercise 19.58 (Helly). Let $\{F_n : n \in N\} \subset \mathfrak{F}$. Prove that there are a function $F : R \to [0, 1]$, which is right continuous on R, and a subsequence $\{F_{n_k} : k \in N\}$ such that $F_{n_k}(x) \to F(x)$ whenever x is a continuity point of F. [Hints. Let $D = \{r_i : i \in I\}$ be a denumerable dense subset of R. As in (6.67), use the Cantor diagonal method to find a subsequence $\{F_{n_k} : k \in N\}$ and a nondecreasing function $G : D \to [0, 1]$ so that $\lim_{k\to\infty} F_{n_k}(r_i) = G(r_i)$, $i \in N$. Put $F(x) = \inf\{G(r) : x < r \in D\}$, $x \in R$.]

Exercise 19.59. Let $F \in \mathfrak{F}$, and assume that $F(a_1x + b_1) = F(a_2x + b_2)$ for all $x \in R$, where $a_1, a_2, b_1, b_2 \in R$ and $a_1 + a_2 \neq 0$. Show that $a_1 = a_2$ and $b_1 = b_2$. [Prove that $a_1x + b_1 = a_2x + b_2$ for any $x \in R$.]

Exercise 19.60. Let $f_n, n \in N$, be random variables, let $F \in \mathfrak{F}$ and let $\{a_n : n \in N\} \subset R$ be such that $|a_n| \to \infty$. If $f_n \xrightarrow{D} F$, then $f_n/a_n \xrightarrow{P} 0$.

Exercise 19.61. For $n \in N$, let $a_n, b_n \in R$ be such that $a_n \to a \in R$ and $b_n \to b \in R$ and let f_n and f be random variables such that $f_n \overset{D}{\to} f$. Use (19.18.c), (19.27) and (19.60) to show that $a_n f_n + b_n \overset{D}{\to} af + b$.

Exercise: Types of distribution functions 19.62. Two distribution functions F and G are said to be of the same *type* if

$$F(ax + b) = G(x), \quad x \in R, \tag{1}$$

for some $a > 0$ and $b \in R$.

Let $\{F_n : n \in N\} \subset \mathfrak{F}$, and let F and G be nondegenerate distribution functions such that $F_n \overset{w}{\to} F$ and $F_n(a_n x + b_n) \to G(x)$, $x \in C(G)$, for some $a_n > 0$ and $b_n \in R$. Prove that F and G are of the same type, i.e. (1) holds, and show that $a_n \to a$ and $b_n \to b$. The following steps may be helpful.

(a) Let a be a limit point of $\{a_n : n \in N\}$, let $\{a_{n_k} : k \in N\}$ be a subsequence such that $a_{n_k} \to a$, let b be a limit point of $\{b_{n_k} : k \in N\}$, and let $\{b_{n_{k_l}} : l \in N\}$ be a subsubsequence such that $b_{n_{k_l}} \to b$. Then $0 < a < \infty$. [Hints. For $n \in N$, let f_n be a random variable whose distribution function is F_n, and set $g_n = (f_n - b_n)/a_n$. Let f and g be random variables whose distribution functions are F and G, respectively. If $a = 0$, then $a_{n_{k_l}} g_{n_{k_l}} \overset{P}{\to} 0$ by (19.60) in view of $g_n \overset{D}{\to} g$. As $f_{n_{k_l}} \overset{D}{\to} f$, this would lead to $b_{n_{k_l}} = f_{n_{k_l}} - a_{n_{k_l}} g_{n_{k_l}} \overset{D}{\to} f$ in view of (19.27), a contradiction. If $a = \infty$, then $f_{n_k}/a_{n_k} \overset{P}{\to} 0$. This would lead to $-b_{n_{k_l}}/a_{n_{k_l}} = g_{n_{k_l}} - f_{n_{k_l}}/a_{n_{k_l}} \overset{D}{\to} g$, a contradiction.

(b) $b \in R$. [Hint. Since $(a_{n_k} - a) g_{n_k} \overset{P}{\to} 0$, $a g_{n_{k_l}} + b_{n_{k_l}} = f_{n_{k_l}} - (a_{n_{k_l}} - a) g_{n_{k_l}} \overset{D}{\to} f$ by (19.27). Then use the fact that $a g_{n_{k_l}} \overset{D}{\to} ag$ (19.18.c).]

(c) $F(ax+b) = G(x)$, $x \in R$. [Hint. $g_{n_{k_l}} \overset{D}{\to} g$ and $g_{n_{k_l}} = (f_{n_{k_l}} - b_{n_{k_l}})/a_{n_{k_l}} \overset{D}{\to} (f - b)/a$ by (19.61). Then apply (19.18.a).]

(d) $\lim_{k\to\infty} b_{n_k} = b$. [Hint. If b' is an alternative limit point of $\{b_{n_k} : k \in N\}$, then $F(ax + b) = F(ax + b')$, $x \in R$, and (19.59) applies.]

(e) $a_n \to a$ and $b_n \to b$. [Hint. If a^* and b^* are alternative limit points of $\{a_n : n \in N\}$ and $\{b_n : n \in N\}$, then $F(ax + b) = F(a^* x + b^*)$, $x \in R$, and (19.59) applies again.]

Exercise 19.63. Let I be a countable set, let $\{F_i : i \in I\} \subset \mathfrak{F}$, let $\{p_i : i \in I\} \subset]0, 1[$ be such that $\sum_{i\in I} p_i = 1$, and set $F = \sum_{i\in I} p_i F_i$. Use (11.25.ii) and (10.8) to show that $F \in \mathfrak{F}$ and $\mu_F = \sum_{i\in I} p_i \mu_{F_i}$.

Exercise 19.64. Let $f_1, \ldots, f_n$ be independent random variables with respective distribution functions $F_1, \ldots, F_n$, let $h : R^n \to R$ be a measurable function, and assume that $h \circ (f_1, \ldots, f_n)$ is P-integrable. Use (18.29) to prove that

$$E[h \circ (f_1, \ldots, f_n)] = \int_{R^n} h(x_1, \ldots, x_n) dF_1(x_1) \ldots dF_n(x_n).$$

Exercise 19.65. Let f_1 and f_2 be independent random variables such that $Ef_2 = 0$, and let $p \geqslant 1$. Show that $E\,|f_1|^p \leqslant E\,|f_1 + f_2|^p$. [Hint. $|x_1|^p = |E[x_1 + f_2]|^p \leqslant$

$E\,|x_1 + f_2|^p$ by (13.5), $x_1 \in R$. Then apply (19.64) with $h(x_1, x_2) = x_1 + x_2$, $(x_1, x_2) \in R^2$.]

Exercise 19.66. Let f and g be independent and P-integrable random variables with a common distribution function F. Prove that $E\,|f - g| = 2\int_R F(x)(1 - F(x))dx$. [Write $E\,|f - g| = 2E(f - g)^+$, and use Fubini's theorem.]

Exercise: Complete convergence 19.67. : Let $\mathfrak{G}$ denote the set of all bounded nondecreasing functions from R to R that are right continuous at any $x \in R$. Let $\{G_n : n \in N\} \subset \mathfrak{G}$ and $G \in \mathfrak{G}$. We say that $\{G_n : n \in N\}$ *converges completely to* G, and we write $G_n \xrightarrow{c} G$, if $G_n(x) \to G(x)$ for any $x \in C(G)$, where $C(G)$ is the set of continuity points of G, $G_n(-\infty) \to G(-\infty)$ and $G_n(\infty) \to G(\infty)$. Prove the following:

(a) If $G_n \xrightarrow{c} G$ and $G_n \xrightarrow{c} G'$, then $G = G'$.

(b) $G_n \xrightarrow{c} G$ if and only if there is a set D dense in R such that $G_n(y) \to G(y)$, $y \in D$, and $G_n(-\infty) \to G(-\infty)$ and $G_n(\infty) \to G(\infty)$.

(c) $G_n \xrightarrow{c} G$ if and only if $\int_R g\,dG_n \to \int_R g\,dF$ for any $g \in \mathfrak{C}_r(R)$. [Hint. Mimic the proof of (19.23). To show that $\mu_{G_n}(R) \to \mu_G(R)$ implies that $G_n(-\infty) \to G(-\infty)$ and $G_n(\infty) \to G(\infty)$, use (19.68.c).]

(d) $G_n \xrightarrow{c} G$ if and only if $\int_R g\,dG_n \to \int_R g\,dF$ for any uniformly continuous $g \in \mathfrak{C}_r(R)$.

(e) If $G_n \xrightarrow{c} G$, then $\int_{-\infty}^x g\,dG_n \to \int_{-\infty}^x g\,dF$ and $\int_x^\infty g\,dG_n \to \int_x^\infty g\,dF$ for any $g \in \mathfrak{C}_r(R)$ and $x \in C(G)$, but the converse statement does not generally hold.

(f) $G_n \xrightarrow{c} G$ if and only if every subsequence of $\{G_n : n \in N\}$ contains a subsubsequence which converges completely to G.

Exercise 19.68. Notation is as in (19.67). Let $\{G_n : n \in N\} \subset \mathfrak{G}$. let $G : R \to R$ be nondecreasing and right continuous on R with $G_n(x) \to G(x)$, $x \in C(G)$, and put $\Delta G_n = G_n(\infty) - G_n(-\infty)$, $n \in N$, and $\Delta G = G(\infty) - G(-\infty)$. Prove the following.

(a) $\limsup_n G_n(-\infty) \leqslant G(-\infty) \leqslant G(\infty) \leqslant \liminf_n G_n(\infty)$.

(b) $\Delta G \leqslant \liminf_n \Delta G_n$.

(c) $G \in \mathfrak{G}$ and $G_n \xrightarrow{c} G$ if and only if $\Delta G_n \to \Delta G < \infty$. [Hint. If $\Delta G_n \to \Delta G < \infty$, then $\limsup_n G_n(\infty) = \Delta G + \limsup_n G_n(-\infty) \leqslant G(\infty) \leqslant \liminf_n G_n(\infty)$ by (a).]

(d) $\Delta G_n \to \Delta G < \infty$ if and only if

$$\lim_{a\to\infty} (\sup_{n\in N}(\Delta G_n - \Delta G_n(a))) = 0, \tag{1}$$

where $\Delta G_n(a) = G_n(a) - G_n(-a)$, $n \in N$, $a \in]0, \infty[$. [Hints. If $\Delta G_n \to \Delta G < \infty$, then, for $\varepsilon > 0$, select $n_1 \in N$ so that $\Delta G_n - \Delta G < \varepsilon$, $n \geqslant n_1$, choose $a > 0$ with $\pm a \in C(G)$ such that $\Delta G - \Delta G(a) < \varepsilon$, where $\Delta G(a) = G(a) - G(-a)$, and then select $n_2 \in N$ so that $\Delta G(a) - \Delta G_n(a) < \varepsilon$, $n \geqslant n_2$. This yields $\Delta G_n - \Delta G_n(a) < 3\varepsilon$, $n \geqslant n_1 \vee n_2$. If (1) holds, then, for $\varepsilon > 0$, choose $a_1 > 0$ such that $\sup_{n\in N}(\Delta G_n - \Delta G_n(a)) < \varepsilon$, $a \geqslant a_1$. Hence

$$\limsup_n \Delta G_n \leqslant \Delta G(a) + \varepsilon < \infty, \quad a \geqslant a_1,\ \pm a \in C(G), \tag{2}$$

and so

$$\limsup_n \Delta G_n \leqslant \Delta G. \tag{3}$$

(2) implies $\Delta G < \infty$ by (b), and (3) implies $\Delta G_n \to \Delta G$ also by (b).]

Exercise 19.69. Let $f_0, f_1 \ldots$ be independent and identically distributed random variables such that $E\,|f_0| < \infty$ and $Ef_0 = \mu$, and assume that ν is a random variable independent of $\sigma(f_0, f_1 \ldots)$ with $P(\nu = k) = p(k; \lambda), k = 0, 1, \ldots$ Use (18.23) to show that $E[\prod_{n=0}^{\nu} f_n] = \mu e^{-\lambda(1-\mu)}$.

Exercise 19.70. Assume that the distribution function of the random variable f is either geometric or exponential. Show that $P(f > x + y | f > x) = P(f > y)$ for any $x, y > 0$.

Exercise 19.71. Let f_1 and f_2 be normal random variables with $\mathrm{Cov}[f_1, f_2] = 0$. Use (19.55.c) and (19.36.a) to prove that f_1 and f_2 are independent.

20 Characteristic Functions

In the preceding section we have seen that most probabilistic properties of random variables can be described in terms of distribution functions. In this section we go further and show that properties of distribution functions can be transferred to their Fourier-Stieltjes transforms called characteristic functions. These transforms have nice properties such as uniform continuity and uniform boundedness, and there is a one-to-one correspondence between distribution functions and characteristic functions. Moreover, there is a one-to-one correspondence between weak convergence of sequences of distribution functions and pointwise convergence of sequences of associated characteristic functions. As there is a parallelism between convolution of distribution functions and multiplication of the corresponding characteristic functions, the study of the asymptotic behavior of sums of independent random variables is mostly indebted to the theory of characteristic functions.

Definition 20.1. Let $(X, \mathcal{A}, P)$ be a probability space, and let f be a random variable on X with distribution function F. The *characteristic function of f* is the function $\varphi_f : R \to C$ defined by

$$\varphi_f(t) = Ee^{itf} = E\cos tf + iE\sin tf, \quad t \in R.$$

In accordance with (19.11.b), φ_f is also the Fourier-Stieltjes transform of μ_F, hereafter denoted by φ_F, having the expression

$$\varphi_F(t) = \int_R e^{itx}\, dF(x) = \int_R \cos tx\, dF(x) + i \int_R \sin tx\, dF(x), \quad t \in R. \tag{1}$$

Since, for fixed $t \in R.$, the function of x with values $\cos tx$ and $\sin tx$ is continuous, the integrals in (1) may be envisaged as Riemann-Stieltjes integrals. In the sequel we shall use either of φ_f, φ_F or simply φ to denote a characteristic function. Throughout $\mathfrak{C}$ will stand for the set of all characteristic functions.

Remarks 20.2. (a) By (11.36.c), it follows that $|\varphi(t)| \leqslant 1 = \varphi(0), t \in R$.

(b) $\varphi_{-f}(t) = \varphi_f(-t) = \overline{\varphi_f(t)}, t \in R$.

(c) φ is uniformly continuous. Actually, for $t, h \in R$, we have $\left|\varphi_f(t+h) - \varphi_f(t)\right| = \left|E[e^{itf}(e^{ihf} - 1)]\right| \leqslant E\left|e^{ihf} - 1\right| \to 0$ as $h \to 0$ by (11.46), since $\left|e^{ihf} - 1\right| \leqslant 2$.

(d) For $a, b \in R$, we have $\varphi_{af+b}(t) = \varphi_f(at)e^{itb}, t \in R$.

(e) If the distribution of f is discrete, say $P(f = a_j) = p_j$ for some countable set $\{a_j : j \in J\} \subset R$, then $\varphi_f(t) = \sum_{j \in J} p_j e^{ita_j}, t \in R$.

(f) If F has density p, then $\varphi_F(t) = \int_R e^{itx} p(x)dx, t \in R$, by (2) in (19.11.b).

Examples 20.3. The characteristic functions below will bear the name of their corresponding random variables or distribution functions.

(a) If F is degenerate at some $a \in R$, then $\varphi_F(t) = e^{iat}$.

(b) If f has a symmetric Bernoulli distribution as in (19.3.b), then $\varphi_f(t) = \cos t$.

(c) If f has a binomial distribution as in (19.3.c), then $\varphi_f(t) = (pe^{it} + 1 - p)^n$.

(d) If f has a Poisson distribution as in (19.3.d), then $\varphi_f(t) = \exp(ita + \lambda(e^{itb} - 1))$.

(e) If f has a geometric distribution as in (19.3.e), then $\varphi_f(t) = p/(1-(1-p)e^{it})$.

(f) If F is the uniform distribution function as in (19.10.a), then $\varphi_F(t) = (e^{itb} - e^{ita})/(b-a)it, t \neq 0$. For $a = -b < 0$, we have $\varphi_F(t) = (\sin bt)/bt, t \neq 0$.

(g) If F is the exponential distribution function with parameter λ, then $\varphi_F(t) = (1 - it/\lambda)^{-1}$.

(h) If F is the normal distribution function as in (19.10.d), then $\varphi_F(t) = e^{imt - \sigma^2 t^2/2}$. This important characteristic function will be derived in (20.19) by making use of differentiability properties.

Theorem (20.8) below shows that there is a one-to-one correspondence between distribution functions and characteristic functions. This situation together with the fact that addition of independent random variables corresponds to multiplication of their characteristic functions, as well as the ease of handling characteristic functions, indicates the crucial importance of characteristic functions in probability theory.

Theorem 20.4. *Let f_1, f_2, F_1 and F_2 be as in* (19.14). *Then* $\varphi_{F_1 * F_2} = \varphi_{f_1 + f_2} = \varphi_{f_1}\varphi_{f_2}$.

Proof. The former equality is a consequence of (19.14), while the latter one follows from (18.42.b), as e^{itf_1} and e^{itf_2} are independent for any $t \in R$ by (18.28.c). □

Corollary 20.5. *If $\varphi_1, \varphi_2 \in \mathfrak{C}$, then $\varphi_1\varphi_2 \in \mathfrak{C}$.*

Proof. Let $F_1, F_2 \in \mathfrak{F}$ be such that $\varphi_1 = \varphi_{F_1}$ and $\varphi_2 = \varphi_{F_2}$. By (18.27), there exist independent random variables f_1 and f_2 such that the distribution of f_i is μ_{F_i} for $i = 1, 2$. □

The following lemmas are steps in the proof of Theorem (20.8).

Lemma 20.6. *Let $F, G \in \mathfrak{F}$ be such that $\varphi_F = \varphi_G$. Then*

(i)
$$\int_R \varphi_H \, dF = \int_R \varphi_H \, dG, \quad H \in \mathfrak{F}.$$

Proof. On account of (11.36) and Fubini's theorem (15.8), we have

$$\int_R \varphi_H \, dF = \int_R \left(\int_R e^{itx} \, d\mu_H(x) \right) d\mu_F(t) = \int_{R^2} e^{itx} \, d(\mu_F \otimes \mu_H)(t, x)$$
$$= \int_R \left(\int_R e^{ixt} \, d\mu_F(t) \right) d\mu_H(x) = \int_R \varphi_F \, dH, \quad H \in \mathfrak{F}.$$

Analogously we get $\int_R \varphi_H \, dG = \int_R \varphi_G \, dH$, $H \in \mathfrak{F}$, and so (i) obtains. □

Put $\mathfrak{C}' = \{\varphi \in \mathfrak{C} : \lim_{t \to -\infty} \varphi(t) = \lim_{t \to \infty} \varphi(t) = 0\}$, and let $\mathfrak{B}$ be the set of all functions of the form $f(t) = \alpha_0 + \sum_{k=1}^{n} \alpha_k \varphi_k(t), t \in R$, where $\alpha_k \in C$ for $k = 0, \ldots, n, \varphi_k \in \mathfrak{C}'$ for $k = 1, \ldots, n$, and $n \in N$. Consider the one-point compactification $R \cup \{\delta\}$ of R (as in (5.94)). To each $f \in \mathfrak{B}$ associate its extension g to $R \cup \{\delta\}$ by setting $g(\delta) = \alpha_0$, and let $\mathfrak{A}$ denote the set of all these extensions.

Lemma 20.7. *$\mathfrak{A}$ is a subalgebra of $\mathfrak{C}_c(R \cup \{\delta\})$ satisfying* (i)–(iii) *in* (6.59).

Proof. If $g \in \mathfrak{A}$, then $g \in \mathfrak{C}_c(R \cup \{\delta\})$ by (20.2.c), (10.50) and (5.47), and so $\mathfrak{A} \subset \mathfrak{C}_c(R \cup \{\delta\})$. Obviously, $\mathfrak{A}$ is a linear subspace of $\mathfrak{C}_c(R \cup \{\delta\})$. Moreover, $g_1, g_2 \in \mathfrak{A}$ implies that $g_1 g_2 \in \mathfrak{A}$ by (20.5). Thus $\mathfrak{A}$ is a subalgebra of $\mathfrak{C}_c(R \cup \{\delta\})$. Plainly, $1 \in \mathfrak{A}$. Now we show that $\mathfrak{A}$ separates points of $R \cup \{\delta\}$. For $t_0 \in R - \{0\}$, choose $b > 0$ so that $\sin bt_0 \neq 0$, and define $g(\delta) = \alpha_0$, $g(0) = \alpha_0 + 1$ and $g(t) = \alpha_0 + (\sin bt)/bt, t \in R - \{0\}$. Then $g \in \mathfrak{A}$, $g(\delta) \neq g(0)$, and $g(\delta) \neq g(t_0)$. Next, for $0 \neq |t_1| < |t_2|$, define g as above with $b = \pi/|t_2|$. Then $g(0) \neq g(t_2)$ and $g(t_1) \neq g(t_2)$. Further, for $t_0 > 0$, define $g(\delta) = \alpha_0$, $g(0) = \alpha_0 + b$, $g(t) = \alpha_0 + e^{iat}(\sin bt)/bt, t \in R - \{0\}$, where $a = b = \pi/2t_0$. Then $g \in \mathfrak{A}$ and $\alpha_0 - i/t_0 = g(-t_0) \neq g(t_0) = \alpha_0 + i/t_0$. Finally, $\overline{g} \in \mathfrak{A}$ whenever $g \in \mathfrak{A}$ by (20.2.b). □

The next theorem settles a one-to-one correspondence between $\mathfrak{F}$ and $\mathfrak{C}$.

Theorem 20.8. *Let $F, G \in \mathfrak{F}$ be such that $\varphi_F = \varphi_G$. Then $F = G$.*

Proof. Let $g \in \mathfrak{A}$, and set $f = g \circ i_R$, where i_R is the inclusion function from R into $R \cup \{\delta\}$. Then $f \in \mathfrak{B}$ and we have, in view of (11.29) and (20.6),

$$\int_{R \cup \{\delta\}} g \, d(\mu_F i_R^{-1}) = \int_R f \, d\mu_F = \int_R f \, d\mu_G = \int_{R \cup \{\delta\}} g \, d(\mu_G i_R^{-1}).$$

On account of (20.7), the complex version of the Stone-Weierstrass theorem (6.59) and (11.26), it follows that

$$\int_{R \cup \{\delta\}} g \, d(\mu_F i_R^{-1}) = \int_{R \cup \{\delta\}} g \, d(\mu_G i_R^{-1}), \quad g \in \mathfrak{C}_c(R \cup \{\delta\}). \tag{1}$$

(1) shows, in view of (6.89) and (11.50.a), that $\mu_F i_R^{-1} = \mu_G i_R^{-1}$. Hence $F(x) = \mu_F(]-\infty, x]) = (\mu_F i_R^{-1})(]-\infty, x]) = (\mu_G i_R^{-1})(]-\infty, x]) = \mu_G(]-\infty, x]) = G(x), x \in R$. □

The following two theorems show that the weak convergence of sequences of distribution functions corresponds to the convergence of the associated sequences of characteristic functions.

Uniform convergence theorem 20.9. *Let $\{F_n : n \in N\} \subset \mathfrak{F}$ and $F \in \mathfrak{F}$ be such that $F_n \xrightarrow{w} F$. Then $\lim_{n\to\infty} \varphi_{F_n}(t) = \varphi_F(t)$ uniformly in $|t| \leqslant T$ for any $T > 0$.*

Proof. Let $\varepsilon > 0$, and choose $M > 0$ with $\pm M \in C(F)$ so that $F(-M) < \varepsilon$ and $1 - F(M) < \varepsilon$. Since $F_n(-M) \to F(-M)$ and $F_n(M) \to F(M)$, determine $n_1 \in N$ such that $F_n(-M) < \varepsilon$ and $1 - F_n(M) < \varepsilon$ for $n \geqslant n_1$. Then, for $t \in R$ and $n \geqslant n_1$,

$$\begin{aligned}
&\left|\varphi_{F_n}(t) - \varphi_F(t)\right| \\
&= \left|\int_R e^{itx}\, dF_n(x) - \int_R e^{itx}\, dF(x)\right| \\
&\leqslant \int_{-\infty}^{-M} \left|e^{itx}\right| dF_n(x) + \int_{-\infty}^{-M} \left|e^{itx}\right| dF(x) + \int_M^{\infty} \left|e^{itx}\right| dF_n(x) \\
&\quad + \int_M^{\infty} \left|e^{itx}\right| dF(x) + \left|\int_{-M}^{M} e^{itx}\, dF_n(x) - \int_{-M}^{M} e^{itx}\, dF(x)\right| \\
&< 4\varepsilon + \left|\int_{-M}^{M} e^{itx}\, dF_n(x) - \int_{-M}^{M} e^{itx}\, dF(x)\right|. \qquad (1)
\end{aligned}$$

Select a division $-M = a_0 < a_1 < \cdots < a_m = M$ of $[-M, M]$ such that $a_j \in C(F)$ and $a_j - a_{j-1} < \varepsilon$, $1 \leqslant j \leqslant m$. Since $\left|e^{itx} - e^{ity}\right| \leqslant |t|\,|x - y| \leqslant T\,|x - y|$ for $|t| \leqslant T$, the reasoning in the proof of (19.22 (with $a = -M$ and $b = M$) ensures that there is $n_2 \in N$ so that, for $n \geqslant n_2$,

$$\left|\int_{-M}^{M} e^{itx}\, dF_n(x) - \int_{-M}^{M} e^{itx}\, dF(x)\right| < 2T\varepsilon + 2\varepsilon, \quad t \in [-T, T]. \qquad (2)$$

From (1) and (2) it follows that $\varphi_{F_n}(t) \to \varphi_F(t)$ uniformly in $|t| \leqslant T$. □

Remark 20.10. Since, for every $t \in R$, the functions $\cos tx$ and $\sin tx$ are continuous and bounded, the convergence $\varphi_{F_n}(t) \to \varphi_F(t)$, apart from the asserted uniformity above, follows at once by the Helly-Bray theorem (19.23).

Theorem 20.11 (Lévy). *Let $\{\varphi_{F_n} : n \in N\} \subset \mathfrak{C}$ be such that $\varphi_{F_n} \to g$ on R, and assume the limit function g is continuous at 0. Then g is the characteristic function of some $F \in \mathfrak{F}$ and $F_n \xrightarrow{w} F$.*

Proof. In view of (19.58), there is a subsequence $\{F_{n_k} : k \in N\}$ and a nondecreasing function $F : R \to [0, 1]$, which is right continuous on R, so that $F_{n_k}(x) \to F(x)$ whenever x is a continuity point of F. Then, for any $\delta > 0$ and $k \in N$,

$$\frac{1}{2\delta}\int_{-\delta}^{\delta} \varphi_{n_k}(t)dt = \frac{1}{2\delta}\int_{0}^{\delta}\left(\int_R e^{itx} + e^{-itx}\right) dF_{n_k}(x))dt$$

$$= \frac{1}{\delta}\int_{0}^{\delta}\left(\int_R \cos txdF_{n_k}(x)\right) dt = \int_R \frac{\sin \delta x}{\delta x}\, dF_{n_k}(x), \tag{1}$$

where $(\sin y)/y$ is to be 1 at $y = 0$. By the dominated convergence theorem (11.25.ii) and (19.57), on letting $k \to \infty$ in (1), we get

$$\frac{1}{2\delta}\int_{-\delta}^{\delta} g(t)dt = \int_R \frac{\sin \delta x}{\delta x} dF(x). \tag{2}$$

Since g is continuous at 0, and again by dominated convergence, as $\delta \to 0$ in (2), we have $1 = g(0) = \mu_F(R)$. Thus $F \in \mathfrak{F}$ and $F_{n_k} \stackrel{w}{\to} F$. By (20.9) $\varphi_{F_{n_k}} \to \varphi_F$, and so $g = \varphi_F$. Now let $\{n_i : i \in N\}$ be an arbitrary subsequence of N. Repeating the argument above with n_i in place of n, we see that $\{F_{n_i} : i \in N\}$ has itself a subsequence converging weakly to some $G \in \mathfrak{F}$ and $\varphi_G = g = \varphi_F$. Then $F = G$ by (20.8), and (19.56) shows that $F_n \stackrel{w}{\to} F$. □

Remarks 20.12. (a) Let $\{F_n : n \in N\} \subset \mathfrak{F}$ and $F \in \mathfrak{F}$. If $\varphi_{F_n} \to \varphi_F$, then $F_n \stackrel{w}{\to} F$ by (20.2.c), (20.11) and (20.8).

(b) Let $\{F_n : n \in N\} \subset \mathfrak{F}$, $\{G_n : n \in N\} \subset \mathfrak{F}$, and $F, G \in \mathfrak{F}$ be such that $F_n \stackrel{w}{\to} F$ and $G_n \stackrel{w}{\to} G$. Taking into account (20.4), (20.9) and (a), it follows that $F_n * G_n \stackrel{w}{\to} F * G$.

(c) Let $\{\varphi_n : n \in N\} \subset \mathfrak{C}$, and $\{t_n : n \in N\} \subset R$ be such that $\varphi_n \to \varphi \in \mathfrak{C}$ and $t_n \to t \in R$. Then $\varphi_n(t_n) \to \varphi(t)$.

(d) For $n \in N$, let F_n be the uniform distribution function on $[-n, n]$. Then $F_n(x) \to 1/2, x \in R$. According to (20.3.f), $\varphi_{F_n}(t) = (\sin nt)/nt$ for $t \neq 0$, so that $\lim_{n\to\infty} \varphi_{F_n} = 1_{\{0\}}$ exists but is not continuous at 0.

The next result provides a special case when we can find the distribution function of a random variable from its characteristic function.

Theorem 20.13. *Assume that $\int_{-\infty}^{\infty} |\varphi_F(t)|\, dt < \infty$, and let λ be the Lebesgue measure on $\mathcal{B}(R)$. Then $\mu_F \ll \lambda$ and $d\mu_F = p\, d\lambda$, i.e. $F(x) = \int_{-\infty}^{x} p(y)dy, x \in R$, where*

(i) $$p(x) = \frac{1}{2\pi}\int_{-\infty}^{\infty} e^{-itx}\varphi_F(t)dt, \quad x \in R,$$

and p is bounded and continuous.

Proof. (I) Suppose first that G_σ is the normal distribution function with density $g_\sigma(x) = \sigma^{-1}(2\pi)^{-1/2}e^{-x^2/2\sigma^2}$, $x \in R$. Then, in accordance with (20.3.h),

$$\varphi_{G_\sigma}(t) = e^{-\sigma^2 t^2/2} = \frac{1}{\sigma\sqrt{2\pi}} \int_{-\infty}^{\infty} e^{itx - x^2/2\sigma^2}\,dx, \quad t \in R. \tag{1}$$

Clearly, $\int_{-\infty}^{\infty} \varphi_{G_\sigma}(t)dt < \infty$. On changing t in $-t$ and σ in σ^{-1} in (1), we have

$$e^{-t^2/2\sigma^2} = \frac{\sigma}{\sqrt{2\pi}} \int_{-\infty}^{\infty} e^{-itx} e^{-\sigma^2 x^2/2}\,dx, \quad t \in R,$$

and so

$$g_\sigma(t) = \frac{1}{2\pi} \int_{-\infty}^{\infty} e^{-ixt} \varphi_{G_\sigma}(x)dx, \quad t \in R. \tag{2}$$

Thus (i) holds in this case.

(II) As $\mu_{G_\sigma} \ll \lambda$, (19.13.b) shows that $\mu_{F*G_\sigma} \ll \lambda$ and $d\mu_{F*G_\sigma} = p_\sigma\, d\lambda$, i.e. $(F * G_\sigma)(x) = \int_{-\infty}^{x} p_\sigma(y)dy$, $x \in R$, with

$$p_\sigma(x) = \int_{-\infty}^{\infty} g_\sigma(x-y)dF(y), \quad x \in R. \tag{3}$$

For $x \in R$, in view of Fubini's theorem (15.10) and (2), we may write

$$\begin{aligned} \frac{1}{2\pi} \int_{-\infty}^{\infty} e^{-itx} \varphi_F(t)\varphi_{G_\sigma}(t)dt &= \frac{1}{2\pi} \int_{-\infty}^{\infty} e^{-itx} \varphi_{G_\sigma}(t) \left(\int_{-\infty}^{\infty} e^{ity} dF(y) \right) dt \\ &= \int_{-\infty}^{\infty} \frac{1}{2\pi} \left(\int_{-\infty}^{\infty} e^{-it(x-y)} \varphi_{G_\sigma}(t)dt \right) dF(y) \\ &= \int_{-\infty}^{\infty} g_\sigma(x-y)dF(y). \end{aligned} \tag{4}$$

From (3) and (4) we find that

$$p_\sigma(x) = \frac{1}{2\pi} \int_{-\infty}^{\infty} e^{-itx} \varphi_F(t)\varphi_{G_\sigma}(t)dt, \quad x \in R.$$

and

$$p_\sigma(x) \leqslant \frac{1}{2\pi} \int_{-\infty}^{\infty} |\varphi_F(t)|\,dt, \quad x \in R. \tag{5}$$

Since $\lim_{\sigma\to 0}\varphi_{G_\sigma}(t) = 1$ for any $t \in R$, we infer by dominated convergence (11.46) that $\lim_{\sigma\to 0} p_\sigma(x) = p(x), x \in R$, where p is as in (i). Boundedness and continuity of p follows from $\int_R |\varphi_F|\, d\lambda < \infty$ and dominated convergence. We see also that $G_\sigma \overset{w}{\to} F_0$ as $\sigma \to 0$ (20.12.a), where F_0 is degenerate at 0, and so $F*G_\sigma \overset{w}{\to} F*F_0 = F$ as $\sigma \to 0$ by (20.12.b). Therefore, for $a, b \in C(F)$ such that $a < b$, once more by dominated convergence and (5), we have

$$\mu_F(]a,b]) = F(b) - F(a) = \lim_{\sigma\to 0}((F*G_\sigma)(b) - (F*G_\sigma)(a))$$
$$= \lim_{\sigma\to 0}\int_a^b p_\sigma(y)dy = \int_a^b p(y)dy. \qquad (6)$$

On letting $a \to -\infty$ and $b \to \infty$ in (6) entails that $\int_R p\, d\lambda = 1$. Since the family of all intervals $]a, b]$ with $a, b \in C(F)$ generates $\mathcal{B}(R)$ and is closed under intersection, (10.8) and (6) yield that $\mu_F(A) = \int_A p\, d\lambda$, $A \in \mathcal{B}(R)$, i.e. $d\mu_F = p\, d\lambda$. □

Common applications of Lévy's theorem (20.11) depend on Taylor expansions of characteristic functions, for which we now study their differentiability properties. We need first the next lemma.

Lemma 20.14. *Let $(X, \mathcal{A}, P)$ be a probability space, let $]a, b[\subset R$, and let f : $]a, b[\times X \to C$ be a function satisfying:*

(i) $|f_t| \in \mathfrak{L}^1(X, \mathcal{A}, P)$ *for any* $t \in]a, b[$, *where* f_t *is the section of* f *at* t;

(ii) *the partial derivative* $f_t'(t, x)$ *of* f *with respect to* t *exists for any* $(t, x) \in]a, b[\times X$;

(iii) *there is* $g \in L^1(X, A, P)$ *such that* $|f_t'(t, x)| \leqslant g(x)$ *for all* $(t, x) \in]a, b[\times X$.

Then, putting $h(t) = \int_X f(t, x)dP(x)$, $t \in]a, b[$, *the derivative* h' *exists and* $h'(t) = \int_X f_t'(t, x)dP(x)$, $t \in]a, b[$.

Proof. For $(t, x) \in]a, b[\times X$ and sufficiently small u, by Lagrange's theorem, we have

$$\left|\frac{f(t+u,x) - f(t,x)}{u}\right| = |f_t'(\xi, x)| \leqslant g(x).$$

Therefore, by dominated convergence,

$$h'(t) = \lim_{u\to 0}\frac{h(t+u) - h(t)}{u} = \lim_{u\to 0}\int_X \frac{f(t+u,x) - f(t,x)}{u} dP(x)$$
$$= \int_X \lim_{u\to 0}\frac{f(t+u,x) - f(t,x)}{u} dP(x) = \int_X f_t'(t,x)dP(x). \quad \square$$

For $k \in N$, let $m^{(k)} = \int_R x^k\, dF(x)$ and $\mu^{(k)} = \int_R |x|^k\, dF(x)$ denote, respectively, the kth moment (if it exists) and the kth absolute moment of a random variable f with distribution function F. Put $m^{(0)} = 1$.

Theorem 20.15. *If $\mu^{(k)} < \infty$ for some $k \in N$, then φ_F has a kth uniformly continuous and bounded derivative $\varphi_F^{(k)}$ given by*

$$\text{(i)} \quad \varphi_F^{(k)}(t) = \int_R (ix)^k e^{itx}\, dF(x), \quad t \in R.$$

Proof. As stated in (17.5), $\mu^{(j)} < \infty$ whenever $j < k$. For $k = 1$, (i) follows from (20.14) with $f(t,x) = e^{itx}$ and $g(x) = |x|$, $(t,x) \in R^2$. The general case follows easily by induction. Uniform continuity of $\varphi_F^{(k)}$ follows by dominated convergence as in (20.2.c), and boundedness of $\varphi_F^{(k)}$ follows from $\left|\varphi_F^{(k)}(t)\right| \leqslant \mu^{(k)}$, $t \in R$. □

Notations 20.16. For $n \in N$, let $a_n, b_n \in C$ with $b_n \neq 0$. We shall write $a_n = O(b_n)$ if there is $M < \infty$ such that $|a_n/b_n| \leqslant M$ for all n, and we shall write $a_n = o(b_n)$ if $a_n/b_n \to 0$. The relation $a_n \sim b_n$ will mean that $a_n/b_n \to 1$.

For $x \in]a,b[$, let $f(x), g(x) \in C$ with $g(x) \neq 0$. Let $x_0 \in [a,b]$. In a similar vein, we shall write $f(x) = O(g(x))$ as $x \to x_0$ if there is $M < \infty$ such that $|f(x)/g(x)| \leqslant M$ for all x in a neighborhood of x_0, and we shall write $f(x) = o(g(x))$ as $x \to x_0$ if $f(x)/g(x) \to 0$ as $x \to x_0$. The relation $f(x) \sim g(x)$ as $x \to x_0$ will mean that $f(x)/g(x) \to 1$ as $x \to x_0$.

Corollary 20.17. *Assume that $\mu^{(n)} < \infty$ for some $n \in N$. Then*

$$\text{(i)} \quad \varphi_F(t) = \sum_{k=0}^{n} \frac{i^k m^{(k)}}{k!} t^k + o(t^n) \text{ as } t \to 0$$

and

$$\text{(ii)} \quad \left|\varphi_F(t) - \sum_{k=0}^{n-1} \frac{i^k m^{(k)}}{k!} t^k\right| \leqslant \frac{\mu^{(n)}}{n!}|t|^n, \quad t \in R.$$

Proof. By (20.15) we get $\varphi_F^{(k)}(0) = i^k m^{(k)}$ (with $\varphi_F^{(0)} = \varphi_F$) for $0 \leqslant k \leqslant n$. Hence, according to a theorem in calculus, we have the Taylor expansion, with the remainder term in the integral form,

$$\begin{aligned}\varphi_F(t) &= \sum_{k=0}^{n-1} \frac{\varphi_F^{(k)}(0)}{k!} t^k + \frac{t^n}{(n-1)!} \int_0^1 \varphi_F^{(n)}(tu)(1-u)^{n-1}\, du \\ &= \sum_{k=0}^{n} \frac{i^k m^{(k)}}{k!} t^k + \frac{t^n}{(n-1)!} \int_0^1 \left(\varphi_F^{(n)}(tu) - \varphi_F^{(n)}(0)\right)(1-u)^{n-1}\, du, \quad t \in R.\end{aligned} \tag{1}$$

Since $\varphi_F^{(n)}$ is continuous at 0, $\int_0^1 (\varphi_F^{(n)}(tu) - \varphi_F^{(n)}(0))(1-u)^{n-1}\, du \to 0$ as $t \to 0$ by dominated convergence. Thus (i) follows from (1). In view of (20.15) and (1), we have also

$$\left|\varphi_F(t) - \sum_{k=0}^{n-1} \frac{i^k m^{(k)}}{k!} t^k\right| \leqslant \frac{|t|^n}{(n-1)!} \int_0^1 \left|\varphi_F^{(n)}(tu)\right| (1-u)^{n-1}\, du$$

$$\leqslant \frac{|t|^n}{(n)!} \mu^{(n)}, \quad t \in R. \ \square$$

Corollary 20.18. *If $\mu^{(n)} < \infty$ for all $k \in N$, then $\varphi_F(t) = \sum_{k=0}^{\infty} \frac{i^k m^{(k)}}{k!} t^k$ in the interval of convergence of the series.*

Example 20.19. Let Φ be the standard normal distribution function. Then $\varphi_\Phi(t) = e^{-t^2/2}$, $t \in R$. Actually, from (19.49) in this case $m^{(2n-1)} = 0$ and $m^{(2n)} = 1{\cdot}3\ldots(2n-1) = (2n)!/2^n n!$, $n \geqslant 1$. In view of (13.5), we have $\mu^{(2n-1)} \leqslant (\mu^{(2n)})^{\frac{2n-1}{2n}} \leqslant \mu^{(2n)} = m^{(2n)}$, $n \geqslant 1$. This shows that $\mu^{(n)}/n! \to 0$ and so, on account of the estimation (ii) in (20.17), we get

$$\varphi_\Phi(t) = \sum_{k=0}^{\infty} \frac{i^k m^{(k)}}{k!} t^k = \sum_{k=0}^{\infty} \frac{(-t^2)^k}{2^k k!} = e^{-t^2/2}, \quad t \in R.$$

Hence, if f has a normal distribution, and g is as in (19.47), then

$$\varphi_f(t) = \varphi_{m+\sigma g}(t) = Ee^{it(m+\sigma g)} = e^{imt} Ee^{i\sigma t g}$$

$$= e^{imt} \varphi_\Phi(\sigma t) = e^{imt - \sigma^2 t^2/2}, \quad t \in R.$$

Theorem 20.20. *If $\varphi_F^{(2k)}(0)$ exists and is finite for some $k \in N$, then $\mu^{(2k)} < \infty$.*

Proof. The result follows by induction.

(I) For $k = 1$, by l'Hospital's rule, we have

$$\varphi_F''(0) = \lim_{h\to 0} \frac{1}{2}\left[\frac{\varphi_F'(h) - \varphi_F'(0)}{h} + \frac{\varphi_F'(-h) - \varphi_F'(0)}{-h}\right]$$

$$= \lim_{h\to 0} \frac{\varphi_F'(h) - \varphi_F'(-h)}{2h}$$

$$= \lim_{h\to 0} \frac{\varphi_F(h) + \varphi_F(-h) - 2}{h^2} = \lim_{h\to 0} \int_R \frac{e^{ihx} + e^{-ihx} - 2}{h^2} dF(x)$$

$$= -2 \lim_{h\to 0} \int_R \frac{1 - \cos hx}{h^2} dF(x).$$

Hence, by Fatou's lemma (11.24), we get

$$\mu^{(2)} = \int_R x^2\, dF(x) = 2 \int_R \lim_{h\to 0} \frac{1 - \cos hx}{h^2} dF(x)$$

$$\leqslant \lim_{h\to 0} 2 \int_R \frac{1 - \cos hx}{h^2} dF(x) = -\varphi_F''(0) < \infty.$$

(II) Suppose the assertion is true for $2k$, and assume that $\varphi_F^{(2k+2)}(0)$ exists and is finite. Then $\varphi_F^{(2k)}(0)$ exists and is finite, so that $\mu^{(2k)} < \infty$ by the induction hypothesis. In view of (20.15), it follows that

$$\int_R e^{itx} x^{2k} \, dF(x) = (-1)^k \varphi_F^{(2k)}(t), \quad t \in R. \tag{1}$$

If $\mu^{(2k)} = 0$, then F is degenerate at 0, and so $\mu^{(2k+2)} = 0$. If $\mu^{(2k)} > 0$, set $G(x) = \int_{-\infty}^{x} y^{2k} dF(y)/\mu^{(2k)}$, $x \in R$. Then $G \in \mathfrak{F}$, and

$$\varphi_G(t) = \int_R e^{itx} \, dG(x) = \int_R e^{itx} x^{2k} \, dF(x) = (-1)^k \varphi_F^{(2k)}(t), \quad t \in R,$$

by (12.8) and (1). Hence $\varphi_G''(0)$ exists and is finite, and $\int_R x^2 \, dG(x) < \infty$ by (I), that is $\int_R x^{2k+2} \, dF(x) < \infty$. □

A characteristic function φ satisfies either $|\varphi(t)| < 1$ for all $t \neq 0$ or $|\varphi(t_0)| = 1$ for some $t_0 \neq 0$. We next examine the latter possibility. For this, some definitions are in order.

Definition 20.21. A random variable f is said to have a *lattice distribution* if there are $a \in R$ and $h > 0$ such that $P(f \in \{a + kh : k \in Z\}) = 1$. The quantity h is called the *span of* f. (Notice that if f is degenerate, then it has a lattice distribution.) Neither a nor h is uniquely determined. Actually, if h' is a divisor of h and $a' = a + h'$, then also $P(f \in \{a' + kh' : k \in Z\}) = 1$. The greatest common divisor of $\{b_2 - b_1 : b_1 < b_2,\ P(f = b_1) > 0,\ P(f = b_2) > 0\}$ is the *maximal span of* f whenever f is not degenerate and has a lattice distribution.

Remark 20.22. If f has a lattice distribution with span h, then $|\varphi_f|$ is periodic of period $2\pi/h$ and $|\varphi_f(2\pi/h)| = 1$. Actually, setting $p_k = P(f = a + kh)$, $k \in Z$, gives

$$\begin{aligned}\varphi_f\left(t + \frac{2\pi}{h}\right) &= \sum_{k \in Z} e^{i(t+2\pi/h)(a+kh)} p_k = e^{i2\pi a/h} \sum_{k \in Z} e^{it(a+kh)} p_k \\ &= e^{i2\pi a/h} \varphi_f(t), \quad t \in R.\end{aligned} \tag{1}$$

Both assertions follows at once from (1).

Theorem 20.23. *Assume there exists $t_0 > 0$ such that $|\varphi_f(t_0)| = 1$. Then f has a lattice distribution with span $2\pi/t_0$.*

Proof. Since $|\varphi_f(t_0)| = 1$, $\varphi_f(t_0) = e^{i\theta}$ for some $\theta \in R$. Then $1 = e^{-i\theta}\varphi_f(t_0) = \int_X e^{i(t_0 f - \theta)} dP = \int_X \cos(t_0 f - \theta) dP$, and so $\int_X (1 - \cos(t_0 f - \theta)) dP = 0$. In view of (11.16), this entails that $\cos(t_0 f - \theta) = 1$ a.s., meaning that $P(t_0 f - \theta \in \{2k\pi : k \in Z\}) = P(f \in \{\theta/t_0 + k2\pi/t_0 : k \in Z\}) = 1$. □

Remark 20.24. The indication $t_0 > 0$ in (20.23) is equivalent to $t_0 \neq 0$, as $|\varphi_f(-t_0)| = |\varphi_f(t_0)|$.

Theorem 20.25. *Assume that* $|\varphi_f(t_0)| = |\varphi_f(\alpha t_0)| = 1$ *for some* $t_0 > 0$ *and* $0 < \alpha \notin Q$. *Then* f *is degenerate.*

Proof. If f is not degenerate, by (20.23) there exist $b_1, b_2 \in R$ so that $P(f = b_1), P(f = b_2) > 0$, and $b_2 - b_1 = 2k\pi/t_0 = 2k'\pi/\alpha t_0$ for some $k, k' \in N$, whence $\alpha k = k'$, a contradiction. □

An important general criterion for deciding whether or not a function φ is a characteristic function is that of nonnegative definiteness that we introduce now.

Definition 20.26. A function $\varphi : R \to C$ is called *nonnegative definite* if for any finite sets $\{t_1, \ldots, t_n\} \subset R$ and $\{z_1, \ldots, z_n\} \subset C, n \in N$,

$$\sum_{j=1}^{n}\sum_{k=1}^{n} \varphi(t_j - t_k) z_j \bar{z}_k \geqslant 0.$$

Theorem 20.27. *Assume* φ *is nonnegative definite and continuous at* 0. *Then the following assertions hold.*

(i) $\varphi(0) \geqslant 0$.

(ii) $\varphi(-t) = \overline{\varphi(t)}, t \in R$.

(iii) $|\varphi(t)| \leqslant \varphi(0)$.

(iv) φ *is uniformly continuous.*

Proof. Taking $n = 1$ and $z_1 = 1$ in the defining relation, we obtain (i). Taking $n = 2, t_1 = t, t_2 = 0$ and $z_1 = z_2 = 1$, we have

$$2\varphi(0) + \varphi(t) + \varphi(-t) \geqslant 0. \tag{1}$$

Taking $n = 2, t_1 = t, t_2 = 0, z_1 = 1$ and $z_2 = i$, we get

$$-\varphi(t)i + \varphi(-t)i \geqslant 0. \tag{2}$$

(1) shows that $\varphi(t) + \varphi(-t) \in R$, while (2) shows that $\varphi(t) - \varphi(-t)$ is pure imaginary, and so $\varphi(-t) = \overline{\varphi(t)}$. With $n = 2, t_1 = t, t_2 = 0, z_1 = \varphi(-t)$ and $z_2 = -|\varphi(t)|$, in view of (ii), we obtain

$$2\varphi(0)|\varphi(t)|^2 - 2|\varphi(t)|^3 \geqslant 0. \tag{3}$$

(iii) follows at once from (3) and (i). According to (iii), if $\varphi(0) = 0$, then $\varphi = 0$ and (iv) is trivial. We exclude this case, and dividing by $\varphi(0)$, we may and will suppose that $\varphi(0) = 1$. Taking $n = 3, t_1 = 0, t_2 = t$ and $t_3 = t = h$, as the determinant of a nonnegative quadratic form is nonnegative, we have

$$0 \leqslant \begin{vmatrix} 1 & \varphi(t) & \varphi(t+h) \\ \varphi(-t) & 1 & \varphi(h) \\ \varphi(-t-h) & \varphi(-h) & 1 \end{vmatrix}$$
$$= 1 - |\varphi(t)|^2 - |\varphi(h)|^2 - |\varphi(t+h)|^2 + 2\mathrm{Re}(\varphi(t)\varphi(h)\overline{\varphi(t+h)}).$$

Therefore, taking into account (iii),

$$
\begin{aligned}
|\varphi(t+h)-\varphi(t)|^2 &= |\varphi(t+h)|^2+|\varphi(t)|^2-2\mathrm{Re}(\varphi(t))\overline{\varphi(t+h)}) \\
&\leqslant 1-|\varphi(h)|^2+2\mathrm{Re}(\varphi(t)(\varphi(h)-1)\overline{\varphi(t+h)}) \\
&\leqslant 2(1-|\varphi(h)|)+2\,|1-\varphi(h)| \leqslant 4\,|1-\varphi(h)|\,. \qquad (4)
\end{aligned}
$$

Since φ is continuous at 0, (4) shows that (iv) holds □

Theorem 20.28 (Bochner). *Let $\varphi : R \to C$ be a function. Then the following are equivalent:*

(i) *φ is a characteristic function;*

(ii) *φ is nonnegative definite and continuous at* 0 *with* $\varphi(0) = 1$.

Proof. If φ is the characteristic function of f, then

$$
\begin{aligned}
\sum_{j=1}^{n}\sum_{k=1}^{n}\varphi(t_j-t_k)z_j\bar{z}_k &= E\left[\sum_{j=1}^{n}\sum_{k=1}^{n}e^{i(t_j-t_k)f}z_j\bar{z}_k\right] \\
&= E\left[\sum_{j=1}^{n}\sum_{k=1}^{n}e^{it_jf}z_j\overline{e^{it_kf}z_k}\right] \\
&= E\left|\sum_{j=1}^{n}e^{it_jf}z_j\right|^2 \geqslant 0
\end{aligned}
$$

for any $\{t_1,\ldots,t_n\}\subset R$ and $\{z_1,\ldots,z_n\}\subset C, n\in N$. Thus (i) implies (ii), taking into account (20.2.a) and (20.2.c). Assume now that (ii) holds. For $m,n\in N$, put

$$
\varphi_m(t)=\begin{cases}\left(1-\frac{|t|}{m}\right)\varphi(t) & \text{if } |t|\leqslant m\\ 0 & \text{if } |t|>m\end{cases},
$$

$$
p_m(x)=\frac{1}{2\pi m}\int_0^m\int_0^m\varphi(s-t)e^{-i(s-t)x}\,ds\,dt,\quad x\in R, \qquad (1)
$$

and

$$
c_m(n)=\int_{-n}^{n}\left(1-\frac{|x|}{n}\right)p_m(x)dx. \qquad (2)
$$

Notice first that $p_m\geqslant 0$, since the integrand of the double integral in (1) is continuous, and so this integral is a limit of nonnegative Riemann sums. On letting $s=t+u$, we have

$$
p_m(x)=\frac{1}{2\pi m}\int_0^m\left(\int_{-t}^{m-t}\varphi(u)e^{-iux}du\right)dt
$$

$$= \frac{1}{2\pi m}\left(\int_{-m}^{0}\left(\int_{-u}^{m}\varphi(u)e^{-iux}dt\right)du + \int_{0}^{m}\left(\int_{0}^{m-u}\varphi(u)e^{-iux}dt\right)du\right)$$

$$= \frac{1}{2\pi}\int_{-m}^{m}\left(1-\frac{|u|}{m}\right)\varphi(u)e^{-iux}\,du$$

$$= \frac{1}{2\pi}\int_{-\infty}^{\infty}\varphi_m(u)e^{-iux}\,du. \tag{3}$$

Since

$$\int_{-n}^{n}\left(1-\frac{|x|}{n}\right)e^{-iux}\,dx = \begin{cases}\frac{2-2\cos nu}{nu^2} & \text{if } u \neq 0\\ n & \text{if } u = 0\end{cases},$$

we have, on account of (3),

$$c_m(n) = \frac{1}{2\pi}\int_{-m}^{m}\left(1-\frac{|u|}{m}\right)\varphi(u)\left(\int_{-n}^{n}\left(1-\frac{|x|}{n}\right)e^{-iux}\,dx\right)du$$

$$= \frac{1}{\pi}\int_{-m}^{m}\left(1-\frac{|u|}{m}\right)\varphi(u)\frac{1-\cos nu}{nu^2}du$$

$$= \frac{1}{\pi}\int_{-mn}^{mn}\left(1-\frac{|u|}{mn}\right)\varphi\left(\frac{u}{n}\right)\frac{1-\cos u}{u^2}du. \tag{4}$$

Then, by dominated convergence, as φ is continuous at 0 with $\varphi(0) = 1$,

$$\lim_n c_m(n) = \frac{1}{\pi}\int_{-mn}^{mn}\frac{1-\cos u}{u^2}du = 1, \quad m \in N. \tag{5}$$

Multiplying the extreme sides of (3) by $\frac{1}{c_m(n)}\left(1-\frac{|x|}{n}\right)e^{itx}$, and integrating with respect to x on $[-n, n]$, we obtain, similarly to (4),

$$\frac{1}{c_m(n)}\int_{-n}^{n}\left(1-\frac{|x|}{n}\right)p_m(x)e^{itx}\,dx$$

$$= \frac{1}{2\pi c_m(n)}\int_{-n}^{n}\left(1-\frac{|x|}{n}\right)\left(\int_{-m}^{m}\left(1-\frac{|u|}{m}\right)\varphi(u)e^{-iux}du\right)e^{itx}\,dx$$

$$= \frac{1}{2\pi c_m(n)}\int_{-m}^{m}\left(1-\frac{|u|}{m}\right)\varphi(u)\left(\int_{-n}^{n}\left(1-\frac{|x|}{n}\right)e^{-i(u-t)x}\,dx\right)du$$

$$= \frac{1}{\pi c_m(n)} \int_{-m}^{m} \left(1 - \frac{|u|}{m}\right) \varphi(u) \frac{1 - \cos n(u-t)}{n(u-t)^2} du$$

$$= \frac{1}{\pi c_m(n)} \int_{-m}^{m} \varphi_m(u) \frac{1 - \cos n(u-t)}{n(u-t)^2} du$$

$$= \frac{1}{\pi c_m(n)} \int_{-\infty}^{\infty} \varphi_m\left(t + \frac{u}{n}\right) \frac{1 - \cos u}{u^2} du. \tag{6}$$

For fixed $m \in N$, the first term in (6) is a characteristic function in view of (2), while the last term in (6) converges to $\varphi_m(t)$ as $n \to \infty$ by dominated convergence and (5). Since φ_m is continuous at 0, (20.11) shows it is a characteristic function. Finally, since $\lim_{m\to\infty} \varphi_m(t) = \varphi(t)$, $t \in R$, and φ is continuous at 0, (20.11) shows that $\varphi \in \mathfrak{C}$. □

Now we define briefly the notion of characteristic function of a random vector.

Definition 20.29. Let $(X, \mathcal{A}, P)$ be a probability space, and let $\mathbf{f} : X \to R^n$ be a random vector with coordinates $f_1, \ldots, f_n$ and distribution function F. The *characteristic function of* $\mathbf{f}$ is the function $\varphi_{\mathbf{f}} : R^n \to C$ defined by

$$\varphi_{\mathbf{f}}(\mathbf{t}) = Ee^{i\langle \mathbf{t}, \mathbf{f}\rangle} = E\left[\exp\left(i\sum_{j=1}^{n} t_j f_j\right)\right], \quad \mathbf{t} = (t_1, \ldots, t_n) \in R^n.$$

According to (19.32.c) and the change of variable formula (11.29), $\varphi_{\mathbf{f}}$ is also expressible as the Fourier-Stieltjes transform of μ_F,

$$\varphi_F(\mathbf{t}) = \int_{R^n} e^{i\langle \mathbf{t}, \mathbf{x}\rangle}\, d\mu_F(\mathbf{x}) = \int_{R^n} \exp\left(i\sum_{j=1}^{n} t_j x_j\right) d\mu_F(x_1, \ldots, x_n),$$
$$\mathbf{t} = (t_1, \ldots, t_n) \in R^n,$$

where $\langle \mathbf{t}, \mathbf{x}\rangle = \sum_{j=1}^{n} t_j x_j$ denotes inner product.

Remarks 20.30. (a) $|\varphi_{\mathbf{f}}(t)| \leqslant 1 = \varphi_{\mathbf{f}}(\mathbf{0})$, $\mathbf{t} \in R^n$.

(b) $\varphi_{-\mathbf{f}}(t_1, \ldots, t_n) = \varphi_{\mathbf{f}}(-t_1, \ldots, -t_n) = \overline{\varphi_{\mathbf{f}}(t_1, \ldots, t_n)}$, $(t_1, \ldots, t_n) \in R^n$.

(c) φ_F is uniformly continuous. Actually, for $\mathbf{u}, \mathbf{v} \in R^n$,

$$|\varphi_F(\mathbf{u}) - \varphi_F(\mathbf{v})| \leqslant \int_{R^n} \left|e^{i\langle \mathbf{u}, \mathbf{x}\rangle} - e^{i\langle \mathbf{v}, \mathbf{x}\rangle}\right| d\mu_F(\mathbf{x})$$
$$= \int_{R^n} \left|e^{i\langle \mathbf{u}-\mathbf{v}, \mathbf{x}\rangle} - 1\right| d\mu_F(\mathbf{x})$$
$$= \int_{R^n} 2\left|\sin\frac{\langle \mathbf{u}-\mathbf{v}, \mathbf{x}\rangle}{2}\right| d\mu_F(\mathbf{x})$$

$$\leqslant \int_{R^n} 2 \wedge |\langle \mathbf{u} - \mathbf{v}, \mathbf{x} \rangle| \, d\mu_F(\mathbf{x})$$

$$\leqslant \int_{R^n} (2 \wedge \|\mathbf{u} - \mathbf{v}\| \, \|\mathbf{x}\|)) d\mu_F(\mathbf{x}), \tag{1}$$

where $\|\mathbf{x}\| = \|(x_1, \ldots, x_n)\| = \left(\sum_{j=1}^n x_j^2\right)^{1/2}$. Since $\int_{R^n} (2 \wedge h \, \|\mathbf{x}\|)) d\mu_F(\mathbf{x}) \to 0$ as $h \to 0$ by dominated convergence, (1) shows that φ_F is uniformly continuous.

(d) If F has density p, then $\varphi_F(\mathbf{t}) = \int_{R^n} e^{i\langle \mathbf{t}, \mathbf{x} \rangle} p(\mathbf{x}) \, d\lambda(\mathbf{x})$, $\mathbf{t} \in R^n$, where λ is the Lebesgue measure on $\mathcal{B}(R^n)$, by (19.34) and (12.8).

Examples 20.31. (a) If F is the uniform distribution function as in (19.35.a), then $\varphi_F(t_1, t_2) = \frac{\sin a_1 t_1}{a_1 t_1} \frac{\sin a_2 t_2}{a_2 t_2}$.

(b) If $\mathbf{f}$ has a multinomial distribution as in (19.30.a), then $\varphi_{\mathbf{f}}(t_1, \ldots, t_n) = \left(\sum_{k=1}^n p_k e^{i t_k}\right)^m$.

(c) If F is the multinormal distribution function as in (19.35.b), then $\varphi_F(\mathbf{t}) = \exp(i \langle \mathbf{m}, \mathbf{t} \rangle - \mathbf{t} V \mathbf{t}^T / 2)$.

Theorem 20.32. *Let $\mathbf{f} = (f_1, \ldots, f_n)$ and $\mathbf{g} = (g_1, \ldots, g_n)$ be independent random vectors with corresponding distribution functions F and G. Then $\varphi_{F*G} = \varphi_{\mathbf{f}+\mathbf{g}} = \varphi_{\mathbf{f}} \varphi_{\mathbf{g}}$.*

Proof. The first equality follows from (19.36.c). For $(t_1, \ldots, t_n) \in R^n$, in view of (7.29) and (7.19), we see that $\sigma(f_1, \ldots, f_n)$ is independent of $\sigma(g_1, \ldots, g_n)$, and so $e^{i(t_1 f_1 + \cdots + t_n f_n)}$ is independent of $e^{i(t_1 g_1 + \cdots + t_n g_n)}$. Hence the last equality follows by (18.42.b) □

Remark 20.33. The statements of (20.5)–(20.8) carry over to R^n and the proofs generalize readily. Thus $\varphi_F = \varphi_G$ implies that $\mu_F = \mu_G$, and so $F = G$.

Theorem 20.34. *Let $\mathbf{f} = (f_1, \ldots, f_n)$ be a random vector. Then the following are equivalent:*

(i) *$f_1, \ldots, f_n$ are independent;*

(ii) *$\varphi_{(f_1, \ldots, f_n)}(t_1, \ldots, t_n) = \prod_{j=1}^n \varphi_{f_j}(t_j)$, $(t_1, \ldots, t_n) \in R^n$.*

Proof. If (i) holds, for $(t_1, \ldots, t_n) \in R^n$, by (18.42.b),

$$\varphi_{(f_1, \ldots, f_n)}(t_1, \ldots, t_n) = E\left[\exp\left(i \sum_{j=1}^n t_j f_j\right)\right] = E\left[\prod_{j=1}^n \exp(i t_j f_j)\right]$$

$$= \prod_{j=1}^n E[\exp(i t_j f_j)] = \prod_{j=1}^n \varphi_{f_j}(t_j).$$

Conversely, assume that (ii) holds, let F be the distribution function of $\mathbf{f}$, and let F_j be the distribution function of f_j, $1 \leqslant j \leqslant n$. Then, on account of Fubini's theorem, we have

$$\int_{R^n} \exp\left(i\sum_{j=1}^{n} t_j x_j\right) dF(x_1, \ldots, x_n) = \varphi_F(t_1, \ldots, t_n)$$
$$= \prod_{j=1}^{n} \varphi_{F_j}(t_j) = \prod_{j=1}^{n} \int_R \exp(it_j x_j) d\mu_{F_j}(x_j)$$
$$= \int_{R^n} \left(\prod_{j=1}^{n} \exp(it_j x_j)\right) d(\bigotimes_{j=1}^{n} \mu_{F_j})(x_1, \ldots, x_n)$$
$$= \int_{R^n} \exp\left(i\sum_{j=1}^{n} t_j x_j\right) d(\bigotimes_{j=1}^{n} \mu_{F_j})(x_1, \ldots, x_n), (t_1, \ldots, t_n) \in R^n.$$

Therefore, $\mu_F = \otimes_{j=1}^{n} \mu_{F_j}$ in view of (20.33), and so $F(x_1, \ldots, x_n) = \prod_{j=1}^{n} F_j(x_j)$, $(x_1, \ldots, x_n) \in R^n$. This leads to (i) by (19.32.a) □

Exercise 20.35. (a) Let I be a countable set, let $\{\varphi_i : i \in I\} \subset \mathfrak{C}$, and let $\{p_i : i \in I\} \subset]0, 1[$ be such that $\sum_{i \in I} p_i = 1$. Use (19.63) and (11.52) to show that $\sum_{i \in I} p_i \varphi_i \in \mathfrak{C}$.

(b) If $\varphi \in \mathfrak{C}$, then $\operatorname{Re} \varphi \in \mathfrak{C}$.

Exercise 20.36. Compute φ_F, where F is the Cauchy distribution function in (19.10.c). Use the result to find an example of random variables f_1 and f_2 that are not independent, but $\varphi_{f_1+f_2} = \varphi_{f_1}\varphi_{f_2}$.

Exercise 20.37. Verify that $\varphi_F(t) = (1 - it/b)^{-a}$, where F is the gamma distribution function as in (19.10.e).

Exercise 20.38. Prove that the following subsets of $\mathfrak{F}$ are closed under convolution: (i) degenerate; (ii) binomial with the same p; (iii) Poisson with the same b; (iv) Cauchy; (v) normal.

Exercise 20.39. Let $f_0, f_1 \ldots$ be independent and identically distributed random variables. Assume that ν is a random variable independent of $\sigma(f_0, f_1 \ldots)$ with $P(\nu = k) = p(k; \lambda)$, $k = 0, 1, \ldots$, and put $g = f_0 + \cdots + f_\nu$. Show that $\varphi_g = \exp(\lambda\varphi_{f_0} - \lambda)$.

Exercise 20.40. Let $f_1, \ldots, f_n$ be independent random variables, each having a normal distribution function with mean 0 and variance σ^2. Prove the following.

(a) $f_1^2 + \cdots + f_n^2$ has a χ^2 distribution function.

(b) The distribution of $(f_1 + \cdots + f_n)/\sqrt{n}$ coincides with the distribution of f_1.

(c) If $\sigma = 1$ and $x_n \to \infty$, then $x_n^{-1} \log P(f_1 + \cdots + f_n \geqslant \sqrt{nx_n}) \to -1/2$. [Use (b) and (19.50).]

Exercise 20.41. Show that any triangular distribution function is a convolution of two uniform distribution functions.

Exercise: Symmetric random variable 20.42. A random variable f is said to be *symmetric* if f and $-f$ have the same distribution, i.e. $Pf^{-1} = P(-f)^{-1}$. Prove the following.

(a) If α is a symmetric Bernoulli random variable and f is an arbitrary random variable independent of α, then αf is symmetric.

(b) If f has an even probability density, then f is symmetric.

(c) If f is symmetric and P-integrable, then $Ef = 0$. [Hint. Write $Ef = \int_{\{f\geqslant 0\}} f\,dP - \int_{\{-f\geqslant 0\}}(-f)dP$, then use the change of variable formula (11.29).]

(d) f is symmetric if and only if $\varphi_f(R) \subset R$. [Hints. If f is symmetric, then $\varphi_f = \varphi_{-f}$. If $\varphi_f(R) \subset R$, then $\int_R e^{itx}\,d(Pf^{-1})(x) = \int_R e^{itx}\,d(P(-f)^{-1})(x)$, $t \in R$, and apply (20.8).]

(e) If f is symmetric, then so is the product $f(h \circ |f|)$ for any measurable function $h : [0, \infty[\to R$; in particular, the *truncated random variable* $f1_{\{|f|\in B\}}$ is symmetric for any $B \in \mathcal{B}(R)$. [Use (11.29) and (d).]

(f) f is symmetric if and only if $P(f^+)^{-1} = P(f^-)^{-1}$.

(g) f is symmetric if and only if f and $f1_{\{f\in B\}} - f1_{\{f\in B^c\}}$ have the same distribution for each symmetric[1] $B \in \mathcal{B}(R)$. [Hints. If f and B are symmetric, use (11.29) to verify that $\varphi_f = \varphi_{f1_{\{f\in B\}}-f1_{\{f\in B^c\}}}$. To prove the converse implication, take $B = \{0\}$ and show that $\varphi_f = \varphi_{-f}$.]

(h) If $f_1, \ldots, f_n$ are independent and symmetric random variables, and $h : R^n \to R$ is a measurable function such that $h(-x_1, \ldots, -x_n) = -h(x_1, \ldots, x_n)$, $(x_1, \ldots, x_n) \in R^n$, then $h \circ (f_1, \ldots, f_n)$ is symmetric; in particular, $f_1 + \cdots + f_n$ is symmetric. [Use (19.32.b).]

(i) If f_n, $n \in N$, are symmetric random variables and $f_n \overset{D}{\to} f$, then f is symmetric.

Exercise 20.43. Assume f has a χ^2 distribution function with n degrees of freedom as in (19.10.e). For $n \in N$, set $g_n = (f - n\sigma^2)/\sigma^2\sqrt{2n}$ and $G_n(x) = P(g_n \leqslant x)$, $x \in R$. Show that $G_n \overset{w}{\to} \Phi$.

Exercise 20.44. Let f and g be independent random variables such that $Pf^{-1} = Pg^{-1}$, $P(f+g)^{-1} = P(f-g)^{-1}$, $Ef = 0$ and $Ef^2 = 1$. Prove that the common distribution function of f and g is Φ. The next steps may be helpful.

(a) $\varphi_f(t) \neq 0$, $t \in R$. [Hints. $\varphi_f(2t) = \varphi_f(t)^3\varphi_f(-t)$, $t \in R$.]

(b) Let $\rho(t) = \varphi_f(t)/\varphi_f(-t)$, $t \in R$. Then $\rho = 1$. [Hint. $\rho(2t) = \rho(t)^2$, and so $\rho(t) = \rho(t/2^m)^{2^m}$, $t \in R$, $m \in N$. Then apply (i) of (20.17) with $n = 1$.]

(c) $\varphi_f(t) = e^{-t^2/2}$, $t \in R$. [Hint. $\varphi_f(2t) = \varphi_f(t)^4$ since $\varphi_f(t) = \varphi_f(-t)$, and so $\varphi_f(t) = \varphi_f(t/2^m)^{4^m}$, $t \in R$, $m \in N$. Then apply (i) of (20.17) with $n = 2$.]

Exercise 20.45 (Riemann-Lebesgue). If F has density p, then $\lim_{t\to-\infty}\varphi_F(t) = \lim_{t\to\infty}\varphi_F(t) = 0$. [Hint. Write $\int_R e^{itx}p(x)dx = \frac{1}{2}\int_R e^{itx}(p(x) - p(x+\frac{\pi}{t}))dx$, $t \in R$, and apply (13.31).]

Exercise 20.46. Let $\varphi \in \mathfrak{C}$ and $t \in R$. Prove the following.

(a) $|1 - \varphi(t)|^2 \leqslant 2(1 - \operatorname{Re}\varphi(t))$.

(b) $1 - \operatorname{Re}\varphi(2t) \leqslant 4(1 - \operatorname{Re}\varphi(t))$.

(c) $1 - |\varphi(2t)|^2 \leqslant 4(1 - |\varphi(t)|^2)$.

(d) $2(\operatorname{Re}\varphi(t))^2 \leqslant 1 + \operatorname{Re}\varphi(2t)$. [Use Schwarz's inequality.]

[1] A set $B \subset R$ is said to be *symmetric* if $-x \in B$ whenever $x \in B$.

(e) $2|\varphi(t)|^2 \leqslant 1 + |\varphi(2t)|$.

(f) $(\operatorname{Im}\varphi(t))^2 \leqslant 2(1 - \operatorname{Re}\varphi(2t))$.

Exercise 20.47. Let f be a random variable such that $0 < Ef^2 < \infty$. Show that there exist $\varepsilon, c > 0$ so that $\left|\varphi_f(t)\right| \leqslant 1 - ct^2$ for $|t| < \varepsilon$.

Exercise 20.48. For $\varphi \in \mathfrak{C}$ and $t, h \in R$, prove that $|\varphi(t+h) - \varphi(t)|^2 \leqslant 2(1 - \operatorname{Re}\varphi(h))$.

Exercise 20.49. Let $\{\varphi_n : n \in N\} \subset \mathfrak{C}$ and $t_0 > 0$. Show that $\varphi_n \to 1$ whenever $\varphi_n(t) \to 1$ for any $t \in [-t_0, t_0]$. [Use (20.46.b) and (20.46.a).]

Exercise 20.50 (Herglotz). Let $g : R \to C$ be a nonnegative definite function with $g(0) = 1$, and let $c > 0$. Prove that there exists a characteristic function $\varphi(t) = \int_{-\pi/c}^{\pi/c} e^{itx}\, dF(x)$ such that $\varphi(mc) = g(mc)$, $m \in Z$. The following steps may be useful.

(a) For $n \in N$, set $\rho_n(x) = \frac{1}{2\pi}\sum_{m=-n+1}^{n-1}\left(1 - \frac{|m|}{n}\right) g(mc)e^{-imx}$ or 0 according as $|x| \leqslant \pi$ or $|x| > \pi$. Then ρ_n is a probability density. [Hint. $\rho_n(x) = \frac{1}{2\pi}\sum_{j=1}^{n}\sum_{k=1}^{n} g((j-k)c)e^{-i(j-k)x}$, $|x| \leqslant \pi$.]

(b) Put $p_n(x) = \frac{1}{c}\rho_n(cx)$, $x \in R$. Then, for $m \in Z$ and $|m| < n$,

$$\int_{-\pi/c}^{\pi/c} e^{imcx}\, p_n(x)dx = (1 - \frac{|m|}{n})g(mc). \tag{1}$$

(c) There exists $F \in \mathfrak{F}$ with $F(-\pi/c) = 0$ and $F(\pi/c) = 1$ such that $\int_{-\pi/c}^{\pi/c} e^{imcx}\, dF(x) = g(mc)$, $m \in Z$. [Hint. Let $F_n \in \mathfrak{F}$ have density p_n. Use (19.58) to find a subsequence $\{F_{n_k} : k \in N\}$ and $F \in \mathfrak{F}$ so that $F_{n_k} \xrightarrow{w} F$, then let $n \to \infty$ in (1) along the subsequence $\{n_k : k \in N\}$.]

Exercise 20.51. For $a > 0$, let $p(x) = (1 - \cos ax)/\pi a x^2$, $x \in R$. Prove the following.

(a) p is a probability density.

(b) If F has density p, then $\varphi_F(t) = \begin{cases} 1 - |t|/a & \text{if } |t| \leqslant a \\ 0 & \text{if } |t| > a \end{cases}$.

Exercise 20.52. Let $\varphi(t) = \begin{cases} 1 - |t| & \text{if } |t| \leqslant 1 \\ 0 & \text{if } |t| > 1 \end{cases}$.

(a) For $m \geqslant 2$, let $0 < a_1 < \cdots < a_m < \infty$, let $p_1, \ldots, p_m \in]0, 1[$ be such that $p_1 + \cdots + p_m = 1$, and define $\psi(t) = p_1\varphi(t/a_1) + \cdots + p_m\varphi(t/a_m)$, $t \in R$, and $\phi(t) = p_1\varphi(t/a_1) + \cdots + p_{m-1}\varphi(t/a_{m-1}) + p_m$, $t \in R$. Show that ψ and ϕ are even characteristic functions whose restrictions to $]0, \infty[$ have convex polygonal descending graphs. [Use (20.2.d) and (20.35).]

(b) Let $\{a_n : n \in N\} \subset]0, \infty[$ be a strictly increasing sequence with $\lim_n a_n = \infty$, let $\{p_n : n \in N\} \subset]0, 1[$ be such that $\sum_{n\geqslant 1} p_n = 1$, and define $\psi(t) = \sum_{n\geqslant 1} p_n\varphi(t/a_n)$, $t \in R$. Prove that ψ is an even characteristic function, its restriction to $]0, \infty[$ being convex, piecewise linear, and strictly decreasing with $\lim_{t\to\infty}\psi(t) = 0$.

Exercise 20.53. Let $g : R \to [0, 1]$ be a function whose graph has the form of that of ψ or ϕ in (20.52). Show that $g \in \mathfrak{C}$.

Exercise 20.54 (Pólya). Let $g : R \to [0, 1]$ be an even function whose restriction to $]0, \infty[$ is convex and nonincreasing with $g(0+) = g(0) = 1$. Prove that $g \in \mathfrak{C}$. [Hint. For $n \in N$, let $g_n \in \mathfrak{C}$ have the form of g in (20.53), and assume that $g_n \to g$ (13.32). Then apply (20.11).]

Exercise 20.55. Let $\varphi(t) = e^{-a|t|^\alpha}$, $t \in R$, where $a > 0$ and $0 < \alpha \leqslant 1$. Show that $\varphi \in \mathfrak{C}$.

Exercise 20.56. Let $f_1, \ldots, f_n$ be independent random variables, each having a characteristic function φ as in (20.55), and put $S_n = f_1 + \cdots + f_n$. Prove that the distribution of $S_n/n^{1/\alpha}$ coincides with the distribution of f_1.

Exercise 20.57. If F has density p as in (19.35.c), then $\varphi_F(t_1, t_2) = \exp\left(-|a|\sqrt{t_1^2 + t_2^2}\right)$, $(t_1, t_2) \in R^2$.

Exercise 20.58. Compute φ_F, where F has density as in (19.35.d).

Exercise 20.59. Let $\mathbf{f}_1, \ldots, \mathbf{f}_n$ be independent random vectors, each having density p as in (19.35.d), and set $\mathbf{S}_n = \mathbf{f}_1 + \cdots + \mathbf{f}_n$. Show that the distribution of $\mathbf{S}_n/n$ coincides with that of $\mathbf{f}_1$.

Exercise 20.60. Assume that $\varphi_F \in \mathfrak{C}$ is an even function and that the graph of φ_F on $[0, \infty[$ is the polygonal line joining in order the points $(j, 1/(j+1))$, $j = 0, 1, \ldots$ Prove that F has density

$$p(x) = \frac{2}{\pi} \sum_{j \geqslant 1} \frac{1}{j(j+1)(j+2)} \frac{1 - \cos jx}{x^2}, \quad x \in R.$$

Exercise 20.61. Let $\{F_{a_n} : n \in N\}$ be a sequence of degenerate distribution functions, and let $F \in \mathfrak{F}$ be such that $F_{a_n} \xrightarrow{w} F$. Prove the following.

(a) F is degenerate at some $a \in R$. [Use (20.3.a) and (20.25).]

(b) $a_n \to a$. [Hints. If $\lim_k a_{n_k} = a' \in R$, then $F_a = F_{a'}$ by (20.12.a) and (19.18.a). $\lim_k a_{n_k} = -\infty$ would imply $0 = F_a(x) = \lim_k F_{a_{n_k}}(x) = 1, x < a$. $\lim_k a_{n_k} = \infty$ would imply $0 = \lim_k F_{a_{n_k}}(x) = F_a(x) = 1, x \geqslant a$.]

Exercise 20.62. Let $\{G_{m_n,\sigma_n} : n \in N\}$ be a sequence of Gaussian distribution functions, and let $G \in \mathfrak{F}$ be such that $G_{m_n,\sigma_n} \xrightarrow{w} G$. Show that there exist $m \in R$ and $\sigma \in [0, \infty[$ such that $m_n \to m$ and $\sigma_n \to \sigma$, and that $G = \begin{cases} G_{m,\sigma} & \text{if } \sigma > 0 \\ F_m & \text{if } \sigma = 0 \end{cases}$. The following steps may be useful.

(a) The sequence $\{\sigma_n : n \in N\}$ is bounded. [Hint. If $\lim_k \sigma_{n_k} = \infty$, then $|\varphi_G(t)| = \lim_k \left|\varphi_{G_{m_{n_k},\sigma_{n_k}}}(t)\right| = \lim_k e^{-\sigma_{n_k}^2 t^2/2} = \begin{cases} 1 \text{ if } t = 0 \\ 0 \text{ if } t > 0 \end{cases}$.]

(b) The sequence $\{m_n : n \in N\}$ is bounded. [Hint. If $\lim_k m_{n_k} = m \in \overline{R}$ and $\lim_k \sigma_{n_k} = \sigma$, then $\lim_k \varphi_{F_{m_{n_k}}}(t) = \lim_k \varphi_{G_{m_{n_k}}, \sigma_{n_k}}(t) e^{\sigma_{n_k}^2 t^2/2} = \varphi_G(t) e^{\sigma^2 t^2/2} = \varphi_F(t), t \in R$, for some $F \in \mathfrak{F}$ by Lévy's theorem (20.11), and so $m \in R$ by (20.61).]

(c) $\{\sigma_n : n \in N\}$ is convergent. [Hint. If σ and σ' are limit points of $\{\sigma_n : n \in N\}$, then there are $m, m' \in R$ such that $e^{imt-\sigma^2 t^2/2} = e^{im't-\sigma'^2 t^2/2} = \varphi_G(t), t \in R$, and so $m = m'$ and $\sigma = \sigma'$ by (19.48).]

(d) $\{m_n : n \in N\}$ is convergent. [Hint. If m and m' are limit points of $\{m_n : n \in N\}$, then there is $\sigma \geqslant 0$ such that $e^{imt-\sigma^2 t^2/2} = e^{im't-\sigma^2 t^2/2} = \varphi_G(t), t \in R$, and so $m = m'$ by (19.48).]

Exercise 20.63. For $n \in N$, let F_n be a Poisson distribution function with parameters $a_n \in R, b \neq 0$, and $\lambda_n > 0$, and let $F \in \mathfrak{F}$ be such that $F_n \xrightarrow{w} F$. Prove that there are $a \in R$ and $\lambda \in [0, \infty[$ so that $a_n \to a$ and $\lambda_n \to \lambda$, and that either F is Poisson with parameters a, b, and λ or F is degenerate at a according as $\lambda > 0$ or $\lambda = 0$. The following steps may be helpful.

(a) The sequence $\{\lambda_n : n \in N\}$ is bounded. [Hint. If $\lim_k \lambda_{n_k} = \infty$, then $|\varphi_F(t)| = \lim_k \left|\varphi_{F_{n_k}}(t)\right| = \lim_k e^{-\lambda_{n_k}(1-\cos bt)} = \begin{cases} 1 \text{ if } t \in \{2k\pi/b : k \in Z\} \\ 0 \text{ if } t \notin \{2k\pi/b : k \in Z\} \end{cases}$ by(20.3.d).]

(b) The sequence $\{a_n : n \in N\}$ is bounded. [Hint. If $\lim_k a_{n_k} = a \in \overline{R}$ and $\lim_k \lambda_{n_k} = \lambda$, then $\lim_k \varphi_{F_{a_{n_k}}}(t) = \lim_k \varphi_{F_{n_k}}(t) \exp(-\lambda_{n_k}(e^{itb} - 1)) = \varphi_F(t) \times \exp(-\lambda(e^{itb} - 1)) = \varphi_G(t), t \in R$, for some $G \in \mathfrak{F}$ by Lévy's theorem (20.11), where $F_{a_{n_k}}$ is the distribution function degenerated at a_{n_k}, and so $a \in R$ by (20.61).]

(c) $\{\lambda_n : n \in N\}$ is convergent. [Hint. If λ and λ' are limit points of $\{\lambda_n : n \in N\}$, then there are $a, a' \in R$ such that $\exp(ita + \lambda(e^{itb} - 1)) = \exp(ita' + \lambda'(e^{itb} - 1)) = \varphi_F(t), t \in R$, and so $\lambda = \lambda'$ by (19.41).]

(d) $\{a_n : n \in N\}$ is convergent. [Hint. If a and a' are limit points of $\{a_n : n \in N\}$, then there is $\lambda \geqslant 0$ such that $\exp(ita + \lambda(e^{itb} - 1)) = \exp(ita' + \lambda(e^{itb} - 1)) = \varphi_F(t), t \in R$, and so $a = a'$ by (19.41).]

Exercise 20.64. Let $f_0, f_1 \ldots$ be independent random variables, each having an exponential distribution function with parameter λ. Assume that ν is a geometric random variable with parameter p, independent of the σ -algebra generated by $\{f_0, f_1 \ldots\}$, and set $g = f_0 + \cdots + f_\nu$. Show that g has an exponential distribution function with parameter λp. [Use (20.3.g).]

Exercise 20.65. If f is a random variable and $T > 0$, then

$$P\left(|f| \geqslant \frac{2}{T}\right) \leqslant 2\left(1 - \frac{1}{2T}\left|\int_{-T}^{T} \varphi_f(t)dt\right|\right).$$

[Hint. Utilizing (1) of (20.11), write

$$\frac{1}{2T}\left|\int_{-T}^{T} \varphi_f(t)dt\right| \leqslant \int_R \left|\frac{\sin Tx}{Tx}\right| dF(x)$$

$$= \int_{\{|x| \geqslant 2/T\}} \left| \frac{\sin Tx}{Tx} \right| dF(x) + \int_{\{|x|<2/T\}} \left| \frac{\sin Tx}{Tx} \right| dF(x),$$

where F is the distribution function of f.]

Exercise 20.66. For $\varphi \in \mathfrak{C}$, use (20.25) to show that the set $\{t \in R : |\varphi(t)| = 1\}$ is either empty or dense in R.

Reference

Natanson, I. P. (1955). *Theory of functions of a real variable.* New York: Frederick Ungar Publishing Co.

8 Probabilities on Metric Spaces

This chapter deals with the theory of probabilities on abstract metric spaces. Throughout it, unless otherwise stated, (X, d) is a metric space, $\mathcal{T} = \mathcal{T}_d$ is the family of all open subsets of X, $\mathcal{F}$ is the family of all closed subsets of X, $\mathcal{C}$ is the family of all compact subsets of X, and probabilities will be defined on $\mathcal{B}(X)$. In Section 21 we mainly study properties of individual probabilities, and in Section 22 we topologize the space of all probabilities on $\mathcal{B}(X)$ and investigate its properties. Convergence of sequences of probabilities with respect to this topology reduces to weak convergence of sequences of distribution functions when $X = R^n$.

21 Probabilities in a Metric Space

The connection between probability theory and topology is illustrated in the first part of this section by several results, culminating in Lusin's theorem, on the approximation of measurable functions by continuous functions. Then we examine the relationship between nonnegative linear functionals and probabilities, obtaining a version of the famous Riesz's representation theorem.

Throughout this section, if not otherwise stated, P is a probability on $\mathcal{B}(X)$, and $\mathfrak{L}^p$ stands for $\mathfrak{L}^p(X, \mathcal{B}(X), P)$. For $A \in \mathcal{B}(X)$, it follows from (16.21) and (16.34) that there are sequences $\{F_n : n \in N\} \subset \mathcal{F}$ and $\{U_n : n \in N\} \subset \mathcal{T}$ so that $1_{F_n} \xrightarrow{\mathfrak{L}^p} 1_A$ and $1_{U_n} \xrightarrow{\mathfrak{L}^p} 1_A$ for $0 < p < \infty$. Moreover, in view of (16.23), if X is complete and separable, then there exists a sequence $\{C_n : n \in N\} \subset \mathcal{C}$ so that $1_{C_n} \xrightarrow{\mathfrak{L}^p} 1_A$. On the other hand, if $F \in \mathcal{F}$, there exists a sequence $\{f_n : n \in N\} \subset \mathfrak{C}_r(X)$ such that $f_n \to 1_F$ (6.76). We deal now with approximation of measurable functions by continuous functions.

Theorem 21.1. *For $0 < p < \infty$, $f \in \mathfrak{L}^p$ and $\varepsilon > 0$, there exists a function $g \in \mathfrak{C}_r(X)$ such that $\|f - g\|_p < \varepsilon$.*

Proof. On account of (13.14), there exists a function $h_1 = \sum_{i=1}^n a_i 1_{A_i}$ such that $\|f - h_1\|_p < \varepsilon/3$. Then, applying (16.21), and making use of (13.7) and (13.23.a), choose $\{F_1, \ldots, F_n\} \subset \mathcal{F}$ such that $\|h_1 - h_2\|_p < \varepsilon/3$, where $h_2 = \sum_{i=1}^n a_i 1_{F_i}$. Further, apply (6.76) and dominated convergence to obtain functions $f_1, \ldots, f_n \in \mathfrak{C}_r(X)$ so that $\left\|h_2 - \sum_{i=1}^n a_i f_i\right\|_p < \varepsilon/3$. Finally, put $g = \sum_{i=1}^n a_i f_i$. □

Theorem 21.2. *If $f : X \to R$ is measurable and $\varepsilon > 0$, then there is a function $g \in \mathfrak{C}_r(X)$ such that $P(f \neq g) < \varepsilon$.*

Analysis and Probability. http://dx.doi.org/10.1016/B978-0-12-401665-1.00008-4

Proof. (I) Assume first that $0 \leqslant f < 1$, and set $f_n = \sum_{k=1}^{n2^n}(k-1) \times 2^{-n} 1_{\{(k-1)2^{-n} \leqslant f < k2^{-n}\}}$, $n \geqslant 1$. Then, according to (8.15), $0 \leqslant f_n \leqslant f$ and $f_n \to f$. Putting $h_0 = 0$ and $h_n = f_n - f_{n-1}$, $n \geqslant 1$, it follows that $h_n(X) \subset \{0, 2^{-n}\}$ and $f = \sum_{n \geqslant 1} h_n$. For $n \geqslant 1$, apply (16.21) and (16.34) to obtain $C_n \in \mathcal{F}$ and $U_n \in \mathcal{T}$ so that $C_n \subset \{h_n = 0\} \subset U_n$ and $P(U_n - C_n) < 2^{-n}\varepsilon$. Apply next (6.37) to obtain a continuous function $g_n : X \to [0, 1]$ such that $g_n(x) = 0$ for $x \in C_n$ and $g_n(x) = 1$ for $x \in U_n^c$. Let $g = \sum_{n \geqslant 1} 2^{-n} g_n \leqslant 1$. Since $\sum_{k=1}^{n} 2^{-k} g_k \stackrel{u}{\to} g$, (6.42) shows that g is continuous on X. For any $n \geqslant 1$, if $x \in C_n$, then $h_n(x) = 2^{-n} g_n(x) = 0$, while if $x \in U_n^c$, then $h_n(x) = 2^{-n} g_n(x) = 2^{-n}$. This entails that $\{f \neq g\} \subset \cup_{n \geqslant 1}(U_n - C_n)$, and so $P(f \neq g) \leqslant \sum_{n \geqslant 1} P(U_n - C_n) < \sum_{n \geqslant 1} 2^{-n}\varepsilon = \varepsilon$.

(II) Assume now that $0 \leqslant f < M < \infty$. Then, in view of (I), there is $g \in \mathfrak{C}_r(X)$ such that $P(f \neq Mg) < \varepsilon$.

(III) Assume next that $0 \leqslant f$ and $\|f\|_u = \infty$. Then choose M so large that $P(f \geqslant M) < \varepsilon/2$. Apply (II) to obtain a function $g \in \mathfrak{C}_r(X)$ such that $P(f 1_{\{f<M\}} \neq g) < \varepsilon/2$. Since $\{f \neq g\} \subset \{f 1_{\{f<M\}} \neq g\} \cup \{f \geqslant M\}$, it follows that $P(f \neq g) < \varepsilon$.

(IV) Finally, consider the general case. Then, according to (III), there are $g_i \in \mathfrak{C}_r(X)$, $i = 1, 2$, such that $P(f^+ \neq g_1) < \varepsilon/2$ and $P(f^- \neq g_2) < \varepsilon/2$. As $\{f \neq g_1 + g_2\} \subset \{f^+ \neq g_1\} \cup \{f^- \neq g_2\}$, it follows that $P(f \neq g) < \varepsilon$. □

Remark 21.3. The function g in (21.2) can be chosen so that $\|g\|_u \leqslant \|f\|_u$. Actually, if $\|f\|_u < \infty$ and g is as in (21.2), then $g_1 = g 1_{\{|g| \leqslant \|f\|_u\}} + \|f\|_u (g/|g|) 1_{\{|g| > \|f\|_u\}} \in \mathfrak{C}_r(X)$ by (6.92). Since $\{f = g\} \subset \{f = g_1\}$, we have $P(f \neq g_1) \leqslant P(f \neq g) < \varepsilon$.

Theorem 21.4. *If $f : X \to R$ is measurable, then there is a sequence $\{f_n : n \in N\} \subset \mathfrak{C}_r(X)$ such that $f_n \stackrel{a.s.}{\to} f$ and $|f_n| \leqslant \|f\|_u$, $n \in N$.*

Proof. For $n \in N$, use (21.2) and (21.3) to select $f_n \in \mathfrak{C}_r(X)$ such that $P(f \neq f_n) < 2^{-n}$ and $|f_n| \leqslant \|f\|_u$. Put $A = \limsup_n \{f \neq f_n\}$. Then $P(A) = 0$ by the Borel-Cantelli lemma (18.24). If $x \notin A$, then there is $n_x \in N$ so that $f(x) = f_n(x)$ for all $n \geqslant n_x$. Thus $f_n \stackrel{a.s.}{\to} f$. □

Theorem 21.5 (Lusin). *If $f : X \to R$ is measurable and $\varepsilon > 0$, there exist $F \in \mathcal{F}$ and $g \in \mathfrak{C}_r(X)$ such that $P(F^c) < \varepsilon$ and $f = g$ on F, i.e. the restriction of f to F is continuous with respect to the relative topology $F \cap \mathcal{T}$ on F.*

Proof. By (21.2), there exists $g \in \mathfrak{C}_r(X)$ such that $P(f \neq g) < \varepsilon/2$. By (16.21), there is $F \in \mathcal{F}$ so that $F \subset \{f \neq g\}$ and $P(F) > P(f = g) - \varepsilon/2$. Hence $P(X - F) = P(f \neq g) + P(\{f \neq g\} - F) < \varepsilon$. □

Remark 21.6. If the metric space X in (21.5) is complete and separable, then F may be taken as compact by (16.23).

We now take up a version of one of the most famous and most important results of modern analysis, namely Riesz's representation theorem. This theorem establishes the relationship between linear functionals and measures. Throughout the rest of this section, up to exercises, X is assumed to be compact.

Definition 21.7. A function $L : \mathfrak{C}_r(X) \to R$ is called a *nonnegative linear functional* if, for all $f, g \in \mathfrak{C}_r(X)$ and $\alpha \in R$:

(i) $L(f+g) = L(f) + L(g)$;

(ii) $L(\alpha f) = \alpha L(f)$;

(iii) $L(f) \geqslant 0$ whenever $f \geqslant 0$.

Notice that $L(f) \leqslant L(g)$ whenever $f \leqslant g$, and so $|L(f)| \leqslant L(|f|)$.

Examples 21.8. (a) For a fixed point $a \in X$, the function L_a defined by $L_a(f) = f(a)$, $f \in \mathfrak{C}_r(X)$, is a nonnegative linear functional called the *evaluation functional.*

(b) Given any finite measure μ on $\mathcal{B}(X)$, the function L_μ defined by $L_\mu(f) = \int_X f\, d\mu$, $f \in \mathfrak{C}_r(X)$, is a nonnegative linear functional, in view of (11.12). If μ is a probability, then $L_\mu(1) = 1$.

(c) Given any finite measure μ on $\mathcal{B}(X)$ and $0 \leqslant h \in \mathfrak{C}_r(X)$ the function $L_{\mu,h}$ defined by $L_{\mu,h}(f) = \int_X fh\, d\mu$, $f \in \mathfrak{C}_r(X)$, is a nonnegative linear functional. $L_{\mu,1} = L_\mu$.

Hereafter, unless otherwise stated, L will denote a fixed nonnegative linear functional on $\mathfrak{C}_r(X)$ with $L(1) = 1$. Starting from it we shall construct a probability P on $\mathcal{B}(X)$ such that $L = L_P$. We need the next lemmas.

Lemma 21.9. *Let $\{f_n : n \in N\} \subset \mathfrak{C}_r(X)$ be a monotone sequence, and let $f \in \mathfrak{C}_r(X)$ be such that $f_n \to f$. Then $L(f_n) \to L(f)$.*

Proof. For $\varepsilon > 0$, according to the proof of Dini's theorem (6.44), there is $n_\varepsilon \in N$ so that $X = \{|f - f_{n_\varepsilon}| < \varepsilon\}$. Consequently, if $n \geqslant n_\varepsilon$, then $|Lf - Lf_n| \leqslant L(|f - f_n|) \leqslant L(|f - f_{n_\varepsilon}|) \leqslant \varepsilon$. □

Lemma 21.10. *Let $\{f_n : n \in N\} \subset \mathfrak{C}_r(X)$ and $\{f_n' : n \in N\} \subset \mathfrak{C}_r(X)$ be nondecreasing sequences of nonnegative functions such that* $\lim_n f_n \leqslant \lim_n f_n'$. *Then* $\lim_n L(f_n) \leqslant \lim_n L(f_n')$.

Proof. For $m \in N$, $\{f_m \wedge f_n' : n \in N\} \subset \mathfrak{C}_r(X)$ by (5.76.b), and $\lim_n(f_m \wedge f_n') = f_m$. Hence, in view of (21.9),

$$L(f_m) = \lim_n L(f_m \wedge f_n') \leqslant \lim_n L(f_n'). \tag{1}$$

Letting $m \to \infty$ in (1) yields the result. □

We now begin the construction of P from L.

Definition 21.11. For $U \in \mathcal{T}$, let $\{f_n : n \in N\} \subset \mathfrak{C}_r(X)$ be a nondecreasing sequence of nonnegative functions with $f_n \to 1_U$. Such a sequence exists by (6.76). Then define $Q(U) = \lim_n L(f_n)$. By virtue of (21.10), Q is well defined on $\mathcal{T}$.

Remark 21.12. If $h \in \mathfrak{C}_r(X)$ and $U, V \in \mathcal{T}$ are such that $1_U \leqslant h \leqslant 1_V$, then $Q(U) \leqslant L(h) \leqslant Q(V)$ by (21.10).

Theorem 21.13.

(i) $Q(\emptyset) = 0$, $Q(X) = 1$ *and* $0 \leqslant Q(U) \leqslant 1$ *for all* $U \in \mathcal{T}$.

(ii) *If* $U, V \in \mathcal{T}$ *and* $U \subset V$, *then* $Q(U) \leqslant Q(V)$.

(iii) *If* $U, V \in \mathcal{T}$, *then* $Q(U) + Q(V) = Q(U \cap V) + Q(U \cup V)$.

(iv) *If* $\{U_n : n \in N\} \subset \mathcal{T}$ *is a nondecreasing sequence, then* $Q(U_n) \to Q(\cup_{n\in N} U_n)$.

Proof. (i) is trivial. To prove (ii) and (iii), let $\{f_n : n \in N\} \subset \mathfrak{C}_r(X)$ and $\{g_n : n \in N\} \subset \mathfrak{C}_r(X)$ be nondecreasing sequences of nonnegative functions such that $f_n \to 1_U$ and $g_n \to 1_V$. If $U \subset V$, then $1_U \leqslant 1_V$, so (ii) follows from (21.10). As for (iii), notice that $f_n \wedge g_n \to 1_{U\cap V}$, $f_n \vee g_n \to 1_{U\cup V}$, and $f_n + g_n = (f_n \wedge g_n) + (f_n \vee g_n)$, $n \in N$. Hence

$$\begin{aligned} L(f_n) + L(g_n) &= L(f_n + g_n) = L((f_n \wedge g_n) + (f_n \vee g_n)) \\ &= L(f_n \wedge g_n) + L(f_n \vee g_n), \quad n \in N. \end{aligned} \tag{1}$$

Letting $n \to \infty$ in (1) yields (iii). To prove (iv), for each $m \in N$, let $\{f_{nm} : n \in N\} \subset \mathfrak{C}_r(X)$ be a nondecreasing sequence of nonnegative functions such that $\lim_n f_{nm} = 1_{U_m}$. Set $h_n = f_{n1} \vee \cdots \vee f_{nn}$, $n \in N$. Then $\{h_n : n \in N\}$ is a nondecreasing sequence of nonnegative functions in $\mathfrak{C}_r(X)$, and

$$f_{nm} \leqslant h_n \leqslant 1_{U_n}, \quad 1 \leqslant m \leqslant n. \tag{2}$$

In (2) letting first $n \to \infty$, then $m \to \infty$, we see that $h_n \to 1_{\cup_{n\in N} U_n}$. From (2) and (21.12) we have

$$L(f_{nm}) \leqslant L(h_n) \leqslant Q(U_n), \quad 1 \leqslant m \leqslant n. \tag{3}$$

Letting $n \to \infty$ in (3), we find $Q(U_m) \leqslant Q(\cup_{n\in N} U_n) \leqslant \lim_n Q(U_n)$, $m \in N$, whence (iv) follows on letting $m \to \infty$. □

We now extend Q to $\mathcal{P}(X)$.

Theorem 21.14. *For* $A \subset X$, *define* $P^*(A) = \inf\{Q(U) : A \subset U \in \mathcal{T}\}$. *Then the following assertions hold.*

(i) *If* $U \in \mathcal{T}$, *then* $P^*(U) = Q(U)$.

(ii) *If* $A_1 \subset A_2$, *then* $P^*(A_1) \leqslant P^*(A_2)$.

(iii) *If* $A_1, A_2 \subset X$, *then* $P^*(A_1) + P^*(A_2) \geqslant P^*(A_1 \cap A_2) + P^*(A_1 \cup A_2)$.

(iv) *If* $\{A_n : n \in N\} \subset \mathcal{P}(X)$ *is a nondecreasing sequence, then* $P^*(A_n) \to P^*(\cup_{n\in N} A_n)$.

Proof. The proof of this theorem is easily obtained, mutatis mutandis, from the proof of (10.4). □

Lemma 21.15. *If* $h \in \mathfrak{C}_r(X)$ *and* $A \subset X$ *are such that* $1_A \leqslant h$, *then* $P^*(A) \leqslant L(h)$.

Proof. Let $0 < \alpha < 1$. Then $A \subset \{h > \alpha\}$ and $1_{\{h>\alpha\}} \leqslant h/\alpha$. Hence, in view of (5.75.d) and (21.12), $P^*(A) \leqslant Q(\{h > \alpha\}) \leqslant L(h/\alpha) = L(h)/\alpha$. Letting $\alpha \to 1$ completes the proof. □

Theorem 21.16. *Let* $\mathcal{M} = \{A \subset X : P^*(A) + P^*(A^c) = 1\}$. *Then the following assertions hold.*

(i) $\mathcal{M} = \{A \subset X : P^*(A) + P^*(A^c) \leqslant 1\}$.

(ii) $\mathcal{T} \subset \mathcal{M}$.

(iii) $\mathcal{M}$ *is a* σ*-algebra, and* P^* *is a probability on* $\mathcal{M}$.

Proof. For $A \subset X$, using (21.14.iii), (21.14.i) and (21.13.i), we get $P^*(A) + P^*(A^c) \geqslant P^*(\emptyset) + P^*(X) = 1$, and so (i) obtains. To prove (ii), let $U \in \mathcal{T}$ and consider $\{f_n : n \in N\} \subset \mathfrak{C}_r(X)$ a nondecreasing sequence of nonnegative functions such that $f_n \to 1_U$. Then $L(f_n) \to Q(U)$ and $1_{U^c} \leqslant 1 - f_n, n \in N$. Therefore, in view of (21.14.i) and (21.15), we have

$$P^*(U) + P^*(U^c) \leqslant Q(U) + L(1 - f_n) = Q(U) + 1 - L(f_n), \quad n \in N. \tag{1}$$

On letting $n \to \infty$ in (1), we find $P^*(U) + P^*(U^c) \leqslant 1$. Thus $U \in \mathcal{M}$ by (i). The proof of (iii) is easily obtained, mutatis mutandis, from the proof of (10.5.iii). □

Since $\mathcal{T} \subset \mathcal{M}$, it follows that $\mathcal{B}(X) \subset \mathcal{M}$. Let P be the restriction of P^* to $\mathcal{B}(X)$. Then $P = Q$ on $\mathcal{T}$ by (21.14.i).

Riesz's representation theorem 21.17. *There exists a unique probability P on $\mathcal{B}(X)$ such that $L(f) = \int_X f\,dP$ for all $f \in \mathfrak{C}_r(X)$.*

Proof. Let P be as above, and let $f \in \mathfrak{C}_r(X)$.

(I) Assume first that $0 \leqslant f \leqslant 1$, and put $U_i = \{f > i/n\}$ for $i = 0, 1, \ldots, n$. Then $U_0 \supset U_1 \supset \cdots \supset U_n = \emptyset$. For $1 \leqslant i \leqslant n$, consider the continuous function $\varphi_i : [0, 1] \to [0, 1]$ which is equal to 0 on $[0, (i-1)/n]$, to 1 on $[i/n, 1]$, and is linear in $](i-1)/n, i/n[$. Since $(\varphi_1(t) + \cdots + \varphi_n(t))/n = t, t \in [0, 1]$, it follows that $(\varphi_1 \circ f + \cdots + \varphi_n \circ f)/n = f$. Therefore, as $\varphi_i \circ f \geqslant 1_{U_i}, 1 \leqslant i \leqslant n$, using (21.12), we have

$$\begin{aligned}
L(f) &= \frac{1}{n}\sum_{i=1}^{n} L(\varphi_i \circ f) \geqslant \frac{1}{n}\sum_{i=1}^{n} Q(U_i) = \sum_{i=1}^{n}\left(\frac{i}{n} - \frac{i-1}{n}\right) P(U_i) \\
&= \sum_{i=1}^{n-1} \frac{i}{n} P(U_i - U_{i+1}) \\
&= \sum_{i=1}^{n-1} \frac{i+1}{n} P(U_i - U_{i+1}) - \frac{P(U_1)}{n} \geqslant \sum_{i=1}^{n-1} \int_{U_i - U_{i+1}} f\,dP - \frac{P(U_1)}{n} \\
&= \int_{U_1} f\,dP - \frac{P(U_1)}{n} \geqslant \int_X f\,dP - \frac{P(U_0)}{n}.
\end{aligned} \tag{1}$$

Letting $n \to \infty$ in (1) yields

$$L(f) \geqslant \int_X f\,dP. \tag{2}$$

Changing f to $1 - f$ in (2), we have $L(1 - f) \geqslant \int_X (1 - f)dP$ or

$$L(f) \leqslant \int_X f\,dP. \tag{3}$$

From (2) and (3) we get $L(f) = \int_X f\,dP$.

(II) Assume next that $0 \leqslant f \leqslant M < \infty$. Then, in view of (I),

$$L(f) = ML\left(\frac{f}{M}\right) = M\int\limits_X \frac{f}{M} dP = \int\limits_X f\,dP.$$

(III) Finally, if f is arbitrary, then $f^+, f^- \in \mathfrak{C}_r(X)$ and

$$L(f) = L(f^+ - f^-) = L(f^+) - L(f^-) = \int\limits_X f^+\,dP - \int\limits_X f^-\,dP = \int\limits_X f\,dP.$$

Suppose now that P' is another probability on $\mathcal{B}(X)$ such that $L(f) = \int_X f\,dP'$ for any $f \in \mathfrak{C}_r(X)$. Then $\int_X f\,dP = \int_X f\,dP'$, $f \in \mathfrak{C}_r(X)$, and so $P = P'$ by (11.50). □

Remarks 21.18. (a) ((21.8.b) and (21.17) show that there is a one-to-one correspondence between the set of all nonnegative linear functionals L on $\mathfrak{C}_r(X)$ with $L(1) = 1$ and the set of all probabilities on $\mathcal{B}(X)$.

(b) If L is a nonnegative linear functional on $\mathfrak{C}_r(X)$ such that $L(1) = 0$, then $L(f) = 0$ for any $f \in \mathfrak{C}_r(X)$. Actually, $L(1)$ implies that $L(\alpha) = 0$ for any $\alpha \in R$, and so $0 \leqslant L(f) \leqslant L(\|f\|_u) = 0$ whenever $0 \leqslant f \in \mathfrak{C}_r(X)$. Consequently, $L(f) = L(f^+) - L(f^-) = 0$, $f \in \mathfrak{C}_r(X)$.

(c) If L is a nonnegative linear functional on $\mathfrak{C}_r(X)$ with $L(1) > 0$, then (21.17) ensures that there exists a unique measure μ on $\mathcal{B}(X)$ with $0 < \mu(X) < \infty$ such that $L(f) = \int_X f\,d\mu$ for any $f \in \mathfrak{C}_r(X)$. This, in conjunction with (b), shows that there is a one-to-one correspondence between the set of all nonnegative linear functionals L on $\mathfrak{C}_r(X)$ and the set of all finite measures on $\mathcal{B}(X)$.

Exercise 21.19. Let X be a complete separable metric space, let $f : X \to R$ be a measurable function, and let $B \subset R$ be such that $f^{-1}(B) \in \mathcal{B}(X)$. Prove that there exists $B_1 \in \mathcal{B}(R)$ with $B_1 \subset B$ and $(Pf^{-1})(B - B_1) = 0$. The following steps may be helpful.

(a) Let $A = f^{-1}(B)$. Use (16.21) to choose a nondecreasing sequence $\{F_n : n \in N\}$ of closed subsets of A such that $P(A - F_n) \to 0$.

(b) Apply (16.22), Lusin's theorem (21.5) and (21.6) to find a nondecreasing sequence $\{C_n : n \in N\} \subset \mathcal{C}$ such that $P(C_n^c) \to 0$ and $f|_{C_n}$ is continuous on C_n.

(c) Write $D_n = F_n \cap C_n, n \in N$, and $B_1 = \cup_{n \in N} f(D_n)$. Show that $B_1 \in \mathcal{B}(R)$. [Use (b) and (5.45).]

(d) $B_1 \subset B$. [$f(F_n) \subset B, n \in N$.]

(e) $(Pf^{-1})(B - B_1) = 0$. [Hint. $f^{-1}(B_1) = f^{-1}(f(\cup_{n \in N} D_n)) \supset \cup_{n \in N} D_n$ by (2.33.b), and so $P(f^{-1}(B) - f^{-1}(B_1)) \leqslant P(A - (\cup_{n \in N} D_n)) = \lim_n P(A - D_n)$.]

Exercise 21.20. Let X, f and B be as in (21.19). Show that there are $B_1, B_2 \in \mathcal{B}(R)$ such that $B_1 \subset B \subset B_2$ and $(Pf^{-1})(B_2 - B_1) = 0$.

Exercise 21.21. Let L_a be as in (21.8.a). Find the probability P corresponding to L_a by Riesz's representation theorem (21.17).

Exercise 21.22. Let $\mathfrak{C}_0(R)$ be the set of all continuous functions $f : R \to R$ such that $f(C_f^c) = \{0\}$ for some compact set $C_f \subset R$, and let $\{a_k : k \in Z\} \subset R$. For $f \in \mathfrak{C}_0(R)$, define $L(f) = \sum_{k \in Z} a_k f(k)$. Prove the following.

(a) $\mathfrak{C}_0(R) \subset \mathfrak{C}_r(R)$.

(b) For all $f, g \in \mathfrak{C}_0(R)$ and $\alpha \in R$, $L(f+g) = L(f)+L(g)$ and $L(\alpha f) = \alpha L(f)$.

(c) If $\{f_n : n \in N\} \subset \mathfrak{C}_0(R)$ is monotone, $f \in \mathfrak{C}_0(R)$ and $f_n \to f$, then $L(f_n) \to L(f)$. [Hint. There exists $[a, b] \subset R$ such that $f_n(x) = f(x)$ for all $x \in [a, b]^c$ and $n \in N$. Then apply Dini's theorem (6.44).]

Exercise 21.23. Let $L : \mathfrak{C}_r(X) \to R$ be a function satisfying conditions (i) and (ii) in (21.7). If L is continuous at some $g_0 \in \mathfrak{C}_r(X)$, then there exists $M \in R$ such that $|L(f)| \leqslant M \|f\|_u$ for all $f \in \mathfrak{C}_r(X)$, and so L is uniformly continuous. [Hint. Let $\delta > 0$ be so that $|L(g) - L(g_0)| \leqslant 1$ whenever $\|g - g_0\|_u \leqslant \delta$. Then, for $f \in \mathfrak{C}_r(X)$ with $\|f\|_u > 0$, write $f = (\|f\|_u / \delta)((\delta f / \|f\|_u + g_0) - g_0)$ and take $M = 1/\delta$.]

Exercise 21.24. Let μ be a signed measure on $\mathcal{B}(X)$ such that $|\mu|(X) < \infty$, and define $L(f) = \int_X f \, d\mu$ for $f \in \mathfrak{C}_r(X)$. Show that L satisfies (i) and (ii) in (21.7), and that L is uniformly continuous.

Exercise 21.25. Let L be as in (21.23), and assume that L is uniformly continuous. For $0 \leqslant f \in \mathfrak{C}_r(X)$, write $L^+(f) = \sup\{L(g) : g \in \mathfrak{C}_r(X), 0 \leqslant g \leqslant f\}$. For $f \in \mathfrak{C}_r(X)$, put $L^+(f) = L^+(f^+) - L^+(f^-)$. Prove that L^+ is a nonnegative linear functional. The following steps may be useful.

(a) $0 \leqslant L^+(f) \leqslant M \|f\|_u$ if $0 \leqslant f \in \mathfrak{C}_r(X)$, where M is as in (21.23).

(b) $L^+(f_1 + f_2) = L^+(f_1) + L^+(f_2)$ whenever $0 \leqslant f_1, f_2 \in \mathfrak{C}_r(X)$. [Hint. If $g \in \mathfrak{C}_r(X), 0 \leqslant g \leqslant f_1 + f_2$, set $g_1 = f_1 \wedge g$ and $g_2 = (g - f_1)^+$. Then $0 \leqslant g_1 \leqslant f_1, 0 \leqslant g_2 \leqslant f_2$, and $g = g_1 + g_2$, and so $L^+(g) \leqslant L^+(f_1) + L^+(f_2)$. Consequently, $L^+(f_1 + f_2) \leqslant L^+(f_1) + L^+(f_2)$.]

(c) $L^+(\alpha f) = \alpha L^+(f)$ if $\alpha \geqslant 0$ and $0 \leqslant f \in \mathfrak{C}_r(X)$.

(d) $L^+(f_1 + f_2) = L^+(f_1) + L^+(f_2)$ for $f_1, f_2 \in \mathfrak{C}_r(X)$. [Hint. Write $(f_1 + f_2)^+ + f_1^- + f_2^- = (f_1 + f_2)^- + f_1^+ + f_2^+$, and use (b).]

(e) $L^+(\alpha f) = \alpha L^+(f)$ for $\alpha \in R$ and $f \in \mathfrak{C}_r(X)$. [$(\alpha f)^+ = -\alpha f^-$ and $(\alpha f)^- = -\alpha f^+$ for $\alpha < 0$.]

Exercise 21.26. Let L and L^+ be as in (21.25). For $0 \leqslant f \in \mathfrak{C}_r(X)$, define $L^-(f) = -\inf\{L(g) : g \in \mathfrak{C}_r(X), 0 \leqslant g \leqslant f\}$. For $f \in \mathfrak{C}_r(X)$, let $L^-(f) = L^-(f^+) - L^-(f^-)$.

(a) Show that L^- is a nonnegative linear functional. [For $0 \leqslant f \in \mathfrak{C}_r(X)$, $L^-(f) = (-L)^+(f)$.]

(b) Prove that $L(f) = L^+(f) - L^-(f)$ for any $f \in \mathfrak{C}_r(X)$.

Exercise 21.27. Let L be as in (21.25). Use (21.26), (21.17) and (11.56) to show that there is a unique signed measure μ on $\mathcal{B}(X)$ with $|\mu|(X) < \infty$ such that $L(f) = \int_X f \, d\mu$ for any $f \in \mathfrak{C}_r(X)$.

22 Topology in the Space of Probabilities

Let P denote the space of all probabilities on $\mathcal{B}(X)$. We shall equip P with a topology $\mathcal{W}$ and study its properties. We shall first investigate convergence problems for sequences in P, and convergence in distribution for sequences of X-valued random elements. Then we proceed to analyze the question of metrization of the topological space $(P, \mathcal{W})$, and prove the important Prohorov's theorems describing the compact subsets of P.

For $P \in$ P, $f \in \mathfrak{C}_r(X)$ and $\varepsilon > 0$, put $W_P(f, \varepsilon) = \{Q \in \mathrm{P} : \left|\int_X f\,dP - \int_X f\,dQ\right| < \varepsilon\}$. For $f \in \mathfrak{C}_r(X)$, define $\psi_f : \mathrm{P} \to R$ by $\psi_f(P) = \int_X f\,dP$, $P \in$ P. Let $\mathcal{W}$ be the topology for P generated by the family $\{\psi_f : f \in \mathfrak{C}_r(X)\}$.

Theorem 22.1.

(i) $\tau(\{W_P(f, \varepsilon) : P \in \mathrm{P},\ f \in \mathfrak{C}_r(X), \varepsilon > 0\}) = \mathcal{W}$.

(ii) (P, $\mathcal{W}$) *is a Hausdorff space.*

(iii) *A sequence* $\{P_n : n \in N\} \subset$ P *converges to* $P \in$ P *if and only if* $\int_X f\,dP_n \to \int_X f\,dP$ *for every* $f \in \mathfrak{C}_r(X)$.

Proof. For $P \in$ P, $f \in \mathfrak{C}_r(X)$, and $\varepsilon > 0$, $W_P(f, \varepsilon) = \psi_f^{-1}(]\int_X f\,dP - \varepsilon, \int_X f\,dP + \varepsilon[)$. This proves (i). If $P_1, P_2 \in$ P and $P_1 \neq P_2$, then there is $f \in \mathfrak{C}_r(X)$ so that $\int_X f\,dP_1 \neq \int_X f\,dP_2$ (11.50). Then $P_1 \in W_{P_1}(f, \varepsilon)$, $P_2 \in W_{P_2}(f, \varepsilon)$ and $W_{P_1}(f, \varepsilon) \cap W_{P_2}(f, \varepsilon) = \emptyset$ whenever $2\varepsilon < \left|\int_X f\,dP_1 - \int_X f\,dP_2\right|$. Thus (ii) holds. To prove (iii), we shall apply (5.6), as $\{W_P(f, \varepsilon) : P \in \mathrm{P},\ f \in \mathfrak{C}_r(X), \varepsilon > 0\}$ is a subbase for $\mathcal{W}$. If $\{P_n : n \in N\}$ converges to P, then, for $f \in \mathfrak{C}_r(X)$ and $\varepsilon > 0$, $P_n \in W_P(f, \varepsilon)$ whenever $n \geqslant n_\varepsilon$, i.e. $\left|\int_X f\,dP - \int_X f\,dP_n\right| < \varepsilon, n \geqslant n_\varepsilon$, and so $\int_X f\,dP_n \to \int_X f\,dP$. Conversely, assume that $P \in W_Q(f, \varepsilon)$ for some $Q \in$ P, $f \in \mathfrak{C}_r(X)$ and $\varepsilon > 0$. Then $\left|\int_X f\,dQ - \int_X f\,dP\right| < \varepsilon$. Since $\int_X f\,dP_n \to \int_X f\,dP$, choose $n_\varepsilon \in N$ so that $\left|\int_X f\,dP - \int_X f\,dP_n\right| < \varepsilon - \left|\int_X f\,dQ - \int_X f\,dP\right|$ for $n \geqslant n_\varepsilon$. Hence $\left|\int_X f\,dQ - \int_X f\,dP_n\right| < \varepsilon, n \geqslant n_\varepsilon$, i.e. $P_n \in W_Q(f, \varepsilon), n \geqslant n_\varepsilon$. In view of (5.6), this means that $\{P_n : n \in N\}$ converges to P. □

Remark 22.2. Inspection of the above proof shows that (i) and (ii) continue to hold if $\mathfrak{C}_r(X)$ is replaced by a subset of it.

Definitions 22.3. The topology $\mathcal{W}$ is called the *weak topology* in P. We say that the sequence $\{P_n : n \in N\} \subset$ P *converges weakly to* $P \in$ P, and we write $P_n \Rightarrow P$, if $\{P_n : n \in N\}$ converges to P with respect to $\mathcal{W}$, i.e. $\int_X f\,dP_n \to \int_X f\,dP$, $f \in \mathfrak{C}_r(X)$.

Remarks 22.4. Let $\{P_n : n \in N\} \subset$ P and $P \in$ P.

(a) According to (5.54), $P_n \Rightarrow P$ if and only if each subsequence $\{P_{n_k} : k \in N\}$ contains a further subsequence $\{P_{n_{k_l}} : l \in N\}$ such that $P_{n_{k_l}} \Rightarrow P$.

(b) Let Y be a metric space, and let $h : X \to Y$ be a continuous function. The change of variable formula (11.29) implies that $P_n h^{-1} \Rightarrow P h^{-1}$ whenever $P_n \Rightarrow P$.

(c) If $P_n h^{-1} \Rightarrow P h^{-1}$ for each $h \in \mathfrak{C}_r(X)$, then $P_n \Rightarrow P$. Actually, $P_n h^{-1} \Rightarrow P h^{-1}$, $h \in \mathfrak{C}_r(X)$, implies that

$$\int_X (f \circ h)\,dP_n = \int_R f\,d(P_n h^{-1}) \to \int_R f\,d(P h^{-1}) = \int_X (f \circ h)\,dP, \quad f \in \mathfrak{C}_r(R). \quad (1)$$

Taking

$$f(t) = \begin{cases} -\|h\|_u & \text{if } t < -\|h\|_u, \\ t & \text{if } -\|h\|_u \leqslant t \leqslant \|h\|_u, \\ \|h\|_u & \text{if } t > \|h\|_u. \end{cases}$$

(1) yields $\int_X h\,dP_n \to \int_X h\,dP$, $h \in \mathfrak{C}_r(X)$.

(d) For $\{x_n : n \in N\} \subset X$ and $x \in X$, (5.37) and (6.94) show that $\varepsilon_{x_n} \Rightarrow \varepsilon_x$ if and only if $x_n \to x$. Hence X is homeomorphic to the set $\{\varepsilon_x : x \in X\} \subset \mathrm{P}$ endowed with the relative topology.

The next theorem yields useful equivalent conditions to weak convergence.

Theorem 22.5. *Let $\{P_n : n \in N\} \subset \mathrm{P}$ and $P \in \mathrm{P}$. Then the following statements are equivalent:*

(i) $P_n \Rightarrow P$;

(ii) $\int_X f\,dP_n \to \int_X f\,dP$ *for any uniformly continuous function* $f \in \mathfrak{C}_r(X)$;

(iii) $\limsup_n P_n(F) \leqslant P(F)$ *for any* $F \in \mathcal{F}$;

(iv) $\liminf_n P_n(U) \geqslant P(U)$ *for any* $U \in \mathcal{T}$;

(v) $\lim_n P_n(A) = P(A)$ *for any* $A \in \mathcal{B}(X)$ *with* $P(\partial(A)) = 0$.

Proof. Clearly, (i) implies (ii). To prove that (ii) implies (iii), let $F \in \mathcal{F}$, and let f_m, $m \in N$, be as in (6.76). Then $1_F \leqslant f_m \leqslant 1$, $m \in N$, and $f_m \to 1_F$. Therefore,

$$P_n(F) \leqslant \int_X f_m\,dP_n, \quad m, n \in N. \tag{1}$$

On account of (5.60.b) and (ii), (1) shows that

$$\limsup_n P_n(F) \leqslant \limsup_n \int_X f_m\,dP_n = \int_X f_m\,dP, \quad m \in N. \tag{2}$$

Letting $m \to \infty$ in (2), we obtain $\limsup_n P_n(F) \leqslant P(F)$ by dominated convergence. The equivalence of (iii) and (iv) follows easily by complementation. If (iii) and (iv) hold and $A \in \mathcal{B}(X)$, since $A^\circ \subset A \subset A^-$, (5.60.b) shows that

$$\begin{aligned} P(A^\circ) &\leqslant \liminf_n P_n(A^\circ) \leqslant \liminf_n P_n(A) \leqslant \limsup_n P_n(A) \\ &\leqslant \limsup_n P_n(A^-) \leqslant P(A^-). \end{aligned} \tag{3}$$

If $P(\partial(A)) = 0$, then $P(A^\circ) = P(A^-)$, as $\partial(A) = A^- - A^\circ$ (5.67.e), and so the extreme terms hence all the terms in (3) coincide. Thus $\lim_n P_n(A) = P(A)$. To complete the proof, assume that (v) holds, let $f \in \mathfrak{C}_r(R)$, and set $\mu = Pf^{-1}$. Suppose that the range of f is included in the bounded interval $]a, b[$. According to (9.37.d) and (9.39.a), $\mu(\{c\}) > 0$ only for a countable set of values of $c \in]a, b[$. Therefore, taking into account (5.69), given any $\varepsilon > 0$, we can select $a = a_0 < a_1 < \cdots < a_m = b$ a division of $[a, b]$ such that $\mu(\{a_j\}) = 0$, $1 \leqslant j < m$, and $a_j - a_{j-1} < \varepsilon$, $1 \leqslant j \leqslant m$.

Put $A_j = f^{-1}([a_{j-1}, a_j[)$, $1 \leqslant j \leqslant m$. Then $\{A_1, \ldots, A_m\}$ is a partition of X. Since f is continuous, we have $f^{-1}(]a_{j-1}, a_j[) \subset A_j^\circ \subset A_j^- \subset f^{-1}([a_{j-1}, a_j])$, and so $\partial(A_j) \subset f^{-1}([a_{j-1}, a_j]) - f^{-1}(]a_{j-1}, a_j[) = f^{-1}(\{a_{j-1}, a_j\})$ for $1 \leqslant j \leqslant m$. Hence $P(\partial(A_j)) \leqslant (Pf^{-1})(\{a_{j-1}, a_j\}) = \mu(\{a_{j-1}\}) + \mu(\{a_j\}) = 0$, $1 \leqslant j \leqslant m$. This entails that

$$\lim_n P_n(A_j) = P(A_j), \quad 1 \leqslant j \leqslant m. \tag{4}$$

Since $0 \leqslant f - \sum_{j=1}^m a_{j-1} 1_{A_j} < \varepsilon$, in view of (4), we can write

$$\begin{aligned}
\limsup_n \left| \int_X f\,dP_n - \int_X f\,dP \right| &\leqslant \limsup_n \left(\int_X f\,dP_n - \int_X \left(\sum_{j=1}^m a_{j-1} 1_{A_j} \right) dP_n \right) \\
&\quad + \limsup_n \left| \int_X \left(\sum_{j=1}^m a_{j-1} 1_{A_j} \right) dP_n - \int_X \left(\sum_{j=1}^m a_{j-1} 1_{A_j} \right) dP \right| \\
&\quad + \limsup_n \left(\int_X f\,dP - \int_X \left(\sum_{j=1}^m a_{j-1} 1_{A_j} \right) dP \right) \\
&\leqslant 2\varepsilon + \limsup_n \sum_{j=1}^m a_{j-1} \left| P_n(A_j) - P(A_j) \right| = 2\varepsilon.
\end{aligned} \tag{5}$$

As ε is arbitrary, (i) follows from (5). □

Example 22.6. Let F_n, $n \in N$, and F be distribution functions on R. Then $F_n \xrightarrow{w} F$ if and only if $\mu_{F_n} \Rightarrow \mu_F$. Actually, if $F_n \xrightarrow{w} F$, then $\mu_{F_n} \Rightarrow \mu_F$ by the Helly-Bray theorem (19.23). Conversely, assume that $\mu_{F_n} \Rightarrow \mu_F$ and that x is a continuity point of F. Then $0 = \mu_F(\{x\}) = \mu_F(\partial(A))$ by (10.36.b), where $A =]-\infty, x]$, and so $F_n(x) = \mu_{F_n}(A) \to \mu_F(A) = F(x)$.

The next general theorem is essential in several concrete situations.

Theorem 22.7. *Let (Y, ρ) be a metric space, let $h : X \to Y$ be a function such that $\mathcal{T} = h^{-1}(\mathcal{T}_\rho)$, and let $\mathcal{V}$ denote the topology for* P *generated by the family $\{\psi_{g \circ h} : g \in \mathfrak{C}_r(Y)\}$. Then $\mathcal{V} = \mathcal{W}$.*

Proof. If $g \in \mathfrak{C}_r(Y)$, then $g \circ h \in \mathfrak{C}_r(X)$, since h is continuous on X (5.40). Hence $\mathcal{V} \subset \mathcal{W}$. To prove the converse inclusion, let $f \in \mathfrak{C}_r(X)$. We shall get the result by showing that ψ_f is continuous on P with respect to the topology $\mathcal{V}$. As f is bounded, choose $c \in R$ such that $f^* = f + c \geqslant 0$. Then $f^* = \sup_{r \in Q_+} r 1_{\{f^* > r\}}$, where $Q_+ = Q \cap]0, \infty[$. The set $\{f^* > r\} = \{f > r - c\}$ belongs to $\mathcal{T}$, and so $\{f^* > r\} = h^{-1}(U_r)$ for some $U_r \in \mathcal{T}_\rho$. Consequently, $f^* = \sup_{r \in Q_+} r 1_{h^{-1}(U_r)} = (\sup_{r \in Q_+} r 1_{U_r}) \circ h$ (2.34). In view of (6.76), for each $r \in Q_+$, there exists a nondecreasing sequence

$\{f_{nr} : n \in N\} \subset \mathfrak{C}_r(Y)$ such that $\lim_n f_{nr} = 1_{U_r}$. Therefore,

$$f^* = \left(\sup_{r\in Q_+} r\left(\sup_{n\in N} f_{nr}\right)\right) \circ h = \left(\sup_{(r,n)\in Q_+\times N} r f_{nr}\right) \circ h. \tag{1}$$

Arranging the functions rf_{nr}, $(r, n) \in Q_+ \times N$, in a sequence $\{f_n : n \in N\}$, and putting $f_n^* = f_1 \vee \cdots \vee f_n$, $n \in N$, (1) becomes

$$f^* = \left(\sup_{n\in N} f_n^*\right) \circ h = \sup_{n\in N}(f_n^* \circ h) \tag{2}$$

with $\{f_n^* : n \in N\} \subset \mathfrak{C}_r(Y)$. Since $0 \leqslant f_n^* \leqslant f_{n+1}^*$, $n \in N$, (2) and the monotone convergence theorem (11.22) imply that

$$\begin{aligned}\psi_{f^*}(P) &= \int_X f^*\,dP = \int_X \lim_n (f_n^* \circ h)dP = \lim_n \int_X (f_n^* \circ h)dP \\ &= \sup_{n\in N} \psi_{f_n^*\circ h}(P), \quad P \in \mathrm{P}.\end{aligned} \tag{3}$$

For any $n \in N$, $\psi_{f_n^*\circ h}$ is continuous on P with respect to $\mathcal{V}$, and so ψ_{f^*} is lower semicontinuous on P by (3) and (5.95.g). Since $\psi_f = \psi_{f^*} - c$, it follows that ψ_f is lower semicontinuous as well. With $-f$ in place of f, this entails that $-\psi_f = \psi_{-f}$ is also semicontinuous. Consequently, ψ_f is continuous on P with respect to $\mathcal{V}$ (5.95.b). □

Corollary 22.8. *Let (Y, ρ) be a metric space, let $h : X \to Y$ be a homeomorphism from X into Y, and let $\{P_n : n \in N\} \subset \mathrm{P}$ and $P \in \mathrm{P}$. Then $P_n \Rightarrow P$ if and only if $P_n h^{-1} \Rightarrow P h^{-1}$.*

Proof. The fact that $P_n \Rightarrow P$ implies $P_n h^{-1} \Rightarrow Ph^{-1}$ has been remarked in (22.4.b).

Conversely, assume that $P_n h^{-1} \Rightarrow Ph^{-1}$. Then, by (11.29), $\int_X (g \circ h)dP_n = \int_Y g\,dP_n h^{-1} \to \int_Y g\,dP\,h^{-1} = \int_X (g \circ h)dP$ for any $g \in \mathfrak{C}_r(Y)$. In accordance with (22.2), this means that the sequence $\{P_n : n \in N\}$ converges to P with respect to the topology $\mathcal{V}$. Since $\mathcal{T} = h^{-1}(\mathcal{T}_\rho)$ (5.86), it follows that $\mathcal{V} = \mathcal{W}$ by (22.7), and so $P_n \Rightarrow P$. □

Other applications of (22.7) will appear in (22.10), (22.12), (22.22) and (22.27).

We shall now study weak convergence of sequences of probabilities on product spaces. Let $\{X_i : i \in I\}$ be a countable family of separable metric spaces, and put $X = \prod_{i\in I} X_i$. Then, on account of (5.71.b) and (6.11), X is a separable metric space. Moreover, in view of (6.12) and (7.28), $\mathcal{B}(X) = \otimes_{i\in I}\mathcal{B}(X_i)$. For $i \in I$, let P_i stand for the set of all probabilities on $\mathcal{B}(X_i)$. Let P be the set of all probabilities on $\mathcal{B}(X)$.

Theorem 22.9. *Notation is as above. For $i \in I$, assume that X_i is compact, and let $\{P_{ni} : n \in N\} \subset \mathrm{P}_i$ and $P_i \in \mathrm{P}_i$. Then $\otimes_{i\in I} P_{ni} \Rightarrow \otimes_{i\in I} P_i$ if and only if $P_{ni} \Rightarrow P_i$ for each $i \in I$.*

Proof. Set $P_n = \otimes_{i\in I} P_{ni}$, $n \in N$, and $P = \otimes_{i\in I} P_i$. Then P_n, $n \in N$, and P belong to P, as X_i, $i \in I$, are separable by (6.19) and (6.18). If $P_n \Rightarrow P$, then, for any

$i \in I$, $P_{ni} = P_n\pi_i^{-1} \Rightarrow P\pi_i^{-1} = P_i$ by (22.4.b), since the projection $\pi_i : X \to X_i$ is continuous.

Conversely, suppose that $P_{ni} \Rightarrow P_i$, $i \in I$. Let J be a nonempty finite subset of I, and let $f_j \in \mathfrak{C}_r(X_j)$, $j \in J$. Then $\prod_{j\in J} f_j \circ \pi_j \in \mathfrak{C}_r(X)$. Since $\prod_{j\in J} f_j \circ \pi_j = (\prod_{j\in J} f_j \circ \pi_j^J) \circ \pi_J$, we can write, on account of (11.29), (16.6) and Fubini's theorem (15.10),

$$\int\limits_X \left(\prod_{j\in J} f_j \circ \pi_j\right) dP_n = \int\limits_{\prod_{j\in J} X_j} \left(\prod_{j\in J} f_j \circ \pi_j^J\right) d\left(\bigotimes_{j\in J} P_{nj}\right)$$

$$= \prod_{j\in J} \int\limits_{X_j} f_j\, dP_{nj} \to \prod_{j\in J} \int\limits_{X_j} f_j\, dP_j = \int\limits_X \left(\prod_{j\in J} f_j \circ \pi_j\right) dP. \tag{1}$$

Now let $\mathfrak{A}$ consist of all finite sums of terms $\prod_{j\in J} f_j \circ \pi_j$ with nonempty finite $J \subset I$ and $f_j \in \mathfrak{C}_r(X_j)$, $j \in J$. Then

$$\int\limits_X g\, dP_n \to \int\limits_X g\, dP, \quad g \in \mathfrak{A}, \tag{2}$$

by (1). Obviously, $\mathfrak{A}$ is a subalgebra of $\mathfrak{C}_r(X)$ and $1 \in \mathfrak{A}$. Moreover, $\mathfrak{A}$ separates points of X. Actually, if $\mathbf{x} = (x_i)_{i\in I}$ and $\mathbf{y} = (y_i)_{i\in I}$ are distinct points in X, then $x_j \neq y_j$ for some $j \in I$. The function $d_{x_j} : X_j \to R$ given by $d_{x_j}(x) = d(x, x_j)$, where d is the metric for X_j, is continuous and bounded by (6.37.ii) and (5.47). Hence $d_{x_j} \circ \pi_j \in \mathfrak{A}$, and $(d_{x_j} \circ \pi_j)(\mathbf{x}) = d_{x_j}(x_j) = 0 \neq d_{x_j}(y_j) = (d_{x_j} \circ \pi_j)(\mathbf{y})$. In view of the Stone-Weierstrass theorem (6.57), it follows that $\mathfrak{A}^- = \mathfrak{C}_r(X)$, since X is compact by Tihonov's theorem (5.28). For $f \in \mathfrak{C}_r(X)$ and $\varepsilon > 0$, choose $g \in \mathfrak{A}$ so that $\|f - g\|_u < \varepsilon$. Next, using (2), choose $n_\varepsilon \in N$ such that $\left|\int_X g\, dP_n - \int_X g\, dP\right| < \varepsilon$ for $n \geqslant n_\varepsilon$. Then we have

$$\left|\int\limits_X f\, dP_n - \int\limits_X f\, dP\right| \leqslant \int\limits_X |f - g|\, dP_n + \left|\int\limits_X g\, dP_n - \int\limits_X g\, dP\right|$$
$$+ \int\limits_X |f - g|\, dP < 3\varepsilon, \quad n \geqslant n_\varepsilon.$$

This shows that $P_n \Rightarrow P$. □

Theorem 22.10. *Notation is as above* (22.9), *i.e.* X_i, $i \in I$ *are separable. For* $i \in I$, *let* $\{P_{ni} : n \in N\} \subset \mathrm{P}_i$ *and* $P_i \in \mathrm{P}_i$. *Then* $\otimes_{i\in I} P_{ni} \Rightarrow \otimes_{i\in I} P_i$ *if and only if* $P_{ni} \Rightarrow P_i$ *for any* $i \in I$.

Proof. If $\otimes_{i\in I} P_{ni} \Rightarrow \otimes_{i\in I} P_i$, then $P_{ni} \Rightarrow P_i$, $i \in I$, as in the proof of (22.9).

Conversely, assume that $P_{ni} \Rightarrow P_i$, $i \in I$. For $i \in I$, by Urysohn's embedding theorem (6.38), there exist a compact metric space Y_i and a homeomorphism h_i from X_i into Y_i. Then, in view of (22.8), $P_{ni}h_i^{-1} \Rightarrow P_i h_i^{-1}$ for each $i \in I$. Consequently,

$$\left(\bigotimes_{i\in I} P_{ni}\right) h^{-1} = \bigotimes_{i\in I} P_{ni}h_i^{-1} \Rightarrow \bigotimes_{i\in I} P_i h_i^{-1} = \left(\bigotimes_{i\in I} P_i\right) h^{-1} \tag{1}$$

by (16.36) and (22.9), where h is as in (5.96). Since h is a homeomorphism from $\prod_{i\in I} X_i$ into $\prod_{i\in I} Y_i$ (5.96), it follows that $\otimes_{i\in I} P_{ni} \Rightarrow \otimes_{i\in I} P_i$ by (1) and (22.8). □

Remark 22.11. The conclusions of (22.9) and (22.10) may be rephrased as follows. If $P_n, n \in N$, and P are product probabilities on $\mathcal{B}(\prod_{i\in I} X_i)$, then $P_n \Rightarrow P$ if and only if $P_n\pi_i^{-1} \Rightarrow P\pi_i^{-1}, i \in I$. If $P_n, n \in N$, and P are arbitrary probabilities on $\mathcal{B}(\prod_{i\in I} X_i)$, then $P_n\pi_i^{-1} \Rightarrow P\pi_i^{-1}, i \in I$, whenever $P_n \Rightarrow P$ by (22.4.b), but the converse statement does not generally hold. Actually, let $X_1 = X_2 = [0, 1]$, let $\mathbb{D}$ be the diagonal of $[0, 1]^2$, and let λ denote the Lebesgue measure on $\mathcal{B}([0, 1])$. Put $P = \lambda \otimes \lambda$, and define the probability Q on $\mathcal{B}([0, 1]^2)$ by $Q(A) = \lambda(\pi_1(A\cap\mathbb{D})) = \lambda(\pi_2(A\cap\mathbb{D}))$, $A \in \mathcal{B}([0, 1]^2)$. Set $P_n = Q, n \in N$. Then, for $i = 1, 2$, $P_n\pi_i^{-1} = P\pi_i^{-1} = \lambda, n \in N$, so that $P_n\pi_i^{-1} \Rightarrow P\pi_i^{-1}$ trivially, but $P_n \Rightarrow Q \neq P$. The following special case contrasts this remark.

Theorem 22.12. *Let X_1 and X_2 be separable metric spaces, let $P_n, n \in N$, be probabilities on $\mathcal{B}(X_1 \times X_2)$, and assume that $P_n\pi_1^{-1} \Rightarrow Q$ and $P_n\pi_2^{-1} \Rightarrow \varepsilon_x$ for some probability Q on $\mathcal{B}(X_1)$ and $x \in X_2$. Then $P_n \Rightarrow Q \otimes \varepsilon_x$.*

Proof. The proof resembles the proof of (22.9) and (22.10).

(I) Suppose first that X_1 and X_2 are compact. Let $f_i \in \mathfrak{C}_r(X_i), i = 1, 2$. Then $f_i \circ \pi_i \in \mathfrak{C}_r(X_1 \times X_2), i = 1, 2$. Since $P_n\pi_2^{-1} \Rightarrow \varepsilon_x$, we see that $\int_{X_2} |f_2 - f_2(x)|\, d(P_n\pi_2^{-1}) \to \int_{X_2} |f_2 - f_2(x)|\, d\varepsilon_x = 0$, and so, by (11.29),

$$\left|\int_{X_1\times X_2} (f_1 \circ \pi_1)(f_2 \circ \pi_2) dP_n - \int_{X_1\times X_2} (f_1 \circ \pi_1) f_2(x) dP_n\right| \leqslant$$
$$\int_{X_1\times X_2} |f_1 \circ \pi_1|\, |f_2 \circ \pi_2 - f_2(x)|\, dP_n \leqslant \|f_1\|_u \int_{X_2} |f_2 - f_2(x)|\, d(P_n\pi_2^{-1}) \to 0. \tag{1}$$

Since $P_n\pi_1^{-1} \Rightarrow Q$, we have

$$\int_{X_1\times X_2} (f_1 \circ \pi_1) f_2(x) dP_n = \int_{X_1\times X_2} f_1 f_2(x) d(P_n\pi_1^{-1}) \to \int_{X_1} f_1 f_2(x) dQ$$
$$= \int_{X_1\times X_2} (f_1 \circ \pi_1)(f_2 \circ \pi_2) d(Q \otimes \varepsilon_x). \tag{2}$$

It follows from (1) and (2) that

$$\int_{X_1\times X_2} (f_1\circ\pi_1)(f_2\circ\pi_2)dP_n \to \int_{X_1\times X_2} (f_1\circ\pi_1)(f_2\circ\pi_2)d(Q\otimes\varepsilon_x),$$

and so $\int_{X_1\times X_2} g\,dP_n \to \int_{X_1\times X_2} g\,d(Q\otimes\varepsilon_x)$, $g\in\mathfrak{A}$, where $\mathfrak{A}$ consists of all finite sums of terms $(f_1\circ\pi_1)(f_2\circ\pi_2)$ with $f_i\in\mathfrak{C}_r(X_i)$, $i=1,2$. $\mathfrak{A}$ is a subalgebra of $\mathfrak{C}_r(X_1\times X_2)$ separating points of $X_1\times X_2$, with $1\in\mathfrak{A}$, and so $\mathfrak{A}^-=\mathfrak{C}_r(X_1\times X_2)$ (6.57). The proof now concludes as in (22.9).

(II) Consider now the general case. For $i=1,2$, let Y_i be a compact metric space, and let h_i be a homeomorphism from X_i into Y_i (6.38). Define $h: X_1\times X_2\to Y_1\times Y_2$ by $h(x_1,x_2)=(h_1(x_1),h_2(x_2))$, $(x_1,x_2)\in X_1\times X_2$. For $i=1,2$, let π_i^X denote the projection from $X_1\times X_2$ onto X_i, and let π_i^Y denote the projection from $Y_1\times Y_2$ onto Y_i. Then

$$\pi_i^Y\circ h = h_i\circ\pi_i^X,\quad i=1,2. \tag{3}$$

Since $P_n(\pi_1^X)^{-1}\Rightarrow Q$ and $P_n(\pi_2^X)^{-1}\Rightarrow\varepsilon_x$, (22.8) shows that $P_n(\pi_1^X)^{-1}h_1^{-1}=P_n(h_1\circ\pi_1^X)^{-1}\Rightarrow Qh_1^{-1}$ and $P_n(\pi_2^X)^{-1}h_2^{-1}=P_n(h_2\circ\pi_1^X)^{-1}\Rightarrow\varepsilon_x h_2^{-1}$. Therefore, in view of (3), $(P_nh^{-1})(\pi_1^Y)^{-1}=P_n(\pi_1^Y\circ h)^{-1}=P_n(h_1\circ\pi_1^X)^{-1}\Rightarrow Qh_1^{-1}$ and $(P_nh^{-1})(\pi_2^Y)^{-1}=P_n(\pi_2^Y\circ h)^{-1}=P_n(h_2\circ\pi_2^X)^{-1}\Rightarrow\varepsilon_xh_2^{-1}$. Hence $P_nh^{-1}\Rightarrow(Qh_1^{-1})\otimes(\varepsilon_xh_2^{-1})=(Q\otimes\varepsilon_x)h^{-1}$ by (I). As h is a homeomorphism from $X_1\times X_2$ into $Y_1\times Y_2$, this entails that $P_n\Rightarrow Q\otimes\varepsilon_x$ by (22.8). □

The result in (22.12) allows a far-reaching generalization of Slutsky's theorem (19.27). For this some definitions are in order.

Definition 22.13. Let $(\Omega,\mathcal{K},\mathbf{P})$ be a probability space, let (X,d) be a metric space, let $f_n:\Omega\to X, n\in N$, be random elements, and let $f:\Omega\to X$ be a random element. We say that $\{f_n : n\in N\}$ *converges in distribution to* f, and we write $f_n\xrightarrow{D}f$, if $\mathbf{P}f_n^{-1}\Rightarrow\mathbf{P}f^{-1}$. If P is a probability on $\mathcal{B}(X)$ and $\mathbf{P}f_n^{-1}\Rightarrow P$, sometimes we will write $f_n\xrightarrow{D}P$.

Definition 22.14. Let $(\Omega,\mathcal{K},\mathbf{P})$, (X,d), $\{f_n : n\in N\}$ and f be as in (22.13) and assume that (X,d) is separable. We say that $\{f_n : n\in N\}$ *converges in probability to* f, and we write $f_n\xrightarrow{\mathbf{P}}f$, if $\mathbf{P}(d(f,f_n)\geqslant\varepsilon)\to 0$ for any $\varepsilon>0$.

Remarks 22.15. (a) The separability assumption in (22.14) ensures the measurability of the function $d\circ(f_n,f)$ via (6.12), (7.28) and (6.74). If $f=x\in X$ is a constant random element, then the function $d\circ(f_n,x)$ is measurable without separability. Actually, for any open set $U\subset X^2$, we have $(f_n,x)^{-1}(U)=f_n^{-1}(U_x)\in\mathcal{K}$, since U_x is an open subset of X (4.36.a). This implies that $(f_n,x)^{-1}(\mathcal{B}(X^2))\subset\mathcal{K}$ by (7.13).

(b) According to (14.7.f), $f_n\xrightarrow{\mathbf{P}}f$ if and only if $\mathbf{E}[d(f,f_n)\wedge 1]\to 0$.

(c) If $X=R$, then the notation $f_n\xrightarrow{D}f$ in (22.13) has the same meaning as the notation $f_n\xrightarrow{D}f$ in (19.17). Actually, $\mathbf{P}f_n^{-1}\Rightarrow\mathbf{P}f^{-1}$ means that $\mu_{F_n}\Rightarrow\mu_F$ by

(19.6.a), where F_n is the distribution function of f_n and F is the distribution function of f. At the same time, $\mu_{F_n} \Rightarrow \mu_F$ is equivalent to $F_n \overset{w}{\to} F$ by (22.6).

Theorem 22.16. *Let* $(\Omega, \mathcal{K}, \mathbf{P})$, (X, d), $\{f_n : n \in N\}$ *and* f *be as in* (22.14) *and assume that* $f_n \overset{\mathbf{P}}{\to} f$. *Then* $f_n \overset{D}{\to} f$.

Proof. For $\emptyset \neq A \in \mathcal{B}(X)$ and $\varepsilon > 0$, we have

$$\mathbf{P}(f_n \in A, f \in A^c) \leqslant \mathbf{P}(d(f_n, f) \geqslant \varepsilon) + \mathbf{P}(d(f, A) < \varepsilon, f \in A^c)$$

and

$$\mathbf{P}(f_n \in A^c, f \in A) \leqslant \mathbf{P}(d(f_n, f) \geqslant \varepsilon) + \mathbf{P}(d(f, A^c) < \varepsilon, f \in A),$$

whence

$$\mathbf{P}(\{f_n \in A\}\Delta\{f \in A\}) \leqslant 2\mathbf{P}(d(f_n, f) \geqslant \varepsilon) + \mathbf{P}(d(f, A) < \varepsilon, f \in A^c)$$
$$+\mathbf{P}(d(f, A^c) < \varepsilon, f \in A).$$

Therefore, as $f_n \overset{\mathbf{P}}{\to} f$, we get

$$\limsup_n \mathbf{P}(\{f_n \in A\}\Delta\{f \in A\}) \leqslant \mathbf{P}(d(f, A) < \varepsilon, f \in A^c)+\mathbf{P}(d(f, A^c) < \varepsilon, f \in A). \quad (1)$$

On letting $\varepsilon \to 0$ in (1), in view of (6.37.i), we obtain

$$\limsup_n \mathbf{P}(\{f_n \in A\}\Delta\{f \in A\}) \leqslant \mathbf{P}(f \in A^-, f \in A^c)$$
$$+\mathbf{P}(f \in (A^c)^-, f \in A) = \mathbf{P}(f \in \partial(A)). \quad (2)$$

If $\mathbf{P}(f \in \partial(A)) = 0$, (2) entails that $\mathbf{P}(\{f_n \in A\}\Delta\{f \in A\}) \to 0$, and so $\mathbf{P}(f_n \in A) \to \mathbf{P}(f \in A)$ by (9.6.iii). It follows from (22.5.v) that $f_n \overset{D}{\to} f$. □

Theorem 22.17. *Let* $(\Omega, \mathcal{K}, \mathbf{P})$, (X, d), $\{f_n : n \in N\}$ *and* f *be as in* (22.14)*and let* $x \in X$. *Then* $f_n \overset{\mathbf{P}}{\to} x$ *if and only if* $f_n \overset{D}{\to} \varepsilon_x$.

Proof. If $f_n \overset{\mathbf{P}}{\to} x$, then $f_n \overset{D}{\to} \varepsilon_x$ by (22.15.a) and (22.16).

Conversely, suppose that $\mathbf{P}f_n^{-1} \Rightarrow \varepsilon_x$, and let $\varepsilon > 0$. Apply (6.37.iii) to obtain a function $g \in \mathfrak{C}_r(X)$ such that $g(x) = 0$ and $g(S(x, \varepsilon)^-) = 1$. Then we have

$$\mathbf{P}(d(f_n, x) \geqslant \varepsilon) \leqslant \int_\Omega (g \circ f_n) d\mathbf{P} = \int_X g\, d(\mathbf{P}f_n^{-1}) \to \int_X g\, d\varepsilon_x = g(x) = 0. \quad \square$$

Theorem 22.18. *Let* $(\Omega, \mathcal{K}, \mathbf{P})$ *be a probability space, let* X_1 *and* X_2 *be separable metric spaces, let* $f_n : \Omega \to X_1, n \in N$, *and* $g_n : \Omega \to X_2, n \in N$, *be random elements, let* $f : \Omega \to X_1$ *be a random element, let* $x \in X_2$, *and assume that* $f_n \overset{D}{\to} f$ *and* $g_n \overset{\mathbf{P}}{\to} x$. *Then* $(f_n, g_n) \overset{D}{\to} (f, x)$.

Proof. For $i = 1, 2$, let π_i stand for the projection from $X_1 \times X_2$ onto X_i. On account of (22.17), $g_n \xrightarrow{D} \varepsilon_x$. For $n \in N$, $\mathbf{P}(f_n, g_n)^{-1}$ is a probability on $\mathcal{B}(X_1 \times X_2)$ with $(\mathbf{P}(f_n, g_n)^{-1})\pi_1^{-1} = \mathbf{P}f_n^{-1} \Rightarrow \mathbf{P}f^{-1}$ and $(\mathbf{P}(f_n, g_n)^{-1})\pi_2^{-1} = \mathbf{P}g_n^{-1} \Rightarrow \varepsilon_x$. Hence, by virtue of (22.12), $\mathbf{P}(f_n, g_n)^{-1} \Rightarrow \mathbf{P}f^{-1} \otimes \varepsilon_x = \mathbf{P}(f, x)^{-1}$, as $(\mathbf{P}f^{-1} \otimes \varepsilon_x)(A_1 \times A_2) = (\mathbf{P}(f, x)^{-1})(A_1 \times A_2) = \mathbf{P}(f^{-1}(A_1))1_{A_2}(x)$, $A_1 \in \mathcal{B}(X_1)$, $A_2 \in \mathcal{B}(X_2)$. □

Remark 22.19. Notation is as in (22.18). Let Y be a metric space, and let $h : X_1 \times X_2 \to Y$ be a continuous function. Then, according to (22.18) and (22.4.b), $h \circ (f_n, g_n) \xrightarrow{D} h \circ (f, x)$. This is a generalization of Slutsky's theorem (19.27) which obtains when $X_1 = X_2 = Y = R$, $x = 0$ and $h(x_1, x_2) = x_1 + x_2$, $x_1, x_2 \in R$.

We now proceed to investigate the problem of metrization of the topological space $(\mathrm{P}, \mathcal{W})$. We begin with the next lemma.

Lemma 22.20. *Let X be an arbitrary nonvoid set, let $\{f_i : i \in I\} \subset R^X$ be a countable family of functions that separates points of X, and let $\mathcal{T}$ denote the topology for X generated by $\{f_i : i \in I\}$. Then the following statements hold.*

(i) *There exists a metric d for X such that $\mathcal{T}_d = \mathcal{T}$.*

(ii) *$(X, \mathcal{T})$ is a separable topological space.*

(iii) *For $\{x_n : n \in N\} \subset X$ and $x \in X$, $x_n \to x$ if and only if $f_i(x_n) \to f_i(x)$ for any $i \in I$.*

Proof. Without loss of generality, suppose that $I = N$.

(I) Assume first that $f_n(X) \subset]0, 1[$, $n \in N$. Define $d(x, y) = \sum_{m \in N} 2^{-m} |f_m(x) - f_m(y)|$ for $x, y \in X$. Since $\{f_n : n \in N\}$ separates points of X, it follows easily that d is a metric for X. For $x \in X$, we have $\sum_{m=1}^{n} 2^{-m} |f_m(x) - f_m| \xrightarrow{u} d_x$, as $|f_m(x) - f_m| < 1$, $m \in N$. Hence, with respect to $\mathcal{T}$, d_x is continuous on X by (5.41) and (6.42). By virtue of (6.77) and (5.41), this entails that $\mathcal{T}_d \subset \mathcal{T}$. For $x, y \in X$ and $n \in N$, we have $|f_n(x) - f_n(y)| \leqslant 2^n d(x, y)$, and so f_n is uniformly continuous with respect to $\mathcal{T}_d$. This implies that $\mathcal{T} \subset \mathcal{T}_d$ by (5.41). Thus $\mathcal{T}_d = \mathcal{T}$. To prove (ii), notice first that R is a separable metric space by (6.7.c) and (5.19.b), and so R has a countable base. Then (4.38.a) shows that there is a countable base for $\mathcal{T}$, and so $(X, \mathcal{T})$ is separable (5.20). Finally, for $\{x_n : n \in N\} \subset X$ and $x \in X$, it is clear that $d(x, x_n) \to 0$ if and only if $\lim_n |f_m(x) - f_m(x_n)| = 0$, $m \in N$.

(II) Consider now the general case. Let $h : R \to]0, 1[$ be a homeomorphism from R onto $]0, 1[$. Then (i)–(iii) follows as in (I) on defining $d(x, y) = \sum_{m \in N} 2^{-m} |h(f_m(x)) - h(f_m(y))|$, $x, y \in X$. □

Theorem 22.21. *X is compact if and only if $(\mathrm{P}, \mathcal{W})$ is metrizable and compact.*

Proof. Assume that X is compact. Then $\mathfrak{C}_r(X)$ is a separable metric space by (6.58), and so there exists a countable set $\{f_i : i \in I\} \subset \mathfrak{C}_r(X)$ such that $\{f_i : i \in I\}^- = \mathfrak{C}_r(X)$. Let $\mathcal{U}$ be the topology for P generated by the family $\{\psi_{f_i} : i \in I\}$. Since $\{\psi_{f_i} : i \in I\}$ separates points of P by (11.50.b), it follows that $(\mathrm{P}, \mathcal{U})$ is metrizable by (22.20.i). As $\{\psi_{f_i} : i \in I\} \subset \{\psi_f : f \in \mathfrak{C}_r(X)\}$, we have $\mathcal{U} \subset \mathcal{W}$. Next, for $f \in \mathfrak{C}_r(X)$ and $\varepsilon > 0$, choose $i \in I$ so that $\|f - f_i\|_u < \varepsilon$. Then $|\psi_f(P) - \psi_{f_i}(P)| = \left|\int_X f\, dP - \int_X f_i\, dP\right| < \varepsilon$, $P \in \mathrm{P}$. By virtue of (6.42), this shows that ψ_f is continuous with respect to $\mathcal{U}$, and so $\mathcal{W} \subset \mathcal{U}$. Hence $\mathcal{U} = \mathcal{W}$, and thus

$(\mathrm{P}, \mathcal{W})$ is metrizable. For the direct implication, it remains to prove that P is compact. Applying (6.19), we will show that P is sequentially compact. So let $\{P_n : n \in N\} \subset \mathrm{P}$. For $i \in I$, choose $c_i \in R$ such that $|c_i f_i| \leqslant 1$. Then, arguing as in (6.67), Cantor's diagonal method ensures the existence of a subsequence $\{P_{n_k} : k \in N\}$ such that the sequence $\{\int_X c_i f_i \, dP_{n_k} : k \in N\}$ has a limit in $[-1, 1]$ for each $i \in I$. Hence $\{\int_X f_i \, dP_{n_k} : k \in N\}$ is a Cauchy sequence for any $i \in I$ (6.21.a). Let $f \in \mathfrak{C}_r(X)$ and $\varepsilon > 0$, and choose $i \in I$ such that $\|f - f_i\|_u < \varepsilon$. Choose next $k_\varepsilon \in N$ so that $\left|\int_X f_i \, dP_{n_j} - \int_X f_i \, dP_{n_k}\right| < \varepsilon$ whenever $j, k \geqslant k_\varepsilon$. Then, for $j, k \geqslant k_\varepsilon$, we have

$$\left|\int_X f \, dP_{n_j} - \int_X f \, dP_{n_k}\right| \leqslant 2\,\|f - f_i\|_u + \left|\int_X f_i \, dP_{n_j} - \int_X f_i \, dP_{n_k}\right| < 3\varepsilon.$$

This shows that $\{\int_X f \, dP_{n_k} : k \in N\}$ is a Cauchy sequence, and so $\lim_k \int_X f \, dP_{n_k}$ exists (6.32). Put $L(f) = \lim_k \int_X f \, dP_{n_k}$, $f \in \mathfrak{C}_r(X)$. Then L is a nonnegative linear functional on $\mathfrak{C}_r(X)$ with $L(1) = 1$. Hence, by Riesz's representation theorem (21.17), there exists a unique probability $P \in \mathrm{P}$ such that $L(f) = \int_X f \, dP$, $f \in \mathfrak{C}_r(X)$. This means that $P_{n_k} \Rightarrow P$, and the proof that $(\mathrm{P}, \mathcal{W})$ is compact is complete.

Conversely, suppose that $(\mathrm{P}, \mathcal{W})$ is metrizable and compact, and let $\{x_n : n \in N\} \subset X$ be such that $x_m \neq x_n$ for $m \neq n$. Then there exist a subsequence $\{x_{n_k} : k \in N\}$ and $P \in \mathrm{P}$ such that $\varepsilon_{x_{n_k}} \Rightarrow P$. To get a contradiction, assume that $\{x_n : n \in N\}$ has no convergent subsequence. Then, in view of (6.72), the sets $A = \{x_{n_1}, x_{n_3}, \ldots\}$ and $B = \{x_{n_2}, x_{n_4}, \ldots\}$ are disjoint and closed. From (22.5.iii) it follows that $1 = \limsup_k \varepsilon_{x_{n_k}}(A) \leqslant P(A)$ and $1 = \limsup_k \varepsilon_{x_{n_k}}(B) \leqslant P(B)$. This would lead to $P(X) \geqslant 2$, a contradiction. □

Theorem 22.22. *X is separable if and only if $(\mathrm{P}, \mathcal{W})$ is metrizable and separable.*

Proof. Suppose that X is separable. By Urysohn's embedding theorem (6.38) there exist a compact metric space (Y, ρ) and a homeomorphism h from X into Y. Since $\mathcal{T} = h^{-1}(\mathcal{T}_\rho)$ (5.86), it follows that $\mathcal{V} = \mathcal{W}$, where $\mathcal{V}$ is as in (22.7). Let $\{g_i : i \in I\} \subset \mathfrak{C}_r(Y)$ be a countable set of functions with $\{g_i : i \in I\}^- = \mathfrak{C}_r(Y)$ (6.58), and let $\mathcal{Z}$ denote the topology for P generated by the family $\{\psi_{g_i \circ h} : i \in I\}$. As $\{\psi_{g_i \circ h} : i \in I\}$ separates points of P by (11.57.b), we see that $(\mathrm{P}, \mathcal{Z})$ is metrizable and separable by (22.20). We have $\mathcal{Z} \subset \mathcal{V}$, since $\{\psi_{g_i \circ h} : i \in I\} \subset \{\psi_{g \circ h} : g \in \mathfrak{C}_r(Y)\}$. For $g \in \mathfrak{C}_r(Y)$ and $\varepsilon > 0$, choose $i \in I$ so that $\|g - g_i\|_u < \varepsilon$. Then $\left|\psi_{g \circ h}(P) - \psi_{g_i \circ h}(P)\right| \leqslant \|g - g_i\|_u < \varepsilon$, $P \in \mathrm{P}$. Hence $\psi_{g \circ h}$ is continuous with respect to $\mathcal{Z}$, and so $\mathcal{V} \subset \mathcal{Z}$. Thus $\mathcal{Z} = \mathcal{V} = \mathcal{W}$.

Conversely, assume that $(\mathrm{P}, \mathcal{W})$ is metrizable and separable. Then, according to (6.70), $\{\varepsilon_x : x \in X\}$ endowed with the relative topology $\{\varepsilon_x : x \in X\} \cap \mathcal{W}$ is a separable metric space. Since X and $\{\varepsilon_x : x \in X\}$ are homeomorphic, it follows that X is separable by (5.50.b). □

When X is separable, the following result yields a countable subset dense in P.

Theorem 22.23. *Assume that (X, d) is separable, and let $\{a_i : i \in I\} \subset X$ be countable with $\{a_i : i \in I\}^- = X$. Then the set $\mathrm{P}_1 = \{\sum_{j \in J} p_j \varepsilon_{a_j} : J$ is finite, $\sum_{j \in J} p_j = 1$ and $0 < p_j \in Q, j \in J\}$ is dense in* P.

Proof. Without loss of generality, suppose that $I = N$. Let $\mathrm{P}_2 = \{\sum_{j=1}^n p_j \varepsilon_{x_j} : \{x_1, \dots, x_n\} \subset X, \sum_{j=1}^n p_j = 1, 0 < p_j \in R$ for $j = 1, \dots, n$, and $n \in N\}$, and put $\mathrm{P}_3 = \{\sum_{j\geqslant 1} p_j \varepsilon_{x_j} : \{x_1, x_2, \dots\} \subset X, \sum_{j\geqslant 1} p_j = 1, 0 < p_j \in R$ for $j \geqslant 1\}$. We shall show that $\mathrm{P}_1^- \supset \mathrm{P}_2, \mathrm{P}_2^- \supset \mathrm{P}_3$, and $\mathrm{P}_3^- = \mathrm{P}$. Let $f \in \mathfrak{C}_r(X)$ and $\varepsilon > 0$.

(I) $\mathrm{P}_1^- \supset \mathrm{P}_2$. Let $P = \sum_{j=1}^n p_j \varepsilon_{x_j} \in \mathrm{P}_2$. For $1 \leqslant j \leqslant n$, choose a_j so close to x_j that $|f(x_j) - f(a_j)| < \varepsilon$. Since $Q^- = R$, choose positive $q_1, \dots, q_n \in Q$ with $\sum_{j=1}^n q_j = 1$ and $\sum_{j=1}^n |p_j - q_j| < \varepsilon$. Set $Q = \sum_{j=1}^n q_j \varepsilon_{a_j} \in \mathrm{P}_1$. Then

$$\left| \int_X f\,dP - \int_X f\,dQ \right| = \left| \sum_{j=1}^n p_j f(x_j) - \sum_{j=1}^n q_j f(a_j) \right|$$
$$\leqslant \sum_{j=1}^n p_j |f(x_j) - f(a_j)| + \sum_{j=1}^n |p_j - q_j|\,|f(a_j)| < (1 + \|f\|_u)\varepsilon.$$

(II) $\mathrm{P}_2^- \supset \mathrm{P}_3$. Let $P = \sum_{j\geqslant 1} p_j \varepsilon_{x_j} \in \mathrm{P}_3$, and put $P_n = s_n^{-1} \sum_{j=1}^n p_j \varepsilon_{x_j}$, where $s_n = \sum_{j=1}^n p_j$ for $n \in N$. Then $P_n \in \mathrm{P}_2$ and

$$\left| \int_X f\,dP - \int_X f\,dP_n \right| = \left| \sum_{j\geqslant 1} p_j f(x_j) - s_n^{-1} \sum_{j=1}^n p_j f(a_j) \right|$$
$$\leqslant (s_n^{-1} - s_n)\,\|f\|_u \to 0.$$

(III) $\mathrm{P}_3^- = \mathrm{P}$. This time assume that f is even uniformly continuous. Let $\delta > 0$ be so that $d(x, y) < \delta$ implies $|f(x) - f(y)| < \varepsilon$. Applying (6.12) and Lindelöf's theorem (5.14), there is a sequence $\{A_n : n \in N\}$ of open balls of radius $\delta/2$ covering X. Define $\{B_n : n \in N\}$ as in the proof of (3.15), and choose $x_n \in B_n$. For $P \in \mathrm{P}$, set $Q = \sum_{n\in N} P(B_n)\varepsilon_{x_n}$. Then $Q \in \mathrm{P}_3$ and

$$\left| \int_X f\,dP - \int_X f\,dQ \right| = \left| \sum_{n\in N} \int_{B_n} (f - f(x_n))dP \right|$$
$$\leqslant \sum_{n\in N} \int_{B_n} |f - f(x_n)|\,dP \leqslant \varepsilon.$$

Thus (22.5.ii) completes the proof. □

We shall now take up a famous result, characterizing the compact subsets of P, due to Yu.V. Prohorov. Its direct statement is Theorem (22.27), and the converse statement is Theorem (22.29). For this, we need the concept of tightness.

Definition 22.24. A set $\mathrm{Q} \subset \mathrm{P}$ is said to be *tight* if for any $\varepsilon > 0$ there exists a compact set $C_\varepsilon \subset X$ such that $Q(C_\varepsilon^c) < \varepsilon$ for all $Q \in \mathrm{Q}$.

Examples 22.25. (a) If X is separable and complete, then any finite subset of P is tight by (5.72) and (16.22).

(b) Let $(\Omega, \mathcal{K}, \mathbf{P})$ be a probability space, let $f_n : \Omega \to X, n \in N$, be random elements, and let F be a distribution function on R such that $f_n \overset{D}{\to} F$. Then $\{\mathbf{P} f_n^{-1} : n \in N\}$ is a tight set of probabilities on $\mathcal{B}(R)$. Actually, for $\varepsilon > 0$, choosing $a, b \in C(F)$ so that $a < b$, $F(a) < \varepsilon/2$ and $1 - F(b) < \varepsilon/2$, we see that

$$(\mathbf{P} f_n^{-1})([a, b]^c) < \varepsilon, \quad n \geqslant n_\varepsilon. \tag{1}$$

By decreasing a and increasing b if necessary, we can ensure that (1) holds for all $n \in N$.

Remark 22.26. Let Y be a metric space, let $h : X \to Y$ be a continuous function, and let $\mathrm{Q} \subset \mathrm{P}$ be tight. Then the set $\{Qh^{-1} : Q \in \mathrm{Q}\}$ is tight. Actually, for $\varepsilon > 0$ choose a compact set $C_\varepsilon \subset X$ so that $Q(C_\varepsilon^c) < \varepsilon, Q \in \mathrm{Q}$. By (5.45), $K_\varepsilon = h(C_\varepsilon)$ is a compact subset of Y, and we have

$$Qh^{-1}(K_\varepsilon^c) = Q(h^{-1}(K_\varepsilon^c)) = Q((h^{-1}(K_\varepsilon))^c) \leqslant Q(C_\varepsilon^c) < \varepsilon, \quad Q \in \mathrm{Q}.$$

Theorem 22.27 (Prohorov). *Suppose that X is separable, and let $\mathrm{Q} \subset \mathrm{P}$ be tight. Then Q^- is compact.*

Proof. Let (Y, ρ) be a metric space, and let h be a homeomorphism from X into Y (6.38). Let $\{Q_n : n \in N\} \subset \mathrm{Q}$. According to (6.66), it suffices to show that $\{Q_n : n \in N\}$ has a convergent subsequence. Since Y is compact, the set of all probabilities on $\mathcal{B}(Y)$ is metrizable and sequentially compact by (22.21) and (6.19). Hence there exist a subsequence $\{Q_{n_k} : k \in N\}$ and a probability R on $\mathcal{B}(Y)$ such that

$$Q_{n_k} h^{-1} \Rightarrow R. \tag{1}$$

For $r \in N$, let $A_r \subset X$ be a compact set such that $Q(A_r) \geqslant 1 - 1/r, Q \in \mathrm{Q}$, and let $B_r = h(A_r)$. Put $A = \cup_{r \in N} A_r$ and $B = \cup_{r \in N} B_r$. Then $B \in \mathcal{B}(Y)$, since each B_r is compact by (5.45), and so closed by (5.27). On account of (22.5.iii), we have $R(B_r) \geqslant \limsup_k (Q_{n_k} h^{-1})(B_r) = \limsup_k Q_{n_k}(A_r) \geqslant 1 - 1/r, r \in N$. This entails that $R(B) = 1$. Let R' denote the restriction of R to $B \cap \mathcal{B}(Y)$. Since h^{-1} is a homeomorphism from $h(X)$ endowed with the relative topology $h(X) \cap \mathcal{T}_\rho$ onto X, (5.86) shows that $h(\mathcal{T}) = h(X) \cap \mathcal{T}_\rho$. Hence, in view of (7.13) and (7.14), $h(\mathcal{B}(X)) = h(X) \cap \mathcal{B}(Y)$. This implies that $h^{-1} \circ i_B$ is $(B \cap \mathcal{B}(Y), \mathcal{B}(X))$-measurable, where i_B stands for the inclusion function from B into $h(X)$. Define $P = R'(h^{-1} \circ i_B)^{-1}$. Then $P \in \mathrm{P}$, as $P(X) = R'(i_B^{-1}(h(X))) = R'(B) = 1$. Moreover, for $C \in \mathcal{B}(Y)$,

$$\begin{aligned}(Ph^{-1})(C) &= (R'(h^{-1} \circ i_B)^{-1})(h^{-1}(C)) = R'(i_B^{-1}(h(h^{-1}(C)))) \\ &= R'(i_B^{-1}(h(X) \cap C)) = R'(B \cap C) = R(C). \end{aligned} \tag{2}$$

From (1) and (2) we infer that $Q_{n_k} h^{-1} \Rightarrow Ph^{-1}$, whence $Q_{n_k} \Rightarrow P$ by (22.8). □

To prove the second part of Prohorov's result, we need the next lemma.

Lemma 22.28. *Assume that X is complete, let $\mathrm{Q} \subset \mathrm{P}$, and suppose that, for every $\varepsilon > 0$ and $r > 0$, there is a finite union U of open balls of radius r with $Q(U^c) < \varepsilon$ for all $Q \in \mathrm{Q}$. Then Q is tight.*

Proof. Let $\varepsilon > 0$. For each $n \in N$, choose a finite union U_n of open balls of radius $1/n$ such that

$$Q(U_n^c) < \varepsilon/2^n, \quad Q \in \mathrm{Q}. \tag{1}$$

Put $A_\varepsilon = \cap_{n\in N} U_n$. Obviously, A_ε is a totally bounded set, and so is A_ε^- by (6.73). Then, in view of (6.25.b), A_ε^- equipped with the topology $A_\varepsilon^- \cap \mathcal{T}$ is complete. Hence A_ε^- is compact by (6.31). Finally, on account of (1), we have $Q((A_\varepsilon^-)^c) \leqslant Q(A_\varepsilon^c) \leqslant \sum_{n\in N} Q(U_n^c) < \sum_{n\in N} \varepsilon/2^n = \varepsilon$, $Q \in \mathrm{Q}$. □

Theorem 22.29 (Prohorov). *Suppose that X is separable and complete, and let $\mathrm{Q} \subset \mathrm{P}$ be such that Q^- is compact. Then Q is tight.*

Proof. Let $\varepsilon > 0$ and $r > 0$. Since X is separable, there exists a sequence $\{A_k : k \in N\}$ of open balls of radius r so that $X = \cup_{k\in N} A_k$ (5.14). Set $U_n = \cup_{k=1}^n A_k$. We claim that there is $n \in N$ such that $Q(U_n^c) < \varepsilon$ for all $Q \in \mathrm{Q}$. Otherwise, for each $n \in N$, there is $Q_n \in \mathrm{Q}$ with $Q_n(U_n) \leqslant 1 - \varepsilon$. As Q^- is compact, there exist a subsequence $\{Q_{n_j} : j \in N\}$ and $P \in \mathrm{P}$ such that $Q_{n_j} \Rightarrow P$. Then, on account of (22.5.iv),

$$P(U_n) \leqslant \liminf_j Q_{n_j}(U_n) \leqslant \liminf_j Q_{n_j}(U_{n_j}) \leqslant 1 - \varepsilon, \quad n \in N. \tag{1}$$

Letting $n \to \infty$ in (1), we get $P(X) \leqslant 1 - \varepsilon$. This is a contradiction, and so the claim is verified. Thus Lemma (22.28) ensures that Q is tight. □

We shall apply Prohorov's theorem (22.27) to prove the version of Lévy's continuity theorem (20.11) for random vectors.

Lemma 22.30. *Let $X = \prod_{i\in I} X_i$, where $\{X_i : i \in I\}$ is a countable family of separable metric spaces, $\mathrm{Q} \subset \mathrm{P}$, and put $\mathrm{Q}_i = \{Q\pi_i^{-1} : Q \in \mathrm{Q}\}$ for $i \in I$. Then Q is tight if and only if Q_i is tight for each $i \in I$.*

Proof. Without loss of generality, assume that $I = N$, and let $\varepsilon > 0$. Suppose that Q is tight, and let C be a compact subset of X with $Q(C^c) < \varepsilon$, $Q \in \mathrm{Q}$. Then, for any $i \in I$, $\pi_i(C)$ is a compact subset of X_i by (5.45), and $(Q\pi_i^{-1})(\pi_i(C)^c) \leqslant Q(C^c) < \varepsilon$, $Q \in \mathrm{Q}$, by (2.33.b). Thus Q_i is tight for every $i \in I$.

Conversely, assume that Q_n $n \in N$, is tight. For $n \in N$, choose a compact subset C_n of X_n so that $(Q\pi_n^{-1})(C_n^c) < \varepsilon/2^n$, $Q \in \mathrm{Q}$. Then $C = \prod_{n\in N} C_n$ is compact by Tihonov's theorem (5.28). Since $C = \cap_{n\in N} \pi_n^{-1}(C_n)$, it follows that $Q(C^c) \leqslant \sum_{n\in N} Q(\pi_n^{-1}(C_n^c)) < \sum_{n\in N} \varepsilon/2^n = \varepsilon$, $Q \in \mathrm{Q}$. Thus Q is tight. □

Theorem 22.31. *Let $(\Omega, \mathcal{K}, \mathbf{P})$ be a probability space, let $\mathbf{f}_k : \Omega \to R^n, k \in N$, be random vectors with coordinates $f_{k1}, \ldots, f_{kn}$, and assume that $\varphi_{\mathbf{f}_k} \to g$ on R^n, where the limit function g is continuous at $\mathbf{0}$. Then g is the characteristic function of some distribution function F on R^n and $\mathbf{f}_k \xrightarrow{D} F$.*

Proof. Since $\lim_k \mathbf{E}[\exp(i\sum_{j=1}^n t_j f_{kj})] = g(t_1, \ldots, t_n)$ for any $(t_1, \ldots, t_n) \in R^n$, it follows that $\lim_k \varphi_{f_{kj}}(t_j) = \lim_k \mathbf{E}[\exp(it_j f_{kj})] = g(\ldots, 0, t_j, 0, \ldots)$ for any $t_j \in R, 1 \leqslant j \leqslant n$. As $g(\ldots, 0, t_j, 0, \ldots)$ is continuous at $t_j = 0$, Lévy's theorem

(20.11) ensures the existence of a distribution function F_j on R such that $f_{kj} \xrightarrow{D} F_j$. Then, according to (22.25.b), the set $\{\mathbf{P} f_{kj}^{-1} : k \in N\}$ is tight for $1 \leqslant j \leqslant n$. Since $(\mathbf{Pf}_k^{-1})\pi_j^{-1} = \mathbf{P}(\pi_j \circ \mathbf{f}_k)^{-1} = \mathbf{P} f_{kj}^{-1}, k \in N, 1 \leqslant j \leqslant n$, (22.30) shows that $\mathrm{Q} = \{\mathbf{Pf}_k^{-1} : k \in N\}$ is tight. Hence, in view of Prohorov's theorem (22.27), Q^- is compact. Let $\{\mathbf{Pf}_{k_l}^{-1} : l \in N\}$ be a subsequence. Then there exist a subsubsequence $\{\mathbf{Pf}_{k_{l_m}}^{-1} : m \in N\}$ and a probability P on $\mathcal{B}(R^n)$ such that

$$\mathbf{Pf}_{k_{l_m}}^{-1} \Rightarrow P. \tag{1}$$

Since, for any $\mathbf{t} \in R^n$, the function $e^{i\langle \mathbf{t}, \mathbf{x} \rangle}$ is bounded and continuous in $\mathbf{x} \in R^n$, (1) and (22.1) lead to

$$\lim_m \varphi_{\mathbf{f}_{k_{l_m}}}(\mathbf{t}) = g(\mathbf{t}) = \int_{R^n} e^{i\langle \mathbf{t}, \mathbf{x} \rangle}\, dP(\mathbf{x}), \quad \mathbf{t} \in R^n. \tag{2}$$

Repeating the above argument with another subsequence, we arrive at the same P by (2) and (20.33). Therefore, on account of (22.4.a), $\mathbf{Pf}_k^{-1} \Rightarrow P$. This can be rewritten as $\mathbf{f}_k \xrightarrow{D} F$, where F is the cumulative distribution function of P. □

Remark 22.32. Let $(\Omega, \mathcal{K}, \mathbf{P})$ and $\mathbf{f}_k, k \in N$, be as in (22.31), and let $\mathbf{f} : \Omega \to R^n$ be a random vector such that $\varphi_{\mathbf{f}_k} \to \varphi_{\mathbf{f}}$. Then $\mathbf{f}_k \xrightarrow{D} \mathbf{f}$ by (20.30.c), (22.31) and (20.33).

An immediate application of (22.32) is the following result, due to H. Cramér and H. Wold, by which problems involving random vectors can be reduced to problems involving random variables.

Cramér-Wold device. 22.33. *Let* $(\Omega, \mathcal{K}, \mathbf{P})$ *and* $\mathbf{f}_k, k \in N$, *be as in* (22.31), *and let* $\mathbf{f} : \Omega \to R^n$ *be a random vector with coordinates* $f_1, \ldots, f_n$. *Then the following are equivalent:*

(i) $\mathbf{f}_k \xrightarrow{D} \mathbf{f}$;

(ii) $\sum_{j=1}^n t_j f_{kj} \xrightarrow{D} \sum_{j=1}^n t_j f_j$ for any $(t_1, \ldots, t_n) \in R^n$.

Proof. If $\mathbf{Pf}_k^{-1} \Rightarrow \mathbf{Pf}^{-1}$, then

$$\begin{aligned}\lim_k \mathbf{E}\left[\exp\left(i\sum_{j=1}^n t_j f_{kj}\right)\right] &= \lim_k \int_{R^n} e^{i\langle \mathbf{t}, \mathbf{x}\rangle}\, d(\mathbf{Pf}_k^{-1})(\mathbf{x}) = \int_{R^n} e^{i\langle \mathbf{t}, \mathbf{x}\rangle}\, d(\mathbf{Pf}^{-1})(\mathbf{x}) \\ &= \mathbf{E}\left[\exp\left(i\sum_{j=1}^n t_j f_j\right)\right], \quad \mathbf{t} = (t_1, \ldots, t_n) \in R^n,\end{aligned}$$

by (11.29) and (20.1). Hence

$$\lim_k \mathbf{E}\left[\exp\left(iu\sum_{j=1}^n t_j f_{kj}\right)\right] = \mathbf{E}\left[\exp\left(iu\sum_{j=1}^n t_j f_j\right)\right], \quad (t_1, \ldots, t_n) \in R^n, \ u \in R, \tag{1}$$

whence (ii) follows by (20.12.a) (see also (22.15.c)). Thus (i) implies (ii). If now (ii) is fulfilled, then (1) holds by (20.9). Taking $u = 1$, we find that

$$\varphi_{\mathbf{f}_k}(\mathbf{t}) = \mathbf{E}\left[\exp\left(i\sum_{j=1}^{n} t_j f_{kj}\right)\right] \to \mathbf{E}\left[\exp\left(i\sum_{j=1}^{n} t_j f_j\right)\right]$$
$$= \varphi_{\mathbf{f}}(\mathbf{t}), \quad \mathbf{t} = (t_1, \ldots, t_n) \in R^n,$$

and so $\mathbf{f}_k \xrightarrow{D} \mathbf{f}$ by (22.32). Thus (ii) implies (i). □

Exercise 22.34. Let $P \in \mathrm{P}$. Prove that $\{A \in \mathcal{B}(X) : P(\partial(A)) = 0\}$ is an algebra, but it may not be a σ-algebra.

Exercise 22.35. Let $x \in X$, $A \subset X$ and $r > 0$, and let $P \in \mathrm{P}$.

(a) Show that $\partial(S(x, r)) \subset \{y \in X : d(x, y) = r\}$, $\partial(\overline{S}(x, r)) \subset \{y \in X : d(x, y) = r\}$, $\partial(\{y \in X : d(y, A) < r\}) \subset \{y \in X : d(y, A) = r\}$ and $\partial(\{y \in X : d(y, A) \leqslant r\}) \subset \{y \in X : d(y, A) = r\}$.

(b) If X is as in (6.2), then the inclusions in (a) may be strict.

(c) Prove that the sets $\{r > 0 : P(\partial(S(x, r))) > 0\}$, $\{r > 0 : P(\partial(\overline{S}(x, r))) > 0\}$, $\{r > 0 : P(\partial(\{y \in X : d(y, A) < r\})) > 0\}$ and $\{r > 0 : P(\partial(\{y \in X : d(y, A) \leqslant r\})) > 0\}$ are countable.

Exercise 22.36. Let $P \in \mathrm{P}$ and $\varepsilon > 0$, and define $W_P^1(F, \varepsilon) = \{Q \in \mathrm{P} : Q(F) < P(F) + \varepsilon\}$ for $F \in \mathcal{F}$, $W_P^2(U, \varepsilon) = \{Q \in \mathrm{P} : Q(U) > P(U) - \varepsilon\}$ for $U \in \mathcal{T}$, and $W_P^3(A, \varepsilon) = \{Q \in \mathrm{P} : |P(A) - Q(A)| < \varepsilon\}$ for $A \in \mathcal{B}(X)$ with $P(\partial(A)) = 0$. Put $\mathcal{W}^1 = \tau(\{W_P^1(F, \varepsilon) : P \in \mathrm{P}, F \in \mathcal{F}, \varepsilon > 0\})$, $\mathcal{W}^2 = \tau(\{W_P^2(U, \varepsilon) : P \in \mathrm{P}, U \in \mathcal{T}, \varepsilon > 0\})$, and $\mathcal{W}^3 = \tau(\{W_P^3(A, \varepsilon) : P \in \mathrm{P}, A \in \mathcal{B}(X)$ with $P(\partial(A)) = 0, \varepsilon > 0\})$. Prove that $\mathcal{W}^1 = \mathcal{W}^2 = \mathcal{W}^3 = \mathcal{W}$. The next steps may be helpful.

(a) $\mathcal{W}^1 = \mathcal{W}^2$. [$W_P^1(F, \varepsilon) = W_P^2(F^c, \varepsilon)$, $F \in \mathcal{F}$.]

(b) $\mathcal{W}^1 = \mathcal{W}^3$. [Hints. To show that $\mathcal{W}^1 \subset \mathcal{W}^3$, notice that $W_P^1(F, 2\varepsilon) \supset W_P^3(F^\delta, \varepsilon)$, $F \in \mathcal{F}$, where $F^\delta = \{x \in X : d(x, F) < \delta\}$ is chosen so that $P(F^\delta) < P(F) + \varepsilon$ and $P(\partial(F^\delta)) = 0$. To show that $\mathcal{W}^3 \subset \mathcal{W}^1$, notice that $W_P^3(A, \varepsilon) \supset W_P^1(A^-, \varepsilon) \cap W_P^1((A^c)^-, \varepsilon)$ for $A \in \mathcal{B}(X)$ with $P(\partial(A)) = 0$.]

(c) $\mathcal{W}^1 \subset \mathcal{W}$. [Hint. For $P \in \mathrm{P}$, $F \in \mathcal{F}$ and $\varepsilon > 0$, choose $\delta > 0$ so that $P(F^\delta) < P(F) + \varepsilon$, and use (6.37) to choose a continuous function $f : X \to [0, 1]$ such that $f(F) = \{1\}$ and $f((F^\delta)^c) = \{0\}$. Then $W_P(f, \varepsilon) \subset W_P^1(F, 2\varepsilon)$.]

(d) $\mathcal{W} \subset \mathcal{W}^1$. [Hints. For $f \in \mathfrak{C}_r(X)$ with $0 < f < 1$ and $\varepsilon > 0$, choose $k \in N$ so that $1/k < \varepsilon$, and put $F_i = \{f \geqslant i/k\}$, $1 \leqslant i \leqslant k$. If $Q \in \cap_{i=1}^k W_P^1(F_i, \varepsilon)$, then

$$\int_X f\,dQ \leqslant \frac{1}{k} + \frac{1}{k}\sum_{i=1}^{k} Q(F_i) < \varepsilon + \frac{1}{k}\sum_{i=1}^{k}(P(F_i) + \varepsilon) = 2\varepsilon + \frac{1}{k}\sum_{i=1}^{k} P(F_i)$$
$$\leqslant 2\varepsilon + \int_X f\,dP,$$

and so $\int_X f\,dQ - \int_X f\,dP < 2\varepsilon$. Similarly, if $Q \in \cap_{i=1}^k W_P^1(F_i', \varepsilon)$, where $F_i' = \{1 - f \geqslant i/k\}$, $1 \leqslant i \leqslant k$, then $\int_X f\,dP - \int_X f\,dQ < 2\varepsilon$. Hence $\cap_{i=1}^k (W_P^1(F_i, \varepsilon) \cap W_P^1(F_i', \varepsilon)) \subset W_P(f, 2\varepsilon)$.]

Exercise 22.37. For $P, Q \in \mathrm{P}$, put $\rho(P, Q) = \inf\{\varepsilon > 0 : Q(A) \leqslant P(A^\varepsilon) + \varepsilon$ and $P(A) \leqslant Q(A^\varepsilon) + \varepsilon$ for all $A \in \mathcal{B}(X)\}$, where $A^\varepsilon = \{x \in X : d(x, A) < \varepsilon\}$. Prove the following.

(a) ρ is a metric for P called the *Prohorov metric*. [Hints. To show that $\rho(P, Q) = 0$ implies $P = Q$, use (6.95.a) and (10.8). To verify the triangle inequality, use (6.95.b).]

(b) $\mathcal{W} \subset \mathcal{T}_\rho$. [Hint. $W_P^1(F, 2\varepsilon) \supset S(P, \delta)$, $F \in \mathcal{F}$, where $\delta > 0$ is chosen so that $\delta < \varepsilon$ and $P(F^\delta) < P(F) + \varepsilon$.]

(c) If X is separable, then $\mathcal{W} = \mathcal{T}_\rho$. [Hints. Let $0 < 3\delta < \varepsilon$. For $P \in \mathrm{P}$, apply (22.35.c) to cover X by open balls S of radius $\delta/2$ with $P(\partial(S)) = 0$, and use Lindelöf's theorem (5.14) to get a denumerable subcover $\{S_n : n \in N\}$. Define $A_1 = S_1$ and $A_{n+1} = S_{n+1} - \cup_{i=1}^n S_i$, $n \geqslant 1$, and select $k \in N$ so that

$$P\left(\bigcup_{i=1}^k A_i\right) > 1 - \delta. \tag{1}$$

Let $\mathcal{A}$ denote the family of all unions of sets in $\{A_1, \ldots, A_n\}$, and apply (22.34) to see that $P(\partial(A)) = 0$, $A \in \mathcal{A}$. Let $Q \in \cap_{A \in \mathcal{A}} W_P^3(A, \delta)$ and $B \in \mathcal{B}(X)$, and let A_B be the union of those sets in $\{A_1, \ldots, A_n\}$ that meet B. Notice that

$$|P(A) - Q(A)| < \delta, \quad A \in \mathcal{A}, \tag{2}$$

and so

$$Q\left(\bigcup_{i=1}^k A_i\right) > 1 - 2\delta. \tag{3}$$

Since $B \subset A_B \cup (\cup_{i=1}^k A_i)^c$ and $A_B \subset B^\delta$, use (1) and (2) to show that

$$P(B) < Q(B^\delta) + 2\delta, \tag{4}$$

and use (2) and (3) to see that

$$Q(B) < P(B^\delta) + 3\delta. \tag{5}$$

From (4) and (5) conclude that $\cap_{A \in \mathcal{A}} W_P^3(A, \delta) \subset S(P, \varepsilon)$.]

(d) If X is separable and complete, then (P, ρ) is a complete metric space. [Hints. Let $\{P_n : n \in N\} \subset \mathrm{P}$ be a Cauchy sequence, let $0 < r < \varepsilon$, and choose $n_r \in N$ so that $\rho(P_{n_r}, P_n) < r, n \geqslant n_r$. Choose open balls $S_1, \ldots, S_k$ of radius r so that

$$P_{n_r}\left(\bigcup_{i=1}^k S_i\right) > 1 - \varepsilon. \tag{6}$$

Let $A_1, \ldots, A_k$ be the open balls with the same centers of radius $2r$. Since $(\cup_{i=1}^k S_i)^r \subset \cup_{i=1}^k A_i$, use (6) to show that

$$P_n\left(\bigcup_{i=1}^k A_i\right) > 1 - 2\varepsilon, \quad n \geqslant n_r. \tag{7}$$

Take then into account (7), (22.28), Prohorov's theorem (22.27) and (6.25.c).]

Exercise 22.38. Define $\varepsilon^* : X \to \mathrm{P}$ by $\varepsilon^*(x) = \varepsilon_x, x \in X$. Prove that ε^* is continuous on X. [Hint. Notice that $\psi_f \circ \varepsilon^* = f, f \in \mathfrak{C}_r(X)$, and use (5.80).]

Exercise 22.39. For $i = 1, 2$, let X^i be a metric space, let P^i denote the set of all probabilities on $\mathcal{B}(X^i)$, and let $\mathcal{W}^i$ be the topology for P^i defined as above (22.1). Let $h : X^1 \to X^2$ be a continuous function, and define $h^* : \mathrm{P}^1 \to \mathrm{P}^2$ by $h^*(P) = Ph^{-1}, P \in \mathrm{P}^1$. Show that h^* is continuous on P^1. [Hint. Notice that $\psi_g \circ h^* = \psi_{g \circ f}, g \in \mathfrak{C}_r(X^2)$, and apply (5.80).]

Exercise 22.40. Let X be an arbitrary topological space, let M denote the space of all finite measures on $\mathcal{B}(X)$, and let M be endowed with the topology $\mathcal{W}_{\mathrm{M}}$ generated by the family $\{\psi_f : f \in \mathfrak{C}_r(X)\}$, where $\psi_f(\mu) = \int_X f\, d\mu, \mu \in \mathrm{M}$.

(a) Let $g : X \to [0, \infty[$ be continuous and bounded, and define $g^* : \mathrm{M} \to \mathrm{M}$ by $g^*(\mu) = g\mu, \mu \in \mathrm{M}$. Use (5.80) and (12.8) to prove that g^* is continuous on M.

(b) Show that the mapping that associates to each pair $(\mu, \nu) \in \mathrm{M}^2$ the measure $\mu + \nu \in \mathrm{M}$ is continuous on M^2.

Exercise 22.41. Let $\mathcal{A} \subset \mathcal{B}(X)$ be such that $A, B \in \mathcal{A}$ implies $A \cap B \in \mathcal{A}$, and assume that each $U \in \mathcal{T}$ is a countable union of sets in $\mathcal{A}$. For $\{P_n : n \in N\} \subset \mathrm{P}$ and $P \in \mathrm{P}$, show that $P_n \Rightarrow P$ whenever $P_n(A) \to P(A), A \in \mathcal{A}$. [Hint. Let $U = \cup_{i \geqslant 1} A_i \in \mathcal{T}$ with $A_i \in \mathcal{A}$. Prove first that

$$\lim_n P_n\left(\bigcup_{i=1}^m A_i\right) = P\left(\bigcup_{i=1}^m A_i\right), \quad m \geqslant 2. \tag{1}$$

Given $\varepsilon > 0$, choose m_ε so that $P(U) - \varepsilon \leqslant P(\cup_{i=1}^{m_\varepsilon} A_i)$, and use (1) to arrive at (22.5.iv).]

Exercise 22.42. Let $\{P_n : n \in N\} \subset \mathrm{P}$ and $P \in \mathrm{P}$, and assume that $P_n \Rightarrow P$. Show that $\liminf_n \int_X f\, dP_n \geqslant \int_X f\, dP$ for any bounded and lower semicontinuous function $f : X \to R$. [Hints. Let $b < \int_X f\, dP$, and choose a nondecreasing sequence $\{f_n : n \in N\} \subset \mathfrak{C}_r(X)$ so that $f_n \to f$ and $|f_n| \leqslant \|f\|_u$ (6.96.b). Select $m_0 \in N$ with

$$\int_X f dP - \int_X f_{m_0}\, dP < \frac{1}{2}\left(\int_X f dP - b\right), \tag{1}$$

and select $n_b \in N$ such that

$$\int_X f_{m_0}\, dP - \int_X f_{m_0}\, dP_n < \frac{1}{2}\left(\int_X f\, dP - b\right), \quad n \geqslant n_b. \tag{2}$$

Use (1) and (2) to prove that $b < \int_X f\,dP_n, n \geqslant n_b$.]

Exercise 22.43. Let $(\Omega, \mathcal{K}, \mathbf{P})$ be a probability space, let (X, d) be a metric space, let $f_n : \Omega \to X$ and $g_n : \Omega \to X, n \in N$, be random elements, and let $f : \Omega \to X$ be a random element such that $f_n \xrightarrow{D} f$ and $d(f_n, g_n) \xrightarrow{\mathbf{P}} 0$. Prove that $g_n \xrightarrow{D} f$.

Exercise 22.44. Let $(\Omega, \mathcal{K}, \mathbf{P})$ be a probability space, let X and Y be metric spaces, let $h : X \to Y$ be a measurable function, and let $f_n : \Omega \to X, n \in N$, be random elements such that $f_n \xrightarrow{\mathbf{P}} x \in X$. Show that $h \circ f_n \xrightarrow{\mathbf{P}} h(x)$ whenever h is continuous at x. [Hint. If F is a closet subset of Y and h is continuous at x, verify that $\limsup_n (\mathbf{P}(h \circ f_n)^{-1})(F) \leqslant \varepsilon_{h(x)}(F)$.]

Exercise 22.45. Let X be a metric space, let M and $\mathcal{W}_{\mathrm{M}}$ be as in (22.40), and let $\{\mu_n : n \in N\} \subset \mathrm{M}$ and $\mu \in \mathrm{M}$. We say that $\{\mu_n : n \in N\}$ *converges weakly to* μ, and we write $\mu_n \Rightarrow \mu$, if $\{\mu_n : n \in N\}$ converges to μ with respect to the topology $\mathcal{W}_{\mathrm{M}}$. Prove that the following assertions are equivalent:

(i) $\mu_n \Rightarrow \mu$;
(ii) $\int_X f\,d\mu_n \to \int_X f\,d\mu$ for any $f \in \mathfrak{C}_r(X)$;
(iii) $\int_X f\,d\mu_n \to \int_X f\,d\mu$ for any uniformly continuous function $f \in \mathfrak{C}_r(X)$;
(iv) $\limsup_n \mu_n(F) \leqslant \mu(F)$ for any $F \in \mathcal{F}$;
(v) $\liminf_n \mu_n(U) \geqslant \mu(U)$ for any $U \in \mathcal{T}$;
(iv) $\lim_n \mu_n(A) = \mu(A)$ for any $A \in \mathcal{B}(X)$ with $\mu(\partial(A)) = 0$.

Exercise 22.46. Notation is as in (22.45). Let Y be a metric space, and let $h : X \to Y$ be a homeomorphism from X into Y. Show that $\mu_n \Rightarrow \mu$ if and only if $\mu_n h^{-1} \Rightarrow \mu h^{-1}$.

Exercise 22.47. Notation is as in (22.45). For $0 < a < \infty$, let $\mathrm{M}_a = \{\mu \in \mathrm{M} : \mu(X) \leqslant a\}$ and put $\mathcal{W}_a = \mathrm{M}_a \cap \mathcal{W}_{\mathrm{M}}$. Prove the following.

(a) M_a is a closed subset of M.
(b) X is compact if and only if $(\mathrm{M}_a, \mathcal{W}_a)$ is metrizable and compact.
(c) X is separable if and only if $(\mathrm{M}_a, \mathcal{W}_a)$ is metrizable and separable.
(d) If X is separable and $\{x_i : i \in I\} \subset X$ is countable with $\{x_i : i \in I\}^- = X$, then the countable set $\{\sum_{j \in J} p_j \varepsilon_{x_j} : J$ is finite, $\sum_{j \in J} p_j \leqslant a$ and $0 < p_j \in Q, j \in J\}$ is dense in M_a.

Exercise 22.48. Notation is as in (22.47). Suppose that X is separable, let $\mathrm{N} \subset \mathrm{M}_a$ for some $a \in]0, \infty[$, and assume that for any $\varepsilon > 0$ there is a compact set $C_\varepsilon \subset X$ such that $\nu(C_\varepsilon^c) < \varepsilon$ for all $\nu \in \mathrm{N}$. Mimic the proof of (22.27) to show that N^- is compact with respect to $\mathcal{W}_a$.

Exercise 22.49. Notation is as in (22.45). Suppose that X is separable and complete, and let $\mathrm{N} \subset \mathrm{M}$ be such that N^- is compact with respect to $\mathcal{W}_{\mathrm{M}}$.

(a) Prove that $\mathrm{N}^- \subset \mathrm{M}_a$ for some $a \in]0, \infty[$. [Use (5.35) and (22.47.a).]
(b) Adapt the arguments in the proofs of (22.28) and (22.29) to show that for each $\varepsilon > 0$ there exists a compact set $C_\varepsilon \subset X$ such that $\nu(C_\varepsilon^c) < \varepsilon$ for all $\nu \in \mathrm{N}$.

9 Central Limit Problem

In this chapter we discuss one of the most important problems in probability theory. The main object is an array of random variables $\{f_{nk} : n \geqslant 1, 1 \leqslant k \leqslant k_n\}$, where $k_n \to \infty$, and $f_{nk}, 1 \leqslant k \leqslant k_n$, are independent for each n, and the problem consists in determining the set of all possible distribution functions F with $\sum_{k=1}^{k_n} f_{nk} \xrightarrow{D} F$, and in seeking conditions for convergence to a specified F. To avoid the trivial situation when any $F \in \mathfrak{F}$ can be a limit of this kind, we impose an additional restriction called the infinite smallness condition. This schema covers the classical problem of searching conditions under which normed sums $(f_1 + \cdots + f_n)/b_n - a_n$, where $f_n, n \geqslant 1$, are independent random variables, $a_n \in R$ and $b_n > 0$, converge in distribution to a degenerate or Gaussian distribution function. (Take $k_n = n$ and $f_{nk} = f_k/b_n - a_n/n, 1 \leqslant k \leqslant n$.) It turns out that the totality of limit distribution functions for sums of the form $\sum_{k=1}^{k_n} f_{nk}$ coincides with a special subset of $\mathfrak{F}$, namely the set of infinitely divisible distribution functions. Section 23 is devoted to a detailed study of the class of infinitely divisible distribution/characteristic functions. Resting on the material of Section 23, in Section 24 we present necessary and sufficient conditions for $\sum_{k=1}^{k_n} f_{nk} \xrightarrow{D} F$ with a given infinitely divisible F, and we specialize the general criterion of convergence obtained to the three basic cases when F is either degenerate, Poisson or Gaussian.

23 Infinitely Divisible Distribution/Characteristic Functions

In the first part of this section we investigate basic properties of infinitely divisible distribution/characteristic functions culminating with the Lévy-Khintchine representation of an infinitely divisible characteristic function. Then we show that the class $\mathfrak{C}_I$ of infinitely divisible characteristic functions coincides with the class of limiting characteristic functions φ_{S_n}, where $S_n = \sum_{k=1}^{k_n} f_{nk}$ and $\{f_{nk} : n \geqslant 1, 1 \leqslant k \leqslant k_n\}$ is as above, and we identify and describe two special subclasses of $\mathfrak{C}_I$, namely the class of self-decomposable characteristic functions and the class of stable characteristic functions.

Definitions 23.1. A distribution function $F \in \mathfrak{F}$ is called *infinitely divisible* if for each $n \in N$ there exists $F_n \in \mathfrak{F}$ such that $F = F_n^{*n}$, where F_n^{*n} denotes the n-fold convolution of F_n. A characteristic function $\varphi \in \mathfrak{C}$ is said to be *infinitely divisible* if

Analysis and Probability. http://dx.doi.org/10.1016/B978-0-12-401665-1.00009-6

for any $n \in N$ there exists $\varphi_n \in \mathfrak{C}$ such that $\varphi = (\varphi_n)^n$. In terms of random variables, this means that for every $n \in N$ there exist a probability space $(X_n, \mathcal{A}_n, P_n)$, a random variable $f : X_n \to R$, and independent random variables $f_{nk} : X_n \to R$, $1 \leqslant k \leqslant n$, such that f has distribution function F, f_{nk} has distribution function F_n for each k, and $f = f_{n1} + \cdots + f_{nn}$. That such a construction is possible follows from (19.8.d). Let $\mathfrak{F}_I$ and $\mathfrak{C}_I$ denote the sets of all infinitely divisible distribution functions and characteristic functions, respectively. Plainly, $F \in \mathfrak{F}_I$ if and only if $\varphi_F \in \mathfrak{C}_I$.

Examples 23.2. (a) The degenerate distribution function $F_a \in \mathfrak{F}_I$, as $F_a = F_{a/n}^{*n}, n \in N$.

(b) The Gaussian distribution function $G_{m,\sigma} \in \mathfrak{F}_I$, since $\varphi_{G_{m,\sigma}} = (\varphi_{G_{m/n,\sigma/\sqrt{n}}})^n, n \in N$, by (20.3.h).

(c) If F is the Poisson distribution function with parameters a, b and λ, then (20.3.d) shows that $F = F_n^{*n}, n \in N$, where F_n is the Poisson distribution function with parameters a/n, b and λ/n and so $F \in \mathfrak{F}_I$.

(d) If φ is as in (20.55), then $\varphi \in \mathfrak{C}_I$.

Remarks 23.3. (a) If $F \in \mathfrak{F}_I$ and $G(x) = F(ax + b), x \in R$, for some $a > 0$ and $b \in R$, then $G \in \mathfrak{F}_I$. Actually, for $n \in N$, let $\varphi_{F,n} \in \mathfrak{C}$ be so that $\varphi_F = (\varphi_{F,n})^n$. If f is a random variable whose distribution function is F, then the distribution function of $(f - b)/a$ is G, and so $\varphi_G(t) = \varphi_F(t/a)e^{-itb/a} = (\varphi_{F,n}(t/a)e^{-itb/na})^n, t \in R, n \in N$.

(b) If $\varphi_1, \varphi_2 \in \mathfrak{C}_I$, then $\varphi_1\varphi_2 \in \mathfrak{C}_I$. Hence $|\varphi|^2 \in \mathfrak{C}_I$ whenever $\varphi \in \mathfrak{C}_I$.

Theorem 23.4. *If $\varphi \in \mathfrak{C}_I$, then $\varphi(t) \neq 0$ for every $t \in R$.*

Proof. Since φ is continuous at 0 with $\varphi(0) = 1$, there exists $t_0 > 0$ such that $\varphi(t) \neq 0$ for $|t| \leqslant t_0$. Let $\varphi = (\varphi_n)^n, n \in N$, with $\varphi_n \in \mathfrak{C}$. Hence $|\varphi_n(t)| = |\varphi(t)|^{1/n} \to 1, |t| \leqslant t_0$. Let $0 < \varepsilon < 1/8$. Then, on account of Dini's theorem (6.44), there is $n_\varepsilon \in N$ so that $|\varphi_n(t)| > 1 - \varepsilon$ whenever $|t| \leqslant t_0$ and $n \geqslant n_\varepsilon$. Therefore, in view of (20.46.b), $1 - |\varphi_n(2t)|^2 \leqslant 4(1 - |\varphi_n(t)|^2) \leqslant 4(2\varepsilon - \varepsilon^2) < 8\varepsilon, |t| \leqslant t_0, n \geqslant n_\varepsilon$. This shows that $\varphi_n(2t) \neq 0$ for $|t| \leqslant t_0$ and $n \geqslant n_\varepsilon$. Thus $\varphi(t) \neq 0$ for $|t| \leqslant 2t_0$, and so $\varphi(t) \neq 0$ for any $t \in R$. □

To get further insight into the structure of $\mathfrak{C}_I$ and the properties of its members, we will need to determine a continuous logarithm and a continuous nth root of a characteristic function. In this connection, notice that $e^{2k\pi i\{t\}} = e^{2k\pi it}, t \in R, k \in Z$, where $\{t\}$ is the fractional part of t. The following lemmas are essential.

Lemma 23.5. *If $h_1 : R \to C$ and $h_2 : R \to C$ are continuous functions such that $h_1(0) = h_2(0) = 0$ and $e^{h_1} = e^{h_2}$, then $h_1 = h_2$.*

Proof. If $e^{h_1(t)} = e^{h_2(t)}, t \in R$, then $h_1(t) - h_2(t) = 2k(t)\pi i, t \in R$, with $k(t) \in Z$. Since the function $k(t)$ is continuous with $k(0) = 0$, it follows that $k(t) = 0, t \in R$, and so $h_1 = h_2$. □

Lemma 23.6. *If $g : R \to C$ is a continuous function such that $g(0) = 1$ and $g(t) \neq 0$ for any $t \in R$, then there exists a unique continuous function $h : R \to C$ with $h(0) = 0$ and $g(t) = e^{h(t)}, t \in R$.*

Proof. Let $T > 0$ and $\rho_T = \inf |g|([-T, T])$. Since $g(0) = 1$ and $g([-T, T])$ is a compact subset of C (5.45) with $0 \notin g([-T, T])$, we have $0 < \rho_T \leqslant 1$. As g is uniformly continuous on $[-T, T]$ (6.62), there is $\delta_T > 0$ such that $t, t' \in [-T, T]$ and $|t - t'| < \delta_T$ imply $|g(t) - g(t')| < \rho_T/2 \leqslant 1/2$. Consider $-T = t_{-m} < \cdots < t_{-1} < t_0 < t_1 < \cdots < t_m = T$ a division of $[-T, T]$ with $t_0 = 0$ and $t_{j+1} - t_j = t_1 - t_0 < \delta_T$, $-m \leqslant j < m$. Define

$$h(t) = \sum_{j \geqslant 1} \frac{(-1)^{j-1}}{j}(g(t) - 1)^j, \quad t \in [-t_1, t_1]. \tag{1}$$

The series in (1) converges uniformly on $[-t_1, t_1]$, since $|g(t) - 1| = |g(t) - g(t_0)| < 1/2$ for $-t_1 \leqslant t \leqslant t_1$. Hence h is continuous on $[-t_1, t_1]$ (6.42) with $h(0) = 0$. According to a theorem in calculus, the power series

$$L(z) = \sum_{j \geqslant 1} \frac{(-1)^{j-1}}{j}(z - 1)^j, \quad z \in C, \ |z - 1| < 1,$$

represents the principal value of $\log z$ vanishing at $z = 1$, and so

$$e^{L(z)} = z, \quad z \in C, \ |z - 1| < 1. \tag{2}$$

Thus $h(t) = L(g(t))$ and $e^{h(t)} = g(t)$, $t \in [-t_1, t_1]$. For $1 \leqslant j \leqslant m$, the definition of h can be extended from $[t_{-j}, t_j]$ to $[t_j, t_{j+1}]$ by putting

$$h(t) = h(t_j) + L(g(t)/g(t_j)), \quad t \in [t_j, t_{j+1}]. \tag{3}$$

Since $|g(t)/g(t_j) - 1| < (\rho_T/2)/\rho_T = 1/2$, $t \in [t_j, t_{j+1}]$, $L(g(t)/g(t_j))$ is defined in (3), and h is continuous on $[t_j, t_{j+1}]$. We have

$$e^{h(t)} = e^{h(t_j)}\, e^{L(g(t)/g(t_j))} = g(t_j)\frac{g(t)}{g(t_j)} = g(t), \quad t \in [t_j, t_{j+1}],$$

by (2). Similar arguments allow the extension of h from $[t_{-j}, t_j]$ to $[t_{-j-1}, t_{-j}]$. Thus the lemma is proved by induction for the interval $[-T, T]$. Now, having defined h on $[-T, T]$, we can extend it to $[T, 2T]$ and $[-2T, -T]$ by the above method, and hence by induction to R. The uniqueness of h follows immediately from (23.5). □

Definitions 23.7. Notation is as in (23.6). The function h is called the *distinguished logarithm of* g and is denoted by $\operatorname{Log} g$. If $|g(t) - 1| < 1$ for $-T \leqslant t \leqslant T$ and some $T > 0$, then $\operatorname{Log} g(t) = L(g(t))$, $-T \leqslant t \leqslant T$. If g_1 and g_2 are as in (23.6), then $\operatorname{Log}(g_1 g_2) = \operatorname{Log} g_1 + \operatorname{Log} g_2$ and $\operatorname{Log}(g_1/g_2) = \operatorname{Log} g_1 - \operatorname{Log} g_2$. For $n \in N$, the function $e^{(1/n)h}$ is called the *distinguished* nth *root of* g and is denoted by $g^{1/n}$. Thus $g^{1/n} = e^{(1/n)\operatorname{Log} g}$.

Lemma 23.8. *Let g and h, and g_n and h_n, $n \in N$, be as in* (23.6), *and assume that $g_n \to g$ uniformly on $[-T, T]$ for some $T > 0$. Then $h_n \to h$ uniformly on $[-T, T]$.*

Proof. We have $|g_n(t)/g(t) - 1| \leqslant |g_n(t) - g(t)| / \rho_T$, $-T \leqslant t \leqslant T$, where ρ_T is as in the proof of (23.6), and so $g_n/g \to 1$ uniformly on $[-T, T]$. Choose $n_0 \in N$ such that $|g_n(t)/g(t) - 1| \leqslant 1/2$ for all $t \in [-T, T]$ and $n \geqslant n_0$. Then we may write

$$h_n(t) - h(t) = \text{Log}(g_n(t)/g(t)) = L(g_n(t)/g(t)), \quad t \in [-T, T], \; n \geqslant n_0. \quad (1)$$

Since L is continuous at 1 with $L(1) = 0$, (1) shows that $h_n \to h$ uniformly on $[-T, T]$. □

Theorem 23.9. *Let $\varphi \in \mathfrak{C}$. Then $\varphi \in \mathfrak{C}_I$ if and only if the distinguished n-th root $\varphi^{1/n}$ exists and $\varphi^{1/n} \in \mathfrak{C}$ for each $n \in N$.*

Proof. If $\varphi \in \mathfrak{C}_I$, let $\varphi = (\varphi_n)^n, n \in N$, with $\varphi_n \in \mathfrak{C}$. Then (23.4) ensures that $\varphi(t) \neq 0, t \in R$, and so $\varphi_n(t) \neq 0, t \in R$. It follows from (23.6) that the distinguished logarithms $\text{Log}\,\varphi$ and $\text{Log}\,\varphi_n$ exist. We have $e^{\text{Log}\,\varphi} = \varphi = (\varphi_n)^n = e^{n\text{Log}\,\varphi_n}$, and so $\text{Log}\,\varphi = n\,\text{Log}\,\varphi_n$ by (23.5). Hence $\varphi^{1/n} = e^{(1/n)\text{Log}\,\varphi} = e^{\text{Log}\,\varphi_n} = \varphi_n \in \mathfrak{C}$.

Conversely, if $\varphi^{1/n}$ is defined and belongs to $\mathfrak{C}$ for every $n \in N$, then $\varphi = e^{\text{Log}\,\varphi} = (e^{(1/n)\text{Log}\,\varphi})^n, n \in N$, i.e. $\varphi \in \mathfrak{C}_I$. □

Remarks 23.10. (a) If $\varphi \in \mathfrak{C}_I$ and $\varphi(R) \subset]0, 1]$, then its distinguished nth root $\varphi^{1/n}$ is just the real positive nth root of φ.

(b) If $\varphi \in \mathfrak{C}_I$, then $\varphi^{1/n} = e^{(1/n)\text{Log}\,\varphi} \to 1$. Since 1 is a characteristic function, (20.9) shows that $\varphi^{1/n}(t) \to 1$ uniformly in $|t| \leqslant t_0$ for any $t_0 > 0$.

Theorem 23.11. *Let $\{\varphi_k : k \in N\} \subset \mathfrak{C}_I$ be such that $\varphi_k \to \varphi \in \mathfrak{C}$. Then $\varphi \in \mathfrak{C}_I$.*

Proof. Write $\psi_k = |\varphi_k|^2$, $k \in N$, and $\psi = |\varphi|^2$. Then $\psi_k \in \mathfrak{C}_I, k \in N, \psi \in \mathfrak{C}_I$ and $\psi_k \to \psi$. Therefore, for $n \in N$,

$$\lim_k \psi_k^{1/n} = \psi^{1/n}, \quad (1)$$

where $\psi_k^{1/n}$ is the distinguished nth root of ψ_k by (23.10.a), and so $\psi_k^{1/n} \in \mathfrak{C}$ by (23.9), while $\psi^{1/n}$ is the distinguished nth root of ψ. Since $\psi^{1/n}$ is continuous at 0, (1) shows that $\psi^{1/n} \in \mathfrak{C}$ by Lévy's theorem (20.11). Thus $\psi = (\psi^{1/n})^n \in \mathfrak{C}_I$, and so ψ and hence φ never vanishes by (23.4). In view of (23.6), φ has a distinguished logarithm $\text{Log}\,\varphi$. Since $\varphi_k \to \varphi$ uniformly on each interval $[-T, T], T > 0$, by (20.12) and (20.9), it follows from (23.8) that $\text{Log}\,\varphi_k \to \text{Log}\,\varphi$. Consequently,

$$\lim_k \varphi_k^{1/n} = \lim_k e^{(1/n)\text{Log}\,\varphi_k} = e^{(1/n)\text{Log}\,\varphi} = \varphi^{1/n}, \quad n \in N. \quad (2)$$

As $\varphi_k^{1/n} \in \mathfrak{C}$ by (23.9), and $\varphi^{1/n}$ is continuous at 0, (1) and (20.11) show that $\varphi^{1/n} \in \mathfrak{C}$. Hence $\varphi \in \mathfrak{C}_I$ by (23.9). □

In what follows we shall establish a canonical representation for infinitely divisible characteristic functions.

Theorem 23.12. *Let $\gamma \in R$, let $G : R \to R$ be nondecreasing and right continuous on R with $G(-\infty) = 0$ and $G(\infty) < \infty$, and define*

$$\chi(t,x) = \begin{cases} \frac{e^{itx}-1-itx}{x^2} + e^{itx} - 1 & \text{if } (t,x) \in R \times (R - \{0\}) \\ \lim\limits_{x\to 0}\left(\frac{e^{itx}-1-itx}{x^2} + e^{itx} - 1\right) = -\frac{t^2}{2} & \text{if } (t,x) \in R \times \{0\} \end{cases},$$

$\psi(t) = i\gamma t + \int_R \chi(t,x)dG(x)$, $t \in R$, and $\varphi(t) = e^{\psi(t)}$, $t \in R$. Then $\varphi \in \mathfrak{C}_I$.

Proof. For $t_0 > 0$, we have

$$\begin{aligned} \sup_{|t|\leqslant t_0, x\in R} |\chi(t,x)| &\leqslant 2 + \sup_{|t|\leqslant t_0, |x|>1} |\chi(t,x)| + \sup_{|t|\leqslant t_0, |x|\leqslant 1} |\chi(t,x)| \\ &\leqslant 4 + t_0 + \sup_{|u|\leqslant t_0} \frac{\left|e^{iu} - 1 - iu\right|}{u^2} t_0^2 \leqslant C < \infty, \end{aligned}$$

since $e^{iu} = 1 + iu + O(u^2)$ as $u \to 0$, where C is a constant depending on t_0. Hence dominated convergence ensures that ψ is continuous on $]-t_0, t_0[$, and so ψ is continuous on R. As $\psi(0) = 0$, we see that $\psi = \operatorname{Log}\varphi$. Clearly, $\chi(t,x)$ is bounded and continuous in x for every fixed t. Now, for $n \in N$, apply (22.47.d) to choose $x_{n1}, \ldots, x_{nj_n} \in Q - \{0\}$ and $p_{n1}, \ldots, p_{nj_n} \in Q_+$ with $\sum_{j=1}^{j_n} p_{nj} \leqslant G(\infty)$ such that $\mu_n = \sum_{j=1}^{j_n} p_{nj}\varepsilon_{x_{nj}} \Rightarrow \mu_G$. Therefore, on account of (22.45), we may write

$$\varphi(t) = e^{i\gamma t} \lim_n \exp\left(\int_R \chi(t,x)d\mu_n(x)\right), \quad t \in R. \tag{1}$$

We have

$$\begin{aligned} \exp\left(\int_R \chi(t,x)d\mu_n(x)\right) &= \exp\left(\sum_{j=1}^{j_n} p_{nj}\chi(t,x_{nj})\right) \\ = \prod_{j=1}^{j_n} \exp\left(\frac{p_{nj}}{x_{nj}^2}(e^{itx_{nj}} - 1)\right) &\exp\left(-it\frac{p_{nj}}{x_{nj}} + p_{nj}(e^{itx_{nj}} - 1)\right), \quad t \in R. \end{aligned} \tag{2}$$

The last term in (2) is a product of characteristic functions of Poisson distribution functions (20.3.d), and so it is an infinite divisible characteristic function by (23.2) and (23.3.b). Since φ is continuous on R, from (1) and (2) it follows that $\varphi \in \mathfrak{C}$ via Lévy's theorem (20.11). From (1), (2) and (22.11) we see that $\varphi \in \mathfrak{C}_I$. □

Notation being as in (23.12), we write $\psi(t) = \psi(t;\gamma,G)$ and $\varphi(t) = \varphi(t;\gamma,G)$ whenever we want to stress that ψ and φ depend on γ and G.

Remarks 23.13. (a) If $\gamma = a$ and $G(x) = 0$, $x \in R$, then $\varphi(t;\gamma,G)$ is the characteristic function of a distribution function degenerated at a.

(b) If $\gamma = m \in R$ and $G(x) = 0$ or σ^2 according as $x < 0$ or $x \geqslant 0$, where $\sigma > 0$, then $\varphi(t; \gamma, G)$ is the characteristic function of a normal distribution function with parameters m and σ.

(c) If $\gamma = a + b\lambda/(1 + b^2)$, where $\lambda > 0, a \in R$ and $b \neq 0$, and $G(x) = 0$ or $b^2\lambda/(1 + b^2)$ according as $x < b$ or $x \geqslant b$, then $\varphi(t; \gamma, G)$ is the characteristic function of a Poisson distribution function with parameters a, b and λ.

(d) $\varphi(t; \gamma_1, G_1)\varphi(t; \gamma_2, G_2) = \varphi(t; \gamma_1 + \gamma_2, G_1 + G_2)$, as $\mu_{G_1} + \mu_{G_2} = \mu_{G_1+G_2}$.

(e) If $\{\gamma_n : n \in N\} \subset R$ and $\gamma_n \to \gamma \in R$, and $\{G_n : n \in N\} \subset \mathfrak{G}$ with $G_n(-\infty) = 0, n \in N$, and $G_n \overset{c}{\to} G \in \mathfrak{G}$, where $\mathfrak{G}$ is as in (19.67), then $\varphi(t; \gamma_n, G_n) \to \varphi(t; \gamma, G)$. This follows from (19.67.c), since $\chi(t, x)$ is continuous and bounded in x for each fixed t.

Theorem 23.14. *If $\varphi(t; \gamma_1, G_1) = \varphi(t; \gamma_2, G_2), t \in R$, then $\gamma_1 = \gamma_2$ and $G_1 = G_2$.*

Proof. If $\psi(t; \gamma, G)$ is as in (23.12) and $h > 0$, write

$$\begin{aligned}
\psi^{(h)}(t; \gamma, G) &= \psi(t + h; \gamma, G) + \psi(t - h; \gamma, G) - 2\psi(t; \gamma, G) \\
&= \int_R \left(\frac{e^{i(t+h)x} + e^{i(t-h)x} - 2\,e^{itx}}{x^2} \right. \\
&\qquad \left. + e^{i(t+h)x} + e^{i(t-h)x} - 2\,e^{itx} \right) dG(x) \\
&= \int_R e^{itx}(e^{ihx} + e^{-ihx} - 2)\frac{1 + x^2}{x^2} dG(x) \\
&= -2 \int_R e^{itx}(1 - \cos hx)\frac{1 + x^2}{x^2} dG(x).
\end{aligned}$$

The integrand in the last integral is bounded and continuous in x, its value at $x = 0$ defined by continuity being h^2. Applying Fubini's theorem, we have

$$\begin{aligned}
\int_0^1 \psi^{(h)}(t; \gamma, G)dh &= -2 \int_R e^{itx} \left(\int_0^1 (1 - \cos hx)dh \right) \frac{1 + x^2}{x^2} dG(x) \\
&= -2 \int_R e^{itx} \left(1 - \frac{\sin x}{x} \right) \frac{1 + x^2}{x^2} dG(x) \\
&= -2 \int_R e^{itx} g(x) dG(x), \qquad (1)
\end{aligned}$$

where $g(x) = (1 - (\sin x)/x)(1 + x^2)/x^2$, $x \in R$, with $g(0)$ defined by continuity to be $1/6$. Clearly, g is bounded and continuous with $\lim_{x\to-\infty} g(x) = \lim_{x\to\infty} g(x) = 1$, g does not vanish on R, and $\inf_{x\in R} g(x) > 0$. This shows that the function $1/g$ is bounded and continuous on R. Now $\varphi(t; \gamma_1, G_1) = \varphi(t; \gamma_2, G_2), t \in R$, implies that

$$\psi(t; \gamma_1, G_1) = \psi(t; \gamma_2, G_2), \quad t \in R, \qquad (2)$$

by (23.5), and so

$$\int_R e^{itx} g(x)dG_1(x) = \int_R e^{itx} g(x)dG_2(x), \quad t \in R, \tag{3}$$

by (1). Taking $t = 0$ in (3) yields $c = \int_R g\,dG_1 = \int_R g\,dG_2$. If $c = 0$, then $0 = \mu_{G_1}(R) = \mu_{G_2}(R)$, whence $G_1 = G_2 = 0$, and hence $\gamma_1 = \gamma_2$ by (2). If $c > 0$ and $i = 1, 2$, put $F_i(x) = c^{-1}\int_{]-\infty,x]} g(y)dG_i(y), x \in R$. Then F_i is a distribution function such that $\mu_{F_i} = c^{-1}g\mu_{G_i}$, and (3) can be rewritten as $\int_R e^{itx}\,dF_1(x) = \int_R e^{itx}\,dF_2(x), t \in R$. In view of (20.8), this entails that $g\mu_{G_1} = g\mu_{G_2}$. Consequently, $\mu_{G_1}(A) = \int_A (1/g)g\,d\mu_{G_1} = \int_A (1/g)d(g\mu_{G_1}) = \int_A (1/g)d(g\mu_{G_2}) = \int_A (1/g)g\,d\mu_{G_2} = \mu_{G_2}(A), A \in \mathcal{B}(R)$, by (12.8), and so $G_1(x) = \mu_{G_1}(]-\infty, x]) = \mu_{G_2}(]-\infty, x]) = G_2(x), x \in R$. Then $\gamma_1 = \gamma_2$ follows from (2). □

Theorem 23.15. *For $n \in N$ and $t \in R$, let $\varphi(t; \gamma_n, G_n)$ be defined as in (23.12), and assume that $\varphi(t; \gamma_n, G_n) \to \xi(t)$ for any $t \in R$, where ξ is continuous at 0. Then there exist $\gamma \in R$ and $G \in \mathfrak{G}$ such that $\xi(t) = \varphi(t; \gamma, G), t \in R$, and $\gamma_n \to \gamma$ and $G_n \xrightarrow{c} G$.*

Proof. It follows from Lévy's theorem (20.11) and (23.11) that $\xi \in \mathfrak{C}_I$, and so $\xi(t) \neq 0, t \in R$, by (23.4). Hence $\operatorname{Log}\xi$ exists by (23.6). Applying the uniform convergence theorem (20.9) and (23.8), we see that $\lim_n \psi(t; \gamma_n, G_n) = \operatorname{Log}\xi(t)$ uniformly in $|t| \leqslant T$ for any $T > 0$. This ensures that $\lim_n \psi^{(h)}(t; \gamma_n, G_n) = \operatorname{Log}\xi(t+h) + \operatorname{Log}\xi(t-h) - 2\operatorname{Log}\xi(t)$ uniformly in $|t| \leqslant T$ and $0 < h \leqslant 1$, and so

$$\lim_n \int_0^1 \psi^{(h)}(t; \gamma_n, G_n)dh = \int_0^1 (\operatorname{Log}\xi(t+h) + \operatorname{Log}\xi(t-h))dh - 2\operatorname{Log}\xi(t), \quad t \in R. \tag{1}$$

If $-2\phi(t)$ denotes the right-hand side of (1), then the continuity of $\operatorname{Log}\xi$ shows that ϕ is continuous. Taking into account (1) of the proof of (23.14), we have

$$\lim_n \int_R e^{itx} g(x)dG_n(x) = \phi(t), \quad t \in R. \tag{2}$$

Taking $t = 0$ in (2) yields $c_n = \int_R g\,dG_n \to \phi(0) \geqslant 0$. If $\phi(0) = 0$, then $G_n(\infty) = \mu_{G_n}(R) \leqslant (\inf_{x\in R} g(x))^{-1}c_n \to 0$, and so $G_n(x) \to 0, x \in R$. Thus $G_n \xrightarrow{c} 0$ and $\gamma_n = (\psi(1; \gamma_n, G_n) - \int_R \chi(1, x)dG_n(x))/i \to \gamma = \operatorname{Log}\xi(1)/i$ by (19.67.c). Then necessarily $\xi(t) = \varphi(t; \gamma, 0), t \in R$, (23.13.e). If $\phi(0) > 0$ and $n \in N$, set $F_n(x) = c_n^{-1}\int_{]-\infty,x]} g(y)dG_n(y), x \in R$. Then F_n is a distribution function such that $\mu_{F_n} = c_n^{-1}g\mu_{G_n}$, and (2) becomes $\lim_n \int_R e^{itx}\,dF_n(x) = \phi(t)/\phi(0), t \in R$. Since ϕ is continuous, Lévy's theorem (20.11) ensures the existence of some $F \in \mathfrak{F}$ so that $F_n \xrightarrow{w} F$. Define $G(x) = \int_{]-\infty,x]} (\phi(0)/g(y))dF(y)$, $x \in R$. Clearly, $G \in \mathfrak{G}$ and $\mu_G = (\phi(0)/g)\mu_F$. Then, applying (12.8), for $f \in \mathfrak{C}_r(R)$, as also $1/g \in \mathfrak{C}_r(R)$, we have

$$\int_R f\,d\mu_{G_n} = \int_R \frac{f}{g} d(g\mu_{G_n}) = c_n \int_R \frac{f}{g} d\mu_{F_n} \to \phi(0) \int_R \frac{f}{g} d\mu_F = \int_R f\,d\mu_G.$$

This shows that $G_n \xrightarrow{c} G$ by (19.67.c). Finally, γ_n tends to a limit $\gamma \in R$ by an argument similar to the case $\phi(0) = 0$, and $\xi(t) = \varphi(t; \gamma, G)$, $t \in R$, by (23.13.e). □

Lévy-Khintchine representation 23.16. *A function $\varphi : R \to C$ is an infinite divisible characteristic function if and only if it admits the representation*

$$\varphi(t) = \exp\left\{ i\gamma t + \int_R \left(\frac{e^{itx} - 1 - itx}{x^2} + e^{itx} - 1 \right) dG(x) \right\}, \quad t \in R, \tag{1}$$

where $\gamma \in R, G \in \mathfrak{G}$ with $G(-\infty) = 0$, and the integrand is equal to $-t^2/2$ for $x = 0$.

Proof. If $\varphi \in \mathfrak{C}_I$, then $\varphi = (\varphi_n)^n$, $n \in N$, with $\varphi_n = e^{(1/n)\mathrm{Log}\,\varphi} \in \mathfrak{C}$. Since $e^z = 1 + z + O(|z|^2)$ as $z \to 0 \in C$, we have

$$\mathrm{Log}\,\varphi(t) = \lim_n n(e^{(1/n)\mathrm{Log}\,\varphi(t)} - 1) = \lim_n n(\varphi_n(t) - 1), \quad t \in R.$$

If F_n is the distribution function corresponding to φ_n, we have further

$$\begin{aligned}
\mathrm{Log}\,\varphi(t) &= \lim_n n \int_R (e^{itx} - 1) dF_n(x) \\
&= \lim_n \Bigg[it \int_R \frac{nx}{1+x^2} dF_n(x) \\
&\qquad + \int_R \left(\frac{e^{itx} - 1 - itx}{x^2} + e^{itx} - 1 \right) \frac{nx^2}{1+x^2} dF_n(x) \Bigg] \\
&= \lim_n \left(it\gamma_n + \int_R \chi(t,x) dG_n(x) \right) = \lim_n \psi(t; \gamma_n, G_n), \quad t \in R,
\end{aligned} \tag{2}$$

where $\gamma_n = \int_R nx/(1+x^2)dF_n(x)$ and $G_n(x) = \int_{]-\infty,x]} ny^2/(1+y^2)dF_n(y)$, $x \in R$. Obviously, $G_n \in \mathfrak{G}$ and $G_n(-\infty) = 0$ for any $n \in N$. From (1) we infer that $\varphi(t) = e^{\mathrm{Log}\,\varphi(t)} = \lim_n e^{\psi(t;\gamma_n,G_n)} = \lim_n \varphi(t; \gamma_n, G_n)$, $t \in R$. Then (23.15) ensures the existence of $\gamma \in R$ and $G \in \mathfrak{G}$ such that $\gamma_n \to \gamma$, $G_n \xrightarrow{c} G$ and $\varphi(t) = \varphi(t; \gamma, G)$, $t \in R$.

If φ has the form (1), then $\varphi \in \mathfrak{C}_I$ by (23.12). □

Now let $\{f_{nk} : n \geqslant 1, 1 \leqslant k \leqslant k_n\}$ be an array of random variables, where $k_n \to \infty$ and f_{nk}, $1 \leqslant k \leqslant k_n$, are independent for each n, and put $S_n = \sum_{k=1}^{k_n} f_{nk}$. Our next purpose is to show that the class of infinitely divisible characteristic functions coincides with the class of limiting characteristic functions φ_{S_n} under the restriction that individual terms in S_n become negligible as $n \to \infty$. This will be done via a rather long sequence of lemmas. We begin with some definitions.

Definition 23.17. Let $(X, \mathcal{A}, P)$ be a probability space, let f be a random variable on X, and let F be its distribution function. The number $m[f]$ is called a *median of* f if $P(f \leqslant m[f]) \geqslant 1/2 \leqslant P(f \geqslant m[f])$, or, equivalently, $F(m[f]-) \leqslant 1/2 \leqslant F(m[f])$. Such a number always exists, since the abscissa of any common point of the line $y = 1/2$ with the graph of $y = F(x)$, completed at its discontinuity points by the segments $(x, F(x-))$ to $(x, F(x))$, is a median of f.

Remarks 23.18. (a) Let M denote the set of all medians of f. Then either M consists of a single point or it is a closed interval in R called the *median segment of* f. Actually, if $m_1, m_2 \in M$ with $m_1 < m_2$, then obviously $[m_1, m_2] \subset M$. Moreover, if $\inf M = m_* < m^* = \sup M$ and $m_* < m < m^*$, we have $P(f \leqslant m_*) = \lim_{\varepsilon \to 0} P(f \leqslant m_* + \varepsilon) \geqslant 1/2$ and $P(f \geqslant m_*) \geqslant P(f \geqslant m) \geqslant 1/2$, and so $m_* \in M$. We also have $P(f \leqslant m^*) \geqslant P(f \leqslant m) \geqslant 1/2$ and $P(f \geqslant m^*) = \lim_{\varepsilon \to 0} P(f \geqslant m^* - \varepsilon) \geqslant 1/2$, and so $m^* \in M$.

(b) If $P(|f| \geqslant a) \leqslant 1/2$ for some $a \in R$, then $|m[f]| \leqslant a$.

(c) $m[cf] = cm[f]$ and $m[f + c] = m[f] + c$ for any $c \in R$.

(d) If $P(f < a) < 1/2$ and $P(f > a) < 1/2$ for some $a \in R$, then a is the unique median of f.

(e) If $m[f] = 0$, then $m[f 1_{\{|f|<c\}}] = 0$ for each $c > 0$. Actually, for $c > 0$, $P(f 1_{\{|f|<c\}} \geqslant 0) = P(f \geqslant 0) + P(f \leqslant -c)$ and $P(f 1_{\{|f|<c\}} \leqslant 0) = P(f \leqslant 0) + P(f \geqslant c)$.

(f) If f is symmetric, then $m[f] = 0$.

A simple but basic idea in the study of sums of independent random variables is the symmetrization procedure we now introduce.

Definitions 23.19. Let $(X, \mathcal{A}, P)$ be a probability space, let $\{f_n : n \in N\}$ be a sequence of random variable on X, and consider the measurable space $(\prod_{n \in N} R_n, \mathcal{B}(\prod_{n \in N} R_n))$, where $R_n = R$ for each $n \in N$. If $J \subset N$ is finite and nonempty, put $J' = \{j \in N : 2j - 1 \in J\}$ and $J'' = \{j \in N : 2j \in J\}$, and define the probability P_J^* on $\mathcal{B}(\prod_{j \in J} R_j)$ by $P_J^* = (P(f_j)_{j \in J'}^{-1}) \otimes (P(f_j)_{j \in J''}^{-1})$. By (16.23) and the Daniell-Kolmogorov extension theorem (16.20), the probabilities P_J^* determine a unique probability P^* on $\mathcal{B}(\prod_{n \in N} R_n)$ such that $P^* \pi_J^{-1} = P_J^*$ for any J. For $n \in N$, set $f_n' = \pi_{2n-1}$ and $f_n'' = \pi_{2n}$. Using (10.8) it follows that $P^*(f_n')_{n \in N}^{-1} = P^*(f_n'')_{n \in N}^{-1} = P(f_n)_{n \in N}^{-1}$, and (18.10) ensures that, with respect to P^*, the σ-algebra generated by the sequence $\{f_n' : n \in N\}$ is independent of the σ-algebra generated by the sequence $\{f_n'' : n \in N\}$. The sequences $\{f_n' : n \in N\}$ and $\{f_n'' : n \in N\}$ are called *independent copies of* $\{f_n : n \in N\}$. The sequence $\{f_n^* : n \in N\}$, where $f_n^* = f_n' - f_n''$, $n \in N$, is called the *symmetrized* $\{f_n : n \in N\}$, and f_n^* is called the *symmetrized* f_n.

Throughout the sequel, whenever a symmetrization method is invoked, P^* will have the above meaning, E^* will designate expectation with respect to P^*, and Var* will designate variance with respect to P^*.

Remarks 23.20. Notation is as in (23.19).

(a) Any finite set of random variables can be symmetrized by the above procedure.

(b) If f is a random variable, and $f^* = f' - f''$ is its associated symmetrized random variable, then $\varphi_{f^*}(t) = \varphi_{f'}(t)\varphi_{f''}(-t) = \left|\varphi_f(t)\right|^2, t \in R$. In view of (20.42.d), this shows that f^* is indeed symmetric, and so $m[f^*] = 0$ (23.18.f).

(c) For $n \in N$, $P^*(f_n')^{-1} = P^*(f_n'')^{-1} = Pf_n^{-1}$, and so we can always choose medians such that $m[f_n'] = m[f_n''] = m[f_n]$.

(d) If $f_n, n \in N$, are P-independent, then $f_n^*, n \in N$, are P^*-independent. Actually, if $f_n, n \in N$, are P-independent, then both $f_n', n \in N$, and $f_n'', n \in N$, are P^*-independent, since $P^*(f_n')_{n\in N}^{-1} = P^*(f_n'')_{n\in N}^{-1} = P(f_n)_{n\in N}^{-1}$. It follows then that the random vectors $(f_n', f_n''), n \in N$, are P^*-independent by (18.14) and (18.28.b). This entails that $f_n^*, n \in N$, are P^*-independent by (18.28.c).

(e) If $f_n, n \in N$, are P-independent and $P(\sum_{j\in J_1} f_j)^{-1} = P\left(\sum_{j\in J_2} f_j\right)^{-1}$ for some finite $J_1, J_2 \subset N$, then $P^*(\sum_{j\in J_1} f_j^*)^{-1} = P^*(\sum_{j\in J_2} f_j^*)^{-1}$. This follows from (20.4), (b), and (20.8).

(f) Let $J_1, J_2 \subset N$ be either denumerable or finite with the same number of elements. If $P(f_j)_{j\in J_1}^{-1} = P(f_j)_{j\in J_2}^{-1}$, then, taking into account that $\sigma(f_n', n \in N)$ and $\sigma(f_n'', n \in N)$ are P^*-independent, and making use of (18.29), it follows that $P^*(f_j^*)_{j\in J_1}^{-1} = P^*(f_j^*)_{j\in J_2}^{-1}$; in particular, $P^*(f_n^*)^{-1} = P^*(f_1^*)^{-1}, n \geqslant 1$, whenever $Pf_n^{-1} = Pf_1^{-1}, n \geqslant 1$.

Definition 23.21. We say that an array of random variables $\{f_{nk} : n \geqslant 1, 1 \leqslant k \leqslant k_n\}$, where $k_n \to \infty$, satisfies the condition of *infinite smallness* if $\sup_{1\leqslant k\leqslant k_n} P(|f_{nk}| \geqslant \varepsilon) \to 0$ for each $\varepsilon > 0$.

Notation 23.22. Throughout this chapter, with respect to the array $\{f_{nk} : n \geqslant 1, 1 \leqslant k \leqslant k_n\}$, we shall denote by F_{nk}, φ_{nk} and m_{nk} the distribution function, the characteristic function and the median of f_{nk}, respectively. To simplify the writing, we shall write $\sup_k$, $\sum_k$ and $\prod_k$ in place of $\sup_{1\leqslant k\leqslant k_n}$, $\sum_{k=1}^{k_n}$ and $\prod_{k=1}^{k_n}$, respectively.

Lemma 23.23. *The condition of infinite smallness implies that* $\sup_k |m_{nk}| \to 0$ *and* $\sup_k \int_{-\tau}^{\tau} |x|^p \, dF_{nk}(x) \to 0$ *for all* $\tau, p > 0$.

Proof. For $\varepsilon > 0$, choose $n_\varepsilon \in N$ such that $\sup_k P(|f_{nk}| \geqslant \varepsilon) < 1/2, n \geqslant n_\varepsilon$. It follows then from (23.18.b) that $\sup_k |m_{nk}| \leqslant \varepsilon, n \geqslant n_\varepsilon$. If $\varepsilon < \tau$, we have also

$$\begin{aligned}\sup_k \int_{-\tau}^{\tau} |x|^p \, dF_{nk}(x) &\leqslant \varepsilon^p + \sup_k \int_{]-\tau,\tau]-]-\varepsilon,\varepsilon]} |x|^p \, dF_{nk}(x) \\ &\leqslant \varepsilon^p + \tau^p \sup_k P(|f_{nk}| \geqslant \varepsilon).\end{aligned}$$

This and the infinite smallness condition entail that $\sup_k \int_{-\tau}^{\tau} |x|^p \, dF_{nk}(x) \to 0$. □

Lemma 23.24. *The following are equivalent:*

(i) $\sup_k P(|f_{nk}| \geqslant \varepsilon) \to 0$ *for any* $\varepsilon > 0$;

(ii) $\sup_k |\varphi_{nk}(t) - 1| \to 0$ *uniformly in* $|t| \leqslant t_0$ *for any* $t_0 > 0$;

(iii) $\sup_k \int_R \frac{x^2}{1+x^2} dF_{nk}(x) \to 0$.

Proof. If (i) holds, and $|t| \leqslant t_0$ and $\varepsilon > 0$, then

$$\sup_k |\varphi_{nk}(t) - 1| \leqslant \sup_k \left| \int_{\{|x|<\varepsilon\}} (e^{itx} - 1) dF_{nk}(x) \right| + \sup_k \left| \int_{\{|x|\geqslant\varepsilon\}} (e^{itx} - 1) dF_{nk}(x) \right|$$
$$\leqslant t_0\varepsilon + 2 \sup_k P(|f_{nk}| \geqslant \varepsilon),$$

since $\left|e^{iy} - 1\right| \leqslant |y|, y \in R$. Hence (i) implies (ii). Assume now that (ii) holds, let $\varepsilon > 0$, and choose first $t_0 > 0$ so that $e^{-t_0} < \varepsilon$ and then $n_\varepsilon \in N$ such that $\sup_k |\varphi_{nk}(t) - 1| < \varepsilon$ whenever $|t| \leqslant t_0$ and $n \geqslant n_\varepsilon$. Since $\int_0^\infty e^{-t}(1 - \cos tx)dt = x^2/(1 + x^2)$, $x \in R$, on account of Fubini's theorem (15.8), we have

$$\int_R \frac{x^2}{1+x^2} dF_{nk}(x) = \int_0^\infty e^{-t} \left(\int_R (1 - \cos tx) dF_{nk}(x) \right) dt$$
$$\leqslant \int_0^\infty e^{-t} |1 - \varphi_{nk}(t)| \, dt \leqslant \int_0^\infty e^{-t} \sup_k |1 - \varphi_{nk}(t)| \, dt.$$

Therefore, for $n \geqslant n_\varepsilon$,

$$\sup_k \int_R \frac{x^2}{1+x^2} dF_{nk}(x) \leqslant \int_0^{t_0} \sup_k |1 - \varphi_{nk}(t)| \, dt + 2 \int_{t_0}^\infty e^{-t} dt \leqslant (t_0 + 2)\varepsilon.$$

This shows that (iii) is fulfilled. As the function $(1 + x^2)/x^2$ is decreasing on $]0, \infty[$, we have

$$\sup_k P(|f_{nk}| \geqslant \varepsilon) \leqslant \frac{1+\varepsilon^2}{\varepsilon^2} \sup_k \int_{\{|x|\geqslant\varepsilon\}} \frac{x^2}{1+x^2} dF_{nk}(x), \quad \varepsilon > 0.$$

Therefore, (iii) implies (i). □

From now on fix $\tau \in]0, \infty[$, and define $a_{nk} = a_{nk}(\tau) = \int_{-\tau}^{\tau} x \, dF_{nk}(x)$, $\tilde{f}_{nk} = f_{nk} - a_{nk}$, $\widetilde{F}_{nk}(x) = P(\tilde{f}_{nk} \leqslant x) = F_{nk}(x + a_{nk})$, $x \in R$, and $\tilde{\varphi}_{nk}(t) = \varphi_{\tilde{f}_{nk}}(t) = e^{-ita_{nk}} \varphi_{nk}(t)$, $t \in R$.

Remark 23.25. If $g : R \to R$ is a measurable function, then $\int_R g(x - a_{nk}) dF_{nk}(x) = \int_R g(x) d\widetilde{F}_{nk}(x)$, in the sense that if one of the integral exists, so does the other, and the two integrals are equal. Actually, setting $t_{a_{nk}}(x) = x - a_{nk}$, $x \in R$, and making use of (11.29) and (11.6.a), we may write

$$\int_R g(x - a_{nk})dF_{nk}(x) = \int_R (g \circ t_{a_{nk}})d(Pf_{nk}^{-1}) = \int_R (g \circ t_{a_{nk}} \circ f_{nk})dP$$
$$= \int_R (g \circ \tilde{f}_{nk})dP = \int_R gd(P\tilde{f}_{nk}^{-1}) = \int_R g(x)d\widetilde{F}_{nk}(x).$$

Lemma 23.26. *If $\{f_{nk} : n \geqslant 1, 1 \leqslant k \leqslant k_n\}$ satisfies the condition of infinite smallness, then $\sup_k |\tilde{\varphi}_{nk}(t) - 1| \to 0$ uniformly in $|t| \leqslant t_0$ for any $t_0 > 0$.*

Proof. From (23.23) it follows that $\sup_k |a_{nk}| \to 0$. Therefore, for $\varepsilon > 0$, if $\sup_k |a_{nk}| \leqslant \varepsilon/2$, then

$$\sup_k P(|\tilde{f}_{nk}| \geqslant \varepsilon) \leqslant \sup_k P(|f_{nk}| + |a_{nk}| \geqslant \varepsilon) \leqslant \sup_k P\left(|f_{nk}| \geqslant \frac{\varepsilon}{2}\right) \to 0.$$

This means that $\{\tilde{f}_{nk} : n \geqslant 1, 1 \leqslant k \leqslant k_n\}$ also satisfies the infinite smallness condition, and the conclusion follows from (23.24). □

Lemma 23.27. *If $\{f_{nk} : n \geqslant 1, 1 \leqslant k \leqslant k_n\}$ satisfies the infinite smallness condition, then, for every $t_0 > 0$, there exists a positive constant $c_1 = c_1(t_0, \tau)$ such that*

$$c_1 \sup_{|t| \leqslant t_0} |\tilde{\varphi}_{nk}(t) - 1| \leqslant \int_R \frac{x^2}{1 + x^2} d\widetilde{F}_{nk}(x), \quad n \geqslant n_\tau, \quad 1 \leqslant k \leqslant k_n.$$

Proof. For $|t| \leqslant t_0$, since $\left|e^{iy} - 1 - iy\right| \leqslant y^2/2$, $y \in R$, we have omitting subscripts

$$|\tilde{\varphi}(t) - 1| \leqslant \left| \int_{]-\tau,\tau]} (e^{it(x-a)} - 1)dF(x) \right| + 2 \int_{]-\tau,\tau]^c} dF(x)$$
$$\leqslant \left| \int_{]-\tau,\tau]} (e^{it(x-a)} - 1 - it(x - a))dF(x) \right| + \left| \int_{]-\tau,\tau]} it(x - a)dF(x) \right|$$
$$+2 \int_{]-\tau,\tau]^c} dF(x) \leqslant \frac{t_0^2}{2} \int_{]-\tau,\tau]} (x - a)^2 dF(x)$$
$$+t_0 \left| \int_{]-\tau,\tau]} (x - a)dF(x) \right| + 2 \int_{]-\tau,\tau]^c} dF(x). \tag{1}$$

We have next

$$\int_{]-\tau,\tau]} (x - a)^2 dF(x) \leqslant (1 + (\tau + |a|)^2) \int_{]-\tau,\tau]} \frac{(x - a)^2}{1 + (x - a)^2} dF(x). \tag{2}$$

Since $\sup_k |a_{nk}| \to 0$ by (23.23), choose $n_\tau \in N$ so that $\sup_k |a_{nk}| < \tau/2$ whenever $n \geqslant n_\tau$. Then, for $n \geqslant n_\tau$, as $|x-a| \geqslant \tau - |a| > \tau/2$ when $|x| \geqslant \tau$, we may write

$$\left|\int\limits_{]-\tau,\tau]} (x-a)dF(x)\right| = \left|a \int\limits_{]-\tau,\tau]^c} dF(x)\right|$$
$$\leqslant \frac{|a|\,(1+(\tau+|a|)^2)}{(\tau-|a|)^2} \int\limits_{]-\tau,\tau]^c} \frac{(x-a)^2}{1+(x-a)^2} dF(x). \quad (3)$$

From (1)–(3) and (23.25) it follows that

$$\sup_{|t|\leqslant t_0} |\tilde{\varphi}_{nk}(t) - 1| \leqslant (1+(\tau+|a_{nk}|)^2)\left(\frac{t_0^2}{2} + \frac{|a_{nk}|\,t_0+2}{(\tau-|a_{nk}|)^2}\right) \int\limits_R \frac{x^2}{1+x^2} d\widetilde{F}_{nk}(x),$$
$$n \geqslant n_\tau,\ 1 \leqslant k \leqslant k_n,$$

whence

$$\sup_{|t|\leqslant t_0} |\tilde{\varphi}_{nk}(t) - 1| \leqslant (1+4\tau^2)\left(\frac{t_0^2}{2} + \frac{4(\tau t_0+2)}{\tau^2}\right) \int\limits_R \frac{x^2}{1+x^2} d\widetilde{F}_{nk}(x)$$
$$= \frac{1}{c_1} \int\limits_R \frac{x^2}{1+x^2} d\widetilde{F}_{nk}(x), \quad n \geqslant n_\tau,\ 1 \leqslant k \leqslant k_n. \qquad \square$$

Lemma 23.28. *If* $\{f_{nk} : n \geqslant 1, 1 \leqslant k \leqslant k_n\}$ *satisfies the condition of infinite smallness, then, for any* $t_0 > 0$, *there exists a positive constant* $c_2 = c_2(t_0, \tau)$ *such that*

$$\int\limits_R \frac{x^2}{1+x^2} d\widetilde{F}_{nk}(x) \leqslant -c_2 \int\limits_0^{t_0} \log|\varphi_{nk}(t)|\, dt, \quad n \geqslant n_\tau,\ 1 \leqslant k \leqslant k_n.$$

Proof. Most part of this proof we omit subscripts. Let m be a median of f, let F^m denote the distribution function of $f - m$, let $f^* = f' - f''$ be the symmetrized f, and denote by F^* the distribution function of f^*. Clearly, m is also a median of both f' and f''. We shall pass from F^* to F^m and then from F^m to $\widetilde{F}$. Let $t_0 > 0$. Since $|\varphi|^2$ is the characteristic function of f^*, applying Fubini's theorem, we have

$$\int\limits_0^{t_0} (1-|\varphi|^2)dt = \int\limits_R \left(\int\limits_0^{t_0} (1-\cos tx)dt\right) dF^*(x) = t_0 \int\limits_R \left(1 - \frac{\sin t_0 x}{t_0 x}\right) dF^*(x)$$
$$\geqslant t_0 c(t_0) \int\limits_R \frac{x^2}{1+x^2} dF^*(x), \quad (1)$$

where $c(t_0) = \inf_{x \in R} (1 - (\sin t_0 x)/t_0 x)(1 + x^2)/x^2 > 0$. To pass from F^* to F^m, notice first that

$$P^*(f^* \geqslant x) \geqslant P^*(f' - m \geqslant x,\, f'' - m \leqslant 0) = P^*(f' - m \geqslant x) P^*(f'' \leqslant m)$$
$$\geqslant P(f - m \geqslant x)/2, \quad x \geqslant 0, \tag{2}$$

where P^* is as in (23.19). Similarly, we have

$$P^*(f^* \leqslant -x) \geqslant P(f - m \leqslant -x)/2, \quad x \geqslant 0. \tag{3}$$

From (2) and (3) we get

$$P^*(|f^*| \geqslant x) \geqslant P(|f - m| \geqslant x)/2, \quad x \geqslant 0. \tag{4}$$

Then, if $h(x) = x^2/(1 + x^2)$, $x \geqslant 0$, we can write applying (15.23.a)

$$\int_R \frac{x^2}{1 + x^2} dF^*(x) = E^*[h \circ |f^*|] = \int_0^1 P^*(h \circ |f^*| \geqslant y) dy.$$

Consequently, putting $y = h(x)$ and using (4), we obtain

$$\int_R \frac{x^2}{1 + x^2} dF^*(x) = \int_0^\infty P^*(|f^*| \geqslant x) h'(x) dx \geqslant \frac{1}{2} \int_0^\infty P(|f - m| \geqslant x) h'(x) dx$$
$$= \frac{1}{2} \int_R \frac{x^2}{1 + x^2} dF^m(x). \tag{5}$$

To pass from F^m to $\widetilde{F}$, we begin with the inequality $(x - a)^2 \leqslant (x - m)^2 + 2(m - a)(x - a)$, $x \in R$. From it and the equality in (3) of the proof of (23.27) we have

$$\int_{]-\tau,\tau]} (x - a)^2 dF(x) \leqslant \int_{]-\tau,\tau]} (x - m)^2 dF(x) + 2(|m| + \tau) \left| \int_{]-\tau,\tau]} (x - a) dF(x) \right|$$
$$\leqslant \int_{]-\tau,\tau]} (x - m)^2 dF(x) + 2\tau(|m| + \tau) \int_{]-\tau,\tau]^c} dF(x),$$

and so

$$\int_R \frac{x^2}{1 + x^2} d\widetilde{F}(x) = \int_R \frac{(x - a)^2}{1 + (x - a)^2} dF(x) \leqslant \int_{]-\tau,\tau]} (x - a)^2 dF(x) + \int_{]-\tau,\tau]^c} dF(x)$$
$$\leqslant \int_{]-\tau,\tau]} (x - m)^2 dF(x) + (1 + 2\tau(|m| + \tau)) \int_{]-\tau,\tau]^c} dF(x). \tag{6}$$

Now take (23.23) and (23.24) into account to choose $n_\tau \in N$ such that both $\sup_k |m_{nk}| < \tau/2$ and $\sup_k |\varphi_{nk}(t)| > 0$ in $|t| \leqslant t_0$ whenever $n \geqslant n_\tau$. Then, for $n \geqslant n_\tau$, we may write

$$\int\limits_{]-\tau,\tau]} (x-m)^2 dF(x) \leqslant (1+(\tau+|m|)^2) \int\limits_{]-\tau,\tau]} \frac{(x-m)^2}{1+(x-m)^2} dF(x)$$
$$\leqslant (1+(\tau+|m|)^2) \int\limits_R \frac{x^2}{1+x^2} dF^m(x) \tag{7}$$

and

$$\int\limits_{]-\tau,\tau]^c} dF(x) \leqslant \frac{1+(\tau+|m|)^2}{(\tau-|m|)^2} \int\limits_{]-\tau,\tau]^c} \frac{(x-m)^2}{1+(x-m)^2} dF(x)$$
$$\leqslant \frac{1+(\tau+|m|)^2}{(\tau-|m|)^2} \int\limits_R \frac{x^2}{1+x^2} dF^m(x). \tag{8}$$

From (6)–(8) it follows that

$$\int\limits_R \frac{x^2}{1+x^2} d\widetilde{F}(x) \leqslant (1+(\tau+|m|)^2)\left(1+\frac{1+2\tau(|m|+\tau)}{(\tau-|m|)^2}\right) \int\limits_R \frac{x^2}{1+x^2} dF^m(x).$$

Combining this with (1) and (5), we obtain

$$\int\limits_R \frac{x^2}{1+x^2} d\widetilde{F}_{nk}(x) \leqslant \frac{2}{t_0 c(t_0)} (1+(\tau+|m_{nk}|)^2)\left(1+\frac{1+2\tau(|m_{nk}|+\tau)}{(\tau-|m_{nk}|)^2}\right)$$
$$\times \int\limits_0^{t_0} (1-|\varphi_{nk}|^2) dt, \quad n \geqslant n_\tau, \ 1 \leqslant k \leqslant k_n. \tag{9}$$

Since $1-y \leqslant -\log y$, $0 < y \leqslant 1$, and $\sup_k |m_{nk}| < \tau/2$, $n \geqslant n_\tau$, (9) leads to

$$\int\limits_R \frac{x^2}{1+x^2} d\widetilde{F}_{nk}(x) \leqslant -\frac{4}{t_0 c(t_0)} (1+4\tau^2)\left(1+\frac{1+4(1+4\tau^2)}{\tau^2}\right) \int\limits_0^{t_0} \log|\varphi_{nk}|\, dt$$
$$= -c_2(t_0,\tau) \int\limits_0^{t_0} \log|\varphi_{nk}(t)|\, dt, \quad n \geqslant n_\tau, \ 1 \leqslant k \leqslant k_n. \qquad \square$$

Lemma 23.29. *Suppose the array $\{f_{nk} : n \geqslant 1, 1 \leqslant k \leqslant k_n\}$ satisfies the infinite smallness condition, and $\prod_k |\varphi_{nk}| \to g$ on R, where g is continuous at 0. Then there exists a positive constant c, depending on τ and the array, such that*

$$\sum_k \int\limits_R \frac{x^2}{1+x^2} d\widetilde{F}_{nk}(x) \leqslant c, \quad n \geqslant 1. \tag{1}$$

Proof. Since g is continuous at 0 with $g(0) = 1$, choose $t_0 > 0$ so that $\sup_{|t| \leqslant t_0} g(t) > 0$. By (23.24) choose next $n_1 \in N$ such that $\sup_k |\varphi_{nk}(t)| > 0$ in $|t| \leqslant t_0$ whenever $n \geqslant n_1$. Since $\prod_k |\varphi_{nk}|^2 \to g^2$ and g^2 is continuous at 0, (20.11) shows that $g^2 \in \mathfrak{C}$, and so $\prod_k |\varphi_{nk}|^2 \to g^2$ uniformly in $|t| \leqslant t_0$. This implies that $\sum_k \log |\varphi_{nk}| \to \log |g|$ uniformly in $|t| \leqslant t_0$. By (23.28) we have

$$\sum_k \int_R \frac{x^2}{1+x^2} d\widetilde{F}_{nk}(x) \leqslant -c_2 \int_0^{t_0} \left(\sum_k \log |\varphi_{nk}(t)| \right) dt, \quad n \geqslant n_1 \vee n_\tau. \tag{2}$$

The right-hand side of (2) has the limit $-c_2 \int_0^{t_0} \log |g(t)| \, dt$ by (11.26), whence (1) follows. □

Lemma 23.30. *Assume the condition of infinite smallness is fulfilled, and there exists a positive constant c with $\sum_k \int_R x^2/(1+x^2) d\widetilde{F}_{nk}(x) \leqslant c$, $n \geqslant 1$. Then*

$$\sum_k (\operatorname{Log} \tilde{\varphi}_{nk}(t) - (\tilde{\varphi}_{nk}(t) - 1)) \to 0, \quad t \in R. \tag{1}$$

Proof. For $t_0 > 0$, apply (23.26) to choose $n_1 \in N$ such that $\sup_k |\tilde{\varphi}_{nk}(t) - 1| \leqslant 1/2$ for $|t| \leqslant t_0$ and $n \geqslant n_1$. According to (23.7), $\operatorname{Log} \tilde{\varphi}_{nk}(t)$ is well defined, and

$$\begin{aligned} \operatorname{Log} \tilde{\varphi}_{nk}(t) &= L(\tilde{\varphi}_{nk}(t)) \\ &= \sum_{j \geqslant 1} \frac{(-1)^{j-1}}{j} (\tilde{\varphi}_{nk}(t) - 1)^j, \quad |t| \leqslant t_0, \ n \geqslant n_1, \ 1 \leqslant k \leqslant k_n. \end{aligned}$$

Since

$$\begin{aligned} |L(z) - (z-1)| &\leqslant |z-1|^2 \sum_{j \geqslant 0} |z-1|^j \\ &= \frac{|z-1|^2}{1 - |z-1|} \leqslant 2\,|z-1|^2, \quad z \in C, \ |z-1| \leqslant \frac{1}{2}, \end{aligned}$$

we have

$$\begin{aligned} |\operatorname{Log} \tilde{\varphi}_{nk}(t) - (\tilde{\varphi}_{nk}(t) - 1)| &\leqslant 2\,|\tilde{\varphi}_{nk}(t) - 1|^2 \\ &\leqslant 2 \left(\sup_k |\tilde{\varphi}_{nk}(t) - 1| \right) |\tilde{\varphi}_{nk}(t) - 1|, \ |t| \leqslant t_0, \ n \geqslant n_1, \ 1 \leqslant k \leqslant k_n. \end{aligned}$$

Therefore, applying (23.27) and (23.26), for $n \geqslant n_1 \vee n_\tau$, we obtain

$$\begin{aligned} \left| \sum_k (\operatorname{Log} \tilde{\varphi}_{nk}(t) - (\tilde{\varphi}_{nk}(t) - 1)) \right| &\leqslant 2c_1^{-1} \sup_k |\tilde{\varphi}_{nk}(t) - 1| \sum_k \int_R \frac{x^2}{1+x^2} d\widetilde{F}_{nk}(x) \\ &\leqslant 2cc_1^{-1} \sup_k |\tilde{\varphi}_{nk}(t) - 1| \to 0, \quad |t| \leqslant t_0. \end{aligned}$$

Since t_0 is arbitrary, (1) follows. □

We are now ready to characterize the class of all infinitely divisible distribution functions.

Theorem 23.31. *Let F be a distribution function. Then the following are equivalent:*

(i) *F is infinitely divisible;*

(ii) *there exists an array $\{f_{nk} : n \geqslant 1, 1 \leqslant k \leqslant k_n\}$ of random variables, where $k_n \to \infty$, $f_{nk}, 1 \leqslant k \leqslant k_n$, are independent for any n, and the infinite smallness condition is satisfied such that $\sum_k f_{nk} \xrightarrow{D} F$.*

Proof. If F is infinitely divisible, let $F = F_n^{*n}$ and $\varphi_F = (\varphi_n)^n, n \in N$, with $F_n \in \mathfrak{F}$ and $\varphi_n \in \mathfrak{C}$. Let $\{f_{nk} : n \geqslant 1, 1 \leqslant k \leqslant n\}$ and f be as in (23.1). Then $\sum_k f_{nk} = f \xrightarrow{D} F$ trivially. Moreover, $\sup_k \left|\varphi_{f_{nk}}(t) - 1\right| = |\varphi_n(t) - 1| = \left|\varphi^{1/n}(t) - 1\right| \to 0$ uniformly in $|t| \leqslant t_0$ for any $t_0 > 0$ by (23.10.b). Hence the condition of infinite smallness is fulfilled by (23.24). Thus (i) implies (ii). Conversely, assume that (ii) holds, and let $t_0 > 0$. Then

$$\prod_k \varphi_{nk} \to \varphi_F \tag{1}$$

by (20.9). From (23.29) and (23.30) it follows that

$$\sum_k (\operatorname{Log} \tilde{\varphi}_{nk}(t) - (\tilde{\varphi}_{nk}(t) - 1)) \to 0, \quad t \in R. \tag{2}$$

If $n \geqslant n_1 \vee n_\tau$ and $1 \leqslant k \leqslant k_n$, where n_1 and n_τ are as in the proof of (23.30), we have

$$\begin{aligned}
\operatorname{Log} \tilde{\varphi}_{nk}(t) - (\tilde{\varphi}_{nk}(t) - 1) &= \operatorname{Log} \tilde{\varphi}_{nk}(t) - ita_{nk} - \int_R (e^{itx} - 1) d\widetilde{F}_{nk}(x) \\
&= \operatorname{Log} \tilde{\varphi}_{nk}(t) - it\left(a_{nk} + \int_R \frac{x}{1+x^2} d\widetilde{F}_{nk}(x)\right) \\
&\quad - \int_R \left(\frac{e^{itx} - 1 - itx}{x^2} + e^{itx} - 1\right) \frac{x^2}{1+x^2} d\widetilde{F}_{nk}(x), \quad |t| \leqslant t_0.
\end{aligned}$$

Therefore, for $n \geqslant n_1 \vee n_\tau$,

$$\begin{aligned}
\sum_k (\operatorname{Log} \tilde{\varphi}_{nk}(t) - (\tilde{\varphi}_{nk}(t) - 1)) &= \operatorname{Log}\left(\prod_k \varphi_{nk}(t)\right) - it\gamma_n - \int_R \chi(t,x) dG_n(x) \\
&= \operatorname{Log}\left(\prod_k \varphi_{nk}(t)\right) - \psi(t; \gamma_n, G_n), \quad |t| \leqslant t_0, \quad (3)
\end{aligned}$$

where $\gamma_n = \sum_k (a_{nk} + \int_R x/(1+x^2) d\widetilde{F}_{nk}(x))$ and $G_n(x) = \sum_k \int_{]-\infty,x]} y^2/(1 + y^2) d\widetilde{F}_{nk}(y), x \in R$. Clearly, G_n is nondecreasing, $G_n(-\infty) = 0$ and $G_n(\infty) \leqslant c$, where c is as in (23.29), for any $n \in N$. For $n \geqslant n_1 \vee n_\tau$ and $|t| \leqslant t_0$, from (1)–(3) and (2) in the proof of (23.6), we have

$$\exp(\psi(t;\gamma_n,G_n)) = \exp\left(\psi(t;\gamma_n,G_n) - \operatorname{Log}\left(\prod_k \varphi_{nk}(t)\right)\right)\prod_k \varphi_{nk}(t) \to \varphi_F(t).$$

Since t_0 is arbitrary, (23.12) and (23.11) show that $\varphi_F \in \mathfrak{C}_I$. □

Corollary 23.32. *Let F be a distribution function. Then the following are equivalent:*
(i) *F is infinitely divisible;*
(ii) *there exists an array $\{f_{nk} : n \geqslant 1, 1 \leqslant k \leqslant k_n\}$ as in* (23.31) *and $\{a_n : n \in N\} \subset R$ such that $\sum_k f_{nk} - a_n \overset{D}{\to} F$.*

Now we introduce and discuss briefly two special classes of infinitely divisible distribution/characteristic functions. A complete description of these classes can be found in *Limit Distributions for Sums of Independent Random Variables* (Gnedenko and Kolmogorov, 1968).

Definition 23.33. Let $f_n, n \in N$, be independent random variables, let $\{a_n : n \in N\} \subset R$ and $\{b_n : n \in N\} \subset]0,\infty[$, and assume that $\sup_{1\leqslant k\leqslant n} P(|f_k| \geqslant \varepsilon b_n) \to 0$ for each $\varepsilon > 0$. Set $S_n = \sum_{k=1}^n f_k, n \in N$. We denote by $\mathfrak{F}_L$ the class of all distribution functions $F \in \mathfrak{F}$ with $S_n/b_n - a_n \overset{D}{\to} F$. Let $\mathfrak{C}_L = \{\varphi_F \in \mathfrak{C} : F \in \mathfrak{F}_L\}$ denote the corresponding class of characteristic functions. If we put $f_{nk} = f_k/b_n, n \geqslant 1, 1 \leqslant k \leqslant n$, it is clear that we are dealing with a special case of array obeying the infinite smallness condition. Since $S_n/b_n - a_n = \sum_{k=1}^n f_{nk} - a_n$, (23.32) shows that actually $\mathfrak{F}_L \subset \mathfrak{F}_I$.

Remarks 23.34. (a) In this special triangular array, according to (23.24), the infinite smallness condition is equivalent to either $\sup_{1\leqslant k\leqslant n} \left|\varphi_{f_k}(t/b_n) - 1\right| \to 0$ uniformly in $|t| \leqslant t_0$ for any $t_0 > 0$ or $\sup_{1\leqslant k\leqslant n} \int_R x^2/(b_n^2 + x^2)dF_k(x) \to 0$, where $F_k(x) = P(f_k \leqslant x), x \in R$.
(b) $F_a \in \mathfrak{F}_L$ for any $a \in R$.

Theorem 23.35. *For $n \in N$, let f_n, a_n, b_n and S_n be as in* (23.33). *If $S_n/b_n - a_n \overset{D}{\to} F$ and F is not degenerate, then $b_n \to \infty$ and $b_{n+1}/b_n \to 1$.*

Proof. Let F_n and φ_n denote the distribution function and characteristic function of $S_n/b_n - a_n$, respectively. Then

$$\varphi_n(t) = e^{-ia_n t}\prod_{k=1}^n \varphi_{f_k}\left(\frac{t}{b_n}\right) \to \varphi_F(t), \quad t \in R. \tag{1}$$

If $b_n \nrightarrow \infty$, then, in view of (5.29) and (6.19), there exist a subsequence $\{b_{n_m} : m \in N\}$ and $b < \infty$ so that $b_{n_m} \to b$. Therefore, for every $k \in N$ and $t \in R$, by (23.34.a), we have $\left|\varphi_{f_k}(t)\right| = \lim_m \left|\varphi_{f_k}(tb_{n_m}/b_{n_m})\right| = 1$. This and (1) imply that $|\varphi_F| = 1$. Hence F is degenerate by (20.25). This contradiction shows that $b_n \to \infty$. Since $f_{n+1}/b_{n+1} \overset{P}{\to} 0$, Slutsky's theorem (19.27) ensures that $S_n/b_{n+1} - a_{n+1} =$

$S_{n+1}/b_{n+1} - a_{n+1} - f_{n+1}/b_{n+1} \xrightarrow{D} F$. Thus

$$P\left(\frac{S_n}{b_{n+1}} - a_{n+1} \leqslant x\right) = F_n\left(\frac{b_{n+1}}{b_n}x + \frac{b_{n+1}}{b_n}a_{n+1} - a_n\right) \to F(x), \quad x \in C(F). \tag{2}$$

As $F_n \xrightarrow{w} F$, (2) and (20.62) show that $b_{n+1}/b_n \to 1$. □

To characterize the class $\mathfrak{C}_L$ we need the next definition.

Definition 23.36. We say that $\varphi \in \mathfrak{C}$ is *self-decomposable* if for each $c \in]0, 1[$ there is $\varphi_c \in \mathfrak{C}$ such that $\varphi(t) = \varphi(ct)\varphi_c(t)$, $t \in R$.

Remarks 23.37. (a) f $\varphi \in \mathfrak{C}$ is degenerate, then φ is self-decomposable.

(b) If $\varphi \in \mathfrak{C}$ is not degenerate and $\varphi(t) = \varphi(ct)\varphi_c(t)$, $t \in R$, for some $c > 0$ and $\varphi_c \in \mathfrak{C} - \{1\}$, then $c < 1$. Actually, $c = 1$ would imply $\varphi_c = 1$, while $c > 1$ would imply $1 \geqslant |\varphi(t)| \geqslant |\varphi(t/c^n)| \to |\varphi(0)| = 1$, leading to φ degenerate by (20.25).

Lemma 23.38. *If φ is self-decomposable, then $\varphi(t) \neq 0$ for every $t \in R$.*

Proof. To get a contradiction, assume that $\{t \geqslant 0 : \varphi(t) = 0\} \neq \emptyset$, and set $2a = \inf\{t \geqslant 0 : \varphi(t) = 0\}$. Since φ is continuous with $\varphi(0) = 1$, it follows that $\varphi(2a) = 0$ and $\varphi(t) \neq 0$ for $0 \leqslant t < 2a$. Moreover, $\lim_{c \nearrow 1} \varphi_c(a) = \lim_{c \nearrow 1} \varphi(a)/\varphi(ca) = 1$. Therefore, in view of (20.46.b), $1 = 1 - |\varphi_c(2a)|^2 \leqslant 4(1 - |\varphi_c(a)|^2) \to 0$ as $c \nearrow 1$. □

Theorem 23.39. *Let $\varphi \in \mathfrak{C}$. Then $\varphi \in \mathfrak{C}_L$ if and only if it is self-decomposable.*

Proof. If $\varphi \in \mathfrak{C}_L$, let f be a random variable such that $\varphi = \varphi_f$. Then $S_n/b_n - a_n \xrightarrow{D} f$, where S_n, a_n and b_n are as in (23.33). We assume that φ is nondegenerate, since degenerate characteristic functions are self-decomposable by (23.37.a). It follows from (23.35) that $b_n \to \infty$ and $b_{n+1}/b_n \to 1$. For $c \in]0, 1[$ and $\{m_n : n \in N\}$ as in (5.97.a), put $T_n = S_{m_n}/b_n - a_{m_n}b_{m_n}/b_n$ and $U_n = (S_n - S_{m_n})/b_n - a_n + a_{m_n}b_{m_n}/b_n$. Then T_n and U_n are independent random variables (18.15) with $S_n/b_n - a_n = T_n + U_n$. Therefore, in view of (20.4),

$$\varphi_{S_n/b_n - a_n} = \varphi_{T_n}\varphi_{U_n}, \quad n \in N. \tag{1}$$

Since $T_n = b_{m_n}/b_n(S_{m_n}/b_{m_n} - a_{m_n}) \xrightarrow{D} cf$ by (19.61), we find that $\varphi_{T_n}(t) \to \varphi_{cf}(t) = \varphi(ct)$, $t \in R$, by (20.9). As $\varphi_{S_n/b_n - a_n} \to \varphi$ and $\varphi(t) \neq 0$ for any $t \in R$ by (23.4), we see on account of (1) that $\varphi_{U_n}(t) \to \varphi(t)/\varphi(ct)$, $t \in R$. Since the function $\varphi(t)/\varphi(ct)$ is continuous, Lévy's theorem (20.11) ensures that $\varphi_c(t) = \varphi(t)/\varphi(ct)$ is a characteristic function. Thus φ is self-decomposable.

Conversely, suppose that φ is self-decomposable. Let $f_1, \ldots, f_n$ be independent random variables such that $\varphi_{f_k}(t) = \varphi_{(k-1)/k}(kt) = \varphi(kt)/\varphi((k-1)t)$, $t \in R$, $1 \leqslant k \leqslant n$ (19.8.d). Then $\varphi_{S_n/n}(t) = \prod_{k=1}^{n} \varphi_{f_k}(t/n) = \varphi(t)$, $t \in R$, where $S_n = f_1 + \cdots + f_n$, so that $\varphi_{S_n/n} \to \varphi$ trivially. Since φ is uniformly continuous (20.2.c) and never vanishes (23.38), it follows that $\sup_{1 \leqslant k \leqslant n} \left|\varphi_{f_k}(t/n) - 1\right| \to 0$ uniformly in $|t| \leqslant t_0$ for any $t_0 > 0$. This and (23.34.a) show that the infinite smallness condition holds. Thus $\varphi \in \mathfrak{C}_L$. □

Examples 23.40. (a) If φ is as in (20.55), then φ is self-decomposable, and all its components φ_c, $0 < c < 1$, have the same form. Hence $\varphi, \varphi_c \in \mathfrak{C}_L$, $0 < c < 1$.

(b) If φ is the characteristic function of the Gaussian distribution function $G_{m,\sigma}$, then (20.3.h) shows that φ is self-decomposable, and the component φ_c is the characteristic function of $G_{(1-c)m,\sqrt{1-c^2}\sigma}$. Thus $G_{m,\sigma} \in \mathfrak{F}_L$.

Theorem 23.41. *If φ is self-decomposable and $c \in]0, 1[$, then its component φ_c is infinite divisible.*

Proof. Let $f_1, \ldots, f_n$ and S_n be as in the second part of the proof of (23.39). Let f be as in the proof of (23.39). Set $T_n = S_{[cn]}/n$ and $U_n = (S_n - S_{[cn]})/n$. Then

$$\varphi = \varphi_{S_n/n} = \varphi_{T_n}\varphi_{U_n}, \quad n \in N. \tag{1}$$

Since $T_n = ([cn]/n)S_{[cn]}/[cn] \xrightarrow{D} cf$, it follows that $\varphi_{T_n}(t) \to \varphi(ct), t \in R$. This in conjunction with (1) and (23.38) shows that $\varphi_{U_n}(t) \to \varphi_c(t), t \in R$. For $n \geqslant 1/c$, we see that $U_n = \sum_{k=1}^{n} f_{nk}$, where $f_{nk} = 0$ or f_k/n according as $1 \leqslant k \leqslant [cn]$ or $[cn] < k \leqslant n$. As above we find that $\sup_{1 \leqslant k \leqslant n} \left|\varphi_{f_{nk}}(t) - 1\right| = \sup_{[cn] < k \leqslant n} \left|\varphi_{f_k}(t/n) - 1\right| \to 0$ uniformly in $|t| \leqslant t_0$ for any $t_0 > 0$, and so the condition of infinite smallness is fulfilled (23.34.a). It follows then that $\varphi_c \in \mathfrak{C}_I$ by (23.31). □

We deal now with a very narrow class of infinitely divisible distribution/characteristic functions.

Definitions 23.42. A distribution function F is said to be *stable* if for every $a_1, a_2 > 0$ and $b_1, b_2 \in R$ there exist $a > 0$ and $b \in R$ such that $F(a_1x + b_1) * F(a_2x + b_2) = F(ax + b), x \in R$. We denote by $\mathfrak{F}_S$ the class of all stable distribution functions.

A characteristic function φ is called *stable* if for any $\alpha_1, \alpha_2 > 0$ there exist $\alpha > 0$ and $\beta \in R$ such that $\varphi(\alpha_1 t)\varphi(\alpha_2 t) = e^{-i\beta t}\varphi(\alpha t), t \in R$. Let $\mathfrak{C}_S$ denote the class of all stable characteristic functions.

Examples 23.43. (a) If φ is degenerate, then $\varphi \in \mathfrak{C}_S$.

(b) If φ is normal, then $\varphi \in \mathfrak{C}_S$.

Remarks 23.44. (a) If $F \in \mathfrak{F}_S$ and $G(x) = F(cx + d), x \in R$, for some $c > 0$ and $d \in R$, then $G \in \mathfrak{F}_S$.

(b) $F \in \mathfrak{F}_S$ if and only if $\varphi_F \in \mathfrak{C}_S$. Actually, if $F \in \mathfrak{F}_S$ and $\alpha_1, \alpha_2 > 0$, then there exist $a > 0$ and $b \in R$ such that $F(\alpha_1^{-1}x) * F(\alpha_2^{-1}x) = F(ax + b), x \in R$, and so $\varphi_F(\alpha_1 t)\varphi_F(\alpha_2 t) = e^{-i(b/a)t}\varphi_F(t/a), t \in R$, by (20.4). Writing $\alpha = 1/a$ and $\beta = b/a$, this shows that $\varphi_F \in \mathfrak{C}_S$. Conversely, if $\varphi_F \in \mathfrak{C}_S$, $a_1, a_2 > 0$ and $b_1, b_2 \in R$, then there exist $\alpha > 0$ and $\beta \in R$ such that

$$\varphi_F(a_1^{-1}t)\varphi_F(a_2^{-1}t) = e^{i\beta t}\varphi_F(\alpha t), \quad t \in R. \tag{1}$$

Putting $a = 1/\alpha$ and $(b_1/a_1) + (b_2/a_2) = \alpha b + \beta$, (1) becomes

$$e^{-i\frac{b_1}{a_1}t}\varphi_F\left(\frac{t}{a_1}\right)e^{-i\frac{b_2}{a_2}t}\varphi_F\left(\frac{t}{a_2}\right) = e^{-i\frac{b}{a}t}\varphi_F\left(\frac{t}{a}\right), \quad t \in R,$$

which is equivalent to $F(a_1x+b_1)*F(a_2x+b_2)=F(ax+b)$, $x\in R$.

(c) If $\varphi\in\mathfrak{C}$ is stable, then there exist $c_1,c_2\in\,]0,1[$ with $c_1\neq c_2$ and $\varphi_{c_i}\in\mathfrak{C}$ such that

$$\varphi(t)=\varphi(c_it)\varphi_{c_i}(t),\quad t\in R,\ i=1,2. \tag{1}$$

Plainly, this property is satisfied whenever φ is degenerate. Otherwise, for $\alpha_1,\alpha_2>0$ with $\alpha_1\neq\alpha_2$, let $\alpha>0$ and $\beta\in R$ be such that $\varphi(\alpha t)=e^{-i\beta t}\varphi(\alpha_1t)\varphi(\alpha_2t)$, $t\in R$. Setting $\alpha t=u$ and $c_i=\alpha_i/\alpha$, $i=1,2$, we obtain $\varphi(u)=\varphi(c_1u)e^{-i(\beta/\alpha)u}\varphi(c_2u)=\varphi(c_1u)\varphi_{c_1}(u)$, $u\in R$, where $\varphi_{c_1}(u)=e^{-i(\beta/\alpha)u}\varphi(c_2u)$ is a characteristic function, and $c_1\in\,]0,1[$ by (23.37.b). Similarly we get (1) for $i=2$.

The next final result of this section shows that $\mathfrak{F}_S$ is indeed a subset of $\mathfrak{F}_I$. In fact, we prove that $\mathfrak{F}_S\subset\mathfrak{F}_L$, the stable elements of $\mathfrak{F}_L$ being limiting distribution functions of normed sums $(f_1+\cdots+f_n)/b_n-a_n$ with $a_n\in R$ and $b_n>0$, under the additional restriction that f_n, $n\in N$, are independent and identically distributed (abbreviated i.i.d.).

Theorem 23.45. *Let F be a distribution function. Then the following are equivalent:*

(i) *F is stable;*

(ii) *there exists a sequence $\{f_n:n\in N\}$ of random variables, $\{a_n:n\in N\}\subset R$ and $\{b_n:n\in N\}\subset\,]0,\infty[$ such that $S_n/b_n-a_n\xrightarrow{D}F$, where $S_n=f_1+\cdots+f_n$, $n\in N$.*

Proof. If F is stable, apply (19.8.d) to provide a sequence $\{f_n:n\in N\}$ of independent random variables such that the distribution function of f_n is F for each $n\in N$. Then, for $n\in N$, $\varphi_{S_n}(t)=(\varphi_F(t))^n=e^{ia_nt}\varphi_F(b_nt)$ for some $a_n\in R$ and $b_n>0$. It follows that $\varphi_{S_n/b_n-a_n}=\varphi_F$, $n\in N$, i.e. $S_n/b_n-a_n\xrightarrow{D}F$ trivially. Suppose now that (ii) holds, and let f be a random variable with distribution function F. We assume that F is nondegenerate, since degenerate distribution functions are stable (23.43.a). This implies that f_1 is not degenerate. Suppose for the moment that there are a subsequence $\{b_{n_k}:k\in N\}$ and $b<\infty$ such that $b_{n_k}\to b$. Then $S_{n_k}-a_{n_k}b_{n_k}\xrightarrow{D}bf$ by (19.61), and so $e^{-ia_{n_k}b_{n_k}t}(\varphi_{f_1}(t))^{n_k}\to\varphi_{bf}(t)$, $t\in R$, by (20.9). This entails that $\varphi_{bf}(t)=0$ whenever $|\varphi_{f_1}(t)|<1$. Since f_1 is not degenerate, (20.64) ensures that $\{t\in R: |\varphi_{f_1}(t)|<1\}$ is dense in R. Hence $\varphi_{bf}(t)=0$, $t\in R$, by continuity, and so f is degenerate. The contradiction we get shows that $b_n\to\infty$, and so $f_1/b_n\xrightarrow{P}0$, i.e. the condition of infinite smallness is satisfied. It follows then that $F\in\mathfrak{F}_L$ and $b_{n+1}/b_n\to1$ (23.35). For $0<\alpha_1<\alpha_2$, let $\{m_n:n\in N\}$ be defined as in (5.97.b) with $c=\alpha_2/\alpha_1$, and put

$$T_n=\alpha_1\left(\frac{S_n}{b_n}-a_n\right),\quad U_n=\alpha_1\frac{b_{m_n}}{b_n}\left(\frac{1}{b_{m_n}}\sum_{k=n+1}^{n+m_n}f_k-a_{m_n}\right),\qquad n\in N.$$

Then T_n and U_n are independent, and $T_n+U_n=V_n$, where

$$V_n=\alpha_{n+m_n}\left(\frac{S_{n+m_n}}{b_{n+m_n}}-a_{n+m_n}\right)-\beta_{n+m_n}$$

with

$$\alpha_{n+m_n} = \alpha_1 \frac{b_{n+m_n}}{b_n}, \quad \beta_{n+m_n} = \alpha_1 \left(a_n + \frac{a_{m_n} b_{m_n}}{b_n} - \frac{a_{n+m_n} b_{n+m_n}}{b_n} \right).$$

Since $\varphi_{T_n}\varphi_{U_n} = \varphi_{V_n}$, and $T_n \xrightarrow{D} \alpha_1 f$ and $U_n \xrightarrow{D} \alpha_2 f$ by (19.61), it follows that

$$\varphi_F(\alpha_1 t)\varphi_F(\alpha_2 t) = \lim_n \varphi_{V_n}(t), \quad t \in R. \tag{1}$$

By (20.11), there is a nondegenerate distribution function G such that $V_n \xrightarrow{D} G$. Then, if F_n stands for the distribution function of $S_n/b_n - a_n$, we have

$$P(V_n \leqslant x) = F_{n+m_n}\left(\alpha_{n+m_n}^{-1} x + \alpha_{n+m_n}^{-1} \beta_{n+m_n}\right) \to G(x), \quad x \in C(G). \tag{2}$$

As $F_n \xrightarrow{w} F$, (2) and (19.62) show that F and G are of the same type, i.e. $G(x) = F(ax+b)$, $x \in R$, for some $a > 0$ and $b \in R$. Hence, on account of (1), $\varphi_F(\alpha_1 t)\varphi_F(\alpha_2 t) = \varphi_G(t) = e^{-i\beta t}\varphi_F(\alpha t)$, $t \in R$, where $\alpha = 1/a$ and $\beta = b/a$, and so $\varphi_F \in \mathfrak{C}_S$, i.e. F is stable (23.44.b). □

Remark 23.46. If $F \in \mathfrak{F}_S$, and $\{f_n : n \in N\}$, $\{a_n : n \in N\}$ and $\{b_n : n \in N\}$ are as in the first part of the proof of (23.45), then $f_1/b_n \xrightarrow{P} 0$ by the second part of the same proof, i.e. the infinite smallness condition is fulfilled. According to (23.33), this means that $F \in \mathfrak{F}_L$. Thus $\mathfrak{F}_S \subset \mathfrak{F}_L$.

Exercise 23.47. If F is the gamma distribution function as in (19.10.e), then F is infinitely divisible.

Exercise 23.48 (Pakshirajan-Chatterjee). If F is an infinitely divisible distribution function such that $F(C) - F(-C) = 1$ for some $C \in]0, \infty[$, then F is degenerate. The following steps may be helpful.

(a) For $n \in N$, let P_n, f and f_{nk}, $1 \leqslant k \leqslant n$, be as in (23.1). Then $|f| \leqslant C$ P_n-a.s. and $|f_{nk}| \leqslant C/n$ P_n-a.s.

(b) $F = F_m$, where $m = \int_R x\,dF(x)$. [Hint. By (17.9) and (a), $0 \leqslant \int_R (x-m)^2 dF(x) = E_n(f - E_n f)^2 = nE_n(f_{n1} - Ef_{n1})^2 \leqslant n(2C/n)^2 \to 0$, where E_n denotes expectation with respect to P_n.]

Exercise 23.49. Let $\varphi(t) = (1-b)/(1-be^{it})$, $t \in R$, where $0 < b < 1$.

(a) Use (23.2.c) and (23.11) to show that $\varphi \in \mathfrak{C}_I$. [Hint. $\varphi(t) = \lim_k \prod_{j=1}^k \exp\left(\frac{b^j}{j}(e^{itj} - 1)\right)$, $t \in R$.]

(b) If $\varphi(t) = \varphi(t; \gamma, G)$, $t \in R$, is the Lévy-Khintchine representation of φ, prove that $\gamma = \sum_{j \geqslant 1} b^j/(1+j^2)$ and $G(x) = \sum_{j\geqslant 1} G_j(x)$, $x \in R$, where $G_j(x) = 0$ or $jb^j/(1+j^2)$ according as $x < j$ or $x \geqslant j$. [Use (23.12.a), (23.12.d) and (23.15).]

Exercise 23.50. (a) If $G \in \mathfrak{G}$ with $G(-\infty) = 0$, then the function $\varphi(t) = \exp\{\int_R (e^{itx} - 1)dG(x)\}$, $t \in R$, is an infinitely divisible characteristic function. [Use (23.16).]

(b) The characteristic function φ_g in (20.39) is infinitely divisible.

Exercise 23.51. Let $\varphi \in \mathfrak{C}$, and assume there exist sequences $\{\varphi_k : k \in N\} \subset \mathfrak{C}$ and $\{n_k : k \in N\} \subset N$ with $n_k \to \infty$ such that $\varphi = (\varphi_k)^{n_k}$, $k \in N$. Prove the following.

(a) $\varphi(t) \neq 0$ for any $t \in R$.

(b) $\varphi^{1/n_k} = \varphi_k$ for any $k \in N$.

(c) φ is infinitely divisible. [Hint. Adapt the proof of (23.16), using (a) and (b).]

Exercise 23.52. If f is a random variable with $E\,|f| < \infty$, then $|m[f] - Ef| \leqslant \sqrt{2\operatorname{Var} f}$. [Use Chebyshev's inequality and (23.18.b).]

Exercise 23.53. Let f_n, $n \in N$, and f be random variables, and let $\{c_n : n \in N\} \subset R$.

(a) If $f_n \xrightarrow{P} 0$, then $m[f_n] \to 0$. [Use (23.18.b).]

(b) If $f_n \xrightarrow{P} f$ and $m[f]$ is unique, then $m[f_n] \to m[f]$.

(c) If $f_n - c_n \xrightarrow{P} 0$, then $f_n - m[f_n] \xrightarrow{P} 0$. [Apply (a), (23.18.c) and (14.23.a).]

Exercise 23.54. Let $\{f_{nk} : n \geqslant 1, 1 \leqslant k \leqslant k_n\}$ be an array of random variables, where $k_n \to \infty$. If $\sup_k E\,|f_{nk}|^p \to 0$ for some $p > 0$, then the condition of infinite smallness is fulfilled.

Exercise 23.55. Let $\{f_{nk} : n \geqslant 1, 1 \leqslant k \leqslant k_n\}$ be as above (23.17). Show that $\sup_k |f_{nk}| \xrightarrow{P} 0$ is equivalent to $\sum_k P(|f_{nk}| \geqslant \varepsilon) \to 0$, $\varepsilon > 0$, and implies the infinite smallness condition. [Hint. $P(\sup_k |f_{nk}| \geqslant \varepsilon) = 1 - \prod_k (1 - P(|f_{nk}| \geqslant \varepsilon)) \geqslant 1 - \exp(-\sum_k P(|f_{nk}| \geqslant \varepsilon))$, $\varepsilon > 0$.]

Exercise 23.56. Let $\{f_{nk} : n \geqslant 1, 1 \leqslant k \leqslant k_n\}$ be as in (23.54). Prove that the following are equivalent:

(i) $\sup_k P(|f_{nk}| \geqslant \varepsilon) \to 0$ for each $\varepsilon > 0$;

(ii) $\lim_n \int_R g\,dF_{nk} = g(0)$ uniformly with respect to $1 \leqslant k \leqslant k_n$ for any $g \in \mathfrak{C}_r(R)$;

(iii) $\lim_n \int_R g\,dF_{nk} = g(0)$ uniformly with respect to $1 \leqslant k \leqslant k_n$ for any uniformly continuous function $g \in \mathfrak{C}_r(R)$.

Exercise 23.57. Let $\{f_n : n \in N\}$ be an arbitrary sequence of random variables, and let $a \in R$. Show that there are $\{a_n : n \in N\} \subset R$ and $\{b_n : n \in N\} \subset]0, \infty[$ such that $f_n/b_n + a_n \xrightarrow{D} a$.

Exercise 23.58. Let $\{f_n : n \in N\}$ be a sequence of i.i.d. random variables, and let $\{a_n : n \in N\} \subset R$ and $\{b_n : n \in N\} \subset]0, \infty[$ be such that $S_n/b_n - a_n \xrightarrow{D} F$, where $S_n = f_1 + \cdots + f_n$, $n \in N$, and $F \in \mathfrak{F}$ is nondegenerate. Prove that $a_{n+1} - a_n \to 0$. [Hint. Show that $f_n/b_n \xrightarrow{P} 0$ as in the proof of (23.45), then apply Slutsky's theorem (19.27) and (19.62).]

Exercise 23.59. Let $p(x) = 0$ or $(e^{1/2x}\sqrt{2\pi x^3})^{-1}$ according as $x \leqslant 0$ or $x > 0$. Prove the following.

(a) p is a probability density on R.

(b) The distribution function corresponding to p is

$$F(x) = \begin{cases} 0 & \text{if } x \leqslant 0 \\ 2(1 - \Phi(x^{-1/2})) & \text{if } x > 0 \end{cases}.$$

(c) $\varphi_F(t) = \begin{cases} e^{-\sqrt{-t}(1+i)} & \text{if } t \leqslant 0 \\ e^{-\sqrt{t}(1-i)} & \text{if } t > 0 \end{cases}$.

(d) φ_F is stable.

Exercise 23.60 (Lévy). Let $\psi(t) = 2\sum_{k\in Z} 2^{-k}(\cos 2^k t - 1), t \in R$. Show that e^ψ is an infinitely divisible characteristic function.

Exercise 23.61. Let $\{f_{nk} : n \geqslant 1, 1 \leqslant k \leqslant k_n\}$ be as in (23.54), and for some $c > 0$ consider the truncated random variables $f_{nk}^c = f_{nk}1_{\{|f_{nk}|<c\}}, n \geqslant 1, 1 \leqslant k \leqslant k_n$. If $\{f_{nk} : n \geqslant 1, 1 \leqslant k \leqslant k_n\}$ satisfies the infinite smallness condition, then $\{f_{nk}^c : n \geqslant 1, 1 \leqslant k \leqslant k_n\}$ satisfies it as well.

24 Convergence to an Infinitely Divisible Distribution/ Characteristic Function

Let $\{f_{nk} : n \geqslant 1, 1 \leqslant k \leqslant k_n\}$ be an array of random variables, where $k_n \to \infty$, $f_{nk}, 1 \leqslant k \leqslant k_n$, are independent for any n, and the condition of infinite smallness is fulfilled. As shown in (23.31), if $\sum_k f_{nk} \xrightarrow{D} F$ for some $F \in \mathfrak{F}$, then F is infinitely divisible. In this section we shall present necessary and sufficient conditions for the convergence in distribution of such sums $\sum_k f_{nk}$ to a given $F \in \mathfrak{F}_I$. We shall apply the general convergence criterion to the three basic limit distribution functions of probability theory, namely the degenerate, the Poisson and the normal distribution function.

Theorem 24.1. *Let $\{f_{nk} : n \geqslant 1, 1 \leqslant k \leqslant k_n\}$ be as above, and let $F \in \mathfrak{F}_I$ with $\varphi_F(t) = \varphi(t; \gamma, G), t \in R$. Then $\sum_k f_{nk} \xrightarrow{D} F$ if and only if $\gamma_n \to \gamma$ and $G_n \xrightarrow{c} G$, where $\gamma_n = \sum_k(a_{nk} + \int_R x/(1 + x^2)d\widetilde{F}_{nk}(x))$ and $G_n(x) = \sum_k \int_{-\infty}^x y^2/(1 + y^2)d\widetilde{F}_{nk}(y), x \in R$.*

Proof. If $\sum_k f_{nk} \xrightarrow{D} F$, then as we showed in the proof of (23.31), $\varphi(t; \gamma_n, G_n) = \exp(\psi(t; \gamma_n, G_n)) \to \varphi_F(t) = \varphi(t; \gamma, G)$ for any $t \in R$. It follows then from (23.14) and (23.15) that $\gamma_n \to \gamma$ and $G_n \xrightarrow{c} G$.

Conversely, if $\gamma_n \to \gamma$ and $G_n \xrightarrow{c} G$, then

$$\varphi(t; \gamma_n, G_n) \to \varphi(t; \gamma, G), \quad t \in R, \tag{1}$$

by (23.13.e). Since $G_n \xrightarrow{c} G$, we have $\sum_k \int_R y^2/(1 + y^2)d\widetilde{F}_{nk}(y) = G_n(\infty) \to G(\infty) < \infty$, and so $\sum_k(\operatorname{Log}\tilde{\varphi}_{nk}(t) - (\tilde{\varphi}_{nk}(t) - 1)) \to 0, t \in R$, by (23.30). As we showed in the proof of (23.31), this implies that

$$\operatorname{Log}\left(\prod_k \varphi_{nk}(t)\right) - \psi(t; \gamma_n, G_n) \to 0, \quad t \in R. \tag{2}$$

From (1) and (2) we infer that $\prod_k \varphi_{nk}(t) \to \varphi(t; \gamma, G), t \in R$, i.e. $\sum_k f_{nk} \xrightarrow{D} F$. □

The conditions in (24.1) are expressed somewhat artificially in terms of the distribution functions $\widetilde{F}_{nk}$. However, for applications it is desirable to have these conditions expressed in terms of the distribution functions F_{nk} themselves. We give the new formulation of the convergence criterion in (24.6). Since the conditions transformation is rather intricate and unpleasant, the proof is broken in several lemmas. Throughout (24.2)–(24.8) $\{f_{nk} : n \geqslant 1, 1 \leqslant k \leqslant k_n\}$ will be as in (24.1).

Lemma 24.2. *Let $\{G_n : n \in N\}$ be as in (24.1), and let $G \in \mathfrak{G}$ with $G(-\infty) = 0$. Then $G_n \xrightarrow{c} G$ if and only if*

(i) $\sum_k \widetilde{F}_{nk}(x) \to \int_{-\infty}^{x} \frac{1+y^2}{y^2} dG(y), 0 > x \in C(G)$,

$\sum_k (1 - \widetilde{F}_{nk}(x)) \to \int_x^{\infty} \frac{1+y^2}{y^2} dG(y), 0 < x \in C(G)$,

and

(ii) $\lim_{\varepsilon \to 0} \limsup_n \sum_k \int_{-\varepsilon}^{\varepsilon} \frac{x^2}{1+x^2} d\widetilde{F}_{nk}(x) = \lim_{\varepsilon \to 0} \liminf_n \sum_k \int_{-\varepsilon}^{\varepsilon} \frac{x^2}{1+x^2} d\widetilde{F}_{nk}(x)$ $= G(0) - G(0-)$.

Proof. Assume that $G_n \xrightarrow{c} G$, and let $x \in C(G)$. If $x < 0$, then (12.8) and (19.67.e) imply that

$$\sum_k \widetilde{F}_{nk}(x) = \sum_k \int_{-\infty}^{x} \frac{1+y^2}{y^2} \frac{y^2}{1+y^2} d\widetilde{F}_{nk}(y)$$
$$= \int_{-\infty}^{x} \frac{1+y^2}{y^2} dG_n(y) \to \int_{-\infty}^{x} \frac{1+y^2}{y^2} dG(y).$$

If $x > 0$, define $g_x(y) = (1 + x^2)/x^2$ or $(1 + y^2)/y^2$ according as $y \leqslant x$ or $y > x$. Then $g_x \in \mathfrak{C}_r(R)$ and

$$\sum_k (1 - \widetilde{F}_{nk}(x)) = \sum_k \int_x^{\infty} \frac{1+y^2}{y^2} \frac{y^2}{1+y^2} d\widetilde{F}_{nk}(y) = \int_x^{\infty} \frac{1+y^2}{y^2} dG_n(y)$$
$$= \int_R g_x\, dG_n - \int_{-\infty}^{x} g_x\, dG_n \to \int_R g_x\, dG - \int_{-\infty}^{x} g_x\, dG$$
$$= \int_x^{\infty} \frac{1+y^2}{y^2} dG(y)$$

by (12.8), (19.67.c) and (19.67.e). Thus (i) holds. Next, for $\varepsilon > 0$ with $\pm\varepsilon \in C(G)$, we have $\sum_k \int_{-\varepsilon}^{\varepsilon} x^2/(1 + x^2) d\widetilde{F}_{nk}(x) = G_n(\varepsilon) - G_n(-\varepsilon), n \geqslant 1$, whence

$$\lim_n \sum_k \int_{-\varepsilon}^{\varepsilon} \frac{x^2}{1+x^2} d\widetilde{F}_{nk}(x) = G(\varepsilon) - G(-\varepsilon). \tag{1}$$

Since G is nondecreasing, (1) ensures that $\lim_{\varepsilon\to 0}\lim_n \sum_k \int_{-\varepsilon}^{\varepsilon} x^2/(1+x^2)d\widetilde{F}_{nk}(x) = G(0) - G(0-)$. Hence (ii) is satisfied.

Suppose now that (i) and (ii) hold, and let $x_0 > 0$ be such that $\pm x_0 \in C(G)$. Define

$$_1\widetilde{F}_{nk}(x) = \begin{cases} \widetilde{F}_{nk}(x) & \text{if } x \leqslant -x_0 \\ \widetilde{F}_{nk}(x_0) & \text{if } x > -x_0 \end{cases}, \quad n \geqslant 1,\ 1 \leqslant k \leqslant k_n,$$

and

$$H_1(x) = \begin{cases} \int_{-\infty}^{x} \frac{1+y^2}{y^2} dG(y) & \text{if } x \leqslant -x_0 \\ \int_{-\infty}^{x_0} \frac{1+y^2}{y^2} dG(y) & \text{if } x > -x_0 \end{cases}.$$

Then $\sum_k ({}_1\widetilde{F}_{nk}) \in \mathfrak{G}$ for any $n \geqslant 1$, $H_1 \in \mathfrak{G}$, $C(G) \cap]-\infty, -x_0] = C(H_1) \cap]-\infty, -x_0]$, and $\sum_k ({}_1\widetilde{F}_{nk}) \overset{c}{\to} H_1$ by (i). Therefore, for $x \leqslant -x_0$ with $x \in C(G)$, in view of (19.67.e) and (12.8), we obtain

$$\begin{aligned} G_n(x) &= \sum_k \int_{-\infty}^{x} \frac{y^2}{1+y^2} d\widetilde{F}_{nk}(y) \\ &= \int_{-\infty}^{x} \frac{y^2}{1+y^2} d\left(\sum_k ({}_1\widetilde{F}_{nk})\right)(y) \to \int_{-\infty}^{x} \frac{y^2}{1+y^2} dH_1(y) \\ &= \int_{-\infty}^{x} \frac{y^2}{1+y^2}\frac{1+y^2}{y^2} dG(y) = G(x). \end{aligned} \tag{2}$$

Define also

$$_2\widetilde{F}_{nk}(x) = \begin{cases} \widetilde{F}_{nk}(x_0) & \text{if } x \leqslant x_0 \\ \widetilde{F}_{nk}(x) & \text{if } x > x_0 \end{cases}, \quad n \geqslant 1,\ 1 \leqslant k \leqslant k_n,$$

and

$$H_2(x) = \begin{cases} -\int_{x_0}^{\infty} \frac{1+y^2}{y^2} dG(y) & \text{if } x \leqslant x_0 \\ -\int_{x}^{\infty} \frac{1+y^2}{y^2} dG(y) & \text{if } x > x_0 \end{cases}.$$

Then $\sum_k ({}_2\widetilde{F}_{nk} - 1) \in \mathfrak{G}$ for any $n \geqslant 1$, $H_2 \in \mathfrak{G}$, $C(G) \cap [x_0, \infty[= C(H_2) \cap [x_0, \infty[$, and $\sum_k ({}_2\widetilde{F}_{nk} - 1) \overset{c}{\to} H_2$ by (i). Consequently, for $x \geqslant x_0$ with $x \in C(G)$, by (19.67.e) and (12.8), we have

$$\begin{aligned}
G_n(\infty) - G_n(x) &= \int_x^\infty \frac{y^2}{1+y^2} d\left(\sum_k \widetilde{F}_{nk}\right)(y) \\
&= \int_x^\infty \frac{y^2}{1+y^2} d\left(\sum_k (\widetilde{F}_{nk} - 1)\right)(y) \\
&= \int_x^\infty \frac{y^2}{1+y^2} d\left(\sum_k (2\widetilde{F}_{nk} - 1)\right)(y) \to \int_x^\infty \frac{y^2}{1+y^2} dH_2(y) \\
&= \int_x^\infty \frac{y^2}{1+y^2} \frac{1+y^2}{y^2} dG(y) = G(\infty) - G(x). \qquad (3)
\end{aligned}$$

From (2) and (3) it follows in fact that

$$G_n(x) \to G(x), \ 0 > x \in C(G), \quad G_n(\infty) - G_n(x) \to G(\infty) - G(x), \\ 0 < x \in C(G). \qquad (4)$$

Next, taking $\pm\varepsilon \in C(G)$, in view of (ii) and (4), we get

$$\begin{aligned}
G(0) - G(0-) &= \lim_{\varepsilon \to 0} \limsup_n (G_n(\varepsilon) - G_n(-\varepsilon)) \\
&= \lim_{\varepsilon \to 0} \limsup_n (G_n(\varepsilon) - G_n(\infty) - G_n(-\varepsilon) + G_n(\infty)) \\
&= \lim_{\varepsilon \to 0} \left(G(\varepsilon) - G(\infty) - G(-\varepsilon) + \limsup_n G_n(\infty) \right) \\
&= G(0) - G(0-) - G(\infty) + \limsup_n G_n(\infty),
\end{aligned}$$

whence $\limsup_n G_n(\infty) = G(\infty)$. Analogously we get $\liminf_n G_n(\infty) = G(\infty)$, so that $G_n(\infty) \to G(\infty)$. This and (4) ensure that $G_n \xrightarrow{c} G$. □

Lemma 24.3. *Notation is as in (24.2). Then the following are equivalent:*

(i) $\sum_k \widetilde{F}_{nk}(x) \to \int_{-\infty}^x \frac{1+y^2}{y^2} dG(y), 0 > x \in C(G),$

$\sum_k (1 - \widetilde{F}_{nk}(x)) \to \int_x^\infty \frac{1+y^2}{y^2} dG(y), 0 < x \in C(G);$

(ii) $\sum_k F_{nk}(x) \to \int_{-\infty}^x \frac{1+y^2}{y^2} dG(y), 0 > x \in C(G),$

$\sum_k (1 - F_{nk}(x)) \to \int_x^\infty \frac{1+y^2}{y^2} dG(y), 0 < x \in C(G).$

Proof. Write $a_n = \sup_k \int_{-\tau}^{\tau} |x| \, dF_{nk}(x), n \geqslant 1$, so that $\sup_k |a_{nk}| \leqslant a_n \to 0$ by (23.23). We have

$$\sum_k \widetilde{F}_{nk}(x - a_n) \leqslant \sum_k F_{nk}(x) \leqslant \sum_k \widetilde{F}_{nk}(x + a_n), \quad x \in R, \ n \geqslant 1, \qquad (1)$$

whence

$$\sum_k F_{nk}(x-a_n) \leqslant \sum_k \widetilde{F}_{nk}(x) \leqslant \sum_k F_{nk}(x+a_n), \quad x \in R, \ n \geqslant 1. \tag{2}$$

For $\varepsilon > 0$, choose $n_\varepsilon \in N$ such that $a_n < \varepsilon, n \geqslant n_\varepsilon$. Then (1) and (2) yield

$$\sum_k \widetilde{F}_{nk}(x-\varepsilon) \leqslant \sum_k F_{nk}(x) \leqslant \sum_k \widetilde{F}_{nk}(x+\varepsilon), \quad x \in R, \ n \geqslant n_\varepsilon, \tag{3}$$

and

$$\sum_k F_{nk}(x-\varepsilon) \leqslant \sum_k \widetilde{F}_{nk}(x) \leqslant \sum_k F_{nk}(x+\varepsilon), \quad x \in R, \ n \geqslant n_\varepsilon, \tag{4}$$

respectively. Let $x+\varepsilon < 0$ with $x, x \pm \varepsilon \in C(G)$, and assume that (i) holds. Then, on account of (3), we have

$$\limsup_n \sum_k F_{nk}(x) \leqslant \limsup_n \sum_k \widetilde{F}_{nk}(x+\varepsilon) = \int_{-\infty}^{x+\varepsilon} \frac{1+y^2}{y^2} dG(y) \tag{5}$$

and

$$\liminf_n \sum_k F_{nk}(x) \geqslant \liminf_n \sum_k \widetilde{F}_{nk}(x-\varepsilon) = \int_{-\infty}^{x-\varepsilon} \frac{1+y^2}{y^2} dG(y). \tag{6}$$

Letting $\varepsilon \to 0$ in (5) and (6), the first limit in (ii) obtains. By making use of (4), the second limit in (ii) obtains similarly. The same arguments work reversely to show that (ii) implies (i). □

Lemma 24.4. *Assume that*

$$\begin{cases} \sum_k F_{nk}(x) \to \int_{-\infty}^{x} \frac{1+y^2}{y^2} dG(y), 0 > x \in C(G) \\ \sum_k (1-F_{nk}(x)) \to \int_{x}^{\infty} \frac{1+y^2}{y^2} dG(y), 0 < x \in C(G) \end{cases}. \tag{1}$$

Then the following are equivalent:

(i) $\lim_{\varepsilon\to 0} \limsup_n \sum_k \int_{-\varepsilon}^{\varepsilon} \frac{x^2}{1+x^2} d\widetilde{F}_{nk}(x) = \lim_{\varepsilon\to 0} \liminf_n \sum_k \int_{-\varepsilon}^{\varepsilon} \frac{x^2}{1+x^2} d\widetilde{F}_{nk}(x) = G(0) - G(0-)$;

(ii) $\lim_{\varepsilon\to 0} \limsup_n \sum_k [\int_{-\varepsilon}^{\varepsilon} x^2\, dF_{nk}(x) - (\int_{-\varepsilon}^{\varepsilon} x\, dF_{nk}(x))^2] = \lim_{\varepsilon\to 0} \liminf_n \sum_k [\int_{-\varepsilon}^{\varepsilon} x^2\, dF_{nk}(x) - \left(\int_{-\varepsilon}^{\varepsilon} x\, dF_{nk}(x)\right)^2] = G(0) - G(0-)$.

Proof. Let $0 < \varepsilon < \tau$ with $\pm\varepsilon \in C(G)$, and choose $n_\varepsilon \in N$ such that $a_n < \varepsilon/2$ for $n \geqslant n_\varepsilon$. It follows from (1) that

$$\sum_k (P(f_{nk} \leqslant -\varepsilon) + P(f_{nk} > \varepsilon)) = \sum_k F_{nk}(-\varepsilon) + \sum_k (1-F_{nk}(\varepsilon))$$

$$\to \int_{-\infty}^{-\varepsilon} \frac{1+y^2}{y^2} dG(y) + \int_{\varepsilon}^{\infty} \frac{1+y^2}{y^2} dG(y),$$

and so there exists $C_\varepsilon < \infty$ such that

$$\sum_k (P(f_{nk} \leqslant -\varepsilon) + P(f_{nk} > \varepsilon)) \leqslant C_\varepsilon, \quad n \geqslant 1. \tag{2}$$

For $n \geqslant 1$, write

$$A_n = \sum_k \int_{-\varepsilon}^{\varepsilon} (x - a_{nk})^2 dF_{nk}(x) - \sum_k \left[\int_{-\varepsilon}^{\varepsilon} x^2 \, dF_{nk}(x) - \left(\int_{-\varepsilon}^{\varepsilon} x \, dF_{nk}(x) \right)^2 \right]$$

and

$$B_n = \sum_k \int_{-\varepsilon}^{\varepsilon} (x - a_{nk})^2 dF_{nk}(x) - \sum_k \int_{-\varepsilon}^{\varepsilon} x^2 \, d\widetilde{F}_{nk}(x).$$

We have

$$\begin{aligned}
A_n &= -2 \sum_k a_{nk} \int_{-\varepsilon}^{\varepsilon} x \, dF_{nk}(x) + \sum_k a_{nk}^2 \int_{-\varepsilon}^{\varepsilon} dF_{nk}(x) + \sum_k \left(\int_{-\varepsilon}^{\varepsilon} x \, dF_{nk}(x) \right)^2 \\
&= \sum_k \left(\int_{-\tau}^{\tau} x \, dF_{nk}(x) - \int_{-\varepsilon}^{\varepsilon} x \, dF_{nk}(x) \right)^2 \\
&\quad - \sum_k a_{nk}^2 (P(f_{nk} \leqslant -\varepsilon) + P(f_{nk} > \varepsilon)) \\
&= \sum_k \left(\int_{]-\tau,-\varepsilon] \cup]\varepsilon,\tau]} x \, dF_{nk}(x) \right)^2 - \sum_k a_{nk}^2 (P(f_{nk} \leqslant -\varepsilon) + P(f_{nk} > \varepsilon)),
\end{aligned}$$

whence $|A_n| \leqslant (a_n \tau + a_n^2) \sum_k a_{nk}^2 (P(f_{nk} \leqslant -\varepsilon) + P(f_{nk} > \varepsilon))$. By virtue of (2), as $a_n \to 0$, this entails that $A_n \to 0$. Taking into account (23.25), we have also

$$\begin{aligned}
B_n &= \sum_k \int_{-\varepsilon}^{\varepsilon} (x - a_{nk})^2 dF_{nk}(x) - \sum_k \int_{a_{nk}-\varepsilon}^{a_{nk}+\varepsilon} (x - a_{nk})^2 dF_{nk}(x) \\
&= \sum_k \int_{]-\varepsilon,\varepsilon] -]a_{nk}-\varepsilon, a_{nk}+\varepsilon]} (x - a_{nk})^2 dF_{nk}(x) \\
&\quad - \sum_k \int_{]a_{nk}-\varepsilon, a_{nk}+\varepsilon] -]-\varepsilon,\varepsilon]} (x - a_{nk})^2 dF_{nk}(x).
\end{aligned}$$

Therefore, for $n \geqslant n_\varepsilon$ and $0 < \eta < \varepsilon/2$, we obtain

$$|B_n| \leqslant 2\sum_k \int\limits_{\{\varepsilon/2<|x|<2\varepsilon\}} (x-a_{nk})^2 dF_{nk}(x) \leqslant 18\varepsilon^2 \sum_k P\left(|f_{nk}| > \frac{\varepsilon}{2}\right)$$

$$\leqslant 18\varepsilon^2 \sum_k (P(f_{nk} \leqslant -\eta) + P(f_{nk} > \eta)). \tag{3}$$

Taking η in (3) such that $\pm\eta \in C(G)$, and applying (2), we see that $B_n \to 0$. Consequently,

$$\sum_k \int_{-\varepsilon}^{\varepsilon} x^2\, d\widetilde{F}_{nk}(x) - \sum_k \left[\int_{-\varepsilon}^{\varepsilon} x^2\, dF_{nk}(x) - \left(\int_{-\varepsilon}^{\varepsilon} x\, dF_{nk}(x)\right)^2\right] = A_n - B_n \to 0. \tag{4}$$

To prove that (i) and (ii) are equivalent, we need another ingredient, namely the obvious inequalities

$$\frac{1}{1+\varepsilon^2}\sum_k \int_{-\varepsilon}^{\varepsilon} x^2\, d\widetilde{F}_{nk}(x) \leqslant \sum_k \int_{-\varepsilon}^{\varepsilon} \frac{x^2}{1+x^2} d\widetilde{F}_{nk}(x)$$

$$\leqslant \sum_k \int_{-\varepsilon}^{\varepsilon} x^2\, d\widetilde{F}_{nk}(x), \quad n \geqslant 1,$$

which ensure that

$$\lim_{\varepsilon\to 0} \limsup_n \sum_k \int_{-\varepsilon}^{\varepsilon} x^2\, d\widetilde{F}_{nk}(x) = \lim_{\varepsilon\to 0} \limsup_n \sum_k \int_{-\varepsilon}^{\varepsilon} \frac{x^2}{1+x^2} d\widetilde{F}_{nk}(x) \tag{5}$$

and

$$\lim_{\varepsilon\to 0} \liminf_n \sum_k \int_{-\varepsilon}^{\varepsilon} x^2\, d\widetilde{F}_{nk}(x) = \lim_{\varepsilon\to 0} \liminf_n \sum_k \int_{-\varepsilon}^{\varepsilon} \frac{x^2}{1+x^2} d\widetilde{F}_{nk}(x). \tag{6}$$

Suppose now that (i) is fulfilled. Then, in view of (4) and (5), we get

$$\lim_{\varepsilon\to 0} \limsup_n \sum_k \left[\int_{-\varepsilon}^{\varepsilon} x^2\, dF_{nk}(x) - \left(\int_{-\varepsilon}^{\varepsilon} x\, dF_{nk}(x)\right)^2\right]$$

$$= \lim_{\varepsilon\to 0} \limsup_n \sum_k \int_{-\varepsilon}^{\varepsilon} x^2\, d\widetilde{F}_{nk}(x)$$

$$= \lim_{\varepsilon\to 0} \limsup_n \sum_k \int_{-\varepsilon}^{\varepsilon} \frac{x^2}{1+x^2} d\widetilde{F}_{nk}(x) = G(0) - G(0-).$$

The second equality in (ii) follows analogously, on account of (4) and (6). The proof that (ii) implies (i) involves the same arguments and is omitted. □

Lemma 24.5. *Let $\{\gamma_n : n \in N\}$ and $\{G_n : n \in N\}$ be as in (24.1), let $\gamma \in R$ and $G \in \mathfrak{G}$, and suppose that $G_n \xrightarrow{c} G$. Then $\gamma_n \to \gamma$ if and only if $\sum_k \int_{-\tau}^{\tau} x\, dF_{nk}(x) \to \gamma + \int_{]-\tau,\tau]} x\, dG(x) - \int_{]-\tau,\tau]^c} 1/x\, dG(x)$ for some $\tau > 0$ such that $\pm\tau \in C(G)$.*

Proof. Clearly, for $\tau > 0$ with $\pm\tau \in C(G)$, it suffices to show that

$$\sum_k \int_R \frac{x}{1+x^2} d\widetilde{F}_{nk}(x) \to \int_{]-\tau,\tau]^c} \frac{1}{x} dG(x) - \int_{]-\tau,\tau]} x\, dG(x). \tag{1}$$

We have

$$\sum_k \int_R \frac{x}{1+x^2} d\widetilde{F}_{nk}(x) = \sum_k \int_{]-\tau,\tau]} x\, d\widetilde{F}_{nk}(x) - \sum_k \int_{]-\tau,\tau]} \frac{x^3}{1+x^2} d\widetilde{F}_{nk}(x)$$
$$+ \sum_k \int_{]-\tau,\tau]^c} \frac{x}{1+x^2} d\widetilde{F}_{nk}(x). \tag{2}$$

Since $G_n \xrightarrow{c} G$, (12.8) and (19.67.e) yield

$$\sum_k \int_{]-\tau,\tau]} \frac{x^3}{1+x^2} d\widetilde{F}_{nk}(x) = \int_{]-\tau,\tau]} x\, dG_n(x) \to \int_{]-\tau,\tau]} x\, dG(x) \tag{3}$$

and

$$\sum_k \int_{]-\tau,\tau]^c} \frac{x}{1+x^2} d\widetilde{F}_{nk}(x) = \int_{]-\tau,\tau]^c} \frac{1}{x} dG_n(x) \to \int_{]-\tau,\tau]^c} \frac{1}{x} dG(x). \tag{4}$$

From (2)–(4) it follows that to verify (1) we need only show that

$$\sum_k \int_{]-\tau,\tau]} x\, d\widetilde{F}_{nk}(x) \to 0. \tag{5}$$

To do this, let $0 < \eta < \tau/2$ with $\tau \pm \eta \in C(G)$, and choose $n_1 \in N$ so that $a_n < \eta$ for $n \geqslant n_1$. Then, using (23.25), we have

$$\int_{]-\tau,\tau]} x\, d\widetilde{F}_{nk}(x) = \int_{]a_{nk}-\tau,a_{nk}+\tau]} (x - a_{nk}) dF_{nk}(x) = \int_{]-\tau,\tau]} (x - a_{nk}) dF_{nk}(x)$$
$$+ \int_{]a_{nk}-\tau,a_{nk}+\tau]-]-\tau,\tau]} (x - a_{nk}) dF_{nk}(x)$$
$$- \int_{]-\tau,\tau]-]a_{nk}-\tau,a_{nk}+\tau]} (x - a_{nk}) dF_{nk}(x),$$

and so

$$\left|\sum_k \int_{]-\tau,\tau]} x\,d\widetilde{F}_{nk}(x)\right| \leqslant \left|\sum_k \int_{]-\tau,\tau]} (x-a_{nk})dF_{nk}(x)\right|$$

$$+\left|\sum_k \int_{]a_{nk}-\tau,a_{nk}+\tau]-]-\tau,\tau]} (x-a_{nk})dF_{nk}(x)\right|$$

$$+\left|\sum_k \int_{]-\tau,\tau]-]a_{nk}-\tau,a_{nk}+\tau]} (x-a_{nk})dF_{nk}(x)\right|$$

$$\leqslant \left|\sum_k a_{nk}(P(f_{nk}\leqslant -\eta)+P(f_{nk}>\eta))\right|$$

$$+\tau\sum_k P(\tau\leqslant |f_{nk}|\leqslant \tau+|a_{nk}|)$$

$$+\sum_k(\tau+|a_{nk}|)P(\tau-|a_{nk}|\leqslant |f_{nk}|\leqslant \tau)$$

$$\leqslant a_n C_\tau+\tau\sum_k P(\tau-a_n\leqslant |f_{nk}|\leqslant \tau+a_n)$$

$$+\sum_k(\tau+a_n)P(\tau-a_n\leqslant |f_{nk}|\leqslant \tau+a_n)$$

$$\leqslant a_n C_\tau+(2\tau+a_n)\sum_k P(\tau-\eta<|f_{nk}|<\tau+\eta), \quad n\geqslant n_1, \qquad (6)$$

where C_τ is as in (24.4). As $a_n \to 0$, (6) shows that to prove (5) it suffices to verify that

$$\sum_k P(\tau-\eta<|f_{nk}|<\tau+\eta)\to 0. \qquad (7)$$

Since $G_n \xrightarrow{c} G$, we see that the assumption in (24.4) is satisfied by (24.2) and (24.3). Therefore, for $n\geqslant n_1$, as in the proof of (24.4) we obtain

$$\sum_k P(\tau-\eta<|f_{nk}|<\tau+\eta)$$

$$\leqslant \sum_k P(f_{nk}\leqslant -\tau+\eta)-\sum_k P(f_{nk}\leqslant -\tau-\eta)+\sum_k P(f_{nk}>\tau-\eta)$$

$$-\sum_k P(f_{nk}>\tau+\eta)\to \int_{-\tau-\eta}^{-\tau+\eta}\frac{1+y^2}{y^2}dG(y)+\int_{\tau-\eta}^{\tau+\eta}\frac{1+y^2}{y^2}dG(y)$$

$$\leqslant \left(1+\frac{4}{\tau^2}\right)(G(-\tau+\eta)-G(-\tau-\eta)+G(\tau+\eta)-G(\tau-\eta)). \qquad (8)$$

Since $G(-\tau+\eta)-G(-\tau-\eta)+G(\tau+\eta)-G(\tau-\eta)\to 0$ as $\eta\to 0$, (8) shows that (7) holds. □

Theorem 24.6. *Let* $\{f_{nk} : n \geqslant 1, 1 \leqslant k \leqslant k_n\}$ *and* F *be as in* (24.1). *Then* $\sum_k f_{nk} \xrightarrow{D} F$ *if and only if*

(i) $\sum_k F_{nk}(x) \to \int_{-\infty}^{x} \frac{1+y^2}{y^2} dG(y), 0 > x \in C(G)$,

$\sum_k (1 - F_{nk}(x)) \to \int_{x}^{\infty} \frac{1+y^2}{y^2} dG(y), 0 < x \in C(G)$,

(ii) $\lim_{\varepsilon\to 0} \limsup_n \sum_k [\int_{-\varepsilon}^{\varepsilon} x^2\, dF_{nk}(x) - (\int_{-\varepsilon}^{\varepsilon} x\, dF_{nk}(x))^2]$
$= \lim_{\varepsilon\to 0} \liminf_n \sum_k [\int_{-\varepsilon}^{\varepsilon} x^2\, dF_{nk}(x) - (\int_{-\varepsilon}^{\varepsilon} x\, dF_{nk}(x))^2] = G(0) - G(0-)$

and

(iii) $\sum_k \int_{-\tau}^{\tau} x\, dF_{nk}(x) \to \gamma + \int_{]-\tau,\tau]} x\, dG(x) - \int_{]-\tau,\tau]^c} 1/x\, dG(x)$ *for some* $\tau > 0$ *such that* $\pm\tau \in C(G)$.

Proof. If $\sum_k f_{nk} \xrightarrow{D} F$, then $\gamma_n \to \gamma$ and $G_n \xrightarrow{c} G$ by (24.1). Then (i) holds on account of (24.2) and (24.3). Furthermore, in view of (24.2), (i) and (24.4), (ii) is fulfilled. Finally, (iii) follows from (24.5).

Suppose now that (i)–(iii) hold. Then (24.2.i) is satisfied by (i) and (24.3), and (24.2.ii) follows from (i), (ii) and (24.4). Thus $G_n \xrightarrow{c} G$ by (24.2), and so $\gamma_n \to \gamma$ by (iii) and (24.5). Hence $\sum_k f_{nk} \xrightarrow{D} F$ by (24.1). □

Remark 24.7. In (24.5) and (24.6.iii) "for some $\tau > 0$" may be replaced by "for every $\tau > 0$".

We specialize now the convergence criterion (24.6) to the case where F is degenerate, Poisson and normal, respectively. For $\varepsilon > 0$, set $\sigma_{nk}^2(\varepsilon) = \int_{-\varepsilon}^{\varepsilon} x^2\, dF_{nk}(x) - (\int_{-\varepsilon}^{\varepsilon} x\, dF_{nk}(x))^2$. Then $\sigma_{nk}^2(\varepsilon) \geqslant 0$ by Schwarz's inequality.

If $F \in \mathfrak{F}$ is degenerate at some $a \in R$, then in the Lévy-Khintchine representation (23.16) of its characteristic function we have $\gamma = a$ and $G = 0$, as remarked in (23.13.a). Hence (24.6) implies the next assertion.

Theorem 24.8. *Let* $\{f_{nk} : n \geqslant 1, 1 \leqslant k \leqslant k_n\}$ *be as in* (24.1), *and let* $a \in R$. *Then* $\sum_k f_{nk} \xrightarrow{D} a$ *if and only if*

(i) $\sum_k P(|f_{nk}| \geqslant \varepsilon) \to 0, \varepsilon > 0$,

(ii) $\lim_{\varepsilon\to 0} \limsup_n \sum_k \sigma_{nk}^2(\varepsilon) = \lim_{\varepsilon\to 0} \liminf_n \sum_k \sigma_{nk}^2(\varepsilon) = 0$

and

(iii) $\sum_k \int_{-\tau}^{\tau} x\, dF_{nk}(x) \to a$ *for some/every* $\tau > 0$.

This theorem allows to get the following result.

Theorem 24.9. *Let* $\{f_{nk} : n \geqslant 1, 1 \leqslant k \leqslant k_n\}$ *be an array of random variables, where* $k_n \to \infty$ *and* f_{nk}, $1 \leqslant k \leqslant k_n$, *are independent for each* n, *and let* $a \in R$. *Then the array satisfies the condition of infinite smallness and* $\sum_k f_{nk} \xrightarrow{D} a$ *if and only if, for every* $\varepsilon > 0$,

(i) $\sum_k P(|f_{nk}| \geqslant \varepsilon) \to 0$,
(ii) $\sum_k [\int_{-\varepsilon}^{\varepsilon} x^2 \, dF_{nk}(x) - (\int_{-\varepsilon}^{\varepsilon} x \, dF_{nk}(x))^2] \to 0$

and

(iii) $\sum_k \int_{-\varepsilon}^{\varepsilon} x \, dF_{nk}(x) \to a$.

Proof. If the infinite smallness condition is fulfilled and $\sum_k f_{nk} \xrightarrow{D} a$, then (24.8.i)–(24.8.iii) hold, whence (i) and (iii) hold. It remains to prove (ii). For this, let $0 < \delta < \varepsilon$. We obtain

$$\int_{-\varepsilon}^{\varepsilon} x^2 \, dF_{nk}(x) - \Big(\int_{-\varepsilon}^{\varepsilon} x \, dF_{nk}(x)\Big)^2 = \int_{-\delta}^{\delta} x^2 \, dF_{nk}(x) - \Big(\int_{-\delta}^{\delta} x \, dF_{nk}(x)\Big)^2$$
$$+ \int_{]-\varepsilon,\varepsilon]-]-\delta,\delta]} x^2 \, dF_{nk}(x) - \Big(\int_{]-\varepsilon,\varepsilon]-]-\delta,\delta]} x \, dF_{nk}(x)\Big)^2$$
$$-2\Big(\int_{-\delta}^{\delta} x \, dF_{nk}(x)\Big) \int_{]-\varepsilon,\varepsilon]-]-\delta,\delta]} x \, dF_{nk}(x),$$

and so

$$\sum_k \sigma_{nk}^2(\varepsilon) = \sum_k \sigma_{nk}^2(\delta) + \sum_k \Big[\int_{]-\varepsilon,\varepsilon]-]-\delta,\delta]} x^2 \, dF_{nk}(x)$$
$$- \Big(\int_{]-\varepsilon,\varepsilon]-]-\delta,\delta]} x \, dF_{nk}(x)\Big)^2\Big]$$
$$-2\sum_k \Big(\int_{-\delta}^{\delta} x \, dF_{nk}(x)\Big) \int_{]-\varepsilon,\varepsilon]-]-\delta,\delta]} x \, dF_{nk}(x). \tag{1}$$

In view of (i), we have

$$0 \leqslant \sum_k \Big[\int_{]-\varepsilon,\varepsilon]-]-\delta,\delta]} x^2 \, dF_{nk}(x) - \Big(\int_{]-\varepsilon,\varepsilon]-]-\delta,\delta]} x \, dF_{nk}(x)\Big)^2\Big]$$
$$\leqslant \varepsilon^2 \sum_k \int_{]-\varepsilon,\varepsilon]-]-\delta,\delta]} dF_{nk}(x)$$
$$\leqslant \varepsilon^2 \sum_k P(\big|f_{nk}\big| \geqslant \delta) \to 0 \tag{2}$$

and

$$\sum_k \left| \int_{-\delta}^{\delta} x\,dF_{nk}(x) \right| \left| \int_{]-\varepsilon,\varepsilon]-]-\delta,\delta]} x\,dF_{nk}(x) \right| \leqslant \delta\varepsilon \sum_k P(|f_{nk}| \geqslant \delta) \to 0. \tag{3}$$

From (1)–(3) it follows that

$$\limsup_n \sum_k \sigma_{nk}^2(\varepsilon) = \limsup_n \sum_k \sigma_{nk}^2(\delta), \quad \liminf_n \sum_k \sigma_{nk}^2(\varepsilon) = \liminf_n \sum_k \sigma_{nk}^2(\delta). \tag{4}$$

Consequently, (24.8.ii) and (4) imply (ii).

Conversely, assume that (i)–(iii) hold. Then (24.8.i)–(24.8.iii) also hold. Moreover, for every $\varepsilon > 0$, $\sup_k P(|f_{nk}| \geqslant \varepsilon) \leqslant \sum_k P(|f_{nk}| \geqslant \varepsilon) \to 0$ by (i), i.e. the condition of infinite smallness is satisfied. Finally, $\sum_k f_{nk} \xrightarrow{D} a$ by (24.8). □

The following version of (24.9) is useful for checking convergence to a degenerate distribution function.

Theorem 24.10. *Let $\{f_{nk} : n \geqslant 1, 1 \leqslant k \leqslant k_n\}$ and a be as in (24.9). Then the array satisfies the condition of infinite smallness and $\sum_k f_{nk} \xrightarrow{D} a$ if and only if, for every $\varepsilon > 0$ and some $\tau > 0$,*

(i) $\sum_k P(|f_{nk}| \geqslant \varepsilon) \to 0$,

(ii) $\sum_k [\int_{-\tau}^{\tau} x^2\,dF_{nk}(x) - (\int_{-\tau}^{\tau} x\,dF_{nk}(x))^2] \to 0$

and

(iii) $\sum_k \int_{-\tau}^{\tau} x\,dF_{nk}(x) \to a$.

Proof. It is sufficient to show that (i)–(iii) imply (24.9.ii) and (24.9.iii). For $0 < \varepsilon < \tau$, according to the proof of (24.9), we have

$$\left| \sum_k \sigma_{nk}^2(\tau) - \sum_k \sigma_{nk}^2(\varepsilon) \right| \leqslant (\tau^2 + \varepsilon\tau) \sum_k P(|f_{nk}| \geqslant \varepsilon) \to 0$$

by (i). Hence (ii) implies (24.9.ii) in this case. For $0 < \tau < \varepsilon$, we have

$$\left| \sum_k \sigma_{nk}^2(\varepsilon) - \sum_k \sigma_{nk}^2(\tau) \right| \leqslant (\varepsilon^2 + \tau\varepsilon) \sum_k P(|f_{nk}| \geqslant \tau) \to 0,$$

and so (ii) implies (24.9.ii) also in this case. Further, for $0 < \varepsilon < \tau$, we obtain

$$\begin{aligned} \left| \sum_k \int_{-\tau}^{\tau} x\,dF_{nk}(x) - \sum_k \int_{-\varepsilon}^{\varepsilon} x\,dF_{nk}(x) \right| &\leqslant \int_{]-\tau,\tau]-]-\varepsilon,\varepsilon]} |x|\,dF_{nk}(x) \\ &\leqslant \tau \sum_k P(|f_{nk}| \geqslant \varepsilon) \to 0 \end{aligned}$$

by (i), and so (iii) implies (24.9.iii) in this case. Similarly, (iii) implies (24.9.iii) in case $0 < \tau < \varepsilon$. □

Corollary 24.11. *Let f_n, $n \geqslant 1$, be independent random variables, let $\{b_n : n \geqslant 1\} \subset]0, \infty[$, and let $a \in R$. For $n \geqslant 1$, set $S_n = \sum_{k=1}^n f_k$ and let F_n denote the distribution function of f_n. Then $\sup_{1 \leqslant k \leqslant n} P(|f_k| \geqslant \varepsilon b_n) \to 0$, $\varepsilon > 0$, and $S_n/b_n \xrightarrow{D} a$ if and only if*

(i) $\sum_{k=1}^n P(|f_k| \geqslant \varepsilon b_n) \to 0$, $\varepsilon > 0$,
(ii) $\frac{1}{b_n^2} \sum_{k=1}^n [\int_{-b_n}^{b_n} x^2\, dF_k(x) - (\int_{-b_n}^{b_n} x\, dF_k(x))^2] \to 0$

and

(iii) $\frac{1}{b_n} \sum_{k=1}^n \int_{-b_n}^{b_n} x\, dF_k(x) \to a$.

Proof. This follows immediately from (24.10) with $f_{nk} = f_k/b_n$, $n \geqslant 1$, $1 \leqslant k \leqslant n$, and $\tau = 1$, if we notice that, by virtue of the change of variable formula (11.29), we have

$$\int_{-1}^{1} x^2\, dF_{nk}(x) = \int_{\{-1 < f_{nk} \leqslant 1\}} f_{nk}^2\, dP = \frac{1}{b_n^2} \int_{\{-b_n < f_k \leqslant b_n\}} f_k^2\, dP$$
$$= \frac{1}{b_n^2} \int_{-b_n}^{b_n} x^2\, dF_k(x)$$

and, similarly, $\int_{-1}^{1} x\, dF_{nk}(x) = (1/b_n) \int_{-b_n}^{b_n} x\, dF_k(x)$. □

An easier way to check convergence to a degenerate distribution function is as follows.

Theorem 24.12. *Let $\{f_{nk} : n \geqslant 1, 1 \leqslant k \leqslant k_n\}$ and a be as in (24.9). Then $\sum_k f_{nk} \xrightarrow{D} a$ if, for some $\tau > 0$,*

(i) $\sum_k P(|f_{nk}| \geqslant \tau) \to 0$,
(ii) $\sum_k [\int_{-\tau}^{\tau} x^2\, dF_{nk}(x) - (\int_{-\tau}^{\tau} x\, dF_{nk}(x))^2] \to 0$

and

(iii) $\sum_k \int_{-\tau}^{\tau} x\, dF_{nk}(x) \to a$.

Proof. Put $f_{nk}^\tau = f_{nk} 1_{\{-\tau < f_{nk} \leqslant \tau\}}$, $n \geqslant 1$, $1 \leqslant k \leqslant k_n$. Then, in view of (11.29), (ii) and (iii) become $\sum_k \operatorname{Var} f_{nk}^\tau \to 0$ and $\sum_k E f_{nk}^\tau \to a$, respectively. Let $\delta > 0$. By Chebyshev's inequality and (18.40.a), we have $P(|\sum_k f_{nk}^\tau - \sum_k E f_{nk}^\tau| \geqslant \delta) \leqslant \delta^{-2} \sum_k \operatorname{Var} f_{nk}^\tau \to 0$. This means that $\sum_k f_{nk}^\tau - \sum_k E f_{nk}^\tau \xrightarrow{P} 0$, and so

$$\sum_k f_{nk}^\tau \xrightarrow{P} a \tag{1}$$

by (14.23.a). On account of (1) and (i), we get

$$\begin{aligned}
P\left(\left|\sum_k f_{nk} - a\right| \geqslant \delta\right) &= P\left(\left|\sum_k f_{nk} - a\right| \geqslant \delta, \sum_k f_{nk} = \sum_k f_{nk}^{\tau}\right) \\
&\quad + P\left(\left|\sum_k f_{nk} - a\right| \geqslant \delta, \sum_k f_{nk} \neq \sum_k f_{nk}^{\tau}\right) \\
&\leqslant P\left(\left|\sum_k f_{nk}^{\tau} - a\right| \geqslant \delta\right) + P\left(\sum_k f_{nk} \neq \sum_k f_{nk}^{\tau}\right) \\
&\leqslant P\left(\left|\sum_k f_{nk}^{\tau} - a\right| \geqslant \delta\right) + \sum_k P(|f_{nk}| \geqslant \tau) \to 0,
\end{aligned}$$

and so $\sum_k f_{nk} \xrightarrow{P} a$. This is equivalent to $\sum_k f_{nk} \xrightarrow{D} a$ by (19.26). □

Corollary 24.13. *For $n \geqslant 1$, let f_n, S_n and F_n be as in* (24.11), *and assume that $0 < b_n \leqslant b_{n+1} \to \infty$. Then $S_n/b_n \xrightarrow{D} 0$ if and only if*

(i) $\sum_{k=1}^{n} P(|f_k| \geqslant b_n) \to 0$,

(ii) $\frac{1}{b_n^2} \sum_{k=1}^{n} [\int_{-b_n}^{b_n} x^2\, dF_k(x) - (\int_{-b_n}^{b_n} x\, dF_k(x))^2] \to 0$

and

(iii) $\frac{1}{b_n} \sum_{k=1}^{n} \int_{-b_n}^{b_n} x\, dF_k(x) \to 0$.

Proof. If (i)–(iii) hold, then $S_n/b_n \xrightarrow{D} 0$ by (24.12).

Conversely, assume that $S_n/b_n \xrightarrow{D} 0$. Then, in view of (24.11), it suffices to prove that $\sup_{1 \leqslant k \leqslant n} P(|f_k| \geqslant 2\varepsilon b_n) \to 0, \varepsilon > 0$. To do this, let $\varepsilon, \delta > 0$, and choose $n_1 \in N$ such that $P(|S_n| \geqslant \varepsilon b_n) < \delta$ for $n \geqslant n_1$. Then, for $n_1 < k \leqslant n$, we have

$$\begin{aligned}
P\left(\frac{|f_k|}{b_n} \geqslant 2\varepsilon\right) &= P\left(\left|\frac{S_k}{b_k}\frac{b_k}{b_n} - \frac{S_{k-1}}{b_{k-1}}\frac{b_{k-1}}{b_n}\right| \geqslant 2\varepsilon\right) \leqslant P\left(\left|\frac{S_k}{b_k}\right| \geqslant \varepsilon\right) \\
&\quad + P\left(\left|\frac{S_{k-1}}{b_{k-1}}\right| \geqslant \varepsilon\right) < 2\delta. \qquad (1)
\end{aligned}$$

Since $b_n \to \infty$, it follows that $f_k/b_n \xrightarrow{a.s.} 0$, hence $f_k/b_n \xrightarrow{P} 0$ for $1 \leqslant k \leqslant n_1$ (14.9). Choose $n_2 \in N$ so that

$$\sup_{1 \leqslant k \leqslant n_1} P(|f_k| \geqslant 2\varepsilon b_n) < 2\delta, \quad n \geqslant n_2. \qquad (2)$$

From (1) and (2) we see that $\sup_{1 \leqslant k \leqslant n} P(|f_k| \geqslant 2\varepsilon b_n) < 2\delta$ for $n > n_1, n_2$. □

Consider now the case that F is a Poisson distribution function with parameters $a \in R, b \neq 0$ and $\lambda > 0$. Then in the Lévy-Khintchine representation of its characteristic function we have $\gamma = a + b\lambda/(1 + b^2)$ and $G(x) = 0$ or $b^2\lambda/(1 + b^2)$ according as $x < b$ or $x \geqslant b$ (23.13.c). In what follows we assume $b > 0$. Applying the convergence criterion (24.6), and following arguments in the proofs of (24.9) and (24.10), the next result obtains easily.

Theorem 24.14. *Let* $\{f_{nk} : n \geqslant 1, 1 \leqslant k \leqslant k_n\}$ *be as in* (24.1) *and let* F *be as above. Then* $\sum_k f_{nk} \xrightarrow{D} F$ *if and only if*

(i) $\sum_k P(f_{nk} \leqslant x) \to 0, x < 0,$
$\sum_k P(f_{nk} > x) \to \lambda, 0 < x < b,$
$\sum_k P(f_{nk} > x) \to 0, x > b,$

(ii) $\sum_k [\int_{-\tau}^{\tau} x^2\, dF_{nk}(x) - (\int_{-\tau}^{\tau} x\, dF_{nk}(x))^2] \to 0$ *for some/every* $\tau \in]0, b[$

and

(iii) $\sum_k \int_{-\tau}^{\tau} x\, dF_{nk}(x) \to a$ *for some/every* $\tau \in]0, b[$.

Finally, we shall discuss the case when F is a Gaussian distribution function $G_{m,\sigma}$ as in (19.10.d). Then, as remarked in (23.13.b), in the Lévy-Khintchine representation of its characteristic function we have $\gamma = m$ and $G(x) = 0$ or σ^2 according as $x < 0$ or $x \geqslant 0$. Upon specializing (24.6), and paralleling the proofs of (24.9) and (24.10), we get the following two theorems.

Theorem 24.15. *Let* $\{f_{nk} : n \geqslant 1, 1 \leqslant k \leqslant k_n\}$ *be as in* (24.9). *Then the array satisfies the condition of infinite smallness and* $\sum_k f_{nk} \xrightarrow{D} G_{m,\sigma}$ *if and only if, for each* $\varepsilon > 0$,

(i) $\sum_k P(|f_{nk}| \geqslant \varepsilon) \to 0,$

(ii) $\sum_k [\int_{-\varepsilon}^{\varepsilon} x^2\, dF_{nk}(x) - (\int_{-\varepsilon}^{\varepsilon} x\, dF_{nk}(x))^2] \to \sigma^2$

and

(iii) $\sum_k \int_{-\varepsilon}^{\varepsilon} x\, dF_{nk}(x) \to m.$

Theorem 24.16. *Let* $\{f_{nk} : n \geqslant 1, 1 \leqslant k \leqslant k_n\}$ *be as in* (24.9). *Then the infinite smallness condition is fulfilled and* $\sum_k f_{nk} \xrightarrow{D} G_{m,\sigma}$ *if and only if, for any* $\varepsilon > 0$ *and some* $\tau > 0$,

(i) $\sum_k P(|f_{nk}| \geqslant \varepsilon) \to 0,$

(ii) $\sum_k [\int_{-\tau}^{\tau} x^2\, dF_{nk}(x) - (\int_{-\tau}^{\tau} x\, dF_{nk}(x))^2] \to \sigma^2$

and

(iii) $\sum_k \int_{-\tau}^{\tau} x\, dF_{nk}(x) \to m.$

Corollary 24.17. *For* $n \geqslant 1$, *let* f_n, S_n, F_n *and* b_n *be as in* (24.11), *and let* $m \in R$ *and* $\sigma > 0$. *Then* $\sup_{1 \leqslant k \leqslant n} P(|f_k| \geqslant \varepsilon b_n) \to 0, \varepsilon > 0,$ *and* $S_n/b_n \xrightarrow{D} G_{m,\sigma}$ *if and only if*

(i) $\sum_{k=1}^{n} P(|f_k| \geqslant \varepsilon b_n) \to 0, \varepsilon > 0,$

(ii) $\frac{1}{b_n^2} \sum_{k=1}^{n} [\int_{-b_n}^{b_n} x^2\, dF_k(x) - (\int_{-b_n}^{b_n} x\, dF_k(x))^2] \to \sigma^2$

and

(iii) $\frac{1}{b_n} \sum_{k=1}^{n} \int_{-b_n}^{b_n} x\, dF_k(x) \to m.$

Proof. This is an immediate consequence of (24.16) for $f_{nk} = f_k/b_n, n \geqslant 1, 1 \leqslant k \leqslant n$, and $\tau = 1$. □

Remarks 24.18. (a) The degenerate distribution function F_a can be considered as a degenerate Gaussian distribution function $G_{a,\sigma}$ with $\sigma = 0$. In this respect, (24.9),

(24.10) and (24.11) can be viewed as special cases of (24.15), (24.16) and (24.17), respectively.

(b) According to (24.15), the conditions (ii) and (iii) in (24.17) can be replaced by $\frac{1}{b_n^2}\sum_{k=1}^n[\int_{-\varepsilon b_n}^{\varepsilon b_n} x^2\,dF_k(x) - (\int_{-\varepsilon b_n}^{\varepsilon b_n} x\,dF_k(x))^2] \to \sigma^2, \varepsilon > 0$, and $\frac{1}{b_n}\sum_{k=1}^n \int_{-\varepsilon b_n}^{\varepsilon b_n} x\,dF_k(x) \to m, \varepsilon > 0$, respectively.

(c) Notation is as in (24.17). If $\sum_{k=1}^n P(|f_k| \geqslant \varepsilon b_n) \to 0, \varepsilon > 0$, $\frac{1}{b_n^2}\sum_{k=1}^n \int_{-b_n}^{b_n} x^2\,dF_k(x) \to \sigma^2$ and $\frac{1}{b_n}\sum_{k=1}^n \left|\int_{-b_n}^{b_n} x\,dF_k(x)\right| \to 0$, then $S_n/b_n \xrightarrow{D} G_{0,\sigma}$.

Theorem 24.19. *Let $\{f_{nk} : n \geqslant 1, 1 \leqslant k \leqslant k_n\}$ be as in (24.9), and suppose that $\sum_k f_{nk} \xrightarrow{D} F \in \mathfrak{F}$, where F is nondegenerate. Then the following are equivalent:*

(i) *F is Gaussian and the infinite smallness condition is satisfied;*

(ii) $\sup_k |f_{nk}| \xrightarrow{P} 0$.

Proof. If (i) holds, then $\sum_k P(|f_{nk}| \geqslant \varepsilon) \to 0$ for any $\varepsilon > 0$ by (24.16), whence $\sup_k |f_{nk}| \xrightarrow{P} 0$ by (23.55). If now (ii) holds, then

$$\sum_k P(|f_{nk}| \geqslant \varepsilon) \to 0, \quad \varepsilon > 0, \tag{1}$$

and the infinite smallness condition is satisfied by (23.55). Let $\varphi(t; \gamma, G)$ be the Lévy-Khintchine representation of $\varphi_F(t)$. From (1) and (24.6.i), as F is nondegenerate, it follows that $G(x) = 0$ or $G(0)$ according as $x < 0$ or $x \geqslant 0$, and that $G(0) - G(0-) = G(0) > 0$. On account of (1), arguing as in the proof of (24.9), (24.6.ii) shows that $\sum_k \sigma_{nk}^2(\tau) \to G(0), \tau > 0$. We see also that (24.6.iii) implies that $\sum_k \int_{-\tau}^{\tau} x\,dF_{nk}(x) \to \gamma$ for some $\tau > 0$. Thus (24.16) ensures that F is Gaussian with parameters γ and $G(0)$. □

Theorem 24.20 (Lindeberg-Feller). *Let $\{f_{nk} : n \geqslant 1, 1 \leqslant k \leqslant k_n\}$ be as in (24.9), and assume that $Ef_{nk} = 0, \sigma_{nk}^2 = Ef_{nk}^2 < \infty, n \geqslant 1, 1 \leqslant k \leqslant k_n$, and $\sum_k \sigma_{nk}^2 = 1, n \geqslant 1$. Then the following are equivalent:*

(i) $\sup_k \sigma_{nk}^2 \to 0$ *and* $\sum_k f_{nk} \xrightarrow{D} \Phi$;

(ii) $\sum_k \int_{\{|x|\geqslant\varepsilon\}} x^2\,dF_{nk}(x) \to 0, \varepsilon > 0$.

Proof. If (i) holds, then $\sup_k P(|f_{nk}| \geqslant \varepsilon) \leqslant \varepsilon^{-2}\sup_k \sigma_{nk}^2 \to 0, \varepsilon > 0$, by Chebyshev's inequality, and so $\sum_k \sigma_{nk}^2(\varepsilon) \to 1, \varepsilon > 0$, by (24.15.ii). Consequently,

$$1 = \sum_k Ef_{nk}^2 \geqslant \sum_k \int_{-\varepsilon}^{\varepsilon} x^2\,dF_{nk}(x) \geqslant \sum_k \sigma_{nk}^2(\varepsilon) \to 1, \quad \varepsilon > 0,$$

whence

$$\sum_k \int_{-\varepsilon}^{\varepsilon} x^2\,dF_{nk}(x) \to 1, \quad \varepsilon > 0. \tag{1}$$

Since $\sum_k Ef_{nk}^2 = 1$, (1) implies (ii). Conversely, assume that (ii) holds. For $\varepsilon > 0$, as $\sigma_{nk}^2 = \int_{\{|x|\geqslant\varepsilon\}} x^2\,dF_{nk}(x) + \int_{\{|x|<\varepsilon\}} x^2\,dF_{nk}(x)$, we have $\sup_k \sigma_{nk}^2 \leqslant \sum_k \int_{\{|x|\geqslant\varepsilon\}} x^2\,dF_{nk}(x) + \varepsilon^2$, and so $\limsup_n(\sup_k \sigma_{nk}^2) \leqslant \varepsilon^2$. Letting $\varepsilon \to 0$, we see that $\sup_k \sigma_{nk}^2 \to 0$. Theorem (24.15.i) is satisfied, as $\sum_k P(|f_{nk}| \geqslant \varepsilon) \leqslant \varepsilon^{-2}\sum_k \int_{\{|x|\geqslant\varepsilon\}} x^2\,dF_{nk}(x) \to 0, \varepsilon > 0$. Notice also that (ii) implies (1). Since $Ef_{nk} = 0$, in view of (ii), we have

$$\sum_k \left(\int_{-\varepsilon}^{\varepsilon} x\,dF_{nk}(x)\right)^2 = \sum_k \left|\int_{]-\varepsilon,\varepsilon]} x\,dF_{nk}(x)\right|\left|\int_{]-\varepsilon,\varepsilon]^c} x\,dF_{nk}(x)\right|$$
$$\leqslant \sum_k \varepsilon \int_{\{|x|\geqslant\varepsilon\}} |x|\,dF_{nk}(x)$$
$$\leqslant \sum_k \int_{\{|x|\geqslant\varepsilon\}} x^2\,dF_{nk}(x) \to 0, \quad \varepsilon > 0. \tag{2}$$

Equations (1) and (2) imply (24.15.ii) with $\sigma = 1$. We have also

$$\left|\sum_k \int_{-\varepsilon}^{\varepsilon} x\,dF_k(x)\right| \leqslant \sum_k \left|\int_{]-\varepsilon,\varepsilon]^c} x\,dF_k(x)\right|$$
$$\leqslant \frac{1}{\varepsilon}\sum_k \int_{\{|x|\geqslant\varepsilon\}} x^2\,dF_{nk}(x) \to 0, \quad \varepsilon > 0,$$

and so (24.15.iii) is fulfilled with $m = 0$. Therefore, on account of (24.15), $\sum_k f_{nk} \xrightarrow{D} \Phi$. □

The condition (ii) in (24.20) is called the *Lindeberg condition.*

Corollary 24.21. *Let* $\{f_{nk} : n \geqslant 1, 1 \leqslant k \leqslant k_n\}$ *be as in* (24.9), *and assume that* $\sum_k \operatorname{Var} f_{nk} = 1$. *Then* $\sup_k \operatorname{Var} f_{nk} \to 0$ *and* $\sum_k (f_{nk} - Ef_{nk}) \xrightarrow{D} \Phi$ *if and only if*

$$\sum_k \int_{\{|x-Ef_{nk}|\geqslant\varepsilon\}} (x - Ef_{nk})^2 dF_{nk}(x) \to 0, \quad \varepsilon > 0.$$

Corollary 24.22. *Let* $f_n, n \geqslant 1$, *be independent random variables with* $Ef_n = 0$ *and* $\sigma_n^2 = Ef_n^2 < \infty$. *For* $n \geqslant 1$, *let* F_n *denote the distribution function of* f_n, *put* $S_n = \sum_{k=1}^n f_k$ *and* $s_n^2 = \sum_k \sigma_k^2$, *and assume that* $s_n > 0$ *for some* n. *Then the following are equivalent:*

(i) $\frac{1}{s_n^2}\sup_{1\leqslant k\leqslant n} \sigma_k^2 \to 0$ *and* $\frac{S_n}{s_n} \xrightarrow{D} \Phi$;

(ii) $\frac{1}{s_n^2}\sum_{k=1}^n \int_{\{|x|\geqslant\varepsilon s_n\}} x^2\,dF_k(x) \to 0, \varepsilon > 0.$

Proof. This follows immediately from (24.20) with $f_{nk} = f_k/s_n$ for n sufficiently large and $1 \leqslant k \leqslant n$. □

Remarks 24.23. Notation is as in (24.22).

(a) The Lindeberg condition (24.22.ii) is equivalent to the condition

$$\frac{1}{s_n^2}\sum_{k=1}^{n}\int_{\{|x|\geqslant \varepsilon s_k\}} x^2\,dF_k(x) \to 0, \quad \varepsilon > 0. \tag{1}$$

Actually, (1) implies (24.22.ii), as $s_k \leqslant s_n$, $1 \leqslant k \leqslant n$. Conversely, if (24.22.ii) holds, then, for $\varepsilon, \delta > 0$ and enough large n, we have

$$\begin{aligned}\frac{1}{s_n^2}\sum_{k=1}^{n}\int_{\{|x|\geqslant \varepsilon s_k\}} x^2\,dF_k(x) &= \frac{1}{s_n^2}\left(\sum_{k:s_k<\delta s_n}\int_{\{|x|\geqslant \varepsilon s_k\}} x^2\,dF_k(x)\right.\\ &\quad \left. + \sum_{k:s_k\geqslant\delta s_n}\int_{\{|x|\geqslant \varepsilon s_k\}} x^2\,dF_k(x)\right)\\ &\leqslant \delta^2 + \frac{1}{s_n^2}\sum_{k=1}^{n}\int_{\{|x|\geqslant \varepsilon\delta s_n\}} x^2\,dF_k(x) \to \delta^2. \end{aligned}\tag{2}$$

Since δ is arbitrary, (2) implies (1).

(b) The Lindeberg condition (24.22.i) implies that $s_n \to \infty$, as $s_n^{-2}\sup_{1\leqslant k\leqslant n}\sigma_k^2 \to 0$ requires that $s_n \to \infty$.

(c) If $f_n, n \geqslant 1$, are i.i.d. random variables with $Ef_1 = 0$ and $\sigma^2 = Ef_1^2 \in]0,\infty[$, then

$$\frac{1}{s_n^2}\sum_{k=1}^{n}\int_{\{|x|\geqslant \varepsilon s_n\}} x^2\,dF_k(x) = \frac{1}{\sigma^2}\int_{\{|x|\geqslant \varepsilon\sigma\sqrt{n}\}} x^2\,dF_1(x) \to 0, \quad \varepsilon > 0,$$

by dominated convergence, and so $\Phi_n(x) = P(S_n/\sigma\sqrt{n} \leqslant x) \to \Phi(x)$, $x \in R$. Moreover, $\Phi_n \overset{u}{\to} \Phi$ by (19.18.d). This result provides the classical so-called central limit theorem in its most common form.

Theorem 24.24 (Liapounov). *Let $\{f_{nk} : n \geqslant 1, 1 \leqslant k \leqslant k_n\}$ and $\{\sigma_{nk} : n \geqslant 1, 1 \leqslant k \leqslant k_n\}$ be as in (24.20), and suppose that $\sum_k E\,|f_{nk}|^{2+\delta} \to 0$ for some $\delta > 0$. Then $\sup_k \sigma_{nk}^2 \to 0$ and $\sum_k f_{nk} \overset{D}{\to} \Phi$.*

Proof. For any $\varepsilon > 0$, we have

$$\begin{aligned}\sum_k \int_{\{|x|\geqslant\varepsilon\}} x^2\,dF_{nk}(x) &\leqslant \varepsilon^{-\delta}\sum_k\int_{\{|x|\geqslant\varepsilon\}} |x|^{2+\delta}\,dF_{nk}(x)\\ &\leqslant \varepsilon^{-\delta}\sum_k\int_R |x|^{2+\delta}\,dF_{nk}(x) = \varepsilon^{-\delta}\sum_k E\,|f_{nk}|^{2+\delta} \to 0,\end{aligned}$$

so that the conclusion follows from the Lindeberg-Feller theorem (24.20). □

Corollary 24.25. *Let* $\{f_{nk} : n \geqslant 1, 1 \leqslant k \leqslant k_n\}$ *and* $\{\sigma_{nk} : n \geqslant 1, 1 \leqslant k \leqslant k_n\}$ *be as in* (24.20)*, and assume that* $|f_{nk}| \leqslant C_{nk} < \infty$ *a.s.,* $n \geqslant 1, 1 \leqslant k \leqslant k_n$*, for some constants* C_{nk} *with* $\sup_k C_{nk} \to 0$*. Then* $\sup_k \sigma_{nk}^2 \to 0$ *and* $\sum_k f_{nk} \xrightarrow{D} \Phi$*.*

Proof. Since $\sum_k \sigma_{nk}^2 = 1, n \geqslant 1$, we have $\sum_k E|f_{nk}|^3 \leqslant (\sup_k C_{nk}) \sum_k Ef_{nk}^2 = \sup_k C_{nk} \to 0$, and so the result holds true on applying the Liapounov theorem (24.24) with $\delta = 1$. □

Exercise 24.26. Let $\{f_{nk} : n \geqslant 1, 1 \leqslant k \leqslant k_n\}$ be as in (24.9), and assume that $Ef_{nk} = 0, n \geqslant 1, 1 \leqslant k \leqslant k_n$, and $\sum_k E|f_{nk}|^{1+\delta} \to 0$ for some $\delta \in]0, 1]$. Use (24.9) to show that the infinite smallness condition is satisfied and $\sum_k f_{nk} \xrightarrow{D} 0$.

Exercise 24.27. Let $\{f_{nk} : n \geqslant 1, 1 \leqslant k \leqslant k_n\}$ be as in (24.9), suppose that $f_{nk} \geqslant 0$ a.s., $n \geqslant 1, 1 \leqslant k \leqslant k_n$, and let $a > 0$. Prove that the infinite smallness condition is satisfied and $\sum_k f_{nk} \xrightarrow{D} a$ if and only if $\sum_k P(|f_{nk}| \geqslant \varepsilon) \to 0, \varepsilon > 0$, and $\sum_k \int_0^\varepsilon x\, dF_{nk}(x) \to a, \varepsilon > 0$. [Hint. For $\varepsilon > 0$ and $M > 1$, write $\sum_k \int_0^\varepsilon x^2\, dF_{nk}(x) = \sum_k \int_0^{\varepsilon/M} x^2\, dF_{nk}(x) + \sum_k \int_{\varepsilon/M}^\varepsilon x^2\, dF_{nk}(x)$, then show that $\sum_k \int_0^\varepsilon x^2\, dF_{nk}(x) \to 0$ and apply (24.9).]

Exercise 24.28. Let $\{f_{nk} : n \geqslant 1, 1 \leqslant k \leqslant k_n\}$ be as in (24.20), and assume that $\sum_k Ef_{nk} = 1, n \geqslant 1$. Show that the infinite smallness condition is fulfilled and $\sum_k f_{nk} \xrightarrow{D} 1$ if and only if $\sum_k \int_0^\varepsilon x\, dF_{nk}(x) \to 1, \varepsilon > 0$.

Exercise 24.29 (Raikov). Let $\{f_{nk} : n \geqslant 1, 1 \leqslant k \leqslant k_n\}$ be as in (24.27), and assume the infinite smallness condition holds. Use (24.15), (24.28) and the Lindeberg-Feller theorem (24.20) to prove that $\sum_k f_{nk} \xrightarrow{D} \Phi$ if and only if $\sum_k f_{nk}^2 \xrightarrow{D} 1$.

Exercise 24.30. Let $f_n, n \geqslant 1$, be independent random variables, and assume that f_n has a uniform distribution function on $[-n, n]$. Verify that the Lindeberg condition holds, and apply (24.22) to prove that $3\sqrt{2}(f_1 + \cdots + f_n)/\sqrt{n(n+1)(2n+1)} \xrightarrow{D} \Phi$.

Exercise 24.31. Let $f_n, n \geqslant 1$, be i.i.d. random variables with $Ef_1 = 0$ and $Ef_1^2 = 1$, let $g_n, n \geqslant 1$, be independent random variables such that $P(g_n = n) = P(g_n = -n) = 1/2n^2$ and $P(g_n = 0) = 1 - 1/n^2$, and assume that $\sigma(f_n, n \geqslant 1)$ is independent of $\sigma(g_n, n \geqslant 1)$. For $n \geqslant 1$, set $h_n = f_n + g_n$, $S_n = \sum_{k=1}^n h_k$ and $s_n^2 = \sum_{k=1}^n Eh_k^2$. Prove the following.

(a) $h_n, n \geqslant 1$, are independent random variables.

(b) $(g_1 + \cdots + g_n)/\sqrt{n} \xrightarrow{a.s.} 0$. [Use (18.24.i).]

(c) $S_n/\sqrt{n} \xrightarrow{D} \Phi$ and $S_n/s_n \xrightarrow{D} G_{0,\sqrt{2}}$.

Exercise 24.32. Let $f_n, n \geqslant 1$, be independent random variables such that $P(f_1 = 1) = P(f_1 = -1) = 1/2$, and $P(f_n = 1) = P(f_n = -1) = (1 - 1/n^2)/2$, $P(f_n = n) = P(f_n = -n) = 1/2n^2, n > 1$. For $n \geqslant 1$, put $S_n = \sum_{k=1}^n f_k$. Prove the following.

(a) $\mathrm{Var}[S_n/\sqrt{n}] \to 2$.

(b) The Lindeberg condition (24.22.ii) does not hold. [Hint. Show that

$$\frac{1}{s_n^2}\sum_{k=1}^{n}\int_{\{|x|\geqslant \varepsilon s_k\}} x^2\,dF_k(x) \to 1, \quad \varepsilon > 0,$$

and use (24.23.a).]

(c) For $M > 1$ and $n \geqslant 1$, set $U_n = \sum_{k=1}^{n} f_k 1_{\{|f_k|<M\}}$. Then $E[U_n^2]/n \to 1$.

(d) $U_n/\sqrt{n} \xrightarrow{D} \Phi$. [Hint. Apply (24.25) to show that $U_n/\sqrt{EU_n^2} \xrightarrow{D} \Phi$, then use (c) and (19.61).]

(e) $S_n/\sqrt{n} \xrightarrow{D} \Phi$. [Hint. For $x \in R$ and $n > M$, $\left|P(S_n \leqslant \sqrt{n}x) - P(U_n \leqslant \sqrt{n}x)\right| \leqslant 2P(S_n \neq U_n) \leqslant 2\sum_{k=1}^{n} P(|f_k| \geqslant M) \leqslant 2\sum_{k\geqslant M} 1/k^2 \to 0$ as $M \to \infty$.]

Exercise 24.33 (De Moivre-Laplace). For $n \geqslant 1$, let S_n be a random variable having a binomial distribution with parameters n and p. Then $(S_n - np)/\sqrt{np(1-p)} \xrightarrow{D} \Phi$. [Use (19.38), (19.37), and (24.23.c).]

Exercise 24.34. For $n \geqslant 1$, let S_n be a random variable having a Poisson distribution with parameters $a = 0$, $b \neq 0$ and $\lambda = n$. Show that $(S_n - nb)/|b|\sqrt{n} \xrightarrow{D} \Phi$.

Exercise 24.35. Let $f_n, n \geqslant 1$, be i.i.d. random variables, and assume that f_1 has density $p(x) = \begin{cases} 1/|x|^3 & \text{if } |x| \geqslant 1 \\ 0 & \text{if } |x| < 1 \end{cases}$. Use (24.17) to prove that $(f_1 + \cdots + f_n)/\sqrt{n\log n} \xrightarrow{D} \Phi$.

Exercise 24.36. Let f be a random variable with $Ef = 0$ and $Ef^2 = 1$, let $b : [0,\infty[\to [0,\infty[$ be a continuous function such that b is strictly increasing on $[x_0,\infty[$ for some $x_0 \geqslant 0$, and let $c : [0,\infty[\to [0,\infty[$ be a function which is constant on $[0, b(x_0)[$ and is the inverse function of b on $[b(x_0),\infty[$. Show that $\sum_{n\geqslant 1}\frac{1}{n}(1 - \mathrm{Var}[f 1_{\{|f|<b(n)\}}]) < \infty$ if and only if $E[f^2\log(1 + c(|f|))] < \infty$; in particular, $\sum_{n\geqslant 1}\frac{1}{n}(1 - \mathrm{Var}[f 1_{\{|f|<\sqrt{n}\}}]) < \infty$ if and only if $E[f^2\log(1+|f|)] < \infty$. [Hint. $E[f^2 1_{\{|f|<b(n)\}}] \leqslant 1 - \mathrm{Var}[f 1_{\{|f|<\sqrt{n}\}}] \leqslant 2E[f^2 1_{\{|f|<b(n)\}}]$, $n \geqslant 1$.]

Reference

Gnedenko, B. V., & Kolmogorov, A. N. (1968). *Limit Distributions for Sums of Independent Random Variables*. Reading: Addison-Wesley.

10 Sums of Independent Random Variables

Sums of independent random variables have already been encountered in the preceding chapter as row sums of arrays of random variables. On the basis of results therefrom, in Section 25 we deepen the study of weak convergence to zero of such sums, leading to the so-called weak laws of large numbers. Starting with Section 26 we investigate almost sure behavior of (stabilized) sums $S_n, n \geqslant 1$, of independent random variables $f_n, n \geqslant 1$. The stochastic process $\{S_n : n \geqslant 1\}$ lies at the heart of classical and modern probability theory. In Section 26 we deal with questions concerning the almost sure convergence of $\{S_n : n \geqslant 1\}$, and obtain results culminating in the Kolmogorov three series theorem and the two series theorem. By making use of results on convergence of random series, Section 27 is devoted to obtaining various so-called strong laws of large numbers, i.e. theorems establishing conditions under which $S_n/b_n - a_n \stackrel{a.s.}{\rightarrow} 0$ for suitable constants $b_n > 0$ and $a_n \in R$. Resting on sharp exponential bounds for probabilities of the form $P(S_n/\sqrt{\mathrm{Var} S_n} \geqslant x)$, in Section 28 we present the fine order of magnitude of the asymptotic fluctuation of $\{S_n : n \geqslant 1\}$ couched in two crowning achievements in classical probability theory, namely the Kolmogorov law of the iterated logarithm and the Hartman-Wintner law of the iterated logarithm.

25 Weak Laws of Large Numbers

In this section we shall deal again with an array $\{f_{nk} : n \geqslant 1, 1 \leqslant k \leqslant k_n\}$ of random variables, where $k_n \to \infty$, and f_{nk}, $1 \leqslant k \leqslant k_n$, are independent for each n, and we shall adopt the notation in (23.22). Resting on methods and results of Section 24, we shall derive conditions under which $\sum_k f_{nk} - a_n \stackrel{P}{\to} 0$ (equivalently $\sum_k f_{nk} - a_n \stackrel{D}{\to} 0$ (19.24), (19.26)) for some sequence $\{a_n : n \in N\} \subset R$. Remark (25.6) motivates why these conditions are sometimes expressed in terms of the array $\{m_{nk} : n \geqslant 1, 1 \leqslant k \leqslant k_n\}$. We begin with some preliminary material of more general interest.

Theorem 25.1. *Let $f_i, i \in I$, be random variables, where either $I = \{1, \ldots, n\}$ or $I = N$, let $\{f_i^* : i \in I\}$ be the symmetrized $\{f_i : i \in I\}$, let $\{a_i : i \in I\} \subset R$, and let $\varepsilon > 0$. Then the following assertions hold.*

(i) $P(\sup_{i \in I}(f_i - m[f_i]) \geqslant \varepsilon) \leqslant 2P^*(\sup_{i \in I} f_i^* \geqslant \varepsilon)$.

(ii) $P(\sup_{i \in I} |f_i - m[f_i]| \geqslant \varepsilon) \leqslant 2P^*(\sup_{i \in I} |f_i^*| \geqslant \varepsilon) \leqslant 4P(\sup_{i \in I} |f_i - a_i| \geqslant \varepsilon/2)$.

Analysis and Probability. http://dx.doi.org/10.1016/B978-0-12-401665-1.00010-2

(iii) $E(\sup_{i\in I}|f_i - m[f_i]|)^p \leqslant 2E^*(\sup_{i\in I}|f_i^*|)^p \leqslant 2^{p+2}E(\sup_{i\in I}|f_i - a_i|)^p$, $p > 0$.

(iv) *If* $\varphi : [\,0,\infty[\, \to [0,\infty[$ *is a differentiable function such that* $\varphi' > 0, \varphi(0) = 0$ *and* $\lim_{x\to\infty}\varphi(x) = \infty$, *then* $E[\varphi\circ(\sup_{i\in I}|f_i - m[f_i]|)] \leqslant 2E^*[\varphi\circ(\sup_{i\in I}|f_i^*|)] \leqslant 4E[\varphi\circ(2\sup_{i\in I}|f_i - a_i|)]$.

Proof. We prove the case $I = \{1,\ldots,n\}$, as the case $I = N$ follows from it on letting $n\to\infty$. Let $\{f_i' : i\in I\}$ and $\{f_i'' : i \in I\}$ be independent copies of $\{f_i : i\in I\}$ so that $f_i^* = f_i' - f_i'', i\in I$. Consider $m_i = m[f_i'] = m[f_i''] = m[f_i], i \in I$, (23.20.c). Put $A_0 = \overline{A}_0 = \emptyset$ and $A_i = \{f_i' - m_i \geqslant \varepsilon\}, \overline{A}_i = \{f_i' - m_i \leqslant -\varepsilon\}, B_i = \{f_i'' \leqslant m_i\}, \overline{B}_i = \{f_i'' \geqslant m_i\}, C_i = \{f_i^* \geqslant \varepsilon\}, D_i = \{|f_i^*| \geqslant \varepsilon\}$ for $1 \leqslant i \leqslant n$. As $C_i \supset A_iB_i$, $1\leqslant i\leqslant n$, we have

$$\begin{aligned} P^*\left(\sup_{1\leqslant i\leqslant n} f_i^* \geqslant \varepsilon\right) &= P^*\left(\bigcup_{i=1}^n C_i\right) \geqslant P^*\left(\bigcup_{i=1}^n A_iB_i\right) \\ &= \sum_{i=1}^n P^*\left(A_iB_i\left(\bigcap_{j=0}^{i-1}(A_jB_j)^c\right)\right) \\ &\geqslant \sum_{i=1}^n P^*\left(A_iB_i\left(\bigcap_{j=0}^{i-1}A_j^c\right)\right). \end{aligned} \tag{1}$$

Since $\sigma(f_i', i\in I)$ is independent of $\sigma(f_i'', i\in I)$ and $P^*(f_i')_{i\in I}^{-1} = P(f_i)_{i\in I}^{-1}$ (23.19), and $P^*(B_i) \geqslant 1/2$, (1) yields

$$\begin{aligned} P^*\left(\sup_{1\leqslant i\leqslant n} f_i^* \geqslant \varepsilon\right) &\geqslant \frac{1}{2}\sum_{i=1}^n P^*\left(A_iB_i\left(\bigcap_{j=0}^{i-1}A_j^c\right)\right) = \frac{1}{2}P^*\left(\bigcup_{i=1}^n A_i\right) \\ &= \frac{1}{2}P^*\left(\sup_{1\leqslant i\leqslant n}(f_i' - m_i) \geqslant \varepsilon\right) \\ &= \frac{1}{2}P\left(\sup_{1\leqslant i\leqslant n}(f_i - m_i)\geqslant\varepsilon\right), \end{aligned}$$

which is (i). By arguments similar to those used in the proof of (i) we obtain

$$\begin{aligned} P^*\left(\sup_{1\leqslant i\leqslant n}|f_i^*| \geqslant \varepsilon\right) &= P^*\left(\bigcup_{i=1}^n D_i\right) \geqslant P^*\left(\bigcup_{i=1}^n A_iB_i\right) \\ &= \sum_{i=1}^n P^*\left((A_iB_i \cup \overline{A}_i\overline{B}_i)\left(\bigcap_{j=0}^{i-1}(A_jB_j\cup\overline{A}_j\overline{B}_j)^c\right)\right) \\ &\geqslant \sum_{i=1}^n P^*\left((A_iB_i\cup\overline{A}_i\overline{B}_i)\left(\bigcap_{j=0}^{i-1}(A_j\cup\overline{A}_j)^c\right)\right) \end{aligned}$$

$$
\begin{aligned}
&= \sum_{i=1}^{n} P^*\left(A_i B_i \left(\bigcap_{j=0}^{i-1}(A_j \cup \overline{A}_j)^c\right)\right) \\
&\quad + \sum_{i=1}^{n} P^*\left(\overline{A}_i \overline{B}_i \left(\bigcap_{j=0}^{i-1}(A_j \cup \overline{A}_j)^c\right)\right) \\
&\geqslant \frac{1}{2}\sum_{i=1}^{n} P^*\left(A_i \left(\bigcap_{j=0}^{i-1}(A_j \cup \overline{A}_j)^c\right)\right) \\
&\quad + \frac{1}{2}\sum_{i=1}^{n} P^*\left(\overline{A}_i \left(\bigcap_{j=0}^{i-1}(A_j \cup \overline{A}_j)^c\right)\right) \\
&= \frac{1}{2}\sum_{i=1}^{n} P^*\left((A_i \cup \overline{A}_i) \left(\bigcap_{j=0}^{i-1}(A_j \cup \overline{A}_j)^c\right)\right) \\
&= \frac{1}{2} P^*\left(\bigcup_{i=1}^{n}(A_i \cup \overline{A}_i)\right) = \frac{1}{2} P^*\left(\sup_{1\leqslant i\leqslant n} |f_i' - m_i| \geqslant \varepsilon\right) \\
&= \frac{1}{2} P\left(\sup_{1\leqslant i\leqslant n} |f_i - m_i| \geqslant \varepsilon\right).
\end{aligned}
$$

This proves the first inequality in (ii). The second inequality in (ii) follows from

$$
\begin{aligned}
P^*\left(\sup_{i\in I} |f_i^*| \geqslant \varepsilon\right) &= P^*\left(\sup_{i\in I} |f_i' - a_i - (f_i'' - a_i)| \geqslant \varepsilon\right) \\
&\leqslant P^*\left(\sup_{i\in I} |f_i' - a_i| \geqslant \frac{\varepsilon}{2}\right) + P^*\left(\sup_{i\in I} |f_i'' - a_i| \geqslant \frac{\varepsilon}{2}\right) \\
&= 2P\left(\sup_{i\in I} |f_i - a_i| \geqslant \frac{\varepsilon}{2}\right),
\end{aligned}
$$

as $P^*(f_i')_{i\in I}^{-1} = P^*(f_i'')_{i\in I}^{-1} = P(f_i)_{i\in I}^{-1}$. Finally, (iii) and (iv) follow from (ii) and (15.23.a). □

Corollary 25.2. *Let $\{f_n : n \in N\}$ be a sequence of random variables, and let $\{a_n : n \in N\} \subset R$. Then the following assertions hold.*

(i) *$f_n - a_n \xrightarrow{P} 0$ if and only if $f_n^* \xrightarrow{P^*} 0$ and $m[f_n] - a_n \to 0$.*

(ii) *$f_n - a_n \xrightarrow{a.s.} 0$ if and only if $f_n^* \xrightarrow{a.s.} 0$ and $m[f_n] - a_n \to 0$.*

Proof. (i) follows via (25.1.ii), (14.22.f) and (14.23.a), and (ii) follows from (25.1.ii), (14.3), (14.9) and (i). □

Theorem 25.3 (Lévy's inequalities). *Let $f_1, \ldots, f_n$ be independent random variables, and set $S_i = f_1 + \cdots + f_i$, $1 \leqslant i \leqslant n$. Then*

(i) *$P(\sup_{1\leqslant i\leqslant n}(S_i - m[S_i - S_n]) \geqslant \varepsilon) \leqslant 2P(S_n \geqslant \varepsilon)$, $\varepsilon \in R$,*

and

(ii) $P(\sup_{1\leqslant i\leqslant n}|S_i - m[S_i - S_n]| \geqslant \varepsilon) \leqslant 2P(|S_n| \geqslant \varepsilon), \varepsilon > 0$.

Proof. Let $\varepsilon \in R$ and put $A_1 = \{S_1 - m[S_1 - S_n] \geqslant \varepsilon\}$, $A_i = \{\sup_{1\leqslant k<i}(S_k - m[S_k - S_n]) < \varepsilon, S_i - m[S_i - S_n] \geqslant \varepsilon\}, 2 \leqslant i \leqslant n$, and $B_i = \{S_n - S_i \geqslant m[S_n - S_i]\}, 1 \leqslant i \leqslant n$. Clearly, A_i is independent of B_i and $P(B_i) \geqslant 1/2$ for $1 \leqslant i \leqslant n$. We can put $m[S_n - S_i] = -m[S_i - S_n]$ by (23.18.c), and so $A_1B_1 \subset \{S_n \geqslant \varepsilon\}$, $A_iB_i \subset \{\sup_{1\leqslant k<i}(S_k - m[S_k - S_n]) < \varepsilon, S_n \geqslant \varepsilon\}, 2 \leqslant i \leqslant n$. Therefore, we have

$$P(S_n \geqslant \varepsilon) \geqslant P\left(\bigcup_{i=1}^{n} A_iB_i\right) = \sum_{i=1}^{n} P(A_iB_i) \geqslant \frac{1}{2}\sum_{i=1}^{n} P(A_i) = \frac{1}{2}P\left(\bigcup_{i=1}^{n} A_i\right)$$
$$= \frac{1}{2}P\left(\sup_{1\leqslant i\leqslant n}(S_i - m[S_i - S_n]) \geqslant \varepsilon\right),$$

which is (i). If we apply it to the random variables $-f_1, \ldots, -f_n$, we get

$$P\left(\inf_{1\leqslant i\leqslant n}(S_i - m[S_i - S_n]) \leqslant -\varepsilon\right) \leqslant 2P(S_n \leqslant -\varepsilon). \tag{1}$$

From (i) and (1) we obtain (ii). □

Corollary 25.4. *If $f_1, \ldots, f_n$ are independent and symmetric random variables with partial sums $S_i = f_1 + \cdots + f_i, 1 \leqslant i \leqslant n$, then $P\left(\sup_{1\leqslant i\leqslant n} S_i \geqslant \varepsilon\right) \leqslant 2P(S_n \geqslant \varepsilon), \varepsilon \in R$, and $P\left(\sup_{1\leqslant i\leqslant n}|S_i| \geqslant \varepsilon\right) \leqslant 2P(|S_n| \geqslant \varepsilon), \varepsilon > 0$.*

Proof. This is an immediate consequence of (25.3) via (20.42.h) and (23.18.f). □

Definition 25.5. The array $\{f_{nk} : n \geqslant 1, 1 \leqslant k \leqslant k_n\}$ is said to obey the *weak laws of large numbers* (WLLN) if there exists a sequence $\{a_n : n \in N\} \subset R$ such that $\sum_k f_{nk} - a_n \xrightarrow{P} 0$.

Remark 25.6. If the array $\{f_{nk} - a_{nk} : n \geqslant 1, 1 \leqslant k \leqslant k_n\}$ satisfies the condition of infinite smallness for some $a_{nk} \subset R, n \geqslant 1, 1 \leqslant k \leqslant k_n$, then the array $\{f_{nk} - m_{nk} : n \geqslant 1, 1 \leqslant k \leqslant k_n\}$ also satisfies this condition. Actually, under the assumption, for $\varepsilon > 0$, choose $n_\varepsilon \in N$ so that $\sup_k P(|f_{nk} - a_{nk}| \geqslant \varepsilon) \leqslant 1/2, n \geqslant n_\varepsilon$. Then, in view of (23.18.b) and (23.18.c), $\sup_k |m_{nk} - a_{nk}| \leqslant \varepsilon$ for any $n \geqslant n_\varepsilon$. Therefore, $\sup_k P(|f_{nk} - m_{nk}| \geqslant 2\varepsilon) \leqslant \sup_k P(|f_{nk} - a_{nk}| + |m_{nk} - a_{nk}| \geqslant 2\varepsilon) \leqslant \sup_k P(|f_{nk} - a_{nk}| \geqslant \varepsilon), n \geqslant n_\varepsilon$. Hence $\sup_k P(|f_{nk} - m_{nk}| \geqslant 2\varepsilon) \to 0$.

Several versions of the WLLN are contained in (24.9)–(24.13). Another form of the WLLN is as follows.

Theorem 25.7. *We have*

(i) $\sum_k f_{nk} - \sum_k(m_{nk} + E[(f_{nk} - m_{nk})1_{\{|f_{nk}-m_{nk}|<1\}}]) \xrightarrow{P} 0$ *if and only if*

(ii) $\sum_k P(|f_{nk} - m_{nk}| \geqslant 1) \to 0$

and

(iii) $\sum_k E[(f_{nk} - m_{nk})^2 1_{\{|f_{nk}-m_{nk}|<1\}}] \to 0$.

Proof. Assume that (i) holds, and write $a_n = \sum_k(m_{nk} + E[(f_{nk} - m_{nk}) 1_{\{|f_{nk}-m_{nk}|<1\}}])$. For $n \geqslant 1$, let $\{f'_{nk} : 1 \leqslant k \leqslant k_n\}$ and $\{f''_{nk} : 1 \leqslant k \leqslant k_n\}$ be independent copies of $\{f_{nk} : 1 \leqslant k \leqslant k_n\}$, and set $f^*_{nk} = f'_{nk} - f''_{nk}$, $1 \leqslant k \leqslant k_n$. Then $\sum_k f'_{nk} - a_n \xrightarrow{P^*} 0$ and $\sum_k f''_{nk} - a_n \xrightarrow{P^*} 0$, and so

$$\sum_k f^*_{nk} \xrightarrow{P^*} 0. \tag{1}$$

by (14.23.a). Consequently, since f^*_{nk}, $1 \leqslant k \leqslant k_n$, are P^*-independent (23.20.d) and symmetric, on applying (25.4), we have

$$\begin{aligned} P^*(\sup_k |f^*_{nk}| \geqslant \varepsilon) &\leqslant P^*\left(\sup_k \left|\sum_{j=1}^{k} f^*_{nj}\right| \geqslant \frac{\varepsilon}{2}\right) \\ &\leqslant 2P^*\left(\left|\sum_k f^*_{nk}\right| \geqslant \frac{\varepsilon}{2}\right) \to 0, \quad \varepsilon > 0. \end{aligned} \tag{2}$$

From (2) and (23.55) we infer that

$$\sum_k P^*(|f^*_{nk}| \geqslant \varepsilon) \to 0, \quad \varepsilon > 0, \tag{3}$$

and so by (25.1) we get

$$\sum_k P(|f_{nk} - m_{nk}| \geqslant \varepsilon) \to 0, \quad \varepsilon > 0. \tag{4}$$

Thus (ii) holds. As $\sum_k (E^*[f^*_{nk} 1_{\{-1 < f^*_{nk} \leqslant 1\}}])^2 = \sum_k P^*(f^*_{nk} = 1)^2 \to 0$ by (20.42.e), (20.42.c) and (3), we obtain via (1), (3) and (24.9) that

$$\sum_k E^*[f^{*2}_{nk} 1_{\{|f^*_{nk}| < 1\}}] \to 0. \tag{5}$$

On account of (15.23.b) and (25.1.ii), we can write

$$\begin{aligned} &\textstyle\sum_k P^*(|f^*_{nk}| \geqslant 1) + \sum_k E^*[f^{*2}_{nk} 1_{\{|f^*_{nk}| < 1\}}] \\ &= \sum_k \int_0^1 P^*\left(f^{*2}_{nk} \geqslant x\right) dx \geqslant \frac{1}{2} \sum_k \int_0^1 P((f_{nk} - m_{nk})^2 \geqslant x) dx \\ &= \frac{1}{2}\left(\sum_k \int_0^1 P(|f_{nk} - m_{nk}| \geqslant 1)\right. \\ &\quad \left. + \sum_k E[(f_{nk} - m_{nk})^2 1_{\{|f_{nk} - m_{nk}| < 1\}}]\right), n \geqslant 1. \end{aligned} \tag{6}$$

From (3)–(6) we see that (iii) holds.

Conversely, assume that (ii) and (iii) are fulfilled. For $\delta > 0$, by Chebyshev's inequality, (18.40.a) and (iii), we have

$$P\left(\left|\sum_k (f_{nk} - m_{nk})1_{\{|f_{nk}-m_{nk}|<1\}} - \sum_k E[(f_{nk} - m_{nk})1_{\{|f_{nk}-m_{nk}|<1\}}]\right| \geqslant \delta\right)$$
$$\leqslant \delta^{-2} \sum_k \mathrm{Var}[(f_{nk} - m_{nk})1_{\{|f_{nk}-m_{nk}|<1\}}]$$
$$\leqslant \delta^{-2} \sum_k E\left[(f_{nk} - m_{nk})^2 1_{\{|f_{nk}-m_{nk}|<1\}}\right] \to 0. \tag{7}$$

On account of (ii), we have also

$$P\left(\left|\sum_k (f_{nk} - m_{nk}) - \sum_k (f_{nk} - m_{nk})1_{\{|f_{nk}-m_{nk}|<1\}}\right| \geqslant \delta\right)$$
$$\leqslant P\left(\sum_k (f_{nk} - m_{nk}) \neq \sum_k (f_{nk} - m_{nk})1_{\{|f_{nk}-m_{nk}|<1\}}\right)$$
$$\leqslant \sum_k P(|f_{nk} - m_{nk}| \geqslant 1) \to 0. \tag{8}$$

In view of (7) and (8), we obtain

$$P\left(\left|\sum_k (f_{nk} - m_{nk}) - \sum_k E[(f_{nk} - m_{nk})1_{\{|f_{nk}-m_{nk}|<1\}}]\right| \geqslant 2\delta\right)$$
$$\leqslant P\left(\left|\sum_k (f_{nk} - m_{nk}) - \sum_k (f_{nk} - m_{nk})1_{\{|f_{nk}-m_{nk}|<1\}}\right| \geqslant \delta\right)$$
$$+ P\left(\left|\sum_k (f_{nk} - m_{nk})1_{\{|f_{nk}-m_{nk}|<1\}} - \sum_k E[(f_{nk} - m_{nk})1_{\{|f_{nk}-m_{nk}|<1\}}]\right| \geqslant \delta\right) \to 0,$$

which is (i). □

Remark 25.8. Inspection of the first part of the above proof shows that if $\sum_k f_{nk} - \alpha_n \xrightarrow{P} 0$ for some sequence $\{\alpha_n : n \in N\} \subset R$, then (25.7.ii) and (25.7.iii) still hold, hence $\sum_k f_{nk} - a_n \xrightarrow{P} 0$, and so $\alpha_n - a_n \to 0$ by (14.22.f). Thus $\sum_k f_{nk} - \alpha_n \xrightarrow{P} 0$ if and only if (25.7.ii) and (25.7.iii) are satisfied and $\alpha_n - a_n \to 0$.

Corollary 25.9. *Let $f_n, n \geqslant 1$, be independent random variables, let $\{b_n : n \geqslant 1\} \subset]0, \infty[$, let m_n be a median of f_n, and set $S_n = \sum_{k=1}^n f_k$. Then*

(i) $\frac{S_n}{b_n} - \frac{1}{b_n} \sum_{k=1}^n (m_k + E[(f_k - m_k)1_{\{|f_k-m_k|<b_n\}}]) \xrightarrow{P} 0$

if and only if

(ii) $\sum_{k=1}^{n} P(|f_k - m_k| \geqslant b_n) \to 0$

and

(iii) $\sum_{k=1}^{n} E[(f_k - m_k)^2 1_{\{|f_k - m_k| < b_n\}}] \to 0.$

Proof. This follows from (25.7) with $f_{nk} = f_k/b_n$ and $m_{nk} = m_k/b_n$ (23.18.c), $n \geqslant 1, 1 \leqslant k \leqslant k_n$. □

To get a new form of the WLLN, we need the next lemma.

Lemma 25.10. *Let f be a random variable, let $c > 0$, and set $f_c = f 1_{\{|f|<c\}}$. Then*

$$\frac{1}{2c^2} E f_c^2 + \frac{1}{2} P(|f| \geqslant c) \leqslant E\left[\frac{f^2}{c^2 + f^2}\right] \leqslant \frac{1}{c^2} E f_c^2 + P(|f| \geqslant c).$$

Proof. If F is the distribution function of f, in view of (11.29), we can write

$$\begin{aligned}
\frac{1}{2c^2} E f_c^2 + \frac{1}{2} P(|f| \geqslant c) &= \frac{1}{2c^2} \int_{\{|x|<c\}} x^2 \, dF(x) + \frac{1}{2} \int_{\{|x| \geqslant c\}} dF(x) \\
&\leqslant \int_{\{|x|<c\}} \frac{x^2}{c^2 + x^2} dF(x) + \int_{\{|x| \geqslant c\}} \frac{x^2}{c^2 + x^2} dF(x) \\
&= E\left[\frac{f^2}{c^2 + f^2}\right] \\
&\leqslant \frac{1}{c^2} \int_{\{|x|<c\}} x^2 \, dF(x) + \int_{\{|x| \geqslant c\}} dF(x) \\
&= \frac{1}{c^2} E f_c^2 + P(|f| \geqslant c). \quad \square
\end{aligned}$$

Theorem 25.11. *We have*

(i) $\sum_k f_{nk} - \sum_k (m_{nk} + E[(f_{nk} - m_{nk}) 1_{\{|f_{nk} - m_{nk}| < 1\}}]) \xrightarrow{P} 0$

if and only if

(ii) $\sum_k E\left[\frac{(f_{nk} - m_{nk})^2}{1 + (f_{nk} - m_{nk})^2}\right] \to 0.$

Proof. By virtue of (25.10), we see that (25.7.ii) and (25.7.iii) are equivalent to (ii). □

Corollary 25.12. *For $n \geqslant 1$, let f_n, b_n, m_n and S_n be as in (25.9). Then*

(i) $\frac{S_n}{b_n} - \frac{1}{b_n} \sum_{k=1}^{n} (m_k + E[(f_k - m_k) 1_{\{|f_k - m_k| < b_n\}}]) \xrightarrow{P} 0$

if and only if

(ii) $\sum_{k=1}^{n} E\left[\frac{(f_k - m_k)^2}{1 + (f_k - m_k)^2}\right] \to 0.$

Theorem 25.13 (Feller). *Let $f_n, n \geqslant 1$, be i.i.d. random variables, and put $S_n = \sum_{k=1}^{n} f_k$. Then*

(i) $\frac{S_n}{n} - E[f_1 1_{\{|f_1| < n\}}] \xrightarrow{P} 0$

if and only if

(ii) $nP(|f_1| \geqslant n) \to 0.$

Proof. If (i) holds, then, arguing as in the first part of the proof of (25.7), we obtain

$$nP(|f_1 - m_1| \geqslant n/2) \to 0, \tag{1}$$

where m_1 is a median of f_1. Since $\{|f_1| \geqslant n\} \subset \{|f_1 - m_1| \geqslant n/2\}$ whenever $n > 2\,|m_1|$, (1) implies (ii).

Conversely, assume that (ii) holds, and let $\delta > 0$. Set $f_{n,i} = f_i 1_{\{|f_i|<n\}}$, $1 \leqslant i \leqslant n$, and $U_n = \sum_{i=1}^n f_{n,i}$. Clearly, $f_{n,i}$, $1 \leqslant i \leqslant n$, are i.i.d. random variables, and so

$$\begin{aligned} P\left(\left|\frac{S_n}{n} - \frac{U_n}{n}\right| \geqslant \delta\right) &\leqslant P(S_n \neq U_n) \leqslant P\left(\bigcup_{i=1}^n \{f_i \neq f_{n,i}\}\right) \\ &\leqslant nP(|f_1| \geqslant n) \to 0. \end{aligned} \tag{2}$$

By Chebyshev's inequality, (18.40.a) and Fubini's theorem (15.8), we have

$$\begin{aligned} &P\left(\left|\frac{U_n}{n} - E\left[f_1 1_{\{|f_1|<n\}}\right]\right| \geqslant \delta\right) \\ &= P\left(\left|\frac{U_n}{n} - E\left[\frac{U_n}{n}\right]\right| \geqslant \delta\right) \leqslant \delta^{-2}\mathrm{Var}\left[\frac{U_n}{n}\right] \\ &= \delta^{-2}\frac{1}{n}\mathrm{Var}\, f_{n,1} \leqslant \delta^{-2}\frac{1}{n}Ef_{n,1}^2 \\ &= \delta^{-2}\frac{1}{n}\sum_{i=1}^n \int_{\{i-1\leqslant|f_1|<i\}} f_1^2\, dP \\ &\leqslant \delta^{-2}\frac{1}{n}\sum_{i=1}^n i^2 P(i-1 \leqslant |f_1| < i) \\ &\leqslant \delta^{-2}\frac{2}{n}\sum_{i=1}^n\sum_{j=1}^i jP(i-1 \leqslant |f_1| < i) \\ &= \delta^{-2}\frac{2}{n}\sum_{j=1}^n\sum_{i=j}^n jP(i-1 \leqslant |f_1| < i) \\ &\leqslant \delta^{-2}\frac{2}{n}\sum_{j=1}^n jP(j-1 \leqslant |f_1|). \end{aligned}$$

In view of (ii), this entails

$$P\left(\left|\frac{U_n}{n} - E[f_1 1_{\{|f_1|<n\}}]\right| \geqslant \delta\right) \to 0. \tag{3}$$

Now (i) follows from (2) and (3). □

Corollary 25.14. *For $n \geqslant 1$, let f_n and S_n be as in* (25.13). *Then $S_n/n \xrightarrow{P} 0$ if and only if $nP(|f_1| \geqslant n) \to 0$ and $E[f_1 1_{\{|f_1|<n\}}] \to 0$.*

Remark 25.15. Notation is as in (25.14). $S_n/n - a_n \xrightarrow{P} 0$ for some sequence $\{a_n : n \in N\} \subset R$ if and only if $nP(|f_1| \geqslant n) \to 0$ and $a_n - E[f_1 1_{\{|f_1|<n\}}] \to 0$.

Corollary (25.14) leads to the next classical WLLN.

Theorem 25.16. *For $n \geqslant 1$, let f_n and S_n be as in* (25.13), *and assume that f_1 is P-integrable. Then $S_n/n \xrightarrow{P} Ef_1$.*

Proof. $f_1 - Ef_1$ is also P-integrable, and so $nP(|f_1 - Ef_1| \geqslant n) \to 0$ by (15.24). Since $E[f_1 - Ef_1] = 0$, we have $\left|E[(f_1 - Ef_1)1_{\{|f_1-Ef_1|<n\}}]\right| = \left|E[(f_1 - Ef_1)1_{\{|f_1-Ef_1|\geqslant n\}}]\right| \leqslant E[|f_1 - Ef_1|\, 1_{\{|f_1-Ef_1|\geqslant n\}}] \to 0$ by dominated convergence. Hence, in view of (25.14), $S_n/n - Ef_1 = \frac{1}{n}\sum_{i=1}^{n}(f_k - Ef_1) \xrightarrow{P} 0$. □

If $Ef_1 = 0$, (25.16) shows that $P(|S_n| \geqslant \varepsilon n) \to 0$ for all $\varepsilon > 0$. The following two theorems provide rates of convergence to 0 of these probabilities.

Theorem 25.17 (Hsu–Robbins–Erdős). *For $n \geqslant 1$, let f_n and S_n be as in* (25.13). *We have*

$$\sum_{n\geqslant 1} P(|S_n| \geqslant \varepsilon n) < \infty, \quad \varepsilon > 0, \tag{1}$$

if and only if $Ef_1^2 < \infty$ and $Ef_1 = 0$.

Proof. Assume first that (1) holds, and let $\{f_n' : n \geqslant\}$ and $\{f_n'' : n \geqslant 1\}$ be independent copies of $\{f_n : n \geqslant 1\}$. For $n \geqslant 1$, set $f_n^* = f_n' - f_n''$, $S_n' = \sum_{k=1}^n f_k'$, $S_n'' = \sum_{k=1}^n f_k''$ and $S_n^* = \sum_{k=1}^n f_k^*$. Then, with respect to P^*, f_n^*, $n \geqslant 1$, are i.i.d. symmetric random variables by (23.20.d) and (23.20.f). By virtue of (1),

$$\begin{aligned}\sum_{n\geqslant 1} P^*\left(|S_n^*| \geqslant 2n\right) &\leqslant \sum_{n\geqslant 1} P^*\left(|S_n'| \geqslant n\right) + \sum_{n\geqslant 1} P^*\left(|S_n''| \geqslant n\right)\\ &= 2\sum_{n\geqslant 1} P\left(|S_n| \geqslant n\right) < \infty.\end{aligned} \tag{2}$$

In view of (25.4), we have

$$\begin{aligned}2P^*(|S_n^*| \geqslant 2n) &\geqslant P^*\left(\sup_{1\leqslant i\leqslant n} |S_i^*| \geqslant 2n\right) \geqslant P^*\left(\sup_{1\leqslant i\leqslant n} |f_i^*| \geqslant 4n\right)\\ &= 1 - \left(P^*\left(|f_1^*| < 4n\right)\right)^n, \quad n \geqslant 1.\end{aligned} \tag{3}$$

On account of (2) and (3), we can write

$$\infty > \sum_{n\geqslant 1}\left(1 - (P^*(|f_1^*| < 4n))^n\right)$$

$$= \sum_{n\geqslant 1} P^*\left(\left|f_1^*\right| \geqslant 4n\right) \sum_{i=0}^{n-1} (P^*(\left|f_1^*\right| < 4n))^i$$

$$\geqslant \sum_{n\geqslant 2} nP^*(\left|f_1^*\right| \geqslant 4n) \cdot \frac{1}{n} \sum_{i=1}^{n-1} (P^*(\left|f_1^*\right| < 4n))^i$$

$$\geqslant \sum_{n\geqslant 2} nP^*(\left|f_1^*\right| \geqslant 4n) \cdot \frac{1}{n} \sum_{i=1}^{n-1} (1 - 2P^*(\left|S_i^*\right| \geqslant 2n)).$$

Therefore, since $\frac{1}{n}\sum_{i=1}^{n-1}(1 - 2P^*(\left|S_i^*\right| \geqslant 2n)) \to 1$ by (2), it follows that $\sum_{n\geqslant 2} nP^*(\left|f_1^*\right| \geqslant 4n) < \infty$, and so $E^* f_1^{*2} < \infty$ by (13.27.a). In view of (25.1.iii), this entails that $Ef_1^2 < \infty$, and so $E\,|f_1| < \infty$ (13.5). To get a contradiction, suppose that $Ef_1 = m \neq 0$. Then $S_n/n \xrightarrow{P} m$ by (25.16). Choose $\delta \in]0, |m|/2[$, then let $n_\delta \in N$ be such that

$$P\left(\left|\frac{S_n}{n} - m\right| < \delta\right) \geqslant \frac{1}{2}, \quad n \geqslant n_\delta. \tag{4}$$

As $\{m-\delta < S_n/n < m+\delta\} \subset \{|S_n| \geqslant \delta n\}$, (4) implies that $\sum_{n\geqslant 1} P(|S_n| \geqslant \delta n) = \infty$, thus contradicting (1).

Conversely, assume that $Ef_1^2 < \infty$ and $Ef_1 = 0$, and let $\varepsilon > 0$. For $n \geqslant 1$, let $f_{n,i}$, $1 \leqslant i \leqslant n$, and U_n be as in the proof of (25.13). Since $Ef_1 = 0$ and $E\,|f_1| < \infty$, we have $EU_n/n = Ef_{n,1} = E[f_1 1_{\{|f_1|<n\}}] = -E[f_1 1_{\{|f_1|\geqslant n\}}] \to 0$ by dominated convergence. Choose $n_\varepsilon \in N$ such that $|EU_n|/n = \left|Ef_{n,1}\right| < \varepsilon \wedge 1, n \geqslant n_\varepsilon$. Then, by making use of Markov's inequality (13.8) and (13.23.b), we obtain

$$P(|U_n - EU_n| \geqslant \varepsilon n) \leqslant \varepsilon^{-4} n^{-4} E(U_n - EU_n)^4 = \varepsilon^{-4} n^{-4}$$
$$\times \left(\sum_{i=1}^{n} E(f_{n,i} - Ef_{n,i})^4 + 6 \sum_{1\leqslant i<j\leqslant n} E(f_{n,i} - Ef_{n,i})^2 E(f_{n,j} - Ef_{n,j})^2\right)$$
$$\leqslant \varepsilon^{-4} n^{-4} \left(8\sum_{i=1}^{n}(Ef_{n,i}^4 + 1) + 6\binom{n}{2}(\mathrm{Var}[f_{n,1}])^2\right)$$
$$\leqslant \varepsilon^{-4} n^{-3}\left(8Ef_{n,1}^4 + 8 + 3n(Ef_1^2)^2\right), \quad n \geqslant n_\varepsilon. \tag{5}$$

We have also

$$\sum_{n\geqslant n_\varepsilon} P(|S_n| \geqslant 2\varepsilon n) \leqslant \sum_{n\geqslant n_\varepsilon} P(S_n \neq U_n) + \sum_{n\geqslant n_\varepsilon} P(|U_n| \geqslant 2\varepsilon n)$$
$$\leqslant \sum_{n\geqslant n_\varepsilon} nP(|f_1| \geqslant n) + \sum_{n\geqslant n_\varepsilon} P(|U_n - EU_n| \geqslant \varepsilon n). \tag{6}$$

Therefore, since $Ef_1^2 < \infty$ implies that $\sum_{n\geqslant n_\varepsilon} nP(|f_1| \geqslant n) < \infty$ by (17.10), according to (25) it suffices to show that $\sum_{n\geqslant 1} n^{-3} Ef_{n,1}^4 < \infty$. By Fubini's theorem (15.8), as $\sum_{n>m} n^{-3} \leqslant m^{-2}/2$, $m \geqslant 1$, we actually have

$$\sum_{n\geqslant 1} \frac{1}{n^3} E f_{n,1}^4 = E\left[\sum_{n\geqslant 1} \frac{1}{n^3} f_{n,1}^4\right] = E\left[f_1^4 \sum_{n\geqslant 1} \frac{1}{n^3} 1_{\{|f_1|<n\}}\right]$$
$$= E\left[f_1^4 \sum_{n>|f_1|} \frac{1}{n^3}\right] \leqslant \frac{1}{2} E f_1^2 < \infty. \quad \square$$

Lemma 25.18. *Let $f_n, n \geqslant 1$, be independent symmetric random variables, put $S_n = \sum_{k=1}^n f_k$, and let $\{b_n : n \geqslant 1\} \subset]0, \infty[$ be a nondecreasing sequence such that $b_n \to \infty$, $b_{2n}/b_n \leqslant C < \infty$, $n \geqslant 1$, and*

$$\sum_{n\geqslant 1} \frac{1}{n} P(|S_n| \geqslant \varepsilon b_n) < \infty, \quad \varepsilon > 0. \tag{1}$$

Then $S_n/b_n \xrightarrow{P} 0$.

Proof. If this convergence in probability does not hold, then there exists $\delta > 0$ such that either $P(S_{n_k} \geqslant \delta b_{n_k}) \geqslant \delta$ or $P(S_{n_k} \leqslant -\delta b_{n_k}) \geqslant \delta$ for infinitely many k. Suppose, for instance, that $P(S_{n_k} \geqslant \delta b_{n_k}) \geqslant \delta$ for infinitely many k and, with no loss of generality, choose $n_{k+1} > 2n_k$. For $n_k < j \leqslant 2n_k$, it follows by (20.42.h) and (23.18.f) that $P\left(\sum_{n_k<i\leqslant j} f_i \geqslant 0\right) \geqslant 1/2$. Therefore, since $\{S_j \geqslant \delta C^{-1} b_j\} \supset \{S_{n_k} \geqslant \delta b_{n_k}, \sum_{n_k<i\leqslant j} f_i \geqslant 0\}$, we obtain $P(S_j \geqslant \delta C^{-1} b_j) \geqslant \delta/2$, $n_k \leqslant j \leqslant 2n_k$. Consequently, as $\sum_{j=n_k}^{2n_k} \frac{1}{j} \geqslant \log 2$, we have

$$\sum_{n\geqslant 1} \frac{1}{n} P(S_n \geqslant \delta C^{-1} b_n) \geqslant \sum_{k\geqslant 1} \sum_{j=n_k}^{2n_k} \frac{1}{j} P(S_j \geqslant \delta C^{-1} b_j) \geqslant \frac{\delta}{2} \sum_{k\geqslant 1} \log 2 = \infty,$$

thus contradicting (1). Thus $S_n/b_n \xrightarrow{P} 0$. $\square$

Theorem 25.19 (Spitzer). *For $n \geqslant 1$, let f_n and S_n be as in (25.13). We have*

$$\sum_{n\geqslant 1} \frac{1}{n} P(|S_n| \geqslant \varepsilon n) < \infty, \quad \varepsilon > 0, \tag{1}$$

if and only if $E\,|f_1| < \infty$ and $E f_1 = 0$.

Proof. Assume that (1) holds, and let f_n^* and S_n^* be as in the proof of (25.17). Then (1) yields

$$\sum_{n\geqslant 1} \frac{1}{n} P^*(|S_n^*| \geqslant \varepsilon n) < \infty, \quad \varepsilon > 0. \tag{2}$$

Hence $S_n^*/n \xrightarrow{P^*} 0$ by (25.18), and so $nP^*(|f_1^*| \geqslant 2n) \to 0$ by (25.14). Choose $n_0 \in N$ such that $nP^*(|f_1^*| \geqslant 2n) \leqslant 1, n \geqslant n_0$. Therefore, since $(1-a)^n \leqslant 1-na+(na)^2/2$ for $0 \leqslant a \leqslant 1$, we have

$$2P^*(|S_n^*| \geqslant n) \geqslant 1-(1-P^*(|f_1^*| \geqslant 2n))^n \geqslant nP^*(|f_1^*| \geqslant 2n)$$
$$-\frac{n^2(P^*(|f_1^*| \geqslant n))^2}{2} \geqslant \frac{nP^*(|f_1^*| \geqslant 2n)}{2}, \quad n \geqslant n_0, \tag{3}$$

where the first inequality is derived like (3) in the proof of (25.17). From (2) and (3) it follows that $\sum_{n\geqslant 1} P^*(|f_1^*| \geqslant 2n) < \infty$, and so $E^* |f_1^*| < \infty$ by (13.27.a). This implies that $E|f_1| < \infty$ by (25.1.iii). Finally, the fact that $Ef_1 = 0$ follows as in the proof of (25.17).

Conversely, assume that $E|f_1| < \infty$ and $Ef_1 = 0$. For $n \geqslant 1$, let $f_{n,i}$, $1 \leqslant i \leqslant n$, and U_n be as in the proof of (25.13). For $\varepsilon > 0$, choose n_ε as in the proof of (25.17). We have

$$P(|U_n - EU_n| \geqslant \varepsilon n) \leqslant \varepsilon^{-2}\frac{1}{n}Ef_{n,1}^2, \quad n \geqslant 1, \tag{4}$$

as in the proof of (25.13). We have also

$$\sum_{n\geqslant n_\varepsilon} \frac{1}{n}P(|S_n|\geqslant 2\varepsilon n) \leqslant \sum_{n\geqslant n_\varepsilon} \frac{1}{n}P(S_n \neq U_n) + \sum_{n\geqslant n_\varepsilon} \frac{1}{n}P(|U_n| \geqslant 2\varepsilon n)$$
$$\leqslant \sum_{n\geqslant n_\varepsilon} P(|f_1| \geqslant n) + \sum_{n\geqslant n_\varepsilon} \frac{1}{n}P(|U_n - EU_n| \geqslant \varepsilon n). \tag{5}$$

Since $E|f_1| < \infty$ implies that $\sum_{n\geqslant n_\varepsilon} P(|f_1| \geqslant n) < \infty$ by (11.37), (4) and (5) show that it suffices to prove that $\sum_{n\geqslant 1} n^{-2}Ef_{n,1}^2 < \infty$. By Fubini's theorem (15.8), as $\sum_{n>m} n^{-2} \leqslant m^{-1}$, $m \geqslant 1$, we obtain

$$\sum_{n\geqslant 1} \frac{1}{n^2}Ef_{n,1}^2 = E\left[f_1^2 \sum_{n\geqslant 1} \frac{1}{n^2} 1_{\{|f_1|<n\}}\right]$$
$$= E\left[f_1^2 \sum_{n>|f_1|} \frac{1}{n^2}\right] \leqslant E|f_1| < \infty. \quad \square$$

In the final part of this section we discuss several general inequalities ((25.21)–(25.31)) due to D. H. Fuk and S. V. Nagaev, which proved to be very sharp in some cases. These inequalities have become a crucial tool in problems devoted to the theory of weakly dependent random variables, to the characterization of the law of the iterated logarithm (see Section 28) in the vector-valued setting, to precise asymptotics concerning tail behavior, to strong laws of large numbers (see Section 27), etc. Throughout (25.21)–(25.30), let $f_1, \ldots, f_n$ be independent random variables with respective distribution functions $F_1, \ldots, F_n$, set $S_n = f_1 + \cdots + f_n$, let $x, y_1, \ldots, y_n > 0$, put $\mathbf{y} = (y_1, \ldots, y_n)$, and let $y \geqslant y_1 \vee \ldots \vee y_n$.

Theorem 25.20. *Assume that* $0 < p < 1$ *and* $A(p; \mathbf{0}, \mathbf{y}) = \sum_{i=1}^{n} E[f_i 1_{\{0<f_i<y_i\}}] > 0$. *Then*

$$P(S_n \geqslant x) \leqslant \sum_{i=1}^{n} P(f_i \geqslant y_i) + \exp\left\{\frac{x}{y} - \frac{x}{y}\log\left(\frac{xy^{p-1}}{A(p; \mathbf{0}, \mathbf{y})} + 1\right)\right\}. \quad (1)$$

Proof. For $n \geqslant 1$, put $g_i = f_i 1_{\{f_i<y_i\}}$, $1 \leqslant i \leqslant n$, and $U_n = g_1 + \cdots + g_n$. Then, for $h > 0$, applying (13.25.b),

$$P(S_n \geqslant x) \leqslant P(S_n \neq U_n) + P(U_n \geqslant x) \leqslant \sum_{i=1}^{n} P(f_i \geqslant y_i) + e^{-hx} E e^{hU_n}. \quad (2)$$

For $1 \leqslant i \leqslant n$ and $h > 0$, we have

$$\begin{aligned} Ee^{hg_i} &\leqslant \int\limits_{]-\infty,0]} dF_i(u) + \int\limits_{]0,y_i[} e^{hu} dF_i(u) + \int\limits_{[y_i,\infty[} dF_i(u) \\ &= 1 + \int\limits_{]0,y_i[} \frac{e^{hu}-1}{u} u\, dF_i(u) \leqslant 1 + \frac{e^{hy_i}-1}{y_i} \int\limits_{]0,y_i[} u\, dF_i(u) \\ &\leqslant 1 + \frac{e^{hy}-1}{y^p} \int\limits_{]0,y_i[} u^p dF_i(u). \end{aligned}$$

Therefore, via independence, as $1 + z \leqslant e^z$, $z \in R$, we get

$$\begin{aligned} Ee^{hU_n} &= \prod_{i=1}^{n} Ee^{hg_i} \leqslant \prod_{i=1}^{n} \left\{1 + \frac{e^{hy}-1}{y^p} \int\limits_{]0,y_i[} u^p dF_i(u)\right\} \\ &\leqslant \prod_{i=1}^{n} \exp\left\{\frac{e^{hy}-1}{y^p} \int\limits_{]0,y_i[} u^p dF_i(u)\right\} \\ &= \exp\left\{\frac{e^{hy}-1}{y^p} \sum_{i=1}^{n} \int\limits_{]0,y_i[} u^p dF_i(u)\right\} = \exp\left\{\frac{e^{hy}-1}{y^p} A(p; \mathbf{0}, \mathbf{y})\right\}, \end{aligned}$$

and so

$$e^{-hx} Ee^{hU_n} \leqslant \exp\left\{\frac{e^{hy}-1}{y^p} A(p; \mathbf{0}, \mathbf{y}) - hx\right\}. \quad (3)$$

Now setting $h = \frac{1}{y}\log\left(\frac{xy^{p-1}}{A(p;\mathbf{0},\mathbf{y})} + 1\right)$ in (3), and using (2), we arrive at (1). □

Remark 25.21. If $0 < p < 1$ and $xy^{p-1} > A(p; \mathbf{0}, \mathbf{y}) > 0$, then (1) of (25.20) can be improved, namely

$$P(S_n \geqslant x) \leqslant \sum_{i=1}^{n} P(f_i \geqslant y_i) + \exp\left\{\frac{x}{y} - \frac{A(p; \mathbf{0}, \mathbf{y})}{y^p} - \frac{x}{y}\log\frac{xy^{p-1}}{A(p; \mathbf{0}, \mathbf{y})}\right\}. \quad (1)$$

Indeed, the right-hand side of (3) in (25.20) attains its minimum at $h = \frac{1}{y}\log\frac{xy^{p-1}}{A(p;\mathbf{0},\mathbf{y})} > 0$. Putting this value in it, and using (2) of (25.20), (1) obtains. As $\frac{1}{z} + \log z \geqslant \log(z+1), z > 0$, it follows that (1) is an improvement over (1) of (25.20).

Theorem 25.22. *Suppose that* $0 < p < 1$ *and* $A(p; -\mathbf{y}, \mathbf{y}) = \sum_{i=1}^{n} E[|f_i|^p 1_{\{|f_i|<y_i\}}] > 0$. *Then*

$$P(|S_n| \geqslant x) \leqslant \sum_{i=1}^{n} P(|f_i| \geqslant y_i) + \exp\left\{\frac{x}{y} - \frac{x}{y}\log\left(\frac{xy^{p-1}}{A(p;-\mathbf{y},\mathbf{y})} + 1\right)\right\}. \quad (1)$$

Proof. For $n \geqslant 1$, put $g_i = f_i 1_{\{|f_i|<y_i\}}$, $1 \leqslant i \leqslant n$, and $U_n = g_1 + \cdots + g_n$. Then, for $h > 0$, we have

$$P(|S_n| \geqslant x) \leqslant \sum_{i=1}^{n} P(|f_i| \geqslant y_i) + e^{-hx} E e^{hU_n}. \quad (2)$$

Further, we have

$$\begin{aligned} E e^{h|U_n|} \leqslant \prod_{i=1}^{n} E e^{h|g_i|} &= \prod_{i=1}^{n} E\left[1 + \frac{e^{h|g_i|} - 1}{|g_i|}|g_i|\right] \\ &\leqslant \prod_{i=1}^{n}\left\{1 + \frac{e^{hy_i} - 1}{y_i}\int_{]-y_i,y_i[} |u|\, dF_i(u)\right\} \\ &\leqslant \prod_{i=1}^{n}\left\{1 + \frac{e^{hy} - 1}{y^p}\int_{]-y_i,y_i[} |u|^p\, dF_i(u)\right\} \\ &\leqslant \exp\left\{\frac{e^{hy} - 1}{y^p}\sum_{i=1}^{n}\int_{]-y_i,y_i[} |u|^p\, dF_i(u)\right\} \\ &= \exp\left\{\frac{e^{hy} - 1}{y^p}A(p;-\mathbf{y},\mathbf{y})\right\}, \end{aligned}$$

and so

$$e^{-hx} E e^{h|U_n|} \leqslant \exp\left\{\frac{e^{hy} - 1}{y^p}A(p;-\mathbf{y},\mathbf{y}) - hx\right\}. \quad (3)$$

By putting $h = \frac{1}{y}\log(\frac{xy^{p-1}}{A(p;-\mathbf{y},\mathbf{y})} + 1)$ in (3), and using (2), we obtain (1). □

Remark 25.23. If $0 < p < 1$ and $xy^{p-1} > A(p; -\mathbf{y}, \mathbf{y}) > 0$, then

$$P(|S_n| \geqslant x) \leqslant \sum_{i=1}^{n} P(|f_i| \geqslant y_i) + \exp\left\{\frac{x}{y} - \frac{A(p;-\mathbf{y},\mathbf{y})}{y^p} - \frac{x}{y}\log\frac{xy^{p-1}}{A(p;-\mathbf{y},\mathbf{y})}\right\}.$$

This is an improvement over (1) of (25.22).

Theorem 25.24. *Assume that* $1 \leqslant p < 2$, *let* $A(p; -\mathbf{y}, \mathbf{y})$ *be as in* (25.22), *and set* $E(-\mathbf{y}, \mathbf{y}) = \sum_{i=1}^{n} E[f_i 1_{\{|f_i|<y_i\}}]$. *Then*

$$P(S_n \geqslant x) \leqslant \sum_{i=1}^{n} P(f_i \geqslant y_i) + \exp\left\{\frac{x}{y} - \left(\frac{x - E(-\mathbf{y}, \mathbf{y})}{y} + \frac{A(p; -\mathbf{y}, \mathbf{y})}{y^p}\right)\right.$$
$$\left. \times \log\left(\frac{xy^{p-1}}{A(p; -\mathbf{y}, \mathbf{y})} + 1\right)\right\}. \qquad (1)$$

Proof. For $n \geqslant 1$, let g_i, $1 \leqslant i \leqslant n$, and U_n be as in the proof of (25.20). Then (2) of (25.20) still holds. For $1 \leqslant i \leqslant n$ and $h > 0$, we have

$$\begin{aligned} Ee^{hg_i} &\leqslant 1 + h \int_{]-y_i, y_i[} u\,dF_i(u) + \int_{]-y_i, y_i[} \frac{e^{hu} - 1 - hu}{u^2} u^2 dF_i(u) \\ &\leqslant 1 + h \int_{]-y_i, y_i[} u\,dF_i(u) + \frac{e^{hy_i} - 1 - hy_i}{y_i^2} \int_{]-y_i, y_i[} u^2 dF_i(u) \\ &\leqslant 1 + h \int_{]-y_i, y_i[} u\,dF_i(u) + \frac{e^{hy} - 1 - hy}{y^p} \int_{]-y_i, y_i[} |u|^p\,dF_i(u). \end{aligned}$$

Consequently, reasoning as in the proof of (25.20), we get

$$\begin{aligned} Ee^{hU_n} &\leqslant \prod_{i=1}^{n} \exp\left\{h \int_{]-y_i, y_i[} u\,dF_i(u) + \frac{e^{hy} - 1 - hy}{y^p} \int_{]-y_i, y_i[} |u|^p\,dF_i(u)\right\} \\ &= \exp\left\{hE(-\mathbf{y}, \mathbf{y}) + \frac{e^{hy} - 1 - hy}{y^p} A(p; -\mathbf{y}, \mathbf{y})\right\}, \end{aligned}$$

and so

$$e^{-hx} Ee^{hU_n} \leqslant \exp\left\{hE(-\mathbf{y}, \mathbf{y}) + \frac{e^{hy} - 1 - hy}{y^p} A(p; -\mathbf{y}, \mathbf{y}) - hx\right\}. \qquad (2)$$

Putting $h = \frac{1}{y} \log\left(\frac{xy^{p-1}}{A(p; -\mathbf{y}, \mathbf{y})} + 1\right)$ in (2), and using (2) of (25.20), (1) obtains. □

Corollary 25.25. *Under the assumptions of* (25.24), *we have*

$$P(S_n \leqslant -x) \leqslant \sum_{i=1}^{n} P(f_i \leqslant -y_i) + \exp\left\{\frac{x}{y} - \left(\frac{x + E(-\mathbf{y}, \mathbf{y})}{y} + \frac{A(p; -\mathbf{y}, \mathbf{y})}{y^p}\right)\right.$$
$$\left. \times \log\left(\frac{xy^{p-1}}{A(p; -\mathbf{y}, \mathbf{y})} + 1\right)\right\}.$$

Remark 25.26. Inspection of the above proofs shows that in (25.23)–(25.26) we may replace $A(p; -\mathbf{y}, \mathbf{y})$ by $\sum_{i=1}^{n} E\,|f_i|^p$ whenever this is finite.

Theorem 25.27. *Assume that* $B^2(-\infty, \mathbf{y}) = \sum_{i=1}^{n} E[f_i^2 1_{\{f_i < y_i\}}] > 0$, *and set* $E(-\infty, \mathbf{y}) = \sum_{i=1}^{n} E[f_i 1_{\{f_i < y_i\}}]$. *If* $p \geqslant 2$, *then*

$$P(S_n \geqslant x) \leqslant \sum_{i=1}^{n} P(f_i \geqslant y_i) + \exp\left\{\frac{x}{y} - \left(\frac{x - E(-\infty, \mathbf{y})}{y} + \frac{B^2(-\infty, \mathbf{y})}{y^2}\right) \times \log\left(\frac{xy}{B^2(-\infty, \mathbf{y})} + 1\right)\right\}. \quad (1)$$

Proof. For $n \geqslant 1$, let g_i, $1 \leqslant i \leqslant n$, and U_n be as in the proof of (25.20). Then (2) of (25.20) still holds. For $1 \leqslant i \leqslant n$ and $h > 0$, we have

$$\begin{aligned} Ee^{hg_i} &\leqslant 1 + h \int_{]-\infty, y_i[} u\,dF_i(u) + \int_{]-\infty, y_i[} \frac{e^{hu} - 1 - hu}{u^2} u^2 dF_i(u) \\ &\leqslant 1 + h \int_{]-\infty, y_i[} u\,dF_i(u) + \frac{e^{hy} - 1 - hy}{y^2} \int_{]-\infty, y_i[} u^2 dF_i(u). \end{aligned}$$

Therefore, we obtain

$$\begin{aligned} Ee^{hU_n} &\leqslant \prod_{i=1}^{n} \exp\left\{h \int_{]-\infty, y_i[} u\,dF_i(u) + \frac{e^{hy} - 1 - hy}{y^2} \int_{]-\infty, y_i[} u^2 dF_i(u)\right\} \\ &= \exp\left\{hE(-\infty, \mathbf{y}) + \frac{e^{hy} - 1 - hy}{y^2} B^2(-\infty, \mathbf{y})\right\}, \end{aligned}$$

and so

$$e^{-hx} Ee^{hU_n} \leqslant \exp\left\{hE(-\infty, \mathbf{y}) + \frac{e^{hy} - 1 - hy}{y^2} B^2(-\infty, \mathbf{y}) - hx\right\}. \quad (2)$$

Putting $h = \frac{1}{y}\log\left(\frac{xy}{B^2(-\infty, \mathbf{y})} + 1\right)$ in (2), and using (2) of (25.20), we have (1). □

Corollary 25.28. *Assume* $Ef_i^2 < \infty$ *and* $Ef_i = 0$ *for* $1 \leqslant i \leqslant n$, *and write* $s_n^2 = \sum_{i=1}^{n} Ef_i^2$. *Then*

$$P(S_n \geqslant x) \leqslant \sum_{i=1}^{n} P(f_i \geqslant y_i) + e^{x/y}\left(\frac{s_n^2}{xy}\right)^{x/y} \quad (1)$$

and

$$P(|S_n| \geqslant x) \leqslant \sum_{i=1}^{n} P(|f_i| \geqslant y_i) + 2e^{x/y}\left(\frac{s_n^2}{xy}\right)^{x/y}. \quad (2)$$

Proof. We have $E(-\infty, \mathbf{y}) \leqslant 0$ and $B^2(-\infty, \mathbf{y}) \leqslant s_n^2$, and so

$$\exp\left\{-\left(\frac{x - E(-\infty, \mathbf{y})}{y} + \frac{B^2(-\infty, \mathbf{y})}{y^2}\right)\log\left(\frac{xy}{B^2(-\infty, \mathbf{y})} + 1\right)\right\}$$

$$= \left(\frac{xy}{B^2(-\infty, \mathbf{y})} + 1\right)^{-\left(\frac{x-E(-\infty,\mathbf{y})}{y} + \frac{B^2(-\infty,\mathbf{y})}{y^2}\right)} \leqslant \left(\frac{s_n^2}{xy}\right)^{x/y}.$$

This in conjunction with (25.27) proves (1), and (2) follows immediately. □

Corollary 25.29 (Benett-Hoeffding). *For $1 \leqslant i \leqslant n$, assume that $Ef_i = 0$ and there is $L_i \in]0, \infty[$ such that $f_i < L_i$. Set $L = L_1 \vee \ldots \vee L_n$ and let s_n^2 be as in (25.28). Then*

$$P(S_n \geqslant x) \leqslant \exp\left\{\frac{x}{L} - \left(\frac{x}{L} + \frac{s_n^2}{L^2}\right)\log\left(\frac{xL}{s_n^2} + 1\right)\right\}.$$

Proof. Put $y_i = L_i, 1 \leqslant i \leqslant n$, and $y = L$ in (1) of (25.28), and notice that $E(-\infty, \mathbf{y}) \leqslant 0$ and $B^2(-\infty, \mathbf{y}) \leqslant s_n^2$. □

The usefulness of the Fuk–Nagaev inequalities has been well demonstrated in studies concerning precise asymptotics, as $\varepsilon \searrow 0$, of series such as those in the Hsu–Robbins–Erdős theorem (25.17) (see (25.32)), in the Spitzer theorem (25.19), in the Davis theorem (25.39) (see (25.42)), etc. For instance, the next result is used to specify the behavior of the series in Spitzer's theorem as $\varepsilon \searrow 0$.

Theorem 25.30. *Let $f_1, \ldots, f_n$ be i.i.d. random variables with $Ef_1 = 0$ and $E\,|f_1|^p < \infty$, where $1 < p \leqslant 2$, and put $S_n = f_1 + \cdots + f_n$. Then, for $x, y > 0$,*

$$P(|S_n| \geqslant x) \leqslant nP(|f_1| \geqslant y) + 2e^{x/y}\left(\frac{nE\,|f_1|^p}{nE\,|f_1|^p + xy^{p-1}}\right)^{x/y}. \tag{1}$$

Proof. By virtue of (25.26), and by specializing (25.24), we have

$$P(S_n \geqslant x) \leqslant nP(f_1 \geqslant y) + \exp\left\{\frac{x}{y} - \left(\frac{x - nE[f_1 1_{\{|f_1|<y\}}]}{y} + \frac{nE\,|f_1|^p}{y^p}\right)\right.$$
$$\left.\times \log\left(\frac{xy^{p-1}}{nE\,|f_1|^p} + 1\right)\right\}. \tag{2}$$

Since $Ef_1 = 0$, we note that

$$\left|E[f_1 1_{\{|f_1|<y\}}]\right| \leqslant E[|f_1|\, 1_{\{|f_1|\geqslant y\}}] \leqslant \frac{E[|f_1|^p\, 1_{\{|f_1|\geqslant y\}}]}{y^{p-1}} \leqslant \frac{E\,|f_1|^p}{y^{p-1}}.$$

Consequently,

$$\frac{x - nE[f_1 1_{\{|f_1|<y\}}]}{y} + \frac{nE\,|f_1|^p}{y^p} \geqslant \frac{x}{y}.$$

Combining this with (2), we see that

$$P(S_n \geqslant x) \leqslant nP(f_1 \geqslant y) + e^{x/y}\left(\frac{nE\,|f_1|^p}{nE\,|f_1|^p + xy^{p-1}}\right)^{x/y}. \tag{3}$$

Adding (3) and the inequality derived from it by writing $-f_i$ instead of f_i, $1 \leqslant i \leqslant n$, leads to (1). □

We shall now illustrate how the inequality (2) of (25.28) helps to provide the exact asymptotic, as $\varepsilon \searrow 0$, of the series in the Hsu–Robbins–Erdős theorem (25.17). We start with the following lemma.

Lemma 25.31. *Let f be a random variable with $Ef^2 < \infty$, and set $b(\varepsilon) = M\varepsilon^{-2}$, where M is a large positive number and $\varepsilon > 0$. Then*

$$\lim_{\varepsilon \searrow 0} \varepsilon^2 \sum_{n > b(\varepsilon)} nP(|f| \geqslant \varepsilon n) = 0. \tag{1}$$

Proof. We have

$$\begin{aligned}
\varepsilon^2 \sum_{n>b(\varepsilon)} nP(|f| \geqslant \varepsilon n) &= \varepsilon^2 \sum_{n>b(\varepsilon)} n \sum_{k \geqslant n} P(\varepsilon k \leqslant |f| < \varepsilon(k+1)) \\
&= \varepsilon^2 \sum_{k>b(\varepsilon)} \left(\sum_{b(\varepsilon)<n\leqslant k} n\right) P(\varepsilon k \leqslant |f| < \varepsilon(k+1)n) \\
&\leqslant \sum_{k \geqslant [b(\varepsilon)]} \varepsilon^2 k^2 P(\varepsilon^2 k^2 \leqslant f^2 < \varepsilon^2 (k+1)^2) \\
&\leqslant E[f^2 1_{\{|f| \geqslant \varepsilon[b(\varepsilon)]\}}].
\end{aligned}$$

Therefore, since $\varepsilon[b(\varepsilon)] \to \infty$ as $\varepsilon \searrow 0$, (1) follows □

Theorem 25.32 (Heyde). *Let f_n, $n \geqslant 1$, be i.i.d. random variables with $\sigma^2 = Ef_1^2 < \infty$ and $Ef_1 = 0$, and put $S_n = f_1 + \cdots + f_n$. Then*

$$\lim_{\varepsilon \searrow 0} \varepsilon^2 \sum_{n \geqslant 1} P(|S_n| \geqslant \varepsilon n) = \sigma^2. \tag{1}$$

Proof. The convergence of the series in (1) is ensured by (25.17).

(I) First we assume that f_1 has the standard normal distribution function Φ. Then $S_n/\sqrt{n}$ has also the distribution function Φ by (20.40.b). Writing $\Psi(x) = 1 - \Phi(x) + \Phi(-x)$, $x \geqslant 0$, and using (15.24), we see that

$$2\int_0^\infty y\Psi(y)dy = 2\int_0^\infty yP(|f_1| \geqslant y)dy = Ef_1^2 = 1. \tag{2}$$

We have $\sum_{n\geqslant 1} P(|S_n| \geqslant \varepsilon n) = \sum_{n\geqslant 1} \Psi(\varepsilon\sqrt{n})$, and so

$$\int_1^\infty \Psi(\varepsilon\sqrt{x})dx \leqslant \sum_{n\geqslant 1} P(|S_n| \geqslant \varepsilon n) \leqslant \int_0^\infty \Psi(\varepsilon\sqrt{x})dx. \tag{3}$$

By putting $y = \varepsilon\sqrt{x}$, (3) becomes

$$2\int_\varepsilon^\infty y\Psi(y)dy \leqslant \varepsilon^2 \sum_{n\geqslant 1} P(|S_n| \geqslant \varepsilon n) \leqslant 2\int_0^\infty y\Psi(y)dy. \tag{4}$$

Letting $\varepsilon \searrow 0$ in (4), and using (2), we obtain (1) in this case.

(II) Assume now that $\sigma^2 = 1$. According to (I), in order to prove (1) it suffices to show that

$$\lim_{\varepsilon\searrow 0} \varepsilon^2 \sum_{n\geqslant 1} \left|P(|S_n| \geqslant \varepsilon n) - \Psi(\varepsilon\sqrt{n})\right| = 0. \tag{5}$$

Write $\Delta_n = \sup_{x\in R} \left|P(|S_n| \geqslant x\sqrt{n}) - \Psi(x)\right|$. Then $\Delta_n \to 0$ by (24.23.c), and so $\frac{1}{m}\sum_{n=1}^m \Delta_n \to 0$. Therefore, we have

$$\begin{aligned}\varepsilon^2 \sum_{n\leqslant b(\varepsilon)} \left|P(|S_n| \geqslant \varepsilon n) - \Psi(\varepsilon\sqrt{n})\right| &\leqslant \varepsilon^2 \sum_{n\leqslant [b(\varepsilon)]} \Delta_n \\ &= \varepsilon^2 [b(\varepsilon)] \cdot \frac{1}{[b(\varepsilon)]} \sum_{n\leqslant [b(\varepsilon)]} \Delta_n \to 0 \text{ as } \varepsilon \searrow 0.\end{aligned}$$

Thus, to get (5), it suffices to prove that

$$\lim_{\varepsilon\searrow 0} \varepsilon^2 \sum_{n>b(\varepsilon)} P(|S_n| \geqslant \varepsilon n) = 0, \tag{6}$$

since $\lim_{\varepsilon\searrow 0} \varepsilon^2 \sum_{n>b(\varepsilon)} \Psi(\varepsilon\sqrt{n}) = 0$ is a particular case of (6). Specializing (2) of (25.28) at $x = \varepsilon n$ and $y_i = x/2$, $1 \leqslant i \leqslant n$, yields

$$P(|S_n| \geqslant \varepsilon n) \leqslant nP(|f_1| \geqslant \varepsilon n/2) + 8e^2\varepsilon^{-4}n^{-2}.$$

Consequently, as $\sum_{n>m} n^{-2} \leqslant m^{-1}$, $m \geqslant 1$, we get

$$\begin{aligned}\varepsilon^2 \sum_{n>b(\varepsilon)} P(|S_n| \geqslant \varepsilon n) &\leqslant \varepsilon^2 \sum_{n>b(\varepsilon)} nP(|2f_1| \geqslant \varepsilon n) + 8e^2\varepsilon^{-2} \sum_{n>b(\varepsilon)} n^{-2} \\ &\leqslant \varepsilon^2 \sum_{n>b(\varepsilon)} nP(|2f_1| \geqslant \varepsilon n) + \frac{8e^2}{M - \varepsilon^2}.\end{aligned} \tag{7}$$

Since M is at our disposal, (6) obtains on account of (7) and (25.31).

(III) Consider now the general case. Applying (II) with f_n replaced by f_n/σ, $n \geqslant 1$, yields

$$\lim_{\varepsilon \searrow 0} \varepsilon^2 \sum_{n \geqslant 1} P(|S_n| \geqslant \varepsilon \sigma n) = 1.$$

Hence replacing $\varepsilon\sigma$ by ε leads to (1). □

Exercise 25.33. Let f_n, $n \geqslant 1$, be i.i.d. random variables. Show that $\sup_{1 \leqslant k \leqslant n} |f_k| \xrightarrow{P} 0$ if and only if $nP(|f_1| \geqslant n) \to 0$. [Apply (23.55).]

Exercise 25.34 (Klass–Teicher). Let f_n, $n \geqslant 1$, and S_n, $n \geqslant 1$, be as in (25.13), and let $\{b_n : n \geqslant 1\} \subset]0, \infty[$ be such that the sequence $\{b_n/n : n \geqslant 1\}$ is nondecreasing. Show that $S_n/b_n - (n/b_n)E[f_1 1_{\{|f_1|<n\}}] \xrightarrow{P} 0$ if and only if $nP(|f_1| \geqslant b_n) \to 0$. [Mimic the proof of (25.13).]

Exercise 25.35. Let f_n, $n \geqslant 1$, and S_n, $n \geqslant 1$, be as in (25.13), and assume that $S_n/n - a_n \xrightarrow{P} 0$ for some sequence $\{a_n : n \in N\} \subset R$. Prove the following.
(a) $E[f_1 1_{\{an \leqslant |f_1| < bn\}}] \to 0$ whenever $0 < a < b < \infty$. [Use (25.15).]
(b) $kE[S_1 1_{\{|S_1|<kn\}}] - E[S_k 1_{\{|S_k|<kn\}}] \to 0$ for any $k \geqslant 1$. [Use (25.15) and (a).]

Exercise 25.36 (Katz). Let f_n, $n \geqslant 1$, and S_n, $n \geqslant 1$, be as in (25.13), and let $1 \leqslant p < 2$. Show that $\sum_{n \geqslant 1} n^{p-2} P(|S_n| \geqslant \varepsilon n) < \infty$, $\varepsilon > 0$, if and only if $E|f|_1^p < \infty$ and $Ef_1 = 0$. [Mimic the proof of (25.19).]

Exercise 25.37 (Baum–Katz). Let f_n, S_n, $n \geqslant 1$, and p be as in (25.36). Prove that

$$\sum_{n \geqslant 1} \frac{1}{n} P(|S_n| \geqslant \varepsilon n^{1/p}) < \infty, \quad \varepsilon > 0, \tag{1}$$

if and only if $E|f_1|^p < \infty$ and $Ef_1 = 0$. Assuming that (1) holds in (a)–(e) below, the following steps may be useful.
(a) Let f_n^* and S_n^* be as in the proof of (25.17). Then $S_n^*/n^{1/p} \xrightarrow{P^*} 0$. [Use (25.18).]
(b) $nP^*\left(|f_1^*| \geqslant n^{1/p}\right) \to 0$. [Use (a) and (24.13).]
(c) $2P^*\left(|S_n^*| \geqslant n^{1/p}\right) \geqslant nP^*(|f_1^*| \geqslant 2n^{1/p})/2$ for n large enough. [Use (25.4) and (b).]
(d) $E|f_1|^p < \infty$. [Use (c), (13.27.a), and (25.1.iii).]
(e) $Ef_1 = 0$. [Hint. $(S_n - ES_n)/n^{1/p} \xrightarrow{P} 0$ by (d), (18.55) and (14.14). Then mimic the corresponding part in the proof of (25.17).]
(f) If $E|f_1|^p < \infty$ and $Ef_1 = 0$, then (1) is satisfied. [Hint. For $n \geqslant 1$, put $f_{n,i} = f_i 1_{\{|f_i|<n\}}$, $1 \leqslant i \leqslant n$, and $U_n = \sum_{i=1}^n f_{n,i}$. Then, utilizing (18.55.a) and (11.37), mimic the second part of the proof of (25.19).]

Exercise 25.38 (Baum–Katz). Let f_n, S_n, $n \geqslant 1$, and p be as in (25.36). Prove that $\sum_{n \geqslant 1} \frac{\log n}{n} P(|S_n| \geqslant \varepsilon n^{1/p}) < \infty$, $\varepsilon > 0$, if and only if $E[|f_1|^p \log(1 + |f_1|)] < \infty$ and $Ef_1 = 0$. [Hint. Adjust the solution to (25.37). To derive the step corresponding to (25.37.d), apply (17.10) with $\varphi(x) = x \log(1 + x)$, $x \geqslant 0$, to get $E^*[|f_1^*|^p \log(1 +$

$|f_1^*|)] < \infty$, then apply (25.1.iv) with $\varphi(x) = x^p \log(1+x), x \geqslant 0$, to obtain $E[|f_1|^p \log(1+|f_1|)] < \infty$.]

Exercise 25.39 (Davis). For $n \geqslant 1$, let f_n and S_n be as in (25.13). Prove that

$$\sum_{n\geqslant 2} \frac{\log n}{n} P(|S_n| \geqslant \varepsilon\sqrt{n \log n}) < \infty, \quad \varepsilon > 0, \tag{1}$$

if and only if $Ef_1^2 < \infty$ and $Ef_1 = 0$. Assuming that (1) holds in (a)–(e) below, while $Ef_1^2 < \infty$ and $Ef_1 = 0$ in (f)–(j) below, the next steps may be helpful.

(a) Let f_n^* and S_n^* be as in the proof of (25.17). Then $S_n^*/\sqrt{n \log n} \xrightarrow{P^*} 0$.

(b) $nP^*(|f_1^*| \geqslant \sqrt{n \log n}) \to 0$. [Use (a) and (24.13).]

(c) $2P^*(|S_n^*| \geqslant \sqrt{n \log n}) \geqslant nP^*(|f_1^*| \geqslant 2\sqrt{n \log n})/2$ for sufficiently large n. [Use (25.4) and (b).]

(d) $Ef_1^2 < \infty$. [Use (c), (17.16.c) and (25.1.iii).]

(e) $Ef_1 = 0$. [Hint. $(S_n - ES_n)/\sqrt{n \log n} \xrightarrow{P} 0$ by (d), (18.56) and (14.14). Then mimic the corresponding part in the proof of (25.17).]

(f) For $n \geqslant 1$, set $f_{n,i} = f_i 1_{\{|f_i| < \sqrt{n \log n}\}}$, $1 \leqslant i \leqslant n$, and $U_n = \sum_{i=1}^n f_{n,i}$. Then $EU_n/\sqrt{n \log n} \to 0$ and $Ef_{n,1} \to 0$.

(g) $\sum_{n\geqslant 2} \frac{1}{n^2(\log n)^2} Ef_{n,1}^4 < \infty$and $\sum_{n\geqslant 2} \frac{1}{n^3(\log n)^2} Ef_{n,1}^6 < \infty$. [Apply Fubini's theorem (15.8), and use (17.16.a) and (17.16.b).]

(h) $\sum_{n\geqslant 2} \frac{1}{n^2(\log n)^2} (E|f_{n,1}|^3)^2 < \infty$. [Hints. By (18.22) and Fubini's theorem (15.8), using (17.16.a) and (17.16.b), write

$$\begin{aligned}
&\sum\nolimits_{n\geqslant 2} \frac{1}{n^2(\log n)^2} (E[|f_{n,1}|^3 1_{\{|f_1|\geqslant e\}}])^2 \\
&= E\left[\sum_{n\geqslant 2} \frac{1}{n^2(\log n)^2} |f_{n,1}|^3 1_{\{|f_1|\geqslant e\}} |f_{n,2}|^3 1_{\{|f_2|\geqslant e\}}\right] \\
&= E\left[|f_1|^3 1_{\{|f_1|\geqslant e\}} |f_2|^3 1_{\{|f_2|\geqslant e\}} \sum_{n > \chi(f_1^2)\vee\chi(f_2^2)} \frac{1}{n^2(\log n)^2}\right] \\
&\leqslant CE\Bigg[|f_1|^3 1_{\{|f_1|\geqslant e\}} |f_2|^3 1_{\{|f_2|\geqslant e\}} \\
&\qquad \times \frac{1}{\chi(f_1^2)\vee\chi(f_2^2)} \cdot \frac{1}{(\log(\chi(f_1^2)\vee\chi(f_2^2)))^2}\Bigg] \\
&\leqslant CE\left[\frac{|f_1|^3}{\sqrt{\chi(f_1^2)\log\chi(f_1^2)}} 1_{\{|f_1|\geqslant e\}} \cdot \frac{|f_2|^3}{\sqrt{\chi(f_2^2)\log\chi(f_2^2)}} 1_{\{|f_2|\geqslant e\}}\right] \\
&\leqslant CE\left[\frac{f_1^2}{\sqrt{\log|f_1|}} 1_{\{|f_1|\geqslant e\}} \cdot \frac{f_2^2}{\sqrt{\log|f_2|}} 1_{\{|f_2|\geqslant e\}}\right] \leqslant C(Ef_1^2)^2,
\end{aligned}$$

where the second inequality comes from $a \vee b \geqslant \sqrt{ab}$ and $\log(a \vee b) \geqslant \sqrt{(\log a)\log b}$, $a, b \geqslant 1$, and C denotes a positive constant, possibly varying from place to place.]

(i) $\sum_{n\geqslant 2} \frac{\log n}{n} P(|U_n - EU_n| \geqslant \varepsilon\sqrt{n\log n}) < \infty, \varepsilon > 0$. [Hint. By Markov's inequality (13.8), write

$$\begin{aligned}P(|U_n - EU_n| \geqslant \varepsilon\sqrt{n\log n}) &\leqslant \varepsilon^{-6}n^{-3}(\log n)^{-3}E(U_n - EU_n)^6 \\ &= \varepsilon^{-6}n^{-3}(\log n)^{-3}[nE(f_{n,1} - Ef_{n,1})^6 + 15n(n-1) \\ &\quad \times E(f_{n,1} - Ef_{n,1})^4 E(f_{n,1} - Ef_{n,1})^2 \\ &\quad + 10n(n-1)(E(f_{n,1} - Ef_{n,1})^3)^2 + 15n(n-1) \\ &\quad \times (n-2)(E(f_{n,1} - Ef_{n,1})^2)^3], \quad n \geqslant 2.\end{aligned}$$

Then utilize (g) and (h), taking into account that $Ef_{n,1} \to 0$.]

(j) $\sum_{n\geqslant 2} \frac{\log n}{n} P(|S_n| \geqslant \varepsilon\sqrt{n\log n}) < \infty, \varepsilon > 0$. [Hint. If $n_\varepsilon \in N$ is so that $|EU_n|/\sqrt{n\log n} < \varepsilon$, $n \geqslant n_\varepsilon$, write

$$\begin{aligned}\sum_{n\geqslant n_\varepsilon} \frac{\log n}{n} P(|S_n| \geqslant 2\varepsilon\sqrt{n\log n}) &\leqslant \sum_{n\geqslant n_\varepsilon} \frac{\log n}{n} P(S_n \neq U_n) \\ &\quad + \sum_{n\geqslant n_\varepsilon} \frac{\log n}{n} P(|U_n - EU_n| \geqslant \varepsilon\sqrt{n\log n}).\end{aligned}$$

Then use (17.16.c) and (i).]

Exercise 25.40. Let $f_n, n \geqslant 1$, be independent random variables, put $S_n = \sum_{k=1}^n f_k$, $n \geqslant 1$, and let $0 < b_n \leqslant b_{n+1} \to \infty$. Prove the following.

(a) For $\varepsilon, \delta > 0$, $P(|S_n| \geqslant \varepsilon) \geqslant (\inf_{1\leqslant i\leqslant n} P(|S_n - S_i| \leqslant \delta))P(\sup_{1\leqslant i\leqslant n} |S_i| \geqslant \varepsilon + \delta)$, $n \geqslant 1$. [Hint. Define the $\overline{R}$-valued random variable $\tau : X \to \{1, \ldots, n\} \cup \{\infty\}$ by $\tau = \inf\{i \leqslant n : |S_i| \geqslant \varepsilon + \delta\}$ (see (3.21.c)). Then $\{\tau = i\} \in \sigma(f_1, \ldots, f_i)$, $\cup_{i=1}^n \{\tau = i\} = \{\sup_{1\leqslant i\leqslant n} |S_i| \geqslant \varepsilon + \delta\}$, and $\cup_{i=1}^n \{\tau = i, |S_n - S_i| < \delta\} \subset \{|S_n| \geqslant \varepsilon\}$.]

(b) For $\varepsilon, \delta \in R$, $P(S_n \geqslant \varepsilon) \geqslant (\inf_{1\leqslant i\leqslant n} P(S_n - S_i \geqslant -\delta))P(\sup_{1\leqslant i\leqslant n} S_i \geqslant \varepsilon + \delta)$, $n \geqslant 1$.

(c) For $\varepsilon, \delta \in R$, $P(S_n \leqslant -\varepsilon) \geqslant (\inf_{1\leqslant i\leqslant n} P(S_n - S_i \leqslant \delta))P(\inf_{1\leqslant i\leqslant n} S_i \leqslant -\varepsilon - \delta)$, $n \geqslant 1$.

(d) If $S_n/b_n \xrightarrow{P} 0$ and $\varepsilon, \delta > 0$, then there is $m_0 = m_0(\delta) \in N$ so that $P(S_n \geqslant \varepsilon b_n) \geqslant \frac{1}{2}P(\sup_{1\leqslant i\leqslant n} S_i \geqslant (\varepsilon + \delta)b_n)$, $n \geqslant m_0$.

The inequalities in (a)–(c) are *Ottaviani's inequalities.* Avoiding consideration of medians, they can replace Lévy's inequality (25.3.ii) in some situations.

Exercise 25.41. Let $f_n, n \geqslant 1$, be i.i.d. random variables with $f_1 \in \mathfrak{L}^p$ for some $p \geqslant 1$, and set $S_n = \sum_{i=1}^n f_i$. Use (13.21.b), (14.28), (25.16) and (14.30) to prove that $E|S_n/n|^p \to |Ef_1|^p$.

Exercise 25.42 (Gut–Spătaru). For $n \geqslant 1$, let f_n and S_n be as in (25.39), and assume that $\sigma^2 = Ef_1^2 < \infty$ and $Ef_1 = 0$. Prove that

$$\lim_{\varepsilon \searrow 0} \varepsilon^4 \sum_{n \geqslant 1} \frac{\log n}{n} P(|S_n| \geqslant \varepsilon\sqrt{n \log n}) = \frac{3}{2}\sigma^4. \tag{1}$$

The following steps may be useful.

(a) Set $b(\varepsilon) = e^{M\varepsilon^{-2}}$, where M is a large positive number and $\varepsilon > 0$. Then

$$\sum_{n > b(\varepsilon)} (\log n) P(|f_1| \geqslant \varepsilon\sqrt{n \log n}) \leqslant \varepsilon^{-2} E f_1^2.$$

(b) If f_1 has the standard normal distribution function Φ, then (1) holds.

(c) Write $\Psi(x) = 1 - \Phi(x) + \Phi(-x)$, $x \geqslant 0$. If $\sigma^2 = 1$, then

$$\lim_{\varepsilon \searrow 0} \varepsilon^4 \sum_{n \leqslant b(\varepsilon)} \frac{\log n}{n} \left| P(|S_n| \geqslant \varepsilon\sqrt{n \log n}) - \Psi(\varepsilon\sqrt{\log n}) \right| = 0.$$

(d) If $\sigma^2 = 1$, then

$$\varepsilon^4 \sum_{n > b(\varepsilon)} \frac{\log n}{n} P(|S_n| \geqslant \varepsilon\sqrt{n \log n}) \leqslant 9\varepsilon^2 + 54e^3 M^{-1}.$$

[Hint. Apply (2) of (25.28) with $x = \varepsilon\sqrt{n \log n}$ and $y_i = x/3$, $1 \leqslant i \leqslant n$, to yield

$$\sum_{n > b(\varepsilon)} \frac{\log n}{n} P(|S_n| \geqslant \varepsilon\sqrt{n \log n}) \leqslant \sum_{n > b(\varepsilon)} (\log n) P(|3 f_1| \geqslant \varepsilon\sqrt{n \log n}) + 54e^3 \varepsilon^{-6} \sum_{n > b(\varepsilon)} \frac{1}{n(\log n)^2}.$$

Then use (a).]

(e) If $\sigma^2 = 1$, then

$$\lim_{\varepsilon \searrow 0} \varepsilon^4 \sum_{n > b(\varepsilon)} \frac{\log n}{n} P(|S_n| \geqslant \varepsilon\sqrt{n \log n}) = 0$$

and

$$\lim_{\varepsilon \searrow 0} \varepsilon^4 \sum_{n > b(\varepsilon)} \frac{\log n}{n} \Psi(\varepsilon\sqrt{\log n}) = 0.$$

[Use (d).]

Exercise 25.43. For $n \geqslant 1$, let f_n and S_n be as in (25.13), and suppose that $Ef_1^2 < \infty$ and $Ef_1 = 0$.

(a) Use (25.28) to show that $\sum_{n\geqslant 1} P(|S_n| \geqslant \varepsilon n) < \infty, \varepsilon > 0$. [Apply (2) of (25.28) with $x = \varepsilon n$ and $y_i = x/2$, $1 \leqslant i \leqslant n$.]

(b) Compare the solution to (a) with the proof of the converse part of the Hsu–Robbins–Erdős theorem (25.17).

Exercise 25.44. Let $f_1, \ldots, f_n$ be independent random variables and write $s_n^2 = Ef_1^2 + \cdots + Ef_n^2$. For $1 \leqslant i \leqslant n$, assume that $Ef_i^2 < \infty$ and $Ef_i = 0$ and set $S_i = f_1 + \cdots + f_i$. Use Lévy's inequality (25.3.i) and (23.52) to prove that $P(\sup_{1\leqslant i\leqslant n} S_i \geqslant \varepsilon) \leqslant 2P(S_n \geqslant \varepsilon - \sqrt{2}s_n), \varepsilon \in R$.

Exercise 25.45. Let $f_1, \ldots, f_n$ be independent and symmetric random variables, and put $S_n = f_1 + \cdots + f_n$. Show that $2P(|S_n| \geqslant \varepsilon n) \geqslant P(\sup_{1\leqslant i\leqslant n} |f_i| \geqslant \varepsilon), \varepsilon > 0$. The next steps may be helpful.

(a) For $1 \leqslant i \leqslant n, \{\tau = i\} \subset \{\tau = i, |S_n| \geqslant \varepsilon\} \cup \{\tau = i, |S_n - 2f_i| \geqslant \varepsilon\}$, where $\tau = \inf\{i \leqslant n : |f_i| \geqslant \varepsilon\}$.

(b) $P(|S_n| \geqslant \varepsilon n) = \sum_{i=1}^n P(\tau = i, |S_n| \geqslant \varepsilon n)$.

(c) For $1 \leqslant i \leqslant n, P(f_1, \ldots, f_n)^{-1} = P(f_1, \ldots, f_{i-1}, -f_i, f_{i+1}, \ldots, f_n)^{-1}$. [Use (18.29).]

(d) $P(|S_n| \geqslant \varepsilon n) = \sum_{i=1}^n P(\tau = i, |S_n - 2f_i| \geqslant \varepsilon n)$. [Use (c).]

(e) $2P(|S_n| \geqslant \varepsilon n) \geqslant P(\sup_{1\leqslant i\leqslant n} |f_i| \geqslant \varepsilon)$. [Use (a), (b) and (d).]

26 Series of Independent Random Variables

Throughout the rest of this chapter, if not otherwise stated, $f_n, n \geqslant 1$, will be non-degenerate independent random variables and $S_n = \sum_{k=1}^n f_k, n \geqslant 1$. The sequence $\{S_n : n \geqslant 1\}$ is traditionally named a *random walk*, and $\{f_n : n \geqslant 1\}$ is called the sequence of its *random steps*. According to (18.21), $P(\sum_{k\geqslant 1} f_k$ converges$) = 0$ or 1. In this section we shall investigate conditions under which $\sum_{k\geqslant 1} f_k$ converges almost surely.

Theorem 26.1 (Lévy). *Let S be a random variable such that $S_n \xrightarrow{P} S$. Then $S_n \xrightarrow{a.s.} S$.*

Proof. Taking into account that $S_n \xrightarrow{P} S$, choose a strictly increasing sequence $\{n_k : k \geqslant 1\} \subset N$ such that $P(|S_{n_k} - S| \geqslant 2^{-k-1}) \leqslant 2^{-k}, k \geqslant 1$, and so

$$\sum_{k\geqslant 1} P(|S_{n_k} - S| \geqslant 2^{-k-1}) < \infty. \tag{1}$$

From (1), via the Borel-Cantelli lemma (18.24) applied to the events $\{|S_{n_k} - S| \geqslant 2^{-k-1}\}, k \geqslant 1$, it follows that

$$S_{n_k} \xrightarrow{a.s.} S. \tag{2}$$

Since $\{|S_{n_k} - S_{n_{k-1}}| \geqslant 2^{-k}\} \subset \{|S_{n_k} - S| \geqslant 2^{-k-1}\} \cup \{|S_{n_{k-1}} - S| \geqslant 2^{-k-2}\}$, $k \geqslant 2$, (1) also implies

$$\sum_{k\geqslant 2} P(|S_{n_k} - S_{n_{k-1}}| \geqslant 2^{-k}) < \infty. \tag{3}$$

By Lévy's inequality (25.3.ii) applied to the random variables $f_{n_{k-1}+1}, \ldots, f_{n_k}$, and by (3), we have

$$\sum_{k\geqslant 2} P\left(\sup_{n_{k-1}<n\leqslant n_k} |S_n - S_{n_{k-1}} - m[S_n - S_{n_k}]| \geqslant 2^{-k}\right)$$
$$\leqslant 2\sum_{k\geqslant 2} P\left(|S_{n_k} - S_{n_{k-1}}| \geqslant 2^{-k}\right) < \infty. \tag{4}$$

This yields

$$\sup_{n_{k-1}<n\leqslant n_k} |S_n - S_{n_{k-1}} - m[S_n - S_{n_k}]| \overset{a.s.}{\to} 0 \tag{5}$$

again by the (18.24). As $S_n \overset{P}{\to} S$, (23.18.b) shows that $\sup\limits_{n_{k-1}<n\leqslant n_k} |m[S_n - S_{n_k}]| \to 0$, and so

$$\sup_{n_{k-1}<n\leqslant n_k} |S_n - S_{n_{k-1}}| \overset{a.s.}{\to} 0 \tag{6}$$

by (5). From (2) and (6) it follows $S_n \overset{a.s.}{\to} S$ by (14.24). □

The following theorem, in conjunction with (26.1), (14.9) and (19.24), shows that for series of independent random variables, convergence in distribution, convergence in probability, and almost sure convergence are equivalent.

Theorem 26.2. *Let F be a distribution function such that $S_n \overset{D}{\to} F$. Then $\{S_n : n \geqslant 1\}$ converges a.s.*

Proof. Let $t_0 > 0$ be such that $\varphi_F(t) \neq 0$ for $|t| \leqslant t_0$. Then $\lim_n \varphi_{S_n}(t)/\varphi_F(t) = 1$ uniformly in $|t| \leqslant t_0$ by the uniform convergence theorem (20.9). By (20.4), this entails that $\sup_{m>n} |1 - \varphi_{S_m - S_n}(t)| = \sup_{m>n} |1 - \varphi_{S_m}(t)/\varphi_{S_n}(t)| \to 0$ uniformly in $|t| \leqslant t_0$. Therefore, on account of (20.48.b), it follows that

$$\sup_{m>n} |1 - \varphi_{S_m - S_n}(t)| \to 0 \tag{1}$$

uniformly in $|t| \leqslant T$ for any $T > 0$. Then, for $m > n$ and $T > 0$, applying (20.65), we have

$$\frac{1}{2}P(|S_m - S_n| \geqslant \frac{2}{T}) \leqslant 1 - \frac{1}{2T}\left|\int_{-T}^{T} \varphi_{S_m - S_n}(t)dt\right|$$
$$\leqslant \left|1 - \frac{1}{2T}\int_{-T}^{T} \varphi_{S_m - S_n}(t)dt\right| \leqslant \frac{1}{2T}\int_{-T}^{T} |1 - \varphi_{S_m - S_n}(t)|\, dt,$$

and so

$$\sup_{m>n} P\left(|S_m - S_n| \geqslant \frac{2}{T}\right) \leqslant \frac{1}{T}\int_{-T}^{T} \sup_{m>n}\left|1 - \varphi_{S_m - S_n}(t)\right| dt, \quad T > 0.$$

Therefore, in view of (1) and (11.26), we see that $\sup_{m>n} P(|S_m - S_n| \geqslant 2/T) \to 0$ for any $T > 0$. This shows that $\{S_n : n \geqslant 1\}$ is Cauchy in probability (14.7.e), hence $\{\widetilde{S_n} : n \geqslant 1\}$ is a Cauchy sequence in (L^0, d_1) (15.25.c), where (L^0, d_1) is defined in (15.25). Since (L^0, d_1) is complete, there exists $S \in \mathfrak{L}^0$ such that $S_n \xrightarrow{P} S$ (see (15.25.d)). Thus $S_n \xrightarrow{a.s.} S$ by (26.1). □

Theorem 26.3 (Khintchine–Kolmogorov). *Assume that* $\sum_{n\geqslant 1} Ef_n^2 < \infty$ *and* $Ef_n = 0, n \geqslant 1$. *Then* $\{S_n : n \geqslant 1\}$ *converges a.s. and in* $\mathfrak{L}^2$, *and* $E(\sum_{n\geqslant 1} f_n)^2 = \sum_{n\geqslant 1} Ef_n^2$.

Proof. Since $E(S_m - S_n)^2 = \sum_{i=n+1}^{m} Ef_i^2, 1 \leqslant n < m$, and $\sum_{i\geqslant 1} Ef_i^2 < \infty$, it follows that $\{\widetilde{S_n} : n \geqslant 1\}$ is a Cauchy sequence in L^2. Hence, as L^2 is a complete metric space (13.16), there is $S \in \mathfrak{L}^2$ such that $S_n \xrightarrow{\mathfrak{L}^2} S$, and so $S_n \xrightarrow{P} S$ by (14.14). In view of Lévy's theorem (26.1), we have $S_n \xrightarrow{a.s.} S$. Then, on applying (14.30), we may write $E(\sum_{i\geqslant 1} f_i)^2 = E(\lim_n S_n)^2 = \lim_n E{S_n}^2 = \lim_n \sum_{i=1}^{n} Ef_i^2 = \sum_{i\geqslant 1} Ef_i^2$. □

The following remarkable inequalities are extremely useful tools in probability theory.

Theorem 26.4 (Kolmogorov's inequalities). *Let* $f_1, \ldots, f_n$ *be independent random variables, and let* $\varepsilon, c > 0$. *For* $1 \leqslant i \leqslant n$, *assume that* $Ef_i^2 < \infty$, $Ef_i = 0$ *and set* $S_i = f_1 + \cdots + f_i$.

(i) $P(\sup_{1\leqslant i\leqslant n} |S_i| \geqslant \varepsilon) \leqslant \varepsilon^{-2} \sum_{i=1}^{n} Ef_i^2$.

(ii) *If* $|f_i| \leqslant c < \infty, 1 \leqslant i \leqslant n$, *then* $P(\sup_{1\leqslant i\leqslant n} |S_i| \geqslant \varepsilon) \geqslant 1 - (\varepsilon + c)^2 / \sum_{i=1}^{n} Ef_i^2$.

Proof. Define the $\overline{R}$-valued random variable $\tau : X \to \{1, \ldots, n\} \cup \{\infty\}$ by $\tau = \inf\{i \leqslant n : |S_i| \geqslant \varepsilon\}$. For $1 \leqslant i \leqslant n$, the event $\{\tau = i\}$ belongs to $\sigma(f_1, \ldots, f_i)$ and $E(S_n - S_i) = 0$, and so $\int_{\{\tau=i\}} S_i(S_n - S_i)dP = 0$ by independence and (18.23.i).

Therefore, since $\{\tau \leqslant n\} = \cup_{i=1}^{n}\{\tau = i\} = \{\sup_{1\leqslant i\leqslant n} |S_i| \geqslant \varepsilon\}$, we have

$$\begin{aligned}
\sum\nolimits_{i=1}^{n} Ef_i^2 = E{S_n}^2 &\geqslant \int_{\{\tau\leqslant n\}} S_n^2 dP = \sum_{i=1}^{n} \int_{\{\tau=i\}} (S_i + (S_n - S_i))^2 dP \\
&= \sum_{i=1}^{n} \int_{\{\tau=i\}} S_i^2 dP + 2\sum_{i=1}^{n} \int_{\{\tau=i\}} S_i(S_n - S_i) dP \\
&\quad + \sum_{i=1}^{n} \int_{\{\tau=i\}} (S_n - S_i)^2 dP \geqslant \sum_{i=1}^{n} \int_{\{\tau=i\}} S_i^2 dP
\end{aligned}$$

$$\geqslant \varepsilon^2 \sum_{i=1}^{n} P(\tau = i) = \varepsilon^2 P(\sup_{1 \leqslant i \leqslant n} |S_i| \geqslant \varepsilon), \tag{1}$$

which is (i). If $|f_i| \leqslant c$, $1 \leqslant i \leqslant n$, then, since $S_i = S_{i-1} + f_i$, $1 \leqslant i \leqslant n$, it follows that $|S_i| < \varepsilon + c$ whenever $\tau = i$. Hence, on account of (1) and (18.23.i), we have

$$\int_{\{\tau \leqslant n\}} S_n^2 dP = \sum_{i=1}^{n} \int_{\{\tau = i\}} S_i^2 dP + \sum_{i=1}^{n} \int_{\{\tau = i\}} (S_n - S_i)^2 dP \leqslant (\varepsilon + c)^2 P(\tau \leqslant n)$$
$$+ \sum_{i=1}^{n} P(\tau = i) \sum_{i < j \leqslant n} Ef_j^2 \leqslant \left((\varepsilon + c)^2 + \sum_{j=1}^{n} Ef_j^2 \right) P(\tau \leqslant n). \tag{2}$$

On the other hand, we have

$$\int_{\{\tau \leqslant n\}} S_n^2 dP = ES_n^2 - \int_{\{\tau = \infty\}} S_n^2 dP \geqslant \sum_{i=1}^{n} Ef_i^2 - \varepsilon^2 P(\tau = \infty)$$
$$= \sum_{i=1}^{n} Ef_i^2 - \varepsilon^2 + \varepsilon^2 P(\tau \leqslant n). \tag{3}$$

Combining (2) with (3), we get $\sum_{i=1}^{n} Ef_i^2 - \varepsilon^2 + \varepsilon^2 P(\tau \leqslant n) \leqslant ((\varepsilon + c)^2 + \sum_{i=1}^{n} Ef_i^2) P(\tau \leqslant n)$, whence

$$P(\tau \leqslant n) \geqslant \frac{\sum_{i=1}^{n} Ef_i^2 - \varepsilon^2}{(\varepsilon + c)^2 + \sum_{i=1}^{n} Ef_i^2 - \varepsilon^2} \geqslant 1 - \frac{(\varepsilon + c)^2}{\sum_{i=1}^{n} Ef_i^2},$$

which is (ii). □

Remark 26.5. Since $\{|S_n| \geqslant \varepsilon\} \subset \{\sup_{1 \leqslant i \leqslant n} |S_i| \geqslant \varepsilon\}$, Kolmogorov's inequality (26.4.i) is an essential improvement of Chebyshev's inequality in this case.

Theorem 26.6. *Suppose that $|f_n| \leqslant c < \infty$ and $Ef_n = 0$ for $n \geqslant 1$, and that $\sum_{n \geqslant 1} f_n$ converges a.s. Then $\sum_{n \geqslant 1} Ef_n^2 < \infty$.*

Proof. For $\varepsilon > 0$, by Kolmogorov's inequality (26.4.ii), we have

$$P\left(\sup_{1 \leqslant i \leqslant n} |S_{m+i} - S_m| \geqslant \varepsilon \right) \geqslant 1 - \frac{(\varepsilon + c)^2}{\sum_{i=1}^{n} Ef_{m+i}^2}, \quad m, n \geqslant 1. \tag{1}$$

Assuming that $\sum_{n \geqslant 1} f_n^2 = \infty$ and letting $n \to \infty$ in (1) implies $\sup_{i \geqslant 1} |S_{m+i} - S_m| \geqslant \varepsilon$ a.s. for any $m \geqslant 1$. However, since $\sum_{n \geqslant 1} f_n$ converges a.s., it follows that $\sup_{i \geqslant 1} |S_{m+i} - S_m| \overset{a.s.}{\to} 0$ as $m \to \infty$. This contradiction proves the theorem. □

Theorem 26.7. *Suppose that $|f_n| \leqslant c < \infty$ for $n \geqslant 1$. Then $\sum_{n \geqslant 1} f_n$ converges a.s. if and only if $\sum_{n \geqslant 1}$ Var f_n and $\sum_{n \geqslant 1} Ef_n$ converge.*

Proof. Assume that $\sum_{n\geqslant 1} f_n$ converges a.s. Let $S(x) = \sum_{n\geqslant 1} f_n(x)$ or 0 according as $\sum_{n\geqslant 1} f_n(x)$ converges or does not converge, and let F denote the distribution function of S. Then $S_n \xrightarrow{D} F$ by (14.9) and (19.24). Let $\{f'_n : n \geqslant\}$ and $\{f''_n : n \geqslant 1\}$ be independent copies of $\{f_n : n \geqslant 1\}$, and set $f^*_n = f'_n - f''_n$, $S'_n = \sum_{k=1}^n f'_k$, $S''_n = \sum_{k=1}^n$, $n \geqslant 1$. Then $f^*_n, n \geqslant 1$, are independent and symmetric with $E^* f^*_n = 0$ by (23.20.d) and (20.42.c). Since $P^*(f'_n)^{-1}_{n\geqslant 1} = P^*(f''_n)^{-1}_{n\geqslant 1} = P(f_n)^{-1}_{n\geqslant 1}$, we see that $|f'_n|, |f''_n| \leqslant c$ P^*-a.s. for any $n \geqslant 1$, and so $|f^*_n| \leqslant 2c$ P^*-a.s. Moreover, as $P^*(S'_n \leqslant x) = P^*(S''_n \leqslant x) = P(S_n \leqslant x) \to F(x), x \in C(F)$, it follows that $S'_n \xrightarrow{D} F$ and $S''_n \xrightarrow{D} F$. In view of (26.2), we see that $\{S'_n : n \geqslant 1\}$ and $\{S''_n : n \geqslant 1\}$ converge P^*-a.s., and so $\sum_{n\geqslant 1} f^*_n = \sum_{n\geqslant 1} f'_n - \sum_{n\geqslant 1} f''_n$ converges P^*-a.s. Since $E^* f^{*2}_n = \text{Var}^* f'_n + \text{Var}^* f''_n = 2\text{Var}\, f_n, n \geqslant 1$, we have $2\sum_{n\geqslant 1} \text{Var}\, f_n = \sum_{n\geqslant 1} E f^{*2}_n < \infty$ by (26.6). Then, on account of the Khintchine–Kolmogorov theorem (26.3), it follows that $\sum_{n\geqslant 1}(f_n - Ef_n)$ converges a.s., and so $\sum_{n\geqslant 1} Ef_n = \sum_{n\geqslant 1} f_n - \sum_{n\geqslant 1}(f_n - Ef_n)$ converges.

Conversely, if $\sum_{n\geqslant 1} E(f_n - Ef_n)^2 < \infty$, and $\sum_{n\geqslant 1} Ef_n$ converges, then $\sum_{n\geqslant 1}(f_n - Ef_n)$ converges a.s. by (26.3), and so $\sum_{n\geqslant 1} f_n$ converges a.s. □

Definition 26.8. Two arbitrary sequences $\{f_n : n \geqslant 1\}$ and $\{g_n : n \geqslant 1\}$ of random variables are said to be *equivalent* if $\sum_{n\geqslant 1} P(f_n \neq g_n) < \infty$.

Lemma 26.9. *If $\{f_n : n \geqslant 1\}$ and $\{g_n : n \geqslant 1\}$ are equivalent sequences of random variables, then the following assertions hold.*

(i) *$\sum_{n\geqslant 1}(f_n - g_n)$ converges a.s.*

(ii) *If $0 < b_n \to \infty$, then $b_n^{-1}\sum_{k=1}^n (f_k - g_k) \xrightarrow{a.s.} 0$.*

(iii) *If $0 < b_n \to \infty$ and f is a random variable such that $b_n^{-1}\sum_{k=1}^n f_k \xrightarrow{P} f$, then $b_n^{-1}\sum_{k=1}^n g_k \xrightarrow{P} f$.*

Proof. Let $A = \limsup_n \{f_n \neq g_n\}$, so that $A^c = \liminf_n \{f_n = g_n\}$ (1.2.40.b). By the Borel-Cantelli lemma (18.24.i), we have $P(A) = 0$, and so $P(A^c) = 1$. This means that if $x \in A^c$, then $f_n(x) = g_n(x)$ for all but finitely many n (depending on x). Hence (i) and (ii) follow. As for (iii), write $b_n^{-1}\sum_{k=1}^n g_k = b_n^{-1}\sum_{k=1}^n (g_k - f_k) + b_n^{-1}\sum_{k=1}^n f_k$, and apply (14.23.a), noticing that $b_n^{-1}\sum_{k=1}^n (g_k - f_k) \xrightarrow{P} 0$ by (ii) and (14.9). □

We are now ready to prove the next important criterion due to A. N. Kolmogorov.

Three series theorem 26.10. *$\sum_{n\geqslant 1} f_n$ converges a.s. if and only if, for every $c > 0$,*

(i) *$\sum_{n\geqslant 1} P(|f_n| \geqslant c) < \infty$,*

(ii) *$\sum_{n\geqslant 1} E[f_n 1_{\{|f_n|<c\}}]$ converges*

and

(iii) *$\sum_{n\geqslant 1} Var[f_n 1_{\{|f_n|<c\}}] < \infty$.*

Proof. If $\sum_{n\geqslant 1} f_n$ converges a.s. and $c > 0$, then $f_n \xrightarrow{a.s.} 0$, and so $P(|f_n| \geqslant c \text{ i.o.}) = 0$ by (14.3). In view of the Borel-Cantelli lemma (18.24.ii), this implies (i). We see also that $\{\sum_{n\geqslant 1} f_n \text{ converges}\} \subset \{\sum_{n\geqslant 1} f_n 1_{\{|f_n|<c\}} \text{ converges}\}$, hence $\sum_{n\geqslant 1} f_n 1_{\{|f_n|<c\}}$ converges a.s. Then (ii) and (iii) follow at once from (26.7).

Conversely, assume that (i)–(iii) hold and let $c > 0$. Then (iii) implies that $\sum_{n\geqslant 1}(f_n 1_{\{|f_n|<c\}} - E[f_n 1_{\{|f_n|<c\}}])$ converges a.s. by the Khintchine-Kolmogorov theorem (26.3). This and (ii) show that $\sum_{n\geqslant 1} f_n 1_{\{|f_n|<c\}}$ converges a.s. According to (i), the sequences $\{f_n : n \geqslant 1\}$ and $\{f_n 1_{\{|f_n|<c\}} : n \geqslant 1\}$ are equivalent, and so $\sum_{n\geqslant 1} f_n 1_{\{|f_n|\geqslant c\}}$ converges a.s. by (26.9.i) Thus $\sum_{n\geqslant 1} f_n$ converges a.s. □

Remark 26.11. The proof of the converse part of the three series theorem shows that in order that $\sum_{n\geqslant 1} f_n$ converge a.s. it suffices that (i)–(iii) hold only for some $c > 0$.

The three series theorem has fruitful consequences as regards strong laws of large numbers to be discussed in the next section. Other consequences of it are presented in what follows.

Theorem 26.12. *If $Ef_n = 0$, $n \geqslant 1$, and $\sum_{n\geqslant 1} E[f_n^2 1_{\{|f_n|<c\}} + |f_n|\, 1_{\{|f_n|\geqslant c\}}] < \infty$ for some $c > 0$, then $\sum_{n\geqslant 1} f_n$ converges a.s.*

Proof. By Markov's inequality (13.8), we get

$$\sum_{n\geqslant 1} P(|f_n| \geqslant c) = \sum_{n\geqslant 1} P(|f_n|\, 1_{\{|f_n|\geqslant c\}} \geqslant c) \leqslant \frac{1}{c} \sum_{n\geqslant 1} E[|f_n|\, 1_{\{|f_n|\geqslant c\}}] < \infty.$$

As $Ef_n = 0$, $n \geqslant 1$, we obtain

$$\sum_{n\geqslant 1} \left|E[f_n 1_{\{|f_n|<c\}}]\right| = \sum_{n\geqslant 1} \left|E[f_n 1_{\{|f_n|\geqslant c\}}]\right| \leqslant \sum_{n\geqslant 1} E[|f_n|\, 1_{\{|f_n|\geqslant c\}}] < \infty.$$

Finally, we have

$$\sum_{n\geqslant 1} \mathrm{Var}[f_n 1_{\{|f_n|<c\}}] \leqslant \sum_{n\geqslant 1} E[f_n^2 1_{\{|f_n|<c\}}] < \infty.$$

Hence the result follows from (26.10). □

Theorem 26.13 (Loève).

(i) *If $\sum_{n\geqslant 1} E\,|f_n|^{r_n} < \infty$, where $0 < r_n \leqslant 1$, then $\sum_{n\geqslant 1} |f_n|$ converges a.s.*

(ii) *If $\sum_{n\geqslant 1} E\,|f_n|^{r_n} < \infty$, where $1 < r_n \leqslant 2$, and $Ef_n = 0$, then $\sum_{n\geqslant 1} f_n$ converges a.s.*

Proof. We always have $\sum_{n\geqslant 1} P(|f_n| \geqslant 1) \leqslant \sum_{n\geqslant 1} E\,|f_n|^{r_n} < \infty$ by (13.8). If $0 < r_n \leqslant 1$, then $f_n^2 1_{\{|f_n|<1\}} + |f_n|\, 1_{\{|f_n|<1\}} \leqslant 2\,|f_n|^{r_n}$, and so $\sum_{n\geqslant 1} E[|f_n|\, 1_{\{|f_n|<1\}}]$ and $\sum_{n\geqslant 1} \mathrm{Var}[f_n 1_{\{|f_n|<1\}}] < \infty$. Thus (i) is a consequence of (26.10). If $1 < r_n \leqslant 2$, then $f_n^2 1_{\{|f_n|<1\}} + |f_n|\, 1_{\{|f_n|\geqslant 1\}} \leqslant |f_n|^{r_n}$, and so (ii) follows from (26.12). □

Theorem 26.14 (Marcinkiewicz–Zygmund). *If f_n, $n \geqslant 1$, are i.i.d. random variables, then the following assertions hold.*

(i) *If $0 < p < 2$ and $E\,|f_1|^p < \infty$, then $\sum_{n\geqslant 1}(f_n - E[f_n 1_{\{|f_n|<n^{1/p}\}}])/n^{1/p}$ converges a.s.*

(ii) *If* $0 < p < 1$ *and* $E\,|f_1|^p < \infty$, *then* $\sum_{n\geqslant 1} f_n/n^{1/p}$ *converges a.s.*
(iii) *If* $E[|f_1|\log(1+|f_1|)] < \infty$ *and* $Ef_1 = 0$, *then* $\sum_{n\geqslant 1} f_n/n$ *converges a.s.*
(iv) *If* $1 < p < 2$, $E\,|f_1|^p < \infty$ *and* $Ef_1 = 0$, *then* $\sum_{n\geqslant 1} f_n/n^{1/p}$ *converges a.s.*

Proof. Assume first $0 < p < 2$ and $E\,|f_1|^p < \infty$, and let $r > p$. Then, on account of Fubini's theorem (15.8) and of the inequality $\sum_{n>m} n^{-\alpha} \leqslant m^{-\alpha+1}/(\alpha-1), \alpha > 1, m \geqslant 1$, we have

$$\begin{aligned}\sum_{n\geqslant 1}\frac{1}{n^{r/p}}E[|f_n|^r\,1_{\{|f_n|<n^{1/p}\}}] &= \sum_{n\geqslant 1}\frac{1}{n^{r/p}}E[|f_1|^r\,1_{\{|f_1|<n^{1/p}\}}]\\ &= E\left[|f_1|^r\sum_{n\geqslant 1}\frac{1}{n^{r/p}}1_{\{|f_1|<n^{1/p}\}}\right]\\ &= E\left[|f_1|^r\sum_{n>|f_1|^p}\frac{1}{n^{r/p}}\right]\\ &\leqslant \frac{p}{r-p}E[|f_1|^r\,(|f_1|^p)^{-r/p+1}]\\ &= \frac{p}{r-p}E[|f_1|^p] < \infty. \qquad (1)\end{aligned}$$

Specializing (1) at $r = 2$, yields

$$\sum_{n\geqslant 1}(f_n 1_{\{|f_n|<n^{1/p}\}} - E[f_n 1_{\{|f_n|<n^{1/p}\}}])/n^{1/p} \text{ converges a.s.} \qquad (2)$$

by the Khintchine–Kolmogorov theorem (26.3). Since $E\,|f_1|^p < \infty$ implies that $\sum_{n\geqslant 1} P(f_n \neq f_n 1_{\{|f_n|<n^{1/p}\}}) = \sum_{n\geqslant 1} P(|f_n| \geqslant n^{1/p}) = \sum_{n\geqslant 1} P(|f_1|^p \geqslant n) < \infty$ by (13.27.a), the sequences $\{f_n/n^{1/p} : n \geqslant 1\}$ and $\{f_n 1_{\{|f_n|<n^{1/p}\}}/n^{1/p} : n \geqslant 1\}$ are equivalent, and so

$$\sum_{n\geqslant 1}(f_n - f_n 1_{\{|f_n|<n^{1/p}\}})/n^{1/p} \text{ converges a.s.} \qquad (3)$$

by (26.9.i). Thus (i) follows from (2) and (3). If $0 < p < 1$ and $E\,|f_1|^p < \infty$, then (1) with $r = 1$ shows that $\sum_{n\geqslant 1} E[f_n 1_{\{|f_n|<n^{1/p}\}}]/n^{1/p}$ converges. This and (i) prove (ii). Assume now that $E[|f_1|\log(1+|f_1|)] < \infty$ and $Ef_1 = 0$. Then, by virtue of (15.8) and of the inequality $\sum_{n=1}^{m}\frac{1}{n} \leqslant 2\log(1+m),\ m \geqslant 1$, we have

$$\begin{aligned}\sum_{n\geqslant 1}\frac{1}{n}\left|E[f_n 1_{\{|f_n|<n\}}]\right| &= \sum_{n\geqslant 1}\frac{1}{n}\left|E[f_1 1_{\{|f_1|\geqslant n\}}]\right| \leqslant E\left[|f_1|\sum_{n\geqslant 1}\frac{1}{n}1_{\{|f_1|\geqslant n\}}\right]\\ &= E\left[|f_1|\,1_{\{|f_1|\geqslant 1\}}\sum_{1\leqslant n\leqslant |f_1|}\frac{1}{n}\right]\\ &\leqslant 2E[|f_1|\log(1+|f_1|)] < \infty. \qquad (4)\end{aligned}$$

Since $E[|f_1|\log(1+|f_1|)] < \infty$ implies $E|f_1| < \infty$, (iii) follows from (i) with $p = 1$ and (4). Finally, assume that $1 < p < 2$, $E|f_1|^p < \infty$ and $Ef_1 = 0$. Similarly, as $\sum_{n=1}^{m} n^{-\alpha} \leqslant m^{1-\alpha}/(1-\alpha)$, $0 < \alpha < 1$, $m \geqslant 1$, we obtain

$$\begin{aligned}\sum_{n\geqslant 1} \frac{1}{n^{1/p}} \left|E\left[f_n 1_{\{|f_n|<n^{1/p}\}}\right]\right| &\leqslant E\left[|f_1|\, 1_{\{|f_1|\geqslant 1\}} \sum_{1\leqslant n\leqslant |f_1|^p} \frac{1}{n^{1/p}}\right] \\ &\leqslant \frac{p-1}{p} E[|f_1|\,(|f_1|^p)^{1-1/p}] \\ &= \frac{p-1}{p} E[|f_1|^p] < \infty. \end{aligned} \tag{5}$$

Now (iv) follows from (i) and (5). □

Theorem 26.15. *For $n \geqslant 1$, let $h_n : R \to [0, \infty[$ be an even function such that $0 < h_n(x_1) \leqslant h_n(x_2)$ and $x_1/h_n(x_1) \leqslant x_2/h_n(x_2)$ whenever $0 < x_1 < x_2$, and let $b_n > 0$ be such that $\sum_{n\geqslant 1} E[h_n \circ f_n]/h_n(b_n) < \infty$. Then $\sum_{n\geqslant 1} f_n/b_n$ converges a.s.*

Proof. By (13.25.c), we have

$$\sum_{n\geqslant 1} P(|f_n| \geqslant b_n) \leqslant \sum_{n\geqslant 1} \frac{E[h_n \circ f_n]}{h_n(b_n)} < \infty. \tag{1}$$

We have also

$$\sum_{n\geqslant 1} E\left[\frac{|f_n|}{b_n} 1_{\{|f_n|<b_n\}}\right] \leqslant \sum_{n\geqslant 1} E\left[\frac{h_n \circ f_n}{h_n(b_n)} 1_{\{|f_n|<b_n\}}\right] \leqslant \sum_{n\geqslant 1} \frac{E[h_n \circ f_n]}{h_n(b_n)} < \infty. \tag{2}$$

If $|x| < b_n$, then $x^2/b_n^2 \leqslant h_n^2(x)/h_n^2(b_n) \leqslant h_n(x)/h_n(b_n)$, and so

$$\sum_{n\geqslant 1} E\left[\frac{f_n^2}{b_n^2} 1_{\{|f_n|<b_n\}}\right] \leqslant \sum_{n\geqslant 1} E\left[\frac{h_n \circ f_n}{h_n(b_n)} 1_{\{|f_n|<b_n\}}\right] \leqslant \sum_{n\geqslant 1} \frac{E[h_n \circ f_n]}{h_n(b_n)} < \infty. \tag{3}$$

From (1)–(3), on account of the three series theorem (26.10), it follows that $\sum_{n\geqslant 1} f_n/b_n$ converges a.s. □

Corollary 26.16. *If $\{b_n : n \geqslant 1\} \subset]0, \infty[$ is such that $\sum_{n\geqslant 1} E|f_n|^p/b_n^p < \infty$ for some $p \in]0, 1]$, then $\sum_{n\geqslant 1} f_n/b_n$ converges a.s.*

Theorem 26.17. *For $n \geqslant 1$, assume that $Ef_n = 0$, let $h_n : R \to [0, \infty[$ be an even function such that $0 < h_n(x_1) \leqslant h_n(x_2)$, $x_1/h_n(x_1) \geqslant x_2/h_n(x_2)$ and $h_n(x_1)/x_1^2 \geqslant h_n(x_2)/x_2^2$ whenever $0 < x_1 < x_2$, and let $b_n > 0$ be such that $\sum_{n\geqslant 1} E[h_n \circ f_n]/h_n(b_n) < \infty$. Then $\sum_{n\geqslant 1} f_n/b_n$ converges a.s.*

Proof. Since $Ef_n = 0, n \geqslant 1$, we obtain

$$\sum_{n\geqslant 1}\left|E\left[\frac{f_n}{b_n}1_{\{|f_n|<b_n\}}\right]\right| \leqslant \sum_{n\geqslant 1} E\left[\frac{|f_n|}{b_n}1_{\{|f_n|\geqslant b_n\}}\right]$$
$$\leqslant \sum_{n\geqslant 1} E\left[\frac{h_n\circ f_n}{h_n(b_n)}1_{\{|f_n|\geqslant b_n\}}\right] \leqslant \sum_{n\geqslant 1}\frac{E[h_n\circ f_n]}{h_n(b_n)} < \infty. \tag{1}$$

If $|x| < b_n$, then $x^2/b_n^2 \leqslant h_n(x)/h_n(b_n)$, and so

$$\sum_{n\geqslant 1} E\left[\frac{f_n^2}{b_n^2}1_{\{|f_n|<b_n\}}\right] < \infty \tag{2}$$

as above. From (1) of (26.15), (1) and (2), in view of (26.10), we see that $\sum_{n\geqslant 1} f_n/b_n$ converges a.s. □

Corollary 26.18. *If $Ef_n = 0, n \geqslant 1$, and $\{b_n : n \geqslant 1\} \subset]0,\infty[$ is such that $\sum_{n\geqslant 1} E|f_n|^p/b_n^p < \infty$ for some $p \in]1,2]$, then $\sum_{n\geqslant 1} f_n/b_n$ converges a.s.*

Remark 26.19. Let $f_n, n \geqslant 1$, be independent random variables such that $P(f_n = 0) = 1 - P(f_n = 1) = n^{-2}, n \geqslant 1$. Then $Ef_n = 1 - n^{-2}$ and $\operatorname{Var} f_n = n^{-2}(1 - n^{-2})$, and so $\sum_{n\geqslant 1} Ef_n = \infty$ and $\sum_{n\geqslant 1}\operatorname{Var} f_n < \infty$. It follows from (26.3) that $\sum_{n\geqslant 1}(f_n - Ef_n)$ converges a.s. At the same time $\sum_{n\geqslant 1} f_n$ is almost surely divergent. This situation motivates the next definition.

Definition 26.20. We say that the series $\sum_{n\geqslant 1} f_n$ is *essentially convergent* if there exists a sequence $\{a_n : n \geqslant 1\} \subset R$ such that $\sum_{n\geqslant 1}(f_n - a_n)$ converges a.s.

Theorem 26.21. *$\sum_{n\geqslant 1} f_n$ is essentially convergent if and only if $\sum_{n\geqslant 1} f_n^*$ converges a.s., where $\{f_n^* : n \geqslant 1\}$ is the symmetrized $\{f_n : n \geqslant 1\}$.*

Proof. If $\sum_{n\geqslant 1} f_n$ is essentially convergent and $\sum_{n\geqslant 1}(f_n - a_n)$ converges a.s. for some constants $a_n \in R, n \geqslant 1$, let $S(x) = \sum_{n\geqslant 1}(f_n(x) - a_n)$ or 0 according as $\sum_{n\geqslant 1}(f_n(x) - a_n)$ converges or does not converge, and let F denote the distribution function of S. Let $\{f_n' : n \geqslant 1\}, \{f_n'' : n \geqslant 1\}, \{S_n' : n \geqslant 1\}$ and $\{S_n'' : n \geqslant 1\}$ be as in the proof of (26.7), and put $\alpha_n = \sum_{k=1}^n a_k, n \geqslant 1$. Since $S_n - \alpha_n \xrightarrow{D} F$ by (14.9) and (19.24), and $P^*(S_n' - \alpha_n \leqslant x) = P^*(S_n'' - \alpha_n \leqslant x) = P(S_n - \alpha_n \leqslant x), x \in R$, it follows that $S_n' - \alpha_n \xrightarrow{D} F$ and $S_n'' - \alpha_n \xrightarrow{D} F$. Hence, by virtue of (26.2), $\{S_n' - \alpha_n : n \geqslant 1\}$ and $\{S_n'' - \alpha_n : n \geqslant 1\}$ converge P^*-a.s., and so $\sum_{n\geqslant 1} f_n^* = \sum_{n\geqslant 1}(f_n' - a_n) - \sum_{n\geqslant 1}(f_n'' - a_n)$ converges P^*-a.s.

Conversely, assume that $\sum_{n\geqslant 1} f_n^*$ converges a.s., and let $c > 0$. Then, by (25.1.ii) and the three series theorem (26.10),

$$\sum_{n\geqslant 1} P(|f_n - m[f_n]| \geqslant c) \leqslant 2P^*(|f_n^*| \geqslant c) < \infty. \tag{1}$$

This entails that

$$\sum_{n \geqslant 1} (f_n - m[f_n]) 1_{\{|f_n - m[f_n]| \geqslant c\}} \text{ converges a.s.} \tag{2}$$

by (26.9.i). Also by (26.10), we have

$$\sum_{n \geqslant 1} E^*[f_n^{*2} 1_{\{|f\_n| < c\}}] = \sum_{n \geqslant 1} \mathrm{Var}^*[f_n^* 1_{\{|f_n| < c\}}] < \infty, \tag{3}$$

since (20.42.e) and (20.42.c) show that $E^*[f_n^* 1_{\{|f_n|<c\}}] = 0, n \geqslant 1$. On account of (15.23.b) and (25.1.ii), we can write

$$\begin{aligned} E[(f_n - m[f_n])^2 1_{\{|f_n - m[f_n]| < c\}}] &\leqslant 2 \int_0^c x P(|f_n - m[f_n]| \geqslant x) dx \\ &\leqslant 4 \int_0^c x P^*(|f_n^*| \geqslant x) dx = 2E^*[f_n^{*2} 1_{\{|f_n|<c\}}] \\ &\quad + 2c^2 P^*(|f_n^*| \geqslant c), \quad n \geqslant 1. \end{aligned} \tag{4}$$

Consequently, in view of (3) and (1), we get

$$\sum_{n \geqslant 1} \mathrm{Var}[(f_n - m[f_n]) 1_{\{|f_n - m[f_n]| < c\}}] \leqslant \sum_{n \geqslant 1} E[(f_n - m[f_n])^2 1_{\{|f_n - m[f_n]| < c\}}] < \infty.$$

This implies, via the Khintchine–Kolmogorov theorem (26.3), that

$$\sum_{n \geqslant 1} ((f_n - m[f_n]) 1_{\{|f_n - m[f_n]| < c\}} - E[(f_n - m[f_n]) 1_{\{|f_n - m[f_n]| < c\}}]) \text{ converges a.s.} \tag{5}$$

From (2) and (5) we see that $\sum_{n \geqslant 1} f_n$ is essentially convergent with centering constants $a_n = m[f_n] + E[(f_n - m[f_n]) 1_{\{|f_n - m[f_n]| < c\}}], n \geqslant 1$. □

The above proof shows in fact we established the following criterion characterizing essential convergence.

Two series theorem 26.22. *$\sum_{n \geqslant 1} f_n$ is essentially convergence if and only if, for any $c > 0$,*

(i) $\sum_{n \geqslant 1} P(|f_n - m[f_n]| \geqslant c) < \infty$

and

(iii) $\sum_{n \geqslant 1} \mathit{Var}[(f_n - m[f_n]) 1_{\{|f_n - m[f_n]| < c\}}] < \infty$.

Remarks 26.23. (a) The proof of the converse part of (29.21) shows that in order that $\sum_{n \geqslant 1} f_n$ be essentially convergent it suffices that (i) and (ii) hold only for some $c > 0$.

(b) If $\sum_{n \geqslant 1} f_n$ is essentially convergent, then $\sum_{n \geqslant 1} (f_n - m[f_n] - E[(f_n - m[f_n]) 1_{\{|f_n - m[f_n]| < c\}}])$ converges a.s. for all $c > 0$.

Exercise 26.24. Assume that $|f_n| \leqslant c$ and $Ef_n = 0$ for $n \geqslant 1$, and that $\{S_n : n \geqslant 1\}$ converges in distribution. Prove that $\{S_n : n \geqslant 1\}$ converges in $\mathfrak{L}^2$. The following steps may be helpful.

(a) $\{S_n : n \geqslant 1\}$ converges in $\mathfrak{L}^2$ if and only if $\sum_{n\geqslant 1} Ef_n^2 < \infty$.

(b) If $\sum_{n\geqslant 1} Ef_n^2 = \infty$, then $P(|S_n| \geqslant a) \to 1$ for any $a > 0$. [Hints. For sufficiently large n and $x \geqslant 0$, let $s_n^2 = \sum_{i=1}^n Ef_i^2$, $\Psi_n(x) = P(|S_n/s_n| \geqslant x)$ and $\Psi(x) = 1 - \Phi(x) + \Phi(-x)$. For $0 < \varepsilon < 1$, since $\Psi_n \xrightarrow{u} \Psi$ by (24.25) and (19.18.d), choose $n_1(\varepsilon)$ such that $\Psi_n > \Psi - \varepsilon/2, n \geqslant n_1(\varepsilon)$. Select $x_\varepsilon > 0$ so that $\Psi(x_\varepsilon) > 1 - \varepsilon/2$, then choose $n_2(\varepsilon)$ such that $s_n x_\varepsilon \geqslant a, n \geqslant n_2(\varepsilon)$. Show that $P(|S_n| \geqslant a) > 1 - \varepsilon, n \geqslant n_1(\varepsilon) \vee n_2(\varepsilon)$.]

(c) Apply (19.18.f).

Exercise 26.25. Suppose that $f_n \geqslant 0$ for any $n \geqslant 1$. Show that $\sum_{n\geqslant 1} f_n$ converges a.s. if and only if $\sum_{n\geqslant 1} E[f_n \wedge 1] < \infty$. [Hint. Using the inequalities $1 - x \leqslant e^{-x} \leqslant 1 - (1 - e^{-1})x, 0 \leqslant x \leqslant 1$, verify that $E[\exp(-\sum_{n\geqslant 1}(f_n \wedge 1))] \leqslant \exp(1 - (1 - e^{-1})\sum_{n\geqslant 1} E[f_n \wedge 1])$.]

Exercise 26.26. Assume that $f_n, n \geqslant 1$, are nonnegative i.i.d. random variables with $Ef_1 < \infty$, and let $\{a_n : n \geqslant 1\} \subset R$ be such that $n\,|a_n| \leqslant C < \infty$ for any $n \geqslant 1$ and $\sum_{n\geqslant 1} a_n$ converges. Prove that $\sum_{n\geqslant 1} a_n f_n$ converges a.s. The next steps may be useful.

(a) The sequences $\{a_n f_n : n \geqslant 1\}$ and $\{a_n f_n 1_{\{f_n<n\}} : n \geqslant 1\}$ are equivalent.

(b) $\sum_{n\geqslant 1} E[a_n f_n 1_{\{f_n<n\}}]$ converges. [Use the monotone convergence theorem (11.22) and Abel's test of convergence.]

(c) $\sum_{n\geqslant 1} \mathrm{Var}[a_n f_n 1_{\{f_n<n\}}] < \infty$. [Hint. By Fubini's theorem (15.8), write

$$\sum_{n\geqslant 1} a_n^2 E\left[f_n^2 1_{\{f_n<n\}}\right] \leqslant C^2 E\left[f_1^2 \sum_{n\geqslant 1} \frac{1}{n^2} 1_{\{f_n<n\}}\right] \leqslant C^2 Ef_1 < \infty$$

as in the proof of (25.19).]

(d) $\sum_{n\geqslant 1} a_n f_n$ converges a.s. [Apply the Khintchine–Kolmogorov theorem (26.3) and (26.9.i).]

Exercise 26.27. Let $f_1, \ldots, f_n$ be independent random variables, and let $\varepsilon > 0$. For $1 \leqslant i \leqslant n$ and $p \geqslant 1$, assume that $E\,|f_i|^p < \infty$, $Ef_i = 0$, and set $S_i = f_1 + \cdots + f_i$. Show that $\varepsilon^p P(A) \leqslant E[|S_n|^p 1_A] \leqslant E\,|S_n|^p$, where $A = \{\sup_{1\leqslant i\leqslant n} |S_i| \geqslant \varepsilon\}$. [Hint. If τ is as in the proof of (26.4), utilizing (18.57), write $E[|S_n|^p 1_A] = \sum_{i=1}^n E[|S_i + (S_n - S_i)|^p 1_{\{\tau=i\}}] \geqslant \sum_{i=1}^n E[|S_i|^p 1_{\{\tau=i\}}] \geqslant \varepsilon^p P(A)$.]

Exercise 26.28. In order that $\sum_{n\geqslant 1} f_n^2 < \infty$ a.s. it is necessary that $\sum_{n\geqslant 1} E[f_n^2/(c^2 + f_n^2)] < \infty$ for every $c > 0$, and it is sufficient that $\sum_{n\geqslant 1} E[f_n^2/(c^2 + f_n^2)] < \infty$ for some $c > 0$. [Use (25.10) and (26.10).]

Exercise 26.29. Prove that $\sum_{n\geqslant 1} f_n$ converges a.s. whenever $\sum_{n\geqslant 1} f_n^2 < \infty$ a.s. and $\sum_{n\geqslant 1} E[f_n 1_{\{f_n<n\}}]$ converges for some $c > 0$.

Exercise 26.30. Assume that $f_n, n \geqslant 1$, are symmetric i.i.d. random variables with $E\,|f_1| < \infty$. Show that $\sum_{n\geqslant 1} f_n/n$ converges a.s.

Exercise 26.31. For $n \geqslant 1$, suppose f_n has an exponential distribution function with parameter λ_n:

(a) If $\sum_{n\geqslant 1} \lambda_n^{-1} < \infty$ (independence of $f_n, n \geqslant 1$, is not assumed), then $\sum_{n\geqslant 1} f_n < \infty$ a.s. [Use (15.8).]

(b) If $\sum_{n\geqslant 1} \lambda_n^{-1} = \infty$ and $f_n, n \geqslant 1$, are independent, then $\sum_{n\geqslant 1} f_n = \infty$ a.s. [Compute $E[\exp(-\sum_{i=1}^{n} f_i)]$.]

27 Strong Laws of Large Numbers

Throughout this section, if not otherwise stated, $b_n, n \geqslant 1$, are positive numbers such that $b_n \leqslant b_{n+1} \to \infty$, and we shall seek conditions under which $S_n/b_n - a_n \overset{a.s.}{\to} 0$ for some sequence $\{a_n : n \geqslant 1\} \subset R$.

Definitions 27.1. The sequence $\{f_n : n \geqslant 1\}$ is said to obey the *strong law of large numbers* (SLLN), with norming constants $b_n, n \geqslant 1$, and centering constants $a_n, n \geqslant 1$, where $\{a_n : n \geqslant 1\} \subset R$, if $S_n/b_n - a_n \overset{a.s.}{\to} 0$. The numbers b_n and a_n are said to *stabilize* S_n.

Remarks 27.2. (a) For an arbitrary sequence $\{S_n : n \geqslant 1\}$ of random variables, there always exist norming constants $b_n, n \geqslant 1$, for which $S_n/b_n \overset{a.s.}{\to} 0$. Actually, for $n \geqslant 1$, choose $b_n > 0$ such that $P(|S_n| \geqslant b_n/n) \leqslant 2^{-n}$. Then it follows from the Borel-Cantelli lemma (18.24.i) that $S_n/b_n \overset{a.s.}{\to} 0$. Consequently, to get more precise SLLNs, it is desirable to find norming constants with small order of magnitude. This problem leads to the law of the iterated logarithm to be studied in Section 28.

(b) For $n \geqslant 1$, let S_n^* be as in the proof of (25.17), and let $a_n \in R$. According to (25.2.ii) and (23.18.c), $S_n/b_n - a_n \overset{a.s.}{\to} 0$ if and only if $S_n^*/b_n \overset{a.s.}{\to} 0$ and $m[S_n]/b_n - a_n \to 0$. Therefore, since $S_n/b_n - a_n \overset{a.s.}{\to} 0$ implies $S_n/b_n - m[S_n]/b_n \overset{a.s.}{\to} 0$, we see that the problem of determining appropriate norming constants b_n may be studied independently of the constants a_n.

To obtain several SLLNs on account of results concerning convergence of series in Section 26, we need the next elementary lemmas.

Toeplitz's lemma 27.3. *Let $\{a_{nk} : n \geqslant 1, 1 \leqslant k \leqslant k_n\}$, where $k_n \to \infty$, be an array of real numbers such that $a_{nk} \to 0$ for each $k \geqslant 1$ and $\sum_{k=1}^{k_n} |a_{nk}| \leqslant c$ for any $n \geqslant 1$, where $c < \infty$, let $\{x_n : n \geqslant 1\} \subset R$, and set $u_n = \sum_{k=1}^{k_n} a_{nk}x_k, n \geqslant 1$. Then the following assertions hold.*

(i) *If $x_n \to 0$, then $u_n \to 0$.*

(ii) *If $\sum_{k=1}^{k_n} a_{nk} \to 1$ and $x_n \to x \in R$, then $u_n \to x$.*

Proof. Let $\varepsilon > 0$. If $x_n \to 0$, choose $n_1(\varepsilon) \in N$ such that $|x_n| < \varepsilon$ for $n \geqslant n_1(\varepsilon)$, then choose $n_2(\varepsilon)$ such that $k_n \geqslant n_1(\varepsilon)$ for $n \geqslant n_2(\varepsilon)$. Putting $n(\varepsilon) = n_1(\varepsilon) \vee n_2(\varepsilon)$, we have $|u_n| \leqslant \sum_{1\leqslant k<n(\varepsilon)} |a_{nk}x_k| + c\varepsilon, n \geqslant n(\varepsilon)$, whence $\limsup_n |u_n| \leqslant c\varepsilon$. This

shows that $u_n \to 0$, and so (i) is proved. If $\sum_{k=1}^{k_n} a_{nk} \to 1$ and $x_n \to x \in R$, then $u_n = \left(\sum_{k=1}^{k_n} a_{nk}\right) x + \sum_{k=1}^{k_n} a_{nk}(x_k - x) \to x$ by (i). □

Corollary 27.4. *Let $\{a_n : n \geqslant 1\} \subset [0, \infty[$, set $b_n = \sum_{k=1}^{n} a_k, n \geqslant 1$, assume that $b_n \to \infty$, and let $\{x_n : n \geqslant 1\} \subset R$ be such that $x_n \to x \in R$. Then $b_n^{-1} \sum_{k=1}^{n} a_k x_k \to x$.*

Proof. This is a consequence of (27.3.ii) for $a_{nk} = a_k/b_n, n \geqslant 1, 1 \leqslant k \leqslant n$. □

Kronecker's lemma 27.5. *If $0 < b_n \leqslant b_{n+1} \to \infty$ and $\{y_n : n \geqslant 1\} \subset R$ is such that $\sum_{n\geqslant 1} y_n$ converges, then $b_n^{-1} \sum_{k=1}^{n} b_k y_k \to 0$.*

Proof. Write $b_0 = x_0 = 0, a_n = b_n - b_{n-1}, n \geqslant 1, x_n = \sum_{k=1}^{n} y_k, n \geqslant 1$, and $x = \sum_{n\geqslant 1} y_n$. For $n \geqslant 1$, we have

$$\sum_{k=1}^{n} b_k y_k = \sum_{k=1}^{n}\sum_{j=1}^{k} a_j y_k = \sum_{j=1}^{n}\sum_{k=j}^{n} a_j y_k = \sum_{j=1}^{n} a_j(x_n - x_{j-1}) = b_n x_n - \sum_{j=1}^{n} a_j x_{j-1},$$

and so $b_n^{-1} \sum_{k=1}^{n} b_k y_k = x_n - b_n^{-1} \sum_{j=1}^{n} a_j x_{j-1} \to x - x = 0$ by (27.4). □

Theorem 27.6 (Loève). *For $n \geqslant 1$, let $0 < r_n \leqslant 2$, and assume that $Ef_n = 0$ whenever $1 < r_n \leqslant 2$. If $\sum_{n\geqslant 1} E\,|f_n|^{r_n} / b_n^{r_n} < \infty$, then $S_n/b_n \overset{a.s.}{\to} 0$.*

Proof. According to (26.13), $\sum_{n\geqslant 1} f_n/b_n$ converges a.s. This entails, on applying Kronecker's lemma with $y_n = f_n/b_n$, that $S_n/b_n \overset{a.s.}{\to} 0$. □

Corollary 27.7 (Kolmogorov). *If $\sum_{n\geqslant 1} \mathrm{Var}[f_n]/n^2 < \infty$, then $(S_n - ES_n)/n \overset{a.s.}{\to} 0$.*

Theorem 27.8. *For each $n \geqslant 1$, let $h_n : R \to [0, \infty[$ satisfy either the conditions of (26.15), or the conditions of (26.17) and $Ef_n = 0$. If $\sum_{n\geqslant 1} E[h_n \circ f_n]/h_n(b_n) < \infty$, then $S_n/b_n \overset{a.s.}{\to} 0$.*

Proof. In view of (26.15) and (26.17), $\sum_{n\geqslant 1} f_n/b_n$ converges a.s. Then $S_n/b_n \overset{a.s.}{\to} 0$ obtains by (27.5). □

Corollary 27.9. (i) *If $\sum_{n\geqslant 1} E\,|f_n|^p / b_n^p < \infty$ for some $p \in]0, 1]$, then $S_n/b_n \overset{a.s.}{\to} 0$.*

(ii) *If $Ef_n = 0, n \geqslant 1$, and $\sum_{n\geqslant 1} E\,|f_n|^p / b_n^p < \infty$ for some $p \in]1, 2]$, then $S_n/b_n \overset{a.s.}{\to} 0$.*

Proof. (i) follows from (26.16) and (27.5), and (ii) follows from (26.18) and (27.5). □

Theorem 27.10. *If $\sum_{n\geqslant 1} E[f_n^2/(b_n^2 + f_n^2)] < \infty$, then $b_n^{-1} \sum_{k=1}^{n} (f_k - E[f_k 1_{\{|f_k|<b_k\}}]) \overset{a.s.}{\to} 0$.*

Proof. By virtue of (25.10), $\sum_{n\geqslant 1} E[f_n^2/(b_n^2 + f_n^2)] < \infty$ is equivalent to

$$\sum_{n\geqslant 1} P(|f_n| \geqslant b_n) < \infty \tag{1}$$

and

$$\sum_{n \geqslant 1} \frac{1}{b_n^2} E[f_n^2 1_{\{|f_n|<b_n\}}] < \infty. \tag{2}$$

It follows from (2) and the Khintchine–Kolmogorov theorem (26.3) that

$$\sum_{n \geqslant 1} \frac{1}{b_n} (f_n 1_{\{|f_n|<b_n\}} - E[f_n 1_{\{|f_n|<b_n\}}]) \text{ converges a.s.} \tag{3}$$

Since the sequences $\{f_n : n \geqslant 1\}$ and $\{f_n 1_{\{|f_n|<b_n\}} : n \geqslant 1\}$ are equivalent by (1), (26.9.i) shows that

$$\sum_{n \geqslant 1} \frac{1}{b_n} (f_n - f_n 1_{\{|f_n|<b_n\}}) \text{ converges a.s.} \tag{4}$$

From (3) and (4) it follows that $\sum_{n \geqslant 1} b_n^{-1}(f_n 1_{\{|f_n|<b_n\}} - E[f_n 1_{\{|f_n|<b_n\}}])$ converges a.s., and so $b_n^{-1} \sum_{k=1}^n (f_k - E[f_k 1_{\{|f_k|<b_k\}}]) \xrightarrow{a.s.} 0$ by Kronecker's lemma (27.5). □

Corollary 27.11. *If* $\sum_{n \geqslant 1} P(|f_n| \geqslant b_n) < \infty$, $b_n^{-1} \sum_{k=1}^n E[f_k 1_{\{|f_k|<b_k\}}] \to 0$ *and* $\sum_{n \geqslant 1} b_n^{-2} E[f_n^2 1_{\{|f_n|<b_n\}}] < \infty$, *then* $S_n/b_n \xrightarrow{a.s.} 0$.

Theorem 27.12. *Let* $b_n, n \geqslant 1$, *be positive numbers such that* $b_{n-1} = O(b_n)$, *and assume that* $S_n/b_n \xrightarrow{a.s.} 0$. *Then* $\sum_{n \geqslant 1} P(|f_n| \geqslant b_n) < \infty$.

Proof. As there is $M < \infty$ with $b_{n-1}/b_n \leqslant M$ for all $n \geqslant 2$, we have

$$\frac{f_n}{b_n} = \frac{S_n}{b_n} - \frac{b_{n-1}}{b_n} \frac{S_{n-1}}{b_{n-1}} \xrightarrow{a.s.} 0. \tag{1}$$

Assuming $\sum_{n \geqslant 1} P(|f_n| \geqslant b_n) = \infty$ implies $P(|f_n| \geqslant b_n \text{ i.o.}) = 1$ by the Borel-Cantelli lemma (18.24.ii), which contradicts (1). □

We now prove some results on SLLN when $f_n, n \geqslant 1$, are i.i.d. random variables.

Theorem 27.13 (Feller). *Assume that* $f_n, n \geqslant 1$, *are i.i.d. with*

$$\sum_{n \geqslant 1} P(|f_1| \geqslant b_n) < \infty, \tag{1}$$

and suppose there is $C < \infty$ *such that*

$$b_n^2 \sum_{k \geqslant n} \frac{1}{b_k^2} \leqslant Cn, \quad n \geqslant 1. \tag{2}$$

If either f_1 *is symmetric or*

$$\frac{b_m}{m} \leqslant C \frac{b_n}{n}, \quad m \geqslant n \geqslant 1, \text{ and } Ef_1 = 0, \tag{3}$$

then $S_n/b_n \xrightarrow{a.s.} 0$.

Proof. Putting $b_0 = 0$, and making use of (2) and (1), we have

$$\begin{aligned}\sum_{n\geqslant 1}\frac{1}{b_n^2}E[f_n^2 1_{\{|f_n|<b_n\}}] &= \sum_{n\geqslant 1}\frac{1}{b_n^2}\sum_{k=1}^{n}E[f_1^2 1_{\{b_{k-1}\leqslant|f_1|<b_k\}}]\\ &= \sum_{k\geqslant 1}E[f_1^2 1_{\{b_{k-1}\leqslant|f_1|<b_k\}}]\sum_{n\geqslant k}\frac{1}{b_n^2}\\ &\leqslant C\sum_{k\geqslant 1}kE\left[\frac{f_1^2}{b_k^2}1_{\{b_{k-1}\leqslant|f_1|<b_k\}}\right]\\ &\leqslant C\sum_{k\geqslant 1}kP(b_{k-1}\leqslant|f_1|<b_k)\\ &= C\sum_{k\geqslant 0}P(|f_1|\geqslant b_k)<\infty. \end{aligned} \tag{4}$$

If f_1 is symmetric, then $\sum_{k=1}^{n}E[f_k 1_{\{|f_k|<b_k\}}] = 0$ by (20.42.e) and (20.42.c). This together with (1) and (4) implies $S_n/b_n \overset{a.s.}{\to} 0$ by (27.11). Otherwise, if (3) holds, then (27.11) shows that to complete the proof it suffices to verify that

$$\frac{1}{b_n}\sum_{k=1}^{n}E[f_k 1_{\{|f_k|<b_k\}}]\to 0. \tag{5}$$

As $Ef_1 = 0$, we obtain

$$\begin{aligned}\frac{1}{b_n}\left|\sum_{k=1}^{n}E[f_k 1_{\{|f_k|<b_k\}}]\right| &= \frac{1}{b_n}\left|\sum_{k=1}^{n}E[f_k 1_{\{|f_k|\geqslant b_k\}}]\right|\\ &\leqslant \frac{1}{b_n}\sum_{k=1}^{n}\sum_{m\geqslant k}E[|f_1|\,1_{\{b_m\leqslant|f_1|<b_{m+1}\}}]\\ &= \frac{1}{b_n}\sum_{m=1}^{n}\sum_{k=1}^{m}E[|f_1|\,1_{\{b_m\leqslant|f_1|<b_{m+1}\}}]\\ &\quad+\frac{1}{b_n}\sum_{m>n}\sum_{k=1}^{n}E[|f_1|\,1_{\{b_m\leqslant|f_1|<b_{m+1}\}}]\\ &\leqslant \frac{1}{b_n}\sum_{m=1}^{n}mb_{m+1}P(b_m\leqslant|f_1|<b_{m+1})\\ &\quad+\frac{1}{b_n}\sum_{m>n}nb_{m+1}P(b_m\leqslant|f_1|<b_{m+1}). \end{aligned} \tag{6}$$

Since

$$\sum_{m\geqslant 1}mP(b_m\leqslant|f_1|<b_{m+1}) = \sum_{m\geqslant 1}P(|f_1|\geqslant b_m)<\infty, \tag{7}$$

on account of (3) and Kronecker's lemma (27.5), we have

$$\frac{1}{b_n}\sum_{m=1}^{n} m b_{m+1} P(b_m \leqslant |f_1| < b_{m+1}) \leqslant 2C\frac{1}{b_n}\sum_{m=1}^{n} b_m m P(b_m \leqslant |f_1| < b_{m+1}) \to 0. \tag{8}$$

By (3) and (7), we have also

$$\frac{1}{b_n}\sum_{m>n} n b_{m+1} P(b_m \leqslant |f_1| < b_{m+1}) \leqslant C\sum_{m>n}(m+1) P(b_m \leqslant |f_1| < b_{m+1}) \to 0. \tag{9}$$

From (6), (8) and (9) we see that (5) is satisfied. □

Theorem 27.14 (Marcinkiewicz–Zygmund). *Suppose that $f_n, n \geqslant 1$, are i.i.d. random variables. Then, for $0 < p < 2$, the following assertions hold.*

(i) *If $E\,|f_1|^p < \infty$, then $(S_n - na)/n^{1/p} \overset{a.s.}{\to} 0$, where a is an arbitrary constant or Ef_1 according as $0 < p < 1$ or $1 \leqslant p < 2$.*

(ii) *If $(S_n - na)/n^{1/p} \overset{a.s.}{\to} 0$ for some $a \in R$, then $E\,|f_1|^p < \infty$ and $a = Ef_1$ when $1 \leqslant p < 2$.*

Proof. If $0 < p < 1$ and $E\,|f_1|^p < \infty$, then $(S_n - na)/n^{1/p} \overset{a.s.}{\to} 0$ for any $a \in R$ by (26.14.ii) and Kronecker's lemma (27.5). If $E\,|f_1| < \infty$, then $E\,|f_1 - Ef_1| < \infty$, and so $\sum_{n\geqslant 1} P(|f_1 - Ef_1| \geqslant n) < \infty$ by (17.10). Applying Feller's theorem (27.13) with $b_n = n$ yields $S_n/n \overset{a.s.}{\to} Ef_1$ in this case. If $1 \leqslant p < 2$ and $E\,|f_1|^p < \infty$, then $E\,|f_1 - Ef_1|^p < \infty$ by (13.5) and (13.23.b), and so $\sum_{n\geqslant 1}(f_n - Ef_1)/n^{1/p}$ converges a.s. by (26.14.iv). This, in conjunction with (27.5), shows that $(S_n - nEf_1)/n^{1/p} \overset{a.s.}{\to} 0$. Thus (i) is proved. If now (ii) holds, then applying (27.12) with $b_n = n^{1/p}$ yields $\sum_{n\geqslant 1} P(|f_1 - a| \geqslant n^{1/p}) = \sum_{n\geqslant 1} P(|f_n - a| \geqslant n^{1/p}) < \infty$. This entails
$E\,|f_1 - a|^p < \infty$ by (17.10), and so $E\,|f_1|^p < \infty$. Consequently, for $1 \leqslant p < 2$, $(S_n - nEf_1)/n^{1/p} \overset{a.s.}{\to} 0$ by (i). Both $(S_n - na)/n^{1/p} \overset{a.s.}{\to} 0$ and $(S_n - nEf_1)/n^{1/p} \overset{a.s.}{\to} 0$ are possible only if $a = Ef_1$. □

Remark 27.15. The fact that $E\,|f_1| < \infty$ implies that $S_n/n \overset{a.s.}{\to} Ef_1$ is the classical SLLN due to A. N. Kolmogorov. There are several different proofs of this famous result available (see (29.84) and (32.19.b)). An elementary proof of Kolmogorov's SLLN, which is more applicable because only pairwise independence is required, is now in order.

Theorem 27.16 (Etemadi). *If $f_n, n \geqslant 1$, are pairwise independent and identically distributed random variables with $E\,|f_1| < \infty$, then $S_n/n \overset{a.s.}{\to} Ef_1$.*

Proof. (I) Assume first that $f_n \geqslant 0, n \geqslant 1$, and let $\varepsilon > 0$ and $\alpha > 1$. For $n \geqslant 1$, set $k_n = [\alpha^n]$ and $U_n = \sum_{i=1}^n f_i 1_{\{f_i < i\}}$. Since $k_n \sim \alpha^n$, it is not difficult to check that

$$\sum_{n \geqslant (\log i)/\log \alpha} \frac{1}{k_n^2} \leqslant C \frac{1}{i^2}, \quad i \geqslant 1,$$

where C is a positive constant depending only on α. Using this, Chebyshev's inequality, (18.40.a) and Fubini's theorem (15.8), we have

$$\sum_{n \geqslant 1} P\left(\frac{|U_{k_n} - EU_{k_n}|}{k_n} \geqslant \varepsilon\right) \leqslant \varepsilon^{-2} \sum_{n \geqslant 1} \frac{\operatorname{Var} U_n}{k_n^2}$$

$$= \varepsilon^{-2} \sum_{n \geqslant 1} \frac{1}{k_n^2} \sum_{i=1}^{k_n} \operatorname{Var}[f_i 1_{\{f_i < i\}}]$$

$$\leqslant \varepsilon^{-2} \sum_{n \geqslant 1} \frac{1}{k_n^2} \sum_{i=1}^{k_n} E[f_1^2 1_{\{f_1 < i\}}] \leqslant \varepsilon^{-2} \sum_{i \geqslant 1} E[f_1^2 1_{\{f_1 < i\}}] \sum_{n \geqslant (\log i)/\log \alpha} \frac{1}{k_n^2}$$

$$\leqslant C\varepsilon^{-2} \sum_{i \geqslant 1} \frac{1}{i^2} E[f_1^2 1_{\{f_1 < i\}}] \leqslant C\varepsilon^{-2} E f_1 < \infty, \tag{1}$$

where the last $\leqslant$ in (1) is obtained as in the last line of the proof of (25.19). From (1) and (18.44) it follows that

$$\frac{U_{k_n}}{k_n} - \frac{EU_{k_n}}{k_n} \xrightarrow{a.s.} 0. \tag{2}$$

As $Ef_1 < \infty$, we have also

$$Ef_1 = \lim_n E[f_1 1_{\{f_1 < n\}}] = \lim_n \frac{EU_{k_n}}{k_n} \tag{3}$$

by dominated convergence. From (2) and (3) we infer that

$$\frac{U_{k_n}}{k_n} \xrightarrow{a.s.} Ef_1. \tag{4}$$

Since $Ef_1 < \infty$ is equivalent to $\sum_{n \geqslant 1} P(f_1 \geqslant n) < \infty$ by (11.37), the sequences $\{f_n : n \geqslant 1\}$ and $\{f_n 1_{\{f_n < n\}} : n \geqslant 1\}$ are equivalent, and so

$$\frac{S_{k_n}}{k_n} \xrightarrow{a.s.} Ef_1 \tag{5}$$

by (4) and (26.9.ii). Now, for $n \geqslant 1$, let $m \geqslant 1$ be such that $k_m \leqslant n < k_{m+1}$. Then, as the sequence $\{S_n : n \geqslant 1\}$ is nondecreasing, we have

$$\frac{k_m}{k_{m+1}} \frac{S_{k_m}}{k_m} \leqslant \frac{S_n}{n} \leqslant \frac{k_{m+1}}{k_m} \frac{S_{k_{m+1}}}{k_{m+1}}.$$

From this and (5) we conclude that

$$\frac{1}{\alpha}Ef_1 \leqslant \liminf_n \frac{S_n}{n} \leqslant \limsup_n \frac{S_n}{n} \leqslant \alpha Ef_1. \tag{6}$$

Letting $\alpha \to 1$ in (6) gives the desired result.

(II) Consider now the general case. By (18.28.c), $f_n^+, n \geqslant 1$, and $f_n^-, n \geqslant 1$, are pairwise independent. Since $\infty > E|f_1| = Ef_1^+ + Ef_1^-$, we can apply (I) to obtain

$$\frac{S_n}{n} = \frac{f_1^+ + \cdots + f_n^+}{n} - \frac{f_1^- + \cdots + f_n^-}{n} \overset{a.s.}{\to} Ef_1^+ - Ef_1^- = Ef_1. \qquad \square$$

We examine now the behavior of S_n/n when either Ef_1 exists but $|Ef_1| = \infty$ or Ef_1 does not exist, i.e. $Ef_1^+ = Ef_1^- = \infty$.

Theorem 27.17. *Assume $f_n, n \geqslant 1$, are i.i.d. random variables.*

(i) *If $f_n \geqslant 0$ for any $n \geqslant 1$ and $Ef_1 = \infty$, then $S_n/n \overset{a.s.}{\to} \infty$.*

(ii) *If Ef_1 exists and $|Ef_1| = \infty$, then $S_n/n \overset{a.s.}{\to} Ef_1$.*

(iii) *If $E|f_1| = \infty$, then $P(|f_n/n| \geqslant M$ i.o.$) = P(|S_n/n| \geqslant M$ i.o.$) = 1$ for every $M > 0$.*

(iv) *If $E|f_1| = \infty$, then* $\limsup_n |f_n|/n = \limsup_n |S_n|/n = \infty$ *a.s.*

Proof. If $f_n \geqslant 0$ for any $n \geqslant 1$, then $\frac{S_n}{n} \geqslant \frac{1}{n}\sum_{i=1}^n f_i 1_{\{f_i<m\}}, m \in N$, whence

$$\liminf_n \frac{S_n}{n} \geqslant E[f_1 1_{\{f_1<m\}}, \quad m \in N, \tag{1}$$

by (18.28.c) and (27.16). Letting $m \to \infty$ in (1) yields the conclusion of (i) by the monotone convergence theorem (11.22). Writing $f_n = f_n^+ - f_n^-, n \geqslant 1$, (ii) follows immediately from (i) and (27.16). If $E|f_1| = \infty$, we have $\sum_{n\geqslant 1} P(|f_n/n| \geqslant M) = \sum_{n\geqslant 1} P(|f_1/n| \geqslant M) = \infty, M > 0$, by (11.37). In view of the Borel-Cantelli lemma (18.24.ii), this amounts to $P(|f_n/n| \geqslant M$ i.o.$) = 1, M > 0$. Since

$$\left|\frac{f_n}{n}\right| = \left|\frac{S_n}{n} - \frac{n-1}{n}\frac{S_{n-1}}{n-1}\right| \leqslant \left|\frac{S_n}{n}\right| + \left|\frac{S_{n-1}}{n-1}\right|, \quad n \geqslant 2,$$

it follows that $\limsup_n\{|f_n|/n \geqslant 2M\} \subset \limsup_n\{|S_n|/n \geqslant M\}, M > 0$, and so $E|f_1| = \infty$ also implies $P(|S_n/n| \geqslant M$ i.o.$) = 1, M > 0$. Thus (iii) is proved, and (iv) follows at once from (iii). $\square$

The classical SLLN has various applications in probability theory and other fields such as number theory, analysis and statistics. In (27.19) we present an interesting application in statistics concerning the problem of estimating an unknown distribution function F.

Definition 27.18. Let $f_1, \ldots, f_n$ be arbitrary random variables. The function $F_n(x) = \frac{1}{n}\sum_{i=1}^n 1_{\{f_i\leqslant x\}}, x \in R$, is called the *empirical [sample] distribution function based on* $f_1, \ldots, f_n$. The value of the random variable $F_n(x)$ at $w \in X$ is $F_n^w(x) = \frac{1}{n}\sum_{i=1}^n 1_{\{f_i\leqslant x\}}(w)$. Note that F_n^w is a distribution function for each $w \in X$.

Theorem 27.19 (Glivenko–Cantelli). *If $f_n, n \geqslant 1$, are pairwise independent random variables with a common distribution function F and $F_n^w(x)$ is as in (27.18), then there is a null set $N \in \mathcal{A}$ such that*

$$\sup_{x \in R} \left| F_n^w(x) - F(x) \right| \to 0, \quad w \in N^c.$$

Proof. For any $x \in R$, $1_{\{f_n \leqslant x\}}, n \geqslant 1$, and $1_{\{f_n < x\}}, n \geqslant 1$, are pairwise independent random variables by (18.28.c), and $E[1_{\{f_1 \leqslant x\}}] = F(x)$ and $E[1_{\{f_1 < x\}}] = F(x-)$. Etemadi's SLLN (27.16) applies and we see that, for each $x \in X$, $F_n(x) \overset{a.s.}{\to} F(x)$ and $F_n(x-) \overset{a.s.}{\to} F(x-)$. This entails that, for every $x \in Q$, there exists a null set $N_1(x)$ such that $F_n^w(x) \to F(x)$, $w \in N_1(x)^c$. Putting $N_1 = \cup_{x \in Q} N_1(x)$, we have $F_n^w(x) \to F(x)$, $x \in Q$, whenever $w \in N_1^c$. In view of (19.19), as Q is dense in R, this implies that

$$F_n^w \overset{w}{\to} F, \quad w \in N_1^c. \tag{1}$$

It follows also that, for every $x \in C(F)^c$, there exists a null set $N_2(x)$ such that $F_n^w(x-) \to F(x-)$, $w \in N_2(x)^c$. Setting $N_2 = \cup_{x \in C(F)^c} N_2(x)$, we have

$$F_n^w(x-) \to F(x-), \quad x \in C(F)^c, w \in N_2^c. \tag{2}$$

Let $N = N_1 \cup N_2$. Since Q and $C(F)^c$ are countable, N is a null set. As (1) and (2) continue to hold with N in place of N_1 and N_2, respectively, the desired result follows by (19.54). □

Theorem 27.20. *Let $\{b_n : n \geqslant 1\} \subset]0, \infty[$ be a nondecreasing sequence such that $b_n \to \infty$ and $b_{2n}/b_n \leqslant C < \infty, n \geqslant 1$, and suppose that $S_n/b_n \overset{P}{\to} 0$ and $S_{2^n}/b_{2^n} \overset{a.s.}{\to} 0$. Then $S_n/b_n \overset{a.s.}{\to} 0$.*

Proof. Set $I(n) = \{k \in N : 2^{n-1} < k \leqslant 2^n\}, n \geqslant 1$. For $\varepsilon > 0, n \geqslant 1$ and $k \in I(n)$, put $A_k = \{|S_k| \geqslant 2\varepsilon b_{2^n}\}$, $B_k = \{|S_{2^n} - S_k| < \varepsilon b_{2^n}\}$, $C_k = \{|S_k| \geqslant 2\varepsilon b_k\}$ and $D_n = \{|S_{2^n}| \geqslant \varepsilon b_{2^n}\}$. Notice that B_k is independent of $A_k A_{k-1}^c \dots A_1^c$ and $A_k \supset C_k$ for all $k \geqslant 2$. Since $D_n \supset \cup_{k \in I(n)} A_k B_k, n \geqslant 1$, on account of (2) in (18.54), we have

$$\begin{aligned} P\left(\bigcup_{n \geqslant m} D_n\right) &\geqslant P\left(\bigcup_{n \geqslant m} \bigcup_{k \in I(n)} A_k B_k\right) = P\left(\bigcup_{k > 2^{m-1}} A_k B_k\right) \\ &\geqslant \left(\inf_{k > 2^{m-1}} P(B_k)\right) P\left(\bigcup_{k > 2^{m-1}} A_k\right) \\ &\geqslant \left(\inf_{k > 2^{m-1}} P(B_k)\right) P\left(\bigcup_{k > 2^{m-1}} C_k\right), \quad m \geqslant 1. \end{aligned} \tag{1}$$

We have also

$$P(B_k^c) \leqslant P\left(|S_{2^n}| \geqslant \frac{\varepsilon}{2} b_{2^n}\right) + P\left(|S_k| \geqslant \frac{\varepsilon}{2} b_k\right) \to 0. \tag{2}$$

From (1), (2) and (14.3) we infer that $\lim_m P(\cup_{k>2^{m-1}} C_k) = 0$. This implies $S_n/b_n \xrightarrow{a.s.} 0$ again by (14.3). □

Theorem 27.21. *Let $\{b_n : n \geqslant 1\} \subset]0, \infty[$ be a nondecreasing sequence such that $b_n \to \infty$ and $1 < c \leqslant b_{2^n}/b_{2^{n-1}} \leqslant C < \infty$, $n \geqslant 1$, and suppose that $S_n/b_n \xrightarrow{P} 0$ and $\sum_{n\geqslant 1} P(|S_{2^n} - S_{2^{n-1}}| \geqslant \varepsilon b_{2^n}) < \infty$, $\varepsilon > 0$. Then $S_n/b_n \xrightarrow{a.s.} 0$.*

Proof. Let $I(n), n \geqslant 1$, be as in the proof of (27.20). Thus for $k \geqslant 2$ there is a unique $n(k) \geqslant 1$ such that $k \in I(n(k))$. For $\varepsilon > 0, n \geqslant 1$ and $k \in I(n)$, put $A_k = \{|S_k - S_{2^{n-1}}| \geqslant 2\varepsilon b_{2^n}\}$, $B_k = \{|S_{2^n} - S_k| < \varepsilon b_{2^n}\}$ and $C_n = \{|S_{2^n} - S_{2^{n-1}}| \geqslant \varepsilon b_{2^n}\}$. Since $C_n \supset \cup_{k\in I(n)} A_k B_k$, $n \geqslant 1$, on account of (2) in (18.54), we have

$$\sum_{n\geqslant m} P\left(|S_{2^n} - S_{2^{n-1}}| \geqslant \varepsilon b_{2^n}\right) \geqslant P\left(\bigcup_{n\geqslant m} C_n\right) \geqslant P\left(\bigcup_{n\geqslant m}\bigcup_{k\in I(n)} A_k B_k\right)$$

$$\geqslant \left(\inf_{k>2^{m-1}} P(B_k)\right) P\left(\bigcup_{k>2^{m-1}} A_k\right), \quad m \geqslant 1. \quad (1)$$

Since $P(B_k^c) \to 0$, (1) and (14.3) show that $\lim_m P(\cup_{k>2^{m-1}} C_k) = 0$, and so $(S_k - S_{2^{n(k)-1}})/b_{2^{n(k)}} \xrightarrow{a.s.} 0$. Let $W \in \mathcal{A}$ with $P(W) = 1$ be the set on which this convergence holds in the ordinary sense. Hence, for $w \in W$ and $\varepsilon > 0$, there exists k_0 such that $|S_k(w) - S_{2^{n(k)-1}}(w)| < \varepsilon b_{2^{n(k)}}$, $k \geqslant k_0$. Then, for $k > 2^{n(k_0)}$, we have

$$|S_k(w)| \leqslant |S_{2^{n(k_0)-1}}(w)| + \sum_{n=n(k_0)}^{n(k)-1} |S_{2^n}(w) - S_{2^{n-1}}(w)| + |S_k(w) - S_{2^{n(k)-1}}(w)|$$

$$< |S_{2^{n(k_0)-1}}(w)| + \varepsilon \sum_{n=n(k_0)}^{n(k)} b_{2^n} \leqslant |S_{2^{n(k_0)-1}}(w)| + \varepsilon b_{2^{n(k)}} \sum_{n\geqslant 0} c^{-n}$$

$$\leqslant |S_{2^{n(k_0)-1}}(w)| + \varepsilon C(1-c)^{-1} b_k.$$

This entails $S_k(w)/b_k \to 0$. □

Exercise 27.22. For $n \geqslant 1$, assume $Ef_n = 0$ and $\sigma_n^2 = Ef_n^2 < \infty$, and write $s_n^2 = \sum_{k=1}^n \sigma_k^2$. If $s_n \to \infty$, prove that $S_n/s_n(\log s_n)^{1/2+\varepsilon} \xrightarrow{a.s.} 0$ for any $\varepsilon > 0$. [Hint. Put $b_n = s_n(\log s_n)^{1/2+\varepsilon}$, n large enough. If $s_{n_0} > 1$, use (15.8.c) to show that $\sum_{n\geqslant n_0} Ef_n^2/b_n^2 < \infty$. Then apply (27.9.ii).]

Exercise 27.23. For $n \geqslant 1$, assume $Ef_n = 0$ and $\sigma_n^2 = Ef_n^2 < \infty$. Use (27.22) to prove the following.

(a) If $\sigma_n^2 = O(1)$, then $S_n/\sqrt{n}(\log n)^{1/2+\varepsilon} \xrightarrow{a.s.} 0$ for each $\varepsilon > 0$.

(b) If $\sigma_n^2 = O(n/(\log n)^{\delta+1})$ for some $\delta > 0$, then $S_n/n(\log n)^{\varepsilon-\delta/2} \xrightarrow{a.s.} 0$, $\varepsilon > 0$.

Exercise 27.24 (Komlos–Revesz). For $n \geqslant 1$, assume $Ef_n = 0$ and $0 < \sigma_n^2 = Ef_n^2 < \infty$. If $\sum_{n\geqslant 1} \sigma_n^{-2} = \infty$ and $\{\sigma_n^2 \sum_{k=1}^n \sigma_k^{-2} : n \geqslant 1\}$ is nondecreasing, then $\sum_{k=1}^n \sigma_k^{-2} f_k / \sum_{k=1}^n \sigma_k^{-2} \xrightarrow{a.s.} 0$. [Use (26.3), (5.98.b), and (27.5).]

Exercise 27.25. For $n \geqslant 1$, assume that $P(f_n = n^\alpha) = P(f_n = -n^\alpha) = 1/2$ for some $\alpha \geqslant 0$. Use (27.9.ii) and (24.13) to prove that $S_n/n \overset{a.s.}{\to} 0$ for $0 \leqslant \alpha < 1/2$, while $\{S_n/n : n \geqslant 1\}$ does not converge in probability to 0 for $\alpha \geqslant 1/2$.

Exercise 27.26. Let S_n be a binomial random variable with $P(S_n = k) = b(k; n, p)$, $0 \leqslant k \leqslant n, 0 < p < 1$. Show that $(S_n - np)/n^\alpha \overset{a.s.}{\to} 0$ for every $\alpha > 1/2$.

Exercise 27.27. Let $\{\sigma_n^2 : n \geqslant 1\} \subset [0, \infty[$ be such that $\sum_{n\in N} \sigma_n^2/b_n^2 = \infty$, put $N_1 = \{n \in N : \sigma_n^2 \leqslant b_n^2\}$, and let $f_n, n \in N$, be independent random variables such that $P(f_n = b_n) = P(f_n = -b_n) = \sigma_n^2/2b_n^2$, $P(f_n = 0) = 1 - \sigma_n^2/b_n^2$, $n \in N_1$, and $P(f_n = b_n) = P(f_n = -b_n) = 1/2, n \in N_1^c$. Prove the following.

(a) $\sum_{n\in N} \mathrm{Var}[f_n/b_n] = \infty$.

(b) $\{S_n/b_n : n \geqslant 1\}$ does not converge a.s. to 0. [Use (27.12).]

Exercise 27.28. Let $\{n_k : k \geqslant 1\} \subset N$ be a strictly increasing sequence such that $1 < c \leqslant b_{n_k}/b_{n_{k-1}} \leqslant C < \infty, k > 1$, and write $T_1 = S_{n_1}/b_{n_1}$ and $T_k = (S_{n_k} - S_{n_{k-1}})/b_{n_k}, k > 1$. Show that the following are equivalent:

(i) $S_n/b_n - m[S_n]/b_n \overset{a.s.}{\to} 0$;

(ii) $T_k - [T_k] \overset{a.s.}{\to} 0$;

(iii) $\sum_{k\geqslant 1} P(|T_k - [T_k]| \geqslant \varepsilon) < \infty, \varepsilon > 0$.

To prove that (ii) implies (i), the next steps may be useful.

(a) Let $S_n^*, n \geqslant 1$, be as in the proof of (25.17), and write $T_1^* = S_{n_1}^*/b_{n_1}$ and $T_k^* = (S_{n_k}^* - S_{n_{k-1}}^*)/b_{n_k}, k > 1$. Then $T_k^* \to 0$ P^*-a.s. [Use (25.2.ii).]

(b) For $k \geqslant 1$ and $1 \leqslant j \leqslant k$, let $a_{kj} = b_{n_j}/b_{n_k}$. Then $\sum_{j=1}^k a_{kj} \leqslant (1 - c)^{-1}$, $k \geqslant 1$.

(c) $S_{n_k}^*/b_{n_k} \to 0$ P^*-a.s. [Write $S_{n_k}^*/b_{n_k} = \sum_{j=1}^k a_{kj} T_j^*, k \geqslant 1$, and apply (27.3.i) using (a) and (b).]

(d) $\sup_{n_{k-1}<n\leqslant n_k} \left|S_n^* - S_{n_{k-1}}^*\right|/b_{n_k} \to 0$ P^*-a.s. [Hint. By (25.4), for $\varepsilon > 0$ and $k > 1$, $P^*\left(\sup_{n_{k-1}<n\leqslant n_k} \left|S_n^* - S_{n_{k-1}}^*\right| \geqslant \varepsilon b_{n_k}\right) \leqslant 2P^*\left(\left|T_k^*\right| \geqslant \varepsilon\right)$. Then use (a), the Borel-Cantelli lemma (18.24) and (18.44).]

(e) $S_n^*/b_n \to 0$ P^*-a.s. [Hint. $\left|S_n^*/b_n\right| \leqslant C \sup_{n_{k-1}<n\leqslant n_k} \left|S_n^* - S_{n_{k-1}}^*\right|/b_{n_k} + \left|S_{n_{k-1}}^*/b_{n_{k-1}}\right|$ for $k > 1$ and $n_{k-1} < n \leqslant n_k$. Then use (c) and (d).]

(f) $S_n/b_n - m[S_n]/b_n \overset{a.s.}{\to} 0$. [Use (25.2.ii).]

Exercise 27.29. Let $\{b_n : n \geqslant 1\}$ satisfy the condition in (27.28). Use (27.28) and (25.53.a) to prove that $S_n/b_n \overset{a.s.}{\to} 0$ if and only if $S_n/b_n \overset{P}{\to} 0$ and $S_{n_k}/b_{n_k} \overset{a.s.}{\to} 0$.

Exercise 27.30. Let $\{b_n : n \geqslant 1\}$ and $\{T_k : k \geqslant 1\}$ be as in (27.28), and assume that $|f_n| < \varepsilon_n b_n$ and $Ef_n = 0$ for each $n \geqslant 1$, where $\varepsilon_n \to 0$. Show that the following are equivalent:

(i) $S_n/b_n \overset{a.s.}{\to} 0$;

(ii) $T_k \overset{a.s.}{\to} 0$;

(iii) $\sum_{k\geqslant 1} P(|T_k| \geqslant \varepsilon) < \infty, \varepsilon > 0$.

To prove that (ii) implies (i), the next steps may be helpful.

(a) Let $f_n^*, n \geqslant 1$, and $S_n^*, n \geqslant 1$, be as in the proof of (25.17), and let $T_k^*, k \geqslant 1$, be as in (27.28). Then $\sup_{n_{k-1}<n\leqslant n_k} \left| S_n^* - S_{n_{k-1}}^* \right| /b_{n_k} \to 0$ P^*-a.s. and $S_n^*/b_n \to 0$ P^*-a.s. [Follow the solution to (27.28).]

(b) $\mathrm{Var}^* T_k^* \to 0$. [Hint. On account of (26.4.ii), for $\varepsilon > 0$ and $k > 1$, $\mathrm{Var}^*\, T_k^* \leqslant (\varepsilon + \sup_{n_{k-1}<n\leqslant n_k} \varepsilon_n)^2 / P^*(\sup_{n_{k-1}<n\leqslant n_k} \left| S_n^* - S_{n_{k-1}}^* \right| < \varepsilon b_{n_k})$. Then use (a).]

(c) $\mathrm{Var}\, T_k \to 0$. [Use (b).]

(d) Let $\{a_{kj} : k \geqslant 1, 1 \leqslant j \leqslant k\}$ be as in (27.28). Then $\sum_{j=1}^k a_{kj}^2 \leqslant (1-c^2)^{-1}$, $k \geqslant 1$.

(e) $\mathrm{Var}[S_n/b_n] \to 0$. [Hint. For $k > 1$ and $n_{k-1} < n \leqslant n_k$, $c^{-2}\mathrm{Var}[S_n/b_n] \leqslant \mathrm{Var}[S_{n_k}/b_{n_k}] = \sum_{j=1}^k a_{kj}^2 \mathrm{Var}\, T_j$. Then apply (27.3.i) using (c) and (d).]

(f) $S_n/b_n \overset{a.s.}{\to} 0$. [Hint. $m[S_n]/b_n \to 0$ by (e) and (23.52). Then use (a) and (25.2.ii).]

Exercise 27.31. Let $\{n_k : k \geqslant 1\} \subset N$ be a strictly increasing sequence such that $n_k/n_{k-1} \leqslant K < \infty, k > 1$, and $b_{n_k}/b_{n_{k-1}} \leqslant C < \infty, k > 1$, and assume that the distribution of $S_{n_{k-1}}$ coincides with the distribution of $S_{n_k} - S_{n_{k-1}}$ for any $k > 1$. If $S_n/b_n \overset{a.s.}{\to} 0$, then $\sum_{n\geqslant 1} \frac{1}{n} P(|S_n| \geqslant \varepsilon b_n) < \infty, \varepsilon > 0$. The next steps may be useful.

(a) Let $S_n^*, n \geqslant 1$, and $T_k^*, k \geqslant 1$, be as in (27.28). Then $T_k^* \to 0$ P^*-a.s. [Use (25.2.ii).]

(b) $\sum_{k\geqslant 1} P^*(\left|T_k^*\right| \geqslant \varepsilon) < \infty, \varepsilon > 0$. [Use (a) and (18.24).]

(c) For $k > 1$ and $n_{k-1} < n \leqslant n_k$, $P^*(\left|S_n^*\right| \geqslant \varepsilon b_n) \leqslant 2P^*(\left|S_{n_k}^*\right| \geqslant C^{-1}\varepsilon b_{n_k}), \varepsilon > 0$. [Apply (25.4).]

(d) $P^*(S_{n_{k-1}}^*)^{-1} = P^*(S_{n_k}^* - S_{n_{k-1}}^*)^{-1}$ for any $k > 1$. [Use (23.20.e).]

(e) $\sum_{n\geq 1} \frac{1}{n} P^*(\left|S_n^*\right| \geq \varepsilon b_n) < \infty$, $\varepsilon > 0$. [Hint. By (c), (d) and (b), write $\sum_{n\geq n_1} \frac{1}{n} P^*(\left|S_n^*\right| \geq \varepsilon b_n) = \sum_{k>1} \sum_{n_{k-1}<n\leq n_k} \frac{1}{n} P^*(\left|S_n^*\right| \geq \varepsilon b_n) \leq 2(\log K) \sum_{k>1} P^*(\left|S_{n_k}^*\right| \geq C^{-1}\varepsilon b_{n_k}) \leq 2(\log K) \sum_{k>1} P^*(\left|T_{k+1}^*\right| \geq C^{-2}\varepsilon) < \infty$, $\varepsilon > 0$.]

(f) $\sum_{n\geqslant 1} \frac{1}{n} P(|S_n| \geqslant \varepsilon b_n) < \infty, \varepsilon > 0$. [Hint. For $\varepsilon > 0$, by (23.53.a) and (25.1.ii), $P(|S_n| \geqslant \varepsilon b_n) \leqslant P(|S_n - m[S_n]| \geqslant \varepsilon b_n/2) \leqslant 2P^*(\left|S_n^*\right| \geqslant \varepsilon b_n/2)$ whenever $m[S_n]/b_n < \varepsilon/2$. Then use (e).]

Exercise 27.32. Let $I = \{0, \ldots, b-1\}$ with $1 < b \in N$, let $X = [0, 1[$, and let λ be the Lebesgue measure on $\mathcal{B}([0, 1[)$. For $x \in [0, 1[$, let $x = \sum_{n\in N} a_n(x)/b^n$ be its b-adic expansion as in (8.23.e).

(a) Use (16.24) and (18.29) to show that $a_n, n \geqslant 1$, are i.i.d. with respect to λ random variables with $\lambda(\{a_n = i\}) = 1/b, n \in N, i \in I$.

(b) Use (a) and (27.14.i) to prove that

(i) $\frac{1}{n} \sum_{k=1}^n 1_{\{i\}} \circ a_k \to \frac{1}{b}$ λ-a.e., $\quad i \in I$.

For $x \in [0, 1[$ and $i \in I$, $\frac{1}{n} \sum_{k=1}^n (1_{\{i\}} \circ a_k)(x)$ is the relative frequency of the digit i in the first n places of the b-adic expansion of x. The number $x \in [0, 1[$ is said to be *normal* (to the base b), if $\frac{1}{n} \sum_{k=1}^n (1_{\{i\}} \circ a_k)(x) \to \frac{1}{b}$ for each $i \in I$. Thus (i) shows that almost every (with respect to λ) number in $[0, 1[$ is normal to the base b.

Throughout (27.33)–(27.37), we consider that $f_n, n \geqslant 1$, are i.i.d. random variables.

Exercise 27.33. Prove that $\frac{\log n}{n}\sum_{k=3}^{n}\frac{f_k}{\log k}\xrightarrow{a.s.}0$ if and only if $E|f_1|<\infty$ and $Ef_1=0$. [Hint. Use (27.11) and (27.12), taking into account that $\frac{\log n}{n}\sum_{k=3}^{n}\frac{1}{\log k}\to 1$.]

Exercise 27.34. (a) If $E|f_1|<\infty$ and $Ef_1=0$, then $\left(\sum_{k=1}^{n}\frac{1}{k}f_k\right)/\sum_{k=1}^{n}\frac{1}{k}\xrightarrow{a.s.}0$. [Hint. Apply (27.11) to the random variables $f_n/n, n\geqslant 1$, noticing that $\sum_{k=1}^{n}\frac{1}{k}\sim\log n$. To show that $\sum_{n\geqslant 1}(\log n)^{-2}E[(f_n/n)^2 1_{\{|f_n/n|<\log n\}}]<\infty$, use Fubini's theorem (15.8) in conjunction with (17.16.a) and (17.16.b).]

(b) If $\left(\sum_{k=1}^{n}\frac{1}{k}f_k\right)/\sum_{k=1}^{n}\frac{1}{k}\xrightarrow{a.s.}0$, then $E[|f_1|/\log(2+|f_1|)]<\infty$. [Use (27.12) and (17.16.a).]

Exercise 27.35. For fixed $k\geqslant 1$, define $g_n=\frac{1}{k}\sum_{j=n}^{n+k-1}f_j, n\geqslant 1$. Assuming $E|f_1|<\infty$, prove that $\frac{1}{n}\sum_{i=1}^{n}g_i\xrightarrow{a.s.}Ef_1$.

Exercise 27.36. Assume $0<p<2$.

(a) If $E|f_1|^p<\infty$ and $Ef_1=0$ for $1\leqslant p<2$, show that $f_n/n^{1/p}\xrightarrow{a.s.}0$ if and only if $S_n/n^{1/p}\xrightarrow{a.s.}0$.

(b) If $E|f_1|^p=\infty$, then $\limsup_n |f_n|/n^{1/p}=\limsup_n |S_n|/n^{1/p}=\infty$ a.s. [Take (27.17.iv) into account.]

Exercise 27.37. Prove that $\sup_{1\leqslant i\leqslant n}|f_i|/n\xrightarrow{a.s.}0$ if and only if $E|f_1|<\infty$. [Hint. $|f_n|\leqslant\sup_{1\leqslant i\leqslant n}|f_i|\leqslant 2\sup_{1\leqslant i\leqslant n}|S_i-iEf_1|+E|f_1|$.]

28 Laws of the Iterated Logarithm

In this section, if not otherwise stated, we make the additional assumption that $0<Ef_n^2<\infty$ and $Ef_n=0$ for all $n\geqslant 1$, write $\sigma_n^2=Ef_n^2$ and $s_n^2=\sum_{i=1}^{n}\sigma_i^2$, and suppose that $s_n\to\infty$. The preceding sections of this chapter were mainly devoted to the problem of finding conditions under which $S_n/b_n-a_n\xrightarrow{P}0$ or $S_n/b_n-a_n\xrightarrow{a.s.}0$ for some sequence $\{a_n:n\geqslant 1\}\subset R$ and some nondecreasing sequence $\{b_n:n\geqslant 1\}\subset\,]0,\infty[$. On account of delicate and complex methods, we shall now give remarkably more precise results about the magnitude of the asymptotic fluctuation of $\{S_n:n\geqslant 1\}$.

Definition 28.1. Let $\{T_n:n\geqslant 1\}$ be a sequence of random variables. We say that a sequence $\{b_n:n\geqslant 1\}\subset R$ belongs to the *upper class* or to the *lower class* of $\{T_n:n\geqslant 1\}$ according as $P(T_n\geqslant b_n \text{ i.o.})=0$ or 1.

Remarks 28.2. (a) There may exist sequences $\{b_n:n\geqslant 1\}\subset R$ which belong to neither of the two classes.

(b) If $b_n\neq 0, n\geqslant 1$, then $\limsup_n T_n/b_n=c\in R$ if and only if, for any $\delta>0$, the sequence $\{(c+\delta)b_n:n\geqslant 1\}$ belongs to the upper class of $\{T_n:n\geqslant 1\}$, while the sequence $\{(c-\delta)b_n:n\geqslant 1\}$ belongs to the lower class of $\{T_n:n\geqslant 1\}$.

(c) If $f_n, n\geqslant 1$, are i.i.d., then any sequence $\{b_n:n\geqslant 1\}\subset R$ belongs to either the upper class or the lower class of $\{S_n:n\geqslant 1\}$. Actually, the event $\{S_n\geqslant b_n \text{ i.o.}\}$

belongs to the exchangeable σ-algebra of $f_1, f_2, \ldots$, which in this case is included in $\mathcal{O}_P$ (18.53), by virtue of the Hewitt-Savage zero-one law (18.52).

In what follows we shall investigate conditions under which $\limsup_n S_n/s_n \sqrt{2\log\log s_n^2} = 1$ a.s. ((18.37.c) ensures that $\limsup_n S_n/s_n\sqrt{2\log\log s_n^2}$ is a.s. constant.) The efforts will culminate in the Kolmogorov law of the iterated logarithm (LIL) for suitably bounded f_n, and in the Hartman–Wintner LIL for i.i.d. f_n. To reach the former law, we need (and present below as lemmas) sharp exponential bounds for $P(S_n/s_n \geqslant x)$, which are referred to as the *Kolmogorov exponential inequalities*. For ease of notation, put $\log_2 x = 1$ (and so $\log x = e$) or $\log\log x$ according as $0 < x \leqslant e$ or $x > e$.

Lemma 28.3. *Assume that* $\sup_{1\leqslant i\leqslant n} |f_i|/s_n \leqslant c_n < \infty$ *a.s. for* $n \geqslant 1$, *and let* $x > 0$.

(i) *If* $xc_n \leqslant 1$, *then* $P(S_n/s_n \geqslant x) \leqslant \exp\{-(x^2/2)(1 - xc_n/2)\}$.

(ii) *If* $xc_n \geqslant 1$, *then* $P(S_n/s_n \geqslant x) \leqslant \exp(-x/4c_n)$.

Proof. Suppose $t > 0$ and $tc_n \leqslant 1$. For $1 \leqslant i \leqslant n$, since $Ef_i = 0$ and $|Ef_i^k| \leqslant \sigma_i^2 s_n^{k-2} c_n^{k-2}$, $k \geqslant 2$, in view of Fubini's theorem (15.10), we have

$$Ee^{tf_i/s_n} = \sum_{k\geqslant 0} \frac{t^k}{s_n^k k!} Ef_i^k \leqslant 1 + \frac{t^2\sigma_i^2}{s_n^2}\left(\frac{1}{2} + \sum_{k\geqslant 3}\frac{(tc_n)^{k-2}}{k!}\right).$$

As $e^y \leqslant 1 + y + y^2/2 + y^3/4$ for $0 \leqslant y \leqslant 1$, and $1 + y \leqslant e^y$, $y \in R$, this entails

$$Ee^{tf_i/s_n} \leqslant 1 + \frac{t^2\sigma_i^2}{2s_n^2}\left(1 + \frac{tc_n}{2}\right) \leqslant \exp\left\{\frac{t^2\sigma_i^2}{2s_n^2}\left(1 + \frac{tc_n}{2}\right)\right\}.$$

Therefore, via (13.25.b) and independence, we obtain

$$P(S_n/s_n \geqslant x) \leqslant e^{-tx} Ee^{tS_n/s_n} = e^{-tx}\prod_{i=1}^n Ee^{tf_i/s_n} \leqslant \exp\left\{-tx + \frac{t^2}{2}\left(1 + \frac{tc_n}{2}\right)\right\}.$$

Setting $t = x$ when $xc_n \leqslant 1$, and $t = c_n^{-1}$ when $xc_n \geqslant 1$, we arrive at the desired inequalities. □

Corollary 28.4. *For* $n \geqslant 1$, *assume that* $|f_n| \leqslant cs_n/\sqrt{\log_2 s_n^2}$ *a.s., where* $c < \infty$. *Then, for all sufficiently large* n,

$$P(S_n/s_n \geqslant \alpha\sqrt{\log_2 s_n^2}) \leqslant (\log s_n^2)^{-(\alpha^2/2)(1-\alpha c/2)}, \quad 0 < \alpha \leqslant c^{-1}.$$

Proof. Since the function $h(x) = x/\log_2 x$, $x > 0$, is eventually nondecreasing with $\lim_{x\to\infty} h(x) = \infty$, there is $n_0 \in N$ such that $\sup_{1\leqslant i\leqslant n} |f_i| \leqslant c\sup_{1\leqslant i\leqslant n} s_i/\sqrt{\log_2 s_i^2} = cs_n/\sqrt{\log_2 s_n^2}$, $n \geqslant n_0$. If $0 < \alpha \leqslant c^{-1}$ and $n \geqslant n_0$, we apply (28.3.i) with $x = \alpha\sqrt{\log_2 s_n^2}$ and $c_n = c/\sqrt{\log_2 s_n^2}$ to obtain

$$P(S_n/s_n \geqslant \alpha\sqrt{\log_2 s_n^2}) \leqslant \exp\left\{-\frac{\alpha^2}{2}\left(1 - \frac{\alpha c}{2}\right)\log_2 s_n^2\right\} = (\log_2 s_n^2)^{-(\alpha^2/2)(1-\alpha c/2)}. \quad \square$$

Lemma 28.5. *For $n \geqslant 1$, let c_n be as in (28.3), and let $x_n > 0$ satisfy $x_n c_n \to 0$ and $x_n \to \infty$. Then, for given $\gamma > 0$ and all sufficiently large n, $P(S_n/s_n \geqslant x_n) \geqslant \exp\{-(x_n^2/2)(1+\gamma)\}$.*

Proof. Let $0 < t_n \leqslant c_n^{-1}$. For $1 \leqslant i \leqslant n$, since $e^y \leqslant 1+y+y^2/2+y^3/4$, $0 \leqslant y \leqslant 1$, and $e^{y-y^2} \leqslant 1+y$, $y \geqslant 0$, we have

$$\begin{aligned} Ee^{t_n f_i/s_n} &= \sum_{k \geqslant 0} \frac{t_n^k}{s_n^k k!} E f_i^k \\ &\geqslant 1 + \frac{t_n^2 \sigma_i^2}{s_n^2}\left(\frac{1}{2} - \sum_{k \geqslant 3} \frac{(t_n c_n)^{k-2}}{k!}\right) \geqslant 1 + \frac{t_n^2 \sigma_i^2}{2 s_n^2}\left(1 - \frac{t_n c_n}{2}\right) \\ &\geqslant \exp\left\{\frac{t_n^2 \sigma_i^2}{2 s_n^2}\left(1 - \frac{t_n c_n}{2} - \frac{t_n^2 \sigma_i^2}{2 s_n^2}\right)\right\} \geqslant \exp\left\{\frac{t_n^2 \sigma_i^2}{2 s_n^2}(1 - t_n c_n)\right\}, \end{aligned} \tag{1}$$

the last inequality coming from the fact that $\sigma_i^2 \leqslant s_n^2 c_n^2$. By independence and (1), we get

$$Ee^{t_n S_n/s_n} = \prod_{i=1}^{n} Ee^{t_n f_i/s_n} \geqslant \exp\left\{\frac{t_n^2}{2}(1 - t_n c_n)\right\}.$$

In view of (15.26), this implies

$$t_n \int_R e^{t_n x} P(S_n/s_n \geqslant x)dx \geqslant \exp\left\{\frac{t_n^2}{2}(1 - t_n c_n)\right\}. \tag{2}$$

Let $0 < \alpha < 1$, choose $n_1 = n_1(\alpha) \in N$ such that

$$x_n c_n < \alpha^2(1-\alpha)/8(1+\alpha)^2, \quad n \geqslant n_1, \tag{3}$$

and set $t_n = x_n/(1-\alpha)$, $n \geqslant n_1$. Then $t_n c_n \leqslant \alpha^2/4$, $n \geqslant n_1$, and (2) yields

$$Ee^{t_n S_n/s_n} = t_n \int_R e^{t_n x} q_n(x)dx \geqslant \exp\left\{\frac{t_n^2}{2}\left(1 - \frac{\alpha^2}{4}\right)\right\}, \quad n \geqslant n_1, \tag{4}$$

where $q_n(x) = P(S_n/s_n \geqslant x)$, $x \in R$. We split R into the intervals $I_1 =]-\infty, 0]$, $I_2 =]0, t_n(1-\alpha)]$, $I_3 =]t_n(1-\alpha), t_n(1+\alpha)]$, $I_4 =]t_n(1+\alpha), 8t_n]$ and $I_5 =]8t_n, \infty[$ and we put $J_i = \int_{I_i} e^{t_n x} q_n(x)dx$, $1 \leqslant i \leqslant 5$. Clearly,

$$t_n J_1 \leqslant t_n \int_{-\infty}^{0} e^{t_n x}\, dx = 1. \tag{5}$$

For $n \geqslant n_1$, we have $8t_n < c_n^{-1}$, and (28.3.i) implies

$$q_n(x) \leqslant \exp\left\{-\frac{x^2}{2}(1 - \frac{x c_n}{2})\right\} \leqslant e^{-x^2/4} \leqslant e^{-2t_n x}, \quad 8t_n < x \leqslant c_n^{-1},$$

while (28.3.ii) implies

$$q_n(x) \leqslant e^{-x/4c_n} \leqslant e^{-2t_n x}, \quad x > c_n^{-1}.$$

Therefore,

$$t_n J_5 = t_n \int_{8t_n}^{\infty} e^{t_n x} q_n(x) dx \leqslant t_n \int_{8t_n}^{\infty} e^{-t_n x}\, dx < 1, \quad n \geqslant n_1. \tag{6}$$

Since $t_n \to \infty$, we select $n_2 = n_2(\alpha) \in N$ such that $\exp\left\{\frac{t_n^2}{2}\left(1 - \frac{\alpha^2}{4}\right)\right\} > 8$ for $n \geqslant n_2$. Then, on account of (4)–(6), we get

$$t_n J_1 + t_n J_5 < 2 < \frac{1}{4} E e^{t_n S_n / s_n}, \quad n \geqslant n_1 \vee n_2. \tag{7}$$

To estimate J_2 and J_4, let $n \geqslant n_1$ and $x \leqslant 8t_n$. Then $x c_n \leqslant 8 t_n c_n < 1$, and (28.3.i) implies

$$\begin{aligned} e^{t_n x} q_n(x) dx &\leqslant \exp\left\{t_n x - \frac{x^2}{2}\left(1 - \frac{x c_n}{2}\right)\right\} \\ &\leqslant \exp\left\{t_n x - \frac{x^2}{2}(1 - 4 t_n c_n)\right\} = e^{g_n(x)}. \end{aligned}$$

The function $g_n(x) = t_n x - \frac{x^2}{2}(1 - 4t_n c_n), x \in R$, attains its maximum at $x_0 = t_n/(1 - 4t_n c_n)$. As $\alpha^2(1-\alpha)/8(1+\alpha)^2 < \alpha(1-\alpha)/4(1+\alpha)$, it is easily checked by (3) that $t_n(1-\alpha) < x_0 < t_n(1+\alpha)$. Therefore, for $x \in I_2$, since $4t_n c_n < \alpha^2/2(1-\alpha)^2$ by (3), we have

$$g_n(x) \leqslant g_n(t_n(1-\alpha)) = \frac{t_n^2}{2}(1-\alpha)(1+\alpha+4t_n c_n(1-\alpha)) \leqslant \frac{t_n^2}{2}\left(1 - \frac{\alpha^2}{2}\right),$$

and so

$$\begin{aligned} t_n J_2 = t_n \int_0^{t_n(1-\alpha)} e^{t_n x} q_n(x) dx &\leqslant t_n \int_0^{t_n(1-\alpha)} e^{g_n(x)}\, dx \\ \leqslant t_n^2 \exp\left\{\frac{t_n^2}{2}\left(1 - \frac{\alpha^2}{2}\right)\right\}. \end{aligned} \tag{8}$$

At the same time, for $x \in I_4$, since $4t_n c_n < \alpha^2/2(1+\alpha)^2$ by (3), we have

$$g_n(x) \leqslant g_n(t_n(1+\alpha)) = \frac{t_n^2}{2}(1+\alpha)(1+\alpha+4t_n c_n(1+\alpha)) \leqslant \frac{t_n^2}{2}\left(1 - \frac{\alpha^2}{2}\right),$$

and so

$$t_n J_4 = t_n \int_{t_n(1+\alpha)}^{8t_n} e^{t_n x} q_n(x) dx \leqslant t_n \int_{t_n(1+\alpha)}^{8t_n} e^{g_n(x)} dx$$

$$\leqslant 8t_n^2 \exp\left\{\frac{t_n^2}{2}\left(1 - \frac{\alpha^2}{2}\right)\right\}. \quad (9)$$

From (8), (9) and (4) we obtain

$$t_n J_2 + t_n J_4 \leqslant 9t_n^2 \exp\left\{\frac{t_n^2}{2}\left(1 - \frac{\alpha^2}{2}\right)\right\}$$

$$= 9t_n^2 \exp\left\{-\frac{t_n^2\alpha^2}{8}\right\} \exp\left\{\frac{t_n^2}{2}\left(1 - \frac{\alpha^2}{4}\right)\right\}.$$

$$\leqslant 9t_n^2 e^{-t_n^2\alpha^2/8} E e^{t_n S_n/s_n}, \quad n \geqslant n_1. \quad (10)$$

Since $t_n \to \infty$, we choose $n_3 = n_3(\alpha) \in N$ such that $t_n^2 e^{-t_n^2\alpha^2/8} < 1/36$ for $n \geqslant n_3$. Then, in view of (10),

$$t_n J_2 + t_n J_4 < \frac{1}{4} E e^{t_n S_n/s_n}, \quad n \geqslant n_1 \vee n_3. \quad (11)$$

For $n \geqslant n_1 \vee n_2 \vee n_3$, on account of (4), (7) and (11), we obtain

$$t_n \int_{t_n(1-\alpha)}^{t_n(1+\alpha)} e^{t_n x} q_n(x) dx = t_n J_3$$

$$= E e^{t_n S_n/s_n} - t_n J_1 - t_n J_5 - t_n J_2 - t_n J_4 > \frac{1}{2} E e^{t_n S_n/s_n}$$

$$\geqslant \frac{1}{2} \exp\left\{\frac{t_n^2}{2}\left(1 - \frac{\alpha^2}{4}\right)\right\},$$

whence

$$2t_n^2 \alpha e^{t_n^2(1+\alpha)} q_n(t_n(1-\alpha)) > \frac{1}{2} \exp\left\{\frac{t_n^2}{2}\left(1 - \frac{\alpha^2}{4}\right)\right\}.$$

This implies

$$P(S_n/s_n \geqslant x_n) > \frac{1}{4t_n^2} \exp\left\{\frac{t_n^2\alpha^2}{8}\right\} \exp\left\{-\frac{t_n^2}{2}\left(1 + 2\alpha + \frac{\alpha^2}{2}\right)\right\}$$

$$> 9 \exp\left\{-\frac{t_n^2}{2}\left(1 + 2\alpha + \frac{\alpha^2}{2}\right)\right\}$$

$$\geqslant 9 \exp\left\{-\frac{t_n^2}{2}\frac{1 + 2\alpha + \alpha^2/2}{(1-\alpha)^2}\right\}, \quad n \geqslant n_1 \vee n_2 \vee n_3.$$

Consequently, if we choose α so that $(1 + 2\alpha + \alpha^2/2)/(1-\alpha)^2 \leqslant 1 + \gamma$, then $P(S_n/s_n \geqslant x_n) \geqslant \exp\{-(x_n^2/2)(1+\gamma)\}$, whenever $n \geqslant n_1 \vee n_2 \vee n_3$. □

Corollary 28.6. *For $n \geqslant 1$, assume that $|f_n| \leqslant \varepsilon_n s_n/\sqrt{\log_2 s_n^2}$ a.s., with $0 < \varepsilon_n \to 0$. Let $\rho > 1$. For each $k \geqslant 1$, let n_k be the smallest integer n such that $s_n > \rho^k$, and put $u_k^2 = s_{n_{k+1}}^2 - s_{n_k}^2$. Then, for given $\gamma > 0$ and all large k,*

$$P((S_{n_{k+1}} - S_{n_k})/u_k \geqslant \alpha\sqrt{\log_2 u_k^2}) \geqslant (\log u_k^2)^{-(\alpha^2/2)(1+\gamma)}, \quad \alpha > 0.$$

Proof. For $n \geqslant 1$, we have

$$\frac{s_{n+1}^2}{s_n^2} = 1 + \frac{\sigma_{n+1}^2}{s_n^2} \leqslant 1 + \frac{s_{n+1}^2}{s_n^2}\frac{\varepsilon_{n+1}^2}{\log_2 s_n^2},$$

and so $s_{n+1}^2/s_n^2 \to 1$ and $(s_{n+m}^2 - s_n^2)/s_n^2 \to 0$ for any $m \geqslant 1$. This entails $s_{n_k}^2/\rho^{2k} \to 1$, $u_k^2/s_{n_k}^2 \to \rho^2 - 1$ and $n_{k+1} - n_k \to \infty$. Therefore, for sufficiently large k,

$$\sup_{n_k<i\leqslant n_{k+1}} |f_i|\frac{1}{u_k} \leqslant \sup_{n_k<i\leqslant n_{k+1}} \varepsilon_i \frac{s_{n_{k+1}}}{\sqrt{\log_2 s_{n_{k+1}}^2}}\frac{1}{u_k}$$
$$\leqslant C\frac{\rho}{\sqrt{\rho^2-1}} \sup_{n_k<i\leqslant n_{k+1}} \varepsilon_i \frac{1}{\sqrt{\log_2 s_{n_{k+1}}^2}},$$

where $1 < C < \infty$. Applying (28.5) with $c_k = C(\rho/\sqrt{\rho^2-1}) \sup_{n_k<i\leqslant n_{k+1}} \varepsilon_i/\sqrt{\log_2 s_{n_{k+1}}^2}$ and $x_k = \alpha\sqrt{\log_2 u_k^2}$, for $\gamma > 0$ and all sufficiently large k, we obtain

$$P((S_{n_{k+1}} - S_{n_k})/u_k \geqslant \alpha\sqrt{\log_2 u_k^2}) \geqslant \exp\left\{-\frac{\alpha^2}{2}(1+\gamma)\log_2 u_k^2\right\}$$
$$= (\log u_k^2)^{-(\alpha^2/2)(1+\gamma)}, \alpha > 0. \quad \square$$

Kolmogorov LIL 28.7. *For $n \geqslant 1$, assume that $|f_n| \leqslant \varepsilon_n s_n/\sqrt{\log_2 s_n^2}$ a.s., with $0 < \varepsilon_n \to 0$. Then*

$$\limsup_n S_n/s_n\sqrt{2\log_2 s_n^2} = 1 \ a.s. \tag{1}$$

Proof. Let ρ, $\{n_k : k \geqslant 1\}$, and $\{u_k : k \geqslant 1\}$ be as in (28.6), and let $0 < \delta' < \delta < 1$. For $n \geqslant 1$, set $v_n^2 = 2\log_2 s_n^2$.

(I) First we show that $\{(1+\delta)s_n v_n : n \geqslant 1\}$ belongs to the upper class of $\{S_n : n \geqslant 1\}$. As δ is at our disposal, this is equivalent to

$$\limsup_n S_n/s_n v_n \leqslant 1 \text{ a.s.} \tag{2}$$

Given δ we select ρ so that $\rho < 1+\delta$, then we choose δ' such that $1+\delta' < (1+\delta)/\rho$. Since $\varepsilon_n \to 0$ and $\limsup_n S_n/s_n v_n$ does not change under modification of any finite number of terms f_i of S_n, without loss of generality we may and do suppose that $\varepsilon_n \leqslant c$

for all $n \geqslant 1$, where $(1+\delta')^2(1-(1+\delta')c/\sqrt{2}) > 1$. Then, applying (28.4) with $\alpha = \sqrt{2}(1+\delta')$, since $s_{n_k}^2 \sim \rho^{2k}$, we get

$$\begin{aligned}
&\sum_{k:n_{k+1}\geqslant n_0} P(S_{n_{k+1}}/s_{n_{k+1}} \geqslant \alpha(1+\delta')v_{n_{k+1}}) \\
&\quad \leqslant \sum_{k:n_{k+1}\geqslant n_0} (\log s_{n_{k+1}}^2)^{-(1+\delta')^2(1-(1+\delta')c/\sqrt{2})} \\
&\quad \leqslant C \sum_{k:n_{k+1}\geqslant n_0} (k+1)^{-(1+\delta')^2(1-(1+\delta')c/\sqrt{2})} < \infty,
\end{aligned} \tag{3}$$

where n_0 is as in the proof of (28.4) and $C < \infty$. Since $(1+\delta)s_{n_k}v_{n_k}/s_{n_{k+1}}v_{n_{k+1}} - \sqrt{2}/v_{n_{k+1}} \to (1+\delta)/\rho > 1+\delta'$, we have

$$\begin{aligned}
&P\left(S_{n_{k+1}}/s_{n_{k+1}} \geqslant (1+\delta)s_{n_k}v_{n_k}/s_{n_{k+1}} - \sqrt{2}\right) \\
&\quad \leqslant P\left(S_{n_{k+1}}/s_{n_{k+1}} \geqslant (1+\delta')v_{n_{k+1}}\right)
\end{aligned}$$

for k sufficiently large, and so (3) implies

$$\sum_{k\geqslant 1} P\left(S_{n_{k+1}}/s_{n_{k+1}} \geqslant (1+\delta)s_{n_k}v_{n_k}/s_{n_{k+1}} - \sqrt{2}\right) < \infty.$$

Therefore, on account of (25.44), we obtain

$$\begin{aligned}
&\sum_{k\geqslant 1} P\left(\sup_{1\leqslant n\leqslant n_{k+1}} S_n \geqslant (1+\delta)s_{n_k}v_{n_k}\right) \\
&\quad \leqslant 2\sum_{k\geqslant 1} P\left(S_{n_{k+1}}/s_{n_{k+1}} \geqslant (1+\delta)s_{n_k}v_{n_k}/s_{n_{k+1}} - \sqrt{2}\right) < \infty.
\end{aligned}$$

In view of the Borel-Cantelli lemma (18.24.i), this shows that

$$P\left(\sup_{1\leqslant n\leqslant n_{k+1}} S_n \geqslant (1+\delta)s_{n_k}v_{n_k} \text{ i.o.}\right) = 0. \tag{4}$$

As $\{S_n \geqslant (1+\delta)s_n v_n \text{ i.o.}\} \subset \{\sup_{1\leqslant n\leqslant n_{k+1}} S_n \geqslant (1+\delta)s_{n_k}v_{n_k} \text{ i.o.}\}$, (4) entails that $\{(1+\delta)s_n v_n : n \geqslant 1\}$ belongs to the upper class of $\{S_n : n \geqslant 1\}$.

(II) Now we prove that $\{(1-\delta)s_n v_n : n \geqslant 1\}$ belongs to the lower class of $\{S_n : n \geqslant 1\}$. Choose $\gamma > 0$ so that $(1-\delta')^2(1+\gamma) < 1$. Then, applying (28.4) with $\alpha = \sqrt{2}(1-\delta')$, since $u_k^2 \sim (\rho^2-1)\rho^{2k}$, there are $k_0 \in N$ and $C < \infty$ such that

$$\begin{aligned}
&\sum\nolimits_{k\geqslant k_0} P(S_{n_{k+1}} - S_{n_k} \geqslant (1-\delta')u_k\sqrt{2\log_2 u_k^2}) \geqslant \sum\nolimits_{k\geqslant k_0} (\log u_k^2)^{-(1-\delta')^2(1+\gamma)} \\
&\quad \geqslant C\sum_{k\geqslant k_0} k^{-(1-\delta')^2(1+\gamma)} = \infty.
\end{aligned}$$

Since $S_{n_{k+1}} - S_{n_k}$, $k \geqslant 1$, are independent, the Borel-Cantelli lemma (18.24.ii) implies that

$$P(S_{n_{k+1}} - S_{n_k} \geqslant (1-\delta')u_k\sqrt{2\log_2 u_k^2} \text{ i.o.}) = 1. \tag{5}$$

If every f_n is replaced by $-f_n$, then (2) yields $\liminf_n S_n/s_n v_n \geqslant -1$ a.s. This means that, for all $x \in X$ except for a set of probability zero,

$$S_n(x) \geqslant -2s_n v_n, \quad n \geqslant n(x). \tag{6}$$

We see that $((1-\delta')u_k\sqrt{2\log_2 u_k^2} - 2s_{n_k}v_{n_k})/s_{n_{k+1}}v_{n_{k+1}} \to (1-\delta')\sqrt{1-\rho^{-2}} - 2\rho^{-1}$, and so, if we take ρ sufficiently large so that $(1-\delta')\sqrt{1-\rho^{-2}} - 2\rho^{-1} > 1-\delta$, then

$$(1-\delta')u_k\sqrt{2\log_2 u_k^2} - 2s_{n_k}v_{n_k} \geqslant (1-\delta)s_{n_{k+1}}v_{n_{k+1}} \tag{7}$$

for all large k. On account of (5)–(7), we find that

$$\begin{aligned} P(S_{n_{k+1}} &\geqslant (1-\delta)s_{n_{k+1}}v_{n_{k+1}} \text{ i.o.}) \\ &\geqslant P(S_{n_{k+1}} \geqslant (1-\delta')u_k\sqrt{2\log_2 u_k^2} - 2s_{n_k}v_{n_k} \text{ i.o.}) \\ &\geqslant P(S_{n_{k+1}} - S_{n_k} \geqslant (1-\delta')u_k\sqrt{2\log_2 u_k^2} \text{ i.o.}) = 1. \end{aligned}$$

This ensures that $\{(1-\delta)s_n v_n : n \geqslant 1\}$ belongs to the lower class of $\{S_n : n \geqslant 1\}$.

From (I), (II) and (28.2.b) we obtain (1). □

Remarks 28.8. Assume the condition in the Kolmogorov LIL is fulfilled.

(a) Substituting the sequence $\{-f_n : n \geqslant 1\}$ for $\{f_n : n \geqslant 1\}$, from (28.7) it follows that $\liminf_n S_n/s_n\sqrt{2\log_2 s_n^2} = -1$ a.s., and so $\limsup_n |S_n|/s_n\sqrt{2\log_2 s_n^2} = 1$ a.s.

(b) Since $-1 = \liminf_n S_n/s_n\sqrt{2\log_2 s_n^2} < \limsup_n S_n/s_n\sqrt{2\log_2 s_n^2} = 1$ a.s., it follows that $P(\{S_n \leqslant 0 \text{ eventually}\} \cup \{S_n \geqslant 0 \text{ eventually}\}) = 0$, i.e. the random walk $\{S_n : n \geqslant 1\}$ crosses zero infinitely often with probability one.

(c) $\liminf_n S_n = -\infty$ a.s., $\limsup_n S_n = \infty$ a.s. and $S_n/s_n^{1+\varepsilon} \xrightarrow{a.s.} 0$ for each $\varepsilon > 0$.

Theorem 28.9. *If* $\limsup_n S_n/s_n\sqrt{2\log_2 s_n^2} = 1$ *a.s., then*

$$\sum_{n\geqslant 1} P(f_n \geqslant \varepsilon s_n\sqrt{\log_2 s_n^2}) < \infty, \quad \varepsilon > \sqrt{2}. \tag{1}$$

Proof. For $n \geqslant 1$, set $g_n = f_n/s_n\sqrt{2\log_2 s_n^2}$ and $h_n = S_{n-1}/s_n\sqrt{2\log_2 s_n^2}$. Clearly, $\sigma(h_n)$ is independent of $\sigma(g_k, k \geqslant n)$ for any $n > 1$. $\limsup_n S_n/s_n\sqrt{2\log_2 s_n^2} = 1$ a.s. implies

$$P(\limsup_n (g_n + h_n) \geqslant \varepsilon) = 0, \quad \varepsilon > 1. \tag{2}$$

For $\delta > 0$, by Chebyshev's inequality, we have $P(|h_n| \geqslant \delta) \leqslant s_{n-1}^2/2\delta^2 s_n^2 \log_2 s_n^2 \leqslant 1/\delta^2 \log_2 s_n^2 \to 0$, and so

$$P(h_n \geqslant -\delta) \to 1. \tag{3}$$

From (2), (3) and (18.58.b) we infer that $P(\limsup_n g_n > \varepsilon + \delta) = 0$ for any $\varepsilon > 1$ and $\delta > 0$. By this and (5.63.b), it follows that $P(\limsup_n \{f_n/s_n\sqrt{2\log_2 s_n^2} \geqslant \varepsilon\}) \leqslant P(\limsup_n f_n/s_n\sqrt{2\log_2 s_n^2} \geqslant \varepsilon) = 0$ for each $\varepsilon > 1$. Hence (1) obtains via the Borel-Cantelli lemma (18.24.ii). □

Corollary 28.10. *If* $\limsup_n S_n/s_n\sqrt{2\log_2 s_n^2} = 1$ *a.s. and* $\liminf_n S_n/s_n\sqrt{2\log_2 s_n^2} = -1$ *a.s., then*

$$\sum_{n\geqslant 1} P(|f_n| \geqslant \varepsilon s_n\sqrt{\log_2 s_n^2}) < \infty, \quad \varepsilon > \sqrt{2}.$$

We embark now on a proof of the Hartman-Wintner LIL and need again some lemmas. For $n \geqslant 1$, we put $L_n = \log_2 n$ and $b_n = \sqrt{2nL_n}$.

Lemma 28.11. *Let* $g_n, n \geqslant 1$, *be independent nondegenerate random variables, and set* $T_n = \sum_{k=1}^n g_k, n \geqslant 1$. *If* $T_n/b_n \xrightarrow{P} 0$ *and* $P(T_n/b_n \geqslant \alpha) \leqslant ce^{-\gamma L_n}, n \geqslant n_0$, *for some* $c < \infty, \alpha > 0, \gamma > 1$ *and* $n_0 \in N$, *then* $\limsup_n T_n/b_n \leqslant \alpha$ *a.s.*

Proof. Let $\delta > 0$ and $1 < \sqrt{\rho} < 1 + \delta/2\alpha$, and put $n_k = [\rho^k], k \geqslant 1$. We shall show that $\{(\alpha+\delta)b_n : n \geqslant 1\}$ belongs to the upper class of $\{T_n : n \geqslant 1\}$. Let $k_0 \in N$ be such that $n_k \geqslant n_0 \vee m_0$ and $b_{n_{k+1}}/b_{n_k} < 1 + \delta/2\alpha$ for $k \geqslant k_0$, where m_0 is as in (25.4.d). Then, in view of (25.40.d), we have

$$\begin{aligned} P\left(\sup_{1\leqslant n\leqslant n_{k+1}} T_n \geqslant (\alpha+\delta)b_{n_k}\right) &\leqslant 2P\left(T_{n_{k+1}} \geqslant \left(\alpha + \frac{\delta}{2}\right) b_{n_k}\right) \\ &= 2P\left(\frac{T_{n_{k+1}}}{b_{n_{k+1}}} \geqslant \left(\alpha+\frac{\delta}{2}\right)\frac{b_{n_k}}{b_{n_{k+1}}}\right) \\ &\leqslant 2P\left(\frac{T_{n_{k+1}}}{b_{n_{k+1}}} \geqslant \alpha\right) \leqslant 2c(\log n_{k+1})^{-\gamma}, \quad k \geqslant k_0. \end{aligned}$$

Therefore, for some $C < \infty$, we obtain

$$\sum_{k\geqslant k_0} P\left(\sup_{1\leqslant n\leqslant n_{k+1}} T_n \geqslant (\alpha+\delta)b_{n_k}\right) \leqslant C\sum_{k\geqslant k_0}(k+1)^{-\gamma} < \infty.$$

Hence, by the Borel-Cantelli lemma (18.24.i),

$$P\left(T_n \geqslant (\alpha+\delta)b_n \text{ i.o.}\right) \leqslant P\left(\sup_{1\leqslant n\leqslant n_{k+1}} T_n \geqslant (\alpha+\delta)b_{n_k} \text{ i.o.}\right) = 0.$$ □

Lemma 28.12. *For $n \geqslant 1$, let g_n and T_n be as in* (28.11), *and assume that $Eg_n = 0$ and $|g_n| \leqslant c\sqrt{n/L_n}$ a.s. for some $c < \infty$. If $\sup_{n\geqslant 1} Eg_n^2 \leqslant b < \infty$, then, for all $n \geqslant 1$ and $\alpha > 0$,*

$$P(T_n/b_n \geqslant \alpha) \leqslant \exp\{-(\alpha^2/b)(2 - e^{\sqrt{2}\alpha c/b})L_n\}. \tag{1}$$

Proof. Let $t > 0$. For $1 \leqslant i \leqslant n$, $|g_i|/b_n \leqslant c\sqrt{i/L_i}/\sqrt{2nL_n} \leqslant c\sqrt{n/L_n}/\sqrt{2nL_n} = c/\sqrt{2}L_n$. Consequently, since $e^y \leqslant 1 + y + (y^2/2)e^{|y|}$, $y \in R$, and $1 + y \leqslant e^y$, $y \in R$, we have

$$Ee^{tg_i/b_n} \leqslant 1 + \frac{t^2 Eg_i^2}{2b_n^2} \exp\frac{tc}{\sqrt{2}L_n} \leqslant \exp\left\{\frac{t^2 b}{4nL_n} \exp\frac{tc}{\sqrt{2}L_n}\right\}.$$

Therefore, via (13.25.b) and independence, we get

$$\begin{aligned} P(T_n/b_n \geqslant \alpha) \leqslant e^{-t\alpha} Ee^{tT_n/b_n} &= e^{-t\alpha} \prod_{i=1}^{n} Ee^{tg_i/b_n} \\ &\leqslant \exp\left\{-t\alpha + \frac{t^2 b}{4L_n} \exp\frac{tc}{\sqrt{2}L_n}\right\}, \quad \alpha > 0. \end{aligned} \tag{2}$$

Setting $t = 2\alpha L_n/b$ in (2) we arrive at (1). □

Lemma 28.13. *Suppose that $f_n, n \geqslant 1$, are i.i.d., let $c > 0$, and define $h_n = f_n 1_{\{|f_n| \geqslant c\sqrt{n/L_n}\}}$ and $U_n = \sum_{i=1}^{n} h_i$ for $n \geqslant 1$. Then $U_n/b_n \overset{a.s.}{\to} 0$.*

Proof. Since $\sum_{n=1}^{m} \frac{1}{b_n} \leqslant C\sqrt{m/L_m}$, $m \geqslant 1$, for some constant C, we have

$$\begin{aligned} \sum_{n\geqslant 1} \frac{1}{b_n} E\,|h_n| &= \sum_{n\geqslant 1} \frac{1}{b_n} E\left[|f_1|\,1_{\{|f_1| \geqslant c\sqrt{n/L_n}\}}\right] \\ &= \sum_{n\geqslant 1} \frac{1}{b_n} \sum_{m\geqslant n} E\left[|f_1|\,1_{\{c\sqrt{m/L_m} \leqslant |f_1| < c\sqrt{(m+1)/L_{m+1}}\}}\right] \\ &= \sum_{m\geqslant 1} \left(\sum_{n=1}^{m} \frac{1}{b_n}\right) E\left[|f_1|\,1_{\{c\sqrt{m/L_m} \leqslant |f_1| < c\sqrt{(m+1)/L_{m+1}}\}}\right] \\ &\leqslant Cc \sum_{m\geqslant 1} \sqrt{m/L_m}\sqrt{(m+1)/L_{m+1}}\, P(c\sqrt{m/L_m} \leqslant |f_1| < c\sqrt{(m+1)/L_{m+1}}) \\ &\leqslant C' \sum_{m\geqslant 1} (m/L_m) P(m/L_m \leqslant f_1^2/c^2 < (m+1)/L_{m+1}) \leqslant (C'/c^2) Ef_1^2 < \infty, \end{aligned}$$

where $C' < \infty$. Hence, on account of (27.9.i), we arrive at the desired result. □

The following lemma is related to (20.40.c).

Lemma 28.14. *Assume that $f_n, n \geqslant 1$, are i.i.d. with $Ef_1^2 = 1$, and let $x_n > 0$ be such that $x_n \to \infty$ and $x_n/n \to 0$. Then $\liminf_n x_n^{-1} \log P(S_n \geqslant \sqrt{nx_n}) \geqslant -1/2$.*

Proof. Put $S_0 = 0$, fix $\delta, t > 0$, and set $l_n = [n/tx_n]$ and $m_n = [tx_n]$ for $n \geqslant 1$. For all sufficiently large n, since $m_n l_n \leqslant n$, we can define $g_i = S_{il_n} - S_{(i-1)l_n}$, $1 \leqslant i \leqslant m_n$. Then $g_1, \ldots, g_{m_n}$ are i.i.d. random variables. We have

$$\begin{aligned}
&P\left(S_n \geqslant (1+\delta)\frac{m_n}{t}\sqrt{\frac{n}{x_n}}\right)\\
&\geqslant P\Bigg(S_{m_n l_n} = g_1 + \cdots + g_{m_n}\\
&\geqslant (1+2\delta)\frac{m_n}{t}\sqrt{\frac{n}{x_n}},\ S_n - S_{m_n l_n} \geqslant -\delta\frac{m_n}{t}\sqrt{\frac{n}{x_n}}\Bigg)\\
&\geqslant P\left(g_i \geqslant \frac{1+2\delta}{t}\sqrt{\frac{n}{x_n}},\ 1 \leqslant i \leqslant m_n\right) P\left(S_n - S_{m_n l_n} \geqslant -\delta\frac{m_n}{t}\sqrt{\frac{n}{x_n}}\right)\\
&= P\left(S_{l_n} = g_1 \geqslant \frac{1+2\delta}{t}\sqrt{\frac{n}{x_n}}\right)^{m_n} P\left(S_n - S_{m_n l_n} \geqslant -\delta\frac{m_n}{t}\sqrt{\frac{n}{x_n}}\right). \qquad (1)
\end{aligned}$$

By Chebyshev's inequality,

$$P\left(\left|S_n - S_{m_n l_n}\right| \geqslant \delta\frac{m_n}{t}\sqrt{\frac{n}{x_n}}\right) \leqslant \frac{t^2 x_n}{\delta^2 n m_n^2}(n - m_n l_n) \leqslant \frac{t^2 x_n}{\delta^2 m_n^2} \sim \frac{1}{\delta^2 x_n} \to 0,$$

and so, for all n large enough,

$$P\left(S_n - S_{m_n l_n} \geqslant -\delta\frac{m_n}{t}\sqrt{\frac{n}{x_n}}\right) \geqslant \frac{1}{2}.$$

This and (1) entail that

$$\liminf_n \frac{1}{m_n}\log P\left(S_n \geqslant (1+\delta)\frac{m_n}{t}\sqrt{\frac{n}{x_n}}\right) \geqslant \liminf_n \log P\left(S_{l_n} \geqslant \frac{1+2\delta}{t}\sqrt{\frac{n}{x_n}}\right). \qquad (2)$$

Since $tl_n x_n/n \to 1$, the central limit theorem (24.23.c), in conjunction with (19.61) and (19.18.e), shows that

$$P\left(S_{l_n} \geqslant \frac{1+2\delta}{t}\sqrt{\frac{n}{x_n}}\right) \to 1 - \Phi\left(\frac{1+2\delta}{\sqrt{t}}\right) \geqslant C\sqrt{t}\exp\left\{-\frac{(1+2\delta)^2}{2t}\right\} \qquad (3)$$

for some constant C and all sufficiently small t (see (19.50)). From (2) and (3), as $m_n \sim tx_n$, we infer that

$$\liminf_n \frac{1}{x_n}\log P\left(S_n \geqslant (1+\delta)\frac{m_n}{t}\sqrt{\frac{n}{x_n}}\right) \geqslant t\left(\log C + \frac{1}{2}\log t\right) - \frac{(1+2\delta)^2}{2}.$$

Since $(m_n/t)\sqrt{n/x_n} \sim \sqrt{nx_n}$, this implies

$$\liminf_n \frac{1}{x_n}\log P(S_n \geqslant \sqrt{nx_n}) \geqslant t\left(\log C + \frac{1}{2}\log t\right) - \frac{(1+2\delta)^2}{2}. \qquad (4)$$

Letting $\delta \to 0$ and $t \to 0$ in (4) proves the result. □

Corollary 28.15. *Let $f_n, n \geqslant 1$, be as in (28.14), and let $\alpha, \gamma > 0$. Then, for all sufficiently large n,*

$$P(S_n \geqslant \sqrt{2\alpha n L_n}) \geqslant (\log n)^{-(\alpha+\gamma)}. \tag{1}$$

Proof. Applying (28.14) with $x_n = 2\alpha L_n$, we obtain $\liminf_n L_n^{-1} \log P(S_n \geqslant \sqrt{2\alpha n L_n}) \geqslant -\alpha$. Consequently, there is $n_0 \in N$ so that $L_n^{-1} \log P(S_n \geqslant \sqrt{2\alpha n L_n}) \geqslant -\alpha - \gamma$ for $n \geqslant n_0$. This immediately implies (1). □

Hartman-Wintner LIL 28.16. *Assume that $f_n, n \geqslant 1$, are i.i.d. with $Ef_1^2 = 1$. Then* $\limsup_n S_n/\sqrt{2nL_n} = 1$ *a.s. and* $\liminf_n S_n/\sqrt{2nL_n} = -1$ *a.s.*

Proof. (I) First we show that $\limsup_n S_n/\sqrt{2nL_n} \leqslant 1$ a.s. Let $\alpha > 1$ and choose c small enough so that $\gamma = \alpha^2(2 - e^{\sqrt{2}\alpha c}) > 1$. Then, applying (28.12) with $g_n = f_n 1_{\{|f_n|<(c/2)\sqrt{n/L_n}\}} - E[f_n 1_{\{|f_n|<(c/2)\sqrt{n/L_n}\}}], n \geqslant 1$, and $b = 1$, we have $P(T_n/b_n \geqslant \alpha) \leqslant \exp\{-\alpha^2(2 - e^{\sqrt{2}\alpha c})L_n\} = e^{-\gamma L_n}, n \geqslant 1$. For $\varepsilon > 0$, by Chebyshev's inequality, we have also $P(|T_n/b_n| \geqslant \varepsilon) \leqslant n/\varepsilon^2 b_n^2 = 1/2\varepsilon^2 L_n \to 0$, i.e. $T_n/b_n \xrightarrow{P} 0$. On account of (28.11), it follows that

$$\limsup_n T_n/b_n \leqslant \alpha \text{ a.s.} \tag{1}$$

For $n \geqslant 1$, write $h_n = f_n 1_{\{|f_n|\geqslant(c/2)\sqrt{n/L_n}\}}$ and $U_n = \sum_{i=1}^n h_i$, and observe that $S_n = T_n + U_n + \sum_{i=1}^n E[f_i 1_{\{|f_i|<(c/2)\sqrt{i/L_i}\}}]$. Since $Ef_1 = 0$, according to the proof of (28.13), we obtain

$$\sum_{n\geqslant 1} \left| E[f_n 1_{\{|f_n|<(c/2)\sqrt{n/L_n}\}}] \right| /b_n \leqslant \sum_{n\geqslant 1} E\,|h_n|\,/b_n < \infty,$$

whence, by Kronecker's lemma (27.5),

$$\frac{1}{b_n} \sum_{i=1}^n \left| E[f_i 1_{\{|f_i|<(c/2)\sqrt{i/L_i}\}}] \right| \to 0. \tag{2}$$

From (1), (28.13) and (2) we get

$$\limsup_n S_n/b_n \leqslant \limsup_n T_n/b_n + \limsup_n U_n/b_n$$
$$+ \lim_n \frac{1}{b_n} \sum_{i=1}^n E[f_i 1_{\{|f_i|<(c/2)\sqrt{i/L_i}\}}] \leqslant \alpha \text{ a.s.}$$

As α is arbitrary, this proves the desired inequality.

(II) Now we prove that $\limsup_n S_n/\sqrt{2nL_n} \geqslant 1$ a.s. Let $0 < \alpha < 1$ and choose $\gamma > 0$ so that $\alpha + \gamma \leqslant 1$. Put $n_k = k^k, k \geqslant 1$, and notice that $n_{k+1} - n_k \sim n_{k+1}$. Since the distribution of $S_{n_{k+1}} - S_{n_k}$ coincides with the distribution of $S_{n_{k+1}-n_k}$, there exist

$k_0 \in N$ and $0 < C < 1$ such that, by (28.15),

$$\sum_{k \geqslant k_0} P(S_{n_{k+1}} - S_{n_k} \geqslant \sqrt{\alpha} b_{n_{k+1}-n_k}) \geqslant \sum_{k \geqslant k_0} (\log(n_{k+1} - n_k))^{-(\alpha+\gamma)}$$
$$\geqslant C \sum_{k \geqslant k_0} ((k+1)\log(k+1))^{-(\alpha+\gamma)} = \infty.$$

Consequently, as $S_{n_{k+1}} - S_{n_k}, k \geqslant 1$, are independent, the Borel-Cantelli lemma (18.24.ii) yields $P(S_{n_{k+1}} - S_{n_k} \geqslant \sqrt{\alpha} b_{n_{k+1}-n_k}$ i.o.$) = 1$. Since α is at our disposal, and $b_{n_{k+1}-n_k} \sim b_{n_{k+1}}$, this shows that

$$\limsup_k (S_{n_{k+1}} - S_{n_k})/b_{n_{k+1}} \geqslant 1 \text{ a.s.} \tag{3}$$

On account of (I), $-1 \leqslant \liminf_n S_n/b_n \leqslant \limsup_n S_n/b_n \leqslant 1$ a.s. Hence, since $b_{n_k}/b_{n_{k+1}} \to 0$, we have

$$S_{n_k}/b_{n_{k+1}} = (S_{n_k}/b_{n_k})(b_{n_k}/b_{n_{k+1}}) \to 0. \tag{4}$$

From (3) and (4) we get

$$\limsup_k S_{n_{k+1}}/b_{n_{k+1}} = \limsup_k (S_{n_{k+1}} - S_{n_k})/b_{n_{k+1}} + \lim_k S_{n_k}/b_{n_{k+1}} \geqslant 1 \text{ a.s.}$$

A fortiori, $\limsup_n S_n/b_n \geqslant 1$ a.s.

From (I) and (II) we obtain $\limsup_n S_n/\sqrt{2nL_n} = 1$ a.s., which in turns implies that $\liminf_n S_n/\sqrt{2nL_n} = -1$ a.s. □

Remarks 28.17. (a) If $f_n, n \geqslant 1$, are i.i.d. with $Ef_1^2 = \sigma^2$, then, substituting $\{f_n/\sigma : n \geqslant 1\}$ for $\{f_n : n \geqslant 1\}$, (28.16) yields $\limsup_n S_n/\sqrt{2nL_n} = \sigma$ a.s. and $\liminf_n S_n/\sqrt{2nL_n} = -\sigma$ a.s., and so $\limsup_n |S_n|/\sqrt{2nL_n} = \sigma$ a.s.

(b) Similarly to (28.8.b) and (28.8.c), the Hartman-Wintner LIL shows that the random walk $\{S_n : n \geqslant 1\}$ crosses zero infinitely often with probability one, and that $\liminf_n S_n = -\infty$ a.s. and $\limsup_n S_n = \infty$ a.s. It also shows that $S_n/n^{1/2+\varepsilon} \overset{a.s.}{\to} 0$ for any $\varepsilon > 0$, and so implies the classical SLLN.

The Hartman-Wintner LIL admits a converse result. To prove it, we need the next lemma.

Lemma 28.18. *Let $(X, \mathcal{A}, P)$ and $(Y, \mathcal{B}, Q)$ be probability spaces. For $n \geqslant 1$, let f_n be a random variable on X, let g_n be a random variable on Y, let $\varphi_n : R^n \to R$ be a measurable function, and let $B_n \in \mathcal{B}(R)$. If $P(f_n)_{n\geqslant 1}^{-1} = Q(g_n)_{n\geqslant 1}^{-1}$, then $P(\varphi_n \circ (f_1, \ldots, f_n) \in B_n$ i.o.$) = Q(\varphi_n \circ (g_1, \ldots, g_n) \in B_n$ i.o.$)$.*

Proof. Put $A_n = \varphi_n^{-1}(B_n)$, $n \geqslant 1$. Notice that $P(f_1, \ldots, f_n)^{-1} = Q(g_1, \ldots, g_n)^{-1}$, $n \geqslant 1$. Then, for $1 \leqslant n < m$, $P(\cup_{k=n}^m \{(f_1, \ldots, f_k) \in A_k\}) = Q(\cup_{k=n}^m \{(g_1, \ldots, g_k) \in$

$A_k\})$ by the Poincaré formula (9.56). Therefore, we have

$$\begin{aligned}
P(\varphi_n \circ (f_1, \dots, f_n) \in B_n \text{ i.o.}) &= P\left(\bigcap_{n \geqslant 1} \bigcup_{k \geqslant n} \{(f_1, \dots, f_k) \in A_k\}\right) \\
&= \lim_n \lim_m P\left(\bigcup_{k=n}^{m} \{(f_1, \dots, f_k) \in A_k\}\right) \\
&= \lim_n \lim_m Q\left(\bigcup_{k=n}^{m} \{(g_1, \dots, g_k) \in A_k\}\right) \\
&= Q(\varphi_n \circ (g_1, \dots, g_n) \in B_n \text{i.o.}). \quad \square
\end{aligned}$$

Corollary 28.19. *For $n \geqslant 1$, let f_n and g_n be as in (28.18), set $S_n = \sum_{k=1}^n f_k$ and $T_n = \sum_{k=1}^n g_k$, and let $b_n \in R - \{0\}$. Assume that $P(f_n)_{n \geqslant 1}^{-1} = Q(g_n)_{n \geqslant 1}^{-1}$, and let $c \in \overline{R}$. Then*

$$\limsup_n S_n/b_n = c \ P\text{-}a.s. \text{ if and only if } \limsup_n T_n/b_n = c \ Q\text{-}a.s. \tag{1}$$

and

$$\liminf_n S_n/b_n = c \ P\text{-}a.s. \text{ if and only if } \liminf_n T_n/b_n = c \ Q\text{-}a.s. \tag{2}$$

Proof. We apply (28.18) with $\varphi_n(x_1, \dots, x_n) = x_1 + \dots + x_n$, $(x_1, \dots, x_n) \in R^n$. If $c \in R$, then (1) follows from (28.2.b) and (28.18) with $B_n = [(c + \delta)b_n, \infty[$ and $B_n = [(c - \delta)b_n, \infty[, \delta > 0$. If $c = \infty$, then (1) follows from (28.18) with $B_n = [Mb_n, \infty[, M > 0$. If $c = -\infty$, then (1) is a consequence of the preceding two situations. Finally, since $\liminf_n S_n/b_n = -\limsup_n S_n/(-b_n)$, (2) obtains by (1). $\square$

Theorem 28.20 (Strassen). *If $f_n, n \geqslant 1$, are i.i.d. arbitrary random variables such that*

$$\limsup_n S_n/\sqrt{2nL_n} = \sigma \text{ a.s.}, \quad \liminf_n S_n/\sqrt{2nL_n} = -\sigma \text{ a.s.} \tag{1}$$

for some $\sigma \in]0, \infty[$, then $Ef_1 = 0$ and $Ef_1^2 = \sigma^2$.

Proof. Observe first that (1) implies $S_n/n \overset{a.s.}{\to} 0$, and so $Ef_1 = 0$ by (27.14.ii). For $n \geqslant 1$, let f_n', f_n'', f_n^*, S_n', S_n'' and S_n^* be as in the proof of (25.17). Then, with respect to P^*, $f_n^*, n \geqslant 1$, are i.i.d. symmetric random variables by (23.20.d) and (23.20.f). Since $P^*(f_n')_{n \geqslant 1}^{-1} = P^*(f_n'')_{n \geqslant 1}^{-1} = P(f_n)_{n \geqslant 1}^{-1}$, we have

$$\limsup_n S_n'/\sqrt{2nL_n} = \sigma \ P^*\text{-a.s.}, \quad \liminf_n S_n''/\sqrt{2nL_n} = -\sigma \ P^*\text{-a.s.}$$

by (28.19). Consequently,

$$\limsup_n S_n^*/\sqrt{2nL_n} = \limsup_n (S_n' - S_n'')/\sqrt{2nL_n}$$

$$\leqslant \limsup_n S'_n/\sqrt{2nL_n} - \liminf_n S''_n/\sqrt{2nL_n} = 2\sigma\, P^*\text{-a.s.} \tag{2}$$

Now, for $M > 0$ and $n \geqslant 1$, define $g_n = f_n^* 1_{\{|f_n^*|<M\}} - f_n^* 1_{\{|f_n^*|\geqslant M\}}$ and set $T_n = \sum_{k=1}^n g_k$. Then, in view of (20.42.g), $P^*(f_n^*)^{-1} = P^* g_n^{-1}$ for each $n \geqslant 1$. Thus $g_n, n \geqslant 1$, are i.i.d. symmetric random variables, and $P^*(f_n^*)_{n\geqslant 1}^{-1} = P^*(g_n)_{n\geqslant 1}^{-1}$ by (18.29). It follows from (28.19) that

$$\limsup_n S_n^*/\sqrt{2nL_n} = \limsup_n T_n/\sqrt{2nL_n}\; P^*\text{-a.s.} \tag{3}$$

Notice that $E^* f_1^{*2} > 0$, since otherwise $f_n, n \geqslant 1$, would be degenerate and (1) would be violated. Notice also that $f_n^* 1_{\{|f_n^*|<M\}}, n \geqslant 1$, are i.i.d. symmetric random variables (20.42.e). Then, on account of (3) and (2), by applying the Hartman-Wintner LIL (28.16) and (28.17.a) for sufficiently large M, we obtain

$$\begin{aligned}(E^*[f_1^{*2} 1_{\{|f_1^*|<M\}}])^{1/2} &= \limsup_n \frac{1}{\sqrt{2nL_n}} \sum_{k=1}^n f_k^* 1_{\{|f_k^*|<M\}} \\ &= \limsup_n \frac{1}{\sqrt{2nL_n}} \frac{S_n^* + T_n}{2} \\ &\leqslant \frac{1}{2}\left(\limsup_n S_n^*/\sqrt{2nL_n} + \limsup_n T_n/\sqrt{2nL_n}\right) \\ &\leqslant 2\sigma\, P^*\text{-a.s.}\end{aligned} \tag{4}$$

Letting $M \to \infty$ in (4), we get $\sqrt{2Ef_1^2} \leqslant 2\sigma$, i.e. $Ef_1^2 \leqslant 2\sigma^2$. In fact, $Ef_1^2 = \sigma^2$ by (28.17.a) and (1). □

Exercise 28.21. If $0 < b_n \to \infty$, does it follow that $P(S_n \geqslant b_n \text{ i.o.}) = 0$ or 1?

Exercise 28.22 (Bernstein). If $\sup_{1\leqslant i\leqslant n} |f_i|/s_n \leqslant c_n < \infty$ a.s., then $P(S_n \geqslant x) \leqslant \exp\{-x^2/2(s_n^2 + xc_n)\}, x > 0$. The next steps may be helpful.

(a) $1 + x \leqslant e^x \leqslant 1 + x + x^2/2 + (|x|^3/6)e^{|x|}, x \in R$.

(b) For $t > 0$ and $1 \leqslant i \leqslant n$, use (a) to show that

$$Ee^{tf_i} \leqslant \exp\left\{\frac{t^2\sigma_i^2}{2}\left(1 + \frac{tc_n e^{tc_n}}{3}\right)\right\}.$$

(c) For $t > 0$, use (b) and (13.25.b) to obtain

$$P(S_n \geqslant x) \leqslant \exp\left\{-tx + \frac{t^2 s_n^2}{2}\left(1 + \frac{tc_n e^{tc_n}}{3}\right)\right\}, \quad x > 0. \tag{1}$$

(d) $P(S_n \geqslant x) \leqslant \exp\{-x^2/2(s_n^2 + xc_n)\}, x > 0$. [Set $t = x/(s_n^2 + xc_n)$ in (1).]

Exercise 28.23. For $n \geqslant 1$, assume that $P(f_n = \sqrt{n}/\log_2 n) = P(f_n = -\sqrt{n}/\log_2 n) = 1/2$. Prove the following.

(a) $\sum_{n\geqslant 1} \mathrm{Var}[f_n]/n^2 = \infty$ (see (27.7)).

(b) The condition in the Kolmogorov LIL is satisfied with $\varepsilon_n = O(\sqrt{(\log_2 n)/n})$.

(c) $S_n/n \overset{a.s.}{\to} 0$.

Exercise 28.24. For $n \geqslant 1$, suppose that f_n has a normal distribution with $\sigma_n/s_n \to 0$.

(a) Show that $(\log_2 s_n^2)^{-1} \log P(S_n/s_n \geqslant \alpha\sqrt{2\log_2 s_n^2}) \to -\alpha^2$. [Do not forget (20.40.c).]

(b) Use (a) to verify that $\limsup_n S_n/s_n\sqrt{2\log_2 s_n^2} = 1$ a.s.

Exercise 28.25. Let $g_n, n \geqslant 1$, be independent Rademacher random variables, let $a_n \in R, n \geqslant 1$, and set $v_n^2 = \sum_{i=1}^n a_i^2, n \geqslant 1$.If $v_n \to \infty$ and $a_n = o(v_n/\log_2 v_n^2)$, verify that $\limsup_n \frac{1}{v_n\sqrt{2\log_2 v_n^2}} \sum_{i=1}^n a_i g_i = 1$ a.s.

Throughout (28.26)–(28.28), we consider that $f_n, n \geqslant 1$, are i.i.d. random variables.

Exercise 28.26. If f_1 has a Bernoulli symmetric distribution, then $P(S_n = 0 \text{ i.o.}) = 1$

Exercise 28.27. If $Ef_1^2 = 1$, then $\liminf_n |S_n|/\sqrt{n} = 0$ a.s. The following steps may be useful.

(a) Let $n_k = k^k$. Then $S_{n_k}/\sqrt{n_{k+1}} \overset{a.s.}{\to} 0$. $[S_{n_k}/\sqrt{n_{k+1}} = (S_{n_k}/\sqrt{2n_k L_{n_k}})/\sqrt{2n_k L_{n_k}/n_{k+1}}$. Use (28.16).]

(b) $(S_{n_{k+1}} - S_{n_k})/\sqrt{n_{k+1}} \overset{D}{\to} \Phi$. [Use (24.23.c), (a) and Slutsky's theorem (19.27).]

(c) $P\left(\left|S_{n_{k+1}} - S_{n_k}\right|/\sqrt{n_{k+1}} < \delta \text{ i.o.}\right) = 1, \delta > 0$. [Hint. For $\delta > 0$, choose $\varepsilon > 0$ so that $\varepsilon < \Phi(\delta) - \Phi(-\delta)$. Then $P(\left|S_{n_{k+1}} - S_{n_k}\right|/\sqrt{n_{k+1}} < \delta) > \Phi(\delta) - \Phi(-\delta) - \varepsilon$ for all sufficiently large k by (b). Apply the Borel-Cantelli lemma (18.24.ii).]

(d) $P(\left|S_{n_{k+1}}\right|/\sqrt{n_{k+1}} < 2\delta \text{ i.o.}) = 1, \delta > 0$. [Hint. In view of (a), $\limsup_k\{\left|S_{n_{k+1}}\right|/\sqrt{n_{k+1}} < 2\delta\} \supset \limsup_k\{\left|S_{n_{k+1}} - S_{n_k}\right|/\sqrt{n_{k+1}} < \delta, \left|S_{n_k}\right|/\sqrt{n_{k+1}} < \delta\} \overset{a.s.}{=} \limsup_k\{\left|S_{n_{k+1}} - S_{n_k}\right|/\sqrt{n_{k+1}} < \delta\}$.]

Exercise 28.28. Put $V_n^2 = \sum_{i=1}^n f_i^2, n \geqslant 1$.

(a) If f_1 is symmetric, then $\limsup_n S_n/V_n\sqrt{2\log_2 V_n^2} = 1$ a.s.

(b) If $g_n, n \geqslant 1$, are as in (28.25), and $\sigma(g_n, n \geqslant 1)$ is independent of $\sigma(f_n, n \geqslant 1)$, then $\limsup_n \frac{1}{V_n\sqrt{2\log_2 V_n^2}} \sum_{i=1}^n f_i g_i = 1$ a.s.

11 Conditioning

While Chapter 10 is devoted to the notion of independence related to sums of random variables, this chapter is concerned with the concept of conditioning with respect to a σ-algebra. This plays an important role in key areas such as Markov processes, semimartingale theory and ergodic theory. Among others, Section 29 includes topics such as regular conditional probability, conditional independence and Markov triple. Applying results of this section, Section 30 contains a basic account of the theory of semimartingales in discrete time.

29 Conditional Expectations, Conditional Probabilities and Conditional Independence

In Section 18 we introduced the notion of conditional probability given a nonnull event. Resting on it, we begin this section by treating the elementary case of conditioning relative to a countable partition of the underlying space. We then introduce the general concepts of conditional expectation and conditional probability with respect to a σ-algebra, subsuming the elementary case, and study their most important properties. We continue with a treatment of regular conditional probabilities and transition probabilities, and present some general conditions for the existence of such probabilities. The last part of the section is devoted to the notions of conditional independence and Markov triple.

Conditional expectation given a partition 29.1. Let $B \in \mathcal{A}$ with $P(B) > 0$, and let $f \in \mathfrak{L}^1(X, \mathcal{A}, P)$. According to (18.2.vii), the *conditional expectation of f* given B, denoted $E_B f$, is

$$E_B f = \int_X f \, dP_B = \frac{1}{P(B)} \int_B f \, dP. \tag{1}$$

(1) can be rewritten as $\int_B E_B f \, dP = \int_B f \, dP$, which makes sense even if $P(B) = 0$, but then $E_B f$ becomes undetermined. If $f = 1_A$ for some $A \in \mathcal{A}$, then (1) gives $E_B 1_A = P_B(A)$. If both $P(B) > 0$ and $P(B^c) > 0$, then the conditional expectation of f given the partition $\{B, B^c\}$ is the two-valued function on X, taking the value $E_B f$ for $x \in B$ and the value $E_{B^c} f$ for $x \in B^c$. In particular, the conditional probability of A given the partition $\{B, B^c\}$ is the two-valued function on X, with values $P_B(A)$ for $x \in B$ and $P_{B^c}(A)$ for $x \in B^c$. More generally, let $\mathcal{P} = \{B_i : i \in I\} \subset \mathcal{A}$ be

Analysis and Probability. http://dx.doi.org/10.1016/B978-0-12-401665-1.00011-4

a countable partition of X. The *conditional expectation of f given $\mathcal{P}$*, denoted $E^{\mathcal{P}}f$, is by definition the elementary function on X, taking the value $E_{B_i}f$ for $x \in B_i$ if $P(B_i) > 0$, while being arbitrary for $x \in B_i$ with $P(B_i) = 0$. In compact form the function $E^{\mathcal{P}}f : X \to R$ is given by

$$E^{\mathcal{P}}f = \sum_{i \in I}(E_{B_i}f)1_{B_i}, \tag{2}$$

where the value $E_{B_i}f$ is undetermined whenever B_i is a null event. Correspondingly, the *conditional probability of A given $\mathcal{P}$*, denoted $P^{\mathcal{P}}(A)$, is obtained from (2) as $P^{\mathcal{P}}(A) = E^{\mathcal{P}}1_A = \sum_{i \in I} P_{B_i}(A)1_{B_i}$, where $P_{B_i}(A)$ is undetermined if B_i is null. Thus both $E^{\mathcal{P}}f$ and $P^{\mathcal{P}}(A)$ are undetermined on the null event which is the union of null B_i. (This situation concerning the possibility of defining conditional expectations/probabilities up to null events will be seen to be a rule.) Clearly, $E^{\mathcal{P}}f$ and $P^{\mathcal{P}}(A)$ are $\sigma(\mathcal{P})$-measurable. Moreover, $E[E^{\mathcal{P}}f] = Ef$ and $E[P^{\mathcal{P}}(A)] = P(A)$ by (1), so that $E^{\mathcal{P}}f, P^{\mathcal{P}}(A) \in \mathfrak{L}^1(X, \sigma(\mathcal{P}), P_{\sigma(\mathcal{P})})$. Let $\Lambda = \cup_{j \in J} B_j$, where $J \subset I$. Then, on account of (2) and (1), we have

$$\int_\Lambda E^{\mathcal{P}}f\, dP_{\sigma(\mathcal{P})} = \sum_{j \in J}(E_{B_i}f)P(B_j) = \sum_{j \in J}(E_{B_i}f)\int_{B_j} f\, dP = \int_\Lambda f\, dP$$

and

$$\int_\Lambda P^{\mathcal{P}}(A) dP_{\sigma(\mathcal{P})} = \sum_{j \in J} P_{B_i}(A)P(B_j) = \sum_{j \in J} P(A \cap B_j) = P(A \cap \Lambda).$$

According to (7.6.a), every event $\Lambda \in \sigma(\mathcal{P})$ is of the above form, and so we may write

$$\int_\Lambda E^{\mathcal{P}}f\, dP_{\sigma(\mathcal{P})} = \int_\Lambda f\, dP, \quad \Lambda \in \sigma(\mathcal{P}), \tag{3}$$

and

$$\int_\Lambda P^{\mathcal{P}}(A) dP_{\sigma(\mathcal{P})} = P(A \cap \Lambda), \quad \Lambda \in \sigma(\mathcal{P}). \tag{4}$$

This has been a constructive definition of the conditional expectation/probability given the countable partition $\mathcal{P}$, in fact given the σ-algebra $\sigma(\mathcal{P})$. For more general conditioning than a countable partition, a constructive definition is not always possible. However, a descriptive definition of the conditional expectation/probability given a σ-algebra $\mathcal{B} \subset \mathcal{A}$ is possible on account of the Radon-Nikodým theorem (12.5). The clue for the descriptive approach is suggested by (3)/(4).

Definitions 29.2. Let $\mathcal{B} \subset \mathcal{A}$ be a σ-algebra, let $P_{\mathcal{B}}$ denote the restriction of P to $\mathcal{B}$, and let $f \in \mathfrak{L}^1(X, \mathcal{A}, P)$ and $A \in \mathcal{A}$. The *conditional expectation of f given $\mathcal{B}$* is any $\mathcal{B}$-measurable function on X, denoted $E^{\mathcal{B}}f$, $E^{\mathcal{B}}[f]$, $E[f|\mathcal{B}]$ or $E_P[f|\mathcal{B}]$, such that

$$\int_B E^{\mathcal{B}}f\, dP_{\mathcal{B}} = \int_B f\, dP, \quad B \in \mathcal{B}.$$

The *conditional probability of A given $\mathcal{B}$* is any $\mathcal{B}$-measurable function on X, denoted $P^{\mathcal{B}}(A)$ or $P(A|\mathcal{B})$, satisfying

$$\int_B P^{\mathcal{B}}(A)dP_{\mathcal{B}} = P(A \cap B), \quad B \in \mathcal{B}.$$

Remarks 29.3. Let $\mathcal{B}$, $P_{\mathcal{B}}$, f and A be as in (29.2).

(a) $E^{\mathcal{B}}1_A = P^{\mathcal{B}}(A)$.

(b) The function $E^{\mathcal{B}}f/P^{\mathcal{B}}(A)$ in (29.2) exists and is unique up to the equivalence relation defined in (11.14.c). Indeed, for $\Lambda \in \mathcal{A}$, set $\nu_f^+(\Lambda) = \int_\Lambda f^+ \, dP$ and $\nu_f^-(\Lambda) = \int_\Lambda f^- \, dP$. Then $\nu_f^+ \ll P$ and $\nu_f^- \ll P$ by (12.2.b), whence $(\nu_f^+)_{\mathcal{B}} \ll P_{\mathcal{B}}$ and $(\nu_f^-)_{\mathcal{B}} \ll P_{\mathcal{B}}$ (12.13.b). Applying the Radon-Nikodým theorem (12.5), write $g_+ = d(\nu_f^+)_{\mathcal{B}}/dP_{\mathcal{B}}$ and $g_- = d(\nu_f^-)_{\mathcal{B}}/dP_{\mathcal{B}}$. g_+ and g_- are $\mathcal{B}$-measurable and $P_{\mathcal{B}}$-integrable, and so they are finite $P_{\mathcal{B}}$-a.s. (12.6.b). They are also unique up to a $P_{\mathcal{B}}$-null event $B_0 \in \mathcal{B}$ by (12.6.a). Define $E^{\mathcal{B}}f$ to be the function that equals $g_+ - g_-$ on B_0^c and is arbitrary on B_0. Then, for $B \in \mathcal{B}$, we actually have

$$\begin{aligned}\int_B E^{\mathcal{B}}f \, dP_{\mathcal{B}} &= \int_B g_+ \, dP_{\mathcal{B}} - \int_B g_- \, dP_{\mathcal{B}} = (\nu_f^+)_{\mathcal{B}}(B) - (\nu_f^-)_{\mathcal{B}}(B) \\ &= \int_B f^+ \, dP - \int_B f^- \, dP = \int_B f \, dP.\end{aligned}$$

(c) It follows from (b) that $E^{\mathcal{B}}f, P^{\mathcal{B}}(A) \in \mathfrak{L}^1(X, \mathcal{B}, P_{\mathcal{B}})$. Any element of the equivalent class $\widetilde{E^{\mathcal{B}}f}, \widetilde{P^{\mathcal{B}}(A)} \in L^1(X, \mathcal{B}, P_{\mathcal{B}})$ is termed a *version* of $E^{\mathcal{B}}f/P^{\mathcal{B}}(A)$.

(d) Since $E^{\mathcal{B}}f/P^{\mathcal{B}}(A)$ is $\mathcal{B}$-measurable, we can write P instead of $P_{\mathcal{B}}$ in (29.2), and properties of $E^{\mathcal{B}}f/P^{\mathcal{B}}(A)$ valid $P_{\mathcal{B}}$-a.s. may be said to hold a.s.

(e) $E^{\mathcal{P}}f = E^{\sigma(\mathcal{P})}f$ and $P^{\mathcal{P}}(A) = P^{\sigma(\mathcal{P})}(A)$, where $\mathcal{P}$ is as in (29.1).

(f) Let $\mathcal{P}$ be as in (29.1). Consider the version $P^{\sigma(\mathcal{P})}(A)(x) = \sum_{i \in I} p_i(x)1_{B_i}(x)$, $x \in X$, where $p_i(x) = P(A|B_i)$ for $x \in B_i$ with $P(B_i) > 0$, while $p_i(x) = P(A)$ for $x \in B_i$ with $P(B_i) = 0$. For each fixed $x \in X$, as a function of A, this is a probability on $\mathcal{A}$, which we denote for a moment by $P_x^{\sigma(\mathcal{P})}$. For any fixed $A \in \mathcal{A}$, as a function of x, this is a $\sigma(\mathcal{P})$-measurable function. For $x_0 \in X$, we can integrate f with respect to $P_{x_0}^{\sigma(\mathcal{P})}$ to obtain

$$\begin{aligned}\int_X f \, dP_{x_0}^{\sigma(\mathcal{P})} &= \sum_{i \in I} \int_{B_i} f \, dP_{x_0}^{\sigma(\mathcal{P})} \\ &= \begin{cases} \sum_{i \in I} \int_{B_i} f \, dP_{B_j} = E_{B_j}f & \text{if } x_0 \in B_j \text{ with } P(B_j) > 0 \\ \sum_{i \in I} \int_{B_i} f \, dP = Ef & \text{if } x_0 \in B_j \text{ with } P(B_j) = 0 \end{cases}.\end{aligned}$$

In view of (2) in (29.1) and (e), this means that $\int_X f\,dP_{x_0}^{\sigma(\mathcal{P})} = (E^{\sigma(\mathcal{P})} f)(x_0)$, $x_0 \in X$. Consequently, in this special case, we may write $E^{\sigma(\mathcal{P})} f = \int_X f\,dP^{\sigma(\mathcal{P})}$ a.s. for any $f \in \mathfrak{L}^1$. However, in general such a representation is impossible.

(g) f has the decomposition $f = E[f|\mathcal{B}] + g$, where $\int_X gh\,dP = 0$ for any bounded $\mathcal{B}$-measurable function h on X. Indeed, g is P-integrable by (c), and it follows from the definition of $E[f|\mathcal{B}]$ that $\int_B g\,dP = 0$, $B \in \mathcal{B}$. Hence $\int_X gh\,dP = 0$ for every simple finite $\mathcal{B}$-measurable h. If h is $\mathcal{B}$-measurable and $0 \leqslant h \leqslant C < \infty$, apply (8.15) to choose a sequence $\{h_n : n \in N\} \subset [0, C]^X$ of simple $\mathcal{B}$-measurable functions such that $h_n \to h$. Since $|gh_n| \leqslant |g|\,C$ and $\int_X gh_n\,dP = 0$ for any $n \in N$, the dominated convergence theorem (11.25.ii) yields $\int_X gh\,dP = 0$. Finally, if h is $\mathcal{B}$-measurable and bounded, then $\int_X gh\,dP = \int_X gh^+\,dP - \int_X gh^-\,dP = 0$.

(h) If $(X, \mathcal{A}, P)$ is a complete probability space, then $E[f|\mathcal{B}] = E[f|\mathcal{B}^P]$ a.s. Actually, for any $B, N \in \mathcal{B}$ with $P(N) = 0$, we have

$$\int\limits_{B\cup N} E[f|\mathcal{B}]dP = \int\limits_{B} E[f|\mathcal{B}]dP = \int\limits_{B} f\,dP = \int\limits_{B\cup N} f\,dP = \int\limits_{B\cup N} E[f|\mathcal{B}^P]dP. \quad (1)$$

Since $E[f|\mathcal{B}]$ and $E[f|\mathcal{B}^P]$ are $\mathcal{B}^P$-measurable, (1) implies that $E[f|\mathcal{B}] = E[f|\mathcal{B}^P]$ a.s. by (11.18).

(i) If B is an atom of $\mathcal{B}$, then

$$E^{\mathcal{B}} f = E_B f \text{ a.s. on } B. \quad (1)$$

Indeed, as $E^{\mathcal{B}} f$ is $\mathcal{B}$-measurable, it is constant a.s. on B by (9.37.g). Consequently, $\int_B E^{\mathcal{B}} f\,dP = \int_B f\,dP$ and (1) of (29.1) yields (1). This is a smoothing property of $E^{\mathcal{B}}$, in the sense that on atoms of $\mathcal{B}$ which are not atoms of $\mathcal{A}$, f is replaced by its average. Thus $E^{\mathcal{B}} f$ generally takes fewer values than f.

Definitions 29.4. Let f be a P-integrable random variable, let $(Y, \mathcal{B})$ be a measurable space, and let $g : X \to Y$ be a random element. The conditional expectation $E[f|g^{-1}(\mathcal{B})]$, denoted $E[f|g]$, is called the *conditional expectation of f given g*. For $A \in \mathcal{A}$, the conditional probability $P(A|g^{-1}(\mathcal{B}))$, denoted $P(A|g)$, is called the *conditional probability of A given g*. If g is an n-dimensional random vector with coordinates $g_1, \ldots, g_n$, then we shall write $E[f|g_1, \ldots, g_n]$ and $P(A|g_1, \ldots, g_n)$ for $E[f|g]$ and $P(A|g)$, respectively; similarly if g is an infinite dimensional random vector with coordinates $g_1, g_2 \ldots$

Theorem 29.5. *Let f, $(Y, \mathcal{B})$ and g be as in* (29.4)*, and let $E[f|g]$ be a version of the conditional expectation of f given g. Then there exists a measurable function $h : Y \to R$ such that $E[f|g] = h \circ g$.*

Proof. This follows immediately from the Doob-Dynkin theorem (8.17) with g in place of f and $E[f|g]$ in place of g. □

Remark 29.6. The above result shows that $E[f|g]$ is constant a.s. on each set on which g is constant. Thus if $y \in g(X)$ and $A_y = g^{-1}(\{y\})$, then $E[f|g](x) = h(g(x)) = h(y)$ a.s. for any $x \in A_y$. By abuse of notation, this is sometimes written as

$E[f|g = y] = h(y)$. If g is elementary, then $\{A_y : y \in g(X)\}$ is a countable partition of X, and $E[f|g]$ coincides with the conditional expectation of f given this partition as constructed in (29.1). If $P(g = y) > 0$, then $E[f|g = y]$ is precisely $E_{\{g=y\}}f$ as given by (1) of (29.1), whereas if $P(g = y) = 0$, then $E[f|g]$ is not well defined on $\{g = y\}$ in the constructive approach of (29.1), but takes the value $h(y)$ according to (29.5).

Theorem 29.7. *Let $(Y, \mathcal{B})$ be a measurable space, let $\mathcal{C} \subset \mathcal{B}$ be a σ-algebra, let $f : X \to Y$ be a measurable function, and let $g \in \mathfrak{L}^1(Y, \mathcal{B}, Pf^{-1})$. Then*

$$E_P[g \circ f|f^{-1}(\mathcal{C})] = E_{Pf^{-1}}[g|\mathcal{C}] \circ f \text{ a.s.} \tag{1}$$

Proof. Notice first that both sides of (1) are $f^{-1}(\mathcal{C})$-measurable functions. Then, according to the change of variable formula (11.29),

$$\begin{aligned}\int\limits_{f^{-1}(C)} E_P[g \circ f|f^{-1}(\mathcal{C})]dP &= \int\limits_{f^{-1}(C)} (g \circ f)dP \\ &= \int\limits_{C} g\,dPf^{-1} = \int\limits_{C} E_{Pf^{-1}}[g|\mathcal{C}]dPf^{-1} \\ &= \int\limits_{f^{-1}(C)} (E_{Pf^{-1}}[g|\mathcal{C}] \circ f)dP, \quad C \in \mathcal{C}.\end{aligned}$$

Thus (1) follows by (11.18). □

We now present some simple consequences of the definition of $E^{\mathcal{B}}/P^{\mathcal{B}}$. To avoid repetitions, throughout (29.8)–(29.32), if not otherwise stated, $\mathcal{B} \subset \mathcal{A}$ is a σ-algebra, $f \in \mathfrak{L}^1$ and $A \in \mathcal{A}$.

Theorem 29.8.

(i) *If $\mathcal{B} = \{\emptyset, X\}$ or $\mathcal{O}_P$, then $E^{\mathcal{B}}f = f$ a.s.; in particular, $P^{\mathcal{B}}(A) = P(A)$ a.s.*

(ii) $E[E^{\mathcal{B}}f] = Ef$.

(iii) *$E^{\mathcal{B}}cf = cE^{\mathcal{B}}f$ a.s. for any $c \in R$.*

(iv) *If f is $\mathcal{B}$-measurable, then $E^{\mathcal{B}}f = f$ a.s.; in particular, if $f = c \in R$, then $E^{\mathcal{B}}c = c$ a.s. and $E^{\mathcal{A}}f = f$ a.s.*

(v) *If $\sigma(f)$ and $\mathcal{B}$ are independent, then $E^{\mathcal{B}}f = Ef$ a.s.*

(vi) *If $f \geqslant 0$, then $E^{\mathcal{B}}f \geqslant 0$ a.s.*

(vii) *If $g \in \mathfrak{L}^1$ and $f \leqslant g$, then $E^{\mathcal{B}}f \leqslant E^{\mathcal{B}}g$ a.s.*

(viii) *$|E^{\mathcal{B}}f| \leqslant E^{\mathcal{B}}|f|$ a.s.*

(ix) *$E^{\mathcal{B}}f \in \mathfrak{L}^1$ and $\|E^{\mathcal{B}}f\|_1 \leqslant \|f\|_1$.*

(x) *If $f = 0$ on some $\Lambda \in \mathcal{B}$, then $E^{\mathcal{B}}f = 0$ a.s. on Λ.*

(xi) *$P^{\mathcal{B}}(A) = 0$ a.s. if and only if $P(A) = 0$, $P^{\mathcal{B}}(A) = 1$ a.s. if and only if $P(A) = 1$, and $0 \leqslant P^{\mathcal{B}}(A) \leqslant 1$ a.s.*

(xii) *If $g \in \mathfrak{L}^1$, then $E^{\mathcal{B}}[f + g] = E^{\mathcal{B}}f + E^{\mathcal{B}}g$ a.s.*

(xiii) *If $B \in \mathcal{A}$ and $A \cap B = \emptyset$, then $P^{\mathcal{B}}(A \cup B) = P^{\mathcal{B}}(A) + P^{\mathcal{B}}(B)$ a.s.*

(xiv) *If* $\mathcal{C} \subset \mathcal{A}$ *is a* σ*-algebra,* $g \in \mathfrak{L}^1$*, and there is* $\Lambda \in \mathcal{B} \cap \mathcal{C}$ *such that* $\Lambda \cap \mathcal{B} = \Lambda \cap \mathcal{C}$ *and* $f = g$ *on* Λ*, then* $E^{\mathcal{B}} f = E^{\mathcal{C}} g$ *a.s. on* Λ.

Proof. The proofs of (i)–(iii) and (ix)–(xiii) are easy and are left to the reader. (iv) follows immediately from (11.18). If $\sigma(f)$ and $\mathcal{B}$ are independent, then

$$\int_B E^{\mathcal{B}} f \, dP = \int_B f \, dP = (Ef)P(B) = \int_B Ef \, dP, \quad B \in \mathcal{B},$$

by (18.23.ii). Hence (v) is a consequence of (11.18). (vi) is immediate from (11.40), (vii) follows from (vi), and (viii) follows from (vii). As for (xiv), observe first that $1_\Lambda E^{\mathcal{B}} f$ and $1_\Lambda E^{\mathcal{C}} g$ are $\mathcal{B} \cap \mathcal{C}$-measurable. Then, by (xii) and (x), we have

$$\begin{aligned}\int_A 1_\Lambda E^{\mathcal{B}} f \, dP &= \int_A 1_\Lambda E^{\mathcal{B}}[f 1_\Lambda] dP = \int_{\Lambda \cap A} f 1_\Lambda \, dP = \int_{\Lambda \cap A} g 1_\Lambda \, dP \\ &= \int_A 1_\Lambda E^{\mathcal{C}}[g 1_\Lambda] dP = \int_A 1_\Lambda E^{\mathcal{C}} g \, dP, \quad A \in \mathcal{B} \cap \mathcal{C}.\end{aligned}$$

In view of (11.18), this shows that $1_\Lambda E^{\mathcal{B}} f = 1_\Lambda E^{\mathcal{C}} g$ a.s. □

The next property of the conditional expectation is very useful and merits to be stated separately.

Theorem 29.9. *If* $\mathcal{B}' \subset \mathcal{B}$ *is a* σ*-algebra, then*

$$E^{\mathcal{B}'} f = E^{\mathcal{B}'}[E^{\mathcal{B}} f] \text{ a.s.} \tag{1}$$

Proof. Both sides of (1) are $\mathcal{B}'$-measurable, and

$$\int_{B'} E^{\mathcal{B}'} f \, dP = \int_{B'} f \, dP = \int_{B'} E^{\mathcal{B}} f \, dP = \int_{B'} E^{\mathcal{B}'}[E^{\mathcal{B}} f] dP, \quad B' \in \mathcal{B}'.$$

Thus (1) ensues from (11.18). □

Corollary 29.10. *If* $\mathcal{B}'$ *and* $\mathcal{B}''$ *are* σ*-algebras with* $\mathcal{B}' \subset \mathcal{B} \subset \mathcal{B}''$ *and* $E^{\mathcal{B}'} f = E^{\mathcal{B}''} f$ *a.s., then* $E^{\mathcal{B}'} f = E^{\mathcal{B}} f$ *a.s.; in particular, if* $(X, \mathcal{A}, P)$ *is a complete probability space, and* $\mathcal{C}$ *is a* σ*-algebra with* $\mathcal{B} \subset \mathcal{C} \subset \mathcal{B}^P$*, then* $E^{\mathcal{B}} f = E^{\mathcal{C}} f$ *a.s.*

Proof. Since $E^{\mathcal{B}'} f$ is $\mathcal{B}$-measurable, we have $E^{\mathcal{B}'} f = E^{\mathcal{B}}[E^{\mathcal{B}'} f] = E^{\mathcal{B}}[E^{\mathcal{B}''} f] = E^{\mathcal{B}} f$ a.s. by (29.8.iv) and (29.9); the particular case follows from the previous assertion and (29.3.h). □

In what follows we expose basic properties of $E^{\mathcal{B}}$ involving a sequence $\{f_n : n \geqslant 1\} \subset \mathfrak{L}^1$. They usually parallel and follow from the unconditional analogues.

Theorem 29.11. *If* $0 \leqslant f_n \leqslant f_{n+1}, n \geqslant 1$, *and* $f_n \overset{a.s.}{\to} f$, *then* $E^{\mathcal{B}} f_n \overset{a.s.}{\to} E^{\mathcal{B}} f$.

Proof. We have $0 \leqslant E^{\mathcal{B}} f_n \leqslant E^{\mathcal{B}} f_{n+1}, n \geqslant 1$, by (29.8.vi) and (29.8.vii), and so $E^{\mathcal{B}} f_n \overset{a.s.}{\to} g$, where g is some $\mathcal{B}$-measurable $\overline{R}$-valued random variable. Then, in view of the monotone convergence theorem (11.22),

$$\int_B E^{\mathcal{B}} f_n \, dP \to \int_B g \, dP, \quad B \in \mathcal{B}, \tag{1}$$

and

$$\int_B E^{\mathcal{B}} f_n \, dP = \int_B f_n \, dP \to \int_B f \, dP = \int_B E^{\mathcal{B}} f \, dP, \quad B \in \mathcal{B}. \tag{2}$$

From (1) and (2) we infer that $\int_B g \, dP = \int_B E^{\mathcal{B}} f \, dP$, $B \in \mathcal{B}$, and so $g = E^{\mathcal{B}} f$ a.s. by (11.18). □

Remarks 29.12. (a) Replacing the sequence $\{f_n : n \geqslant 1\}$ in (29.11) by $\{f_n - f_1 : n \geqslant 2\}$, and making use of (29.8.xii) and (29.8.iii), show that the condition $0 \leqslant f_n, n \geqslant 1$, therefrom may be dropped.

(b) If $f_n \geqslant f_{n+1}, n \geqslant 1$, and $f_n \overset{a.s.}{\to} f$, then $E^{\mathcal{B}} f_n \overset{a.s.}{\to} E^{\mathcal{B}} f$. This follows from (29.11) and (a) on substituting $\{-f_n : n \geqslant 1\}$ for $\{f_n : n \geqslant 1\}$.

Corollary 29.13. *If* $f_n \geqslant 0, n \geqslant 1$, *and* $\sum_{n \geqslant 1} f_n \in \mathfrak{L}^1$, *then* $E^{\mathcal{B}}[\sum_{n \geqslant 1} f_n] = \sum_{n \geqslant 1} E^{\mathcal{B}} f_n$ *a.s.; in particular,* $P^{\mathcal{B}}(\cup_{n \in N} A_n) = \sum_{n \geqslant 1} P^{\mathcal{B}}(A_n)$ *a.s. whenever* $\{A_n : n \geqslant 1\} \subset \mathcal{A}$ *is such that* $A_m \cap A_n = \emptyset$ *for* $m \neq n$.

Theorem 29.14.

(i) *If there is* $g \in \mathfrak{L}^1$ *such that* $g \leqslant f_n, n \geqslant 1$, *and* $\liminf_n f_n \in \mathfrak{L}^1$, *then* $E^{\mathcal{B}}[\liminf_n f_n] \leqslant \liminf_n E^{\mathcal{B}} f_n$ *a.s.*

(ii) *If there is* $h \in \mathfrak{L}^1$ *such that* $f_n \leqslant h, n \geqslant 1$, *and* $\limsup_n f_n \in \mathfrak{L}^1$, *then* $\limsup_n E^{\mathcal{B}} f_n \leqslant E^{\mathcal{B}}[\limsup_n f_n]$ *a.s.*

Proof. To prove (i), set $g_n = \inf_{k \geqslant n} f_k - g$. Then $0 \leqslant g_n \leqslant g_{n+1} \to \liminf_n f_n - g$ and $g_n \leqslant f_n - g, n \geqslant 1$. By (29.8.xii), (29.11) and (29.8.vii), we obtain $E^{\mathcal{B}}[\liminf_n f_n] - E^{\mathcal{B}} g = E^{\mathcal{B}}[\liminf_n f_n - g] = \lim_n E^{\mathcal{B}} g_n \leqslant \liminf_n E^{\mathcal{B}}[f_n - g] = \liminf_n E^{\mathcal{B}} f_n - E^{\mathcal{B}} g$ a.s. Since $E^{\mathcal{B}} g$ is finite a.s., the result follows. (ii) is deduced from (i) by considering $\{-f_n : n \geqslant 1\}$ in place of $\{f_n : n \geqslant 1\}$. □

Theorem 29.15. *If* $f_n \overset{a.s.}{\to} f$ *and there is* $g \in \mathfrak{L}^1$ *such that* $|f_n| \leqslant g, n \geqslant 1$, *then* $E^{\mathcal{B}} f_n \overset{a.s.}{\to} E^{\mathcal{B}} f$ *and* $E^{\mathcal{B}} f_n \overset{\mathfrak{L}^1}{\to} E^{\mathcal{B}} f$.

Proof. Since $-g \leqslant \liminf_n f_n \leqslant \limsup_n f_n \leqslant g$, we may apply (29.14) to obtain

$$E^{\mathcal{B}}[\liminf_n f_n] \leqslant \liminf_n E^{\mathcal{B}} f_n \leqslant \limsup_n E^{\mathcal{B}} f_n \leqslant E^{\mathcal{B}}[\limsup_n f_n] \text{ a.s.} \tag{1}$$

As $\liminf_n f_n = \limsup_n f_n = f$ a.s., the first statement follows from (1). Since $\|E^{\mathcal{B}} f - E^{\mathcal{B}} f_n\|_1 = \|E^{\mathcal{B}}[f - f_n]\|_1 \leqslant \|f - f_n\|_1 \to 0$ by (29.8.xii), (29.8.ix) and the dominated convergence theorem (11.25.ii), the second assertion is proved. □

Theorem (29.11) enables us to establish the next result which is of great importance in applications.

Theorem 29.16. *If $g : X \to R$ is $\mathcal{B}$-measurable and $fg \in \mathfrak{L}^1$, then*

$$E^{\mathcal{B}}[fg] = gE^{\mathcal{B}}f \ a.s. \tag{1}$$

Proof. If $g = 1_\Lambda$ with $\Lambda \in \mathcal{B}$, we have

$$\int_B E^{\mathcal{B}}[fg]dP = \int_B fg\,dP = \int_{B\cap\Lambda} f\,dP = \int_{B\cap\Lambda} E^{\mathcal{B}}f\,dP$$
$$= \int_B gE^{\mathcal{B}}f\,dP, \quad B \in \mathcal{B},$$

so that (1) holds in this case by (11.18). Therefore, in view of (29.8.xii), (1) is satisfied by any simple $\mathcal{B}$-measurable g. If $f, g \geqslant 0$, choose a nondecreasing sequence $\{g_n : n \geqslant 1\} \subset [0, \infty[^X$ of simple $\mathcal{B}$-measurable functions such that $g_n \to g$ (8.15). Then, on account of (29.11), $E^{\mathcal{B}}[fg] = \lim_n E^{\mathcal{B}}[fg_n] = \lim_n g_n E^{\mathcal{B}}f = gE^{\mathcal{B}}f$ a.s. Finally, writing $f = f^+ - f^-$ and $g = g^+ - g^-$, and using (29.8.xii), we arrive at (1) by the preceding step. □

We next establish some classical inequalities for conditional expectations. The following theorem is a generalization of Hölder's inequality (13.4).

Theorem 29.17. *Assume that $f \in \mathfrak{L}^p$ and $g \in \mathfrak{L}^{p'}$, where $1 < p < \infty$ and $\frac{1}{p} + \frac{1}{p'} = 1$. Then*

$$E^{\mathcal{B}}|fg| \leqslant (E^{\mathcal{B}}|f|^p)^{1/p}(E^{\mathcal{B}}|g|^{p'})^{1/p'} \ a.s. \tag{1}$$

Proof. (I) Suppose first that f and g are bounded. Put $B_f = \{E^{\mathcal{B}}|f|^p = 0\} \in \mathcal{B}$ and $B_g = \{E^{\mathcal{B}}|g|^{p'} = 0\} \in \mathcal{B}$. Then $0 = \int_{B_f} E^{\mathcal{B}}|f|^p\,dP = \int_{B_f}|f|^p\,dP$ and $0 = \int_{B_g} E^{\mathcal{B}}|g|^{p'}\,dP = \int_{B_g}|g|^{p'}\,dP$, and so $|fg|\,1_{B_f\cup B_g} = 0$ a.s. by (11.16). Clearly, this entails that

$$E^{\mathcal{B}}|fg| = 0 \text{ a.s. on } B_f \cup B_g. \tag{2}$$

Since $(E^{\mathcal{B}}|f|^p)(E^{\mathcal{B}}|g|^{p'}) > 0$ on $(B_f \cup B_g)^c$, applying (13.3) with $a = |f|/(E^{\mathcal{B}}|f|^p)^{1/p}$ and $b = |g|/(E^{\mathcal{B}}|g|^{p'})^{1/p'}$, we obtain

$$\frac{|fg|}{(E^{\mathcal{B}}|f|^p)^{1/p}(E^{\mathcal{B}}|g|^{p'})^{1/p'}}1_{(B_f\cup B_g)^c} \leqslant \frac{|f|^p}{pE^{\mathcal{B}}|f|^p}1_{(B_f\cup B_g)^c}$$
$$+\frac{|g|^{p'}}{p'E^{\mathcal{B}}|g|^{p'}}1_{(B_f\cup B_g)^c}.$$

Taking now conditional expectations given $\mathcal{B}$, and using (29.8.vii), (29.8.xii) and (29.16), we have

$$\frac{E^{\mathcal{B}}|fg|}{(E^{\mathcal{B}}|f|^p)^{1/p}(E^{\mathcal{B}}|g|^{p'})^{1/p'}}1_{(B_f\cup B_g)^c}\leqslant\frac{1}{p}1_{(B_f\cup B_g)^c}+\frac{1}{p'}1_{(B_f\cup B_g)^c}\text{ a.s.,}$$

whence

$$E^{\mathcal{B}}|fg| \leqslant (E^{\mathcal{B}}|f|^p)^{1/p}(E^{\mathcal{B}}|g|^{p'})^{1/p'} \text{ a.s. on } (B_f \cup B_g)^c. \tag{3}$$

From (2) and (3) we get (1) in this case.

(II) Consider now the general case. For $n \geqslant 1$, set $f_n = f1_{\{|f|<n\}}$ and $g_n = g1_{\{|g|<n\}}$.

Then, in view of (I),

$$E^{\mathcal{B}}|f_n g_n| \leqslant (E^{\mathcal{B}}|f_n|^p)^{1/p}(E^{\mathcal{B}}||g_n|^{p'})^{1/p'} \text{ a.s.,} \quad n \geqslant 1. \tag{4}$$

Letting $n \to \infty$ in (4), and utilizing (29.11), we arrive at (1). □

The next theorem is a generalization of Minkowski's inequality (13.7).

Theorem 29.18. *If $f, g \in \mathfrak{L}^p$ with $1 \leqslant p < \infty$, then*

$$(E^{\mathcal{B}}|f+g|^p)^{1/p} \leqslant (E^{\mathcal{B}}|f|^p)^{1/p} + (E^{\mathcal{B}}|g|^p)^{1/p} \text{ a.s.} \tag{1}$$

Proof. (1) holds if $p = 1$ because of the triangle inequality. If $p > 1$, let $p' = p/(p-1)$. As $f + g \in \mathfrak{L}^p$, it follows that $|f+g|^{p-1} \in \mathfrak{L}^{p'}$. Therefore, on account of (29.17), we have

$$\begin{aligned}
E^{\mathcal{B}}|f+g|^p &\leqslant E^{\mathcal{B}}[|f+g|^{p-1}(|f|+|g|)] \\
&= E^{\mathcal{B}}[|f+g|^{p-1}|f|] + E^{\mathcal{B}}[|f+g|^{p-1}|g|] \\
&\leqslant (E^{\mathcal{B}}[|f+g|^{(p-1)p'})^{1/p'}((E^{\mathcal{B}}|f|^p)^{1/p} + (E^{\mathcal{B}}|g|^p)^{1/p}) \text{ a.s.}
\end{aligned}$$

Since $(p-1)p' = 1$, this leads to (1). □

The following result is a generalization of Jensen's inequality (13.13).

Theorem 29.19. *Let $\varphi :]a, b[\to R$ be a convex function, and assume that $a < f < b$ and $\varphi \circ f \in \mathfrak{L}^1$. Then*

$$E^{\mathcal{B}}[\varphi \circ f] \geqslant \varphi \circ E^{\mathcal{B}} f \; a.s.$$

Proof. Notice first that $a < E^{\mathcal{B}} f < b$ a.s. For $a \leqslant E^{\mathcal{B}} f \leqslant b$ a.s. by (29.8.vii) and (29.8.iv), and $\int_{\{E^{\mathcal{B}} f=a\}}(f-a)dP = \int_{\{E^{\mathcal{B}} f=a\}}(E^{\mathcal{B}} f - a)dP = 0$ whenever $-\infty < a$, whence $P(E^{\mathcal{B}} f = a) = 0$ by (11.16). On account of (13.12.vi), there are $\{a_n : n \in N\} \subset R$ and $\{b_n : n \in N\} \subset R$ such that $\varphi(y) = \sup_{n\in N}(a_n y + b_n)$, $y \in]a, b[$. Consequently, $\varphi \circ f \geqslant a_n f + b_n, n \in N$, and so

$$E^{\mathcal{B}}[\varphi \circ f] \geqslant a_n E^{\mathcal{B}} f + b_n \text{ a.s.} \tag{1}$$

for any $n \in N$. On taking the pointwise supremum on the denumerable collection of the right-hand side of (1) gives (1). □

Corollary 29.20. *If $f \in \mathfrak{L}^p$ with $1 \leqslant p < \infty$, then $\left|E^{\mathcal{B}} f\right|^p \leqslant E^{\mathcal{B}} |f|^p$ a.s., $E^{\mathcal{B}} f \in \mathfrak{L}^p$ and $\left\|E^{\mathcal{B}} f\right\|_p \leqslant \|f\|_p$.*

Proof. Take $\varphi(y) = |y|^p$, $y \in R$, and observe that φ is convex. Then (29.19) implies $\left|E^{\mathcal{B}} f\right|^p \leqslant E^{\mathcal{B}} |f|^p$ a.s., whence $\left\|E^{\mathcal{B}} f\right\|_p^p = \int_X \left|E^{\mathcal{B}} f\right|^p dP \leqslant \int_X E^{\mathcal{B}} |f|^p \, dP = \int_X |f|^p \, dP = \|f\|_p^p$. □

Theorem 29.21. *For $1 \leqslant p < \infty$, assume that $f \in \mathfrak{L}^p, \{f_n : n \geqslant 1\} \subset \mathfrak{L}^p$ and $f_n \overset{\mathfrak{L}^p}{\to} f$. Then $E^{\mathcal{B}} f_n \overset{\mathfrak{L}^p}{\to} E^{\mathcal{B}} f$.*

Proof. We have $\left\|E^{\mathcal{B}} f - E^{\mathcal{B}} f_n\right\|_p = \left\|E^{\mathcal{B}}[f - f_n]\right\|_p \leqslant \|f - f_n\|_p \to 0$ by (29.20). □

Corollary 29.22. *Let $\{f_n : n \geqslant 1\} \subset \mathfrak{L}^p$ with $1 \leqslant p < \infty$, let f be a random variable such that $f_n \overset{P}{\to} f$, and suppose that $\{|f_n|^p : n \geqslant 1\}$ is uniformly integrable. Then $E^{\mathcal{B}} f \in \mathfrak{L}^p$ and $E^{\mathcal{B}} f_n \overset{\mathfrak{L}^p}{\to} E^{\mathcal{B}} f$.*

Proof. According to (14.30), $f \in \mathfrak{L}^p$ and $f_n \overset{\mathfrak{L}^p}{\to} f$. Hence $E^{\mathcal{B}} f_n \overset{\mathfrak{L}^p}{\to} E^{\mathcal{B}} f$ by (29.21). □

To get the following theorem, we need to introduce a new concept.

Definition 29.23. A sequence $\{f_n : n \geqslant 1\} \subset \mathfrak{L}^1$ is said to be *conditionally uniformly integrable given $\mathcal{B}$* if $\lim_k \sup_{n \geqslant 1} E^{\mathcal{B}}[|f_n| \, 1_{\{|f_n| \geqslant k\}}] = 0$ a.s.

Remarks 29.24. (a) If $\{f_n : n \geqslant 1\}$ is conditionally uniformly integrable given $\mathcal{B}$, and $g_n, n \geqslant 1$, are random variables such that $|g_n| \leqslant |f_n|, n \geqslant 1$, then $\{g_n : n \geqslant 1\}$ is conditionally uniformly integrable given $\mathcal{B}$.

(b) If $f_n, n \geqslant 1$, are random variables such that $|f_n| \leqslant g, n \geqslant 1$, for some $g \in \mathfrak{L}^1$, then $\{f_n : n \geqslant 1\}$ is conditionally uniformly integrable given $\mathcal{B}$.

(c) If $\mathcal{B} = \{\emptyset, X\}$, then (29.8.i) shows that the above concept reduces to the ordinary uniform integrability.

Theorem 29.25. *Let $\{f_n : n \geqslant 1\} \subset \mathfrak{L}^1$ be conditionally uniformly integrable given $\mathcal{B}$, and assume that $f_n \overset{a.s.}{\to} f$. Then $E^{\mathcal{B}} f_n \overset{a.s.}{\to} E^{\mathcal{B}} f$.*

Proof. For $k \geqslant 1$, set $g_k = \sup_{n \geqslant 1} E^{\mathcal{B}}[f_n^- 1_{\{f_n^- \geqslant k\}}]$. Then $g_k \overset{a.s.}{\to} 0$ by (29.24.a). For $n, k \geqslant 1$, we have

$$E^{\mathcal{B}} f_n = E^{\mathcal{B}} f_n^+ - E^{\mathcal{B}}[f_n^- 1_{\{f_n^- < k\}}] - E^{\mathcal{B}}[f_n^- 1_{\{f_n^- \geqslant k\}}] \geqslant E^{\mathcal{B}}[f_n^+ 1_{\{f_n^- < k\}}] \\ - E^{\mathcal{B}}[f_n^- 1_{\{f_n^- < k\}}] - g_k = E^{\mathcal{B}}[f_n 1_{\{f_n^- < k\}}] - g_k \text{ a.s.} \tag{1}$$

Since $-k < f_n 1_{\{f_n^- < k\}}$, $n \geqslant 1$, and $f_n \overset{a.s.}{\to} f$, on account of (1) and (29.14.i), we obtain

$$\liminf_n E^{\mathcal{B}} f_n \geqslant \liminf_n E^{\mathcal{B}}[f_n 1_{\{f_n^- < k\}}] - g_k \geqslant E^{\mathcal{B}}[\liminf_n f_n 1_{\{f_n^- < k\}}] - g_k \\ \geqslant E^{\mathcal{B}}[\liminf_n f_n] - g_k \text{ a.s.} \tag{2}$$

for any $k \geqslant 1$. Letting $k \to \infty$ in (2) gives

$$\liminf_n E^{\mathcal{B}} f_n \geqslant E^{\mathcal{B}}[\liminf_n f_n] \text{ a.s.} \tag{3}$$

If every f_n is replaced by $-f_n$, then (3) yields

$$\limsup_n E^{\mathcal{B}} f_n \leqslant E^{\mathcal{B}}[\limsup_n f_n] \text{ a.s.} \tag{4}$$

From (3) and (4), as in the proof of (29.15), we conclude that $E^{\mathcal{B}} f_n \overset{a.s.}{\to} E^{\mathcal{B}} f$. □

To prove the following two conditional expectation continuity theorems, due to B. Jessen and P. Lévy, we need the next lemma.

Lemma 29.26 (Doob). *Let I be a nonempty set, and for each $i \in I$ let $\mathcal{B}_i \subset \mathcal{A}$ be a σ-algebra. Then $\{E^{\mathcal{B}_i} f : i \in I\}$ is uniformly integrable.*

Proof. Let $\varepsilon > 0$, and choose $M > 0$ such that $\int_{\{f^+ \geqslant M\}} f^+ \, dP < \varepsilon/2$. Then select $n_\varepsilon \geqslant 1$ so that $MEf^+/n < \varepsilon/2$ for $n \geqslant n_\varepsilon$. By Markov's inequality (13.8) and (29.8.ii), we have $P(E^{\mathcal{B}_i} f^+ \geqslant n) \leqslant E[E^{\mathcal{B}_i} f^+]/n = Ef^+/n, i \in I$. Therefore, for $n \geqslant n_\varepsilon$ and $i \in I$,

$$\begin{aligned}
\int_{\{E^{\mathcal{B}_i} f^+ \geqslant n\}} E^{\mathcal{B}_i} f^+ \, dP &= \int_{\{E^{\mathcal{B}_i} f^+ \geqslant n\}} f^+ \, dP \\
&= \int_{\{E^{\mathcal{B}_i} f^+ \geqslant n, f^+ < M\}} f^+ \, dP + \int_{\{E^{\mathcal{B}_i} f^+ \geqslant n, f^+ \geqslant M\}} f^+ \, dP \\
&\leqslant M P(E^{\mathcal{B}_i} f^+ \geqslant n) + \int_{\{f^+ \geqslant M\}} f^+ \, dP < \frac{MEf^+}{n} + \frac{\varepsilon}{2} < \varepsilon.
\end{aligned}$$

This means that $\{E^{\mathcal{B}_i} f^+ : i \in I\}$ is uniformly integrable. The fact that $\{E^{\mathcal{B}_i} f^- : i \in I\}$ is uniformly integrable is similarly obtained. Thus the result follows by (14.29). □

Theorem 29.27. *Let $\{\mathcal{B}_n : n \geqslant 1\}$ be a nondecreasing sequence of σ-algebras, where $\mathcal{B}_n \subset \mathcal{A}$, put $\mathcal{B} = \vee_{n \geqslant 1} \mathcal{B}_n$, and assume that $f \in \mathfrak{L}^p$ with $1 \leqslant p < \infty$. Then $E^{\mathcal{B}_n} f \overset{a.s.}{\to} E^{\mathcal{B}} f$ and $E^{\mathcal{B}_n} f \overset{\mathfrak{L}^p}{\to} E^{\mathcal{B}} f$.*

Proof. To simplify the notation, write $f_n = E^{\mathcal{B}_n} f, n \geqslant 1$. Let $b, c \in Q$ with $b < c$, let $\{b_n : n \geqslant 1\} \subset]b, \infty[$ be a nonincreasing sequence with $b_n \to b$, and set $B = \{\liminf_n f_n \leqslant b\}$ and $C = \{\limsup_n f_n \geqslant c\}$. For $k \geqslant n$, put $B_{nk} = \{\inf_{n \leqslant i \leqslant k} f_i < b_n\}$. Then $B = \cap_{n \geqslant 1} \cup_{k \geqslant n} B_{nk}$. Define $\tau = \inf\{i \in \{n, \dots, k\} : f_i < b_n\}$ and $B'_{ni} = B_{nk} \cap \{\tau = i\}, n \leqslant i \leqslant k$. Then $B'_{ni} \cap B'_{nj} = \emptyset$ for $i \neq j$, $\cup_{i=n}^k B'_{ni} = B_{nk}$ and $B'_{ni} \in \mathcal{B}_i$. Consequently, for $A \in \mathcal{B}_{n_0}$ and $n_0 \leqslant n \leqslant k$, we have

$$\begin{aligned}
\int_{A \cap B_{nk}} f \, dP &= \sum_{i=n}^k \int_{A \cap B'_{ni}} f \, dP = \sum_{i=n}^k \int_{A \cap B'_{ni}} f_i \, dP \leqslant b_n \sum_{i=n}^k P(A \cap B'_{ni}) \\
&= b_n P(A \cap B_{nk}).
\end{aligned} \tag{1}$$

Letting first $k \to \infty$, then $n \to \infty$ in (1), and using the dominated convergence theorem (11.25.ii), yields $\int_{A\cap B} f\,dP \leqslant bP(A \cap B)$, $A \in \cup_{n\geqslant 1}\mathcal{B}_n$. The family of all events in $\mathcal{A}$ satisfying this inequality is a monotone class that contains the algebra $\cup_{n\geqslant 1}\mathcal{B}_n$. Hence

$$\int\limits_{A\cap B} f\,dP \leqslant bP(A\cap B), \quad A \in \mathcal{B}, \tag{2}$$

by (7.69). If the triple f, b, c is replaced by $-f, -c, -b$ in (2), we get

$$\int\limits_{A\cap C} f\,dP \geqslant cP(A\cap C), \quad A \in \mathcal{B}. \tag{3}$$

Taking $A = B\cap C$ in (2) and (3), we obtain $cP(B\cap C) \leqslant \int_{B\cap C} f\,dP \leqslant bP(B\cap C)$, whence $P(B\cap C) = 0$. Taking $A = X$, $b < 0$ and $c > 0$ in (2) and (3) gives $P(B) \leqslant E\,|f|\,/\,|b|$ and $P(C) \leqslant E\,|f|\,/c$, and so $P(\liminf_n f_n = -\infty) = P(\limsup_n f_n = \infty) = 0$. In view of this, since the event $\{\liminf_n f_n \neq \limsup_n f_n\}$ is a denumerable union of events of the form $B \cap C$, we conclude that there exists a $\mathcal{B}$-measurable function $g : X \to R$ such that $f_n \overset{a.s.}{\to} g$. For $n \geqslant 1$, $|f_n|^p \leqslant E^{\mathcal{B}_n}|f|^p$ by (29.20), and so $\{|f_n|^p : n \geqslant 1\}$ is uniformly integrable by (29.26) and (14.27.a). According to (14.30), it follows that $g \in \mathfrak{L}^p$ and $f_n \overset{\mathfrak{L}^p}{\to} g$. To complete the proof, we have only to show that

$$E^{\mathcal{B}} f = g \text{ a.s.} \tag{4}$$

Observe first that $\{f_n : n \geqslant 1\} \subset \mathfrak{L}^1$, $g \in \mathfrak{L}^1$, and $f_n \overset{\mathfrak{L}^1}{\to} g$ by (13.5) and (14.13.d). If $\Lambda \in \cup_{n\geqslant 1}\mathcal{B}_n$, then $\Lambda \in \mathcal{B}_n$, $n \geqslant n_0$, for some $n_0 \geqslant 1$, and so

$$\int\limits_{\Lambda} E^{\mathcal{B}} f\,dP = \int\limits_{\Lambda} f\,dP = \int\limits_{\Lambda} f_n\,dP, \quad n \geqslant n_0. \tag{5}$$

Letting $n \to \infty$ in (5), in view of the $\mathfrak{L}^1$ convergence, we find $\int_\Lambda E^{\mathcal{B}} f\,dP = \int_\Lambda g\,dP$, $\Lambda \in \cup_{n\geqslant 1}\mathcal{B}_n$. This and (11.39) establish (4). □

Corollary 29.28. *If f in (29.27) is $\mathcal{B}$-measurable, then $E^{\mathcal{B}_n} f \overset{a.s.}{\to} f$ and $E^{\mathcal{B}_n} f \overset{\mathfrak{L}^p}{\to} f$.*

Theorem 29.29. *Let $\{\mathcal{B}_n : n \geqslant 1\}$ be a nonincreasing sequence of σ-algebras with $\mathcal{B}_1 \subset \mathcal{A}$, put $\mathcal{B} = \cap_{n\geqslant 1}\mathcal{B}_n$, and assume that $f \in \mathfrak{L}^p$ with $1 \leqslant p < \infty$. Then $E^{\mathcal{B}_n} f \overset{a.s.}{\to} E^{\mathcal{B}} f$ and $E^{\mathcal{B}_n} f \overset{\mathfrak{L}^p}{\to} E^{\mathcal{B}} f$.*

Proof. Let f_n, $n \geqslant 1$, b, c, B and C be as in the proof of (29.27), and consider a nondecreasing sequence $\{c_n : n \geqslant 1\} \subset]-\infty, c[$ such that $c_n \to c$. For $k \geqslant n$, put $C_{nk} = \{\sup_{n\leqslant i\leqslant k} f_i > c_n\}$. Then $C = \cap_{n\geqslant 1} \cup_{k\geqslant n} C_{nk}$. Define $\tau = \sup\{i \in \{n, \ldots, k\} : f_i > c_n\}$ and $C'_{ni} = C_{nk} \cap \{\tau = i\}$, $n \leqslant i \leqslant k$. Then $C'_{ni} \cap C'_{nj} = \emptyset$ for

$i \neq j$, $\cup_{i=n}^{k} C'_{ni} = C_{nk}$, and $C'_{ni} \in \mathcal{B}_i$. For $A \in \mathcal{B}$, we have

$$\int_{A\cap C_{nk}} f\,dP = \sum_{i=n}^{k} \int_{A\cap C'_{ni}} f\,dP = \sum_{i=n}^{k} \int_{A\cap C'_{ni}} f_i\,dP \geqslant c_n \sum_{i=n}^{k} P(A \cap C'_{ni})$$
$$= c_n P(A \cap C_{nk}). \tag{1}$$

Letting successively $k \to \infty$ and $n \to \infty$ in (1) yields

$$\int_{A\cap C} f\,dP \geqslant cP(A \cap C), \quad A \in \mathcal{B}. \tag{2}$$

Replacing f, b, c by $-f, -c, -b$ in (2), we find

$$\int_{A\cap C} f\,dP \leqslant bP(A \cap C), \quad A \in \mathcal{B}. \tag{3}$$

Since $B, C \in \mathcal{B}$, we may take $A = B \cap C$ in (2) and (3) to obtain $P(B \cap C) = 0$ as in the preceding proof. Arguing as above, since $\liminf_n f_n$ and $\limsup_n f_n$ are $\mathcal{B}$-measurable, it follows that there is a $\mathcal{B}$-measurable function $g \in \mathfrak{L}^p$ such that $f_n \overset{a.s.}{\to} g$, $f_n \overset{\mathfrak{L}^p}{\to} g$ and $f_n \overset{\mathfrak{L}^1}{\to} g$. If $\Lambda \in \mathcal{B}$, then

$$\int_{\Lambda} E^{\mathcal{B}} f\,dP = \int_{\Lambda} f\,dP = \int_{\Lambda} f_n\,dP, \quad n \geqslant 1. \tag{4}$$

Letting $n \to \infty$ in (4), in view of the $\mathfrak{L}^1$ convergence, implies $\int_{\Lambda} E^{\mathcal{B}} f\,dP = \int_{\Lambda} g\,dP$, $\Lambda \in \mathcal{B}$. This and (11.18) show that $E^{\mathcal{B}} f = g$ a.s. □

Corollary 29.30. *If f in (29.29) is $\mathcal{B}$-measurable, then $E^{\mathcal{B}_n} f \overset{a.s.}{\to} f$ and $E^{\mathcal{B}_n} f \overset{\mathfrak{L}^p}{\to} f$.*

On account of (29.27) and (29.29), we can now present more general conditional analogs of Fatou's lemma and of the dominated convergence theorem.

Theorem 29.31. *Let $\{\mathcal{B}_n : n \geqslant 1\}$ be a monotone sequence of σ-algebras with $\mathcal{B}_n \subset \mathcal{A}$, put $\mathcal{B} = \vee_{n\geqslant 1}\mathcal{B}_n$ or $\cap_{n\geqslant 1}\mathcal{B}_n$ according as $\{\mathcal{B}_n : n \geqslant 1\}$ is nondecreasing or nonincreasing, and assume that $0 \leqslant f_n \in \mathfrak{L}^1$, $n \geqslant 1$, and $\liminf_n f_n \in \mathfrak{L}^1$. Then*

$$E^{\mathcal{B}}[\liminf_n f_n] \leqslant \liminf_n E^{\mathcal{B}_n} f_n \text{ a.s.} \tag{1}$$

Proof. For $k \geqslant 1$, set $g_k = \inf_{n\geqslant k} f_k$. Then $0 \leqslant g_k \leqslant g_{k+1} \to \liminf_n f_n$ and $g_k \leqslant f_n$, $n \geqslant k$. By virtue of (29.27) or (29.29), we have

$$E^{\mathcal{B}} g_k = \lim_n E^{\mathcal{B}_n} g_k \leqslant \liminf_n E^{\mathcal{B}_n} f_n \text{ a.s.} \tag{2}$$

Letting $k \to \infty$ in (2) yields (1) by (29.11). □

Theorem 29.32. *Let $\{\mathcal{B}_n : n \geqslant 1\}$ and $\mathcal{B}$ be as in (29.31), and let $\{f_n : n \geqslant 1\}$ be a sequence of random variables such that $f_n \overset{a.s.}{\to} f$ and $|f_n| \leqslant g, n \geqslant 1$, for some $g \in \mathfrak{L}^1$. Then $E^{\mathcal{B}_n} f_n \overset{a.s.}{\to} E^{\mathcal{B}} f$ and $E^{\mathcal{B}_n} f_n \overset{\mathfrak{L}^1}{\to} E^{\mathcal{B}} f$.*

Proof. Applying (29.31) to the sequences $\{g + f_n : n \geqslant 1\}$ and $\{g - f_n : n \geqslant 1\}$, and using (29.27) or (29.29), we have

$$E^{\mathcal{B}} g + E^{\mathcal{B}} f = E^{\mathcal{B}}[g + f] \leqslant \liminf_n E^{\mathcal{B}_n}[g + f_n] = E^{\mathcal{B}} g + \liminf_n E^{\mathcal{B}_n} f_n \text{ a.s.} \quad (1)$$

and

$$E^{\mathcal{B}} g - E^{\mathcal{B}} f = E^{\mathcal{B}}[g - f] \leqslant \liminf_n E^{\mathcal{B}_n}[g f_n] = E^{\mathcal{B}} g - \limsup_n E^{\mathcal{B}_n} f_n \text{ a.s.} \quad (2)$$

From (1) and (2) we infer $E^{\mathcal{B}} f \leqslant \liminf_n E^{\mathcal{B}_n} f_n \leqslant \limsup_n E^{\mathcal{B}_n} f_n \leqslant E^{\mathcal{B}} f$ a.s., and so $E^{\mathcal{B}_n} f_n \overset{a.s.}{\to} E^{\mathcal{B}} f$. Since $\left\| E^{\mathcal{B}} f - E^{\mathcal{B}_n} f_n \right\|_1 \leqslant \left\| E^{\mathcal{B}} f - E^{\mathcal{B}_n} f \right\|_1 + \left\| E^{\mathcal{B}_n} f - E^{\mathcal{B}_n} f_n \right\|_1 \leqslant \left\| E^{\mathcal{B}} f - E^{\mathcal{B}_n} f \right\|_1 + \|f - f_n\|_1 \to 0$ by (29.8.ix), (29.27) or (29.29), and dominated convergence, the second statement follows. □

We now return and look more closely to the notion of conditional probability. Observe first that, in view of (29.8.xi) and (29.13), $P^{\mathcal{B}}$ seems to be a genuine probability. However, the σ-additivity property of $P^{\mathcal{B}}$ in (29.13) holds only a.s., and the exceptional null event therefrom may depend on the sequence $\{A_n : n \geqslant 1\} \subset \mathcal{A}$. Consequently, for $x \in X$, writing $P_x^{\mathcal{B}}(A)$ in place of $P^{\mathcal{B}}(A)(x)$ as in (29.3.f), it is desirable for $P_x^{\mathcal{B}}$ to behave as a probability on $\mathcal{A}$ for each $x \in N^c$ with $P(N) = 0$ in order that classical integration methods and results can be freely utilized. Unfortunately, not every conditional probability satisfies this regularity property, and so we have to consider a special class of conditional probabilities.

Definition 29.33. Let $\mathcal{B}$ and $\mathcal{C}$ be σ-algebras included in $\mathcal{A}$. A *regular conditional probability on $\mathcal{C}$ given $\mathcal{B}$* is a function $P^{\mathcal{B}} : X \times \mathcal{C} \to [0, 1]$ such that:

(i) the section of $P^{\mathcal{B}}$ at any $x \in X$ is a probability on $\mathcal{C}$;

(ii) the section of $P^{\mathcal{B}}$ at each $C \in \mathcal{C}$ is $\mathcal{B}$-measurable;

(iii) $\int_B P^{\mathcal{B}}(x, C) dP(x) = P(B \cap C)$ for every $B \in \mathcal{B}$ and $C \in \mathcal{C}$.

Remarks 29.34. (a) If there is a regular conditional probability $P^{\mathcal{B}}$ on $\mathcal{C}$ given $\mathcal{B}$, then, for $C \in \mathcal{C}$, there exists some null event $N_C \in \mathcal{B}$ so that $P^{\mathcal{B}}(x, C) = P_x^{\mathcal{B}}(C), x \in N_C^c$.

(b) If $P^{\mathcal{B}}$ is a conditional probability as defined in (29.2), having the property that the versions $P^{\mathcal{B}}(A)$ can be chosen so that $P_x^{\mathcal{B}}$ be probabilities on $\mathcal{A}$ except for points x belonging to a null event $N \in \mathcal{B}$, then, putting $P^{\mathcal{B}}(x, A) = P_x^{\mathcal{B}}(A)$ or $P(A)$ according as $x \in N^c$ or $x \in N$, we see that $P^{\mathcal{B}}(x, A)$ is a regular conditional probability on $\mathcal{A}$ given $\mathcal{B}$. In this case we shall simply say that $P^{\mathcal{B}}$ is *regular*.

Example 29.35. Let $\mathcal{P}$ be as in (29.1). For $x \in X$ and $A \in \mathcal{A}$, define $P^{\sigma(\mathcal{P})}(x, A) = \sum_{i \in I} p_i(x) 1_{B_i}(x)$, where $p_i(x) = P(A|B_i)$ for $x \in B_i$ with $P(B_i) > 0$, and $p_i(x) = P(A)$ for $x \in B_i$ with $P(B_i) = 0$. Then, according to (29.3.f), $P^{\sigma(\mathcal{P})}$ is a regular conditional probability on $\mathcal{A}$ given $\sigma(\mathcal{P})$.

As expected, conditional expectations may be envisaged as expectations with respect to regular conditional probabilities.

Theorem 29.36. *Let $\mathcal{B}, \mathcal{C}$ and $P^{\mathcal{B}}$ be as in (29.33), let $P_x^{\mathcal{B}}$ denote the section of $P^{\mathcal{B}}$ at $x \in X$, and let $f \in \mathfrak{L}^1(X, \mathcal{C}, P_{\mathcal{C}})$. Then there exists a null event $N \in \mathcal{B}$ such that*

$$(E^{\mathcal{B}} f)(x) = \int_X f\,dP_x^{\mathcal{B}}, \quad x \in N^c. \tag{1}$$

Proof. If $f = 1_C$ with $C \in \mathcal{C}$, then $(E^{\mathcal{B}} f)(x) = P^{\mathcal{B}}(C)(x)$ and $\int_X f\,dP_x^{\mathcal{B}} = P(x, C)$ for $x \in X$. Then, in view of (29.33.iii),

$$\int_B P^{\mathcal{B}}(C)(x)dP(x) = P(B \cap C) = \int_B P^{\mathcal{B}}(x, C)dP(x), \quad B \in \mathcal{B},$$

so that (1) holds by (11.18). If $f = \sum_{i=1}^n c_i 1_{C_i}$, where $\{C_1, \ldots, C_n\} \subset \mathcal{C}$ is a partition of X, then, since $P^{\mathcal{B}}(C_i)(x) = P^{\mathcal{B}}(x, C_i)$ for almost every $x \in X$ by the first step, (1) follows on account of (29.8.xii). If $f \geqslant 0$ and $\{f_n : n \geqslant 1\} \subset [0, \infty[^X$ is a nondecreasing sequence of simple $\mathcal{C}$-measurable functions such that $f_n \to f$, then $(E^{\mathcal{B}} f_n)(x) = \int_X f_n\,dP_x^{\mathcal{B}}$ for almost every $x \in X$, whence (1) follows by (29.11) and (11.22). In the general case (1) holds for f^+ and f^- and hence for f by (29.8.xii). □

Since the supposition that $P^{\mathcal{B}}$ is regular for each $\mathcal{B}$ is not valid, we proceed to examine sufficient conditions for regularity that are fundamental for theoretical developments. We begin with a definition.

Definition 29.37. Let $(X, \mathcal{A})$ and $(Y, \mathcal{B})$ be measurable spaces. A function $Q : X \times \mathcal{B} \to [0, 1]$ is called a *transition probability from* $(X, \mathcal{A})$ *to* $(Y, \mathcal{B})$ if:
(i) the section of Q at any $x \in X$ is a probability on $\mathcal{B}$;
(ii) the section of Q at each $B \in \mathcal{B}$ is $\mathcal{A}$-measurable.

Example 29.38. Notation is as in (29.33). Then $P^{\mathcal{B}}$ is a transition probability from $(X, \mathcal{B})$ to $(X, \mathcal{C})$.

Remarks 29.39. (a) If $\mathcal{B}$ and $\mathcal{C}$ are as in (29.33), and Q is a transition probability from $(X, \mathcal{B})$ to $(X, \mathcal{C})$ such that $\int_B Q(x, C)dP(x) = P(B \cap C)$, $B \in \mathcal{B}$, $C \in \mathcal{C}$, then Q is a regular conditional probability on $\mathcal{C}$ given $\mathcal{B}$.

(b) Let $(X, \mathcal{A})$ and $(Y, \mathcal{B})$ be measurable spaces, and let $\mathcal{M} \subset \mathcal{B}$ be such that $\sigma(\mathcal{M}) = \mathcal{B}$ and $M_1, M_2 \in \mathcal{M}$ implies $M_1 \cap M_2 \in \mathcal{M}$. If $Q : X \times \mathcal{B} \to [0, 1]$ satisfies (29.37.i) and (29.37.ii) with $\mathcal{M}$ in place of $\mathcal{B}$, then Q is a transition probability from $(X, \mathcal{A})$ to $(Y, \mathcal{B})$. Actually, $\{B \in \mathcal{B}$: the section of Q at each $B \in \mathcal{B}$ is $\mathcal{A}$-measurable$\}$ is a Dynkin system containing $\mathcal{M}$, and (7.36) concludes the argument.

(c) If P is a probability on $\mathcal{A}$, and Q is a transition probability from $(X, \mathcal{A})$ to $(Y, \mathcal{B})$, then, according to (15.7), there exists a unique probability on $\mathcal{A} \otimes \mathcal{B}$, which we denote $P \otimes Q$, such that $(P \otimes Q)(A \times B) = \int_A Q(x, B)dP(x)$, $A \in \mathcal{A}$, $B \in \mathcal{B}$. This probability is defined by $(P \otimes Q)(\Lambda) = \int_X Q(x, \Lambda_x)dP(x)$, $\Lambda \in \mathcal{A} \otimes \mathcal{B}$, where Λ_x stands for the section of Λ at x. If $Q(x, B)$ is constant in $x \in X$ for any $B \in \mathcal{B}$, then Q reduces to an ordinary probability on $\mathcal{B}$, and so $P \otimes Q$

becomes the usual product of two probabilities. Notice that $(P \otimes Q)\pi_X^{-1} = P$, where π_X denotes the projection from $X \times Y$ onto X.

(d) Let $\mathcal{B}$, $\mathcal{C}$, and $P^{\mathcal{B}}$ be as in (29.33), and consider the measurable space $(X \times X, \mathcal{B} \otimes \mathcal{C})$ and the $(\mathcal{A}, \mathcal{B}\otimes\mathcal{C})$-measurable mapping $d : X \to X \times X$ given by $d(x) = (x, x)$, $x \in X$. Then $Pd^{-1} = P_{\mathcal{B}} \otimes P^{\mathcal{B}}$. Indeed, Pd^{-1} and $P_{\mathcal{B}} \otimes P^{\mathcal{B}}$ are probabilities on $\mathcal{B} \otimes \mathcal{C}$. For $B \in \mathcal{B}$ and $C \in \mathcal{C}$, in view of (29.33.iii), we have $(Pd^{-1})(B \times C) = P(B \cap C) = \int_B P^{\mathcal{B}}(x, C) dP_{\mathcal{B}}(x) = (P_{\mathcal{B}} \otimes P^{\mathcal{B}})(B \times C)$. Since $\{B \times C : B \in \mathcal{B}, C \in \mathcal{C}\}$ is closed under intersection and generates $\mathcal{B} \otimes \mathcal{C}$, the assertions follow from (10.8).

Theorem 29.40. *Let $(F, \mathcal{F})$ and $(G, \mathcal{G})$ be measurable spaces, let $f : X \to F$ and $g : X \to G$ be measurable functions, and let Q be a transition probability from $(F, \mathcal{F})$ to $(G, \mathcal{G})$. Then the following statements are equivalent:*

(i) $P(f, g)^{-1} = (Pf^{-1}) \otimes Q$;

(ii) $P^{f^{-1}(\mathcal{F})}(g^{-1}(B)) = Q(f, B)$ (Pf^{-1})*-a.s. for any* $B \in \mathcal{G}$.

Proof. Observe first that $P(f, g)^{-1}$ and $(Pf^{-1}) \otimes Q$ are probabilities on $\mathcal{F} \otimes \mathcal{G}$. Let $A \in \mathcal{F}$ and $B \in \mathcal{G}$. If (i) holds, then

$$\int_{f^{-1}(A)} 1_{g^{-1}(B)} dP = P(f^{-1}(A) \cap g^{-1}(B)) = \int_A Q(x, B) d(Pf^{-1})(x)$$
$$= \int_{f^{-1}(A)} Q(f, B) dP$$

by the change of variable formula (11.29). This implies (ii). If now (ii) is satisfied, using (11.29), we have

$$\begin{aligned}(P(f, g)^{-1})(A \times B) &= P(f^{-1}(A) \cap g^{-1}(B)) \\ &= \int_{f^{-1}(A)} P^{f^{-1}(\mathcal{F})}(g^{-1}(B)) dP = \int_{f^{-1}(A)} Q(f, B) dP \\ &= \int_A Q(x, B) d(Pf^{-1})(x) = ((Pf^{-1}) \otimes Q)(A \times B). \quad (1)\end{aligned}$$

As $\{A \times B : A \in \mathcal{F}, B \in \mathcal{G}\}$ is closed under intersection and generates $\mathcal{F} \otimes \mathcal{G}$, (i) follows from (1) via (10.8). □

Corollary 29.41. *Let $(F, \mathcal{F})$, $(G, \mathcal{G})$, f, g and Q be as in* (29.40), *assume that $P(f, g)^{-1} = (Pf^{-1}) \otimes Q$, and let $\varphi \in \mathfrak{L}^1(G, \mathcal{G}, Pg^{-1})$. Then there is a null event $N \in f^{-1}(\mathcal{F})$ such that*

$$E^{f^{-1}(\mathcal{F})}[\varphi \circ g](x) = \int_G \varphi(s) Q(f(x), ds), \quad x \in N^c. \quad (1)$$

Proof. (1) holds for $\varphi = 1_B$ with $B \in \mathcal{G}$ by (29.40). Then the demonstration continues by the indicator function method as in the proof of (29.36). □

Remarks 29.42. (a) Let $(F, \mathcal{F}), (G, \mathcal{G}), f, g$ and Q be as in (29.40), assume that $S = g(X) \in \mathcal{G}$, and suppose that $P(f, g)^{-1} = (Pf^{-1}) \otimes Q$. Then, according to (29.40), there exists a null event $N \in f^{-1}(\mathcal{F})$ so that

$$Q(f(x), S) = P_x^{f^{-1}(\mathcal{F})}(X) = 1, \quad x \in N^c. \tag{1}$$

Let $x_0 \in N^c$ and $\Lambda = g^{-1}(B)$ with $B \in \mathcal{G}$. Define

$$P^{f^{-1}(\mathcal{F})}(x, \Lambda) = \begin{cases} Q(f(x), B) & \text{if } x \in N^c \\ Q(f(x_0), B) & \text{if } x \in N \end{cases}.$$

$P^{f^{-1}(\mathcal{F})}$ is well defined on $X \times g^{-1}(\mathcal{G})$. Indeed, if there are $B_1, B_2 \in \mathcal{G}$ such that $\Lambda = g^{-1}(B_1) = g^{-1}(B_2)$, then $B_1 \Delta B_2 \subset S^c$, and so, for $x \in N^c$, each $Q(f(x), B_1)$ and $Q(f(x), B_2)$ equals $Q(f(x), B_1 \cap B_2)$ by (1). Taking (1) into account, it is easily verified that $P^{f^{-1}(\mathcal{F})}(x, \Lambda)$ is a probability in $\Lambda \in g^{-1}(\mathcal{G})$ for fixed $x \in X$. $P^{f^{-1}(\mathcal{F})}(x, \Lambda)$ is also $f^{-1}(\mathcal{F})$-measurable in $x \in X$ for fixed $\Lambda \in g^{-1}(\mathcal{G})$. Thus $P^{f^{-1}(\mathcal{F})}$ is a transition probability from $(X, f^{-1}(\mathcal{F}))$ to $(X, g^{-1}(\mathcal{G}))$. In fact, $P^{f^{-1}(\mathcal{F})}$ is a regular conditional probability on $g^{-1}(\mathcal{G})$ given $f^{-1}(\mathcal{F})$. Actually, let $A \in \mathcal{F}$, $\Lambda = g^{-1}(B)$ with $B \in \mathcal{G}$, and $N = f^{-1}(C)$ with $C \in \mathcal{F}$. Then

$$\begin{aligned} \int_{f^{-1}(A)} P^{f^{-1}(\mathcal{F})}(x, \Lambda) dP(x) &= \int_{f^{-1}(A) \cap N^c} Q(f(x), B) dP(x) \\ &= \int_{A \cap C^c} Q(y, B) d(Pf^{-1})(y) \\ &= ((Pf^{-1}) \otimes Q)(A \times B) = (P(f, g)^{-1})(A \times B) \\ &= P(f^{-1}(A) \cap \Lambda). \end{aligned}$$

(b) Let $(F, \mathcal{F}), (G, \mathcal{G}), f$ and g be as in (29.40), assume that $\mathcal{G} = \sigma(\mathcal{M})$ for some countable family $\mathcal{M} \subset \mathcal{P}(G)$, and suppose that $P(f, g)^{-1} = (Pf^{-1}) \otimes Q_1 = (Pf^{-1}) \otimes Q_2$ for two transition probabilities Q_1 and Q_2 from $(F, \mathcal{F})$ to $(G, \mathcal{G})$. Then there is an event $N \in \mathcal{F}$ with $(Pf^{-1})(N) = 0$ such that, for all $B \in \mathcal{G}$,

$$Q_1(y, B) = Q_2(y, B), \quad y \in N^c. \tag{1}$$

Indeed, according to (7.5.c), we may and do assume that $\mathcal{M}$ is closed under intersection. Then, for $A \in \mathcal{F}$ and $B \in \mathcal{G}$, we have

$$\begin{aligned} \int_A Q_1(y, B) d(Pf^{-1})(y) &= ((Pf^{-1}) \otimes Q_1)(A \times B) \\ &= ((Pf^{-1}) \otimes Q_2)(A \times B) \\ &= \int_A Q_2(y, B) d(Pf^{-1})(y). \end{aligned}$$

In view of (11.18), this implies that $Q_1(y, B) = Q_2(y, B)$ for every $y \in F$ except for points belonging to a (Pf^{-1})-null event $N_B \in \mathcal{F}$. Put $N = \cup_{B \in \mathcal{M}} N_B$. Since $\mathcal{M}$ is countable, we have $N \in \mathcal{F}$ and $(Pf^{-1})(N) = 0$. Thus $Q_1(y, B) = Q_2(y, B)$ for all $y \in N^c$ and $B \in \mathcal{M}$, and so (1) follows by (10.8).

Now we have to consider an associativity property relative to $\otimes$ for transition probabilities.

Lemma 29.43. *Let $(X, \mathcal{A})$, $(Y, \mathcal{B})$ and $(Z, \mathcal{C})$ be measurable spaces, let Q be a transition probability from $(X, \mathcal{A})$ to $(Y, \mathcal{B})$, and let R be a transition probability from $(X \times Y, \mathcal{A} \otimes \mathcal{B})$ to $(Z, \mathcal{C})$. For $x \in X$ and $\Lambda \in \mathcal{B} \otimes \mathcal{C}$, write*

$$(Q \otimes R)(x, \Lambda) = \int_Y R(x, y, \Lambda_y) Q(x, dy). \tag{1}$$

Then $Q \otimes R$ defines a transition probability from $(X, \mathcal{A})$ to $(Y \times Z, \mathcal{B} \otimes \mathcal{C})$.

Proof. For fixed $x \in X$ and $\Lambda = B \times C$ with $B \in \mathcal{B}$ and $C \in \mathcal{C}$, the function $y \to R(x, y, \Lambda_y) = R(x, y, C) 1_B(y)$ is $\mathcal{B}$-measurable by (15.2.i). As in (29.39.b), on account of (7.36), this implies that the function $y \to R(x, y, \Lambda_y)$ is $\mathcal{B}$-measurable for any $\Lambda \in \mathcal{B} \otimes \mathcal{C}$. Consequently, for fixed $\Lambda \in \mathcal{B} \otimes \mathcal{C}$, the function $x \to \int_Y R(x, y, \Lambda_y) Q(x, dy)$ is $\mathcal{A}$-measurable by (15.11.i). The fact that $(Q \otimes R)(x, \Lambda)$ as defined by (1) is a probability in $\Lambda \in \mathcal{B} \otimes \mathcal{C}$ for fixed $x \in X$ follows easily from the fact that the section of R at any $(x, y) \in X \times Y$ is a probability on $\mathcal{C}$. □

Theorem 29.44. *Let $(X, \mathcal{A})$, $(Y, \mathcal{B})$, $(Z, \mathcal{C})$, Q, R and $Q \otimes R$ be as in* (29.43), *and consider the usual probability P on $\mathcal{A}$. Then*

$$P \otimes (Q \otimes R) = (P \otimes Q) \otimes R. \tag{1}$$

Proof. $P \otimes (Q \otimes R)$ is a probability on $\mathcal{A} \otimes (\mathcal{B} \otimes \mathcal{C})$, and $(P \otimes Q) \otimes R$ is a probability on $(\mathcal{A} \otimes) \mathcal{B} \otimes \mathcal{C}$. By virtue of (2.28.a) and (7.25), we identify $(X \times (Y \times Z), \mathcal{A} \otimes (\mathcal{B} \otimes \mathcal{C}))$ and $((X \times Y) \times Z, (\mathcal{A} \otimes) \mathcal{B} \otimes \mathcal{C})$ with $(X \times Y \times Z, \mathcal{A} \otimes \mathcal{B} \otimes \mathcal{C})$. Let $A \in \mathcal{A}$, $B \in \mathcal{B}$ and $C \in \mathcal{C}$. Then

$$\begin{aligned}(P \otimes (Q \otimes R))(A \times B \times C) &= \int_A (Q \otimes R)(x, B \times C) dP(x) \\ &= \int_A \left(\int_B R(x, y, C) Q(x, dy) \right) dP(x) \end{aligned} \tag{2}$$

by (29.43), and

$$\begin{aligned}((P \otimes Q) \otimes R)(A \times B \times C) &= \int_{A \times B} R(x, y, C) d(P \otimes Q)(x, y) \\ &= \int_A \left(\int_B R(x, y, C) Q(x, dy) \right) dP(x) \end{aligned} \tag{3}$$

by (15.11.ii). From (2) and (3), in view of (10.8), (1) follows. □

Theorem 29.45 (Doob). *Let $(X, \mathcal{A})$ be a measurable space, consider the measurable space $(R, \mathcal{B}(R))$, and let P be a probability on $\mathcal{A}\otimes\mathcal{B}(R)$. Then there exists a transition probability Q from $(X, \mathcal{A})$ to $(R, \mathcal{B}(R))$ such that $P = (P\pi_X^{-1}) \otimes Q$, where π_X is the projection from $X \times R$ onto X.*

Proof. For every $r \in Q$, choose a version of $P(X\times\,]-\infty, r]|\pi_X)$. Then, on account of (29.5), there exists a measurable function $h_r : X \to [0, 1]$ such that $P(X\times\,]-\infty, r]|\pi_X) = h_r \circ \pi_X$. For $r, s \in Q$ with $r < s$, in view of (29.8.vii), $h_r \circ \pi_X \leqslant h_s \circ \pi_X$ P-a.s., and so there is an event $N_{rs} \in \mathcal{A}$ with $(P\pi_X^{-1})(N_{rs}) = 0$ such that $h_r(x) \leqslant h_s(x), x \in N_{rs}^c$. For $r \in Q$ and $n \geqslant 1$, according to (29.12.b) and (29.8.iv), $h_{r+1/n} \circ \pi_X \to h_r \circ \pi_X$ P-a.s., and so there is an event $N_r \in \mathcal{A}$ with $(P\pi_X^{-1})(N_r) = 0$ such that $h_{r+1/n}(x) \to h_r(x), x \in N_r^c$. Similarly, there are events $N_{-\infty}, N_\infty \in \mathcal{A}$ with $(P\pi_X^{-1})(N_{-\infty}) = (P\pi_X^{-1})(N_\infty) = 0$ such that $h_n(x) \to 0, x \in N_{-\infty}^c$, and $h_n(x) \to 1, x \in N_\infty^c$. Put $N = (\cup_{r,s\in Q, r<s} N_{rs}) \cup (\cup_{r\in Q} N_r) \cup N_{-\infty} \cup N_\infty$, and observe that $(P\pi_X^{-1})(N) = 0$, whence $P(N^c \times R) = 1$. For $x \in R$, define

$$F^x(t) = \left(\inf_{t<r\in Q} h_r(x)\right) 1_{N^c}(x) + F_0(t)1_N(x), \quad t \in R,$$

where F_0 is the distribution function degenerated at 0, and note that F^x is a distribution function on R. Then, by virtue of (10.15), there exists a probability Q^x on $\mathcal{B}(R)$ such that $Q^x(]-\infty, t]) = F^x(t), t \in R$. Define $Q : X \times \mathcal{B}(R) \to [0, 1]$ by $Q(x, B) = Q^x(B), x \in X, B \in \mathcal{B}(R)$. For fixed $r \in Q$, $Q(x,]-\infty, r]) = h_r(x))1_{N^c}(x) + F_0(r)1_N(x)$ is $\mathcal{A}$-measurable in $x \in X$. Since the family $\{]-\infty, r] : r \in Q\}$ is closed under intersection and generates $\mathcal{B}(R)$, (29.39.b) shows that Q is a transition probability from $(X, \mathcal{A})$ to $(R, \mathcal{B}(R))$. Now, for $A \in \mathcal{A}$ and $r \in Q$, we have

$$\begin{aligned}
P(A\times]-\infty, r]) &= P((A\times]-\infty, r]) \cap (N^c \times R)) = P((A \cap N^c)\times]-\infty, r]) \\
&= P((X\times]-\infty, r]) \cap ((A \cap N^c) \times R)) \\
&= \int_{(A\cap N^c)\times R} P(X\times]-\infty, r]|\pi_X)dP = \int_{\pi^{-1}(A\cap N^c)} (h_r \circ \pi_X)dP \\
&= \int_{A\cap N^c} h_r(x)d(P\pi_X^{-1})(x) = \int_A Q(x,]-\infty, r])d(P\pi_X^{-1})(x) \\
&= ((P\pi_X^{-1}) \otimes Q)(A\times]-\infty, r]). \qquad (1)
\end{aligned}$$

As $\{A\times]-\infty, r] : A \in \mathcal{A}, r \in Q\}$ is closed under intersection and generates $\mathcal{A}\otimes\mathcal{B}(R)$, (1) and (10.8) imply $P = (P\pi_X^{-1}) \otimes Q$. □

Remark 29.46. The conclusion of (29.45) means that (i) of (29.40) with $(F, \mathcal{F}) = (X, \mathcal{A}), (G, \mathcal{G}) = (R, \mathcal{B}(R)), f = \pi_X$ and $g = \pi_R$, where π_R denotes the projection from $X \times R$ onto R, is fulfilled. This implies that $P(X \times B|\pi_X) = Q(\pi_X, B)$ $(P\pi_X^{-1})$-a.s. for any $B \in \mathcal{B}(R)$. Furthermore, as $\mathcal{B}(R)$ is denumerably generated (7.10.x), if Q' is another transition probability from $(X, \mathcal{A})$ to $(R, \mathcal{B}(R))$ such that $P = (P\pi_X^{-1}) \otimes Q'$, then there is an event $N \in \mathcal{A}$ with $(P\pi_X^{-1})(N) = 0$, so that, for all $B \in \mathcal{B}(R)$, $Q(x, B) = Q'(x, B), x \in N^c$.

Corollary 29.47. *Let $(X, \mathcal{A})$ be a measurable space, consider the measurable space $(R^n, \mathcal{B}(R^n))$, and let P be a probability on $\mathcal{A} \otimes \mathcal{B}(R^n)$. Then there exists a transition probability Q from $(X, \mathcal{A})$ to $(R^n, \mathcal{B}(R^n))$ such that $P = (P\pi_n^{-1}) \otimes Q$, where π_n is the projection from $X \times R^n$ onto X.*

Proof. We prove the assertion inductively. For $n = 1$ it holds by (29.45). Assuming that the stated property holds for $1, \ldots, n$, we shall verify it for $n+1$. Let π_n^{n+1} denote the projection from $X \times R^{n+1}$ onto $X \times R^n$. On account of the associativity properties discussed in (2.28.a) and (7.25), and by (7.28), we identify $(X \times R^{n+1}, \mathcal{A} \otimes \mathcal{B}(R^{n+1}))$ with $((X \times R^n) \times R, (\mathcal{A} \otimes \mathcal{B}(R^n)) \otimes \mathcal{B}(R))$. Then, applying (29.45) with $n = 1$ yields the existence of a transition probability Q_1 from $(X \times R^n, \mathcal{A} \otimes \mathcal{B}(R^n))$ to $(R, \mathcal{B}(R))$ such that

$$P = (P(\pi_n^{n+1})^{-1}) \otimes Q_1. \tag{1}$$

Now $P(\pi_n^{n+1})^{-1}$ is a probability on $\mathcal{A} \otimes \mathcal{B}(R^n)$, and again (29.45) and the induction hypothesis ensure the existence of a transition probability Q_2 from $(X, \mathcal{A})$ to $(R^n, \mathcal{B}(R^n))$ such that

$$P(\pi_n^{n+1})^{-1} = ((P(\pi_n^{n+1})^{-1})\pi_n^{-1}) \otimes Q_2 = (P\pi_{n+1}^{-1}) \otimes Q_2. \tag{2}$$

From (1) and (2), in view of (29.44), we obtain $P = ((P\pi_{n+1}^{-1}) \otimes Q_2) \otimes Q_1 = (P\pi_{n+1}^{-1}) \otimes (Q_2 \otimes Q_1)$. Putting $Q = Q_2 \otimes Q_1$, (29.43) shows that Q is a transition probability from $(X, \mathcal{A})$ to $(R^{n+1}, \mathcal{B}(R^{n+1}))$. □

Corollary 29.48. *Let $(X, \mathcal{A})$ be a measurable space, let $S \in \mathcal{B}(R^n)$, and let P be a probability on $\mathcal{A} \otimes (S \cap \mathcal{B}(R^n))$. Then there exists a transition probability Q from $(X, \mathcal{A})$ to $(S, S \cap \mathcal{B}(R^n))$ such that $P = (P\pi_X^{-1}) \otimes Q$, where π_X is the projection from $X \times S$ onto X.*

Proof. Set $P' = Pi_{X\times S}^{-1}$, where $i_{X\times S}$ is the inclusion function from $X \times S$ into $X \times R^n$. Then, by (29.47), there exists a transition probability Q' from $(X, \mathcal{A})$ to $(R^n, \mathcal{B}(R^n))$ such that $P' = (P'\pi_n^{-1}) \otimes Q'$, where π_n is as in (29.47). Since $\pi_n \circ i_{X\times S} = \pi_X$, we have $P'\pi_n^{-1} = P\pi_X^{-1}$, and so $1 = P(X \times S) = P'(X \times S) = \int_X Q'(x, S) d(P\pi_X^{-1})(x)$. Consequently, there is $N \in \mathcal{A}$ with $(P\pi_X^{-1})(N) = (P'\pi_n^{-1})(N) = 0$ such that $Q'(x, S) < 1$ for $x \in N$. Let $x_0 \in N^c$. For $B \in S \cap \mathcal{B}(R^n)$, define $Q(x, B) = Q'(x, B)1_{N^c}(x) + Q'(x_0, B)1_N(x)$, $x \in X$. Clearly, Q is a transition probability from $(X, \mathcal{A})$ to $(S, S \cap \mathcal{B}(R^n))$. For $A \in \mathcal{A}$ and $B \in S \cap \mathcal{B}(R^n)$, we obtain

$$P(A \times B) = P'(A \times B) = ((P'\pi_n^{-1}) \otimes Q')(A \times B) = \int_{A\cap N^c} Q'(x, B) d(P'\pi_n^{-1})(x)$$

$$= \int_{A\cap N^c} Q(x, B) d(P\pi_X^{-1})(x) = ((P\pi_X^{-1}) \otimes Q)(A \times B). \tag{1}$$

From (1), by the familiar Dynkin system argument, it follows that $P = (P\pi_X^{-1}) \otimes Q$. □

Corollary 29.49. *Let $(X, \mathcal{A})$ be a measurable space, consider the measurable space $(R^\infty, \mathcal{B}(R^\infty))$, and let P be a probability on $\mathcal{A}\otimes\mathcal{B}(R^\infty)$. Then there exists a transition probability Q from $(X, \mathcal{A})$ to $(R^\infty, \mathcal{B}(R^\infty))$ such that $P = (P\pi_X^{-1}) \otimes Q$, where π_X is the projection from $X \times R^\infty$ onto X.*

Proof. For $n \geqslant 1$, let $\pi_{X,n}^{X,\infty}, \pi_{X,n}^{X,n+1}, \pi_X^{X,n}$ denote the projections from $X \times R^\infty$ onto $X \times R^n$, $X \times R^{n+1}$ onto $X \times R^n$, $X \times R^n$ onto X, respectively. Note that

$$\pi_X^{X,n} \circ \pi_{X,n}^{X,\infty} = \pi_X, \quad n \geqslant 1, \tag{1}$$

and

$$\pi_{X,n}^{X,n+1} \circ \pi_{X,n+1}^{X,\infty} = \pi_{X,n}^{X,\infty}, \quad n \geqslant 1. \tag{2}$$

For $n \geqslant 1$, on account of (29.47) and (1), it follows that there exists a transition probability Q_n from $(X, \mathcal{A})$ to $(R^n, \mathcal{B}(R^n))$ such that

$$P\left(\pi_{X,n}^{X,\infty}\right)^{-1} = (P\pi_X^{-1}) \otimes Q_n. \tag{3}$$

From (2) and (3) we see that

$$((P\pi_X^{-1}) \otimes Q_{n+1})(\pi_{X,n}^{X,n+1})^{-1} = (P\pi_X^{-1}) \otimes Q_n, \quad n \geqslant 1. \tag{4}$$

For $n \geqslant 1$ and $\mathbf{r} = (r_1, \ldots, r_n) \in Q^n$, write $B_\mathbf{r} = \prod_{i=1}^n]-\infty, r_i]$. Then, in view of (4), we have

$$\int_A Q_{n+1}(x, B_\mathbf{r} \times R)d(P\pi_X^{-1})(x) = \int_A Q_n(x, B_\mathbf{r})d(P\pi_X^{-1})(x), \quad A \in \mathcal{A}.$$

Therefore, according to (11.18), there is $N_\mathbf{r} \in \mathcal{A}$ with $(P\pi_X^{-1})(N_\mathbf{r}) = 0$ such that $Q_{n+1}(x, B_\mathbf{r} \times R) = Q_n(x, B_\mathbf{r}), x \in N_\mathbf{r}^c$. Put $N_n = \cup_{\mathbf{r}\in Q^n} N_\mathbf{r}$. Hence

$$Q_{n+1}(x, B_\mathbf{r} \times R) = Q_n(x, B_\mathbf{r}), \quad x \in N_n^c, \ \mathbf{r} \in Q^n.$$

Since $\{B_\mathbf{r} : \mathbf{r} \in Q^n\}$ is closed under intersection and generates $\mathcal{B}(R^n)$, applying (10.8), we find that

$$Q_{n+1}(x, B_n \times R) = Q_n(x, B_n), \quad x \in N_n^c, \ B_n \in \mathcal{B}(R^n). \tag{5}$$

Set $N = \cup_{n\geqslant 1} N_n$. Then $N \in \mathcal{A}$ and $(P\pi_X^{-1})(N) = 0$. For $x \in N^c$, in view of (5) and (16.39), it follows that there is a probability $Q(x, \cdot)$ on $\mathcal{B}(R^\infty)$ such that

$$Q(x, B_n \times R \times R \times \cdots) = Q_n(x, B_n), \quad n \geqslant 1, \ B_n \in \mathcal{B}(R^n). \tag{6}$$

For $x \in N$, define $Q(x, B) = Q(x_0, B)$, $B \in \mathcal{B}(R^\infty)$, where $x_0 \in N^c$ is arbitrary. Thus $Q(x, B)$ is a probability in $B \in \mathcal{B}(R^\infty)$ for fixed $x \in X$. Whatever $n \geqslant 1$, $Q(x, B_n \times R \times R \times \cdots)$ is $\mathcal{A}$-measurable in $x \in X$ for fixed $B_n \in \mathcal{B}(R^n)$ by (6). Consequently, by virtue of (29.37.b), $Q(x, B)$ is a transition probability from $(X, \mathcal{A})$ to $(R^\infty, \mathcal{B}(R^\infty))$. For $A \in \mathcal{A}$ and $B_n \in \mathcal{B}(R^n)$, using (3) and (6), we have

$$P(A \times B_n \times R \times R \times \cdots) = (P(\pi_{X,n}^{X,\infty})^{-1})(A \times B_n)$$
$$= \int_A Q_n(x, B_n) d(P\pi_X^{-1})(x) = \int_A Q(x, B_n \times R \times R \times \cdots) d(P\pi_X^{-1})(x)$$
$$= ((P\pi_X^{-1}) \otimes Q)(A \times B_n \times R \times R \times \cdots). \tag{7}$$

Since the family $\cup_{n \geqslant 1}\{A \times B_n \times R \times R \times \cdots : A \in \mathcal{A}, B_n \in \mathcal{B}(R^n)\}$ is closed under intersection and generates $\mathcal{A} \otimes \mathcal{B}(R^\infty)$, (7) implies $P = (P\pi_X^{-1}) \otimes Q$ by the usual Dynkin system argument. □

Remark 29.50. (29.48) remains valid if $\mathcal{B}(R^n)$ is replaced by $\mathcal{B}(R^\infty)$.

Theorem 29.51. *Let $(X, \mathcal{A}, P)$ be the usual probability space, let $\mathcal{F} \subset \mathcal{A}$ be a σ-algebra, and let $g : X \to R^n$ be a measurable function such that $g(X) \in \mathcal{B}(R^n)$. Then there exists a regular conditional probability on $g^{-1}(\mathcal{B}(R^n))$ given $\mathcal{F}$.*

Proof. Consider the identity function $i_X : X \to X$ which is $(\mathcal{A}, \mathcal{F})$-measurable. In view of (8.4), $P(i_X, g)^{-1}$ is a probability on $\mathcal{F} \otimes \mathcal{B}(R^n)$. Then, according to (29.47), there is a transition probability Q from $(X, \mathcal{F})$ to $(R^n, \mathcal{B}(R^n))$ such that $P(i_X, g)^{-1} = ((P(i_X, g)^{-1})\pi_n^{-1}) \otimes Q = (Pi_X^{-1}) \otimes Q = P_{\mathcal{F}} \otimes Q$. Applying (29.42.a) with $(F, \mathcal{F}) = (X, \mathcal{F})$, $(G, \mathcal{G}) = (R^n, \mathcal{B}(R^n))$, and $f = i_X$, it follows that there exists a regular conditional probability $P^{\mathcal{F}}$ on $g^{-1}(\mathcal{B}(R^n))$ given $\mathcal{F}$. More precisely, $P^{\mathcal{F}}$ can be defined by

$$P^{\mathcal{F}}(x, g^{-1}(B)) = Q(x, B)1_{N^c}(x) + Q(x_0, B)1_N(x), \quad x \in X, \; B \in \mathcal{B}(R^n), \tag{1}$$

where $N \in \mathcal{F}$ is a null event and $x_0 \in N^c$ is arbitrary. □

Remarks 29.52. (a) Let Q' be another transition probability from $(X, \mathcal{F})$ to $(R^n, \mathcal{B}(R^n))$ satisfying $P(i_X, g)^{-1} = P_{\mathcal{F}} \otimes Q'$. Then, since $\mathcal{B}(R^n)$ is denumerably generated, (29.42.b) shows that there is $N' \in \mathcal{F}$ with $P(N') = 0$ such that $Q(x, B) = Q'(x, B)$ for all $x \in (N')^c$ and $B \in \mathcal{B}(R^n)$. Hence $P^{\mathcal{F}}$ given by (1) of (29.51) is essentially unique.

(b) Taking (29.49) into account, we see that (29.51) remains true if R^n is replaced by R^∞. Moreover, the corresponding $P^{\mathcal{F}}$ is also essentially unique, as $\mathcal{B}(R^\infty)$ is denumerably generated (7.29).

In the final part of this section we treat the subject of conditional independence, which turns out to be useful for important topics in probability theory such as Markov processes, exchangeable sequences, or stochastic differential equations. As usual, we fix a probability space $(X, \mathcal{A}, P)$. If not otherwise stated, the σ-algebras we consider are included in $\mathcal{A}$. Throughout (29.53)–(29.58), $\mathcal{B}$ is also a fixed σ-algebra.

Definitions 29.53. Let $I \neq \emptyset$, let $\{\mathcal{A}_i : i \in I\}$ be a set of σ-algebras, let $\{A_i : i \in I\} \subset \mathcal{A}$, and let $\{(Y_i, \mathcal{B}_i) : i \in I\}$ be a collection of measurable spaces. For $i \in I$, let $f_i : X \to Y_i$ be an $(\mathcal{A}, \mathcal{B}_i)$-measurable function. We say that $\mathcal{A}_i, i \in I$, are *conditionally independent given* $\mathcal{B}$ if, for every finite set $\{i_1, \ldots, i_n\} \subset I$ and $A_1 \in \mathcal{A}_{i_1}, \ldots, A_n \in \mathcal{A}_{i_n}$, $P(A_1 \cap \cdots \cap A_n|\mathcal{B}) = P(A_1|\mathcal{B}) \cdots P(A_n|\mathcal{B})$ a.s. We say that $A_i, i \in I$, are *conditionally independent given* $\mathcal{B}$ if $\sigma(\{A_i\}), i \in I$, are conditionally independent given $\mathcal{B}$. We say that $f_i, i \in I$, are *conditionally independent given* $\mathcal{B}$ if $f_i^{-1}(\mathcal{B}_i), i \in I$, are conditionally independent given $\mathcal{B}$.

Remarks 29.54. (a) Most statements and formulas involving ordinary independence in Section 18 can be transposed to the case of conditional independence.

(b) The smaller the σ-algebra $\mathcal{B}$ is, the more stringent the defining condition of conditional independence becomes. Thus, if $\{\emptyset, X\} \subset \mathcal{B} \subset \mathcal{O}_P$, then conditional independence given $\mathcal{B}$ coalesces to ordinary independence, while $\mathcal{A}_i, i \in I$, are always conditionally independent given $\mathcal{B}$ if $\mathcal{A}_i \subset \mathcal{B}, i \in I$.

(c) Assume that $\vee_{i\in I}\mathcal{A}_i$ is independent of $\mathcal{B}$. Then, in view of (29.8.v), $\mathcal{A}_i, i \in I$, are conditionally independent given $\mathcal{B}$ if and only if $\mathcal{A}_i, i \in I$, are independent.

(d) Independence may be lost under conditioning. Indeed, let f_1 and f_2 be independent random variables with $P(f_1 = 0) = P(f_2 = 0) = p$ and $P(f_1 = 1) = P(f_2 = 1) = 1 - p$ for some $p \in]0, 1[$. Then $P(f_1 = 0, f_2 = 0|f_1 + f_2) = 1_{\{f_1+f_2=0\}}$, whereas $P(f_1 = 0|f_1 + f_2)P(f_2 = 0|f_1 + f_2) = 1_{\{f_1+f_2=0\}} + \frac{1}{4}1_{\{f_1+f_2=1\}}$.

(e) Dependent σ-algebras may gain independence under conditioning. Indeed, for $n \geqslant 1$, let f_n be a symmetric Bernoulli random variable and set $S_n = f_1 + \cdots + f_n$. Plainly, $S_n, n \geqslant 1$, are not independent. Nevertheless, it is easily verified that, whatever $n \geqslant 2$ and $m \geqslant 1$,

$$\begin{aligned} &P((S_1, \ldots, S_{n-1}) = (i_1, \ldots, i_{n-1}), (S_{n+1}, \ldots, S_{n+m}) = (i_{n+1}, \ldots, i_{n+m})|S_n) \\ &\quad = P((S_1, \ldots, S_{n-1}) = (i_1, \ldots, i_{n-1})|S_n)P((S_{n+1}, \ldots, S_{n+m}) \\ &\quad = (i_{n+1}, \ldots, i_{n+m})|S_n) \end{aligned}$$

for all possible vectors $(i_1, \ldots, i_{n-1})$ and $(i_{n+1}, \ldots, i_{n+m})$.

(f) $\mathcal{A}_i, i \in I$, are conditionally independent given $\mathcal{B}$ if and only if $\mathcal{A}_i, i \in J$, are conditionally independent given $\mathcal{B}$ for any finite set $J \subset I$.

(g) If $\{\mathcal{B}_i : i \in I\}$ is a set of σ-algebras such that $\mathcal{B}_i \subset \mathcal{A}_i, i \in I$, and $\mathcal{A}_i, i \in I$, are conditionally independent given $\mathcal{B}$, then $\mathcal{B}_i, i \in I$, are conditionally independent given $\mathcal{B}$.

(h) Let $\{\mathcal{B}_a : a \in A\}$ be a set of σ-algebras such that $\mathcal{B}_a, a \in A$, are conditionally independent given $\mathcal{B}$. If for each $a \in A$ there exists a set of σ-algebras $\{\mathcal{A}_i : i \in I_a\}$ such that $\mathcal{A}_i \subset \mathcal{B}_a, i \in I_a$, and $\mathcal{A}_i, i \in I_a$, are conditionally independent given $\mathcal{B}$, then $\mathcal{A}_i, i \in \cup_{a\in A} I_a$, are conditionally independent given $\mathcal{B}$.

The following result is an ingredient of many proofs.

Theorem 29.55. *For $i \in I$, let $\mathcal{M}_i$ be as in* (18.10). *If for any finite set $\{i_1, \ldots, i_n\} \subset I$ and $A_1 \in \mathcal{M}_{i_1}, \ldots, A_n \in \mathcal{M}_{i_n}$, $P(A_1 \cap \cdots \cap A_n|\mathcal{B}) = P(A_1|\mathcal{B}) \ldots P(A_n|\mathcal{B})$ a.s., then $\sigma(\mathcal{M}_i), i \in I$, are conditionally independent given $\mathcal{B}$.*

Proof. We may and do assume that $X \in \mathcal{M}_i$ for any $i \in I$. For fixed $A_1 \in \mathcal{M}_{i_1}, \ldots, A_{n-1} \in \mathcal{M}_{i_{n-1}}$, put

$$\mathcal{D} = \{A \in \mathcal{A} : P(A_1 \cap \cdots \cap A_{n-1} \cap A|\mathcal{B}) = P(A_1|\mathcal{B}) \ldots P(A_{n-1}|\mathcal{B})P(A|\mathcal{B}) \text{ a.s.}\}.$$

Then $\mathcal{D}$ is a Dynkin system containing $\mathcal{M}_{i_n}$. On account of (7.36), it follows that

$$P(A_1 \cap \cdots \cap A_{n-1} \cap A_n|\mathcal{B}) = P(A_1|\mathcal{B}) \ldots P(A_{n-1}|\mathcal{B})P(A_n|\mathcal{B}) \text{ a.s., } A_n \in \sigma(\mathcal{M}_{i_n}).$$

Applying the above argument n times proves the statement. □

Corollary 29.56. *Let $\mathcal{A}_i, i \in I$, be conditionally independent given $\mathcal{B}$ σ-algebras, and let $\{I_a : a \in A\}$ be a partition of I with $I_a \neq \emptyset, a \in A$. Then $\vee_{i \in I_a} \mathcal{A}_i, a \in A$, are conditionally independent given $\mathcal{B}$.*

Proof. Reasoning as in the proof of (18.15), this is a consequence of (29.55). □

Corollary 29.57. *For σ-algebras $\mathcal{A}_i, i \in I$, the following are equivalent:*
(i) *$\mathcal{A}_i, i \in I$, are conditionally independent given $\mathcal{B}$;*
(ii) *$\mathcal{A}_{i_0}$ and $\vee_{i \in I - \{i_0\}} \mathcal{A}_i$ are conditionally independent given $\mathcal{B}$ for each $i_0 \in I$.*

Proof. (i) implies (ii) by (29.56). If now (ii) holds and $i_1, \ldots, i_n$ are distinct indices in $I - \{i_0\}$, then $\mathcal{A}_{i_0}$ and $\vee_{j=1}^{n} \mathcal{A}_{i_j}$ are conditionally independent given $\mathcal{B}$ by (29.54.g), and so $\mathcal{A}_{i_0}, \mathcal{A}_{i_1}, \ldots, \mathcal{A}_{i_n}$ are conditionally independent given $\mathcal{B}$ by (29.54.h). This implies (i) by (29.54.f). □

We now give a generalization of the Kolmogorov zero-one law (18.20).

Theorem 29.58. *Assume that $\mathcal{A}_n, n \geqslant 1$, are conditionally independent σ-algebras given $\mathcal{B}$, and put $\mathcal{T} = \cap_{n \geqslant 1} (\vee_{k \geqslant n} \mathcal{A}_k)$. Then $\mathcal{T}$ is essentially contained in $\mathcal{B}$, in the sense that for any $A \in \mathcal{T}$ there is $B \in \mathcal{B}$ with $P(A \Delta B) = 0$.*

Proof. Paralleling the proof of (18.20), we arrive at the conclusion that $\mathcal{T}$ is conditionally independent of itself given $\mathcal{B}$. Hence, for $A \in \mathcal{T}$, $P(A|\mathcal{B}) = P(A \cap A|\mathcal{B}) = P(A|\mathcal{B})^2$ a.s. Thus $P(A|\mathcal{B}) = 1_B$ a.s. for some $B \in \mathcal{B}$. We have $P(A - B) = \int_{B^c} P(A|\mathcal{B})dP = \int_{B^c} 1_B\, dP = 0$ and $P(A \cap B) = \int_B P(A|\mathcal{B})dP = \int_B 1_B\, dP = P(B)$. These equalities show that $P(A \Delta B) = 0$. □

Remark 29.59. If $\mathcal{B} = \{\emptyset, X\}$, then (29.58) yields the classical result (18.20). Actually, in this case $\mathcal{A}_n, n \geqslant 1$, are independent by (29.54.b) and, for every $A \in \mathcal{T}$, either $A \sim \emptyset$ or $A \sim X$ where $\sim$ is the equivalent relation on $\mathcal{A}$ defined in (9.36.b). Thus $P(A) = 0$ or 1 for any $A \in \mathcal{T}$.

To further investigate properties of conditional independence, we introduce a new concept, which is extremely useful in the theory of Markov processes.

Definition 29.60. Let $\mathcal{B}_1, \mathcal{B}_2, \mathcal{B}_3$ be σ-algebras. We say that $(\mathcal{B}_1, \mathcal{B}_2, \mathcal{B}_3)$ is a *Markov triple* if $P(B|\mathcal{B}_1 \vee \mathcal{B}_2) = P(B|\mathcal{B}_2)$ a.s. for any $B \in \mathcal{B}_3$.

Theorem 29.61 (Doob). *Let $\mathcal{B}_1, \mathcal{B}_2, \mathcal{B}_3$ be σ-algebras. The following are equivalent:*
(i) *$(\mathcal{B}_1, \mathcal{B}_2, \mathcal{B}_3)$ is a Markov triple;*
(ii) *$E[f|\mathcal{B}_1 \vee \mathcal{B}_2) = E(f|\mathcal{B}_2)$ a.s. for each $f \in \mathfrak{L}^1(X, \mathcal{B}_3, P)$;*
(iii) *$\mathcal{B}_1$ is conditionally independent of $\mathcal{B}_3$ given $\mathcal{B}_2$.*

Proof. (i) coincides with (ii) for $f = 1_B$ and $B \in \mathcal{B}_3$. Utilizing the indicator function method, it follows that (i) implies (ii). Thus (i) $\iff$ (ii). Assume now that (i) holds and let $B_1 \in \mathcal{B}_1$, $B_3 \in \mathcal{B}_3$. Then, by virtue of (29.9) and (29.16), we have

$$\begin{aligned} P(B_1 \cap B_3|\mathcal{B}_2) &= E[P(B_1 \cap B_3|\mathcal{B}_1 \vee \mathcal{B}_2)|\mathcal{B}_2] = E[1_{B_1} P(B_3|\mathcal{B}_1 \vee \mathcal{B}_2)|\mathcal{B}_2] \\ &= E[1_{B_1} P(B_3|\mathcal{B}_2)|\mathcal{B}_2] = P(B_1|\mathcal{B}_2)P(B_3|\mathcal{B}_2) \text{ a.s.} \end{aligned} \quad (1)$$

Hence (i) $\Rightarrow$ (iii). Conversely, assume that $P(B_1 \cap B_3|\mathcal{B}_2) = P(B_1|\mathcal{B}_2)P(B_3|\mathcal{B}_2)$ a.s. for any $B_1 \in \mathcal{B}_1$, $B_3 \in \mathcal{B}_3$. Then, in view of the last equality in (1), for $B_1 \in \mathcal{B}_1$, $B_2 \in \mathcal{B}_2$, $B_3 \in \mathcal{B}_3$, we obtain

$$\begin{aligned} \int_{B_1 \cap B_2} P(B_3|\mathcal{B}_1 \vee \mathcal{B}_2)dP &= \int_{B_1 \cap B_2} 1_{B_3}\, dP = \int_{B_2} P(B_1 \cap B_3|\mathcal{B}_2)dP \\ &= \int_{B_2} E[1_{B_1} P(B_3|\mathcal{B}_2)|\mathcal{B}_2]dP = \int_{B_1 \cap B_2} P(B_3|\mathcal{B}_2)dP. \end{aligned} \quad (2)$$

Since the family $\{B_1 \cap B_2 : B_1 \in \mathcal{B}_1, B_2 \in \mathcal{B}_2\}$ is closed under intersection and generates $\mathcal{B}_1 \vee \mathcal{B}_2$, (2) implies $P(B_3|\mathcal{B}_1 \vee \mathcal{B}_2) = P(B_3|\mathcal{B}_2)$ a.s. by (11.39). Thus (iii) $\Rightarrow$ (i). □

Besides the fundamental characterization in (29.61), Markov triples have many other useful properties we present now.

Theorem 29.62. *Let $\mathcal{B}_1, \mathcal{B}_2, \mathcal{B}_3, \mathcal{B}_4$ be σ-algebras. Then the following statements hold.*

(i) *If $(\mathcal{B}_1, \mathcal{B}_2, \mathcal{B}_3)$ is a Markov triple, then $(\mathcal{B}_1 \vee \mathcal{B}_2, \mathcal{B}_2, \mathcal{B}_3)$ is a Markov triple.*

(ii) *If $(\mathcal{B}_1, \mathcal{B}_2, \mathcal{B}_3)$ is a Markov triple, and $\mathcal{B}$ is a σ-algebra such that $\mathcal{B}_2 \subset \mathcal{B} \subset \mathcal{B}_1 \vee \mathcal{B}_2$, then $(\mathcal{B}_1, \mathcal{B}, \mathcal{B}_3)$ is a Markov triple.*

(iii) *$(\mathcal{B}_1, \mathcal{B}_2, \mathcal{B}_3)$ is a Markov triple if and only if $(\mathcal{B}_3, \mathcal{B}_2, \mathcal{B}_1)$ is a Markov triple.*

(iv) *If $(\mathcal{B}_1, \mathcal{B}_2, \mathcal{B}_3)$ is a Markov triple, and $\mathcal{B}'_1 \subset \mathcal{B}_1$, $\mathcal{B}'_3 \subset \mathcal{B}_3$ are σ-algebras, then $(\mathcal{B}'_1, \mathcal{B}_2, \mathcal{B}'_3)$ is a Markov triple.*

(v) *If $\mathcal{M} \subset \mathcal{A}$ is a family closed under intersection, and $P(B|\mathcal{B}_1 \vee \mathcal{B}_2) = P(B|\mathcal{B}_2)$ a.s. for any $B \in \mathcal{M}$, then $(\mathcal{B}_1, \mathcal{B}_2, \sigma(\mathcal{M}))$ is a Markov triple.*

(vi) *$(\mathcal{B}_1, \mathcal{B}_2, \mathcal{B}_3)$ and $(\mathcal{B}_1, \mathcal{B}_2 \vee \mathcal{B}_3, \mathcal{B}_4)$ are Markov triples if and only if $(\mathcal{B}_1, \mathcal{B}_2, \mathcal{B}_3 \vee \mathcal{B}_4)$ is a Markov triple.*

(vii) *If $(\mathcal{B}_1, \mathcal{B}_2, \mathcal{B}_3)$ and $(\mathcal{B}_1 \vee \mathcal{B}_2, \mathcal{B}_3, \mathcal{B}_4)$ are Markov triples, then $(\mathcal{B}_1, \mathcal{B}_2, \mathcal{B}_3 \vee \mathcal{B}_4)$ is a Markov triple.*

Proof. (i) is obvious. (ii) is immediate by (29.10). (iii) and (iv) are easy consequences of (29.61). To prove (v), let $\mathcal{D} = \{B \in \mathcal{A} : P(B|\mathcal{B}_1 \vee \mathcal{B}_2) = P(B|\mathcal{B}_2 \text{ a.s.})\}$. Then $\mathcal{D}$ is a Dynkin system and $\mathcal{D} \supset \mathcal{M}$. On account of (7.36), this implies that $(\mathcal{B}_1, \mathcal{B}_2, \sigma(\mathcal{M}))$ is a Markov triple. Assume now that $(\mathcal{B}_1, \mathcal{B}_2, \mathcal{B}_3)$ and $(\mathcal{B}_1, \mathcal{B}_2 \vee \mathcal{B}_3, \mathcal{B}_4)$ are Markov triples, and let $B \in \mathcal{B}_1$. Then, in view of (iii), we have $P(B|\mathcal{B}_2 \vee (\mathcal{B}_3 \vee \mathcal{B}_4)) = P(B|\mathcal{B}_2 \vee \mathcal{B}_3) = P(B|\mathcal{B}_2)$ a.s. This means that $(\mathcal{B}_3 \vee \mathcal{B}_4, \mathcal{B}_2, \mathcal{B}_1)$ is a Markov triple, and so $(\mathcal{B}_1, \mathcal{B}_2, \mathcal{B}_3 \vee \mathcal{B}_4)$ is a Markov triple by (iii). Conversely, if $(\mathcal{B}_1, \mathcal{B}_2, \mathcal{B}_3 \vee \mathcal{B}_4)$ is a Markov triple, then $(\mathcal{B}_1, \mathcal{B}_2, \mathcal{B}_3)$ is a Markov triple by (iv), and $(\mathcal{B}_3 \vee \mathcal{B}_4, \mathcal{B}_2, \mathcal{B}_1)$ is a Markov triple by (iii). Since $\mathcal{B}_2 \subset \mathcal{B}_2 \vee \mathcal{B}_3 \subset \mathcal{B}_2 \vee \mathcal{B}_3 \vee \mathcal{B}_4$, it follows that $(\mathcal{B}_3 \vee \mathcal{B}_4, \mathcal{B}_2 \vee \mathcal{B}_3, \mathcal{B}_1)$ is a Markov triple by (ii). Hence $(\mathcal{B}_1, \mathcal{B}_2 \vee \mathcal{B}_3, \mathcal{B}_4)$ is a Markov triple by (iii) and (iv).

Thus (vi) is proved. Finally, assume that $(\mathcal{B}_1, \mathcal{B}_2, \mathcal{B}_3)$ and $(\mathcal{B}_1 \vee \mathcal{B}_2, \mathcal{B}_3, \mathcal{B}_4)$ are Markov triples. Since $\mathcal{B}_3 \subset \mathcal{B}_2 \vee \mathcal{B}_3 \subset \mathcal{B}_1 \vee \mathcal{B}_2 \vee \mathcal{B}_3$, it follows that $(\mathcal{B}_1 \vee \mathcal{B}_2, \mathcal{B}_2 \vee \mathcal{B}_3, \mathcal{B}_4)$ is a Markov triple by (ii), and so $(\mathcal{B}_1, \mathcal{B}_2 \vee \mathcal{B}_3, \mathcal{B}_4)$ is a Markov triple by (iv). Then $(\mathcal{B}_1, \mathcal{B}_2, \mathcal{B}_3 \vee \mathcal{B}_4)$ is a Markov triple by (vi). □

Corollary 29.63. *Let $\mathcal{C}, \mathcal{B}$ and $\mathcal{A}_1, \mathcal{A}_2, \ldots$ be σ-algebras. Then the following are equivalent:*

(i) *$(\mathcal{C}, \mathcal{B}, \vee_{n\geqslant 1}\mathcal{A}_n)$ is a Markov triple;*

(ii) *$(\mathcal{C}, \mathcal{B}, \mathcal{A}_1)$ and $(\mathcal{C}, \mathcal{B} \vee (\vee_{i=1}^n \mathcal{A}_i, \mathcal{A}_{n+1}), n \geqslant 1$, are Markov triples.*

Proof. If (i) holds, then $(\mathcal{C}, \mathcal{B}, \mathcal{A}_1)$ and $(\mathcal{C}, \mathcal{B}, \vee_{i=1}^{n+1}\mathcal{A}_i), n \geqslant 1$, are Markov triples by (29.62.iv). Then $(\mathcal{C}, \mathcal{B} \vee (\vee_{i=1}^n \mathcal{A}_i), \mathcal{A}_{n+1}), n \geqslant 1$, are Markov triples by (18.12) and (29.62.vi). If now (ii) holds, then it follows inductively by (29.62.vi) that $(\mathcal{C}, \mathcal{B}, \vee_{i=1}^n \mathcal{A}_i), n \geqslant 1$, are Markov triples. Applying (29.62.v) with $\mathcal{M} = \cup_{n\geqslant 1} (\vee_{i=1}^n \mathcal{A}_i)$, this implies (i). □

Corollary 29.64. *Let $\mathcal{C}$ and $\mathcal{A}_1, \mathcal{A}_2, \ldots$ be σ-algebras. Then $\mathcal{C}$ is independent of $\vee_{n\geqslant 1}\mathcal{A}_n$ if and only if $\mathcal{C}$ is independent of $\mathcal{A}_1$ and, for each $n \geqslant 1$, $\mathcal{C}$ is independent of $\mathcal{A}_{n+1}$ given $\vee_{i=1}^n \mathcal{A}_i$.*

Proof. Take $\mathcal{B} = \{\emptyset, X\}$ in (29.63). □

Theorem 29.65. *Let $\mathcal{B}_1, \mathcal{B}_2, \mathcal{B}_3, \mathcal{B}_4$ be σ-algebras. Then the following are equivalent:*

(i) *$(\mathcal{B}_1, \mathcal{B}_2, \mathcal{B}_3)$ is a Markov triple;*

(ii) *$E[f|\mathcal{B}_1] = E[E[f|\mathcal{B}_2]|\mathcal{B}_1]$ a.s. for each $f \in \mathfrak{L}^1(X, \mathcal{B}_2 \vee \mathcal{B}_3, P)$;*

(iii) *$P(B|\mathcal{B}_1) = E[P(B|\mathcal{B}_2)|\mathcal{B}_1]$ a.s. for each $B \in \mathcal{B}_2 \vee \mathcal{B}_3$.*

Proof. The proof that (i) $\Longleftrightarrow$ (ii) is similar to the corresponding proof of (29.61). Assume now that (i) holds and let $B_2 \in \mathcal{B}_2$, $B_3 \in \mathcal{B}_3$. Then, in view of (29.9) and (29.16), we have

$$\begin{aligned} P(B_2 \cap B_3|\mathcal{B}_1) &= E[P(B_2 \cap B_3|\mathcal{B}_1 \vee \mathcal{B}_2)|\mathcal{B}_1] = E[1_{B_2} P(B_3|\mathcal{B}_1 \vee \mathcal{B}_2)|\mathcal{B}_1] \\ &= E[1_{B_2} P(B_3|\mathcal{B}_2)|\mathcal{B}_1] = E[P(B_2 \cap B_3|\mathcal{B}_2)|\mathcal{B}_1] \text{ a.s.} \end{aligned} \tag{1}$$

This shows that (iii) is satisfied for $B = B_2 \cap B_3$ with $B_2 \in \mathcal{B}_2$, $B_3 \in \mathcal{B}_3$. As $\{B_2 \cap B_3 : B_2 \in \mathcal{B}_2, B_3 \in \mathcal{B}_3\}$ is closed under intersection and generates $\mathcal{B}_2 \vee \mathcal{B}_3$, (1) and the customary Dynkin system argument entail (iii). Thus (i) $\Rightarrow$ (iii). Conversely, suppose that (iii) holds, and let $B_1 \in \mathcal{B}_1$, $B_2 \in \mathcal{B}_2$, $B_3 \in \mathcal{B}_3$. Then

$$\begin{aligned} \int_{B_1 \cap B_2} P(B_3|\mathcal{B}_1 \vee \mathcal{B}_2) dP &= P(B_1 \cap B_2 \cap B_3) = \int_{B_1} P(B_2 \cap B_3|\mathcal{B}_1) dP \\ &= \int_{B_1} P(B_2 \cap B_3|\mathcal{B}_2) dP = \int_{B_1 \cap B_2} P(B_3|\mathcal{B}_2) dP. \end{aligned}$$

This leads to (i) by the argument at the end of the proof of (29.61). Thus (iii) $\Rightarrow$ (i). □

Theorem 29.66. *Let $\mathcal{B}_1, \mathcal{B}_2, \mathcal{B}_3, \mathcal{B}_4, \mathcal{B}_5$ be σ-algebras such that $(\mathcal{B}_1, \mathcal{B}_2, \mathcal{B}_3 \vee \mathcal{B}_4 \vee \mathcal{B}_5)$ and $(\mathcal{B}_1 \vee \mathcal{B}_2 \vee \mathcal{B}_3, \mathcal{B}_4, \mathcal{B}_5)$ are Markov triples. Then the following hold.*

(i) $\mathcal{B}_1, \mathcal{B}_3, \mathcal{B}_5$ *are conditional independent given* $\mathcal{B}_2 \vee \mathcal{B}_4$.
(ii) $(\mathcal{B}_1 \vee \mathcal{B}_5, \mathcal{B}_2 \vee \mathcal{B}_4, \mathcal{B}_3)$ *is a Markov triple.*

Proof. Let $B_1 \in \mathcal{B}_1$, $B_3 \in \mathcal{B}_3$, $B_5 \in \mathcal{B}_5$. By (29.62.iii) and (29.62.iv), we have $P(B_1|\mathcal{B}_2 \vee \mathcal{B}_3 \vee \mathcal{B}_4) = P(B_1|\mathcal{B}_2) = P(B_1|\mathcal{B}_2 \vee \mathcal{B}_4)$ a.s., and so

$$P(B_1 \cap B_3|\mathcal{B}_2 \vee \mathcal{B}_4) = E[1_{B_3} P(B_1|\mathcal{B}_2 \vee \mathcal{B}_3 \vee \mathcal{B}_4)|\mathcal{B}_2 \vee \mathcal{B}_4]$$
$$= P(B_1|\mathcal{B}_2 \vee \mathcal{B}_4) P(B_3|\mathcal{B}_2 \vee \mathcal{B}_4) \text{ a.s.} \tag{1}$$

Similarly, we have

$$P(B_5|\mathcal{B}_1 \vee \mathcal{B}_2 \vee \mathcal{B}_3 \vee \mathcal{B}_4) = P(B_5|\mathcal{B}_4) = P(B_5|\mathcal{B}_2 \vee \mathcal{B}_4) \text{ a.s.} \tag{2}$$

From (1) and (2) we get

$$P(B_1 \cap B_3 \cap B_5|\mathcal{B}_2 \vee \mathcal{B}_4) = E[1_{B_1 \cap B_3} P(B_5|\mathcal{B}_1 \vee \mathcal{B}_2 \vee \mathcal{B}_3 \vee \mathcal{B}_4)|\mathcal{B}_2 \vee \mathcal{B}_4]$$
$$= P(B_1|\mathcal{B}_2 \vee \mathcal{B}_4) P(B_3|\mathcal{B}_2 \vee \mathcal{B}_4) P(B_5|\mathcal{B}_2 \vee \mathcal{B}_4) \text{ a.s.}$$

This proves (i). From (i) and (29.56) it follows that $\mathcal{B}_1 \vee \mathcal{B}_5$ is independent of $\mathcal{B}_3$ given $\mathcal{B}_2 \vee \mathcal{B}_4$. This entails (ii) by (29.61). □

Remark 29.67. Inspection of the above proof reveals that the assumption that $(\mathcal{B}_1, \mathcal{B}_2, \mathcal{B}_3 \vee \mathcal{B}_4 \vee \mathcal{B}_5)$ is a Markov triple may be relaxed to the statement that $(\mathcal{B}_1, \mathcal{B}_2, \mathcal{B}_3 \vee \mathcal{B}_4)$ is a Markov triple. However, if $(\mathcal{B}_1 \vee \mathcal{B}_2 \vee \mathcal{B}_3, \mathcal{B}_4, \mathcal{B}_5)$ is a Markov triple, then $(\mathcal{B}_1 \vee \mathcal{B}_2, \mathcal{B}_3 \vee \mathcal{B}_4, \mathcal{B}_5)$ is a Markov triple by (29.62.vi). Hence, in view of (29.62.vii), $(\mathcal{B}_1, \mathcal{B}_2, \mathcal{B}_3 \vee \mathcal{B}_4 \vee \mathcal{B}_5)$ is a Markov triple whenever $(\mathcal{B}_1, \mathcal{B}_2, \mathcal{B}_3 \vee \mathcal{B}_4)$ is a Markov triple.

Throughout (29.68)–(29.78), if present, $\mathcal{B} \subset \mathcal{A}$ is a σ-algebra and $f \in \mathfrak{L}^1$.

Exercise 29.68. Let $\mathcal{C} \subset \mathcal{A}$ be a σ-algebra. Show that the following are equivalent:
(i) $E^{\mathcal{B}} f = E^{\mathcal{C}} f$ a.s. for all $f \in \mathfrak{L}^1$;
(ii) $\mathcal{B}$ essentially coincides with $\mathcal{C}$, in the sense that for any $B \in \mathcal{B}$ there is $C \in \mathcal{C}$ with $P(B \Delta C) = 0$, and for any $C \in \mathcal{C}$ there is $B \in \mathcal{B}$ with $P(B \Delta C) = 0$.

Exercise 29.69. If g is a bounded random variable, then $E[f E^{\mathcal{B}} g] = E[(E^{\mathcal{B}} f) g]$. [Use (29.16).]

Exercise 29.70. For $f \in \mathfrak{L}^2$, put $\operatorname{Var}^{\mathcal{B}} f = E^{\mathcal{B}}(f - E^{\mathcal{B}} f)^2$. Then $\operatorname{Var}^{\mathcal{B}} f$ is termed the *conditional variance of* f *given* $\mathcal{B}$. Show that $\operatorname{Var}[E^{\mathcal{B}} f] \leqslant \operatorname{Var}[E^{\mathcal{B}} f] + E[\operatorname{Var}^{\mathcal{B}} f] = \operatorname{Var} f$.

Exercise 29.71 (Wald). Let $f_1, f_2, \ldots$ be random variables with $P f_n^{-1} = P f_1^{-1}, n \geqslant 1$, let $\nu : X \to N$ be a random variable that is independent of $\sigma(f_n, n \geqslant 1)$, and put $g = f_1 + \cdots + f_\nu$. Prove the following.
(a) $E[g|\nu] = \nu E f$ a.s. and $E g = (E\nu) E f$.
(b) If $f_n, n \geqslant 1$, are independent, $E f_1^2 < \infty$ and $E\nu^2 < \infty$, then $\operatorname{Var}^{\sigma(\nu)} g = \nu \operatorname{Var} f$ a.s.

(c) Under the assumptions of (b), $\mathrm{Var}\, g = (\mathrm{Var}\nu)Ef^2 + (E\nu)\mathrm{Var}\, f$. [Use (29.70), (a) and (b).]

Exercise 29.72. Let $\{A_n : n \geqslant 1\} \subset \mathcal{A}$ be such that $A_m \cap A_n = \emptyset$ for $m \neq n$. Show that $\sum_{k=1}^{n} P^{\mathcal{B}}(A_k) \overset{\mathfrak{L}^p}{\to} \sum_{k\geqslant 1} P^{\mathcal{B}}(A_k)$ for $1 \leqslant p < \infty$. [Use (29.21).]

Exercise 29.73. Let $\{f_n : n \geqslant 1\} \subset \mathfrak{L}^p$ with $1 \leqslant p < \infty$, and assume that $f_n \overset{P}{\to} f$ and $\sup_{n\geqslant 1} E^{\mathcal{B}} |f_n|^p \leqslant C$ a.s. for some $C < \infty$. Prove that $E^{\mathcal{B}} f \in \mathfrak{L}^p$ and $E^{\mathcal{B}} f_n \overset{\mathfrak{L}^p}{\to} E^{\mathcal{B}} f$. [Use (29.22).]

Exercise 29.74. Let $\{f_n : n \geqslant 1\} \subset \mathfrak{L}^p$ with $1 < p < \infty$, and suppose that $f_n \overset{a.s.}{\to} f$ and $\sup_{n\geqslant 1} E^{\mathcal{B}} |f_n|^p \leqslant C$ a.s. for some $C < \infty$. Prove the following.

(a) $E\,|f|^p \leqslant C$.

(b) $E^{\mathcal{B}} f_n \overset{a.s.}{\to} E^{\mathcal{B}} f$. [Use (29.25).]

Exercise 29.75. For $g \in \mathfrak{L}^1$, show that $E^{\mathcal{B}}[f \vee g] \geqslant (E^{\mathcal{B}} f) \vee E^{\mathcal{B}} g$ and $E^{\mathcal{B}}[f \wedge g] \leqslant (E^{\mathcal{B}} f) \wedge E^{\mathcal{B}} g$.

Exercise 29.76. If $\mathcal{C} \subset \mathcal{A}$ is a σ-algebra such that $\sigma(f) \vee \mathcal{B}$ is independent of $\mathcal{C}$, then $E^{\mathcal{B}\vee\mathcal{C}} f = E^{\mathcal{B}} f$.

Exercise 29.77. Let Q be a probability on $\mathcal{B}$ such that $P \ll Q$, and let $g = dP/dQ$. Prove that $dP_{\mathcal{B}}/dQ_{\mathcal{B}} = E_Q[g|\mathcal{B}]$ Q-a.s.

Exercise 29.78. For $n \geqslant 1$, let $0 \leqslant f_n \in \mathfrak{L}^1$ and let $\mathcal{B}_n \subset \mathcal{A}$ be a σ-algebra. If $E^{\mathcal{B}_n} f_n \overset{P}{\to} 0$, then $f_n \overset{P}{\to} 0$. [Use (14.7.f) and (29.75).]

Exercise 29.79. Notation is as in (29.5). Assuming that $f \geqslant 0$, and putting $\mu(B) = \int_{g^{-1}(B)} f\, dP$, $B \in \mathcal{B}$, prove that $\mu \ll Pg^{-1}$ and $h = d\mu/d(Pg^{-1})$.

Exercise 29.80 (Doob). Let $\{\mathcal{B}_n : n \geqslant 1\}$, $\mathcal{B}$ and f be as in (29.27), suppose that f is $\mathcal{B}$-measurable, and write $f_n = E^{\mathcal{B}_n} f$, $n \geqslant 1$. Show that $E^{\mathcal{B}_n} |f_n - f| \overset{a.s.}{\to} 0$ and $E^{\mathcal{B}_n} |f_n - f| \overset{\mathfrak{L}^p}{\to} 0$.

Exercise 29.81. Let $\{\mathcal{B}_n : n \geqslant 1\}$ and $\mathcal{B}$ be as in (29.31). Prove the following.

(a) If $\{f_n : n \geqslant 1\} \subset \mathfrak{L}^p$ with $1 \leqslant p < \infty$ is such that $f_n \overset{\mathfrak{L}^p}{\to} f$ for some $f \in \mathfrak{L}^p$, then $E^{\mathcal{B}_n} f_n \overset{\mathfrak{L}^p}{\to} E^{\mathcal{B}} f$.

(b) If $\{f_n : n \geqslant 1\} \subset \mathfrak{L}^1$ is such that $f_n \leqslant f_{n+1}$, $n \geqslant 1$, and $f_n \overset{a.s.}{\to} f$ for some $f \in \mathfrak{L}^1$, then $E^{\mathcal{B}_n} f_n \overset{a.s.}{\to} E^{\mathcal{B}} f$ and $E^{\mathcal{B}_n} f_n \overset{\mathfrak{L}^1}{\to} E^{\mathcal{B}} f$.

Exercise 29.82. Let P be a probability on $\mathcal{B}(R)$, consider the probability space $(R^\infty, \mathcal{B}(R^\infty), \mathbf{P})$ with $\mathbf{P} = \otimes_{i\geqslant 1}\mathsf{P}_i$, where $\mathsf{P}_i = \mathsf{P}$, $i \geqslant 1$, let π_i denote the projection from R^∞ onto R_i, where $R_i = R$, $i \geqslant 1$, and fix $n \geqslant 2$ and $1 \leqslant j \leqslant n$. For $(x_1, x_2, \ldots) \in R^\infty$, define $u(x_1, x_2, \ldots) = (x_j, x_2, \ldots, x_{j-1}, x_1, x_{j+1}, \ldots)$ and $s_m(x_1, x_2, \ldots) = x_1 + \cdots + x_m$, $m \geqslant n$. Prove the following.

(a) $\mathbf{P}u^{-1} = \mathsf{P}$, $\pi_1 \circ u = \pi_j$ and $s_m \circ u = s_m$, $m \geqslant n$.

(b) For $m \geqslant n$, s_m is $\mathcal{B}(R^\infty)$-measurable.

(c) $u^{-1}(B) = B$, $B \in \sigma(s_m, m \geqslant n)$, and so $u^{-1}(\sigma(s_m, m \geqslant n)) = \sigma(s_m, m \geqslant n)$. [Hint. For $m \geqslant n$, and $B_n, \ldots, B_m \in \mathcal{B}(R)$, $u^{-1}(\cap_{i=n}^m s_i^{-1}(B_i)) = \cap_{i=n}^m s_i^{-1}(B_i)$. Then apply (7.36).]

(d) If $h : R^\infty \to R$ is $\sigma(s_m, m \geqslant n)$-measurable, then $h \circ u = h$. [Use (c) and the indicator function method.]

(e) If $\int_R |x|\, d\mathsf{P}(x) < \infty$, then $E_{\mathbf{P}}[\pi_j|\sigma(s_m, m \geqslant n)] = E_{\mathbf{P}}[\pi_1|\sigma(s_m, m \geqslant n)]$ $\mathbf{P}$-a.s. [Hint. Apply (29.7) with $f = u$, $g = \pi_1$ and $\mathcal{C} = \sigma(s_m, m \geqslant n)$, utilizing (a), (c) and (d).]

(f) If $\int_R x^2\, d\mathsf{P}(x) < \infty$ and $a = \int_R x\, d\mathsf{P}(x)$, then $E_{\mathbf{P}}[(\pi_j - \frac{s_n}{n})^2|\sigma(s_m, m \geqslant n)] = E_{\mathbf{P}}[(\pi_1 - \frac{s_n}{n})^2|\sigma(s_m, m \geqslant n)] = \frac{n-1}{n} E_{\mathbf{P}}[(\pi_1 - a)^2|\sigma(s_m, m \geqslant n)]$ $\mathbf{P}$-a.s.

Exercise 29.83. Let $f_n, n \geqslant 1$, be i.i.d. random variables with $E|f_1| < \infty$ and $a = Ef_1$, fix $n \geqslant 2$ and $1 \leqslant j \leqslant n$, and set $S_n = f_1 + \cdots + f_n$.

(a) Show that $E[f_j|\sigma(S_m, m \geqslant n)] = E[f_1|\sigma(S_m, m \geqslant n)]$ a.s. [Hints. Let π_1, π_j, and $s_m, m \geqslant n$, be as in (29.82). Write $\mathbf{f} = (f_1, f_2, \ldots)$ and $\mathbf{P} = P\mathbf{f}^{-1}$. By (18.29), $\mathbf{P} = \otimes_{i \geqslant 1} \mathsf{P}_i$, where $\mathsf{P}_i = Pf_1^{-1}, i \geqslant 1$. In view of (29.7), $E[f_j|\sigma(S_m, m \geqslant n)] = E_P[\pi_j \circ \mathbf{f}|\mathbf{f}^{-1}(\sigma(S_m, m \geqslant n))] = E_{\mathbf{P}}[\pi_j|\sigma(s_m, m \geqslant n)] \circ \mathbf{f}$ a.s.]

(b) If $Ef_1^2 < \infty$, prove that $E[(f_j - \frac{S_n}{n})^2|\sigma(S_m, m \geqslant n)] = E[(f_1 - \frac{S_n}{n})^2|\sigma(S_m, m \geqslant n)] = \frac{n-1}{n} E[(f_1 - Ef_1)^2|\sigma(S_m, m \geqslant n)]$ a.s.

Exercise 29.84. Let f_n and $S_n, n \geqslant 1$, be as in (29.83). Use (29.29) to prove Kolmogorov's SLLN, i.e. $S_n/n \xrightarrow{a.s.} Ef_1$. [Hints. Let $\mathcal{T}$ and $\mathcal{E}$ denote the tail σ-algebras of $\{f_n : n \geqslant 1\}$ and $\{S_n : n \geqslant 1\}$, respectively, and put $g = E[f_1|\mathcal{E}]$. By (29.83) and (29.29), $\frac{S_n}{n} = \frac{1}{n}E[S_n|\sigma(S_m, m \geqslant n)] = E[f_1|\sigma(S_m, m \geqslant n)] \xrightarrow{a.s.} g$ and $\frac{S_n}{n} \xrightarrow{\mathfrak{L}^1} g$. By (7.72), g is $\mathcal{T}$-measurable, hence $\mathcal{O}_P$-measurable by the Kolmogorov zero-one law (18.20). Then use (18.19) and (14.13.e).]

Exercise 29.85. Let $\{\mathcal{B}_n : n \geqslant 1\}$ and $\mathcal{B}$ be as in (29.27), let Q be a probability on $\mathcal{B}$ such that $P_{\mathcal{B}_n} \ll Q_{\mathcal{B}_n}, n \geqslant 1$, let $g_n = dP_{\mathcal{B}_n}/Q_{\mathcal{B}_n}, n \geqslant 1$, and set $R = (P_{\mathcal{B}} + Q)/2$. Prove the following.

(a) If $P_{\mathcal{B}} \ll Q$ and $g = dP_{\mathcal{B}}/dQ$, then $E_Q[g|\mathcal{B}_n] = g_n$ Q-a.s., $n \geqslant 1$, $g_n \xrightarrow{a.s.} g$, and $g_n \xrightarrow{\mathfrak{L}^1} g$ (relative to Q).

(b) $P_{\mathcal{B}} \ll R$ and $P_{\mathcal{B}_n} \ll R_{\mathcal{B}_n}, n \geqslant 1$.

(c) If $h = dP_{\mathcal{B}}/dR$ and $h_n = dP_{\mathcal{B}_n}/dR_{\mathcal{B}_n}, n \geqslant 1$, then $h \leqslant 2$ R-a.s., and $h_n = 2g_n/(1 + g_n)$ Q-a.s., $n \geqslant 1$. [Use (11.41) for the first assertion.]

(d) $h_n \xrightarrow{a.s.} h$ and $h_n \xrightarrow{\mathfrak{L}^1} h$ (relative to R, hence relative to Q).

(e) $g_n \xrightarrow{a.s.} g$ (relative to Q), where $g = \begin{cases} h/(2-h) & \text{if } h < 2 \\ \infty & \text{if } h = 2 \end{cases}$.

(f) Define $\nu_1(B) = \int_{B \cap \{g < \infty\}} g\, dQ$, $B \in \mathcal{B}$, and $\nu_2(B) = P(B \cap \{g = \infty\})$, $B \in \mathcal{B}$. Then $P_{\mathcal{B}} = \nu_1 + \nu_2$ is the Lebesgue decomposition of $P_{\mathcal{B}}$ with respect to Q.

Exercise 29.86. Let $f_i, i = 1, 2$, be random variables, let $(Y, \mathcal{B})$ be a measurable space, let $g_i : X \to Y, i = 1, 2$, be random elements, and assume that $P(f_1, g_1)^{-1} = P(f_2, g_2)^{-1}$ and $f_1 \in \mathfrak{L}^1$.

(a) For $i = 1, 2$, let $E[f_i|g_i]$ be a version of the conditional expectation of f_i given g_i. Show that the distributions of $E[f_1|g_1]$ and $E[f_2|g_2]$ coincide. [Hint. For $i = 1, 2$, let $h_i : Y \to R$ be a measurable function such that $E[f_i|g_i] = h_i \circ g_i$ a.s. Then $\int_B h_1\, d(Pg_1^{-1}) = \int_B h_2\, d(Pg_2^{-1})$, $B \in \mathcal{B}$.]

(b) Let $\varphi : R \to R$ and $\psi : Y \to Y$ be measurable functions, and suppose that $\varphi \circ f_1 \in \mathfrak{L}^1$. For $i = 1, 2$, let $E[\varphi \circ f_i|\psi \circ g_i]$ be a version of the conditional expectation of $\varphi \circ f_i$ given $\psi \circ g_i$. Prove that the distributions of $E[\varphi \circ f_1|\psi \circ g_1]$ and $E[\varphi \circ f_2|\psi \circ g_2]$ coincide.

Exercise 29.87. Let $(X, \mathcal{A})$, $(Y, \mathcal{B})$ and $(Z, \mathcal{C})$ be measurable spaces, let $\varphi : X \times Y \to Z$ be a measurable function, write $\psi(x, y) = (x, \varphi(x, y))$, $(x, y) \in X \times Y$, let Q be a transition probability from $(X, \mathcal{A})$ to $(Y, \mathcal{B})$, and define $(Q\varphi^{-1})(x, C) = Q(x, \varphi_x^{-1}(C))$, $x \in X$, $C \in \mathcal{C}$. Prove the following.

(a) ψ is a measurable function.

(b) $Q\varphi^{-1}$ is a transition probability from $(X, \mathcal{A})$ to $(Z, \mathcal{C})$.

(c) $P \otimes (Q\varphi^{-1}) = (P \otimes Q)\psi^{-1}$.

Exercise 29.88. Let $(X, \mathcal{A})$, $(Y, \mathcal{B})$ and $(Z, \mathcal{C})$ be measurable spaces, let $\varphi : X \to Y$ be a measurable function, write $\psi(x, z) = (\varphi(x), z)$, $(x, z) \in X \times Z$, let Q be a transition probability from $(Y, \mathcal{B})$ to $(Z, \mathcal{C})$, and define $(Q\varphi)(x, C) = Q(\varphi(x), C)$, $x \in X$, $C \in \mathcal{C}$. Prove the following.

(a) ψ is a measurable function.

(b) $Q\varphi$ is a transition probability from $(X, \mathcal{A})$ to $(Z, \mathcal{C})$.

(c) $(P\varphi^{-1}) \otimes Q = (P \otimes (Q\varphi))\psi^{-1}$.

Exercise 29.89. Let $(F, \mathcal{F})$, $(G, \mathcal{G})$, f, g and Q be as in (29.40), let $(H, \mathcal{H})$ be a measurable space, let $\varphi : F \times G \to H$ be a measurable function, and assume that $P(f, g)^{-1} = (Pf^{-1}) \otimes Q$. Show that $P(f, \varphi \circ (f, g))^{-1} = (Pf^{-1}) \otimes (Q\varphi^{-1})$.

Exercise 29.90. Let $(F, \mathcal{F})$, $(G, \mathcal{G})$ and $(H, \mathcal{H})$ be measurable spaces, let $f : X \to F$ and $h : X \to H$ be measurable functions, let $\varphi : F \to G$ be a measurable function, let Q be a transition probability from $(G, \mathcal{G})$ to $(H, \mathcal{H})$, and assume that $P(f, h)^{-1} = (Pf^{-1}) \otimes (Q\varphi)$. Prove that $P(\varphi \circ f, h)^{-1} = (P(\varphi \circ f)^{-1}) \otimes Q$.

Exercise 29.91. For σ-algebras $\mathcal{A}_i$, $i \in I$, the following are equivalent:

(i) $\mathcal{A}_i$, $i \in I$, are independent;

(ii) for each $i_0 \in I$ and $A \in \mathcal{A}_{i_0}$, $P(A|\vee_{i \in I - \{i_0\}} \mathcal{A}_i) = P(A)$ a.s.;

(iii) for each $i_0 \in I$ and $f \in \mathfrak{L}^1(X, \mathcal{A}_{i_0}, P)$, $E[f|\vee_{i \in I - \{i_0\}} \mathcal{A}_i) = Ef$ a.s.

Exercise 29.92. Let $\mathcal{A}_n$, $n \geqslant 1$, and $\mathcal{B}$ be σ-algebras. Show that $\mathcal{A}_n$, $n \geqslant 1$, are conditionally independent given $\mathcal{B}$ if and only if, for each $n \geqslant 1$, $\vee_{i=1}^n \mathcal{A}_i$ is conditionally independent of $\mathcal{A}_{n+1}$ given $\mathcal{B}$.

Exercise 29.93. Let $f \in \mathfrak{L}^1$, let $(Y, \mathcal{B})$ be a measurable space, let $g_i : X \to Y$, $i = 1, 2$, be random elements such that $g_1^{-1}(\mathcal{B}) \subset g_2^{-1}(\mathcal{B})$, and assume that $P(f, g_1)^{-1} = P(f, g_2)^{-1}$. Prove that $(f^{-1}(\mathcal{B}(R)), g_1^{-1}(\mathcal{B}), g_2^{-1}(\mathcal{B}))$ is a Markov triple. [Hint. For $B \in \mathcal{B}(R)$ and $i = 1, 2$, let $P(f \in B|g_i)$ be a version of the conditional probability of $\{f \in B\}$ given g_i. Use (29.86.b) to show that the distributions of $P(f \in B|g_1)$ and $P(f \in B|g_2)$ coincide.]

Exercise 29.94. Let $\mathcal{B}_1, \mathcal{B}_2, \mathcal{B}_3, \mathcal{B}_4$ be σ-algebras such that $(\mathcal{B}_1, \mathcal{B}_2, \mathcal{B}_3)$ is a Markov triple and $\mathcal{B}_1 \vee \mathcal{B}_2 \vee \mathcal{B}_3$ is independent of $\mathcal{B}_4$. Verify that $(\mathcal{B}_1, \mathcal{B}_2, \mathcal{B}_3 \vee \mathcal{B}_4)$ and $(\mathcal{B}_1, \mathcal{B}_2 \vee \mathcal{B}_4, \mathcal{B}_3)$ are Markov triples.

Exercise 29.95 (Buhr-Carrière). Let $\mathcal{A}_i, i \in I$, be σ-algebras, let $i_0, i_1, \dots, i_n \in I$ be distinct, let $\mathcal{B}$ and $\mathcal{C}$ be σ-algebras with $\mathcal{C} \subset \mathcal{A}_{i_0} \vee \mathcal{B}$, and assume that $\mathcal{A}_i, i \in I$, are conditionally independent given $\mathcal{B}$.

(a) Use (29.57) and (29.61) to prove that $P(A_1 \cap \cdots \cap A_n | \mathcal{A}_{i_0} \vee \mathcal{B}) = P(A_1 | \mathcal{B} \vee \mathcal{C}) \cdots P(A_n | \mathcal{B} \vee \mathcal{C})$ a.s. whenever $A_1 \in \mathcal{A}_{i_1}, \dots, A_n \in A_{i_n}$.

(b) Write $\mathcal{A}^{(i_0)} = \vee_{i \in I - \{i_0\}} \mathcal{A}_i$. Show that $P(A | \mathcal{A}^{(i_0)} \vee \mathcal{B} \vee \mathcal{C}) = P(A | \mathcal{B} \vee \mathcal{C})$ a.s. for any $A \in \mathcal{A}_{i_0} \vee \mathcal{B}$. [Use (29.55), (29.57), (29.61) and (a).]

Exercise 29.96. Let $\mathcal{A}_1$ and $\mathcal{A}_2$ be σ-algebras, and let $\mathcal{B}_n, n \geqslant 1$, and $\mathcal{B}$ be as in (29.27). If $(\mathcal{A}_1, \mathcal{B}_n, \mathcal{A}_2)$ is a Markov triple for each $n \geqslant 1$, then $(\mathcal{A}_1, \mathcal{B}, \mathcal{A}_2)$ is also a Markov triple.

Exercise 29.97. Let $\mathcal{B}$ be a σ-algebra, and let f be a random variable such that $|f| \leqslant 1$ and $E^{\mathcal{B}} f = 0$ a.s. Show that $E^{\mathcal{B}} e^{tf} \leqslant e^{t^2/2}$ a.s. for any $t \in R$. [Hint. For $|y| \leqslant 1$ and $t \in R$, $e^{ty} \leqslant \operatorname{ch} t + y \operatorname{sh} t$.]

30 Stopping Times and Semimartingales

In this section we apply large parts of the content in Section 29 to one of the most important topics in probability theory, namely the mathematics of martingales and semimartingales. We develop the basic theory of these processes in discrete time, commencing the section with the closely related notion of stopping time. This concept is useful in streamlining the treatment and plays the role similar to the truncation method in the study of sums of independent random variables in Chapter 10. Our systematic exposition includes powerful general results such as the optional sampling theorem, supremum inequalities, the submartingale convergence theorem, allowing an extension of the Borel-Cantelli lemma to sums of conditional probabilities, Wald's equations, etc.

Throughout this section, if not otherwise stated, we fix a nondecreasing sequence of σ-algebras $\{\mathcal{B}_n : n \in N\}$, where $\mathcal{B}_n \subset \mathcal{A}$, we set $\mathcal{B}_0 = \{\emptyset, X\}$ and $\mathcal{B}_\infty = \vee_{n \in N} \mathcal{B}_n$, and we also fix a sequence $\{f_n : n \in N\} \subset \mathfrak{L}^1$. (When finiteness of $Ef_n, n \in N$, is not necessary, this inclusion may be dropped.)

Definition 30.1. We say that $\{f_n : n \in N\}$ is *adapted* [$\{\mathcal{B}_n\}$*–adapted*] if f_n is $\mathcal{B}_n$–measurable for each $n \in N$. Notice that $\{f_n : n \in N\}$ is always $\{\sigma(f_1, \dots, f_n)\}$–adapted.

Remark 30.2. For $n \in N$, let $g_n : R^n \to R$ be measurable. Then $\{g_n \circ (f_1, \dots, f_n) : n \in N\}$ is adapted whenever $\{f_n : n \in N\}$ is adapted.

Definition 30.3. An $\overline{R}$-valued random variable $\tau : X \to N \cup \{\infty\}$ is called a *stopping time* [*optional time*, $\{\mathcal{B}_n\}$*–time*] if $\{\tau = n\} \in \mathcal{B}_n$ for any $n \in N$. To any stopping time τ we assign the family $\mathcal{B}_\tau = \{B \in \mathcal{B}_\infty : B \cap \{\tau = n\} \in \mathcal{B}_n, n \in N\}$.

Examples 30.4. (a) If $\tau = m$ for some $m \in N$, then τ is a stopping time and $\mathcal{B}_\tau = \mathcal{B}_m$. In this case we say that τ is *degenerate* (at m).

(b) Let $\{B_n : n \in N\}$ be a partition of X such that $\emptyset \neq B_n \in \mathcal{B}_n$ for each $n \in N$, and define $\tau : X \to N$ by $\tau(x) = n$ if $x \in B_n$. Then τ is a stopping time.

(c) Assume that $\{f_n : n \in N\}$ is adapted, let $C \in \mathcal{B}(R)$, and set $\tau_C = \inf\{n \in N : f_n \in C\}$. Then τ_C is a stopping time called the *hitting time of C*. We already encountered such hitting times in (25.41) and (26.4) in connection with Ottaviani's and Kolmogorov's inequalities.

Remarks 30.5. Throughout (b)–(j) τ is a stopping time.

(a) Let $\tau : X \to N \cup \{\infty\}$ be a function. Then τ is a stopping time $\iff \{\tau \leqslant n\} \in \mathcal{B}_n$ for any $n \in N \iff \{\tau > n\} \in \mathcal{B}_n$ for any $n \in N$. If τ is a stopping time, then $\{\tau \geqslant n\}$ and $\{\tau < n\}$ belong to $\mathcal{B}_{n-1}$ for all $n \in N$.

(b) $\mathcal{B}_\tau = \{B \in \mathcal{B}_\infty : B \cap \{\tau \leqslant n\} \in \mathcal{B}_n, n \in N\}$.

(c) $\mathcal{B}_\tau$ is a σ-algebra called the *pre-τ σ-algebra*, and τ is $\mathcal{B}_\tau$-measurable.

(d) $\{\tau = n\} \cap \mathcal{B}_\tau = \{\tau = n\} \cap \mathcal{B}_n, n \in N$, and $\mathcal{B}_\tau \subset \mathcal{B}_m$ if $\tau \leqslant m$ for some $m \in N$.

(e) $\tau \wedge m$ is $\mathcal{B}_m$-measurable for each $m \in N$.

(f) If $B \in \mathcal{B}_\tau$, then $\tau_B = \tau 1_B + \infty \cdot 1_{B^c}$ is a stopping time such that $\tau \leqslant \tau_B$ and $B \in \mathcal{B}_{\tau_B}$.

(g) $\tau + m$ is a stopping time for any $m \in N$. Moreover, $\mathcal{B}_m, \mathcal{B}_\tau \subset \mathcal{B}_{\tau+m}, m \in N$, and so $\vee_{n\in N}\mathcal{B}_n = \vee_{n\in N}\mathcal{B}_{\tau+n}$. Notice also that $\{\tau = n\} \cap \mathcal{B}_{\tau+m} = \{\tau = n\} \cap \mathcal{B}_{n+m}, m, n \in N$.

(h) If $\tau' : X \to N \cup \{\infty\}$ is $\mathcal{B}_\tau$-measurable and $\tau' \geqslant \tau$, then τ' is a stopping time. Indeed, for any $n \in N$, we have $\{\tau' \leqslant n\} = \{\tau' \leqslant n\} \cap \{\tau \leqslant n\} \in \mathcal{B}_n$ by (a) and (b).

(i) Define $f_\tau : X \to \overline{R}$ by $f_\tau(x) = \begin{cases} f_{\tau(x)}(x) & \text{if } \tau(x) < \infty \\ \limsup_n f_n(x) & \text{if } \tau(x) = \infty \end{cases}$. If $\{f_n : n \in N\}$ is adapted, then f_τ is $\mathcal{B}_\tau$-measurable.

(j) If τ is bounded, then $f_\tau \in \mathfrak{L}^1$.

The following two theorems list commonly used properties of stopping times.

Theorem 30.6. *Let τ and τ' be stopping times, and let $f \in \mathfrak{L}^1$. Then the following assertions hold.*

(i) *$\tau + \tau'$, $\tau \vee \tau'$, and $\tau \wedge \tau'$ are stopping times.*

(ii) *If $\tau \leqslant \tau'$, then $\mathcal{B}_\tau \subset \mathcal{B}_{\tau'}$.*

(iii) *$\mathcal{B}_\tau \cap \mathcal{B}_{\tau'} = \mathcal{B}_{\tau\wedge\tau'}$.*

(iv) *$\{\tau = \tau'\} \cap \mathcal{B}_\tau \subset \mathcal{B}_{\tau\wedge\tau'}$, $\{\tau \leqslant \tau'\} \cap \mathcal{B}_\tau \subset \mathcal{B}_{\tau\wedge\tau'}$, and $\{\tau < \tau'\} \cap \mathcal{B}_\tau \subset \mathcal{B}_{\tau\wedge\tau'}$.*

(v) *$\{\tau = \tau'\}$, $\{\tau \leqslant \tau'\}$, and $\{\tau < \tau'\}$ belong to $\mathcal{B}_{\tau\wedge\tau'}$.*

(vi) *If g is a $\mathcal{B}_\tau$-measurable random variable, then $1_{\{\tau=\tau'\}}g$, $1_{\{\tau\leqslant\tau'\}}g$, and $1_{\{\tau<\tau'\}}g$ are $\mathcal{B}_{\tau\wedge\tau'}$-measurable.*

(vii) *If g is $\mathcal{B}_{\tau\vee\tau'}$-measurable, then $1_{\{\tau\leqslant\tau'\}}g$ is measurable relative to $\mathcal{B}_{\tau'}$, and $1_{\{\tau>\tau'\}}g$ is measurable relative to $\mathcal{B}_\tau$.*

(viii) *$E^{\mathcal{B}_{\tau\wedge\tau'}} f + E^{\mathcal{B}_{\tau\vee\tau'}} f = E^{\mathcal{B}_\tau} f + E^{\mathcal{B}_{\tau'}} f$ a.s.*

Proof. For $n \geqslant 2$, we have $\{\tau + \tau' = n\} = \cup_{i=1}^{n-1}\{\tau = i, \tau' = n - i\} \in \mathcal{B}_n$. The rest of (i) and (ii) obtains immediately by (30.5.a) and (30.5.b). (iii) follows from (ii), since $B \in \mathcal{B}_\tau \cap \mathcal{B}_{\tau'}$ implies $B \cap \{\tau \wedge \tau' \leqslant n\} = B \cap (\{\tau \leqslant n\} \cup \{\tau' \leqslant n\}) = (B \cap \{\tau \leqslant n\}) \cup (B \cap \{\tau' \leqslant n\}) \in \mathcal{B}_n, n \in N$. If $B \in \mathcal{B}_\tau$, then $B \cap \{\tau =$

$\tau'\} \cap \{\tau' = n\} = B \cap \{\tau = n\} \cap \{\tau' = n\} \in \mathcal{B}_n, n \in N$, and so $B \cap \{\tau = \tau'\} \in \mathcal{B}_{\tau'}$. Thus $\{\tau = \tau'\} \cap \mathcal{B}_\tau \subset \mathcal{B}_{\tau'}$, whence $\{\tau = \tau'\} \cap \mathcal{B}_\tau \subset \{\tau = \tau'\} \cap \mathcal{B}_{\tau'} \subset \mathcal{B}_\tau$ by interchanging τ with τ'. Therefore, $\{\tau = \tau'\} \cap \mathcal{B}_\tau \subset \mathcal{B}_\tau \cap \mathcal{B}_{\tau'} = \mathcal{B}_{\tau\wedge\tau'}$ by (iii). Similarly, we have $B \cap \{\tau \leqslant \tau'\} \cap \{\tau' = n\} = B \cap \{\tau \leqslant n\} \cap \{\tau' = n\} \in \mathcal{B}_n, n \in N$, by (30.5.b), and so $B \cap \{\tau = \tau'\} \in \mathcal{B}_{\tau'}$. Consequently, $\{\tau \leqslant \tau'\} \cap \mathcal{B}_\tau \subset \mathcal{B}_{\tau'}$. Replacing τ' by $\tau \wedge \tau'$, we obtain $\{\tau \leqslant \tau'\} \cap \mathcal{B}_\tau = \{\tau \leqslant \tau \wedge \tau'\} \cap \mathcal{B}_\tau \subset \mathcal{B}_{\tau\wedge\tau'}$. Hence the first two inclusions in (iv) are verified, and the third one follows from them, as $\{\tau < \tau'\} = \{\tau \leqslant \tau'\} - \{\tau = \tau'\}$. Plainly, (v) is an easy consequence of (iv). As for (vi), if $g = 1_B$ with $B \in \mathcal{B}_\tau$, then $1_{\{\tau=\tau'\}}g = 1_{\{\tau=\tau'\}\cap B}$, $1_{\{\tau\leqslant\tau'\}}g = 1_{\{\tau\leqslant\tau'\}\cap B}$ and $1_{\{\tau<\tau'\}}g = 1_{\{\tau<\tau'\}\cap B}$ are $\mathcal{B}_{\tau\wedge\tau'}$-measurable by (iv). The assertion now follows via the indicator function method. Next, if g is $\mathcal{B}_{\tau\wedge\tau'}$-measurable, applying (vi), (v) and (ii), we see that $1_{\{\tau\leqslant\tau'\}}g = 1_{\{\tau\leqslant\tau'\}}(1_{\{\tau\vee\tau'\leqslant\tau'\}}g)$ is measurable relative to $\mathcal{B}_{(\tau\vee\tau')\wedge\tau'} = \mathcal{B}_{\tau'}$, and $1_{\{\tau>\tau'\}}g = 1_{\{\tau>\tau'\}}(1_{\{\tau\vee\tau'=\tau\}}g)$ is measurable relative to $\mathcal{B}_{(\tau\vee\tau')\wedge\tau} = \mathcal{B}_\tau$. To verify (viii), observe first that in view of (vi) and (v),

$$1_{\{\tau\leqslant\tau'\}}E^{\mathcal{B}_\tau}f = E^{\mathcal{B}_{\tau\wedge\tau'}}[1_{\{\tau\leqslant\tau'\}}E^{\mathcal{B}_\tau}f] = 1_{\{\tau\leqslant\tau'\}}E^{\mathcal{B}_{\tau\wedge\tau'}}f \text{ a.s.}$$

and

$$1_{\{\tau>\tau'\}}E^{\mathcal{B}_{\tau'}}f = E^{\mathcal{B}_{\tau\wedge\tau'}}[1_{\{\tau>\tau'\}}E^{\mathcal{B}_{\tau'}}f] = 1_{\{\tau>\tau'\}}E^{\mathcal{B}_{\tau\wedge\tau'}}f \text{ a.s.,}$$

whence

$$E^{\mathcal{B}_{\tau\wedge\tau'}}f = 1_{\{\tau\leqslant\tau'\}}E^{\mathcal{B}_\tau}f + 1_{\{\tau>\tau'\}}E^{\mathcal{B}_{\tau'}}f \text{ a.s.} \tag{1}$$

Then, in view of (vii) and (v),

$$1_{\{\tau\leqslant\tau'\}}E^{\mathcal{B}_{\tau\vee\tau'}}f = E^{\mathcal{B}_{\tau'}}[1_{\{\tau\leqslant\tau'\}}E^{\mathcal{B}_{\tau\vee\tau'}}f] = 1_{\{\tau\leqslant\tau'\}}E^{\mathcal{B}_{\tau'}}f \text{ a.s.}$$

and

$$1_{\{\tau>\tau'\}}E^{\mathcal{B}_{\tau\vee\tau'}}f = E^{\mathcal{B}_\tau}[1_{\{\tau>\tau'\}}E^{\mathcal{B}_{\tau\vee\tau'}}f] = 1_{\{\tau>\tau'\}}E^{\mathcal{B}_\tau}f \text{ a.s.,}$$

whence

$$E^{\mathcal{B}_{\tau\vee\tau'}}f = 1_{\{\tau\leqslant\tau'\}}E^{\mathcal{B}_{\tau'}}f + 1_{\{\tau>\tau'\}}E^{\mathcal{B}_\tau}f \text{ a.s.} \tag{2}$$

Adding (1) and (2) yields (viii). □

Corollary 30.7. *Let τ_B be as in* (30.5.f). *Then* $\mathcal{B}_{\tau_B} = \{C_1 \cup C_2 : C_1 \in B \cap \mathcal{B}_\tau, C_2 \in B^c \cap \mathcal{B}_\infty\}$.

Proof. Write $\mathcal{B} = \{C_1 \cup C_2 : C_1 \in B \cap \mathcal{B}_\tau, C_2 \in B^c \cap \mathcal{B}_\infty\}$, and let $C \in \mathcal{B}_{\tau_B}$. We have $C = (B \cap C) \cup (B^c \cap C)$, where $B \cap C \in \mathcal{B}_{\tau_B}$ by (30.5.f), and $B^c \cap C \in B^c \cap \mathcal{B}_\infty$. For $n \in N$, $B \cap C \cap \{\tau = n\} = B \cap C \cap \{\tau_B = n\} \in \mathcal{B}_n$, and so $B \cap C \in \mathcal{B}_\tau$. Hence $\mathcal{B}_{\tau_B} \subset \mathcal{B}$. In the other direction, since $\mathcal{B}_\tau \subset \mathcal{B}_{\tau_B}$ by (30.5.f), we have $C_1 \in \mathcal{B}_{\tau_B}$. For $n \in N$, we have also $C_2 \cap \{\tau = n\} = C_2 \cap B^c \cap \{\tau = n\} = \emptyset \in \mathcal{B}_n$, and so $C_2 \in \mathcal{B}_{\tau_B}$. Thus $\mathcal{B} \subset \mathcal{B}_{\tau_B}$. □

Remark 30.8. Taking $B = \{\tau \leqslant n\}$ in (30.7), and putting $\tau_n = n1_{\{\tau\leqslant n\}} + \infty \cdot 1_{\{\tau>n\}}$, we find that $\mathcal{B}_{\tau_n} = \{C_{1n} \cup C_{2n} : C_{1n} \in \{\tau \leqslant n\} \cap \mathcal{B}_\tau, C_{2n} \in \{\tau > n\} \cap \mathcal{B}_\infty\}$.

Theorem 30.9. *Let $\{\tau_n : n \in N\}$ be a sequence of stopping times. Then the following assertions hold.*

(i) $\vee_{n\in N}\tau_n$ *is a stopping time and* $\mathcal{B}_{\vee_{n\in N}\tau_n} \supset \vee_{n\in N}\mathcal{B}_{\tau_n}$.

(ii) *Put* $\tau = \vee_{n\in N}\tau_n$ *and assume that, for each* $x \in X$, *there is* $n(x) \in N$ *such that* $\tau(x) = \tau_{n(x)}(x)$. *Then* $\mathcal{B}_\tau = \vee_{n\in N}\mathcal{B}_{\tau_n}$.

(iii) $\wedge_{n\in N}\tau_n$ *is a stopping time and* $\mathcal{B}_{\wedge_{n\in N}\tau_n} = \cap_{n\in N}\mathcal{B}_{\tau_n}$.

(iv) $\limsup_n \tau_n$ *and* $\liminf_n \tau_n$ *are stopping times.*

Proof. For $m \in N$, we have $\{\vee_{n\in N}\tau_n \leqslant m\} = \cap_{n\in N}\{\tau_n \leqslant m\} \in \mathcal{B}_m$, hence $\vee_{n\in N}\tau_n$ is a stopping time. Since $\vee_{n\in N}\tau_n \geqslant \tau_n, n \in N$, the second assertion in (i) follows by (30.6.ii). To prove (ii), let $B \in \mathcal{B}_\tau$. Then $B \cap \{\tau = \tau_n\} \in \mathcal{B}_{\tau_n} \subset \vee_{n\in N}\mathcal{B}_{\tau_n}, n \in N$, by (30.6.iv), and so $B = \cup_{n\in N}(B \cap \{\tau = \tau_n\}) \in \vee_{n\in N}\mathcal{B}_{\tau_n}$. This in conjunction with (i) establishes the result. For $m \in N$, we have $\{\wedge_{n\in N}\tau_n \leqslant m\} = \cup_{n\in N}\{\tau_n \leqslant m\} \in \mathcal{B}_m$, hence $\wedge_{n\in N}\tau_n$ is a stopping time. As $\wedge_{n\in N}\tau_n \leqslant \tau_n, n \in N$, (30.6.ii) shows that $\mathcal{B}_{\wedge_{n\in N}\tau_n} \subset \cap_{n\in N}\mathcal{B}_{\tau_n}$. In the other direction, if $B \in \cap_{n\in N}\mathcal{B}_{\tau_n}$, then, for each $n \in N$, $B \cap \{\tau_n \leqslant m\} \in \mathcal{B}_m, m \in N$. Therefore, $B \cap \{\wedge_{n\in N}\tau_n \leqslant m\} = \cup_{n\in N}(B \cap \{\tau_n \leqslant m\}) \in \mathcal{B}_m, m \in N$, so that $B \in \mathcal{B}_{\wedge_{n\in N}\tau_n}$. Thus (iii) holds. The assertion (iv) follows at once from (i) and (iii). □

Theorem 30.10. *Let* $h : \Omega \to X$ *be a function, write* $\mathcal{A}_n = h^{-1}(\mathcal{B}_n), n \in N$, *let* $\mathfrak{T}$ *denote the set of all* $\{\mathcal{B}_n\}$*-times, and let* $\mathfrak{S}$ *denote the set of all* $\{\mathcal{A}_n\}$*-times. Then the following assertions hold.*

(i) $\mathfrak{S} = \{\tau \circ h : \tau \in \mathfrak{T}\}$.

(ii) $\mathcal{A}_{\tau\circ h} = h^{-1}(\mathcal{B}_\tau)$ *for any* $\tau \in \mathfrak{T}$.

Proof. If $\tau \in \mathfrak{T}$, then $\{\tau \circ h = n\} = h^{-1}(\{\tau = n\}) \in \mathcal{A}_n, n \in N$, hence $\tau \circ h \in \mathfrak{S}$. Conversely, if $\sigma \in \mathfrak{S}$, set $\sigma_n = n1_{\{\sigma=n\}} + \infty \cdot 1_{\{\sigma\neq n\}}, n \in N$. Then $\sigma_n \in \mathfrak{S}$ and $\sigma = \wedge_{n\in N}\sigma_n$. For $n \in N$, there is $B_n \in \mathcal{B}_n$ so that $\{\sigma = n\} = h^{-1}(B_n)$. Define $\tau_n = n1_{B_n} + \infty \cdot 1_{B_n^c}, n \in N$, and $\tau = \wedge_{n\in N}\tau_n$. Then $\tau \in \mathfrak{T}$ by (30.9.iii). Observe that $\sigma_n = \tau_n \circ h, n \in N$. Therefore, in view of (5.59), $\sigma = \wedge_{n\in N}(\tau_n \circ h) = (\wedge_{n\in N}\tau_n) \circ h = \tau \circ h$. This establishes (i). To prove (ii), let $\tau \in \mathfrak{T}$, and let $\tau_n, n \in N$, be as in (30.8). Then, applying (30.8), for $n \in N$, we have

$$\begin{aligned} h^{-1}(\mathcal{B}_{\tau_n}) &= \{h^{-1}(C_{1n}) \cup h^{-1}(C_{2n}) : h^{-1}(C_{1n}) \in h^{-1}(\{\tau \leqslant n\}) \cap h^{-1}(\mathcal{B}_n),\\ &\qquad h^{-1}(C_{2n}) \in h^{-1}(\{\tau > n\}) \cap h^{-1}(\mathcal{B}_\infty)\}\\ &= \{h^{-1}(C_{1n}) \cup h^{-1}(C_{2n}) : h^{-1}(C_{1n}) \in \{\tau \circ h \leqslant n\} \cap h^{-1}(\mathcal{B}_n) \cap \mathcal{A}_n,\\ &\qquad h^{-1}(C_{2n}) \in \{\tau \circ h > n\} \cap h^{-1}(\mathcal{A}_\infty)\} = \mathcal{A}_{(\tau\circ h)_n}, \end{aligned} \tag{1}$$

where $\mathcal{A}_\infty = \vee_{n\in N}\mathcal{A}_n$ and $(\tau \circ h)_n = (\tau \circ h)1_{\{\tau\circ h\leqslant n\}} + \infty \cdot 1_{\{\tau\circ h>n\}}, n \in N$. Since $\{\tau_n : n \in N\}$ and $\{(\tau\circ h)_n : n \in N\}$ are nonincreasing sequences such that $\tau = \wedge_{n\in N}\tau_n$ and $\tau \circ h = \wedge_{n\in N}(\tau \circ h)_n$, on account of (30.9.iii), (30.6.ii), (7.71) and (1), we get

$$h^{-1}(\mathcal{B}_\tau) = h^{-1}\left(\bigcap_{n\in N}\mathcal{B}_{\tau_n}\right) = \bigcap_{n\in N} h^{-1}(\mathcal{B}_{\tau_n}) = \bigcap_{n\in N}\mathcal{A}_{(\tau\circ h)_n} = \mathcal{A}_{\tau\circ h}.$$

Thus (ii) holds. □

Theorem 30.11. *Let $(Z, \mathcal{C})$ be a measurable space, let $f_n : X \to Z, n \geqslant 1$, be independent random elements such that $Pf_n^{-1} = Pf_1^{-1}, n \geqslant 1$, and let τ be a finite $\{\mathcal{B}_n\}$-time, where $\mathcal{B}_n = \sigma(f_1, \ldots, f_n), n \geqslant 1$. Then the following assertions hold.*

(i) *$\mathcal{B}_\tau$ is independent of $\sigma(f_{\tau+n}, n \geqslant 1)$.*

(ii) *$f_{\tau+n}, n \geqslant 1$, are independent with $Pf_{\tau+n}^{-1} = Pf_1^{-1}, n \geqslant 1$.*

Proof. For $B \in \mathcal{B}_\tau$ and $C_1, \ldots, C_n \in \mathcal{C}$, since $B \cap \{\tau = j\} \in \mathcal{B}_j$ and $\mathcal{B}_j$ is independent of $\sigma(f_{j+n}, n \geqslant 1)$ by (18.15), we have

$$\begin{aligned} P\left(B \cap \bigcap_{i=1}^n \{f_{\tau+i} \in C_i\}\right) &= \sum_{j \geqslant 1} P\left(B \cap \{\tau = j\} \cap \bigcap_{i=1}^n \{f_{j+i} \in C_i\}\right) \\ &= \sum_{j \geqslant 1} P(B \cap \{\tau = j\}) P\left(\bigcap_{i=1}^n \{f_{j+i} \in C_i\}\right) \\ &= \sum_{j \geqslant 1} P(B \cap \{\tau = j\}) \prod_{i=1}^n P(f_1 \in C_i) \\ &= P(B) \prod_{i=1}^n P(f_1 \in C_i). \end{aligned} \tag{1}$$

Taking $B = X$ in (1) yields

$$P\left(\bigcap_{i=1}^n \{f_{\tau+i} \in C_i\}\right) = \prod_{i=1}^n P(f_1 \in C_i), \tag{2}$$

whence $P(B \cap \cap_{i=1}^n \{f_{\tau+i} \in C_i\}) = P(B)P(\cap_{i=1}^n \{f_{\tau+i} \in C_i\})$. Since the family $\cup_{n \geqslant 1} \{\cap_{i=1}^n \{f_{\tau+i} \in C_i\}) : C_1, \ldots, C_n \in \mathcal{C}\}$ is closed under intersection and generates $\sigma(f_{\tau+n}, n \geqslant 1)$, this in conjunction with (18.10) proves (i). From (2) it follows immediately that $P(f_{\tau+n} \in C) = P(f_1 \in C)$ for any $n \geqslant 1$ and $C \in \mathcal{C}$, and so $P(\cap_{i=1}^n \{f_{\tau+i} \in C_i\}) = \prod_{i=1}^n P(f_{\tau+i} \in C_i)$. Hence (ii) holds. □

Theorem 30.12 (Wald). *For $n \geqslant 1$, set $S_n = f_1 + \cdots + f_n$ and $\mathcal{B}_n = \sigma(f_1, \ldots, f_n)$. If $f_n, n \geqslant 1$, are i.i.d. random variables, and τ is a finite $\{\mathcal{B}_n\}$-time with $E\tau < \infty$, then $S_\tau \in \mathfrak{L}^1$ and $ES_\tau = (Ef_1)E\tau$.*

Proof. Observe that $S_\tau = \sum_{n \geqslant 1} f_n 1_{\{\tau \geqslant n\}}$. Then, by (11.23), (30.5.a) and (18.22), $E|S_\tau| \leqslant \sum_{n \geqslant 1} E[|f_n| 1_{\{\tau \geqslant n\}}] = \sum_{n \geqslant 1} (E|f_n|)P(\tau \geqslant n) = (E|f_1|)E\tau < \infty$, and so $S_\tau \in \mathfrak{L}^1$. Consequently, we may write $ES_\tau = \sum_{n \geqslant 1} E[f_n 1_{\{\tau \geqslant n\}}] = \sum_{n \geqslant 1} (Ef_n)P(\tau \geqslant n) = (Ef_1)E\tau$. □

Theorem 30.13 (Wald). *Let $f_n, S_n, \mathcal{B}_n, n \geqslant 1$, and τ be as in* (30.12), *and assume that $Ef_1 = 0$ and $Ef_1^2 = \sigma^2 < \infty$. Then $S_\tau \in \mathfrak{L}^2$ and $ES_\tau^2 = \sigma^2 E\tau$.*

Proof. For $n \geqslant 1$, put $\tau_n = \tau \wedge n$. Then τ_n is a $\mathcal{B}_n$-measurable stopping time with $n \geqslant \tau_n \to \tau$ and $S_{\tau_n} = \sum_{i=1}^n f_i 1_{\{\tau \geqslant i\}} = S_{\tau_{n-1}} + f_n 1_{\{\tau \geqslant n\}}, n \geqslant 2$. Consequently, for

$n \geqslant 2$, since $S_{\tau_{n-1}} 1_{\{\tau \geqslant n\}}$ is $\mathcal{B}_{n-1}$-measurable, we have

$$\begin{aligned} ES_{\tau_n}^2 &= ES_{\tau_{n-1}}^2 + 2E[f_n S_{\tau_{n-1}} 1_{\{\tau \geqslant n\}}] + E[f_n^2 1_{\{\tau \geqslant n\}}] \\ &= ES_{\tau_{n-1}}^2 + 2(Ef_n)E[S_{\tau_{n-1}} 1_{\{\tau \geqslant n\}}] + \sigma^2 P(\tau \geqslant n) = ES_{\tau_{n-1}}^2 + \sigma^2 P(\tau \geqslant n). \end{aligned}$$

This shows that

$$ES_{\tau_n}^2 = \sigma^2 \sum_{i=1}^{n} P(\tau \geqslant i) \to \sigma^2 E\tau. \tag{1}$$

For $1 \leqslant n < m$, as $f_i 1_{\{\tau \geqslant i\}} 1_{\{\tau \geqslant j\}}$ is $\mathcal{B}_{j-1}$-measurable whenever $n < i < j \leqslant m$, we obtain

$$\begin{aligned} E(S_{\tau_m}^2 - S_{\tau_n}^2) &= E\left[\sum_{i=n+1}^{m} f_i 1_{\{\tau \geqslant i\}}\right] = \sum_{i=n+1}^{m} E\left[f_i^2 1_{\{\tau \geqslant i\}}\right] \\ &\quad + 2 \sum_{n<i<j\leqslant m} E\left[f_i 1_{\{\tau \geqslant i\}} 1_{\{\tau \geqslant j\}} f_j\right] = \sigma^2 \sum_{i=n+1}^{m} P(\tau \geqslant i), \end{aligned}$$

whence it follows that $\{\widetilde{S_{\tau_n}} : n \geqslant 1\}$ is a Cauchy sequence in L^2. Since L^2 is a complete metric space (13.16), there is $S \in \mathfrak{L}^2$ such that $S_{\tau_n} \xrightarrow{\mathfrak{L}^2} S$, and so $S_{\tau_n} \xrightarrow{P} S$ by (14.14). Then, in view of (14.30) and (1), we have

$$ES^2 = \lim_n ES_{\tau_n}^2 = \sigma^2 E\tau. \tag{2}$$

According to Riesz's theorem (14.8), there exists a subsequence $\{S_{\tau_{n_k}} : k \geqslant 1\}$ such that $S_{\tau_{n_k}} \xrightarrow{a.s.} S$. As $\tau_{n_k} \to \tau$, this implies that $S_\tau = S$ a.s. Thus (2) shows that $S_\tau \in \mathfrak{L}^2$ and $ES_\tau^2 = \sigma^2 E\tau$. □

Remark 30.14. If $\tau : X \to N$ is a random variable that is independent of $\sigma(f_n, n \geqslant 1)$ but not necessarily a stopping time, then, in view of (29.71), the conclusions of (30.12) and (30.13) still hold.

We now embark on the study of an extremely important topic in probability theory, namely martingale theory, which refers to the mathematics of semimartingale as well as martingales. Though we restrict ourselves to semimartingales in discrete time, we present powerful general results whose high potential can be fruitfully used in various branches of probability theory and mathematical analysis. The concept of a (semi)martingale originates in games of chances. We may think of f_n as the fortune of a gambler after n trials in a succession of games. The expected fortune after the $(n+1)$st trial, having known the previous n outcomes $f_1, \ldots, f_n$, is $E[f_{n+1}|f_1, \ldots, f_n]$. If this equals f_n, the game is considered to be fair. If $E[f_{n+1}|f_1, \ldots, f_n] \geqslant f_n$, the game is favorable, while if $E[f_{n+1}|f_1, \ldots, f_n] \leqslant f_n$, the game is called unfavorable.

Definitions 30.15. Assume the sequence $\{f_n : n \in N\}$ is adapted. Then $\{f_n : n \in N\}$ is said to be a *martingale* [$\{\mathcal{B}_n\}$*−martingale*] if $E^{\mathcal{B}_n} f_{n+1} = f_n$ a.s. for all $n \in N$, a *submartingale* [$\{\mathcal{B}_n\}$*−submartingale*] if $E^{\mathcal{B}_n} f_{n+1} \geqslant f_n$ a.s. for all $n \in N$, and a *supermartingale* [$\{\mathcal{B}_n\}$*−supermartingale*] if $E^{\mathcal{B}_n} f_{n+1} \leqslant f_n$ a.s. for all $n \in N$. We say that $\{f_n : n \in N\}$ is a *semimartingale* [$\{\mathcal{B}_n\}$*−semimartingale*] if it is either a submartingale [$\{\mathcal{B}_n\}$-submartingale] or a supermartingale [$\{\mathcal{B}_n\}$-supermartingale]. If $I = \{m+1, m+2, \ldots\}$ or $\{m+1, \ldots, m+p\}$ for some $m, p \in N$ with $p \geqslant 2$, then the above definitions extend mutatis mutandis to I.

Examples 30.16. (a) Let $(X, \mathcal{A}, P) = ([0,1[, \mathcal{B}([0,1[), \lambda)$, where λ is the Lebesgue measure. For $n \geqslant 1$, define $f_n = 2^n 1_{[0,2^{-n}[}$ and $\mathcal{B}_n = \sigma(f_1, \ldots, f_n)$. Then $\mathcal{B}_n = \sigma(\mathcal{P}_n)$, where $\mathcal{P}_n = \{[0, 2^{-n}[\} \cup \{[2^{-i}, 2^{-i+1}[: 1 \leqslant i \leqslant n\}$ is a partition of $[0,1[$. Since $f_{n+1} = 2f_n - 2^{n+1} 1_{[2^{-n-1}, 2^{-n}[}$, $n \geqslant 1$, in view of (29.1), we have

$$\begin{aligned} E^{\mathcal{B}_n} f_{n+1} &= 2f_n - 2^{n+1} P^{\mathcal{B}_n}([2^{-n-1}, 2^{-n}[) \\ &= 2f_n - 2^{n+1} P_{[0,2^{-n}[}([2^{-n-1}, 2^{-n}[) 1_{[0,2^{-n}[} \\ &= 2f_n - 2^n 1_{[0,2^{-n}[} = f_n \text{ a.s.}, \quad n \geqslant 1. \end{aligned}$$

(b) If $\{f_n : n \in N\}$ is adapted, then it is a submartingale whenever $f_{n+1} \geqslant f_n, n \in N$, and a supermartingale whenever $f_{n+1} \leqslant f_n, n \in N$.

(c) Let $f \in \mathfrak{L}^1$ and write $f_n = E^{\mathcal{B}_n} f, n \in N$. Then, for $n \in N$, $f_n \in \mathfrak{L}^1$ and $E^{\mathcal{B}_n} f_{n+1} = E^{\mathcal{B}_n}[E^{\mathcal{B}_{n+1}} f] = E^{\mathcal{B}_n} f = f_n$ a.s. by (29.9). Hence $\{f_n : n \in N\}$ is a martingale.

(d) For $n \in N$, put $S_n = f_1 + \cdots + f_n$ and $\mathcal{B}_n = \sigma(f_1, \ldots, f_n)$. Then $E^{\mathcal{B}_n} S_{n+1} = E^{\mathcal{B}_n} S_n + E^{\mathcal{B}_n} f_{n+1} = S_n + E^{\mathcal{B}_n} f_{n+1}$ a.s., $n \in N$. Therefore, $\{S_n : n \in N\}$ is a martingale, a submartingale or a supermartingale according as $E^{\mathcal{B}_n} f_{n+1} = 0, \geqslant 0$ or $\leqslant 0$ a.s. for any $n \in N$. Hence, if $f_n, n \in N$, are independent, then $\{S_n : n \in N\}$ is a martingale, a submartingale or a supermartingale according as $Ef_{n+1} = 0, \geqslant 0$ or $\leqslant 0$ for any $n \in N$. If $\{\mathcal{B}_n : n \in N\}$ is a general nondecreasing sequence of σ-algebras, and $E^{\mathcal{B}_n} f_{n+1} = 0$ a.s. for all $n \in N$, then $\{f_n : n \in N\}$ is called a *martingale difference sequence*.

(e) For $n \geqslant 2$, set $g_n = \sum_{1 \leqslant i < j \leqslant n} f_i f_j$ and let S_n and $\mathcal{B}_n$ be as in (d). Since $g_{n+1} = g_n + S_n f_{n+1}$, we have $E^{\mathcal{B}_n} g_{n+1} = g_n + S_n E^{\mathcal{B}_n} f_{n+1}$ a.s., $n \geqslant 2$. Consequently, if either $f_n, n \geqslant 1$, are independent with $Ef_{n+1} = 0, n \geqslant 2$, or $\{f_n : n \geqslant 1\} \subset \mathfrak{L}^2$ with $E^{\mathcal{B}_n} f_{n+1} = 0$ a.s., $n \geqslant 2$, then $\{g_n : n \geqslant 2\}$ is a martingale.

(f) For $n \in N$, assume that $m_n = Ef_n \neq 0$, put $g_n = \prod_{i=1}^n \frac{f_i}{m_i}$, and let $\mathcal{B}_n = \sigma(f_1, \ldots, f_n)$. If $f_n, n \in N$, are independent, then $\{g_n : n \in N\}$ is a martingale.

(g) Let μ be a signed measure on $\mathcal{A}$ with $|\mu|(X) < \infty$. For $n \in N$, let $\mathcal{B}_n = \sigma(\mathcal{P}_n)$, where $\mathcal{P}_n = \{B_{ni} : i \in I_n\} \subset \mathcal{A}$ is a countable partition of X with $P(B_{ni}) > 0$ for all $i \in I_n$. Suppose that $\mathcal{P}_{n+1}$ refines $\mathcal{P}_n$, and define $f_n = \sum_{i \in I_n} (\mu(B_{ni})/P(B_{ni})) 1_{B_{ni}}, n \in N$. Then $\{f_n : n \in N\}$ is a martingale.

(h) Assume that $\{f_n : n \in N\}$ is adapted, and define $g_n = \sum_{i=1}^n (f_i - E^{\mathcal{B}_{i-1}} f_i), n \in N$. Then $\{g_n : n \in N\}$ is a martingale.

Remarks 30.17. Suppose that $\{f_n : n \in N\}$ is adapted.

(a) $\{f_n : n \in N\}$ is a martingale, a submartingale or a supermartingale if and only if, for all $m, n \in N$ with $m > n$, $E^{\mathcal{B}_n} f_m = f_n$ a.s., $E^{\mathcal{B}_n} f_m \geqslant f_n$ a.s. or $E^{\mathcal{B}_n} f_m \leqslant f_n$ a.s. respectively.

(b) $\{f_n : n \in N\}$ is a martingale, a submartingale or a supermartingale if and only if, for all $m, n \in N$ with $m > n$ and $\Lambda \in \mathcal{B}_n$, $\int_\Lambda f_m\,dP = \int_\Lambda f_n\,dP$, $\int_\Lambda f_m\,dP \geqslant \int_\Lambda f_n\,dP$ or $\int_\Lambda f_m\,dP \leqslant \int_\Lambda f_n\,dP$, respectively.

(c) If $\{f_n : n \in N\}$ is a martingale, a submartingale or a supermartingale then $\{Ef_n : n \in N\}$ is a constant, a nondecreasing or a nonincreasing sequence, respectively.

(d) If $\{f_n : n \in N\}$ is a $\{\mathcal{B}_n\}$-martingale or a $\{\mathcal{B}_n\}$-semimartingale, $\mathcal{A}_n \subset \mathcal{B}_n$ is a σ-algebra for each $n \in N$, and $\{f_n : n \in N\}$ is $\{\mathcal{A}_n\}$-adapted, then it follows by (29.9) that $\{f_n : n \in N\}$ is also an $\{\mathcal{A}_n\}$-martingale or an $\{\mathcal{A}_n\}$-semimartingale, respectively; in particular $\{f_n : n \in N\}$ is a $\{\sigma(f_1, \ldots, f_n)\}$-martingale or a $\{\sigma(f_1, \ldots, f_n)\}$-semimartingale, respectively.

(e) $\{f_n : n \in N\}$ is a submartingale if and only if $\{-f_n : n \in N\}$ is a supermartingale.

(f) $\{f_n : n \in N\}$ is a martingale if and only if it is both a submartingale and a supermartingale.

(g) If $\{f_n : n \in N\}$ is a semimartingale such that $Ef_n = Ef_1$, $n \in N$, then according to (11.16), $\{f_n : n \in N\}$ is a martingale.

(h) If $\{f_n : n \in N\}$ and $\{g_n : n \in N\}$ are martingales, then $\{af_n + bg_n + c : n \in N\}$ is a martingale for any $a, b, c \in R$; in particular, $\{-f_n : n \in N\}$ is a martingale.

(i) If $\{f_n : n \in N\}$ and $\{g_n : n \in N\}$ are submartingales, then $\{af_n + bg_n + c : n \in N\}$ is a submartingale for any $a, b \geqslant 0$ and $c \in R$.

(j) If $\{f_n : n \in N\}$ and $\{g_n : n \in N\}$ are submartingales, then $\{f_n \vee g_n : n \in N\}$ is a submartingale, and so if $\{f_n : n \in N\}$ and $\{g_n : n \in N\}$ are supermartingales, then $\{f_n \wedge g_n : n \in N\}$ is a supermartingale.

Theorem 30.18. *Let $\{f_n : n \in N\}$ be a $\{\mathcal{B}_n\}$-martingale, and let $\{g_n : n \in N\}$ be a $\{\mathcal{C}_n\}$-martingale, where $\mathcal{B}_n$ is independent of $\mathcal{C}_n$ for each $n \in N$. Then $\{f_n + g_n : n \in N\}$ and $\{f_n g_n : n \in N\}$ are $\{\mathcal{B}_n \vee \mathcal{C}_n\}$-martingales.*

Proof. For $n \in N$, $B \in \mathcal{B}_n$ and $C \in \mathcal{C}_n$, since $f_{n+1}1_B$ is independent of 1_C and $g_{n+1}1_C$ is independent of 1_B, we have by (30.17.b)

$$\int\limits_{B\cap C} (f_{n+1} + g_{n+1})dP = P(C)\int f_{n+1}\,dP + P(B)\int g_{n+1}\,dP$$

$$= P(C)\int\limits_B f_n\,dP + P(B)\int\limits_C g_n\,dP = \int\limits_{B\cap C} (f_n + g_n)dP.$$

Let $\mathcal{D} = \{\Lambda \in \mathcal{A} : \int_\Lambda (f_{n+1} + g_{n+1})dP = \int_\Lambda (f_n + g_n)dP\}$. As $f_n, f_{n+1}, g_n, g_{n+1} \in \mathfrak{L}^1$, $Ef_{n+1} = Ef_n$ and $Eg_{n+1} = Eg_n$, it follows that $\mathcal{D}$ is a Dynkin system containing the family $\mathcal{M} = \{B \cap C : B \in \mathcal{B}_n, C \in \mathcal{C}_n\}$. Since $\mathcal{M}$ is closed under intersection and generates $\mathcal{B}_n \vee \mathcal{C}_n$, we see that $\mathcal{D} \supset \mathcal{B}_n \vee \mathcal{C}_n$. According to (30.17.b), this proves the first assertion. The proof of the second assertion is similar and is omitted. □

In view of (30.17.e), since properties of submartingales yield corresponding properties of supermartingales, in what follows we shall mostly treat martingales and submartingales.

Doob decomposition 30.19. *Let $\{f_n : n \in N\}$ be a submartingale. Then the following assertions hold.*

(i) *There exist a martingale $\{f_n' : n \in N\}$ and a nondecreasing sequence $\{f_n'' : n \in N\}$ of nonnegative random variables such that $f_n = f_n' + f_n''$, $n \in N$.*

(ii) *For $n \in N$, $E\left|f_n'\right| \leqslant 2E\left|f_n\right| + E|f_1|$ and $Ef_n'' \leqslant E\left|f_n\right| + E\left|f_1\right|$.*

Proof. Fix versions of $E^{\mathcal{B}_n} f_{n+1}, n \in N$. Set $f_1' = f_1, f_1'' = 0$, and define recursively $f_{n+1}' = f_n' + f_{n+1} - E^{\mathcal{B}_n} f_{n+1}$, $f_{n+1}'' = f_n'' + E^{\mathcal{B}_n} f_{n+1} - f_n$ for $n \in N$. It follows inductively that f_n' and f_n'' are $\mathcal{B}_n$-measurable and $f_n = f_n' + f_n''$ for any $n \in N$. We have $E^{\mathcal{B}_n} f_{n+1}' = f_n' + E^{\mathcal{B}_n}[f_{n+1} - E^{\mathcal{B}_n} f_{n+1}] = f_n'$ a.s., $n \in N$, and $f_{n+1}'' \geqslant f_n''$ a.s., $n \in N$. Hence (i) holds. Moreover, $Ef_{n+1}'' - Ef_n'' = Ef_{n+1} - Ef_n, n \in N$. Summing we obtain $Ef_n'' = Ef_n - Ef_1 \leqslant Ef_n'' \leqslant E\left|f_n\right| + E\left|f_1\right|, n \in N$, whence $E\left|f_n'\right| \leqslant E\left|f_n\right| + Ef_n'' \leqslant 2E\left|f_n\right| + E\left|f_1\right|, n \in N$. □

Remarks 30.20. (a) Observe that $f_1'' = 0$ and f_{n+1}'' is $\mathcal{B}_n$-measurable for any $n \in N$. The Doob decomposition is unique, in the sense that if $f_n = g_n' + g_n''$, $n \in N$, where $\{g_n' : n \in N\}$ is a martingale and $g_1'' = 0$, g_{n+1}'' is $\mathcal{B}_n$-measurable for any $n \in N$, then $f_n' = g_n'$ a.s., $n \in N$, and $f_n'' = g_n''$ a.s., $n \in N$. Indeed, $f_{n+1}'' - f_n'' = g_{n+1}'' - g_n'' = E^{\mathcal{B}_n} f_{n+1} - f_n$ a.s., $n \in N$.

(b) There is a converse to (30.19.i), namely if $\{f_n' : n \in N\}$ is a martingale, and $\{f_n'' : n \in N\} \subset \mathfrak{L}^1$ is adapted and nondecreasing, then $\{f_n' + f_n'' : n \in N\}$ is a submartingale.

We now present the effect of various transformations on martingales and submartingales.

Theorem 30.21. *Let $\varphi :]a, b[\to R$ be a convex function, and assume that $\{f_n : n \in N\}$ is adapted with $a < f_n < b$ and $\varphi \circ f_n \in \mathfrak{L}^1$ for all $n \in N$. If either $\{f_n : n \in N\}$ is a martingale or φ is nondecreasing and $\{f_n : n \in N\}$ is a submartingale, then $\{\varphi \circ f_n : n \in N\}$ is a submartingale.*

Proof. In both cases, on account of (29.19), we have $E^{\mathcal{B}_n}[\varphi \circ f_{n+1}] \geqslant \varphi \circ f_n$ a.s., $n \in N$. □

Corollary 30.22.

(i) *If $\{f_n : n \in N\}$ is a martingale, then $\{|f_n| : n \in N\}$ and $\{f_n^- : n \in N\}$ are submartingales.*

(ii) *If $\{f_n : n \in N\} \subset \mathfrak{L}^p$ with $1 \leqslant p < \infty$ is a martingale, then $\{|f_n|^p : n \in N\}$ is a submartingale.*

(iii) *If $\{f_n : n \in N\}$ is a submartingale, then $\{f_n \vee c : n \in N\}$ is a submartingale for every $c \in R$; in particular, $\{f_n^+ : n \in N\}$ is a submartingale.*

The following basic result implies preservation of martingale/submartingale property under a random time change, and yields also very useful inequalities.

Optional sampling theorem 30.23 (Doob). *Let τ and τ' be stopping times, where $\tau' \leqslant m$ for some $m \in N$.*

(i) *If* $\{f_n : n \in N\}$ *is a martingale, then* $E^{\mathcal{B}_\tau} f_{\tau'} = f_{\tau \wedge \tau'}$ *a.s.*

(ii) *If* $\{f_n : n \in N\}$ *is a submartingale, then* $E^{\mathcal{B}_\tau} f_{\tau'} \geqslant f_{\tau \wedge \tau'}$ *a.s.*

Proof. For $1 \leqslant n \leqslant m$, using (30.5.d), and applying (29.8.xiv) with $\Lambda = \{\tau' = n\}, \mathcal{B} = \mathcal{B}_{\tau'}$ and $\mathcal{C} = \mathcal{B}_n$, we have $1_{\{\tau'=n\}} E^{\mathcal{B}_{\tau'}} f_m = 1_{\{\tau'=n\}} E^{\mathcal{B}_n} f_m = 1_{\{\tau'=n\}} f_n = 1_{\{\tau'=n\}} f_{\tau'}$ a.s. Hence $E^{\mathcal{B}_{\tau'}} f_m = f_{\tau'}$ a.s. Replacing τ' by $\tau \wedge \tau'$, we get $E^{\mathcal{B}_{\tau\wedge\tau'}} f_m = f_{\tau\wedge\tau'}$ a.s. Therefore, in view of (30.6.ii), $E^{\mathcal{B}_{\tau\wedge\tau'}} f_{\tau'} = E^{\mathcal{B}_{\tau\wedge\tau'}}[E^{\mathcal{B}_{\tau'}} f_m] = E^{\mathcal{B}_{\tau\wedge\tau'}} f_m = f_{\tau\wedge\tau'}$ a.s. Consequently, making use of (30.6.v), (30.6.ii) and (30.6.iv), and applying again (29.8.xiv) with $\Lambda = \{\tau \leqslant \tau'\}, \mathcal{B} = \mathcal{B}_\tau$ and $\mathcal{C} = \mathcal{B}_{\tau\wedge\tau'}$, we obtain

$$1_{\{\tau\leqslant\tau'\}} E^{\mathcal{B}_\tau} f_{\tau'} = 1_{\{\tau\leqslant\tau'\}} E^{\mathcal{B}_{\tau\wedge\tau'}} f_{\tau'} = 1_{\{\tau\leqslant\tau'\}} f_{\tau\wedge\tau'} \text{ a.s.} \tag{1}$$

On the other hand, by (30.6.v), (30.5.i), (30.6.ii) and (29.8.iv), we have

$$1_{\{\tau>\tau'\}} E^{\mathcal{B}_\tau} f_{\tau'} = 1_{\{\tau>\tau'\}} E^{\mathcal{B}_\tau} f_{\tau\wedge\tau'} = 1_{\{\tau>\tau'\}} f_{\tau\wedge\tau'} \text{ a.s.} \tag{2}$$

Now (i) follows from (1) and (2). To prove (ii), assume for a moment that $\tau \leqslant \tau'$ and let $B \in \mathcal{B}_\tau$. Then

$$\begin{aligned}\int_B (f_{\tau'} - f_\tau)dP &= \sum_{n=1}^m \int_{B\cap\{\tau=n\}} (f_{\tau'} - f_\tau)dP = \sum_{n=1}^m \int_{B\cap\{\tau=n\}\cap\{\tau'>n\}} (f_{\tau'} - f_\tau)dP \\ &= \sum_{n=1}^m \int_{B\cap\{\tau=n\}\cap\{\tau'>n\}} \left(\sum_{i=n}^{\tau'-1} (f_{i+1} - f_i)\right) dP.\end{aligned}$$

Since $1_{\{\tau'>n\}} \sum_{i=n}^{\tau'-1}(f_{i+1} - f_i) = \sum_{i=n}^{m-1}(f_{i+1} - f_i)1_{\{\tau'>i\}}$, and $B \cap \{\tau = n\} \cap \{\tau' > i\} \in \mathcal{B}_i$ for $n \leqslant i \leqslant m - 1$, we have further

$$\int_B (f_{\tau'} - f_\tau)dP = \sum_{n=1}^m \sum_{i=n}^{m-1} \int_{B\cap\{\tau=n\}\cap\{\tau'>i\}} (f_{i+1} - f_i) \geqslant 0$$

by (30.17.b). Hence $E^{\mathcal{B}_\tau} f_{\tau'} \geqslant f_\tau$ a.s. by (11.40). In the general case, this implies that $E^{\mathcal{B}_{\tau\wedge\tau'}} f_{\tau'} \geqslant f_{\tau\wedge\tau'}$ a.s. The proof now proceeds as in the proof of (i) with the last "=" in (1) replaced by "$\geqslant$". □

Corollary 30.24. *Let* $\{\tau_n : n \in N\}$ *be a nondecreasing sequence of bounded stopping times.*

(i) *If* $\{f_n : n \in N\}$ *is a* $\{\mathcal{B}_n\}$*-martingale, then* $\{f_{\tau_n} : n \in N\}$ *is a* $\{\mathcal{B}_{\tau_n}\}$*-martingale.*

(ii) *If* $\{f_n : n \in N\}$ *is a* $\{\mathcal{B}_n\}$*-submartingale, then* $\{f_{\tau_n} : n \in N\}$ *is a* $\{\mathcal{B}_{\tau_n}\}$*-submartingale.*

Proof. The sequence $\{\mathcal{B}_{\tau_n} : n \in N\}$ is nondecreasing by (30.6.ii), $f_{\tau_n} \in \mathfrak{L}^1$ for each $n \in N$ by (30.5.j), and $\{f_{\tau_n} : n \in N\}$ is $\{\mathcal{B}_{\tau_n}\}$-adapted by (30.5.i). The result follows from (30.23) applied to every pair (τ_n, τ_{n+1}). □

Remark 30.25. Let τ be a stopping time. Then, in view of (30.24), $\{f_{\tau\wedge n} : n \in N\}$ is a $\{\mathcal{B}_{\tau\wedge n}\}$-martingale or a $\{\mathcal{B}_{\tau\wedge n}\}$-submartingale according as $\{f_n : n \in N\}$ is a $\{\mathcal{B}_n\}$-martingale or a $\{\mathcal{B}_n\}$-submartingale. Even a stronger statement holds, namely that $\{f_{\tau\wedge n} : n \in N\}$ is a $\{\mathcal{B}_n\}$-martingale or a $\{\mathcal{B}_n\}$-submartingale according as $\{f_n : n \in N\}$ is a $\{\mathcal{B}_n\}$-martingale or a $\{\mathcal{B}_n\}$-submartingale. (Notice that $\mathcal{B}_{\tau\wedge n} \subset \mathcal{B}_n, n \in N$, by (30.5.d), and recall (30.17.d).) Actually, for $n \in N$, $f_{\tau\wedge(n+1)} = f_{\tau\wedge n}$ on $\{\tau \leqslant n\}$, while $f_{\tau\wedge(n+1)}$ and $f_{\tau\wedge n}$ are f_{n+1} and f_n, respectively, on $\{\tau > n\}$. Therefore, in view of (30.17.b), for $\Lambda \in \mathcal{B}_n$, we have $\int_{\Lambda\cap\{\tau\leqslant n\}} f_{\tau\wedge(n+1)}\,dP = \int_{\Lambda\cap\{\tau\leqslant n\}} f_{\tau\wedge n}\,dP$, and $\int_{\Lambda\cap\{\tau>n\}} f_{\tau\wedge(n+1)}\,dP = \int_{\Lambda\cap\{\tau>n\}} f_{n+1}\,dP = \int_{\Lambda\cap\{\tau>n\}} f_n\,dP$ or $\geqslant \int_{\Lambda\cap\{\tau>n\}} f_n\,dP$ according as $\{f_n : n \in N\}$ is a $\{\mathcal{B}_n\}$-martingale or a $\{\mathcal{B}_n\}$-submartingale, whence the desired conclusion follows by addition.

Corollary 30.26. *Suppose that $\{f_n : n \in N\}$ is adapted.*

(i) *$\{f_n : n \in N\}$ is a martingale if and only if $Ef_\tau = Ef_{\tau'}$ for any stopping times τ and τ' that take at most two values.*

(ii) *$\{f_n : n \in N\}$ is a submartingale if and only if $Ef_\tau \leqslant Ef_{\tau'}$ for any stopping times τ and τ' with $\tau \leqslant \tau'$ that take at most two values.*

Proof. If $\{f_n : n \in N\}$ is a martingale, then $Ef_\tau = Ef_{\tau'} = Ef_{\tau\wedge\tau'}$ for any bounded stopping times τ and τ' by (30.23.i). Conversely, if the stated property holds, for $n \in N$ and $B \in \mathcal{B}_n$, consider the stopping times $\tau = n$ and $\tau' = n1_{B^c} + (n+1)1_B$. Then we have $\int_B f_{n+1}\,dP = Ef_{\tau'} - \int_{B^c} f_n\,dP = Ef_n - \int_{B^c} f_n\,dP = \int_B f_n\,dP$, which proves (i). The proof of (ii) is similar. □

Corollary 30.27. *Assume that $\{f_n : n \in N\}$ is a martingale, let τ be a stopping time, let $h : X \to R$ be a bounded $\mathcal{B}_\tau$-measurable function, and put $g_n = h(f_n - f_{\tau\wedge n}), n \in N$. Then $\{g_n : n \in N\}$ is a martingale.*

Proof. Observe first that $\{g_n : n \in N\} \subset \mathfrak{L}^1$ by (30.5.j). For $n \in N, g_n = 1_{\{\tau\leqslant n\}}hf_n - 1_{\{\tau\leqslant n\}}hf_\tau$ is $\mathcal{B}_n$-measurable by (30.6.vi). Let τ' be a bounded stopping time. Then, by virtue of (29.16) and (30.23.i), since $f_{\tau\wedge\tau'}$ is $\mathcal{B}_\tau$-measurable, we get $E^{\mathcal{B}_\tau} g_{\tau'} = hE^{\mathcal{B}_\tau}[f_{\tau'} - f_{\tau\wedge\tau'}] = h(f_{\tau\wedge\tau'} - f_{\tau\wedge\tau'}) = 0$ a.s., and so $Eg_{\tau'} = 0$. In view of (30.26.i), this shows that $\{g_n : n \in N\}$ is a martingale. □

The statements in the optional sampling theorem (30.23) can be extended to unbounded stopping times, provided that additional uniform integrability conditions are imposed. We need the next lemma.

Lemma 30.28. *Let $\{f_n : n \in N\}$ be a submartingale such that $\sup_{n\in N} Ef_n^+ = M < \infty$, and let τ be a stopping time. Then $E[f_\tau^+ 1_{\{\tau<\infty\}}] \leqslant M$ and $E[|f_\tau|\,1_{\{\tau<\infty\}}] \leqslant 2M - Ef_1$.*

Proof. As $\{f_n^+ : n \in N\}$ is a submartingale by (30.22.iii), applying (30.23.ii) with $\tau' = n$, we get $E[f_{\tau\wedge n}^+ 1_{\{\tau<\infty\}}] \leqslant Ef_{\tau\wedge n}^+ \leqslant Ef_n^+ \leqslant M, n \in N$. Consequently, since $f_{\tau\wedge n}^+ 1_{\{\tau<\infty\}} \to f_\tau^+ 1_{\{\tau<\infty\}}$, the first assertion follows on account of Fatou's lemma (11.24). Next, as $\{f_{\tau\wedge n} : n \in N\}$ is a submartingale by (30.25), applying (30.17.c), we have $E[|f_{\tau\wedge n}|\,1_{\{\tau<\infty\}}] \leqslant E\,|f_{\tau\wedge n}| = 2Ef_{\tau\wedge n}^+ - Ef_{\tau\wedge n} \leqslant 2M - Ef_1, n \in N$, whence the second assertion follows again by Fatou's lemma. □

Theorem 30.29. *Let τ and τ' be finite stopping times.*

(i) *If $\{f_n : n \in N\}$ is a uniformly integrable martingale, then $E^{\mathcal{B}_\tau} f_{\tau'} = f_{\tau\wedge\tau'}$ a.s.*

(ii) *If $\{f_n : n \in N\}$ is a submartingale and $\{f_n^+ : n \in N\}$ is uniformly integrable, then $E^{\mathcal{B}_\tau} f_{\tau'} \geqslant f_{\tau\wedge\tau'}$ a.s.*

Proof. We prove only (ii), as the proof of (i) is simpler and easier. We begin with the observation that $\sup_{n\in N} Ef_n^+ < \infty$ by (14.28), and so $f_{\tau'} \in \mathfrak{L}^1$ by (30.28). Since $f_{\tau'\wedge n}^+ \to f_{\tau'}^+$, and the uniform integrability of $\{f_n^+ : n \in N\}$ implies the uniform integrability of $\{f_{\tau'\wedge n}^+ : n \in N\}$, it follows that $E^{\mathcal{B}_\tau} f_{\tau'\wedge n}^+ \overset{\mathfrak{L}^1}{\to} E^{\mathcal{B}_\tau} f_{\tau'}^+$ by (29.22). Hence, according to (14.14) and (14.8), $E^{\mathcal{B}_\tau} f_{\tau'\wedge n_k}^+ \overset{a.s.}{\to} E^{\mathcal{B}_\tau} f_{\tau'}$ for some subsequence $\{n_k : k \in N\} \subset N$. Therefore, on account of (29.14.i) and (30.23.ii),

$$E^{\mathcal{B}_\tau} f_{\tau'} = E^{\mathcal{B}_\tau} f_{\tau'}^+ - E^{\mathcal{B}_\tau} f_{\tau'}^- \geqslant \lim_k E^{\mathcal{B}_\tau} f_{\tau'\wedge n_k}^+ - \liminf_k E^{\mathcal{B}_\tau} f_{\tau'\wedge n_k}^-$$
$$\geqslant \limsup_k E^{\mathcal{B}_\tau} f_{\tau'\wedge n_k} \geqslant \limsup_k f_{\tau\wedge\tau'\wedge n_k} = f_{\tau\wedge\tau'} \text{ a.s.} \qquad \square$$

Corollary 30.30. *Let $\{\tau_n : n \in N\}$ be a nondecreasing sequence of finite stopping times. Then $\{f_{\tau_n} : n \in N\}$ is a $\{\mathcal{B}_{\tau_n}\}$-martingale or a $\{\mathcal{B}_{\tau_n}\}$-submartingale according as $\{f_n : n \in N\}$ is a $\{\mathcal{B}_n\}$-martingale or a $\{\mathcal{B}_n\}$-submartingale as in* (30.29).

Remark 30.31. For a nonpositive submartingale the conclusions of (30.23.ii) and (30.29.ii), as well as the conclusions of their corollaries (30.24.ii) and (30.30), hold without additional assumptions such as boundedness or uniform integrability. In fact, if $\{f_n : n \in N\}$ is a nonnegative supermartingale, and τ and τ' are finite stopping times, then

$$E^{\mathcal{B}_\tau} f_{\tau'} \leqslant f_{\tau\wedge\tau'} \text{ a.s.} \tag{1}$$

Indeed, observe first that, in view of (30.17.e) and (30.23.ii), for $n \in N$, we have $f_{\tau\wedge n} \leqslant E^{\mathcal{B}_\tau} f_n$ a.s., and so $0 \leqslant Ef_{\tau\wedge n} \leqslant Ef_n \leqslant Ef_1$, whence $0 \leqslant Ef_\tau \leqslant Ef_1$ by Fatou's lemma. Next, once more by (30.23.ii), $E^{\mathcal{B}_\tau} f_{\tau'\wedge n} \leqslant f_{\tau\wedge\tau'\wedge n}$ a.s., $n \in N$, and (1) follows by (29.14.i).

Our next aim is to establish several important inequalities, which apply to main convergence theorems for martingales and semimartingales, and to other problems.

Theorem 30.32 (Bernstein-Lévy). *Let $\{f_n : n \in N\}$ be a submartingale, and let $\varepsilon > 0$.*

(i) $\varepsilon P(\sup_{1\leqslant i\leqslant n} f_i \geqslant \varepsilon) \leqslant \int_{\{\sup_{1\leqslant i\leqslant n} f_i \geqslant \varepsilon\}} f_n \, dP \leqslant Ef_n^+, n \in N.$

(ii) $\varepsilon P(\sup_{1\leqslant i\leqslant n} |f_i| \geqslant \varepsilon) \leqslant 2Ef_n^+ - Ef_1, n \in N.$

Proof. Let τ be the hitting time of $[\varepsilon, \infty[$, write $B = \{\sup_{1\leqslant i\leqslant n} f_i \geqslant \varepsilon\}$, and notice that $B = \{\tau \leqslant n\}$. Since $B \in \mathcal{B}_\tau$ and $f_\tau \geqslant \varepsilon$ on B, applying (30.23) with $\tau' = n$, we get $\varepsilon P(B) \leqslant \int_B f_\tau \, dP = \int_B f_{\tau\wedge n} \, dP \leqslant \int_B f_n \, dP \leqslant Ef_n^+, n \in N$, which is (i). For $n \in N$, let $f_n = f_n' + f_n''$ be the Doob decomposition of f_n (30.19.i). Then, in view of

(30.17.h) and (30.17.f), as $f_1' = f_1$, we have

$$\varepsilon P\left(\inf_{1\leqslant i\leqslant n} f_i \leqslant -\varepsilon\right) \leqslant \varepsilon P\left(\inf_{1\leqslant i\leqslant n} f_i' \leqslant -\varepsilon\right) \leqslant Ef_n'^- = Ef_n'^+ - Ef_n'$$
$$\leqslant Ef_n^+ - Ef_1, \quad n \in N. \tag{1}$$

Now (ii) obtains from (i) and (1). □

Remark 30.33. If $\{f_n : n \in N\} \subset \mathfrak{L}^p$, where $1 \leqslant p < \infty$, is a martingale, then $\{|f_n|^p : n \in N\}$ is a submartingale by (30.22.ii), and so, for any $\varepsilon > 0$, (30.32.i) yields $P(\sup_{1\leqslant i\leqslant n} |f_i| \geqslant \varepsilon) = P(\sup_{1\leqslant i\leqslant n} |f_i|^p \geqslant \varepsilon^p) \leqslant \varepsilon^{-p} E\,|f_n|^p$, $n \in N$. For $n = 2$, applying this inequality to the martingale $\{S_n : n \in N\}$ in (30.16.d), we obtain, for $\varepsilon > 0$, $P(\sup_{1\leqslant i\leqslant n} |S_i| \geqslant \varepsilon) \leqslant \varepsilon^{-2} E S_n^2$, $n \in N$, which is precisely Kolmogorov's inequality (26.4.i). (See also (26.27).)

Corollary 30.34. *Let* $\{f_n : n \in N\}$ *and* ε *be as in* (30.32), *and put* $B = \{\sup_{k\leqslant i\leqslant n} f_i \geqslant \varepsilon\}$, *where* $1 \leqslant k \leqslant n$. *Then*

$$\varepsilon P(B|\mathcal{B}_k) \leqslant E[1_B f_n|\mathcal{B}_k] \; a.s., \quad n \geqslant k. \tag{1}$$

Proof. For $\Lambda \in \mathcal{B}_k$, consider the submartingale $\{1_\Lambda f_n : n \geqslant k\}$. Then, according to (30.32.i), for $n \geqslant k$,

$$\varepsilon P(\Lambda \cap B) = \varepsilon P(\sup_{k\leqslant i\leqslant n} 1_\Lambda f_i \geqslant \varepsilon) \leqslant \int\limits_{\{\sup_{k\leqslant i\leqslant n} 1_\Lambda f_i \geqslant \varepsilon\}} 1_\Lambda f_n \, dP = \int\limits_\Lambda 1_B f_n \, dP.$$

This proves (1). □

Theorem 30.35. *Let* $\{f_n : n \in N\}$ *be a nonnegative submartingale.*

(i) $\left\|\sup_{1\leqslant i\leqslant n} f_i\right\|_1 \leqslant \frac{e}{e-1}\left(1 + \left\| f_n \log^+ f_n\right\|_1\right)$, $n \in N$, *where* $\log^+ x = \log(x \vee 1)$, $x \geqslant 0$.

(ii) *If* $\{f_n : n \in N\} \subset \mathfrak{L}^p$ *with* $1 < p < \infty$, *then* $\left\|\sup_{1\leqslant i\leqslant n} f_i\right\|_p \leqslant \frac{p}{p-1} \left\| f_n\right\|_p$, $n \in N$.

Proof. To ease the writing, put $f_n^* = \sup_{1\leqslant i\leqslant n} f_i$, $n \in N$. Using (15.24), (30.32.i) and Fubini's theorem (15.8), we have

$$Ef_n^* \leqslant 1 + \int_1^\infty P(f_n^* \geqslant x)dx \leqslant 1 + \int_1^\infty \left(\frac{1}{x}\int_X f_n 1_{\{f_n^* \geqslant x\}}\, dP\right) dx$$
$$= 1 + \int_X \left(f_n 1_{\{f_n^* \geqslant 1\}} \int_1^{f_n^*} \frac{dx}{x}\right) dP$$
$$= 1 + E[f_n 1_{\{f_n^* \geqslant 1\}} \log f_n^*] = 1 + E[f_n \log^+ f_n^*].$$

Next, since $a \log b \leqslant a \log a + b/e$, and so $a \log^+ b \leqslant a \log^+ a + b/e$, for $a, b \geqslant 0$, we obtain $Ef_n^* \leqslant 1 + E[f_n \log^+ f_n] + (Ef_n^*)/e$, from which (i) follows. Now, if

$1 < p < \infty$, once more on account of (15.24), (30.32.i) and Fubini's theorem, we have

$$E(f_n^*)^p = p\int_0^\infty x^{p-1}P(f_n^* \geqslant x)dx \leqslant p\int_0^\infty \left(x^{p-2}\int_X f_n 1_{\{f_n^* \geqslant x\}}dP\right)dx$$

$$= p\int_X \left(f_n \int_0^{f_n^*} x^{p-2}\right)dP = \frac{p}{p-1}E[f_n(f_n^*)^{p-1}].$$

Therefore, by virtue of Hölder's inequality (13.4), we get $E(f_n^*)^p \leqslant \frac{p}{p-1}\,\|f_n\|_p\,(E(f_n^*)^p)^{1-\frac{1}{p}}$, whence (ii) obtains. □

Definitions 30.36. Let $\{f_n : n \geqslant 1\} \subset R^X$ be an arbitrary sequence of functions, and let $a, b \in R$ with $a < b$. Set $\tau_0 = 0$ and define successively the *crossing times* $\tau_k, k \geqslant 1$, by

$$\tau_k = \begin{cases} \inf\{n > \tau_{k-1} : f_n \leqslant a\} \text{ if } k \text{ is odd} \\ \inf\{n > \tau_{k-1} : f_n \geqslant b\} \text{ if } k \text{ is even} \end{cases}.$$

As usual, we take $\inf\emptyset = \infty$ here. Let $U_{a,b} = \sup\{k \geqslant 0 : \tau_{2k} < \infty\}$. Then $U_{a,b}$ represents the number of *upcrossings* of $[a, b]$ by $\{f_n : n \geqslant 1\}$. For $m \geqslant 2$, let $U_{a,b}^{(m)}$ denote the number of upcrossings of $[a, b]$ by $\{f_n : 1 \leqslant n \leqslant m\}$, i.e. $U_{a,b}^{(m)} = \sup\{0 \leqslant k \leqslant m/2 : \tau_{2k} < \infty\}$.

Remarks 30.37. (a) $\{\liminf_n f_n < a < b < \limsup_n f_n\} \subset \{U_{a,b} = \infty\} \subset \{\liminf_n f_n \leqslant a < b \leqslant \limsup_n f_n\}$, and so $\{\lim_n f_n \text{ exists}\} = \bigcap_{\substack{a,b\in R\\ a<b}}\{U_{a,b} < \infty\} = \bigcap_{\substack{a,b\in Q\\ a<b}}\{U_{a,b} < \infty\}$.

(b) Assume that $\{f_n : n \geqslant 1\}$ is an adapted sequence of random variables. For each $k \geqslant 1$, the event $\{\tau_k = n\}$ is determined by the random variables $f_1, \ldots, f_n$. Consequently, $\{\tau_k : k \geqslant 1\}$ is a sequence of stopping times. Since $\{U_{a,b} \geqslant k\} = \{\tau_{2k} < \infty\}, k \geqslant 0$, we see that $U_{a,b}$ is a random variable.

(c) If $\{f_n : n \geqslant 1\}$ is as in (b) and $f_n \geqslant a$ for all $n \geqslant 1$, then, since $f_{\tau_k} = a$ whenever τ_k is odd and finite, while $f_{\tau_k} \geqslant b$ whenever $\tau_k \neq 0$ is even and finite, it follows that $\sum_{\substack{k=2\\ k\,\text{even}}}^{m}(f_{\tau_k\wedge m} - f_{\tau_{k-1}\wedge m}) \geqslant (b-a)U_{a,b}^{(m)}$.

Theorem 30.38 (Upcrossing inequality). *Let $\{f_n : n \geqslant 1\}$ be a submartingale, and let $U_{a,b}^{(m)}$ be as in (30.36). Then*

$$EU_{a,b}^{(m)} \leqslant \frac{Ef_m^+ + |a|}{b-a}. \tag{1}$$

Proof. (I) Consider first the case where $f_n \geqslant a$ for all $n \geqslant 1$. We have

$$f_m - f_{\tau_1 \wedge m} = \sum_{k=2}^{m} \left(f_{\tau_k \wedge m} - f_{\tau_{k-1} \wedge m}\right)$$
$$= \sum_{\substack{k=2 \\ k \text{ even}}}^{m} (f_{\tau_k \wedge m} - f_{\tau_{k-1} \wedge m}) + \sum_{\substack{k=2 \\ k \text{ odd}}}^{m} (f_{\tau_k \wedge m} - f_{\tau_{k-1} \wedge m}).$$

In view of (30.37.c),

$$E\left[\sum_{\substack{k=2 \\ k \text{ even}}}^{m} (f_{\tau_k \wedge m} - f_{\tau_{k-1} \wedge m})\right] \geqslant (b-a)U_{a,b}^{(m)}. \tag{2}$$

Since $\{f_{\tau_k \wedge m} : k \geqslant 1\}$ is a submartingale by (30.24.ii), on account of (30.17.c), we get

$$E\left[\sum_{\substack{k=2 \\ k \text{ odd}}}^{m} (f_{\tau_k \wedge m} - f_{\tau_{k-1} \wedge m})\right] = \sum_{\substack{k=2 \\ k \text{ odd}}}^{m} (Ef_{\tau_k \wedge m} - Ef_{\tau_{k-1} \wedge m}) \geqslant 0. \tag{3}$$

Adding (2) and (3), we have

$$Ef_m - a \geqslant Ef_m - Ef_{\tau_1 \wedge m} \geqslant (b-a)U_{a,b}^{(m)}, \tag{4}$$

whence (1) obtains in this case.

(II) Consider now the general case. According to (30.22.ii), $\{f_n \vee a : n \geqslant 1\}$ is a submartingale. Observe also that $U_{a,b}^{(m)}$ is exactly the number of upcrossings of $[a, b]$ by $\{f_n \vee a : 1 \leqslant n \leqslant m\}$. Therefore, applying (4), we get $(b-a)U_{a,b}^{(m)} \leqslant E[f_m \vee a - a] \leqslant Ef_m^+ + |a|$, since $f_m \vee a - a \leqslant f_m^+ + |a|$. □

We are now in the position to explore basic convergence results for martingales and submartingales. First, a definition is in order.

Definition 30.39. Let $\{f_n : n \in N\}$ be a martingale or a submartingale. It is said to be *closed* (by f) if there exists $f \in \mathfrak{L}^1$ such that, for all $n \in N$, $E^{\mathcal{B}_n} f = f_n$ a.s. or $E^{\mathcal{B}_n} f \geqslant f_n$ a.s., respectively.

Remarks 30.40. (a) The martingale $\{E^{\mathcal{B}_n} f : n \in N]$ in (30.16.c) is closed by f.

(b) Let φ and $\{f_n : n \in N\}$ be as in (30.21). If f closes $\{f_n : n \in N\}$, then $\varphi \circ f$ closes $\{\varphi \circ f_n : n \in N\}$.

(c) If $\{f_n : n \in N\}$ is a martingale or a nonnegative submartingale closed by $f \in \mathfrak{L}^p$ with $1 \leqslant p < \infty$, then $\{f_n : n \in N\} \subset \mathfrak{L}^p$ and $\{|f_n|^p : n \in N\}$ is uniformly integrable. The first statement follows from (29.20). Since $|f_n|^p \leqslant E^{\mathcal{B}_n} |f|^p$ a.s., $n \in N$, the second statement follows from (29.26) and (14.27.a).

(d) If $\{f_n : n \in N\}$ is a submartingale closed by f, then $\{f_n \vee c : n \in N\}$ is uniformly integrable for every $c \in R$. Indeed, for $n \in N$, $0 \leqslant f_n \vee c - c \leqslant E^{\mathcal{B}_n}[f \vee c] - c \leqslant E^{\mathcal{B}_n} |f \vee c| + |c|$ a.s. The assertion now follows from (29.26) and (14.27.a).

Theorem 30.41. *Let $\{f_n : n \in N\}$ be a submartingale.*

(i) *If $\sup_{n\in N} Ef_n^+ < \infty$, then there exists $f_\infty \in \mathfrak{L}^1$ such that $f_n \overset{a.s.}{\to} f_\infty$.*

(ii) *If $\{f_n^+ : n \in N\}$ is uniformly integrable, then there exists f_∞ as in* (i), *and $\{f_n : n \in N\}$ is closed by f_∞.*

(iii) *If $\{f_n : n \in N\}$ is closed by some g, then there exists f_∞ as in* (ii), *and $E^{\mathcal{B}_\infty} g \geqslant f_\infty$ a.s.*

Proof. For $a, b \in Q$ with $a < b$ and $m \geqslant 2$, let $U_{a,b}$ and $U_{a,b}^{(m)}$ be as in (30.36). If $\sup_{n\in N} Ef_n^+ < \infty$, since $U_{a,b}^{(m)} \to U_{a,b}$, it follows by the monotone convergence theorem (11.22) and (30.38) that $EU_{a,b} < \infty$, and so $U_{a,b} < \infty$ a.s. In view of (30.37.a), this ensures the existence of a measurable function f_∞ such that $f_n \overset{a.s.}{\to} f_\infty$. Moreover, by Fatou's lemma and (30.17.c), $E\,|f_\infty| \leqslant \liminf_n E\,|f_n| = \liminf_n (2Ef_n^+ - Ef_n) \leqslant \liminf_n (2Ef_n^+ - Ef_1) < \infty$. Thus (i) holds. Next, if $\{f_n^+ : n \in N\}$ is uniformly integrable, then $\sup_{n\in N} Ef_n^+ < \infty$ by (14.28), and so there exists f_∞ as in (i). To finish the proof of (ii), fix $n \in N$. Then $E^{\mathcal{B}_n} f_m^+ \overset{\mathfrak{L}^1}{\to} E^{\mathcal{B}_n} f_\infty^+$ by (29.22), and so $E^{\mathcal{B}_n} f_{m_k}^+ \overset{a.s.}{\to} E^{\mathcal{B}_n} f_\infty^+$ for some subsequence $\{m_k : k \in N\}$. Consequently, on account of (29.14.i),

$$E^{\mathcal{B}_n} f_\infty = E^{\mathcal{B}_n} f_\infty^+ - E^{\mathcal{B}_n} f_\infty^- \geqslant \lim_k E^{\mathcal{B}_n} f_{m_k}^+ - \liminf_k E^{\mathcal{B}_n} f_{m_k}^-$$

$$\geqslant \limsup_k E^{\mathcal{B}_n} f_{m_k} \geqslant f_n \text{ a.s.}$$

Thus $\{f_n : n \in N\}$ is closed by f_∞. Apropos of (iii), the hypothesis implies that g^+ closes $\{f_n^+ : n \in N\}$ by (30.40.b), and so $\{f_n^+ : n \in N\}$ is uniformly integrable by (30.40.d). Hence there exists f_∞ as in (ii). Finally, from $E^{\mathcal{B}_n} g \geqslant f_n$ a.s., $n \in N$, it follows that $E^{\mathcal{B}_\infty} g \geqslant f_\infty$ a.s. by (29.27). □

Corollary 30.42.

(i) *If $\{f_n : n \in N\}$ is a nonpositive submartingale, then $f_n \overset{a.s.}{\to} f_\infty$ for some $f_\infty \in \mathfrak{L}^1$.*

(ii) *If $\{f_n : n \in N\}$ is a nonnegative or nonpositive martingale, then $f_n \overset{a.s.}{\to} f_\infty$ for some $f_\infty \in \mathfrak{L}^1$.*

Proof. Since $\{f_n : n \in N\}$ is closed by 0, (i) follows by (30.41.iii). (ii) follows from (i) via (30.17.f) and (30.17.e). □

Theorem 30.43. *Let $\{f_n : n \in N\}$ be a martingale.*

(i) *If $\{f_n : n \in N\}$ is uniformly integrable, then there exists $f_\infty \in \mathfrak{L}^1$ such that $f_n \overset{a.s.}{\to} f_\infty$, and $\{f_n : n \in N\}$ is closed by f_∞.*

(ii) *If $\{f_n : n \in N\}$ is closed by some g, then there exists f_∞ as in* (i), *and $E^{\mathcal{B}_\infty} g = f_\infty$ a.s.*

Proof. If $\{f_n : n \in N\}$ is uniformly integrable, then both $\{f_n^+ : n \in N\}$ and $\{f_n^- : n \in N\}$ are uniformly integrable. The existence of $f_\infty \in \mathfrak{L}^1$ with $f_n \overset{a.s.}{\to} f_\infty$ and $E^{\mathcal{B}_n} f_\infty \geqslant f_n$ a.s., $n \in N$, follows from (30.41.ii). Since $\{-f_n : n \in N\}$ is a submartingale by (30.17.f) and (30.17.e), and $\{(-f_n)^+ : n \in N\} = \{f_n^- : n \in N\}$, we have also $E^{\mathcal{B}_n} f_\infty \leqslant f_n$ a.s., $n \in N$. This finishes the proof of (i). If now g closes

$\{f_n : n \in N\}$, then $\{f_n : n \in N\}$ is uniformly integrable by (30.40.c), so that there exists f_∞ as in (i). As $E^{\mathcal{B}_n} g = f_n$ a.s., $n \in N$, it follows that $E^{\mathcal{B}_\infty} g = f_\infty$ a.s. □

We now look at the convergence in $\mathfrak{L}^p$ of martingales and submartingales.

Theorem 30.44. *Let $\{f_n : n \in N\} \subset \mathfrak{L}^p$ be a submartingale, where $1 \leqslant p < \infty$. The following are equivalent:*

(i) *$\{|f_n|^p : n \in N\}$ is uniformly integrable;*

(ii) *there is $f_\infty \in \mathfrak{L}^p$ such that $f_n \overset{\mathfrak{L}^p}{\to} f_\infty$;*

(iii) *there is $f_\infty \in \mathfrak{L}^p$ such that $f_n \overset{a.s.}{\to} f_\infty$ and $E\,|f_n|^p \to E\,|f_\infty|^p$.*

Proof. If (i) holds, then $\sup_{n\in N} Ef_n^+ < \infty$ by (14.28), and so there is $f_\infty \in \mathfrak{L}^1$ with $f_n \overset{a.s.}{\to} f_\infty$ by (30.41.i). In view of (14.30), it follows that $f_\infty \in \mathfrak{L}^p$ and $f_n \overset{\mathfrak{L}^p}{\to} f_\infty$. Thus (i) $\Longrightarrow$ (ii). If (ii) holds, then $\sup_{n\in N} Ef_n^+ \leqslant \sup_{n\in N} \|f_n\|_p < \infty$, and so there is $f_\infty \in \mathfrak{L}^p$ with $f_n \overset{a.s.}{\to} f_\infty$ by the above argument. Moreover, $E\,|f_n|^p \to E\,|f_\infty|^p$ by (14.13.c). Hence (ii) $\Longrightarrow$ (iii). Invoking again (14.30), we see that (iii) $\Longrightarrow$ (i). □

Remarks 30.45. (a) Let $\{f_n : n \in N\}$ and f_∞ (see (14.15.c)) be as in (30.44). Then f_∞ closes $\{f_n : n \in N\}$, and if g also closes $\{f_n : n \in N\}$, then $E^{\mathcal{B}_\infty} g \geqslant f_\infty$ a.s.; if $\{f_n : n \in N\}$ is a martingale, then $E^{\mathcal{B}_\infty} g = f_\infty$ a.s.

(b) The conditional expectation continuity theorem (29.27) can be proved by making use of the martingale theory. Indeed, let $f \in \mathfrak{L}^p$ with $1 \leqslant p < \infty$, and write $f_n = E^{\mathcal{B}_n} f$, $n \geqslant 1$. Then $\{|f_n|^p : n \geqslant 1\}$ is uniformly integrable by (30.40.c). On account of (30.43.i) and (30.44), there exists $f_\infty \in \mathfrak{L}^p$ such that $f_n \overset{a.s.}{\to} f_\infty$ and $f_n \overset{\mathfrak{L}^p}{\to} f_\infty$. Now (29.22) helps to identify the limit, namely $f_\infty = E^{\mathcal{B}_\infty} f$ a.s.

Here is the place for an interlude about limiting properties of reverse martingales and semimartingales. Throughout (30.46)–(30.52), $\{\mathcal{B}_n : n \in N\}$ is a nonincreasing sequence of σ-algebras with $\mathcal{B}_1 \subset \mathcal{A}$, $\mathcal{B} = \cap_{n\in N} \mathcal{B}_n$, and $\{f_n : n \in N\} \subset \mathfrak{L}^1$ is adapted [$\{\mathcal{B}_n\}$-adapted], i.e. f_n is $\mathcal{B}_n$-measurable for any $n \in N$.

Definitions 30.46. $\{f_n : n \in N\}$ is said to be a *reverse martingale* if $E^{\mathcal{B}_{n+1}} f_n = f_{n+1}$ a.s. for all $n \in N$, a *reverse submartingale* if $E^{\mathcal{B}_{n+1}} f_n \geqslant f_{n+1}$ a.s. for all $n \in N$, and a *reverse supermartingale* if $E^{\mathcal{B}_{n+1}} f_n \leqslant f_{n+1}$ a.s. for all $n \in N$. We say that $\{f_n : n \in N\}$ is a *reverse semimartingale* if it is either a reverse submartingale or a reverse supermartingale. These definitions extend in an obvious manner to an index set of the form $\{m+1, m+2, \ldots\}$ or $\{m+1, \ldots, m+p\}$ for some $m, p \in N$ with $p \geqslant 2$.

Examples 30.47. (a) $\{f_n : n \in N\}$ is a reverse submartingale whenever $f_n \geqslant f_{n+1}$, $n \in N$, and a reverse supermartingale whenever $f_n \leqslant f_{n+1}$, $n \in N$.

(b) Let $f \in \mathfrak{L}^1$ and write $f_n = E^{\mathcal{B}_n} f$, $n \in N$. Then $E^{\mathcal{B}_{n+1}} f_n = E^{\mathcal{B}_{n+1}}[E^{\mathcal{B}_n} f] = E^{\mathcal{B}_{n+1}} f$ a.s., $n \in N$. Hence $\{f_n : n \in N\}$ is a reverse martingale.

(c) Let $f_n, n \geqslant 1$, and $S_n, n \geqslant 1$, be as in (29.83). Write $\mathcal{B}_n = \sigma(S_m, m \geqslant n)$, $n \geqslant 1$. If $f_1 \in \mathfrak{L}^1$, then, according to (29.83.a), $E^{\mathcal{B}_{n+1}} f_1 = E^{\mathcal{B}_{n+1}}[S_n/n] = E^{\mathcal{B}_{n+1}}[S_{n+1}/(n+1)] = S_{n+1}/(n+1)$ a.s. $n \geqslant 1$. Consequently, $\{S_n/n : n \geqslant 1\}$ is a reverse martingale.

Remarks 30.48. (a) If $\{f_n : n \in N\}$ is a reverse martingale, then it follows by induction that $f_n = E^{\mathcal{B}_n} f_1$ a.s. for any $n \in N$. Therefore, according to (29.29), $f_n \overset{a.s.}{\to} E^{\mathcal{B}} f_1$. Moreover, if $f_1 \in \mathfrak{L}^p$ for some $p \in [1, \infty[$, then $\{f_n : n \in N\} \subset \mathfrak{L}^p$ and $f_n \overset{\mathfrak{L}^p}{\to} E^{\mathcal{B}} f_1$. Notice also that $\{f_n : n \in N\}$ is uniformly integrable by (29.26).

(b) All properties of ordinary martingales and semimartingales in (30.17), as well as other useful properties, can be transferred with possible obvious changes to reverse martingales and semimartingales.

(c) (30.21) and (30.22) hold with "martingale" replaced by "reverse martingale" and "submartingale" replaced by "reverse submartingale".

(d) Consider the set $-N = \{\ldots, -2, -1\}$ with its usual ordering. If we write $\mathcal{C}_{-n} = \mathcal{B}_n$ and $g_{-n} = f_n$ for $n \in N$, then $\mathcal{C}_i \subset \mathcal{C}_j$ whenever $i, j \in -N$ with $i < j$, and $\{g_i : i \in -N\}$ is $\{\mathcal{C}_i\}$-adapted. Furthermore, if $\{f_n : n \in N\}$ is a reverse martingale, a reverse submartingale or a reverse supermartingale, then, for all $i, j \in -N$ with $i < j$, $E^{\mathcal{B}_j} g_i = g_j$ a.s., $E^{\mathcal{B}_j} g_i \geqslant g_j$ a.s. or $E^{\mathcal{B}_j} g_i \leqslant g_j$ a.s., respectively, i.e. $\{g_i : i \in -N\}$ fulfills the defining relations for an ordinary martingale or semimartingale.

(e) Let $\{f_n : n \geqslant 1\}$ be a reverse submartingale. For $m \geqslant 2$, write $\mathcal{C}_i = \mathcal{B}_{m-i+1}$, $1 \leqslant i \leqslant m$, and $g_i = f_{m-i+1}$, $1 \leqslant i \leqslant m$. Then $\{g_i : 1 \leqslant i \leqslant m\}$ is a $\{\mathcal{C}_i\}$-submartingale. Now let $a, b \in R$ with $a < b$, let $U_{a,b}^{(m)}$ denote the number of upcrossings of $[a, b]$ by $\{g_i : 1 \leqslant i \leqslant m\}$, and let $D_{a,b}^{(m)}$ denote the number of downcrossings of $[a, b]$ by $\{f_i : 1 \leqslant i \leqslant m\}$. Notice that there is a one-to-one correspondence between the set of upcrossings and the set of downcrossings, more precisely the first upcrossing of $[a, b]$ by $\{g_i : 1 \leqslant i \leqslant m\}$ corresponds to the last downcrossing of $[a, b]$ by $\{f_i : 1 \leqslant i \leqslant m\}, \ldots$, the last upcrossing of $[a, b]$ by $\{g_i : 1 \leqslant i \leqslant m\}$ corresponds to the first downcrossing of $[a, b]$ by $\{f_i : 1 \leqslant i \leqslant m\}$. Hence $U_{a,b}^{(m)} = D_{a,b}^{(m)}$. Consequently, according to (30.38), $E D_{a,b}^{(m)} = E U_{a,b}^{(m)} \leqslant (E g_m^+ + |a|)/(b-a) = (E f_1^+ + |a|)/(b-a)$. Let $D_{a,b} = \lim_m D_{a,b}^{(m)}$. Then $\{\liminf_n f_n < a < b < \limsup_n f_n\} \subset \{D_{a,b} = \infty\} \subset \{\liminf_n f_n \leqslant a < b \leqslant \limsup_n f_n\}$, and so $\{\lim_n f_n \text{ exists}\} = \bigcap_{\substack{a,b \in R \\ a<b}} \{D_{a,b} < \infty\} = \bigcap_{\substack{a,b \in Q \\ a<b}} \{D_{a,b} < \infty\}$.

Definition 30.49. Let $\{f_n : n \in N\}$ be a reverse martingale or a reverse submartingale. It is said to be *closed* (by f) if there exists $f \in \mathfrak{L}^1$ such that, for all $n \in N$, $E^{\mathcal{B}} f_n = f$ a.s. or $E^{\mathcal{B}} f_n \geqslant f$ a.s., respectively.

Theorem 30.50. *Let $\{f_n : n \in N\}$ be a reverse submartingale.*

(i) *$f = \lim_n f_n$ exists a.s. with $Ef^+ \leqslant Ef_1^+$.*

(ii) *If $\inf_{n \in N} Ef_n > -\infty$, then $\sup_{n \in N} E|f_n| < \infty$, there exists $f \in \mathfrak{L}^1$ such that $f_n \overset{a.s.}{\to} f$, and $\{f_n : n \in N\}$ is closed by f.*

Proof. For $a, b \in Q$ with $a < b$, it follows by (30.48.e) and the monotone convergence theorem that $E D_{a,b} \leqslant (E f_1^+ + |a|)/(b - a)$, and so $D_{a,b} < \infty$ a.s. Once more (30.48.e) ensures the existence of a measurable function f such that $f_n \overset{a.s.}{\to} f$. Moreover, since $\{f_n^+ : n \in N\}$ is a reverse submartingale by (30.48.c), and so $\{E f_n^+ : n \in N\}$ is a nonincreasing sequence, we have $Ef^+ \leqslant \liminf_n E f_n^+ \leqslant E f_1^+$ by Fatou's lemma.

Thus (i) holds. Next if $\inf_{n\in N} Ef_n = L > -\infty$, we get $E|f_n| = 2Ef_n^+ - Ef_n \leqslant 2Ef_1^+ - L, n \in N$, and so $\sup_{n\in N} E|f_n| < \infty$. Consequently, if f is as in (i), once more by Fatou's lemma, $E|f| \leqslant \liminf_n E|f_n| < \infty$, and so $f \in \mathfrak{L}^1$. To finish the proof, fix $m \in N$. As $E^{\mathcal{B}_n} f_m \geqslant f_n$ a.s. for all $n > m$, it follows that $E^{\mathcal{B}} f_m \geqslant f$ a.s. by (29.29). This means that f closes $\{f_n : n \in N\}$. □

Theorem 30.51. *Let $\{f_n : n \geqslant 1\}$ be a reverse submartingale. Then the following are equivalent:*

(i) $\inf_{n\geqslant 1} Ef_n > -\infty$;

(ii) $\{f_n : n \geqslant 1\}$ *is uniformly integrable;*

(iii) *there is $f \in \mathfrak{L}^1$ such that $f_n \overset{\mathfrak{L}^1}{\to} f$;*

(iv) *there is $f \in \mathfrak{L}^1$ that closes $\{f_n : n \geqslant 1\}$.*

Proof. Assume (i) and let $h = \sum_{i\geqslant 2}(E^{\mathcal{B}_i} f_{i-1} - f_i)$. Then $h \geqslant 0$ a.s. and $Eh = \sum_{i\geqslant 2}(Ef_{i-1} - Ef_i) = Ef_1 - \inf_{i\geqslant 1} Ef_i < \infty$, so that $h \in \mathfrak{L}^1$. For $n \geqslant 1$, write $g_n = E^{\mathcal{B}_n} f_1 - f_n$. We have

$$0 \leqslant g_n = E^{\mathcal{B}_n}[f_1 - f_n] = \sum_{i=2}^{n} E^{\mathcal{B}_n}[f_{i-1} - f_i] = \sum_{i=2}^{n} E^{\mathcal{B}_n}[E^{\mathcal{B}_i}[f_{i-1} - f_i]]$$

$$= E^{\mathcal{B}_n}[\sum_{i=2}^{n}(E^{\mathcal{B}_i} f_{i-1} - f_i)] \leqslant E^{\mathcal{B}_n} h \text{ a.s.}, \quad n \geqslant 2. \tag{1}$$

In view of (29.26), $\{E^{\mathcal{B}_n} h : n \geqslant 1\}$ is uniformly integrable, and so $\{g_n : n \geqslant 1\}$ is uniformly integrable by (1) and (14.27.a). Since $\{E^{\mathcal{B}_n} f_1 : n \geqslant 1\}$ is also uniformly integrable by (29.26), (ii) is verified. If (ii) holds, then $\inf_{n\geqslant 1} Ef_n > -\infty$ by (14.28), and so there is $f \in \mathfrak{L}^1$ with $f_n \overset{a.s.}{\to} f$ by (30.50.ii). It follows that $f_n \overset{\mathfrak{L}^1}{\to} f$ by (14.30). Thus (ii) $\Longrightarrow$ (iii). Next, if f is as in (iii), then $f_{n_k} \overset{a.s.}{\to} f$ for some subsequence $\{n_k : k \geqslant 1\} \subset N$. For $m \in N$, as $E^{\mathcal{B}_{n_k}} f_m \geqslant f_{n_k}$ a.s. whenever $n_k > m$, we see that $E^{\mathcal{B}} f_m \geqslant f$ a.s. by (29.29). Hence (iii) $\Longrightarrow$ (iv). Finally, if (iv) holds, then $-\infty < Ef \leqslant Ef_m, m \geqslant 1$, so that (i) is fulfilled. □

Remarks 30.52. (a) Taking $f_n = -n, n \in N$, we see that f in (30.50.i) may be identically $-\infty$.

(b) If $\{f_n : n \in N\}$ is a reverse submartingale such that $\{|f_n|^p : n \in N\}$ is uniformly integrable, where $1 \leqslant p < \infty$, then there is $f \in \mathfrak{L}^p$ with $f_n \overset{\mathfrak{L}^p}{\to} f$, f closes $\{f_n : n \geqslant 1\}$, and f is the nearest of the closing random variables, in the sense that if $\{f_n : n \geqslant 1\}$ is closed by some g, then $E^{\mathcal{B}} f \geqslant g$ a.s. Actually, we have $\sup_{n\in N} E|f_n| < \infty$ by (14.28), and so there is $f \in \mathfrak{L}^1$ closing $\{f_n : n \in N\}$ with $f_n \overset{a.s.}{\to} f$ by (30.50.ii). In view of (14.30), it follows that $f \in \mathfrak{L}^p$ and $f_n \overset{\mathfrak{L}^p}{\to} f$. Moreover, on account of (29.22), $E^{\mathcal{B}} f_n \overset{\mathfrak{L}^p}{\to} E^{\mathcal{B}} f$, and so $E^{\mathcal{B}} f_{n_k} \overset{a.s.}{\to} E^{\mathcal{B}} f$ for some subsequence $\{n_k : k \geqslant 1\} \subset N$. Consequently, if $E^{\mathcal{B}} f_n \geqslant g$ a.s., $n \in N$, for some $g \in \mathfrak{L}^1$, then $E^{\mathcal{B}} f \geqslant g$ a.s.; if $\{f_n : n \in N\}$ is a reverse martingale closed by g, then $E^{\mathcal{B}} f = g$ a.s.

We devote the next part of this section to applications of the basic convergence theorems for martingales and submartingales. We revert to the assumptions about $\{\mathcal{B}_n : n \in N\}$ and $\{f_n : n \in N\}$ at the beginning of the section, and denote $d_n = f_n - f_{n-1}, n \in N$, where $f_0 = 0$.

Theorem 30.53. *Let $\{f_n : n \in N\}$ be a submartingale with $E[\sup_{n\in N} d_n^+] < \infty$. Then $\{\{f_n : n \in N\}$ converges$\} \overset{a.s.}{=} \{\sup_{n\in N} f_n < \infty\}$.*

Proof. For $m \in N$, introduce the stopping time $\tau_m = \inf\{n \in N : f_n \geqslant m\}$. Then, according to (30.25), $\{f_{\tau_m \wedge n} : n \in N\}$ is also a submartingale. Since $f_{\tau_m\wedge n} = f_{\tau_m - 1} + d_{\tau_m} < m + d_{\tau_m}^+$ if $\tau_m \leqslant n$, while $f_{\tau_m \wedge n} < m$ if $\tau_m > n$, we see that $f_{\tau_m\wedge n}^+ < m + \sup_{n\in N} d_n^+$. Therefore, in view of (30.41.i), $\{f_{\tau_m\wedge n} : n \in N\}$ converges a.s. As $\{\sup_{n\in N} f_n < \infty\} = \cup_{m\in N}\{\tau_m = \infty\} = \cup_{m\in N}\{f_n = f_{\tau_m\wedge n}$ for all $n \in N\}$, it follows that $\{f_n : n \in N\}$ converges a.s. on $\{\sup_{n\in N} f_n < \infty\}$. The converse inclusion is trivial, since $\sup_{n\in N} f_n < \infty$ whenever $\{f_n : n \in N\}$ converges. □

This last result allows the following useful extension of the Borel-Cantelli lemma (18.24) due to P. Lévy.

Corollary 30.54. *For $n \geqslant 1$, let $B_n \in \mathcal{B}_n$. Then $\limsup_n B_n \overset{a.s.}{=} \left\{\sum_{n\geqslant 1} P^{\mathcal{B}_{n-1}}(B_n) = \infty\right\}$.*

Proof. Fix versions of $P^{\mathcal{B}_{n-1}}(B_n), n \geqslant 1$, and put $g_n = \sum_{i=1}^n (1_{B_i} - P^{\mathcal{B}_{i-1}}(B_i))$, $n \geqslant 1$. Then $\{g_n : n \geqslant 1\}$ is a martingale by (30.16.h) with $|g_n - g_{n-1}| \leqslant 1, n \geqslant 1$, where $g_0 = 0$. On account of (30.53), we have

$$\begin{aligned}(\limsup_n B_n)^c &= \left\{\sum_{n\geqslant 1} 1_{B_n} < \infty\right\} = \{\sup_{n\geqslant 1} g_n < \infty\} \cap \left\{\sum_{n\geqslant 1} 1_{B_n} < \infty\right\} \\ &\overset{a.s.}{=} \left\{\sum_{n\geqslant 1}(1_{B_n} - P^{\mathcal{B}_{n-1}}(B_n)) \text{ converges}\right\} \cap \left\{\sum_{n\geqslant 1} 1_{B_n} < \infty\right\} \\ &\subset \left\{\sum_{n\geqslant 1} P^{\mathcal{B}_{n-1}}(B_n) < \infty\right\}. \end{aligned} \tag{1}$$

Since $\{-g_n : n \geqslant 1\}$ is again a martingale, in view of (30.53), we have also

$$\begin{aligned}\left\{\sum_{n\geqslant 1} P^{\mathcal{B}_{n-1}}(B_n) < \infty\right\} &= \{\sup_{n\geqslant 1}(-g_n) < \infty\} \cap \left\{\sum_{n\geqslant 1} P^{\mathcal{B}_{n-1}}(B_n) < \infty\right\} \\ &\overset{a.s.}{=} \left\{\sum_{n\geqslant 1}(1_{B_n} - P^{\mathcal{B}_{n-1}}(B_n)) \text{ converges}\right\} \cap \left\{\sum_{n\geqslant 1} 1_{B_n} < \infty\right\} \\ &\subset \left\{\sum_{n\geqslant 1} 1_{B_n} < \infty\right\} = (\limsup_n B_n)^c. \end{aligned} \tag{2}$$

The assertion follows from (1) and (2). □

Lemma 30.55. *Let $\{f_n : n \in N\} \subset \mathfrak{L}^2$ be a martingale, let $i, j \in N$ with $i < j$, and let $g \in \mathfrak{L}^2(X, \mathcal{B}_i, P)$. Then $E[gd_j] = 0$; in particular, $E[d_i d_j] = 0$.*

Proof. We have $E[gd_j] = E[E^{\mathcal{B}_i}[gd_j]] = E[gE^{\mathcal{B}_i}[f_j - f_{j-1}]] = E[g(f_i - f_i)] = 0$. □

Theorem 30.56. *Let $\{f_n : n \geqslant 1\} \subset \mathfrak{L}^2$ be a martingale.*

(i) $\left\{\sum_{n\geqslant 1} E^{\mathcal{B}_{n-1}} d_n^2 < \infty\right\} \overset{a.s.}{\subset} \{\{f_n : n \geqslant 1\}\ converges\}$.

(ii) *If* $E[\sup_{n\geqslant 1} d_n^2] < \infty$, *then* $\{\{f_n : n \geqslant 1\}\ converges\} \overset{a.s.}{\subset} \left\{\sum_{n\geqslant 1} E^{\mathcal{B}_{n-1}} d_n^2 < \infty\right\}$.

Proof. Fix versions of $E^{\mathcal{B}_{n-1}} d_n^2, n \geqslant 1$. For $m \in N$, define the stopping time $\tau_m = \inf\{n \geqslant 1 : \sum_{i=1}^{n+1} E^{\mathcal{B}_{i-1}} d_i^2 \geqslant m\}$. Then $\{f_{\tau_m \wedge n} : n \geqslant 1\}$ is a martingale by (30.25), and $f_{\tau_m \wedge n} = \sum_{i=1}^{n} d_i 1_{\{\tau_m \geqslant i\}}$. Applying (30.55), we have

$$\begin{aligned} Ef_{\tau_m \wedge n}^2 &= \sum_{i=1}^{n} E\left[d_i^2 1_{\{\tau_m \geqslant i\}}\right] = \sum_{i=1}^{n} E\left[E^{\mathcal{B}_{i-1}}[d_i^2 1_{\{\tau_m \geqslant i\}}]\right] \\ &= E\left[\sum_{i=1}^{n} 1_{\{\tau_m \geqslant i\}} E^{\mathcal{B}_{i-1}} d_i^2\right] = E\left[\sum_{i=1}^{\tau_m \wedge n} E^{\mathcal{B}_{i-1}} d_i^2\right] \leqslant m. \end{aligned}$$

Hence $\sup_{n\geqslant 1} E\left|f_{\tau_m \wedge n}\right| \leqslant \sqrt{m}$ by Schwarz's inequality, and so $\{f_{\tau_m \wedge n} : n \geqslant 1\}$ converges a.s. by (30.41.i). Since $\{\sum_{n\geqslant 1} E^{\mathcal{B}_{n-1}} d_n^2 < \infty\} = \cup_{m\in N}\{\tau_m = \infty\} = \cup_{m\in N}\{f_n = f_{\tau_m \wedge n}$ for all $n \geqslant 1\}$, it follows that $\{f_n : n \geqslant 1\}$ converges a.s. on $\{\sum_{n\geqslant 1} E^{\mathcal{B}_{n-1}} d_n^2 < \infty\}$. This proves (i). Next, for $m \in N$, consider the stopping time $\sigma_m = \inf\{n \geqslant 1 : |f_n| \geqslant m\}$. Then $\{f_{\sigma_m \wedge n} : n \geqslant 1\}$ is a martingale, $f_{\sigma_m \wedge n} = \sum_{i=1}^{n} d_i 1_{\{\sigma_m \geqslant i\}}$ and $\left|f_{\sigma_m \wedge n}\right| < m + \sup_{n\geqslant 1} |d_n|$. Hence $\{f_{\sigma_m \wedge n}^2 : n \geqslant 1\}$ is uniformly integrable. Therefore, in view of (30.44), there is $f_{(m)} \in \mathfrak{L}^2$ such that $f_{\sigma_m \wedge n} \overset{a.s.}{\to} f_{(m)}$ and $Ef_{\sigma_m \wedge n}^2 \to Ef_{(m)}^2 < \infty$. As $Ef_{\sigma_m \wedge n}^2 = E[\sum_{i=1}^{n} 1_{\{\sigma_m \geqslant i\}} E^{\mathcal{B}_{i-1}} d_i^2]$ by (30.55), we see that $E[\sum_{i\geqslant 1} 1_{\{\sigma_m \geqslant i\}} E^{\mathcal{B}_{i-1}} d_i^2] < \infty$, and so $\sum_{i\geqslant 1} 1_{\{\sigma_m \geqslant i\}} E^{\mathcal{B}_{i-1}} d_i^2 < \infty$ a.s. Since $\{\{f_n : n \geqslant 1\}$ converges$\} \subset \cup_{m\in N}\{\sigma_m = \infty\}$, it follows that $\sum_{i\geqslant 1} E^{\mathcal{B}_{i-1}} d_i^2 < \infty$ a.s. on $\{\{f_n : n \geqslant 1\}$ converges$\}$. □

Corollary 30.57. *Let $\{f_n : n \geqslant 1\}$ be as in (30.56).*

(i) $\left\{\sum_{n\geqslant 1} d_n^2 < \infty\right\} \overset{a.s.}{\subset} \{\{f_n : n \geqslant 1\}\ converges\}$.

(ii) *If* $E[\sup_{n\geqslant 1} d_n^2] < \infty$, *then* $\{\{f_n : n \geqslant 1\}\ converges\} \overset{a.s.}{\subset} \left\{\sum_{n\geqslant 1} d_n^2 < \infty\right\}$.

Proof. For $m \in N$, putting $\tau_m = \inf\{n \geqslant 1 : \sum_{i=1}^{n} d_i^2 \geqslant m\}$ and $\sigma_m = \inf\{n \geqslant 1 : |f|_n \geqslant m\}$, the proof parallels closely the proof of (30.56). □

Remarks 30.58. Let $d_n, n \geqslant 1$, be independent random variables with $Ed_n = 0, n \geqslant 1$, and put $f_n = d_1 + \cdots + d_n, n \geqslant 1$. Then $\{f_n : n \geqslant 1\}$ is a $\{\sigma(d_1, \ldots, d_n)\}$-martingale by (30.16.d).

(a) The Khintchine-Kolmogorov theorem (26.3) asserts that $\{f_n : n \geqslant 1\}$ converges a.s. whenever $\sum_{n\geqslant 1} Ed_n^2 < \infty$. This result follows as well from (30.57.i), since $\sum_{n\geqslant 1} Ed_n^2 < \infty$ implies $\sum_{n\geqslant 1} d_n^2 < \infty$ a.s.

(b) An improvement over (26.7) is possible when $E[\sup_{n\geqslant 1} d_n^2] < \infty$, namely that $\sum_{n\geqslant 1} d_n$ converges a.s. if and only if $\sum_{n\geqslant 1} \operatorname{Var} d_n$ and $\sum_{n\geqslant 1} Ed_n$ converge. Actually, if $E[\sup_{n\geqslant 1} d_n^2] < \infty$, and $\sum_{n\geqslant 1} d_n$ converges a.s., then $\sum_{n\geqslant 1} \operatorname{Var} d_n \leqslant \sum_{n\geqslant 1} Ed_n^2 = \sum_{n\geqslant 1} E^{\mathcal{B}_{n-1}} d_n^2 < \infty$ a.s. by (29.8.v) and (30.56.ii). Consequently, $\sum_{n\geqslant 1}(d_n - Ed_n)$ converges a.s. by (26.3), and so $\sum_{n\geqslant 1} Ed_n$ converges. The converse implication does not make use of the condition $E[\sup_{n\geqslant 1} d_n^2] < \infty$ and is proved as in (26.7).

In order to obtain the conditional version of half of the three series Theorem (26.10), we need an auxiliary result.

Lemma 30.59. *Assume that $\{f_n : n \geqslant 1\}$ is adapted, let $\{c_n : n \geqslant 1\} \subset]0, \infty]$, and put $A = \left\{\sum_{n\geqslant 1} P^{\mathcal{B}_{n-1}}(|f_n| \geqslant c_n) < \infty\right\}$. Then*

$$\left\{\sum_{n\geqslant 1} f_n converges\right\} \cap A \overset{a.s.}{=} \left\{\sum_{n\geqslant 1} f_n 1_{\{|f_n|<c_n\}} converges\right\} \cap A. \tag{1}$$

Proof. According to (30.54), $A \overset{a.s.}{=} \liminf_n\{|f_n| < c_n\} \subset \liminf_n\{f_n = f_n 1_{|f_n|<c_n}\}$, which implies (1). □

Theorem 30.60. *Assume that $\{f_n : n \geqslant 1\}$ is adapted , let $c \in]0, \infty[$, and put $A = \left\{\sum_{n\geqslant 1} P^{\mathcal{B}_{n-1}}(|f_n| \geqslant c) < \infty\right\}$, $B = \left\{\sum_{n\geqslant 1} E^{\mathcal{B}_{n-1}}[f_n 1_{\{|f_n|<c\}}] \; converges\right\}$, $C = \left\{\sum_{n\geqslant 1} \operatorname{Var}^{\mathcal{B}_{n-1}}[f_n 1_{\{|f_n|<c\}}] \; converges\right\}$. Then $A\cap B\cap C \overset{a.s.}{\subset} \left\{\sum_{n\geqslant 1} f_n \; converges\right\}$.*

Proof. For $n \geqslant 1$, set $d_n = f_n 1_{\{|f_n|<c\}} - E^{\mathcal{B}_{n-1}}[f_n 1_{\{|f_n|<c\}}]$ and $g_n = \sum_{i=1}^n d_i$. Then $|d_n| \leqslant 2c, n \geqslant 1, \{g_n : n \geqslant 1\}$ is a martingale by (30.16.h), $\left\{\sum_{n\geqslant 1} f_n 1_{\{|f_n|<c_n\}} \; converges\right\} \cap B = \left\{\sum_{n\geqslant 1} d_n \; converges\right\} \cap B$ and $C = \left\{\sum_{n\geqslant 1} E^{\mathcal{B}_{n-1}} d_n^2 < \infty\right\}$. Therefore, on account of (30.59) and (30.56), we get

$$\left\{\sum_{n\geqslant 1} f_n \text{converges}\right\} \cap A \cap B \cap C = \left\{\sum_{n\geqslant 1} f_n 1_{\{|f_n|<c\}} \text{converges}\right\} \cap A \cap B \cap C$$

$$= \left\{\sum_{n\geqslant 1} d_n \text{converges}\right\} \cap A \cap B \cap C \overset{a.s.}{=} A \cap B \cap C,$$

whence the desired inclusion follows. □

Remark 30.61. Comparing (26.10) with (30.60) raises the question whether $\left\{\sum_{n\geqslant 1} f_n \text{ converges}\right\} \overset{a.s.}{\subset} A \cap B \cap C$. For any $c \in]0, \infty[$, according to (30.54), we always have $\left\{\sum_{n\geqslant 1} f_n \text{ converges}\right\} \subset (\limsup_n\{|f_n| \geqslant c\})^c \overset{a.s.}{=}$

$\left\{\sum_{n\geqslant 1} P^{\mathcal{B}_{n-1}}(|f_n| \geqslant c) < \infty\right\} = A$. However, as the next example shows, the inclusion $\left\{\sum_{n\geqslant 1} f_n \text{converges}\right\} \overset{a.s.}{\subset} B \cap C$ does not hold in general. Let $h_n, n \geqslant 0$, be independent random variables with $h_0 = 0$ and $P(h_n = 1/\sqrt{n}) = P(h_n = -1/\sqrt{n}) = 1/2, n \geqslant 1$, define $f_n = h_n - h_{n-1}, n \geqslant 1$, and let $\mathcal{B}_n = \sigma(f_1, \ldots, f_n) = \sigma(h_1, \ldots, h_n), n \geqslant 1$. Then $\left|\sum_{i=1}^n f_i\right| = |h_n| \overset{a.s.}{\to} 0$, and so $\sum_{n\geqslant 1} f_n = 0$ a.s. For $n \geqslant 1$, $E^{\mathcal{B}_{n-1}} h_n = 0$ a.s. and $E^{\mathcal{B}_{n-1}} h_n^2 = E h_n^2 = 1/n$ a.s., and so $E^{\mathcal{B}_{n-1}}[f_n 1_{\{|f_n|<2\}}] = E^{\mathcal{B}_{n-1}} f_n = -h_{n-1}$ a.s. and $\operatorname{Var}^{\mathcal{B}_{n-1}}[f_n 1_{\{|f_n|<2\}}] = E^{\mathcal{B}_{n-1}} f_n^2 - (E^{\mathcal{B}_{n-1}} f_n)^2 = E^{\mathcal{B}_{n-1}} h_n^2 = 1/n$ a.s. Consequently, $\sum_{n\geqslant 1} \operatorname{Var}^{\mathcal{B}_{n-1}}[f_n 1_{\{|f_n|<2\}}] = \infty$ a.s., i.e. $C \overset{a.s.}{=} \emptyset$ in this case. Moreover, $\sum_{n\geqslant 1} \operatorname{Var} h_n = \infty$, and so $\sum_{n\geqslant 1} E^{\mathcal{B}_{n-1}}[f_n 1_{\{|f_n|<2\}}] = -\sum_{n\geqslant 1} h_n$ a.s. is divergent almost surely by (26.7), i.e. also $B \overset{a.s.}{=} \emptyset$ in this case.

Corollary 30.62 (Chow). *Let $\{f_n : n \geqslant 1\} \subset \mathfrak{L}^p$ be a martingale, where $1 \leqslant p \leqslant 2$. Then $\left\{\sum_{n\geqslant 1} E^{\mathcal{B}_{n-1}} |d_n|^p < \infty\right\} \overset{a.s.}{\subset} \{\{f_n : n \geqslant 1\}$ converges$\}$.*

Proof. Write $\Lambda = \{\sum_{n\geqslant 1} E^{\mathcal{B}_{n-1}} |d_n|^p < \infty\}$, let $c \in]0, \infty[$, and let A, B, C be as in (30.60) with f_n replaced by d_n. We have

$$P^{\mathcal{B}_{n-1}}(|d_n| \geqslant c) = E^{\mathcal{B}_{n-1}} 1_{\{|d_n|\geqslant c\}} \leqslant E^{\mathcal{B}_{n-1}}[|d_n|\,|^p 1_{\{|d_n|\geqslant c\}}]/c^p$$
$$\leqslant E^{\mathcal{B}_{n-1}} |d_n|^p / c^p \text{ a.s.,} \quad n \geqslant 1,$$

and so $\Lambda \overset{a.s.}{\subset} A$. Next, for $n \geqslant 1$, as $0 = E^{\mathcal{B}_{n-1}} d_n = E^{\mathcal{B}_{n-1}}[d_n 1_{\{|d_n|<c\}}] + E^{\mathcal{B}_{n-1}}[d_n 1_{\{|d_n|\geqslant c\}}]$ a.s. and $p \geqslant 1$, we get

$$\left|E^{\mathcal{B}_{n-1}}[d_n 1_{\{|d_n|<c\}}]\right| = \left|E^{\mathcal{B}_{n-1}}[d_n 1_{\{|d_n|\geqslant c\}}]\right| \leqslant E^{\mathcal{B}_{n-1}}[|d_n|\, 1_{\{|d_n|\geqslant c\}}]$$
$$\leqslant E^{\mathcal{B}_{n-1}} |d_n|^p / c^{p-1} \text{ a.s.,}$$

whence $\Lambda \overset{a.s.}{\subset} B$. Finally, since $p \leqslant 2$,

$$\operatorname{Var}^{\mathcal{B}_{n-1}}[d_n 1_{\{|d_n|<c\}}] \leqslant E^{\mathcal{B}_{n-1}}[d_n^2 1_{\{|d_n|<c\}}] \leqslant E^{\mathcal{B}_{n-1}} |d_n|^p / c^{p-2} \text{ a.s.,} \quad n \geqslant 1,$$

and so $\Lambda \overset{a.s.}{\subset} C$. Consequently, according to (30.60), $\Lambda \overset{a.s.}{\subset} \{\{f_n : n \geqslant 1\}$ converges$\}$. □

Remark 30.63. Let $\{f_n : n \geqslant 1\} \subset \mathfrak{L}^p$ be a martingale, where $p \in]2, \infty[$. The plausible conjecture that $\left\{\sum_{n\geqslant 1} E^{\mathcal{B}_{n-1}} |d_n|^p < \infty\right\} \overset{a.s.}{\subset} \{\{f_n : n \geqslant 1\}$ converges$\}$ is false. Actually, let $d_n, n \geqslant 1$, be independent random variables with $P(d_n = 1/\sqrt{n}) = P(d_n = -1/\sqrt{n}) = 1/2, n \geqslant 1$, set $f_n = d_1 + \cdots + d_n, n \geqslant 1$, and let $\mathcal{B}_n = \sigma(d_1, \ldots, d_n), n \geqslant 1$. Then $\sum_{n\geqslant 1} E^{\mathcal{B}_{n-1}} |d_n|^p = \sum_{n\geqslant 1} E |d_n|^p = \sum_{n\geqslant 1} \frac{1}{n^{p/2}} < \infty$ a.s., and so $\left\{\sum_{n\geqslant 1} E^{\mathcal{B}_{n-1}} |d_n|^p < \infty\right\} \overset{a.s.}{=} X$. On the other hand, $E[\sup_{n\geqslant 1} d_n^2] = 1$ and $\sum_{n\geqslant 1} E^{\mathcal{B}_{n-1}} d_n^2 = \sum_{n\geqslant 1} \frac{1}{n} = \infty$ a.s., whence $\{\{f_n : n \geqslant 1\}$ converges$\} \overset{a.s.}{=} \emptyset$ by (30.56.ii).

In Section 26 we saw that convergence in distribution, convergence in probability, and a.s. convergence are equivalent for series of independent random variables. While convergence in $\mathfrak{L}^p$ always implies convergence in probability by (14.14), the following theorem completes the picture. (See also (26.24).)

Theorem 30.64 (Marcinkiewicz). *Let $f_n, n \geqslant 1$, be independent random variables, assume that $Ef_n = 0$ and $f_n \in \mathfrak{L}^p$, where $1 \leqslant p < \infty$, put $S_n = f_1 + \cdots + f_n$, and let $S \in \mathfrak{L}^p$. Then $S_n \overset{\mathfrak{L}^p}{\to} S$ if and only if $S_n \overset{a.s.}{\to} S$.*

Proof. Recall that $\{S_n : n \geqslant 1\}$ is a $\{\sigma(f_1, \ldots, f_n)\}$-martingale (30.16.d). If $S_n \overset{\mathfrak{L}^p}{\to} S$, then $S_n \overset{a.s.}{\to} S$ by (30.44).

Conversely, suppose that $S_n \overset{a.s.}{\to} S$. We may and do assume that $S = \limsup_n S_n$. Then, for fixed $m \geqslant 1$, $S - S_m = \limsup_n S_{m+n} - S_m = \limsup_n (S_{m+n} - S_m)$. As $S_{m+n} - S_m$ is $\sigma(f_{m+i}, i \geqslant 1)$-measurable for each $n \geqslant 1$, it follows that $\sigma(S - S_m)$ and $\mathcal{B}_m$ are independent by (8.12.i). Therefore,

$$E^{\mathcal{B}_m} S = E^{\mathcal{B}_m}[S - S_m] + E^{\mathcal{B}_m} S_m = E[S - S_m] + S_m = ES + S_m \text{ a.s.}, \quad m \geqslant 1. \quad (1)$$

Letting $m \to \infty$ in (1), and using (29.28), we find $ES = 0$, and so $E^{\mathcal{B}_m} S = S_m$ a.s., $m \geqslant 1$. Moreover, $S_m \overset{\mathfrak{L}^p}{\to} S$ once more by (29.28). □

In the final part of this section we generalize Wald's theorems (30.12) and (30.13). For this, we first derive a version of the optional sampling theorem under conditions different from those in (30.29).

Theorem 30.65. *Let τ and τ' be finite stopping times.*

(i) *If $\{f_n : n \geqslant 1\}$ is a martingale with $f_{\tau'} \in \mathfrak{L}^1$ and $\liminf_n \int_{\{\tau' > n\}} |f_n| \, dP = 0$, then $E^{\mathcal{B}_\tau} f_{\tau'} = f_{\tau \wedge \tau'}$ a.s.*

(ii) *If $\{f_n : n \geqslant 1\}$ is a submartingale with $f_{\tau'} \in \mathfrak{L}^1$ and $\liminf_n \int_{\{\tau' > n\}} f_n^+ \, dP = 0$, then $E^{\mathcal{B}_\tau} f_{\tau'} \geqslant f_{\tau \wedge \tau'}$ a.s.*

Proof. We prove only (ii), as the proof of (i) is similar. For $l \geqslant 1$ and $B \in \mathcal{B}_l$, we have

$$\begin{aligned} \int_{B \cap \{\tau' \geqslant l\}} f_l \, dP &= \int_{B \cap \{\tau' = l\}} f_l \, dP + \int_{B \cap \{\tau' > l\}} f_l \, dP \leqslant \int_{B \cap \{\tau' = l\}} f_l \, dP \\ &\quad + \int_{B \cap \{\tau' > l\}} E^{\mathcal{B}_l} f_{l+1} \, dP = \int_{B \cap \{\tau' = l\}} f_{\tau'} \, dP + \int_{B \cap \{\tau' \geqslant l+1\}} f_{l+1} \, dP, \end{aligned}$$

whence we inductively obtain, since $B \in \mathcal{B}_m$ for all $m \geqslant l$,

$$\int_{B \cap \{\tau' \geqslant l\}} f_l \, dP \leqslant \int_{B \cap \{l \leqslant \tau' \leqslant n\}} f_{\tau'} \, dP + \int_{B \cap \{\tau' > n\}} f_n \, dP, \quad n \geqslant l. \quad (1)$$

Since $\int_X |f_{\tau'}| \, dP < \infty$ and $\int_{B \cap \{\tau' > n\}} f_n \, dP \leqslant \int_{B \cap \{\tau' > n\}} f_n^+ \, dP$, letting $n \to \infty$ through a subsequence in (1), and using the dominated convergence theorem (11.25.ii), we obtain

$$\int_{B\cap\{\tau'\geqslant l\}} f_l\,dP \leqslant \int_{B\cap\{\tau'\geqslant l\}} f_{\tau'}\,dP. \tag{2}$$

Replacing B in (2) by $B \cap \{\tau = l\}$, where $B \in \mathcal{B}_\tau$, we get

$$\begin{aligned}\int_{B\cap\{\tau=l\}\cap\{\tau'\geqslant\tau\}} f_{\tau'}\,dP &= \int_{B\cap\{\tau=l\}\cap\{\tau'\geqslant l\}} f_{\tau'}\,dP \geqslant \int_{B\cap\{\tau=l\}\cap\{\tau'\geqslant l\}} f_l\,dP \\ &= \int_{B\cap\{\tau=l\}\cap\{\tau'\geqslant\tau\}} f_\tau\,dP, \quad l \geqslant 1. \end{aligned}\tag{3}$$

Since $\{f_n^+ : n \geqslant 1\}$ is a submartingale by (30.22.iii), we also have

$$\int_{B\cap\{\tau=l\}\cap\{\tau'\geqslant\tau\}} f_{\tau'}^+\,dP \geqslant \int_{B\cap\{\tau=l\}\cap\{\tau'\geqslant\tau\}} f_\tau^+\,dP, \quad l \geqslant 1.$$

This, in conjunction with (11.28), permits summation on l in (3) to obtain $\int_{B\cap\{\tau'\geqslant\tau\}} f_{\tau'}\,dP \geqslant \int_{B\cap\{\tau'\geqslant\tau\}} f_\tau\,dP = \int_{B\cap\{\tau'\geqslant\tau\}} f_{\tau\wedge\tau'}\,dP$, which, in view of (30.6.v) and (30.6.ii), shows that

$$1_{\{\tau'\geqslant\tau\}} E^{\mathcal{B}_\tau} f_{\tau'} \geqslant 1_{\{\tau'\geqslant\tau\}} f_{\tau\wedge\tau'} \text{ a.s.} \tag{4}$$

On the other hand, we know that

$$1_{\{\tau'<\tau\}} E^{\mathcal{B}_\tau} f_{\tau'} = 1_{\{\tau>\tau'\}} E^{\mathcal{B}_\tau} f_{\tau\wedge\tau'} = 1_{\{\tau'<\tau\}} f_{\tau\wedge\tau'} \text{ a.s.} \tag{5}$$

by (2) in the proof of (30.23). Adding (4) and (5) yields the result. □

Corollary 30.66. *Let τ and τ' be finite stopping times, and let $\{f_n : n \geqslant 1\}$ be a submartingale with $\sum_{j=1}^{\tau'} E^{\mathcal{B}_{j-1}} |d_j| \in \mathfrak{L}^1$. Then $E^{\mathcal{B}_\tau} f_{\tau'} \geqslant f_{\tau\wedge\tau'}$ a.s., and $E^{\mathcal{B}_\tau} f_{\tau'} = f_{\tau\wedge\tau'}$ a.s. if $\{f_n : n \geqslant 1\}$ is a martingale.*

Proof. To prove the second assertion, on account of (11.23) and (29.16), we write

$$\begin{aligned} E\,|f_{\tau'}| &\leqslant E\left[\sum_{j=1}^{\tau'} |d_j|\right] = E\left[\sum_{j\geqslant 1} |d_j|\, 1_{\{\tau'\geqslant j\}}\right] \\ &= \sum_{j\geqslant 1} E[|d_j|\, 1_{\{\tau'\geqslant j\}}] = \sum_{j\geqslant 1} E[E^{\mathcal{B}_{j-1}}[|d_j|\, 1_{\{\tau'\geqslant j\}}]] \\ &= E\left[\sum_{j\geqslant 1} 1_{\{\tau'\geqslant j\}} E^{\mathcal{B}_{j-1}} |d_j|\right] = \sum_{j=1}^{\tau'} E^{\mathcal{B}_{j-1}} |d_j|, \end{aligned}$$

and so $f_{\tau'} \in \mathfrak{L}^1$ and $\sum_{j=1}^{\tau'} |d_j| \in \mathfrak{L}^1$. Consequently,

$$\int\limits_{\{\tau'>n\}} |f_n|\, dP \leqslant \int\limits_{\{\tau'>n\}} \left(\sum_{j=1}^{n} |d_j|\right) dP \leqslant \int\limits_{\{\tau'>n\}} \left(\sum_{j=1}^{\tau'} |d_j|\right) dP \to 0$$

by dominated convergence. Hence the result follows by (30.65.i). The proof of the first assertion is similar. □

The next remark should be related to Wald's theorem (30.12).

Remark 30.67. Let τ' be a finite stopping time, and let $\{f_n : n \geqslant 1\}$ be a martingale. If either $f_{\tau'} \in \mathfrak{L}^1$ and $\liminf_n \int_{\{\tau'>n\}} |f_n|\, dP = 0$, or $E[\sum_{j=1}^{\tau'} |d_j|] < \infty$, then $E^{\mathcal{B}_n} f_{\tau'} = f_{\tau' \wedge n}$ a.s., $n \geqslant 1$, and $E f_{\tau'} = E f_1$. In particular, if $d_n, n \geqslant 1$, are independent with $E d_n = 0, n \geqslant 1$, and $E[\sum_{j=1}^{\tau'} |d_j|] < \infty$, then $E[\sum_{j=1}^{\tau'} d_j] = 0$.

Lemma 30.68. *Let τ' be a finite stopping time, let $\{f_n : n \geqslant 1\} \subset \mathfrak{L}^2$ be a martingale, and consider the following conditions:* (i) $E[\sum_{j=1}^{\tau'} d_j^2] < \infty$, (ii) $\liminf_n \int_{\{\tau'>n\}} f_n^2 dP < \infty$, (iii) $\liminf_n \int_{\{\tau'>n\}} |f_n|\, dP = 0$, (iv) $E[\sum_{j=1}^{\tau'} |d_j|] < \infty$. *Then* (i) $\Longrightarrow$ (ii) $\Longrightarrow$ (iii) *and* (iv) $\Longrightarrow$ (iii).

Proof. If (i) holds, then, for $n \geqslant 1$,

$$\begin{aligned}
\int\limits_{\{\tau'>n\}} f_n^2\, dP &= \sum_{j=1}^{n} \Bigg(\int\limits_{\{\tau'>j\}} f_j^2\, dP - \int\limits_{\{\tau'>j-1\}} f_{j-1}^2\, dP \Bigg) \\
&\leqslant \sum_{j=1}^{n} \int\limits_{\{\tau'\geqslant j\}} (f_j^2 - f_{j-1}^2) dP = \sum_{j=1}^{n} \int\limits_{\{\tau'\geqslant j\}} d_j^2\, dP \\
&\leqslant E\Bigg[\sum_{j\geqslant 1} d_j^2 1_{\{\tau'\geqslant j\}} \Bigg] = E\left[\sum_{j=1}^{\tau'} d_j^2 \right] < \infty,
\end{aligned}$$

as $E[f_{j-1} 1_{\{\tau'\geqslant j\}} d_j] = 0, 1 \leqslant j \leqslant n$, by (30.55). Hence (i) $\Longrightarrow$ (ii). Next, to get a contradiction, suppose that $\liminf_n \int_{\{\tau'>n\}} |f_n|\, dP = a > 0$, and let $b > 0$. Then, since $\int_{\{\tau'>n, |f_n|<b\}} |f_n|\, dP \leqslant b P(\tau' > n) \to 0$, we have

$$\liminf_n \int\limits_{\{\tau'>n\}} f_n^2 dP \geqslant b \liminf_n \int\limits_{\{\tau'>n, |f_n|\geqslant b\}} |f_n|\, dP = ab \to \infty \text{ as } b \to \infty.$$

Thus (ii) $\Longrightarrow$ (iii). In the proof of (30.66) we already verified that (iv) $\Longrightarrow$ $\lim_n \int_{\{\tau'>n\}} |f_n|\, dP = 0$. □

Theorem 30.69. *Let τ' and $\{f_n : n \geqslant 1\}$ be as in* (30.68), *and assume that any of the conditions* (i)–(iv) *in* (30.68) *holds. Then*

$$Ef_{\tau'}^2 = E\left[\sum_{j=1}^{\tau'} d_j^2\right]. \tag{1}$$

Proof. For $n \geqslant 1$, we have $f_{\tau' \wedge n} = \sum_{j=1}^{n} d_j 1_{\{\tau' \geqslant j\}}$, whence

$$Ef_{\tau' \wedge n}^2 = \sum_{j=1}^{n} E[d_j^2 1_{\{\tau' \geqslant j\}}] = E\left[\sum_{j=1}^{\tau' \wedge n} d_j^2\right] \tag{2}$$

by (30.55). Therefore, as $\tau' \wedge n \to \tau'$, we get

$$Ef_{\tau'}^2 \leqslant E\left[\sum_{j=1}^{\tau'} d_j^2\right]. \tag{3}$$

Consequently, if $Ef_{\tau'}^2 = \infty$, then (1) is verified. If $Ef_{\tau'}^2 < \infty$, then, since $\liminf_n \int_{\{\tau' > n\}} |f_n|\, dP = 0$ by (30.68), we have $E^{\mathcal{B}_n} f_{\tau'} = f_{\tau' \wedge n}$ a.s., $n \geqslant 1$, by (30.65.ii). Hence, in view of (29.20), $E^{\mathcal{B}_n} f_{\tau'}^2 \geqslant (E^{\mathcal{B}_n} f_{\tau'})^2 = f_{\tau' \wedge n}^2$ a.s., $n \geqslant 1$, whence, using (2),

$$Ef_{\tau'}^2 \geqslant Ef_{\tau' \wedge n}^2 = E\left[\sum_{j=1}^{\tau' \wedge n} d_j^2\right], \quad n \geqslant 1. \tag{4}$$

Letting $n \to \infty$ in (4), by (11.22), we obtain

$$Ef_{\tau'}^2 \geqslant E\left[\sum_{j=1}^{\tau'} d_j^2\right]. \tag{5}$$

Finally, (1) follows from (3), (4) and (5). □

Theorem (30.69) as well as the next remark should be compared with Wald's theorem (30.13).

Remark 30.70. Let d_n, $n \geqslant 1$, be independent random variables with $Ed_n^2 < \infty$ and $Ed_n = 0$, and let τ' be a finite $\{\sigma(d_1, \ldots, d_n)\}$-time. If either $E[\sum_{j=1}^{\tau'} d_j^2] < \infty$ or $E[\sum_{j=1}^{\tau'} |d_j|] < \infty$, then $E(\sum_{j=1}^{\tau'} d_j)^2 = E[\sum_{j=1}^{\tau'} d_j^2]$.

Exercise 30.71. Let τ be a stopping time, let $\{n_k : k \in N\} \subset N$ be such that $n_k \geqslant k$ for any $k \in N$, and define $n_\tau = n_k 1_{\{\tau = k\}} + \infty \cdot 1_{\{\tau = \infty\}}$. Show that n_τ is a $\mathcal{B}_\tau$-measurable stopping time.

Exercise 30.72. Let $(X, \mathcal{A}) = (]0,1], \mathcal{B}(]0,1]))$, consider $\mathcal{B}_1 = \{\emptyset, X\}$ and $\mathcal{B}_n = \sigma(\{]0, \frac{1}{n}],]\frac{1}{n}, \frac{1}{n-1}], \ldots,]\frac{1}{2}, 1]\}, n \geqslant 2$, define $\tau = \sum_{n\geqslant 2} n 1_{]\frac{1}{n}, \frac{1}{n-1}]}$, and let $B = \bigcup_{k\geqslant 1}\{\tau = 2k\}$. Prove the following.

(a) $\{\mathcal{B}_n : n \geqslant 1\}$ is a nondecreasing sequence.

(b) τ is a $\{\mathcal{B}_n\}$-time and $B \in \mathcal{B}_\tau$.

(c) Whatever $n \geqslant 1$, $B \cap \{\tau > n\} \notin \mathcal{B}_n$.

(d) Contrast (c) with (30.5.a).

Exercise 30.73. Let τ be a $\{\mathcal{B}_n\}$-time, and let τ' be a $\{\mathcal{C}_n\}$-time, where $\mathcal{C}_n = \mathcal{B}_{\tau+n}$, $n \in N$.

(a) Show that $\tau + \tau'$ is a $\{\mathcal{B}_n\}$-time.

(b) Verify that $\mathcal{B}_{\tau+\tau'} = \mathcal{C}_{\tau'}$.

Exercise 30.74. Let τ and τ' be $\{\mathcal{B}_n\}$-times with $\tau < \tau'$, and set $\sigma = (\tau' - \tau)1_{\{\tau<\infty\}} + \infty \cdot 1_{\{\tau=\infty\}}$. Prove that σ is a $\{\mathcal{B}_{\tau+n}\}$-time.

Exercise 30.75. Let τ be a $\{\mathcal{B}_n\}$-time, and let $\tau' : X \to N \cup \{\infty\}$ be an $\overline{R}$-valued random variable such that $\tau + \tau'$ is a $\{\mathcal{B}_n\}$-time. Show that τ' is a $\{\mathcal{B}_{\tau+n}\}$-time.

Exercise 30.76. Let τ and τ' be stopping times. Use (30.6.iv) to prove that $\mathcal{B}_{\tau\vee\tau'} = \{C_1 \cup C_2 : C_1 \in \{\tau \leqslant \tau'\} \cap \mathcal{B}_{\tau'}, C_2 \in \{\tau > \tau'\} \cap \mathcal{B}_\tau\}$ and $\mathcal{B}_{\tau\wedge\tau'} = \{C_1 \cup C_2 : C_1 \in \{\tau \leqslant \tau'\} \cap \mathcal{B}_\tau, C_2 \in \{\tau > \tau'\} \cap \mathcal{B}_{\tau'}\}$. [Hint. If $C \in \mathcal{B}_{\tau\vee\tau'}$ or $C \in \mathcal{B}_{\tau\wedge\tau'}$, write $C = \{\tau \leqslant \tau'\}C \cup \{\tau > \tau'\}C$.]

Exercise 30.77. Let τ, τ' and f be as in (30.6). Show that $E^{\mathcal{B}_\tau}[E^{\mathcal{B}_{\tau'}} f] = E^{\mathcal{B}_{\tau'}}[E^{\mathcal{B}_\tau} f] = E^{\mathcal{B}_{\tau\wedge\tau'}} f$ a.s.

Exercise 30.78. Let τ and τ' be bounded stopping times with $\tau \leqslant \tau'$. Prove that there exist finitely many stopping times $\tau = \tau_0 \leqslant \cdots \leqslant \tau_n = \tau'$ such that τ_i is $\mathcal{B}_{\tau_{i-1}}$-measurable for $1 \leqslant i \leqslant n$.

Exercise 30.79. Let $(Z, \mathcal{C}, \mathsf{P})$ be a probability space, put $(Z_i, \mathcal{C}_i, \mathsf{P}_i) = (Z, \mathcal{C}, \mathsf{P})$, $i \geqslant 1$, let $(X, \mathcal{A}, P) = (\prod_{i\geqslant 1} Z_i, \otimes_{i\geqslant 1}\mathcal{C}_i, \otimes_{i\geqslant 1}\mathsf{P}_i)$, let $\mathcal{B}_n = \sigma(\pi_1, \ldots, \pi_n), n \geqslant 1$, where π_i denotes the projection from X onto Z_i, let τ be a finite stopping time, and let $f = (\pi_{\tau+i})_{i\geqslant 1}$. Prove the following.

(a) f is measurable.

(b) $\mathcal{B}_\tau$ is independent of $f^{-1}(\mathcal{A})$, and $Pf^{-1} = P$. [Use (30.11) and (18.29).]

(c) $f_n^{-1}(\mathcal{B}_\tau), n \geqslant 0$, are independent, where $f_0 = i_X$ and $f_{n+1} = f_n \circ f, n \geqslant 0$. [Apply (18.61.c).]

(d) Let $(Y, \mathcal{B})$ be a measurable space, and let $g : X \to Y$ be a $\mathcal{B}_\tau$-measurable function. Then $g \circ f_n, n \geqslant 0$, are independent and identically distributed random elements.

(e) For $n \geqslant 1$, $f_n = (\pi_{n\tau+i})_{i\geqslant 1}$.

(f) For $m, n \geqslant 1$, $f_m^{-1}(\mathcal{B}_n) \subset \mathcal{B}_{m\tau+n}$.

Exercise 30.80. Assume that $f_n, n \geqslant 1$, are independent with $|f_n| \leqslant M < \infty, n \geqslant 1$, set $S_n = f_1 + \cdots + f_n, n \geqslant 1$, and let τ be a finite $\{\sigma(f_1, \ldots, f_n)\}$-time. Then $E|S_\tau| \leqslant ME\tau$.

Exercise 30.81. Let $\mathcal{B} \subset \mathcal{A}$ be a σ-algebra, and assume that $f_n, n \in N$, are conditionally independent given $\mathcal{B}$. For $n \in N$, put $S_n = f_1 + \cdots + f_n$ and $\mathcal{B}_n = \mathcal{B} \vee \sigma(f_1, \ldots, f_n)$. Use (29.56) and (29.61.ii) to show that $\{S_n : n \in N\}$ is a martingale, a submartingale or a supermartingale according as $E^{\mathcal{B}} f_{n+1} = 0, \geqslant 0$ or $\leqslant 0$ a.s. for any $n \in N$.

Exercise 30.82. For each $m \in N$, suppose that $\{f_n(m) : n \in N\} \subset \mathfrak{L}^1$ and is adapted. For $n \in N$, assume that $g_n = \sup_{m \in N} f_n(m) \in \mathfrak{L}^1$ and $h_n = \liminf_m f_n(m) \in \mathfrak{L}^1$. Prove the following.

(a) If $\{f_n(m) : n \in N\}$ is a submartingale for any $m \in N$, then $\{g_n : n \in N\}$ is a submartingale. [Use (30.17.j).]

(b) If $f_n(m) \leqslant f_n(m+1)$ for all $m, n \in N$, then $\{g_n : n \in N\}$ is a martingale or a supermartingale according as, for each $m \in N$, $\{f_n(m) : n \in N\}$ is a martingale or a supermartingale.

(c) If $f_n(m) \geqslant 0$ for all $m, n \in N$, and $\{f_n(m) : n \in N\}$ is a supermartingale for any $m \in N$, then $\{h_n : n \in N\}$ is a supermartingale.

(d) If $\{f_n(m) : n \in N\}$ is a submartingale for any $m \in N$ and $\sup_{m \in N} |f_n(m)| \in \mathfrak{L}^1, n \in N$, then $\{\limsup_m f_n(m) : n \in N\}$ is a submartingale.

Exercise 30.83. For $n \in N$, let S_n and $\mathcal{B}_n$ be as in (30.16.d) and suppose that $E^{\mathcal{B}_n} f_{n+1} = S_n/n$ a.s. Show that $\{S_n/n : n \in N\}$ is a martingale.

Exercise 30.84 (Halmos). Assume that $\{f_n : n \geqslant 1\}$ is adapted, put $f_0 = 0$ and $\varepsilon_1 = 1$, let $\varepsilon_n : X \to [0, 1]$ be $\mathcal{B}_{n-1}$-measurable for $n \geqslant 2$, and define $g_n = \sum_{i=1}^{n} \varepsilon_i (f_i - f_{i-1}), n \geqslant 1$.

(a) If $\{f_n : n \geqslant 1\}$ is a martingale, a submartingale or a supermartingale, then so is $\{g_n : n \geqslant 1\}$.

(b) For $n \geqslant 1$, $Eg_n = Ef_n$, $\leqslant Ef_n$ or $\geqslant Ef_n$ according as $\{f_n : n \geqslant 1\}$ is a martingale, a submartingale or a supermartingale. [Hint. $f_{n+1} - g_{n+1} = (1 - \varepsilon_{n+1})(f_{n+1} - f_n) + f_n - g_n, n \geqslant 1$.]

Exercise 30.85. Suppose that $f_n, n \in N$, are independent with $P(f_n = 1) = 1 - P(f_n = -1) = p, n \in N$, where $0 < p < 1$. For $n \in N$, let S_n and $\mathcal{B}_n$ be as in (30.16.d). Prove that $\{((1-p)/p)^{S_n} : n \in N\}$ is a martingale.

Exercise 30.86. Let $\{\mathcal{C}_n : n \in N\}$ be a nondecreasing sequence of σ-algebras with $\mathcal{C}_n \subset \mathcal{A}$. Show that every $\{\mathcal{B}_n\}$-martingale is a $\{\mathcal{C}_n\}$-martingale if and only if, for any $n \in N$, $\mathcal{B}_n$ is essentially contained in $\mathcal{C}_n$ and $(\mathcal{B}_\infty, \mathcal{B}_n, \mathcal{C}_n)$ is a Markov triple. [Hint. For the necessity, consider martingales of the form $\{P^{\mathcal{B}_n}(B) : n \in N\}$ with $B \in \cup_{n \in N} \mathcal{B}_n$, and apply (29.62.v).]

Exercise 30.87. Let $\{f_n : n \in N\}$ and $\{g_n : n \in N\}$ be submartingales, and let τ be a stopping time such that $f_\tau \leqslant g_\tau$ on $\{\tau < \infty\}$. Prove that $\{1_{\{\tau > n\}} f_n + 1_{\{\tau \leqslant n\}} g_n : n \in N\}$ is a submartingale.

Exercise 30.88. Suppose that $\{f_n : n \in N\}$ is adapted, let τ be a finite stopping time such that $f_\tau \in \mathfrak{L}^1$, and set $g_n = E^{\mathcal{B}_n} f_{\tau \wedge n}, n \in N$. If $\{f_n : n \in N\}$ is a martingale, a submartingale or a supermartingale, then so is $\{g_n : n \in N\}$.

Exercise 30.89. Let $\varphi : R \to [0, \infty[$ be a nondecreasing convex function, let $\{f_n : n \in N\}$ be submartingales, let $a, \varepsilon > 0$, and assume that $\varphi \circ (af_n) \in \mathfrak{L}^1$ for all $n \in N$. Show that $P(\sup_{1 \leqslant i \leqslant n} f_i \geqslant \varepsilon) \leqslant E[\varphi \circ (af_n)]/\varphi(a\varepsilon), n \in N$.

Exercise 30.90. Suppose that $f_n, n \in N$, are independent with $f_n \geqslant 1$, and let $a > 0$ and $\varepsilon > 1$. Prove that $\varepsilon^a P\big(\sup_{1 \leqslant i \leqslant n} \prod_{j=1}^{i} f_j \geqslant \varepsilon\big) \leqslant \prod_{i=1}^{n} Ef_i^a, n \in N$. [Apply (30.89) with $\varphi(x) = e^x, x \in R$.]

Exercise 30.91 (Azuma). Let $\{f_n : n \in N\}$ be a martingale with $Ef_1 = 0$, set $f_0 = 0$, assume that $|d_n| = |f_n - f_{n-1}| \leqslant c_n \in]0, \infty[, n \in N$, and put $C_n^2 = \sum_{i=1}^{n} c_i^2$. For $\varepsilon > 0$, show that $P(|f_n| \geqslant \varepsilon) \leqslant 2e^{-\varepsilon^2/2C_n^2}, n \in N$. The following steps may be helpful.

(a) For $1 \leqslant i \leqslant n$, $E^{\mathcal{B}_{i-1}} e^{td_i} \leqslant e^{t^2 c_i^2/2}$ a.s., $t > 0$. [Apply (29.97).]

(b) $Ee^{tf_n} \leqslant e^{t^2 C_n^2/2}, t > 0$. [Hint. Write $Ee^{tf_n} = E[e^{tf_{n-1}} E^{\mathcal{B}_{n-1}} e^{td_n}], t > 0$, and use (a) and backward iteration.]

(c) $P(f_n \geqslant \varepsilon) \leqslant e^{-\varepsilon^2/2C_n^2}$. [Hint. $P(f_n \geqslant \varepsilon) \leqslant \exp(-t\varepsilon + t^2 C_n^2/2), t > 0$, by (13.25.b) and (b). Choose the optimal value $t = \varepsilon/C_n^2$.]

Exercise 30.92. Let $\{f_n : n \geqslant 1\}$ be as in (30.16.a). Use (30.40.c) and (14.30) to prove that $\{f_n : n \geqslant 1\}$ is not closable.

Exercise 30.93. Let $\{f_n : n \in N\}$ and $\{g_n : n \in N\}$ be as in (30.16.f), and assume that $P(f_n = 1) = 1 - P(f_n = 0) = p, n \in N$, where $0 < p < 1$. Show that $g_n \overset{a.s.}{\to} 0$ and $\{g_n : n \geqslant 1\}$ is not closable.

Exercise 30.94. Let $\{f_n : n \in N\}$ be a submartingale with $\sup_{n \in N} Ef_n^+ < \infty$, and let $f_n = f_n' + f_n'', n \in N$, be the Doob decomposition of f_n in (30.19.i). Prove the following.

(a) $f_n' \overset{a.s.}{\to} f_\infty'$ for some $\overline{R}$-valued random variable f_∞'.

(b) $\{f_n' : n \in N\}$ is closed by f_∞' if and only if $\{f_n : n \in N\}$ is uniformly integrable. [Hints. For the necessity, use (29.26). For the sufficiency, show first that $f_\infty' \in \mathfrak{L}^1$ by Fatou's lemma and (30.19.ii), then apply (30.43.i).]

Exercise 30.95. Let $\{f_n : n \in N\}$ be a martingale or a nonnegative submartingale with $\sup_{n \in N} E|f_n|^p < \infty$, where $1 < p < \infty$. Show that there is $f_\infty \in \mathfrak{L}^p$ such that $f_n \overset{a.s.}{\to} f_\infty$ and $f_n \overset{\mathfrak{L}^p}{\to} f_\infty$. The following steps may be useful.

(a) There is $f_\infty \in \mathfrak{L}^p$ such that $f_n \overset{a.s.}{\to} f_\infty$. [Use (30.41.i) and Fatou's lemma.]

(b) $\{f_n : n \in N\}$ is closed by f_∞. [Use (14.27.c) and (30.43.i) or (30.41.ii).]

(c) $f_n \overset{\mathfrak{L}^p}{\to} f_\infty$. [Use (30.40.c) and (30.44).]

Exercise 30.96. Let $f_n, n \geqslant 1$, and $S_n, n \geqslant 1$, be as in (29.83), and assume that $f_1 \in \mathfrak{L}^2$. For $n \geqslant 2$, put $T_n = f_1^2 + \cdots + f_n^2$, $U_n = \frac{1}{n-1}\sum_{i=1}^{n}(f_i - \frac{S_n}{n})^2$ and $\mathcal{B}_n = \sigma((S_m, T_m), m \geqslant n)$. The random variable U_n is called the *sample variance.* Prove the following:

(a) $\{U_n : n \geqslant 2\}$ is $\{\mathcal{B}_n\}$-adapted.

(b) For $n \geqslant 2$ and $1 \leqslant i < j \leqslant n$, $E^{\mathcal{B}_n}(f_i - f_j)^2 = E^{\mathcal{B}_n}(f_1 - f_2)^2$ a.s. [Hint. First state and solve an analog of (29.82), then imitate the solution of (29.83.a).]

(c) For $n \geqslant 2, U_n = \frac{1}{2}E^{\mathcal{B}_n}(f_1 - f_2)^2$ a.s. [Hint. Show that $U_n = \frac{1}{n(n-1)}\sum_{1\leqslant i<j\leqslant n}(f_i - f_j)^2$, and use (a) and (b).]

(d) $\{U_n : n \geqslant 2\}$ is a reverse martingale. [Use (c).]

Exercise 30.97. Let $\{f_n : n \in N\}$ be a reverse submartingale with $\inf_{m\in N} Ef_m > -\infty$, fix versions of $E^{\mathcal{B}_{n+1}} f_n, n \in N$, and write

$$f_n = f_n' + f_n'', \quad n \in N, \tag{1}$$

where $f_n'' = \sum_{k\geqslant n}(E^{\mathcal{B}_{k+1}} f_k - f_{k+1})$. Prove the following.

(a) Almost surely $\{f_n'' : n \in N\}$ is a nonincreasing sequence of nonnegative and finite random variables such that $f_n'' \to 0$. [Hint. For $n \in N, Ef_n'' = Ef_n - \inf_{m\in N} Ef_m$.]

(b) f_n'' is $\mathcal{B}_{n+1}$-measurable for any $n \in N$.

(c) $\{f_n' : n \in N\}$ is a reverse martingale.

(d) The decomposition (1) is unique, in the sense that if $f_n = g_n' + g_n'', n \in N$, where $\{g_n' : n \in N\}$ is a reverse martingale, and g_n'' is $\mathcal{B}_{n+1}$-measurable for any $n \in N$ with $g_n'' \overset{a.s.}{\to} 0$, then $f_n' = g_n'$ a.s., $n \in N$, and $f_n'' = g_n''$ a.s., $n \in N$. [Hint. For $n \in N$, $f_{n+1}'' - f_n'' = g_{n+1}'' - g_n'' = f_{n+1} - E^{\mathcal{B}_{n+1}} f_n$ a.s.]

Exercise 30.98. Let $\{f_n : n \geqslant 1\}$ be a martingale such that $\sum_{n\geqslant 1} E\,|d_n|^p < \infty$ for some $p \in]0, 2]$, where $d_n = f_n - f_{n-1}, n \geqslant 1$, with $f_0 = 0$. Use (30.60) to show that $\{f_n : n \geqslant 1\}$ converges a.s.

Exercise 30.99 (Chow). For $n \geqslant 1$, let $a_n : X \to]0, \infty[$ be a $\mathcal{B}_{n-1}$-measurable random variable such that $a_n \overset{a.s.}{\to} \infty$. Let $\{f_n : n \geqslant 1\}$ be a martingale such that the series $\sum_{n\geqslant 1} E^{\mathcal{B}_{n-1}} |d_n|^p / a_n^p$ converges a.s. for some $p \in]0, 2]$, where $d_n, n \geqslant 1$, are as in (30.98) with $d_n/a_n \in \mathfrak{L}^p, n \geqslant 1$. Prove the following.

(a) $\{d_n/a_n : n \geqslant 1\}$ is a martingale difference sequence.

(b) $f_n/a_n \overset{a.s.}{\to} 0$. [Hint. Take into account the reasoning to solve (30.98) and apply Kronecker's lemma (27.5).]

Exercise 30.100 (Lévy). Let $\{f_n : n \geqslant 1\} \subset \mathfrak{L}^2$ be a martingale, set $t_n^2 = \sum_{i=1}^n E^{\mathcal{B}_{i-1}} d_i^2, n \geqslant 1$, where $d_n = f_n - f_{n-1}, n \geqslant 1$, with $f_0 = 0$, and assume that $t_n^2 \overset{a.s.}{\to} \infty$. Let $\varphi : [0, \infty[\to]0, \infty[$ be a nondecreasing function such that $\int_0^\infty \frac{dt}{\varphi^2(t)} < \infty$.

(a) $\{d_n/(\varphi \circ t_n^2) : n \geqslant 1\}$ is a martingale difference sequence.

(b) $f_n/(\varphi \circ t_n^2) \overset{a.s.}{\to} 0$. [Use (30.56.i) and (27.5).]

12 Ergodicity, Mixing, and Stationarity

The classical ergodic theorems for measure-preserving transformations and their role in the theory of stationary sequences are the ultimate objectives of this chapter. To this end, a remarkable mathematical machinery is first built up, and a number of preliminary and adjacent results are established. Some topics on measure-preserving transformations such as weak-mixing and strong-mixing are also discussed.

31 Ergodicity and Mixing

The word "ergodic" hints at the origins in statistical mechanics. The main purpose of this section is to formulate and prove the most important classical ergodic theorems. To this end, we first introduce and discuss basic concepts such as measure-preserving transformation, invariant σ-algebra, invariant random variable and ergodic transformation. The final part of the section is devoted to results concerning the related notions of weak-mixing and strong-mixing.

Throughout this section, we consider a probability space $(X, \mathcal{A}, P)$ as usual, let $T : X \to X$ be an $(\mathcal{A}, \mathcal{A})$-measurable transformation, set $T^0 = i_X$, and define $T^n = T \circ T^{n-1}, n \geqslant 1$. For $n \geqslant 0$, write $PT^{-n} = P(T^n)^{-1}$ and $T^{-n}(A) = (T^n)^{-1}(A)$, $A \in \mathcal{A}$.

Definition 31.1. We say that T is *measure-preserving* [*P-preserving*] (or that T preserves P) if $PT^{-1} = P$. If T is P-preserving, then $PT^{-n} = P, n \geqslant 0$.

Remarks 31.2. (a) If $T_1 : X \to X$ and $T_2 : X \to X$ are measure-preserving transformations, then so is $T_1 \circ T_2$.

(b) If $\{P_i : i \in I\}$ is a countable set of probabilities on $\mathcal{A}$, each preserved by T, and $\{a_i : i \in I\} \subset [0, 1]$ is such that $\sum_{i \in I} a_i = 1$, then $\sum_{i \in I} a_i P_i$ is preserved by T.

(c) If T is P-preserving and $T(X) \subset A \in \mathcal{A}$, then $T^{-1}(A) = X$, and so $P(A) = P(T^{-1}(A)) = 1$.

(d) If T is P-preserving and $A \in \mathcal{A}$, then

$$P(T^{-1}(A)\Delta A) = 2P(T^{-1}(A) - A) = 2P(A - T^{-1}(A)); \tag{1}$$

in particular, $P(T^{-1}(A)\Delta A) = 0$ whenever either $T^{-1}(A) \overset{a.s.}{\subset} A$ or $A \overset{a.s.}{\subset} T^{-1}(A)$. Indeed, $P(T^{-1}(A) - A) = P(T^{-1}(A) - (T^{-1}(A) \cap A)) = P(A) - P(T^{-1}(A) \cap A) = P(A - T^{-1}(A))$, whence (1) follows.

Analysis and Probability. http://dx.doi.org/10.1016/B978-0-12-401665-1.00012-6

(e) If T is P-preserving, then T is an $(\mathcal{A}^P, \mathcal{A}^P)$-measurable transformation and $\overline{P}T^{-1} = \overline{P}$.

(f) If $\mathcal{M} \subset \mathcal{A}$ is closed under intersection and generates $\mathcal{A}$, and $P(T^{-1}(M)) = P(M)$, $M \in \mathcal{M}$, then T is P-preserving by (10.8).

Examples 31.3. (a) Let $X = \{x_1, \ldots, x_m\}$ with $m \geqslant 2$, put $\mathcal{A} = \mathcal{P}(X)$ and $P(\{x_i\}) = 1/m$, $1 \leqslant i \leqslant m$, and define $T(x_i) = x_{i+1}$, $1 \leqslant i < m$ and $T(x_m) = x_1$. Then T is P-preserving.

(b) Assume that $X = \{x_1, \ldots, x_m, x'_1, \ldots, x'_n\}$ with $m, n \geqslant 2$, put $\mathcal{A} = \mathcal{P}(X)$, let $P(\{x_i\}) = p$, $1 \leqslant i \leqslant m$, and $P(\{x'_j\}) = q$, $1 \leqslant j \leqslant n$, where $mp + nq = 1$, and define $T(x_i) = x_{i+1}$, $1 \leqslant i < m$, and $T(x_m) = x_1$, and $T(x'_j) = x_{j+1}$, $1 \leqslant i < n$ and $T(x'_n) = x'_1$. Then T is P-preserving.

(c) Let $I = N$ or Z, put $(X_i, \mathcal{A}_i, P_i) = (X, \mathcal{A}, P)$ for any $i \in I$, consider the probability space $(\prod_{i \in I} X_i, \otimes_{i \in I} \mathcal{A}_i, \otimes_{i \in I} P_i)$, and define $T : \prod_{i \in I} X_i \to \prod_{i \in I} X_i$ by $T((x_i)_{i \in I}) = (y_i)_{i \in I}$, where $y_i = x_{i+1}$ for any $i \in I$, i.e. $\pi_i \circ T = \pi_{i+1}$, $i \in I$. Consequently, according to (7.18), we have $T^{-1}(\otimes_{i \in I} \mathcal{A}_i) = \sigma(\pi_i \circ T, i \in I) = \sigma(\pi_{i+1}, i \in I) \subset \sigma(\pi_i, i \in I) = \otimes_{i \in I} \mathcal{A}_i$, which shows that T is measurable. Let $J \subset I$ be finite and nonempty. Since we may identify the probability spaces $(\prod_{j \in J} X_j, \otimes_{j \in J} \mathcal{A}_j, \otimes_{j \in J} P_j)$ and $(\prod_{j \in J+1} X_j, \otimes_{j \in J+1} \mathcal{A}_j, \otimes_{j \in J+1} P_j)$, we have $((\otimes_{i \in I} P_i) T^{-1})\pi_J^{-1} = (\otimes_{i \in I} P_i)(\pi_J \circ T)^{-1} = (\otimes_{i \in I} P_i)\pi_{J+1}^{-1} = \otimes_{j \in J+1} P_j = \otimes_{j \in J} P_j = (\otimes_{i \in I} P_i)\pi_J^{-1}$. By (16.6), this ensures that T preserves $\otimes_{i \in I} P_i$. If $I = N$, T is called the *one-sided Bernoulli shift*, while if $I = Z$, T is called the *two-sided Bernoulli shift*.

(d) Notation and assumptions are as in (16.40). Suppose that $\sum_{x_1 \in X} p_1(x_1) = 1$, and define $T : \prod_{i \in I} X_i \to \prod_{i \in I} X_i$ as in (c). We have already shown that T is measurable. For $J \in \mathcal{J}$, identifying the measurable spaces $(X^J, \mathcal{P}(X^J))$ and $(X^{J+1}, \mathcal{P}(X^{J+1}))$, we see that $P_J = P_{J+1}$. Therefore, $(PT^{-1})\pi_J^{-1} = P(\pi_J \circ T)^{-1} = P\pi_{J+1}^{-1} = P_{J+1} = P_J = P\pi_J^{-1}$. The uniqueness assertion in (16.40) shows that $PT^{-1} = P$. If we take $p_l(x_1, \ldots, x_l) = p_{x_1} \cdots p_{x_l}$, where p_x, $x \in X$, are nonnegative numbers with $\sum_{x \in X} p_x = 1$, then the projections π_i, $i \in I$, become independent and identically distributed random elements, and so we obtain a special case of (c).

(e) A special case of (d) is as follows. Assume we are given a vector $\mathbf{p} = (p_x)_{x \in X}$, where p_x, $x \in X$, are as in (d), and a *stochastic matrix* $\Pi = (p_{xx'})_{x,x' \in X}$, i.e. a matrix satisfying $p_{xx'} \geqslant 0$ and $\sum_{x' \in X} p_{xx'} = 1$ for all $x, x' \in X$, such that $\mathbf{p}\Pi = \mathbf{p}$. If we take $p_1(x_1) = p_{x_1}$ and $p_l(x_1, \ldots, x_l) = p_{x_1} p_{x_1 x_2} \cdots p_{x_{l-1} x_l}$ for $l \geqslant 2$, we see that conditions (i)–(iii) in (16.40) are fulfilled. In this case T is called the *one-sided Markov shift* or the two-sided Markov shift according as $I = N$ or $I = Z$. If each row of Π coincides with $\mathbf{p}$, then the Markov shift becomes a Bernoulli shift.

According to (16.40), for any $i \in I$ and $k \geqslant 0$, $P(\pi_i = x, \pi_{i+k} = x') = p_x p_{xx'}^{(k)}$, where $p_{xx'}^{(k)}$ is the (x, x')th entry of the matrix Π^k.

(f) This example is related to the notion of f-expansion, so notation is as in (8.23). Denote $a = (a_n)_{n \geqslant 1}$ and $\mathcal{B} = \otimes_{n \geqslant 1} \mathcal{P}(I_n)$. In case that $1 < b \in N$, consider the denumerable set $E = \{(i_n)_{n \geqslant 1} \in I^N : \text{there is } n \geqslant 1 \text{ such that } i_m = b - 1 \text{ for all } m \geqslant n\}$. In case that $I = N$, put $E^c = I^N$. Assume that a is injective, so that a maps

$[0, 1[$ onto E^c. Let $\mathcal{M}$ stand for the family of all sets of the form $B_l(i_1, \ldots, i_l)$ with $l \geqslant 1$ and $i_1, \ldots, i_l \in I$, where $B_l(i_1, \ldots, i_l) = \{a_1 = i_1, \ldots, a_l = i_l\} = [\underline{x}_l(i_1, \ldots, i_l), \bar{x}_l(i_1, \ldots, i_l)[$. Then, in view of (8.23.g), $\sigma(\mathcal{M}) = \sigma(a_n, n \geqslant 1) = a^{-1}(\mathcal{B}) = \mathcal{B}([0, 1[)$. For any $l \geqslant 1$ and $i_1, \ldots, i_l \in I$, suppose a nonnegative number $p_l(i_1, \ldots, i_l)$ is given such that:

(i) $\sum_{i \in I} p_1(i) = 1$;

(ii) $\sum_{i \in I} p_{l+1}(i_1, \ldots, i_l, i) = p_l(i_1, \ldots, i_l)$;

(iii) $\sum_{i \in I} p_{l+1}(i, i_1, \ldots, i_l) = p_l(i_1, \ldots, i_l)$.

If $1 < b \in N$, suppose furthermore that

(iv) $\lim_{m \to \infty} p_{l+m}(i_1, \ldots, i_l, \underbrace{b-1, \ldots, b-1}_{m}) = 0.$

Then, according to (16.40), there exists a unique probability P on $\mathcal{B}$ such that $P(\pi_{k+1} = i_1, \ldots, \pi_{k+l} = i_l) = p_l(i_1, \ldots, i_l)$ for each $k \geqslant 0$. Moreover, $P(E^c) = 1$ (by (iv) in case $1 < b \in N$). Define a probability λ on $\mathcal{B}([0, 1[)$ by setting $\lambda(A) = P(B)$ if $A = a^{-1}(B)$ with $B \in \mathcal{B}$. λ is well defined, as $P(E) = 0$. From (8.23.c) we see that, for $n \geqslant 1, r_n = r_1 \circ r_{n-1}$ and $a_n = a_1 \circ r_{n-1}$, and so $a_{n+1} = a_1 \circ r_n = a_n \circ r_1$. Put $r_1 = T$. Then $T^{-1}(\mathcal{B}([0, 1[)) = r_1^{-1}(\sigma(a_n, n \geqslant 1)) = \sigma(a_n \circ r_1, n \geqslant 1) = \sigma(a_{n+1}, n \geqslant 1) = \sigma(a_n, n \geqslant 1)$, since I is countable. Hence T is measurable. Finally, for $B_l(i_1, \ldots, i_l) \in \mathcal{M}$, we have $\lambda(B_l(i_1, \ldots, i_l)) = \lambda(a^{-1}(\{(i_1, \ldots, i_l)\} \times \prod_{j>l} I_j)) = P(\{(i_1, \ldots, i_l)\} \times \prod_{j>l} I_j) = p_l(i_1, \ldots, i_l)$. Consequently,

$$
\begin{aligned}
(\lambda T^{-1})(B_l(i_1, \ldots, i_l)) &= \lambda(a_1 \circ r_1 = i_1, \ldots, a_l \circ r_1 = i_l) \\
&= \lambda(a_2 = i_1, \ldots, a_{l+1} = i_l) \\
&= \sum_{i \in I} \lambda(B_{l+1}(i, i_1, \ldots, i_l)) \\
&= \sum_{i \in I} p_{l+1}(i, i_1, \ldots, i_l) = p_l(i_1, \ldots, i_l) \\
&= \lambda(B_l(i_1, \ldots, i_l))
\end{aligned}
$$

by (iii). In view of (31.2.f), it follows that T is λ-preserving. Suppose now that $p_l(i_1, \ldots, i_l) = p_{i_1} \cdots p_{i_l}$, where $p_i \geqslant 0, i \in I$, and $\sum_{i \in I} p_i = 1$. If $1 < b \in N$, suppose furthermore that $p_{b-1} < 1$. Then conditions (i)–(iv) are satisfied, and $\lambda(a_n = i) = p_i$ for all $n \geqslant 1$ and $i \in I$, and so $a_n, n \geqslant 1$, become i.i.d. random variables. If $1 < b \in N$ and $p_i = 1/b, i \in I$, then, according to (16.24), λ is in fact the Lebesgue measure on $\mathcal{B}([0, 1[)$. In this case T is called the *b-adic transformation.*

(g) Let X be a compact metric space, let $T : X \to X$ be a continuous transformation on X, let $\{Q_n : n \in N\} \subset$ P, where P is as in Section 22, and denote $P_n = \frac{1}{n} \sum_{i=0}^{n-1} Q_n T^{-i}, n \in N$. Since P is sequentially compact by (22.21) and (6.19), there is $P \in$ P such that $P_{n_k} \Rightarrow P$ for some subsequence $\{n_k : k \in N\}$. Then T is P-preserving. Actually, for $f \in \mathfrak{C}_r(X)$, by the change of variable formula (11.29), we obtain

$$\left|\int_X (f\circ T)dP - \int_X f\,dP\right| = \lim_k \left|\int_X (f\circ T)dP_{n_k} - \int_X f\,dP_{n_k}\right|$$

$$= \lim_k \left|\frac{1}{n_k}\sum_{i=0}^{n_k-1}\int_X (f\circ T)d(Q_{n_k}T^{-i}) - \frac{1}{n_k}\sum_{i=0}^{n_k-1}\int_X f\,d(Q_{n_k}T^{-i})\right|$$

$$= \lim_k \frac{1}{n_k}\left|\int_X (f\circ T^{n_k})dQ_{n_k} - \int_X f\,dQ_{n_k}\right| \leqslant \lim_k \frac{2\,\|f\|_u}{n_k} = 0.$$

Consequently, $\int_X (f\circ T)dP = \int_X f\,dP$, whence $PT^{-1} = P$ by (11.50).

The next result describes a way of constructing a measure-preserving transformation from given ones.

Theorem 31.4. *For $i \in I \neq \emptyset$, let $(X_i, \mathcal{A}_i, P_i)$ be a probability space, and let $T_i : X_i \to X_i$ be a measurable transformation. Let P stand for the product of $P_i, i \in I$, and define $T : \prod_{i\in I} X_i \to \prod_{i\in I} X_i$ by $T = (T_i \circ \pi_i)_{i\in I}$. Then T is P-preserving if and only if T_i is P_i-preserving for each $i \in I$.*

Proof. Observe first that T is measurable by (8.5). Suppose that T is P-preserving, and let $i \in I$. Then we get

$$P_iT_i^{-1} = (P\pi_i^{-1})T_i^{-1} = P(T_i\circ\pi_i)^{-1} = P(\pi_i\circ T)^{-1}$$
$$= (PT^{-1})\pi_i^{-1} = P\pi_i^{-1} = P_i,$$

which shows that T_i is P_i-preserving.

Conversely, assume that $P_iT_i^{-1} = P_i, i \in I$, and let $J \subset I$ be finite and nonempty. Since $T_i \circ \pi_i, i \in I$, are independent by (18.27) and (18.28.c), applying (18.29) we have

$$(PT^{-1})\pi_J^{-1} = P(\pi_J\circ T)^{-1} = P(T_j\circ\pi_j)_{j\in J}^{-1}$$
$$= \bigotimes_{j\in J} P(T_j\circ\pi_j)^{-1} = \bigotimes_{j\in J} P_jT_j^{-1} = \bigotimes_{j\in J} P_j = P_J.$$

Therefore, the uniqueness assertion in (16.11) shows that $PT^{-1} = P$. □

Definition 31.5. Let T be a measure-preserving transformation. For $x \in X$, the set $O_T(x) = \{T^n(x) : n \geqslant 0\}$ is called the *orbit* [*T-orbit*] *of* x.

Anticipating central facts in ergodic theory, to be later presented, the following two theorems indicate important properties of orbits from measure-theoretical and topological perspectives.

Poincaré recurrence theorem 31.6. *Suppose that T is P-preserving, let $A \in \mathcal{A}$, and denote $B = \{x \in A : T^n(x) \in A$ for infinitely many $n\}$. Then $B \in \mathcal{A}$ and $P(B) = P(A)$.*

Proof. Clearly, $B = A \cap \limsup_n T^{-n}(A)$, so that $B \in \mathcal{A}$. For $n \geqslant 0$, put $B_n = \cup_{k\geqslant n} T^{-k}(A)$. Since $T^{-1}(B_n) = B_{n+1}$, we have $P(B_n) = P(B_{n+1})$, and so $P(B_n) = P(B_0)$, $n \geqslant 0$. As $B = \cap_{n\geqslant 0}(A \cap B_n)$ and $\{A \cap B_n : n \geqslant 0\}$ is nondecreasing, and $B_0 \supset A$, it follows that $P(B) = P(A \cap B_0) \geqslant P(A)$. □

Theorem 31.7. *Let X be a topological space that has a countable base $\{U_i : i \in I\}$, let $T : X \to X$ be a continuous transformation on X, and let P be a probability on $\mathcal{B}(X)$ preserved by T such that:*

(i) *$P(U) > 0$ whenever $U \neq \emptyset$ is an open set;*

(ii) *$A \in \mathcal{O}_P$ whenever $A \in A$ and $P(T^{-1}(A)\Delta A) = 0$.*

Then the set $A = \{x \in X : O_T(x)$ is dense in $X\} \in \mathcal{A}$ and $P(A) = 1$.

Proof. Assume $U_i \neq \emptyset$, $i \in I$. We have

$$A = \bigcap_{i\in I} \bigcup_{n\geqslant 0} T^{-n}(U_i). \tag{1}$$

Indeed, if $x \in \cup_{i\in I} \cap_{n\geqslant 0} T^{-n}(U_i^c)$, then there exists an open set U_{i_0} such that $O_T(x) \cap U_{i_0} = \emptyset$, and so $x \notin A$. On the other hand, let $x \in \cap_{i\in I} \cup_{n\geqslant 0} T^{-n}(U_i)$ and $y \in X$, and assume that $y \in U_{i_y}$ for some $i_y \in I$. Then $O_T(x) \cap U_{i_y} \neq \emptyset$, and so $x \in A$. Thus (1) is established, which shows that $A \in \mathcal{A}$. For $i \in I$, put $A_i = \cup_{n\geqslant 0} T^{-n}(U_i)$. Since T is P-preserving, and $T^{-1}(A_i) \subset A_i$, we have $P(T^{-1}(A_i)\Delta A_i) = 0$ by (31.2.d), and so $P(A_i) = 0$ or 1 by (ii). As $A_i \neq \emptyset$ and is open, it follows that $P(A_i) = 1$ by (i). Hence $P(A) = 1$ by (1) □

To accomplish the main aim of this section for presenting the most useful classical ergodic theorems, we need to first discuss the concept of invariance.

Definitions 31.8. Let $A \in \mathcal{A}$ and let $f : X \to R$ be a measurable function. We say that A is *invariant* [*T-invariant*] if $T^{-1}(A) = A$, and we say that A is a.s. invariant [*T-a.s. invariant*] if $T^{-1}(A) \overset{a.s.}{=} A$, i.e. $P(T^{-1}(A)\Delta A) = 0$. We say that f is *invariant* [*T-invariant*] if $f \circ T = f$, and we say that f is a.s. invariant [*T-a.s. invariant*] if $f \circ T = f$ a.s.

Examples 31.9. (a) $\emptyset$ and X are invariant, so that the family of invariant sets is not empty.

(b) The set $\{x_1, \ldots, x_m\}$ and $\{x_1', \ldots, x_n'\}$ in (31.3.b) are invariant.

(c) For any $A \in \mathcal{A}$, the sets $\liminf_n T^{-n}(A)$ and $\limsup_n T^{-n}(A)$ are invariant.

(d) The set A in (31.7) is a.s. invariant, since $T^{-1}(A) \subset A$ and (31.2.d) applies.

(e) For any measurable function $f : X \to R$, $\liminf_n(f \circ T^n)$ and $\limsup_n(f \circ T^n)$ are invariant.

(f) For f as in (e) and $n \geqslant 1$, set $S_n(f) = \frac{1}{n}\sum_{i=0}^{n-1} f \circ T^i$. Since $S_n(f \circ T) = \frac{n+1}{n} S_{n+1}(f) - \frac{f}{n}$, $n \geqslant 1$, it follows that the functions $\liminf_n S_n(f)$ and $\limsup_n S_n(f)$ are invariant.

Remarks 31.10. (a) Invariant sets and functions are a.s. invariant sets and functions, respectively.

(b) A is invariant if and only if 1_A is invariant, and A is a.s. invariant if and only if 1_A is a.s. invariant.

(c) If A is a.s. invariant, then $P(T^{-1}(A)) = P(A)$ by (9.6.iii).

(d) If $T^{-1}(A) \overset{a.s.}{=} A$, this does not imply that $T^{-2}(A) \overset{a.s.}{=} A$. For instance, let $X = \{a, b, c\}$, $\mathcal{A} = \mathcal{P}(X)$, $P = \varepsilon_c$, $A = \{a\}$, $T(a) = T(b) = a$, and $T(c) = b$. Then A is a.s. invariant, but $P(T^{-2}(A)\Delta A) = \varepsilon_c(A^c) = 1$. In general, if T is P-preserving, then $T^{-n}(A) \overset{a.s.}{=} A$ for any $n \geqslant 1$. Actually, observe first that $A, B, C \in \mathcal{A}$ with $A \overset{a.s.}{=} B$ and $B \overset{a.s.}{=} C$ imply $A \overset{a.s.}{=} C$ by (9.36.d). Next, if $P(T^{-n}(A)\Delta A) = 0$, then $P(T^{-n-1}(A)\Delta T^{-1}(A)) = (PT^{-1})(T^{-n}(A)\Delta A) = P(T^{-n}(A)\Delta A) = 0$, and so $T^{-n-1}(A) \overset{a.s.}{=} T^{-1}(A)$. Hence $T^{-n-1}(A) \overset{a.s.}{=} A$, and the assertion follows by induction.

(e) If T is P-preserving, $A, B \in \mathcal{A}$ with $A \overset{a.s.}{=} B$ and B is a.s. invariant, then so is A. Indeed, $T^{-1}(A) \overset{a.s.}{=} T^{-1}(B) \overset{a.s.}{=} A$.

(f) If A is a.s. invariant and T is P-preserving, then, according to (d) and (17.17), $A \overset{a.s.}{=} \cup_{n\geqslant 0} T^{-n}(A) \overset{a.s.}{=} \cap_{n\geqslant 0} T^{-n}(A) \overset{a.s.}{=} \liminf_n T^{-n}(A) \overset{a.s.}{=} \limsup_n T^{-n}(A)$. Therefore, in view of (31.9.c), A equals almost surely the invariant sets $\liminf_n T^{-n}(A)$ and $\limsup_n T^{-n}(A)$.

(g) If f is invariant and $\varphi : R \to R$ is measurable, then $\varphi \circ f$ is invariant.

(h) If f is a.s. invariant and T is P-preserving, then $f \circ T^n = f$ a.s. for any $n \geqslant 1$. Moreover, according to (31.9.e), f equals a.s. the invariant functions $\liminf_n (f \circ T^n)$ and $\limsup_n (f \circ T^n)$.

Theorem 31.11. *Let $\mathcal{I}$ $[\mathcal{I}_T]$ and $\mathcal{I}'$ $[\mathcal{I}'_T]$ denote the family of all T-invariant and T-a.s. invariant sets in $\mathcal{A}$, respectively, and let $f : X \to R$ be a measurable function. Then the following assertions hold.*

(i) *$\mathcal{I}$ and $\mathcal{I}'$ are σ-algebras. ($\mathcal{I}$ and $\mathcal{I}'$ are called the invariant σ-algebra and the a.s. invariant σ-algebra, respectively.)*

(ii) *If T is P-preserving, then $\mathcal{I}'$ is the P-completion of $\mathcal{I}$ in $\mathcal{A}$.*

(iii) *f is invariant if and only if f is $\mathcal{I}$-measurable.*

(iv) *f is a.s. invariant if and only if f is $\mathcal{I}'$-measurable.*

Proof. Since T^{-1} preserves the basic set operations, $\mathcal{I}$ is obviously a σ-algebra. Next $A^c \in \mathcal{I}'$ whenever $A \in \mathcal{I}'$, because $T^{-1}(A^c)\Delta A^c = T^{-1}(A)\Delta A$. Let $\{A_n : n \geqslant 1\} \subset \mathcal{I}'$, so that $P(T^{-1}(A_n)\Delta A_n) = 0$, $n \geqslant 1$. We have

$$P\left(T^{-1}\left(\bigcup_{n\geqslant 1} A_n\right)\Delta\left(\bigcup_{n\geqslant 1} A_n\right)\right) = P\left(\left(\bigcup_{n\geqslant 1} T^{-1}(A_n)\right)\Delta\left(\bigcup_{n\geqslant 1} A_n\right)\right)$$
$$\leqslant P\left(\bigcup_{n\geqslant 1}(T^{-1}(A_n)\Delta A_n)\right) = 0.$$

Hence $\cup_{n\geqslant 1} A_n \in \mathcal{I}'$ and the proof of (i) is finished. To verify (ii), let $\mathcal{I}''$ denote the P-completion of $\mathcal{I}$ in $\mathcal{A}$. If $A \in \mathcal{I}'$, then $A \overset{a.s.}{=} \liminf_n T^{-n}(A) \in \mathcal{I}$ by (31.10.f). In view of (9.14.d), this amounts to $A \in \mathcal{I}''$. Conversely, if $A \in \mathcal{I}''$, there exists some $B \in \mathcal{I}$ with $A \overset{a.s.}{=} B$. Then, according to (31.10.a) and (31.10.e), we see that $A \in \mathcal{I}'$. Thus $\mathcal{I}' = \mathcal{I}''$.

Further, if $f \circ T = f$ and $B \in \mathcal{B}(R)$, then $T^{-1}(f^{-1}(B)) = (f \circ T)^{-1}(B) = f^{-1}(B)$, and so $f^{-1}(B) \in \mathcal{I}$. Hence f is $\mathcal{I}$-measurable. Conversely, if f is $\mathcal{I}$-measurable and $f = 1_A$ with $A \in \mathcal{I}$, then $1_A \circ T = 1_{T^{-1}(A)} = 1_A$, i.e. f is invariant. This property carries over to the general f by the indicator function method, using also (31.10.f). Thus (iii) is established. Finally, if $f \circ T = f$ a.s. and $B \in \mathcal{B}(R)$, then $T^{-1}(f^{-1}(B))\Delta f^{-1}(B) = (f \circ T)^{-1}(B)\Delta f^{-1}(B) \subset \{f \circ T \neq f\}$. This shows that f is $\mathcal{I}'$-measurable. Conversely, if f is $\mathcal{I}'$-measurable and $f = 1_A$ with $A \in \mathcal{I}'$, then $T^{-1}(A) \overset{a.s.}{=} A$, and so $1_A \circ T = 1_{T^{-1}(A)} = 1_A$ a.s., i.e. f is a.s. invariant. The general case follows by the indicator function method. □

Theorem 31.12. *Let $(X, \mathcal{A})$ and $(Y, \mathcal{B})$ be measurable spaces, let $T_X : X \to X$ and $T_Y : Y \to Y$ be measurable transformations, and let $f : X \to Y$ be a measurable function such that $f \circ T_X = T_Y \circ f$. Then the following hold.*

(i) $\mathcal{I}_{T_X} \supset f^{-1}(\mathcal{I}_{T_Y})$.

(ii) *If $\mathcal{A} = f^{-1}(\mathcal{B})$, then $\mathcal{I}_{T_X} = f^{-1}(\mathcal{I}_{T_Y})$.*

Proof. If $B \in \mathcal{I}_{T_Y}$, then $T_X^{-1}(f^{-1}(B)) = (f \circ T_X)^{-1}(B) = f^{-1}(T_Y^{-1}(B)) = f^{-1}(B)$, and so $f^{-1}(B) \in \mathcal{I}_{T_X}$. This proves (i). Next, observe that $f \circ T_X^n = T_Y^n \circ f$, $n \geqslant 1$. If $\mathcal{A} = f^{-1}(\mathcal{B})$ and $A \in \mathcal{I}_{T_X}$, then $A = f^{-1}(B)$ for some $B \in \mathcal{B}$. Therefore, for $n \geqslant 1$, $A = T_X^{-n}(f^{-1}(B)) = (f \circ T_X^n)^{-1}(B) = f^{-1}(T_Y^{-n}(B)) = \liminf_n f^{-1}(T_Y^{-n}(B)) = f^{-1}(\liminf_n T_Y^{-n}(B))$. Hence, according to (31.9.c), $A \in f^{-1}(\mathcal{I}_{T_Y})$. In view of (i), this proves (ii). □

Another basic concept in this section is that of an ergodic transformation.

Definition 31.13. We say that T is *ergodic* [P-*ergodic*] if $\mathcal{I} \subset \mathcal{O}_P$, where $\mathcal{I}$ is as in (31.11).

Remarks 31.14. (a) The assumption (ii) in (31.7) means that $\mathcal{I}' \subset \mathcal{O}_P$. As $\mathcal{I} \subset \mathcal{I}'$, this implies that T therefrom is ergodic. In (31.16) we shall show that $\mathcal{I}' \subset \mathcal{O}_P$ is in fact equivalent to the seemingly weaker condition $\mathcal{I} \subset \mathcal{O}_P$ if T is P-preserving.

(b) If T is P-ergodic, and Q is a probability on $\mathcal{A}$ with $Q \ll P$, then T is also Q-ergodic.

Examples 31.15. (a) T in (31.3.a) is ergodic, but T in (31.3.b) is not ergodic if $pq \neq 0$.

(b) Both the one-sided and two-sided Bernoulli shifts are **P**-ergodic, where $\mathbf{P} = \otimes_{i \in I} P_i$. Actually, let $A \in \mathcal{I}$ and $\varepsilon > 0$. Then, according to (7.21) and (10.33), there exist a finite and nonempty set $J \subset I$ and $B \in \otimes_{j \in J} \mathcal{A}_j$ such that

$$\left|\mathbf{P}(A) - \mathbf{P}(\pi_J^{-1}(B))\right| \leqslant \mathbf{P}(A \Delta \pi_J^{-1}(B)) < \varepsilon, \tag{1}$$

where the first inequality comes from (9.6.iii). Choose n so large that $J \cap (J + n) = \emptyset$. Since π_J and π_{J+n} are independent, we have

$$\mathbf{P}(\pi_J^{-1}(B) \cap \pi_{J+n}^{-1}(B)) = \mathbf{P}(\pi_J^{-1}(B))\mathbf{P}(\pi_{J+n}^{-1}(B)) = \mathbf{P}(\pi_J^{-1}(B))^2. \tag{2}$$

As $\pi_J \circ T^n = \pi_{J+n}$, we get $\mathbf{P}(A\Delta\pi_{J+n}^{-1}(B)) = \mathbf{P}(T^{-n}(A)\Delta T^{-n}(\pi_J^{-1}(B))) = \mathbf{P}(A\Delta\pi_J^{-1}(B)) < \varepsilon$. Therefore, by virtue of (9.6.iii) and (1.8.j),

$$\begin{aligned}\left|\mathbf{P}(A) - \mathbf{P}(\pi_J^{-1}(B) \cap \pi_{J+n}^{-1}(B))\right| &\leqslant \mathbf{P}(A\Delta(\pi_J^{-1}(B) \cap \pi_{J+n}^{-1}(B)))\\ &\leqslant \mathbf{P}(A\Delta\pi_J^{-1}(B)) + \mathbf{P}(A\Delta\pi_{J+n}^{-1}(B)) < 2\varepsilon.\end{aligned} \tag{3}$$

From (1)–(3) we obtain

$$\begin{aligned}\left|\mathbf{P}(A) - \mathbf{P}(A)^2\right| &\leqslant \left|\mathbf{P}(A) - \mathbf{P}(\pi_J^{-1}(B) \cap \pi_{J+n}^{-1}(B))\right|\\ &+ \left|\mathbf{P}(\pi_J^{-1}(B) \cap \pi_{J+n}^{-1}(B)) - \mathbf{P}(A)^2\right| < 2\varepsilon\\ &+ \left|\mathbf{P}(\pi_J^{-1}(B))^2 - \mathbf{P}(A)^2\right| < 4\varepsilon.\end{aligned} \tag{4}$$

As ε is arbitrary, it follows that $\mathbf{P}(A) = \mathbf{P}(A)^2$, and so $A \in \mathcal{O}_{\mathbf{P}}$.

(c) The b-adic transformation is ergodic. Indeed, let $a = (a_n)_{n\geqslant 1}$, T and P be as in (31.3.f), and let $T' : \prod_{n\geqslant 1} I_n \to \prod_{n\geqslant 1} I_n$ denote the one-sided Bernoulli shift, where $I_n = \{0, \ldots, b-1\}$, $n \geqslant 1$. Since $a_n \circ T = a_{n+1}$, $n \geqslant 1$, it follows that $a \circ T = T' \circ a$. We have also $\mathcal{B}([0, 1[) = a^{-1}(\otimes_{n\geqslant 1}\mathcal{P}(I_n))$ and $\lambda a^{-1} = P$ (16.24), where λ is the Lebesgue measure on $\mathcal{B}([0, 1[)$, so that $a^{-1}(\mathcal{O}_P) \subset \mathcal{O}_\lambda$. As T' is P-ergodic by (b), applying (31.12.ii), we obtain $\mathcal{I}_T = a^{-1}(\mathcal{I}_{T'}) \subset a^{-1}(\mathcal{O}_P) \subset \mathcal{O}_\lambda$. Thus T is λ-ergodic.

The question whether T is ergodic is sometimes very deep, and the proof of the ergodicity exploits the specific nature of the transformation. Some general necessary and sufficient conditions for ergodicity are presented in the next two theorems as well as in (31.23), (31.28) and (31.52).

Theorem 31.16. *If T is P-preserving then the following assertions are equivalent:*
(i) *T is ergodic;*
(ii) $\mathcal{I}' \subset \mathcal{O}_P$;
(iii) $P(\cup_{n\geqslant 1}T^{-n}(A)) = 1$ *whenever* $A \in \mathcal{A}$ *with* $P(A) > 0$;
(iv) *for* $A, B \in \mathcal{A}$ *with* $P(A)P(B) > 0$ *there is* $n \geqslant 1$ *such that* $P(T^{-n}(A) \cap B) > 0$.

Proof. Assume that $\mathcal{I} \subset \mathcal{O}_P$, let $A \in \mathcal{I}'$, and put $B = \liminf_n T^{-n}(A)$. Then $A \overset{a.s.}{=} B$ by (31.10.f). Since $B \in \mathcal{I}$, we have $P(B) = 0$ or 1, and so $P(A) = 0$ or 1. Thus (i) $\Longrightarrow$ (ii). Suppose now that $\mathcal{I}' \subset \mathcal{O}_P$, let $A \in \mathcal{A}$ with $P(A) > 0$, and put $C = \cup_{n\geqslant 1}T^{-n}(A)$. Then $P(C) \geqslant P(T^{-1}(A)) = P(A) > 0$ and $T^{-1}(C) \subset C$. In view of (31.2.d), we have $C \in \mathcal{I}'$, and so $P(C) = 1$. Hence (ii) $\Longrightarrow$ (iii). If (iii) holds and $A, B \in \mathcal{A}$ with $P(A)P(B) > 0$, then we obtain $0 < P(B) = P((\cup_{n\geqslant 1}T^{-n}(A)) \cap B) \leqslant \sum_{n\geqslant 1} P(T^{-n}(A) \cap B)$, and so $P(T^{-n}(A) \cap B) > 0$ for some $n \geqslant 1$. Thus (iii) $\Longrightarrow$ (iv). Finally, assume that (iv) holds, and let $A \in \mathcal{I}$. If $0 < P(A) < 1$, then $P(T^{-n}(A) \cap A^c) > 0$ for some $n \geqslant 1$. But $P(T^{-n}(A) \cap A^c) = P(A \cap A^c) = 0$, a contradiction. Consequently, (iv) $\Longrightarrow$ (i). □

Theorem 31.17. *Suppose that T is P-preserving, and let $f : X \to R$ be a measurable function. Then the following assertions are equivalent:*

(i) *T is ergodic;*
(ii) *if $f \circ T \geqslant f$ a.s., then f is constant a.s.;*
(iii) *if $f \circ T = f$, then f is constant a.s.*

Proof. Assume that T is ergodic, $f \circ T \geqslant f$ a.s., but f is not constant a.s. Then there is $a \in R$ such that, putting $A = \{f \geqslant a\}, 0 < P(A) < 1$. As $A \overset{a.s.}{\subset} T^{-1}(A)$, it follows that $P(T^{-1}(A)\Delta A) = 0$ by (31.2.d), and so $P(A) = 0$ or 1 by (31.16). This contradiction shows that (i) $\Longrightarrow$ (ii). Trivially, (ii) $\Longrightarrow$ (iii). Suppose now that (iii) holds, and let A be invariant. Then 1_A is invariant by (31.10.b), and so 1_A is constant a.s. Hence $1_A = 0$ or 1 a.s., i.e. $P(A) = 0$ or 1. □

The necessary mathematical machinery is now built up to formulate and prove the fundamental ergodic theorems. The ingenious proof of the next theorem is due to A. Garsia.

Maximal ergodic theorem 31.18. *Assume that T is P-preserving, let $f \in \mathfrak{L}^1$, and put $U_n = \sum_{i=0}^{n-1} f \circ T^i$, $M_n = U_1 \vee \cdots \vee U_n$ and $N_n = 0 \vee M_n$ for $n \geqslant 1$. Then*

$$\int_{\{M_n>0\}} f\,dP \geqslant 0. \tag{1}$$

Proof. We have $N_n \circ T \geqslant U_1 - f$ and $N_n \circ T \geqslant U_i \circ T = U_{i+1} - f, 1 \leqslant i \leqslant n$, and so $N_n \circ T \geqslant M_n - f$. Therefore,

$$\begin{aligned}\int_{\{N_n>0\}} f dP &\geqslant \int_{\{N_n>0\}} M_n\,dP - \int_{\{N_n>0\}} (N_n \circ T) dP \\ &= \int_{\{N_n>0\}} N_n\,dP - \int_{\{N_n>0\}} (N_n \circ T) dP.\end{aligned} \tag{2}$$

Since $\int_{\{N_n>0\}} N_n\,dP = \int_X N_n\,dP$, $\int_{\{N_n>0\}} (N_n \circ T) dP \leqslant \int_X (N_n \circ T) dP = \int_X N_n\,dP$ by (11.29), and $\int_{\{M_n>0\}} f\,dP = \int_{\{N_n>0\}} f\,dP$, (1) follows from (2). □

Corollary 31.19. *Let T, f, U_n and M_n be as in* (31.18), *suppose that A is invariant, let $S_n(f)$ be as in* (31.9.f), *and denote $B_\varepsilon = \{\sup_{n\geqslant 1} S_n(f) > \varepsilon\}, \varepsilon \in R$. Then $\int_{A\cap B_\varepsilon} f\,dP \geqslant \varepsilon P(A \cap B_\varepsilon)$.*

Proof. Write $U_n = U_n(f)$ to stress the dependence of U_n on f. Then, as $T^{-1}(A) = A$, we see that $U_n(f 1_A) = U_n(f) 1_A$, so that applying (31.18) to $f 1_A$ yields $\int_{A\cap\{M_n>0\}} f\,dP \geqslant 0$, and so $\int_{A\cap\{\sup_{n\geqslant 1} U_n>0\}} f\,dP \geqslant 0$ by dominated convergence. Finally, as $S_n(f) = U_n(f)/n$, replacing f by $f - \varepsilon$ we get the result. □

The following famous theorem is sometimes called the pointwise ergodic theorem.

Birkhoff ergodic theorem 31.20. *If T is P-preserving and $f \in \mathfrak{L}^1$, then*

$$\frac{1}{n}\sum_{i=0}^{n-1} f \circ T^i \overset{a.s.}{\to} E[f|\mathcal{I}]. \tag{1}$$

Proof. (I) Assume first that $E[f|\mathcal{I}] = 0$ a.s., let $\varepsilon > 0$, put $A_\varepsilon = \{\limsup_n S_n(f) > \varepsilon\}$ and let B_ε be as in (31.19). Then $A_\varepsilon \cap B_\varepsilon = A_\varepsilon$ and $A_\varepsilon \in \mathcal{I}$ by (31.9.f) and (31.11.iii). Applying (31.19) with $A = A_\varepsilon$, we get

$$0 \leqslant \int_{A_\varepsilon \cap B_\varepsilon} (f - \varepsilon)dP = \int_{A_\varepsilon} (f - \varepsilon)dP = \int_{A_\varepsilon} E[(f - \varepsilon)|\mathcal{I}]dP = -\varepsilon P(A_\varepsilon),$$

which implies $P(A_\varepsilon) = 0$, and so $\limsup_n S_n(f) \leqslant \varepsilon$ a.s. As ε is arbitrary, we have $\limsup_n S_n(f) \leqslant 0$ a.s. Consequently, $\liminf_n S_n(f) = -\limsup_n S_n(-f) \geqslant 0$ a.s. Hence (1) holds in this case.

(II) Consider now the general case, fix a version of $E[f|\mathcal{I}]$, and set $g = f - E[f|\mathcal{I}]$. Since $E[f|\mathcal{I}] \circ T = E[f|\mathcal{I}]$ by (31.11.iii), we see that $S_n(g) = S_n(f) - E[f|\mathcal{I}]$, $n \geqslant 1$. Therefore, as $E[g|\mathcal{I}] = 0$ a.s., we obtain $S_n(f) \overset{a.s.}{\to} E[f|\mathcal{I}]$ by (I). □

Remarks 31.21. (a) If T is P-preserving and ergodic, and $f \in \mathfrak{L}^1$, then $\frac{1}{n}\sum_{i=0}^{n-1} f \circ T^i \overset{a.s.}{\to} Ef$. Indeed, $E[f|\mathcal{I}]$ is invariant by (31.11.iii), and so constant a.s. by (31.17.iii). Thus $\frac{1}{n}\sum_{i=0}^{n-1} f \circ T^i \overset{a.s.}{\to} E[E[f|\mathcal{I}]] = Ef$ by (31.20).

(b) Taking $f = 1_A$ with $A \in \mathcal{A}$ in (a) yields $\frac{1}{n}\sum_{i=0}^{n-1} 1_A(T^i(x)) \to P(A)$ for almost all $x \in X$. Since the proportion of number of elements of $\{x, \dots, T^{n-1}(x)\}$ in A equals $\frac{1}{n}\sum_{i=0}^{n-1} 1_A(T^i(x))$, this shows that the orbit of almost every $x \in X$ enters A with asymptotic relative frequency $P(A)$.

(c) Assume that T is P-preserving. If f is a bounded random variable and $g \in \mathfrak{L}^1$, then $\frac{1}{n}\sum_{i=0}^{n-1} E[(f \circ T^i)g] \to E[E[f|\mathcal{I}]E[g|\mathcal{I}]]$; in particular, if $A, B \in \mathcal{A}$, then $\frac{1}{n}\sum_{i=0}^{n-1} P(T^{-i}(A) \cap B) \to E[P(A|\mathcal{I})P(B|\mathcal{I})]$. If T is also ergodic, then $\frac{1}{n}\sum_{i=0}^{n-1} E[(f \circ T^i)g] \to (Ef)Eg$; in particular, $\frac{1}{n}\sum_{i=0}^{n-1} P(T^{-i}(A) \cap B) \to P(A)P(B)$. Actually, $\frac{1}{n}\sum_{i=0}^{n-1} f \circ T^i \overset{a.s.}{\to} E[f|\mathcal{I}]$ by (31.20). Multiplying by g, and using the bounded convergence theorem (11.25.ii), we obtain $\frac{1}{n}\sum_{i=0}^{n-1} E[(f \circ T^i)g] \to E[E[f|\mathcal{I}]g] = E[E[f|\mathcal{I}]E[g|\mathcal{I}]]$.

Example 31.22. Consider the b-adic transformation $T = r_1$ in (31.3.f). Then T preserves the Lebesgue measure λ on $\mathcal{B}([0, 1[)$ and is ergodic by (31.3.f) and (31.15.c). The b-adic expansion of $x \in [0, 1[$ is $x = \sum_{n \geqslant 1} a_n(x)/b^n$, where $r_n = r_1 \circ r_{n-1}$ and $a_n = a_1 \circ r_{n-1}$ (with $r_0(x) = x, x \in [0, 1[$). Taking $A = \{a_1 = i\}$ with $i \in \{0, \dots, b-1\}$ in (31.21.b), we have $\frac{1}{n}\sum_{k=0}^{n-1} 1_{\{a_1 = i\}} \circ r_k = \frac{1}{n}\sum_{k=0}^{n-1} 1_{\{a_{k+1} = i\}} = \frac{1}{n}\sum_{k=1}^{n} 1_{\{a_k = i\}} \overset{a.s.}{\to} \lambda(\{a_1 = i\}) = 1/b$. This proves that almost all (with respect to λ) numbers in $[0, 1[$ are normal to the base b (see also (27.35)).

The next theorem is sometimes used to establish ergodicity of a transformation.

Theorem 31.23. *Assume that T is P-preserving, and let $\mathcal{A}_0$ be an algebra that generates $\mathcal{A}$. Then the following statements are equivalent:*

(i) T *is ergodic;*

(ii) $\frac{1}{n}\sum_{i=0}^{n-1} 1_A \circ T^i \overset{a.s.}{\to} P(A)$ *for any* $A \in \mathcal{A}$;

(iii) $\frac{1}{n}\sum_{i=0}^{n-1} 1_A \circ T^i \overset{a.s.}{\to} P(A)$ *for any* $A \in \mathcal{A}_0$;

(iv) $\frac{1}{n}\sum_{i=0}^{n-1} P(T^{-i}(A) \cap B) \to P(A)P(B)$ *for any* $A, B \in \mathcal{A}_0$;

(v) $\frac{1}{n}\sum_{i=0}^{n-1} P(T^{-i}(A) \cap B) \to P(A)P(B)$ *for any* $A, B \in \mathcal{A}$.

Proof. (i) $\Longrightarrow$ (ii) by (31.21.b), and (ii) $\Longrightarrow$ (iii) trivially. If (iii) holds and $B \in \mathcal{A}_0$, then $\frac{1}{n}\sum_{i=0}^{n-1}(1_A \circ T^i)1_B \overset{a.s.}{\to} P(A)1_B$, whence (iv) follows by dominated convergence. Suppose now that (iv) holds, let $A, B \in \mathcal{A}$ and $\varepsilon > 0$, and use (10.33) to choose $A_0, B_0 \in \mathcal{A}_0$ such that $P(A \Delta A_0) < \varepsilon$ and $P(B \Delta B_0) < \varepsilon$. We have

$$\begin{aligned} |P(A)P(B) - P(A_0)P(B_0)| &\leqslant |P(A) - P(A_0)| + |P(B) - P(B_0)| \\ &\leqslant P(A \Delta A_0) + P(B \Delta B_0) < 2\varepsilon. \end{aligned} \tag{1}$$

Also, as $PT^{-1} = P$, for $0 \leqslant i \leqslant n-1$,

$$\begin{aligned} \left|P(T^{-i}(A) \cap B) - P(T^{-i}(A_0) \cap B_0)\right| &\leqslant P((T^{-i}(A) \cap B)\Delta(T^{-i}(A_0) \cap B_0)) \\ &\leqslant P(T^{-i}(A)\Delta T^{-i}(A_0)) \\ &\quad + P(B \Delta B_0) < 2\varepsilon. \end{aligned}$$

Hence

$$\left|\frac{1}{n}\sum_{i=0}^{n-1} P(T^{-i}(A) \cap B) - \frac{1}{n}\sum_{i=0}^{n-1} P(T^{-i}(A_0) \cap B_0)\right| < 2\varepsilon. \tag{2}$$

Since ε is arbitrary, (1) and (2) entail that $\frac{1}{n}\sum_{i=0}^{n-1} P(T^{-i}(A) \cap B) \to P(A)P(B)$. Thus (iv) $\Longrightarrow$ (v). Finally, if (v) holds and $T^{-1}(A) = A$, then, taking $B = A$, we get $\frac{1}{n}\sum_{i=0}^{n-1} P(A) \to P(A)^2$, which is possible only if $P(A) = 0$ or 1. Thus (v) $\Longrightarrow$ (i). □

Remark 31.24. The property of T in (31.23) of being P-preserving is used only to show that (iv) $\Longrightarrow$ (v). Since the proof of (ii) $\Longrightarrow$ (v) is similar to that of (iii) $\Longrightarrow$ (iv), we see that T is ergodic whenever (ii) or (v) holds, without the assumption $PT^{-1} = P$.

Example 31.25. In (31.15.b) we verified that both the one-sided and two-sided Bernoulli shifts are **P**-ergodic. This can now be easily proved on account of (31.23). Indeed, T is **P**-preserving, where **P** is as in (31.15.b), and $\cup_{J \in \mathcal{J}} \mathcal{A}_J$ is an algebra that generates $\otimes_{i \in I} \mathcal{A}_i$, where $\cup_{J \in \mathcal{J}} \mathcal{A}_J$ is as in (7.21). If $A \in \mathcal{A}_{\mathcal{J}_A}$, $B \in \mathcal{A}_{\mathcal{J}_B}$, and n is so large that $(J_A + n) \cap J_B = \emptyset$, then $\mathbf{P}(T^{-n}(A) \cap B) = \mathbf{P}(T^{-n}(A))\mathbf{P}(B) = \mathbf{P}(A)\mathbf{P}(B)$ by independence. Consequently, $\frac{1}{n}\sum_{i=0}^{n-1} \mathbf{P}(T^{-i}(A) \cap B) \to \mathbf{P}(A)\mathbf{P}(B)$.

Another example of application of (31.23) is the following theorem concerning both the one-sided and two-sided Markov shifts. We need first an auxiliary result.

Lemma 31.26. *Let T, $\mathbf{p}$, Π and P be as in* (31.3.e), *and assume that X is finite and $p_x > 0$ for all $x \in X$. Then the matrix $Q = (q_{xx'})_{x,x' \in X} = \lim_n \frac{1}{n}\sum_{k=0}^{n-1} \Pi^k$ exists, Q is a stochastic matrix, $Q\Pi = \Pi Q$ and $Q^2 = Q$.*

Proof. Let $x, x' \in X$. Taking $A = \{\pi_0 = x'\}$ and $B = \{\pi_0 = x\}$ in (31.21.c), and using (31.3.e), we see that $\lim_n \frac{1}{n}\sum_{k=0}^{n-1} P(T^{-k}(A) \cap B) = \lim_n \frac{1}{n}\sum_{k=0}^{n-1} P(\pi_0 = x, \pi_k = x') = p_x \lim_n \frac{1}{n}\sum_{k=0}^{n-1} p_{xx'}^{(k)}$ exists, and so $\lim_n \frac{1}{n}\sum_{k=0}^{n-1} p_{xx'}^{(k)}$ exists. Setting $q_{xx'} = \lim_n \frac{1}{n}\sum_{k=0}^{n-1} p_{xx'}^{(k)}$, the first assertion is proved. In matrix form, we may write $Q = \lim_n \frac{1}{n}\sum_{k=0}^{n-1} \Pi^k$, whence the following two assertions follow at once. As for the last one, notice that $Q\Pi^k = \Pi^k Q = Q$, $k \geqslant 0$, and so $Q^2 = \lim_n \frac{1}{n}\sum_{k=0}^{n-1} Q\Pi^k = Q$. □

Theorem 31.27. *Notation and assumptions are as in* (31.26). *The following statements are equivalent:*

(i) *T is ergodic;*

(ii) *all rows of Q are identical;*

(iii) *each entry in Q is positive;*

(iv) *Π is irreducible, i.e. for any $x, x' \in X$, there is $k = k(x, x') \geqslant 1$ such that $p_{xx'}^{(k)} > 0$.*

Proof. We shall show (i) $\Longrightarrow$ (ii) $\Longrightarrow$ (iii) $\Longrightarrow$ (iv) $\Longrightarrow$ (iii) $\Longrightarrow$ (ii) $\Longrightarrow$ (i). Let $x, x' \in X$. If T is ergodic, then, according to (31.23) and the proof of (31.26) with $A = \{\pi_0 = x'\}$ and $B = \{\pi_0 = x\}$, we have $p_x q_{xx'} = P(\pi_0 = x')P(\pi_0 = x) = p_x p_{x'}$, and so $q_{xx'} = p_{x'}$. Thus (i) $\Longrightarrow$ (ii). As $\mathbf{p}\Pi^k = \mathbf{p}$, $k \geqslant 0$, we get $\mathbf{p}Q = \lim_n \frac{1}{n}\sum_{k=0}^{n-1} \mathbf{p}\Pi^k = \mathbf{p}$. Therefore, if (ii) holds, then $q_{xx'} = p_{x'} > 0$, i.e. (iii) is verified. If (iii) holds, then $0 < q_{xx'} = \lim_n \frac{1}{n}\sum_{k=0}^{n-1} p_{xx'}^{(k)}$, and so $p_{xx'}^{(k)} > 0$ for some $k \geqslant 1$, i.e. (iv) holds. Suppose now that Π is irreducible. Since Q is stochastic, there is $y \in X$ with $q_{xy} > 0$. As Π is irreducible, there is $k \geqslant 1$ such that $p_{yx'}^{(k)} > 0$. Since $Q = Q\Pi^k$, it follows that $q_{xx'} \geqslant q_{xy} p_{yx'}^{(k)} > 0$. Hence (iv) $\Longrightarrow$ (iii). If (iii) holds, assume there is $z \in X$ with $q_{zx'} < q_{x'} = \sup_{y \in X} q_{yx'}$. Then, as $Q^2 = Q$, $q_{xx'} = \sum_{y \in X} q_{xy} q_{yx'} < \sum_{y \in X} q_{xy} q_{x'} = q_{x'}$ for all $x \in X$. This contradiction shows that (iii) $\Longrightarrow$ (ii). Finally, suppose that (ii) holds, and consider the sets $A = \{\pi_{i_1} = x_1, \ldots, \pi_{i_l} = x_l\}$ and $B = \{\pi_{j_1} = y_1, \ldots, \pi_{j_m} = y_m\}$, where the sets $\{i_1, \ldots, i_l\} \subset I$ and $\{j_1, \ldots, j_m\} \subset I$ consist of consecutive integers. Then A and B belong to the algebra $\mathcal{A}_0$ in (16.40) that generates $\otimes_{i \in I}\mathcal{P}(X_i)$, and $\lim_n \frac{1}{n}\sum_{k=0}^{n-1} P(T^{-k}(A) \cap B)$ exists by (31.21.c). To prove that T is ergodic, in view of (31.23), it suffices to verify that this limit is $P(A)P(B)$. If $j_m < i_1 + k$, then

$$\begin{aligned} P(T^{-k}(A) \cap B) &= P(\{\pi_{j_1} = y_1, \ldots, \pi_{j_m} = y_m\} \cap \{\pi_{i_1+k} = x_1, \ldots, \pi_{i_l+k} = x_l\}) \\ &= p_{y_1} p_{y_1 y_2} \cdots p_{y_{m-1} y_m} p_{y_m x_1}^{(i_1+k-j_m)} p_{x_1 x_2} \cdots p_{x_{l-1} x_l}. \qquad (1) \end{aligned}$$

We know that (ii) implies $q_{y_m x_1} = p_{x_1}$. Therefore, (1) entails that

$$\lim_n \frac{1}{n} \sum_{k=0}^{n-1} P(T^{-k}(A) \cap B)$$
$$= p_{y_1} p_{y_1 y_2} \cdots p_{y_{m-1} y_m} p_{x_1} p_{x_1 x_2} \cdots p_{x_{l-1} x_l} = P(A)P(B).$$

Thus (ii) $\Longrightarrow$ (i). □

Theorem 31.28.
(i) *If T is P-preserving and ergodic, then*

$$\frac{1}{n} \sum_{i=0}^{n-1} E[(f \circ T^i)g] \to (Ef)Eg, \quad f, g \in \mathfrak{L}^2. \tag{1}$$

(ii) *If (1) holds, then T is ergodic.*

Proof. Suppose that T is P-preserving and ergodic, let $f, g \in \mathfrak{L}^2$, and put $S_n(f, g) = \frac{1}{n} \sum_{i=0}^{n-1} E[(f \circ T^i)g]$, $n \geqslant 1$. Then, on account of Schwarz's inequality, $|S_n(f, g)| \leqslant \frac{1}{n} \sum_{i=0}^{n-1} \|f \circ T^i\|_2 \|g\|_2 = \|f\|_2 \|g\|_2$, $n \geqslant 1$. For $\varepsilon > 0$, use (13.14) to choose a simple function $h \in \mathfrak{L}^2$ with $\|f - h\|_2 < \varepsilon$. Then, applying (31.21.c), choose $n_\varepsilon \in N$ such that $|S_n(h, g) - (Eh)Eg| < \varepsilon$, $n \geqslant n_\varepsilon$. For $n \geqslant n_\varepsilon$, we have

$$|S_n(f, g) - (Ef)Eg| \leqslant |S_n(f, g) - S_n(h, g)| + |S_n(h, g) - (Eh)Eg|$$
$$+ |(Eh)Eg - (Ef)Eg| < |S_n(f - h, g)| + \varepsilon + |E[h - f]|\,|Eg|$$
$$\leqslant 2\|f - h\|_2 \|g\|_2 + \varepsilon \leqslant 2\varepsilon \|g\|_2 + \varepsilon,$$

which proves (1). If (1) holds, then, taking $f = 1_A$ and $g = 1_B$ with $A, B \in \mathcal{A}$, we get $\frac{1}{n} \sum_{i=0}^{n-1} P(T^{-i}(A) \cap B) \to P(A)P(B)$. This implies that T is ergodic by (31.23). □

Remark 31.29. If X is a metric space, T is P-preserving, and $\frac{1}{n} \sum_{i=0}^{n-1} E[(h \circ T^i)g] \to (Eh)Eg$ for any $h \in \mathfrak{C}_r(X)$ and $g \in \mathfrak{L}^2(X, \mathcal{B}(X), P)$, then T is ergodic. Actually, for $f \in \mathfrak{L}^2(X, \mathcal{B}(X), P)$ and $\varepsilon > 0$, use (21.1) to choose a function $h \in \mathfrak{C}_r(X)$ with $\|f - h\|_2 < \varepsilon$. Then, following the reasoning in the proof of (31.28), we see that (1) therein holds, and so T is ergodic.

Theorem 31.30. *Assume that X is a metric space and T is P-preserving, and let* P *be as in Section 22. Then T is ergodic if and only if $\frac{1}{n} \sum_{i=0}^{n-1} QT^{-i} \Rightarrow P$ for any $Q \in$* P *with $Q \ll P$.*

Proof. Suppose that T is P-preserving and ergodic, let $Q \in$ P with $Q \ll P$, let $g = dQ/dP$ and let $f \in \mathfrak{C}_r(X)$. Then, in view of (31.21.c),

$$\int_X f\, d\left(\frac{1}{n} \sum_{i=0}^{n-1} QT^{-i}\right) = \frac{1}{n} \sum_{i=0}^{n-1} \int_X f\, d(QT^{-i}) = \frac{1}{n} \sum_{i=0}^{n-1} \int_X (f \circ T^i) dQ$$
$$= \frac{1}{n} \sum_{i=0}^{n-1} \int_X (f \circ T^i) g\, dP \to (Ef)Eg = \int_X f\, dP.$$

Hence $\frac{1}{n}\sum_{i=0}^{n-1} QT^{-i} \Rightarrow P$.

Conversely, assume that the convergence condition holds, let $h \in \mathfrak{C}_r(X)$ and $g \in \mathfrak{L}^2(X, \mathcal{B}(X), P)$, and define $Q^+(A) = \int_A g^+\,dP$, $A \in \mathcal{B}(X)$, and $Q^-(A) = \int_A g^-\,dP$, $A \in \mathcal{B}(X)$. Then

$$\frac{1}{n}\sum_{i=0}^{n-1}\int_X (h\circ T^i)g^+\,dP = \int_X hd\left(\frac{1}{n}\sum_{i=0}^{n-1} Q^+T^{-i}\right) \to \int_X h\,dP = (Eh)Eg^+. \quad (1)$$

Equation (1) holds as well with g^- in place of g^+, and so $\frac{1}{n}\sum_{i=0}^{n-1} E[(h\circ T^i)g] \to (Eh)Eg$. Therefore, T is ergodic by (31.29). □

Theorem 31.31. *Assume that X is a compact metric space and T is continuous on X. Then the following statements are equivalent:*

(i) *for any $f \in \mathfrak{C}_r(X)$, $\frac{1}{n}\sum_{i=0}^{n-1} f\circ T^i \xrightarrow{u} c$ for some $c \in R$;*

(ii) *for any $f \in \mathfrak{C}_r(X)$, $\frac{1}{n}\sum_{i=0}^{n-1} f\circ T^i \to c$ for some $c \in R$;*

(iii) *there exists a probability P on $\mathcal{B}(X)$ which is preserved by T such that $\frac{1}{n}\sum_{i=0}^{n-1} f\circ T^i \to \int_X f\,dP$ for all $f \in \mathfrak{C}_r(X)$;*

(iv) *there is only one probability P on $\mathcal{B}(X)$ such that $PT^{-1} = P$, and T is P-ergodic.*

Proof. (i) $\Longrightarrow$ (ii) trivially. If (ii) holds, define $L : \mathfrak{C}_r(X) \to R$ by $L(f) = \lim_n \frac{1}{n}\sum_{i=0}^{n-1} f\circ T^i$, $f \in \mathfrak{C}_r(X)$. Then L is a nonnegative linear functional with $L(1) = 1$. According to Riesz's representation theorem (21.17), there exists a unique probability P on $\mathcal{B}(X)$ such that $L(f) = \int_X f dP$, $f \in \mathfrak{C}_r(X)$. For $f \in \mathfrak{C}_r(X)$, we have $L(f\circ T) = L(f)$, and so $\int_X fd(PT^{-1}) = \int_X f\,dP$. Hence $PT^{-1} = P$ by (11.50). Thus (ii) $\Longrightarrow$ (iii). Suppose that (iii) holds, and that Q is a probability on $\mathcal{B}(X)$ with $QT^{-1} = Q$. Then $\int_X \left(\frac{1}{n}\sum_{i=0}^{n-1} f\circ T^i\right) dQ = \frac{1}{n}\sum_{i=0}^{n-1}\int_X (f\circ T^i)dQ = \int_X f\,dQ \to \int_X f\,dP$, $f \in \mathfrak{C}_r(X)$, by the bounded convergence theorem. This shows that $Q = P$ by (11.50). Hence P is the unique probability on $\mathcal{B}(X)$ that is preserved by T. To get a contradiction, assume T is not P-ergodic. Then there is $B \in \mathcal{B}(X)$ such that $T^{-1}(B) = B$ and $0 < P(B) < 1$. Since $P_B T^{-1} = P_B$ and $P_B \neq P$, the above uniqueness assertion is contradicted. Thus (iii) $\Longrightarrow$ (iv). If (iv) holds, then $\frac{1}{n}\sum_{i=0}^{n-1} f\circ T^i \xrightarrow{a.s.} \int_X f\,dP$, $f \in \mathfrak{C}_r(X)$, by (31.21.a). Suppose that $\frac{1}{n}\sum_{i=0}^{n-1} f\circ T^i \xrightarrow{u} \int_X f\,dP$, $f \in \mathfrak{C}_r(X)$, is false. Then there are $g \in \mathfrak{C}_r(X)$, $\varepsilon > 0$ and $\{x_k : k \in N\} \subset X$ such that $\left|\frac{1}{n_k}\sum_{i=0}^{n_k-1} g(T^i(x_k)) - \int_X g\,dP\right| \geqslant \varepsilon$ for some subsequence $\{n_k : k \in N\}$, i.e. $\left|\int_X g\,dP_{n_k} - \int_X g\,dP\right| \geqslant \varepsilon$, $k \in N$, where $P_{n_k} = \frac{1}{n_k}\sum_{i=0}^{n_k-1} \varepsilon_{x_k} T^{-i}$. Following (31.3.g), there exists a probability P' on $\mathcal{B}(X)$ with $P'T^{-1} = P'$ such that $P_{n_{k_l}} \Rightarrow P'$ for some subsequence $\{k_l : l \in N\}$. This implies $\left|\int_X g\,dP' - \int_X g\,dP\right| \geqslant \varepsilon$, and so $P' \neq P$, a contradiction. Thus (iv) $\Longrightarrow$ (i). □

von Neumann ergodic theorem 31.32. *If T is P-preserving and $f \in \mathfrak{L}^p$ with $1 \leqslant p < \infty$, then $\frac{1}{n}\sum_{i=0}^{n-1} f\circ T^i \xrightarrow{\mathfrak{L}^p} E[f|\mathcal{I}]$.*

Proof. Let $S_n(f), n \geqslant 1$, be as in (31.9.f). Then, on account of Minkowski's inequality, $\|S_n(f)\|_p \leqslant \frac{1}{n}\sum_{i=0}^{n-1}\|f \circ T^i\|_p = \|f\|_p, n \geqslant 1$. For $\varepsilon > 0$, choose a simple function $h \in \mathfrak{L}^p$ with $\|f-h\|_p < \varepsilon$ (13.14). Then $S_n(h) - E[h|\mathcal{I}] \overset{a.s.}{\to} 0$ by (31.20), and so $\|S_n(h) - E[h|\mathcal{I}]\|_p \to 0$ by dominated convergence. Choose $n_\varepsilon \in N$ such that $\|S_n(h) - E[h|\mathcal{I}]\|_p < \varepsilon, n \geqslant n_\varepsilon$. For $n \geqslant n_\varepsilon$, using (29.20), we get

$$\begin{aligned}\|S_n(f) - E[f|\mathcal{I}]\|_p &\leqslant \|S_n(f) - S_n(h)\|_p + \|S_n(h) - E[h|\mathcal{I}]\|_p \\ &+ \|E[h|\mathcal{I}] - E[f|\mathcal{I}]\|_p < \|S_n(f-h)\|_p + \varepsilon \\ &+ \|E[f-h|\mathcal{I}]\|_p \leqslant 2\|f-h\|_p + \varepsilon < 3\varepsilon,\end{aligned}$$

which shows that $S_n(f) \overset{\mathfrak{L}^p}{\to} E[f|\mathcal{I}]$. □

Before passing to the related notion of mixing, we take up the following remarkable dichotomy property.

Theorem 31.33. *Let P and Q be probabilities on $\mathcal{A}$, and assume that T is both P-preserving and P-ergodic and Q-preserving and Q-ergodic. Then either $P = Q$ or $P \perp Q$.*

Proof. If $P \neq Q$, there is $A \in \mathcal{A}$ with $P(A) \neq Q(A)$. Then, by (31.21.b), there exist sets $B_1, B_2 \in \mathcal{A}$ with $P(B_1) = Q(B_2) = 1$ such that $\frac{1}{n}\sum_{i=0}^{n-1} 1_A(T^i(x)) \to P(A), x \in B_1$, and $\frac{1}{n}\sum_{i=0}^{n-1} 1_A(T^i(x)) \to Q(A), x \in B_2$. This is possible only if $B_1 \cap B_2 = \emptyset$. As $P(B_1) = 1$ and $Q(B_1^c) \geqslant Q(B_2) = 1$, it follows that $P \perp Q$. □

Measure-preserving ergodic transformations possess also the next extremal property.

Theorem 31.34. *If T is P-preserving, then the following are equivalent:*

(i) *T is P-ergodic;*

(ii) *P cannot be represented in the form $P = tP_1 + (1-t)P_2$ with $0 < t < 1$, where P_1 and P_2 are probabilities on $\mathcal{A}$, $P_1T^{-1} = P_1$, $P_2T^{-1} = P_2$ and $P_1 \neq P_2$.*

Proof. Suppose that T is P-ergodic, but (ii) is false. Then $P_1 \ll P$, and so T is also P_1-ergodic by (31.14.b). As $P_1 \neq P_2$, (31.33) implies that $P_1 \perp P$, a contradiction. Thus (i) $\Longrightarrow$ (ii). If now T is not P-ergodic, then there is $A \in \mathcal{A}$ such that $T^{-1}(A) = A$ and $0 < P(A) < 1$. Since $P_AT^{-1} = P_A$, $P_{A^c}T^{-1} = P_{A^c}$, $P_A \neq P_{A^c}$ and $P = P(A)P_A + (1-P(A))P_{A^c}$, (ii) does not hold. Thus (ii) $\Longrightarrow$ (i). □

A notion closed to ergodicity is that of mixing we now discuss.

Definitions 31.35. We say that the transformation T is *weak-mixing* [*P-weak-mixing*] if $\frac{1}{n}\sum_{i=0}^{n-1}\left|P(T^{-i}(A)\cap B) - P(A)P(B)\right| \to 0$ for any $A, B \in \mathcal{A}$ and we say that T is *strong-mixing* [*P-strong-mixing*] if $P(T^{-n}(A)\cap B) \to P(A)P(B)$ for any $A, B \in \mathcal{A}$.

Remarks 31.36. (a) Any strong-mixing transformation is weak-mixing, and any weak-mixing transformation is ergodic (by (31.23) and (31.24)).

(b) Assume that T is P-preserving. Then, since $P(T^{-n}(A)) = P(A), n \geqslant 1$, strong-mixing may be interpreted as asymptotic independence of the events $T^{-n}(A)$ and B for any $A, B \in \mathcal{A}$, and ergodicity means intuitively that $T^{-n}(A)$ becomes independent of B on the average for any $A, B \in \mathcal{A}$.

(c) Many results involving ergodicity have correspondent statements concerning mixing.

(d) Let the transformation T and $\mathcal{A}_0$ be as in (31.23). Then T is weak-mixing if and only if $\frac{1}{n}\sum_{i=0}^{n-1}\left|P(T^{-i}(A)\cap B) - P(A)P(B)\right| \to 0$ for any $A, B \in \mathcal{A}_0$, and T is strong-mixing if and only if $P(T^{-n}(A) \cap B) \to P(A)P(B)$ for any $A, B \in \mathcal{A}_0$. The proof of these statements is similar to that of (31.23).

Examples 31.37. (a) T in (31.3.a) is ergodic (31.15.a), but it is not strong-mixing. Indeed, take $A = \{x_k\}$ and $B = \{x_l\}$ with $k \neq l$.

(b) Both the one-sided and two-sided Bernoulli shifts are strong-mixing. This can be shown on account of (31.36.d), following (31.25).

(c) Notation and assumptions are as in (31.26). Then the Markov shift T is strong-mixing if and only if $p_{xx'}^{(n)} \to p_{x'}$ for any $x, x' \in X$. Actually, if T is strong-mixing and $x, x' \in X$, then, taking $A = \{\pi_0 = x'\}$ and $B = \{\pi_0 = x\}$, we see that $P(T^{-n}(A) \cap B) = P(\pi_0 = x, \pi_n = x') = p_x p_{xx'}^{(n)} \to P(A)P(B) = p_{x'}p_x$, and so $p_{xx'}^{(n)} \to p_{x'}$. Conversely, if $p_{xx'}^{(n)} \to p_{x'}, x, x' \in X$, then, taking $A = \{\pi_{i_1} = x_1, \dots, \pi_{i_l} = x_l\}$ and $B = \{\pi_{j_1} = y_1, \dots, \pi_{j_m} = y_m\}$, where $\{i_1, \dots, i_l\} \subset I$ and $\{j_1, \dots, j_m\} \subset I$ consist of consecutive integers, on account of (1) in the proof of (31.27), we have for large n

$$\begin{aligned}P(T^{-n}(A) \cap B) &= p_{y_1}p_{y_1y_2}\cdots p_{y_{m-1}y_m}p_{y_mx_1}^{(i_1+n-j_m)}p_{x_1x_2}\cdots p_{x_{l-1}x_l}\\ &\to p_{y_1}p_{y_1y_2}\cdots p_{y_{m-1}y_m}p_{x_1}p_{x_1x_2}\cdots p_{x_{l-1}x_l} = P(A)P(B).\end{aligned}$$

As $A, B \in \mathcal{A}_0$, where $\mathcal{A}_0$ is the algebra in (16.40) generating $\otimes_{i\in I}\mathcal{P}(X_i)$, this and (31.36.d) show that T is strong-mixing.

The next theorem provides a way of checking the strong-mixing property of a transformation, given that another transformation has this property.

Theorem 31.38. *Let $(X, \mathcal{A}, P)$ and $(Y, \mathcal{B}, Q)$ be probability spaces, let $T_X : X \to X$ and $T_Y : Y \to Y$ be measurable transformations, and let $f : X \to Y$ be a function such that $f^{-1}(\mathcal{B}) = \mathcal{A}$, $Pf^{-1} = Q$ and $f \circ T_X = T_Y \circ f$. Then T_X is P-strong-mixing if and only if T_Y is Q-strong-mixing.*

Proof. Observe first that $f \circ T_X^n = T_Y^n \circ f$, $n \geqslant 1$. Suppose that T_X is P-strong-mixing and let $C, D \in \mathcal{B}$. Then we have

$$\begin{aligned}Q(T_Y^{-n}(C) \cap D) &= P(f^{-1}(T_Y^{-n}(C)) \cap f^{-1}(D)) = P(T_X^{-n}(f^{-1}(C)) \cap f^{-1}(D))\\ &\to P(f^{-1}(C))P(f^{-1}(D)) = Q(C)Q(D),\end{aligned}$$

which shows that T_Y is Q-strong-mixing.

Conversely, assume that T_Y is Q-strong-mixing, and let $A, B \in \mathcal{A}$. Then $A = f^{-1}(C)$ and $B = f^{-1}(D)$ for some $C, D \in \mathcal{B}$, and we obtain

$$\begin{aligned}P(T_X^{-n}(A) \cap B) &= P(T_X^{-n}(f^{-1}(C)) \cap f^{-1}(D)) = Q(T_Y^{-n}(C) \cap D)\\ &\to Q(C)Q(D) = P(f^{-1}(C))P(f^{-1}(D)) = P(A)P(B),\end{aligned}$$

which proves that T_X is P-strong-mixing. □

Remark 31.39. The above proof reveals that it suffices to assume only $f^{-1}(\mathcal{B}) \subset \mathcal{A}$ to ensure that T_Y is Q-strong-mixing whenever T_X is P-strong-mixing.

Corollary 31.40. *The b-adic transformation is strong-mixing.*

Proof. Let a, T and P be as in (31.3.f), and let T' be as in (31.15.c). Then $\lambda a^{-1} = P$, where λ is the Lebesgue measure on $\mathcal{B}([0, 1[)$, and $a \circ T = T' \circ a$. Since T' is P-strong-mixing by (31.37.b), it follows that T is λ-strong-mixing by (31.38). □

Theorem 31.41. *Notation is as in* (31.4).

(i) *If T is P-strong-mixing, then T_i is P_i-strong-mixing for each $i \in I$.*

(ii) *If T_i is P_i-preserving and P_i-strong-mixing for any $i \in I$, then T is P-strong-mixing.*

Proof. Set $\mathcal{A} = \otimes_{i \in I}\mathcal{A}_i$. Suppose that T is P-strong-mixing, and let $i \in I$. Applying (31.38) and (31.39) with $(Y, \mathcal{B}, Q) = (X_i, \mathcal{A}_i, P_i)$ and $f = \pi_i$, it follows that T_i is P_i-strong-mixing. Thus (i) holds. Assume now that $P_i T_i^{-1} = P_i$ and T_i is P_i-strong-mixing for any $i \in I$, let J be a nonempty finite subset of I, and let $A = \pi_J^{-1}(\prod_{j \in J} A_j)$ and $B = \pi_J^{-1}(\prod_{j \in J} B_j)$ be product cylinders in $\mathcal{A}$. For $n \geqslant 1$, we have $T^n = (T_i^n \circ \pi_i)_{i \in I}$, and so $\pi_J \circ T^n = (T_j^n \circ \pi_j)_{j \in J}$. Since

$$\begin{aligned} T^{-n}(A) &= T^{-n}\left(\pi_J^{-1}\left(\prod_{j \in J} A_j\right)\right) \\ &= (T_j^n \circ \pi_j)_{j \in J}^{-1}\left(\prod_{j \in J} A_j\right) = \bigcap_{j \in J}(T_j^n \circ \pi_j)^{-1}(A_j) \\ &= \bigcap_{j \in J} \pi_j^{-1}(T_j^{-n}(A_j)), \quad n \geqslant 1, \end{aligned}$$

$B = \cap_{j \in J}\pi_j^{-1}(B_j)$, and π_j, $j \in J$, are P-independent, we get

$$\begin{aligned} P(T^{-n}(A) \cap B) &= P\left(\bigcap_{j \in J} \pi_j^{-1}(T_j^{-n}(A_j) \cap B_j)\right) \\ &= \prod_{j \in J} P_j(T_j^{-n}(A_j) \cap B_j) \to \prod_{j \in J} P_j(A_j)P_j(B_j) \\ &= \left(\prod_{j \in J} P_j(A_j)\right) \prod_{j \in J} P_j(B_j) = P(A)P(B). \end{aligned}$$

Now if $A = \pi_{J_A}^{-1}(\prod_{j \in J_A} A_j)$ and $B = \pi_{J_B}^{-1}(\prod_{j \in J_B} B_j)$ are product cylinders in $\mathcal{A}$, let $J = J_A \cup J_B$, and write $A = \pi_J^{-1}(\prod_{j \in J} A'_j)$, where $A'_j = \begin{cases} A_j \text{ if } j \in J_A \\ X_j \text{ if } j \in J - J_A \end{cases}$, and $B = \pi_J^{-1}(\prod_{j \in J} B'_j)$, where $B'_j = \begin{cases} B_j \text{ if } j \in J_B \\ X_j \text{ if } j \in J - J_B \end{cases}$. Then, repeating the above argument, we obtain $P(T^{-n}(A) \cap B) \to P(A)P(B)$. Therefore, we have $P(T^{-n}(A) \cap$

$B) \to P(A)P(B)$ for any $A, B \in \mathcal{A}_0$, where $\mathcal{A}_0$ is as in (7.17.b), since each element of $\mathcal{A}_0$ is a finite union of pairwise product cylinders. Consequently, T is P-strong-mixing by (31.4) and (31.36.d). □

The next theorem is useful for testing whether a transformation has the weak-mixing property.

Theorem 31.42. *If T is P-preserving, then the following assertions are equivalent:*
(i) *T is weak-mixing;*
(ii) $\frac{1}{n}\sum_{i=0}^{n-1}\left|E[(f\circ T^i)f]-(Ef)^2\right| \to 0$ *for any* $f \in \mathfrak{L}^2$;
(iii) $\frac{1}{n}\sum_{i=0}^{n-1}\left|E[(f\circ T^i)g]-(Ef)Eg\right| \to 0$ *for any* $f, g \in \mathfrak{L}^2$.

Proof. Assume that T is weak-mixing, let $f \in \mathfrak{L}^2$, and let $A, B \in \mathcal{A}$. Then $\frac{1}{n}\sum_{i=0}^{n-1}|E[(1_A\circ T^i)1_B]-(E1_A)E1_B| \to 0$. Fixing B, it follows that $\frac{1}{n}\sum_{i=0}^{n-1}\left|E[(h\circ T^i)1_B]-(Eh)E1_B\right| \to 0$ for any finite simple function h. Then, fixing h, we see that $\frac{1}{n}\sum_{i=0}^{n-1}\left|E[(h\circ T^i)h]-(Eh)^2\right| \to 0$. For $\varepsilon > 0$, select a simple function $h \in \mathfrak{L}^2$ with $\|f-h\|_2 < \varepsilon$, then choose $n_\varepsilon \in N$ such that $\frac{1}{n}\sum_{i=0}^{n-1}\left|E[(h\circ T^i)h]-(Eh)^2\right| < \varepsilon, n \geqslant n_\varepsilon$. For $n \geqslant n_\varepsilon$, using Schwarz's inequality and (11.29), we have

$$\begin{aligned}
\frac{1}{n}\sum_{i=0}^{n-1}\left|E[(f\circ T^i)f]-(Ef)^2\right| &\leqslant \frac{1}{n}\sum_{i=0}^{n-1}\left|E[((f-h)\circ T^i)f]\right| \\
&+\frac{1}{n}\sum_{i=0}^{n-1}\left|E[(h\circ T^i)(f-h)]\right| + \frac{1}{n}\sum_{i=0}^{n-1}\left|E[(h\circ T^i)h]-(Eh)^2\right| \\
&+\frac{1}{n}\sum_{i=0}^{n-1}|(Ef)E[h-f]| < \frac{1}{n}\sum_{i=0}^{n-1}\|f-h\|_2\,\|f\|_2 \\
&+\frac{1}{n}\sum_{i=0}^{n-1}\|h\|_2\,\|f-h\|_2+\varepsilon \\
&+\|h\|_2\,\|f-h\|_2+\|f\|_2\,\|f-h\|_2 \leqslant 2\varepsilon(\|f\|_2+\|h\|_2)+\varepsilon,
\end{aligned}$$

which shows that (i) $\Longrightarrow$ (ii). Suppose now that (ii) holds, let $f \in \mathfrak{L}^2$, and write $M_f = \{\tilde{g} \in L^2 : \frac{1}{n}\sum_{i=0}^{n-1}\left|E[(f\circ T^i)g]-(Ef)Eg\right| \to 0\}$. Clearly, M_f is a linear subspace of L^2. Observe that $\tilde{f} \in M_f$ and $\tilde{1} \in M_f$ (as $E[f\circ T^n] = Ef, n \geqslant 0$). Moreover, if $\tilde{g} \in M_f$, then $\widetilde{g\circ T} \in M_f$, because

$$\begin{aligned}
&\left|E[(f\circ T^n)(g\circ T)]-(Ef)E(g\circ T)-(E[(f\circ T^n)g]-(Ef)Eg)\right| \\
&\quad= \left|E[(f\circ T^{n-1})g]-E[(f\circ T^n)g]\right| = \left|E[((f\circ T^{n-1})\right. \\
&\left.-(f\circ T^n))g]\right| \leqslant 2\,\|f\|_2\,\|g\|_2\,, \quad n \geqslant 1,
\end{aligned}$$

by (11.29) and Schwarz's inequality. M_f is a closed subset of L^2. Indeed, let $\{\widetilde{g_k} : k \in N\} \subset M_f$ and $\tilde{g} \in L^2$ be such that $\|g-g_k\|_2 \to 0$. For $\varepsilon > 0$, select $k \in N$ with

$\|g - g_k\|_2 < \varepsilon$, then choose $n_\varepsilon \in N$ such that $\frac{1}{n}\sum_{i=0}^{n-1}\left|E[(f\circ T^i)g_k] - (Ef)Eg_k\right| < \varepsilon, n \geqslant n_\varepsilon$. For $n \geqslant n_\varepsilon$, we obtain

$$\begin{aligned}\frac{1}{n}\sum_{i=0}^{n-1}\left|E[(f\circ T^i)g] - (Ef)Eg\right| &\leqslant \frac{1}{n}\sum_{i=0}^{n-1}\left|E[(f\circ T^i)(g - g_k)\right| \\ &+\frac{1}{n}\sum_{i=0}^{n-1}\left|E[(f\circ T^i)g_k] - (Ef)Eg_k\right| + \frac{1}{n}\sum_{i=0}^{n-1}|(Ef)E[g_k - g]| \\ &< \|f\|_2\,\|g - g_k\|_2 + \varepsilon + \|f\|_2\,\|g - g_k\|_2 \leqslant 2\varepsilon\,\|f\|_2 + \varepsilon, \end{aligned} \tag{1}$$

which shows that $\tilde{g} \in M_f$. Thus M_f is closed by (6.15). Next, let $g \in \mathfrak{L}^2$, and consider the decomposition $g = g_0 + h_0$ a.s. in (13.33), where $\widetilde{g_0} \in M_f$ and $E[g'h_0] = 0$ whenever $\widetilde{g'} \in M_f$. Therefore, since $\widetilde{f\circ T^n} \in M_f, n \geqslant 0$, and $\tilde{1} \in M_f$, we have $\frac{1}{n}\sum_{i=0}^{n-1}\left|E[(f\circ T^i)h_0] - (Ef)Eh_0\right| = 0, n \geqslant 1$, and so $\widetilde{h_0} \in M_f$. Hence $Eh_0^2 = 0$, which implies that $g = g_0 + 0 = g_0$ a.s., and so $\tilde{g} \in M_f$. Thus (ii) $\Longrightarrow$ (iii). If (iii) holds, then, putting $f = 1_A$ and $g = 1_B$ with $A, B \in \mathcal{A}$, (i) obtains. □

Remarks 31.43. (a) If T is P-preserving, then T is strong-mixing $\Longleftrightarrow E[(f\circ T^n)f] \to (Ef)^2, f \in \mathfrak{L}^2, \Longleftrightarrow E[(f\circ T^n)g] \to (Ef)Eg, f, g \in \mathfrak{L}^2$. The proof of these equivalences is similar to the proof of (31.42).

(b) T is weak-mixing whenever (iii) in (31.42) holds, without the assumption $PT^{-1} = P$; similarly, T is strong-mixing whenever $E[(f\circ T^n)g] \to (Ef)Eg, f, g \in \mathfrak{L}^2$, without assuming $PT^{-1} = P$.

Corollary 31.44. *If T is P-preserving and weak-mixing, then $\frac{1}{n}\sum_{i=0}^{n-1}|E[(f\circ T^i)g] - (Ef)Eg| \to 0$ for any bounded random variable f and $g \in \mathfrak{L}^1$.*

Proof. For $\varepsilon > 0$, use (13.14) to choose a simple function g_ε with $\|g - g_\varepsilon\|_1 < \varepsilon$. Then, applying (31.42), choose $n_\varepsilon \in N$ such that $\frac{1}{n}\sum_{i=0}^{n-1}\left|E[(f\circ T^i)g_\varepsilon] - (Ef)Eg_\varepsilon\right| < \varepsilon, n \geqslant n_\varepsilon$. Paralleling (1) in the proof of (31.42), we find $\frac{1}{n}\sum_{i=0}^{n-1}|E[(f\circ T^i)g] - (Ef)Eg| \leqslant 2\varepsilon\,\|f\|_u + \varepsilon, n \geqslant n_\varepsilon$. □

Corollary 31.45. *If X is a metric space, T is P-preserving, and $\frac{1}{n}\sum_{i=0}^{n-1}|E[(h\circ T^i)g] - (Eh)Eg| \to 0$ for any $h \in \mathfrak{C}_r(X)$ and $g \in \mathfrak{L}^2(X, \mathcal{B}(X), P)$, then T is weak-mixing.*

Proof. Let $f, g \in \mathfrak{L}^2(X, \mathcal{B}(X), P)$. For $\varepsilon > 0$, use (21.1) to choose a function $h \in \mathfrak{C}_r(X)$ with $\|f - h\|_2 < \varepsilon$, then select $n_\varepsilon \in N$ such that $\frac{1}{n}\sum_{i=0}^{n-1}|E[(h\circ T^i)g] -$

$(Eh)Eg| < \varepsilon, n \geqslant n_\varepsilon$. For $n \geqslant n_\varepsilon$, we get

$$\begin{aligned}\frac{1}{n}\sum_{i=0}^{n-1}\left|E[(f\circ T^i)g]-(Ef)Eg\right| &\leqslant \frac{1}{n}\sum_{i=0}^{n-1}\left|E[((f-h)\circ T^i)g]\right|\\ &+\frac{1}{n}\sum_{i=0}^{n-1}|E[(h\circ T^i)g]-(Eh)Eg|\\ &+\frac{1}{n}\sum_{i=0}^{n-1}|E[h-f]Eg| < \|f-h\|_2\,\|g\|_2\\ &+\varepsilon+\|f-h\|_2\,\|g\|_2 \leqslant 2\varepsilon\,\|g\|_2+\varepsilon,\end{aligned}$$

which shows that $\frac{1}{n}\sum_{i=0}^{n-1}|E[(f\circ T^i)g]-(Ef)Eg| \to 0$. Hence T is weak-mixing by (31.42). □

Remarks 31.46. Suppose that T is P-preserving.

(a) If T is strong-mixing, then $E[(f\circ T^n)g] \to (Ef)Eg$ for any bounded random variable f and $g \in \mathfrak{L}^1$.

(b) If X is a metric space, and $E[(h\circ T^n)g] \to (Eh)Eg$ for any $h \in \mathfrak{C}_r(X)$ and $g \in \mathfrak{L}^2(X, \mathcal{B}(X), P)$, then T is weak-mixing.

(c) Assume that X is a metric space, and let P be as in Section 22. Then T is strong-mixing if and only if $QT^{-n} \Rightarrow P$ for any $Q \in$ P with $Q \ll P$. The proof of this assertion is similar to that of (31.30), resting on (a) and (b).

Exercise 31.47. Define $T : [0, 1[\to [0, 1[$ by $T(x) = x^p$, where $p > 1$. Show that there is no probability P on $\mathcal{B}([0, 1[)$ such that $P(\{0\}) < 1$ and $PT^{-1} = P$. [Hint. If $PT^{-1} = P$, then $P([a, a^{1/p}[) = 0, 0 < a < 1$.]

Exercise 31.48. Let $c \in [0, 1[$, and define $T : [0, 1[\to [0, 1[$ by $T(x) = x + c \pmod 1$. Prove the following.

(a) T preserves the Lebesgue measure λ on $\mathcal{B}([0, 1[)$.

(b) If $c \in Q$, then T is periodic, and the orbit $O_T(x)$ is finite and has the same number of points for each $x \in [0, 1[$.

(c) If $c \notin Q$, then $O_T(x)$ is dense in $[0, 1[$ for any $x \in [0, 1[$.

(d) If $c \in Q$, then T is not λ-ergodic. [Hint. If $c = l/m$ with l, m relatively prime, put $A = \cup_{k=0}^{m-1}[\frac{2k}{2m}, \frac{2k+1}{2m}[$. Then $T^{-1}(A) = A$.]

(e) If $c \notin Q$, then T is λ-ergodic, but fails to be λ-strong-mixing.

(f) If $c \notin Q$ and $f : [0, 1[\to R$ is continuous on $[0, 1[$ with $\lim_{x\to 1} f(x) = f(0)$, then $\frac{1}{n}\sum_{i=0}^{n-1} f\circ T^i \xrightarrow{u} \int_0^1 f(x)dx$. [Hint. The interval $[0, 1[$ may be viewed as a compact metric space by joining its ends and forming a circle, and T may be viewed as a rotation. Apply (e), (31.20), and (31.31).]

Exercise 31.49. Notation is as in (31.4). Prove the following.

(a) $T^{-1}(\otimes_{i\in I}\mathcal{I}_{T_i}) = \otimes_{i\in I}\mathcal{I}_{T_i} \subset \mathcal{I}_T$. [Hint. For $i \in I, \pi_i \circ T = T_i \circ \pi_i$.]

(b) If T is P-preserving, then $T^{-1}(\otimes_{i\in I}\mathcal{I}'_{T_i}) \subset \otimes_{i\in I}\mathcal{I}'_{T_i} \subset \mathcal{I}'_T$.

(c) If T is P-ergodic, then T_i is P_i-ergodic for each $i \in I$. [Hint. For $i \in I$, $\pi_i^{-1}(\mathcal{I}_{T_i}) \subset \mathcal{I}_T$.]

Exercise 31.50. Assume that T is P-preserving and ergodic, and let $f \geqslant 0$ be a random variable with $Ef = \infty$. Prove that $\frac{1}{n}\sum_{i=0}^{n-1} f \circ T^i \to \infty$ a.s. [Hint. For each $k \in N$, apply the Birkhoff ergodic theorem (31.20) to $f 1_{\{f<k\}}$.]

Exercise 31.51. Notation and assumptions are as in (31.26). Use (31.27) to show that the following assertions are equivalent:

(i) T is ergodic;

(ii) the equation $\mathbf{y}Q = \mathbf{y}$ has a unique (up to a multiplicative constant) solution;

(iii) the equation $Q\mathbf{z}^T = \mathbf{z}^T$ has a unique solution;

(iv) the equation $\mathbf{y}\Pi = \mathbf{y}$ has a unique solution;

(v) the equation $\Pi\mathbf{z}^T = \mathbf{z}^T$ has a unique solution.

Exercise 31.52. Suppose that T preserves P. Prove that $\frac{1}{n}\sum_{i=0}^{n-1} E[(f \circ T^i)f] \to (Ef)^2$, $f \in \mathfrak{L}^2$, if and only if $\frac{1}{n}\sum_{i=0}^{n-1} E[(f \circ T^i)g] \to (Ef)Eg$, $f, g \in \mathfrak{L}^2$.

Exercise 31.53. If X is a metric space, $PT^{-1} = P$ and $\frac{1}{n}\sum_{i=0}^{n-1} E[(f \circ T^i)h] \to (Ef)Eh$ for any $f \in \mathfrak{L}^2(X, \mathcal{B}(X), P)$ and $h \in \mathfrak{C}_r(X)$, then T is ergodic.

Exercise 31.54. Assume that X is a compact metric space and T is P-preserving and ergodic. Then there is $A \in \mathcal{B}(X)$ with $P(A) = 1$ such that $\frac{1}{n}\sum_{i=0}^{n-1} f(T^i(x)) \to Ef$ for all $x \in A$ and $f \in \mathfrak{C}_r(X)$. [Use (6.58) and (31.21.a).]

Exercise 31.55. Suppose that X is a compact metric space and $PT^{-1} = P$. Show that T is ergodic if and only if there is $A \in \mathcal{B}(X)$ with $P(A) = 1$ such that $\frac{1}{n}\sum_{i=0}^{n-1} \varepsilon_x T^{-i} \Rightarrow P$ for all $x \in A$. [Use (31.54) and (31.29).]

Exercise 31.56. Assume that T is P-preserving and P-ergodic, and let μ be a finite measure on $\mathcal{A}$ such that $\mu \ll P$ and $\mu T^{-1} \leqslant \mu$. Prove that $\mu T^{-1} = \mu$ and $\mu = aP$ for some $a > 0$. The next steps may be helpful.

(a) $\mu - \mu T^{-1} \ll P$.

(b) If $A \in \mathcal{A}$ satisfies $\mu(A) > \mu(T^{-1}(A))$, then $\mu(\cap_{n\geqslant 0} T^{-n}(A^c)) = 0$. [Hint. $P(\cap_{n\geqslant 0} T^{-n}(A^c)) = 0$ by (a) and (31.16).]

(c) Put $A_0 = A$ and define $A_n = T^{-1}(A_{n-1}) \cap A$ and $B_n = T^{-1}(A_{n-1}) \cap A^c$ for $n \geqslant 1$. Then $\mu(A_n) \to 0$. [Use (b).]

(d) $\mu T^{-1} = \mu$. [Hint. For $n \geqslant 1$, $\mu(T^{-1}(A)) \geqslant \mu(A_n) + \sum_{i=1}^{n} \mu(B_i) \to \sum_{i\geqslant 1} \mu(B_i) = \mu(A)$ by (c).]

(e) $\mu = \mu(X)P$. [Use (31.14.b) and (31.33).]

Exercise 31.57. Notation and assumptions are as in (31.38). Show that T_X is P-weak-mixing if and only if T_Y is Q-weak-mixing.

Exercise 31.58. Notation is as in (31.4). Prove the following.

(a) If T is P-weak-mixing, then T_i is P_i-weak-mixing for each $i \in I$.

(b) If T_i is P_i-preserving and P_i-weak-mixing for any $i \in I$, then T is P-weak-mixing. [Hint. Mimic the proof of (31.41.ii), noticing that $\left|\prod_{l=1}^{m} a_l - \prod_{l=1}^{m} b_l\right| \leqslant \sum_{l=1}^{m} |a_l - b_l|$ whenever $0 \leqslant a_l, b_l \leqslant 1, l = 1, \ldots, m$.]

Exercise 31.59 (Rényi). Suppose that T is P-preserving. Show that T is strong-mixing if and only if $P(T^{-n}(A) \cap A) \to P(A)^2$ for any $A \in \mathcal{A}$. [Remember the proof of (31.42).]

Exercise 31.60. If T is P-preserving and weak-mixing, then $\frac{1}{n}\sum_{i=0}^{n-1} |E[(f \circ T^i)g] - (Ef)Eg| \to 0$ for any $f \in \mathfrak{L}^1$ and any bounded random variable g.

Exercise 31.61. Notation is as in (31.3.c). Assume that X is a topological space and $\prod_{i \in I} X_i$ is endowed with the product topology. Prove that the one-sided Bernoulli shift is continuous on $\prod_{i \in I} X_i$, and that the two-sided Bernoulli shift is a homeomorphism from $\prod_{i \in I} X_i$ onto $\prod_{i \in I} X_i$. [Use (5.44) and (5.41).]

Exercise 31.62. If T is P-preserving and $f \in \mathfrak{L}^1$, then $E[f|\mathcal{I}_T] = E[f \circ T|\mathcal{I}_T]$.

Exercise 31.63. Notation is as in (8.23) and (31.3.f), and assume that a is injective. For any $n \geqslant 1$, suppose numbers $p_i^{(n)} \geqslant 0$, $i \in I$, are given such that $\sum_{i \in I} p_i^{(n)} = 1$ and $\prod_{n \geqslant 1} p_{b-1}^{(n)} = 0$ in case $1 < b \in N$.

(a) Construct a probability λ on $\mathcal{B}([0, 1[)$ making the digits a_n, $n \geqslant 1$, λ-independent random variables with $\lambda(a_n = i) = p_i^{(n)}$, $n \geqslant 1$, $i \in I$.

(b) λ is either atomic or nonatomic. [Hint. If λ is not nonatomic, there exists $x \in [0, 1[$ such that $\lambda(\{x\}) > 0$ by (9.39.c). The set $\liminf_n \{a_n = a_n(x)\}$ is denumerable, and $\{x\} \subset \liminf_n \{a_n = a_n(x)\} \in \mathcal{O}_\lambda$].

(c) λ is nonatomic if and only if $\prod_{n \geqslant 1} \sup_{i \in I} p_i^{(n)} = 0$. [For $x \in [0, 1[$, $\lambda(\{x\}) = \prod_{n \geqslant 1} p_{a_n(x)}^{(n)}$.]

(d) Define $F(x) = \sum_{m \geqslant 1} (\prod_{n=1}^{m-1} p_{a_n(x)}^{(n)}) s_{a_m(x)}^{(m)}$, $x \in [0, 1[$, where $s_i^{(n)} = 1 - \sum_{l \geqslant i} p_l^{(n)}$, $n \geqslant 1$, $i \in I$. Show that $F(0) = 0$ and $F(x) = \lambda([0, x[)$, $x \in]0, 1[$.

(e) F is strictly increasing if and only if $p_i^{(n)} > 0$ for all $n \geqslant 1$ and $i \in I$.

Exercise 31.64. Let λ be a probability on $\mathcal{B}([0, 1[)$ such that the function $F : [0, 1] \to [0, 1]$ defined by $F(0) = 0$, $F(1) = 1$ and $F(x) = \lambda([0, x[)$, $x \in]0, 1[$, is strictly increasing and continuous, let b be as in (8.23), and let I denote the set of all nonnegative integers that are less than b. For $n \geqslant 1$, suppose numbers $p_i^{(n)} > 0$, $i \in I$, are given such that $\sum_{i \in I} p_i^{(n)} = 1$ and $\prod_{n \geqslant 1} \sup_{i \in I} p_i^{(n)} = 0$. Construct a function f as in (8.23) such that the associated function $a = (a_n)_{n \geqslant 1}$ in (31.3.f) is injective, and the digits a_n, $n \geqslant 1$, are λ-independent random variables with $\lambda(a_n = i) = p_i^{(n)}$, $n \geqslant 1$, $i \in I$. The next steps may be helpful.

(a) For $i_1, i_2, \ldots \in I$ and $r \geqslant 1$, put

$$\underline{y}_r(i_1, \ldots, i_r) = \sum_{m=1}^{r} \left(\prod_{n=1}^{m-1} p_{i_n}^{(n)} \right) s_{i_m}^{(m)},$$

$$\bar{y}_r(i_1, \ldots, i_r) = \underline{y}_r(i_1, \ldots, i_r) + \prod_{n=1}^{r} p_{i_n}^{(n)},$$

where $s_i^{(n)} = 1 - \sum_{l \geqslant i} p_l^{(n)}$, $n \geqslant 1$, $i \in I$. Then

$$\underline{y}_r(i_1, \ldots, i_r) \leqslant \underline{y}_{r+1}(i_1, \ldots, i_r, i_{r+1}) \leqslant \bar{y}_{r+1}(i_1, \ldots, i_r) \leqslant \bar{y}_r(i_1, \ldots, i_r).$$

(b) $\underline{y}(i_1, i_2, \ldots) = \lim_r \underline{y}_r(i_1, \ldots, i_r)$ and $\bar{y}(i_1, i_2, \ldots) = \lim_r \bar{y}_r(i_1, \ldots, i_r)$ exist, and $\underline{y}(i_1, i_2, \ldots) \leqslant \bar{y}(i_1, i_2, \ldots)$.

(c) If $b = \infty$, then $\bar{y}(i_1, i_2, \ldots) < \bar{y}_r(i_1, \ldots, i_r)$, $r \geqslant 1$, while if $1 < b \in N$, then $\bar{y}(i_1, i_2, \ldots) = \bar{y}_k(i_1, \ldots, i_k)$ if and only if $i_n = b - 1$ for $n > k$.

(d) For $x \in [0, 1[$, define recursively the sequence $\{i_n(x) : n \geqslant 1\}$ by

$$F^{-1}(\underline{y}_r(i_1(x), \ldots, i_r(x))) \leqslant x < F^{-1}(\bar{y}_r(i_1(x), \ldots, i_r(x))), \quad r \geqslant 1. \quad (1)$$

Then, for $i \in I$,

$$\bar{y}(i, i_1(x), i_2(x), \ldots) < \bar{y}_r(i, i_1(x), \ldots, i_r(x)), \quad r \geqslant 1.$$

[Hint. Letting $r \to \infty$ in (1), we get

$$F^{-1}(\underline{y}(i_1(x), i_2(x), \ldots)) = x = F^{-1}(\bar{y}(i_1(x), i_2(x), \ldots)) \quad (2)$$

by (b), and so

$$\bar{y}(i_1(x), i_2(x), \ldots) < \bar{y}_r(i_1(x), \ldots, i_r(x)), \quad r \geqslant 1.$$

Then use (c).]

(e) Define $f : [0, b[\to [0, 1[$ by

$$f(i + x) = F^{-1}(\underline{y}(i, i_1(x), i_2(x), \ldots)), \quad i \in I, \; x \in [0, 1[. \quad (3)$$

Then f is strictly increasing with $f(0) = 0$. [Hint. If $0 \leqslant x < x' < 1$, let $k = \inf\{n : i_n(x) \neq i_n(x')\}$. Then $i_k(x) < i_k(x')$ and

$$\begin{aligned} f(i + x) &< F^{-1}(\bar{y}_{k+1}(i, i_1(x), \ldots, i_k(x))) \\ &\leqslant F^{-1}(\underline{y}_{k+1}(i, i_1(x'), \ldots, i_k(x'))) \leqslant f(i + x') \end{aligned}$$

by (d).]

(f) Write $\Delta_1 = I$, $\Delta_{r+1} = I + f(\Delta_r)$, $r \geqslant 1$, and $\Delta = \cup_{r \geqslant 1} \Delta_r$. Then $f(\Delta)$ is dense in $[0, 1[$. [Hint. If $1 < b \in N$, set $f(b) = 1$. For $r \geqslant 1$ and $t_1, \ldots, t_r \in [0, b[$, let $\underline{x}_r(t_1, \ldots, t_r)$ and $\bar{x}_r(t_1, \ldots, t_r)$ be as in (8.23.b). Iterating (3) and using (2), we have

$$\underline{x}_r(i_1, \ldots, i_r) = F^{-1}(\underline{y}_r(i_1, \ldots, i_r)), \; \bar{x}_r(i_1, \ldots, i_r) = F^{-1}(\bar{y}_r(i_1, \ldots, i_r)), \quad (4)$$

and so

$$x = \lim_r F^{-1}(\underline{y}_r(i_1(x), \ldots, i_r(x))) = \lim_r \underline{x}_r(i_1(x), \ldots, i_r(x))$$

by (2) and (b).]

(g) f is continuous on $[0, b[$ with $\lim_{t \to b} f(t) = 1$. [Use (f).]

(h) For $x \in [0, 1[$, consider the sequence $\{a_n(x) : n \geqslant 1\}$ of digits of x as in (8.23.c), and put $a = (a_n)_{n \geqslant 1}$. Then a is injective. [Hints. For $r \geqslant 1$ and $i_1, \ldots, i_r \in I$, we have $i_n(x) = i_n$, $1 \leqslant n \leqslant r$, if and only if

$$F^{-1}(\underline{y}_r(i_1, \ldots, i_r)) \leqslant x < F^{-1}(\bar{y}_r(i_1, \ldots, i_r)),$$

and so, according to (4), if and only if

$$\underline{x}_r(i_1, \ldots, i_r) \leqslant x < \bar{x}_r(i_1, \ldots, i_r).$$

In view of (8.23.c), this means that $a_n(x) = i_n(x), 1 \leqslant n \leqslant r$. Thus $a = i$, where $i(x) = (i_n(x))_{n \geqslant 1}, x \in [0, 1[$. But i is injective by (2).]

(i) $a_n, n \geqslant 1$, are λ-independent. [Hint. For $r \geqslant 1$ and $i_1, \ldots, i_r \in I$,

$$\begin{aligned}\lambda(a_1 = i_1, \ldots, a_r = i_r) &= \lambda([\underline{x}_r(i_1, \ldots, i_r), \bar{x}_r(i_1, \ldots, i_r)[) \\ &= F(\bar{x}_r(i_1, \ldots, i_r)) - F(\underline{x}_r(i_1, \ldots, i_r)) \\ &= \bar{y}_r(i_1, \ldots, i_r) - \underline{y}_r(i_1, \ldots, i_r) = \prod_{n=1}^{r} p_{i_n}^{(n)}\end{aligned}$$

by (4).]

Exercise 31.65. Let λ be the Lebesgue measure on $\mathcal{B}([0, 1[)$, and let b and I be as in (31.64). For $n \geqslant 1$, suppose numbers $p_i^{(n)} > 0, i \in I$, are given such that $\sum_{i \in I} p_i^{(n)} = 1$ and $p_i^{(k+n)} = p_i^{(n)}, i \in I$, for some $k \geqslant 1$. Let f be the corresponding function constructed as in (31.64), and let Δ_k be as in (31.64.f). Show that the graph of f is the polygonal line joining in order the points $(t, f(t)), t \in \Delta_k$. [Hint. For $r \geqslant 1$ and $t_1, \ldots, t_r \in [0, b[$, let $\underline{x}_r(t_1, \ldots, t_r)$ be as in (8.23.b). If $k = 1$, then $f(i + x) - f(i) = p_i^{(1)} x, i \in I, x \in [0, 1[$, while if $k > 1$ and $i_1, \ldots, i_k \in I$, then

$$\begin{aligned}&\underline{x}_k(i_1, i_2, \ldots, i_k + x) - \underline{x}_k(i_1, i_2, \ldots, i_k) = \left(\left(\prod_{n=1}^{k} p_{i_n}^{(n)} \right) \Big/ \prod_{n=1}^{k-1} p_{i_{n+1}}^{(n)} \right) \\ &\quad \times (\underline{x}_{k-1}(i_2, \ldots, i_k + x) - \underline{x}_{k-1}(i_2, \ldots, i_k))\end{aligned}$$

for any $x \in [0, 1[$.]

32 Stationary Sequences

The aim of this section is to present some of the most useful results for general stationary sequences of random elements, which are sequences whose distributions do not depend on the placement of the time origin. Beside martingales and semimartingales studied in Section 30, stationary sequences encompass a dependence structure of great importance in probability theory. Basic results included here are the Birkhoff ergodic theorem and the von Neumann ergodic theorem, spelled out for stationary sequences of random variables. The former theorem contains strictly Kolmogorov's SLLN.

Throughout this section, if not otherwise stated, we consider a measurable space $(Y, \mathcal{B})$, put $(Y_i, \mathcal{B}_i) = (Y, \mathcal{B})$ for any $i \in Z$, let $\pi_i, i \in Z$, denote the projection from Y^∞ onto Y_i, let $f_i : X \to Y_i, i \in Z$, be random elements, and set $\mathbf{f} = (f_n)_{n \geqslant 0}$. As usual, for any $J \subset Z$ and $k \geqslant 1$, we identify the measurable spaces $(\prod_{j \in J} Y_j, \otimes_{j \in J} \mathcal{B}_j)$ and $(\prod_{j \in J+k} Y_j, \otimes_{j \in J+k} \mathcal{B}_j)$.

Definition 32.1. The sequence $\{f_n : n \geqslant 0\}$ is said to be *stationary* [P-stationary] if $P(f_{n_1}, \ldots, f_{n_m})^{-1} = P(f_{n_1+k}, \ldots, f_{n_m+k})^{-1}$ whenever $0 \leqslant n_1 < \cdots < n_m$ and $m, k \geqslant 1$. This definition extends in a obvious way to $\{f_i : i \in Z\}$ and $\{f_n : n \geqslant r\}$ with $r \in Z$.

Remarks 32.2. (a) Assume that $\mathcal{M} \subset \mathcal{B}$ is closed under intersection and generates $\mathcal{B}$. Then $\{f_n : n \geqslant 0\}$ is stationary if and only if $P(f_{n_1} \in M_1, \ldots, f_{n_m} \in M_m) = P(f_{n_1+k} \in M_1, \ldots, f_{n_m+k} \in M_m)$ whenever $0 \leqslant n_1 < \cdots < n_m, m, k \geqslant 1$ and $M_1, \ldots, M_m \in \mathcal{M}$; in particular, if $(Y, \mathcal{B}) = (R, \mathcal{B}(R))$, then $\{f_n : n \geqslant 0\}$ is stationary if and only if the distribution function of $(f_{n_1}, \ldots, f_{n_m})$ coincides with the distribution function of $(f_{n_1+k}, \ldots, f_{n_m+k})$. Indeed, if the stated condition holds, then the probabilities $P(f_{n_1}, \ldots, f_{n_m})^{-1}$ and $P(f_{n_1+k}, \ldots, f_{n_m+k})^{-1}$ agree on $\mathcal{C} = \{M_1 \times \cdots \times M_m : M_1, \ldots, M_m \in \mathcal{M}\}$. Since $\mathcal{C}$ is closed under intersection, and generates $\otimes_{i=1}^m \mathcal{B}_{n_i} = \otimes_{i=1}^m \mathcal{B}_{n_i+k}$ by (7.24), it follows from (10.8) that the two probabilities are equal.

(b) $\{f_n : n \geqslant 0\}$ is stationary if and only if

$$P(f_0, \ldots, f_{m-1})^{-1} = P(f_k, \ldots, f_{m-1+k})^{-1}, \quad m, k \geqslant 1. \tag{1}$$

Actually, stationarity of $\{f_n : n \geqslant 0\}$ implies (1) trivially. On the other direction, if $0 \leqslant n_1 < \cdots < n_m$ and $m, k \geqslant 1$, then $P(f_{n_1} \in B_1, \ldots, f_{n_m} \in B_m) = P(f_{n_1+k} \in B_1, \ldots, f_{n_m+k} \in B_m)$ for all $B_1, \ldots, B_m \in \mathcal{B}$. This implies stationarity of $\{f_n : n \geqslant 0\}$ by (a) with $\mathcal{M} = \mathcal{B}$.

(c) Suppose that $\mathcal{M}$ is as in (a). Then $\{f_n : n \geqslant 0\}$ is stationary if and only if $P(f_0 \in M_1, \ldots, f_{m-1} \in M_m) = P(f_k \in M_1, \ldots, f_{m-1+k} \in M_m)$ whenever $m, k \geqslant 1$ and $M_1, \ldots, M_m \in \mathcal{M}$.

(d) From (b) and (c) it follows that $\{f_n : n \geqslant 0\}$ is stationary if and only if, for any $m \geqslant 1$, $P(f_0, \ldots, f_{m-1})^{-1} = P(f_1, \ldots, f_m)^{-1}$.

(e) Let $\mathcal{M}$ be as in (a). Then $\{f_n : n \geqslant 0\}$ is stationary if and only if, for any $m \geqslant 1$, $P(f_0 \in M_1, \ldots, f_{m-1} \in M_m) = P(f_1 \in M_1, \ldots, f_m \in M_m)$ whenever $M_1, \ldots, M_m \in \mathcal{M}$.

(f) $\{f_n : n \geqslant 0\}$ is stationary if and only if $P(f_0, f_1, \ldots)^{-1} = P(f_k, f_{k+1}, \ldots)^{-1}$, $k \geqslant 1$. Indeed, if $\{f_n : n \geqslant 0\}$ is stationary and $k \geqslant 1$, then the probabilities $P(f_0, f_1, \ldots)^{-1}$ and $P(f_k, f_{k+1}, \ldots)^{-1}$ agree on the algebra $\mathcal{A}$ in (7.21) (with $I = \{0, 1, \ldots\}$). By (10.8), this ensures that they are equal.

(g) From (f) and (d) it follows that $\{f_n : n \geqslant 0\}$ is stationary if and only if $P(f_0, f_1, \ldots)^{-1} = P(f_1, f_2, \ldots)^{-1}$.

(h) $\{f_n : n \geqslant 0\}$ is stationary if and only if $\{(f_n, \ldots, f_{n+l-1}) : n \geqslant 0\}$ is stationary for each $l \geqslant 1$. Actually, if $\{f_n : n \geqslant 0\}$ is stationary and $l, m \geqslant 1$, let $B_1 \in \otimes_{i=0}^{l-1} \mathcal{B}_i, \ldots, B_m \in \otimes_{i=0}^{l-1} \mathcal{B}_{m-1+i}$ be product cylinders. Then $P((f_0, \ldots, f_{l-1}) \in$

$B_1, \ldots, (f_{m-1}, \ldots, f_{m-1+l-1} \in B_m)) = P((f_1, \ldots, f_l) \in B_1, \ldots, (f_m, \ldots, f_{m+l-1} \in B_m))$, and the implication follows from (e).

(i) Define the shift transformation $\theta : \prod_{n \geqslant 0} Y_n \to \prod_{n \geqslant 0} Y_n$ by $\theta(y_0, y_1, \ldots) = (y_1, y_2, \ldots)$, i.e. $\pi_n \circ \theta = \pi_{n+1}, n \geqslant 0$. We have seen in (31.3.c) that θ is measurable. Since $\theta \circ (f_0, f_1, \ldots) = (f_1, f_2, \ldots)$, (g) shows that $\{f_n : n \geqslant 0\}$ is stationary if and only if θ is $P\mathbf{f}^{-1}$-preserving. (Throughout this section, if not otherwise stated, θ will denote this transformation.)

(j) If $\{f_n : n \geqslant 0\}$ is stationary, then $f_n, n \geqslant 0$, are identically distributed random elements.

Examples 32.3. (a) If $f_n, n \geqslant 0$, are independent and identically distributed, then, in view of (18.29), $\{f_n : n \geqslant 0\}$ is stationary.

(b) Notation and assumptions are as in (16.40). For $i \in I$, let π_i denote the projection from X^I onto X_i. Then $\{\pi_i : i \in I\}$ is P-stationary; in particular, $\{\pi_i : i \in I\}$ is stationary with respect to the probability P defining the Markov shift in (31.3.e).

Starting with any stationary sequence, other stationary sequences can be derived.

Theorem 32.4. *Let $\{f_n : n \geqslant 0\}$ be a stationary sequence, let $(Z, \mathcal{C})$ be a measurable space, and let $\varphi : Y^\infty \to Z$ be a measurable function. Then $\{\varphi \circ (f_n, f_{n+1}, \ldots) : n \geqslant 0\}$ is stationary.*

Proof. By (32.2.g), we have

$$\begin{aligned} P(\varphi \circ (f_0, f_1, \ldots))^{-1} &= (P(f_0, f_1, \ldots)^{-1})\varphi^{-1} \\ &= (P(f_1, f_2, \ldots)^{-1})\varphi^{-1} = P(\varphi \circ (f_1, f_2, \ldots))^{-1}, \end{aligned}$$

which proves the assertion. □

Corollary 32.5. *Let $\{f_n : n \geqslant 0\}$ and $(Z, \mathcal{C})$ be as in (32.4), fix $l \geqslant 1$, and let $\varphi_l : Y^l \to Z$ be a measurable function. Then $\{\varphi_l \circ (f_n, \ldots, f_{n+l-1}) : n \geqslant 0\}$ is stationary.*

Proof. Define $\varphi : Y^l \times Y^l \times \cdots \to Z \times Z \times \cdots$ by $\varphi(\mathbf{y}_0, \mathbf{y}_1, \ldots) = (\varphi_l(\mathbf{y}_0), \varphi_l(\mathbf{y}_1), \ldots)$. Then φ is measurable by (8.4), and

$$(\varphi_l \circ (f_0, \ldots, f_{l-1}), \varphi_l \circ (f_1, \ldots, f_l), \ldots) = \varphi \circ ((f_0, \ldots, f_{l-1}), (f_1, \ldots, f_l), \ldots).$$

Stationarity of $\{\varphi_l \circ (f_n, \ldots, f_{n+l-1}) : n \geqslant 0\}$ now follows from (32.2.h) and (32.4). □

Example 32.6. If $Y = R$ and $\{f_n : n \geqslant 0\}$ is stationary, then the sequence $\{\frac{1}{l}(f_n + \ldots + f_{n+l-1}) : n \geqslant 0\}$ is stationary. The random variables $g_n = \frac{1}{l}(f_n + \ldots + f_{n+l-1}), n \geqslant 0$, are called *moving averages*. In general $g_n, n \geqslant 0$, are dependent even if $f_n, n \geqslant 0$, are independent.

Starting from measure-preserving transformations, many stationary sequences can be constructed.

Theorem 32.7. *Let $T : X \to X$ be a P-preserving transformation, and let $f : X \to Y$ be a measurable function. Then $\{f \circ T^n : n \geqslant 0\}$ is stationary.*

Proof. We have $P(f \circ T^0, f \circ T^1, \ldots)^{-1} = (PT^{-1})(f \circ T^0, f \circ T^1, \ldots)^{-1} = P((f \circ T^0, f \circ T^1, \ldots) \circ T)^{-1} = P(f \circ T^1, f \circ T^2, \ldots)^{-1}$. □

Remark 32.8. In terms of distribution, every stationary sequence $\{f_n : n \geqslant 0\}$ can be generated by a measure-preserving transformation. Indeed, consider the probability space $(\prod_{n \geqslant 0} Y_n, \otimes_{n \geqslant 0} \mathcal{B}_n, P\mathbf{f}^{-1})$. Then the sequence of projections $\{\pi_n : n \geqslant 0\}$ is stationary, since both $(f_0, f_1, \ldots)$ and $(\pi_0, \pi_1, \ldots)$ have distribution $P\mathbf{f}^{-1}$. Moreover, $\pi_n = \pi_0 \circ \theta^n$, $n \geqslant 0$, where θ is $P\mathbf{f}^{-1}$-preserving by (32.2.i).

The stationary $\{f_i : i \in Z\}$ indexed by Z is simpler than the stationary $\{f_n : n \geqslant 0\}$, because the associate shift transformation θ is invertible, and the set $\{\theta^i : i \in Z\}$ forms a group, in the bilateral case, while $\{\theta^n : n \geqslant 0\}$ forms merely a semigroup in the unilateral case. If Y satisfies mild regularity conditions, the next result shows that the two cases are essentially equivalent.

Theorem 32.9. *Assume that Y is a complete separable metric space, $\mathcal{B} = \mathcal{B}(Y)$ and $\{f_n : n \geqslant 0\}$ is P-stationary. Then there exist a probability space $(\Omega, \mathcal{K}, \mu)$ and random elements $g_i : \Omega \to Y, i \in Z$, such that $\{g_i : i \in Z\}$ is μ-stationary and $\mu(g_0, g_1, \ldots)^{-1} = P(f_0, f_1, \ldots)^{-1}$.*

Proof. Take $(\Omega, \mathcal{K}) = (\prod_{i \in Z} Y_i, \otimes_{i \in Z} \mathcal{B}_i)$ and, for $i \in Z$, let $g_i = \pi_i$. For $J \in \mathcal{J}$, where $\mathcal{J}$ is as in (7.21), choose $r \geqslant 0$ so that $J + r \subset \{0, 1, \ldots\}$, and put $\mu_J = P(f_j)^{-1}_{j \in J+r}$. As $\{f_n : n \geqslant 0\}$ is P-stationary, μ_J is a well-defined probability on $\otimes_{j \in J} \mathcal{B}_j$. Moreover, $\mu_J(\pi^J_K)^{-1} = \mu_K$ whenever $\emptyset \neq K \subset J$. Then, according to (16.23) and the Daniell-Kolmogorov extension theorem (16.20), there exists a unique probability μ on $\otimes_{i \in Z} \mathcal{B}_i$ such that $\mu \pi_J^{-1} = \mu_J$, $J \in \mathcal{J}$. For $i_1, \ldots, i_m \in Z$ with $i_1 < \cdots < i_m$, select $r \geqslant 0$ so that $i_1 + r \geqslant 0$. Then, for $k \geqslant 1$, we have

$$\begin{aligned}\mu(g_{i_1}, \ldots, g_{i_m})^{-1} &= \mu_{\{i_1, \ldots, i_m\}} = P(f_{i_1+r}, \ldots, f_{i_m+r})^{-1} \\ &= P(f_{i_1+r+k}, \ldots, f_{i_m+r+k})^{-1} \\ &= \mu_{\{i_1+k, \ldots, i_m+k\}} = \mu(g_{i_1+k}, \ldots, g_{i_m+k})^{-1}.\end{aligned}$$

This shows that $\{g_i : i \in Z\}$ is μ-stationary. Finally, for any $n \geqslant 0$ and $B_0, \ldots, B_n \in \mathcal{B}$, $\mu(g_0 \in B_0, \ldots, g_n \in B_n) = \mu_{\{0, \ldots, n\}}(B_0 \times \cdots \times B_n) = P(f_0 \in B_0, \ldots, f_n \in B_n)$, which shows that $\mu(g_0, g_1, \ldots)^{-1} = P(f_0, f_1, \ldots)^{-1}$ by (10.8). □

It will now be desirable to define the concepts of invariance and ergodicity relative to the sequence $\{f_n : n \geqslant 0\}$. Since these concepts involve the shift transformation θ, it is appropriate to have more information about it.

Remarks 32.10. (a) If $B \in \mathcal{I}_\theta$, then $(y_0, y_1, \ldots) \in B$ implies $(y_n, y_{n+1}, \ldots) \in B$ for any $n \geqslant 1$. Conversely, if $(y_n, y_{n+1}, \ldots) \in B$, $n \geqslant 1$, whenever $(y_0, y_1, \ldots) \in B$, then $B \subset \cup_{n \geqslant 1} \theta^{-n}(B)$.

(b) Let $\mathcal{T} = \cap_{n \geqslant 0} \sigma(\pi_n, \pi_{n+1}, \ldots)$ be the tail σ-algebra of $\{\pi_n : n \geqslant 0\}$. Then $\mathcal{I}_\theta \subset \mathcal{T}$. Actually, as $\pi_i \circ \theta^n = \pi_{n+i}$, $i, n \geqslant 0$, if $B \in \mathcal{I}_\theta$, then $B \in \sigma(\pi_0, \pi_1, \ldots)$, and so $B = \theta^{-n}(B) \in \theta^{-n}(\sigma(\pi_0, \pi_1, \ldots)) = \sigma(\pi_n, \pi_{n+1}, \ldots)$ for any $n \geqslant 0$.

(c) The inclusion $\mathcal{I}_\theta \subset \mathcal{T}$ is strict if Y has at least two elements. Indeed, for $y, y' \in Y$ with $y \neq y'$, let $B = \limsup_n \{\pi_{2n} = y\}$. Then $B \in \mathcal{T}$. However, $(y, y', y, y', \ldots) \in B$, but $(y', y, y', y, \ldots) \notin B$, and so $B \notin \mathcal{I}_\theta$ by (a).

(d) On $\otimes_{n\geqslant 0}\mathcal{B}_n$ we consider the probability $P\mathbf{f}^{-1}$ with respect to which θ-a.s. invariance is defined. Thus $\mathcal{I}'_\theta = \{B \in \otimes_{n\geqslant 0}\mathcal{B}_n : (P\mathbf{f}^{-1})(\theta^{-1}(B)\Delta B) = 0\}$.

Definitions 32.11. The sets in $\mathbf{f}^{-1}(\mathcal{I}_\theta)$ are called $\mathbf{f}$-*invariant*, and the sets in $\mathbf{f}^{-1}(\mathcal{I}'_\theta)$ are called *a.s.* $\mathbf{f}$-*invariant*.

Let $(Z, \mathcal{C})$ be a measurable space. A random element $g : X \to Z$ is said to be $\mathbf{f}$-*invariant* if there is a measurable function $\varphi : Y^\infty \to Z$ such that $g = \varphi \circ (f_n, f_{n+1}, \ldots)$ for all $n \geqslant 0$.

Remarks 32.12. (a) A set $A \in \mathcal{A}$ is $\mathbf{f}$-invariant if and only if there exists $B \in \otimes_{n\geqslant 0}\mathcal{B}_n$ such that

$$A = (f_n, f_{n+1}, \ldots)^{-1}(B), \quad n \geqslant 0. \tag{1}$$

Actually, if A is $\mathbf{f}$-invariant, then $A = \mathbf{f}^{-1}(B)$ for some $B \in \mathcal{I}_\theta$. For $n \geqslant 0$, since $B = \theta^{-n}(B)$ and $\theta^n \circ \mathbf{f} = (f_n, f_{n+1}, \ldots)$, we have $A = \mathbf{f}^{-1}(\theta^{-n}(B)) = (f_n, f_{n+1}, \ldots)^{-1}(B)$, which is (1). Conversely, assume that (1) holds. Then $A = \cup_{n\geqslant 0}\cap_{k\geqslant n}(f_k, f_{k+1}, \ldots)^{-1}(B) = \cup_{n\geqslant 0}\cap_{k\geqslant n}\mathbf{f}^{-1}(\theta^{-k}(B)) = \mathbf{f}^{-1}(\liminf_n \theta^{-n}(B)) \in \mathbf{f}^{-1}(\mathcal{I}_\theta)$ by (31.9.c).

(b) If A is a.s. $\mathbf{f}$-invariant and θ is $P\mathbf{f}^{-1}$-preserving, then there exists $B \in \mathcal{I}_\theta$ such that

$$A \overset{a.s.}{=} (f_n, f_{n+1}, \ldots)^{-1}(B), \quad n \geqslant 0. \tag{1}$$

Indeed, if A is a.s. $\mathbf{f}$-invariant, then $A = \mathbf{f}^{-1}(C)$ for some $C \in \mathcal{I}'_\theta$. According to (31.10.f) and (31.9.c), $(P\mathbf{f}^{-1})((\liminf_k \theta^{-k}(C))\Delta C) = 0$, where $\liminf_k \theta^{-k}(C) \in \mathcal{I}_\theta$. Consequently, for $n \geqslant 0$, we get

$$\begin{aligned}
&P\left((f_n, f_{n+1}, \ldots)^{-1}\left(\liminf_k \theta^{-k}(C)\right)\Delta A\right)\\
&\quad = P\left(\mathbf{f}^{-1}\left(\theta^{-n}\left(\liminf_k \theta^{-k}(C)\right)\right)\Delta \mathbf{f}^{-1}(C)\right)\\
&\quad = (P\mathbf{f}^{-1})\left(\left(\liminf_k \theta^{-k}(C)\right)\Delta C\right) = 0,
\end{aligned}$$

which establishes (1) with $B = \liminf_k \theta^{-k}(C)$.

(c) If $A \in \mathcal{A}$ is such that $A \overset{a.s.}{=} (f_n, f_{n+1}, \ldots)^{-1}(B)$, $n \geqslant 0$, for some $B \in \otimes_{n\geqslant 0}\mathcal{B}_n$, then there is $C \in \mathcal{I}_\theta$ with $A \overset{a.s.}{=} \mathbf{f}^{-1}(C)$. Actually, if $P((f_n, f_{n+1}, \ldots)^{-1}(B)\Delta A) = P(\mathbf{f}^{-1}(\theta^{-n}(B))\Delta A) = 0, n \geqslant 0$, for some $B \in \otimes_{n\geqslant 0}\mathcal{B}_n$, then $P(\mathbf{f}^{-1}(\liminf_n \theta^{-n}(B))\Delta A) = 0$ by (2.41.g) and (2.41.h), and $C = \liminf_n \theta^{-n}(B) \in \mathcal{I}_\theta$.

(d) Notation is as in (32.11). If g is $\mathbf{f}$-invariant, then it is $\mathbf{f}^{-1}(\mathcal{I}_\theta)$-measurable. Indeed, for $C \in \mathcal{C}, g^{-1}(C) = (f_n, f_{n+1}, \ldots)^{-1}(\varphi^{-1}(C)), n \geqslant 0$. Hence $g^{-1}(C) \in \mathbf{f}^{-1}(\mathcal{I}_\theta)$ by (a).

(e) If $g : X \to R$ is $\mathbf{f}^{-1}(\mathcal{I}_\theta)$-measurable, then it is $\mathbf{f}$-invariant. Actually, on account of (8.17), there is an $\mathcal{I}_\theta$-measurable function $\varphi : Y^\infty \to R$ such that $g = \varphi \circ (f_0, f_1, \ldots)$. But $\varphi \circ \theta^n = \varphi, n \geqslant 0$, by (31.11.iii), and so $g = \varphi \circ (f_n, f_{n+1}, \ldots)$, $n \geqslant 0$.

Definition 32.13. We say that the sequence $\{f_n : n \geqslant 0\}$ is *ergodic* [P-*ergodic*] if $\mathbf{f}^{-1}(\mathcal{I}_\theta) \subset \mathcal{O}_P$. The definition extends in an obvious way to $\{f_i : i \in Z\}$ and $\{f_n : n \geqslant r\}$ with $r \in Z$.

Remarks 32.14. (a) Since $\mathbf{f}^{-1}(\mathcal{I}_\theta) \subset \mathcal{O}_P$ is equivalent to $\mathcal{I}_\theta \subset \mathcal{O}_{P\mathbf{f}^{-1}}$, it follows that $\{f_n : n \geqslant 0\}$ is ergodic if and only if θ is $P\mathbf{f}^{-1}$-ergodic.

(b) Let $(\Omega, \mathcal{K}, \mathbf{P})$ be a probability space, and let $h : \Omega \to X$ be a random element with distribution P. If $\{f_n : n \geqslant 0\}$ is P-ergodic, then $\{f_n \circ h : n \geqslant 0\}$ is $\mathbf{P}$-ergodic. Indeed, putting $g_n = f_n \circ h, n \geqslant 0$, and $\mathbf{g} = (g_n)_{n\geqslant 0}$, we have $\mathbf{g} = \mathbf{f} \circ h$, and so $\mathbf{g}^{-1}(\mathcal{I}_\theta) = h^{-1}(\mathbf{f}^{-1}(\mathcal{I}_\theta)) \subset h^{-1}(\mathcal{O}_P) \subset \mathcal{O}_{\mathbf{P}}$, as $\mathbf{P}h^{-1} = P$.

(c) $\{f_n : n \geqslant 0\}$ is ergodic if and only if $\{(f_n, \ldots, f_{n+l-1}) : n \geqslant 0\}$ is ergodic for each $l \geqslant 1$. Actually, for $l \geqslant 1$, consider the shift transformation $\theta_l : Y^l \times Y^l \times \cdots \to Y^l \times Y^l \times \cdots$ defined by $(\pi_n, \ldots, \pi_{n+l-1}) \circ \theta_l = (\pi_{n+1}, \ldots, \pi_{n+l}), n \geqslant 0$, and set $\psi = ((\pi_0, \ldots, \pi_{l-1}), (\pi_1, \ldots, \pi_l), \ldots)$. Then ψ is a measurable function from $Y \times Y \times \cdots$ onto $Y^l \times Y^l \times \cdots$ satisfying $\psi \circ \theta = \theta_l \circ \psi$. Applying (31.12.i), it follows that $\psi^{-1}(\mathcal{I}_{\theta_l}) \subset \mathcal{I}_\theta$. We have $((f_0, \ldots, f_{l-1}), (f_1, \ldots, f_l), \ldots) = \psi \circ \mathbf{f}$, and so $(\psi \circ \mathbf{f})^{-1}(\mathcal{I}_{\theta_l}) = \mathbf{f}^{-1}(\psi^{-1}(\mathcal{I}_{\theta_l})) \subset \mathbf{f}^{-1}(\mathcal{I}_\theta) \subset \mathcal{O}_P$, which establishes the assertion.

An ergodic transformation generates many ergodic sequences.

Theorem 32.15. *Let $T : X \to X$ be an ergodic transformation, and let $f : X \to Y$ be a measurable function. Then $\{f \circ T^n : n \geqslant 0\}$ is ergodic.*

Proof. Set $f_n = f \circ T^n, n \geqslant 0$, and $\mathbf{f} = (f_n)_{n\geqslant 0}$. Since $\mathbf{f} \circ T = \theta \circ \mathbf{f}$, applying (31.12.i), we see that $\mathbf{f}^{-1}(\mathcal{I}_\theta) \subset \mathcal{I}_T \subset \mathcal{O}_P$. □

Theorem 32.16. *Let $\{f_n : n \geqslant 0\}$ be an ergodic sequence, and let $(Z, \mathcal{C})$ and φ be as in* (32.4). *Then $\{\varphi \circ (f_n, f_{n+1}, \ldots) : n \geqslant 0\}$ is ergodic.*

Proof. For $n \geqslant 0$, we have $\varphi \circ (f_n, f_{n+1}, \ldots) = \varphi \circ \theta^n \circ \mathbf{f}$. According to (32.14.a) and (32.15), $\{\varphi \circ \theta^n : n \geqslant 0\}$ is $P\mathbf{f}^{-1}$-ergodic. Then $\{\varphi \circ \theta^n \circ \mathbf{f} : n \geqslant 0\}$ is P-ergodic by (32.14.b). □

Corollary 32.17. *Let $\{f_n : n \geqslant 0\}$ and $(Z, \mathcal{C})$ be as in* (32.16), *fix $l \geqslant 1$, and let $\varphi_l : Y^l \to Z$ be a measurable function. Then $\{\varphi_l \circ (f_n, \ldots, f_{n+l-1}) : n \geqslant 0\}$ is ergodic.*

Proof. Define φ as in the proof of (32.5). Since

$$(\varphi_l \circ (f_0, \ldots, f_{l-1}), \varphi_l \circ (f_1, \ldots, f_l), \ldots) = \varphi \circ ((f_0, \ldots, f_{l-1}), (f_1, \ldots, f_l), \ldots),$$

ergodicity of $\{\varphi_l \circ (f_n, \ldots, f_{n+l-1}) : n \geqslant 0\}\}$ follows from (32.14.c) and (32.16). □

We can now translate the Birkhoff ergodic theorem (31.20) and the von Neumann ergodic theorem (31.32) for stationary sequences of random variables.

Theorem 32.18. *If $\{f_n : n \geqslant 0\}$ is a stationary sequence of random variables with $f_0 \in \mathfrak{L}^1$, then $\frac{1}{n}\sum_{i=0}^{n-1} f_i \overset{a.s.}{\to} E[f_0|\mathbf{f}^{-1}(\mathcal{I}_\theta)]$.*

Proof. Observe first that θ is $P\mathbf{f}^{-1}$-preserving by (32.2.i), and $\pi_0 \in \mathfrak{L}^1(R^\infty, \mathcal{B}(R^\infty), P\mathbf{f}^{-1})$ by (11.29). Applying (31.20) with $f = \pi_0$ and $T = \theta$, we find that

$\frac{1}{n}\sum_{i=0}^{n-1}\pi_i \quad\to\quad E_{P\mathbf{f}^{-1}}[\pi_0|\mathcal{I}_\theta]\;\; P\mathbf{f}^{-1}$-a.s. Then, according to (29.7), we have $\frac{1}{n}\sum_{i=0}^{n-1} f_i = (\frac{1}{n}\sum_{i=0}^{n-1}\pi_i)\circ\mathbf{f} \to E_{P\mathbf{f}^{-1}}[\pi_0|\mathcal{I}_\theta]\circ\mathbf{f} = E_P[f_0|\mathbf{f}^{-1}(\mathcal{I}_\theta)]\; P$-a.s. □

Remarks 32.19. (a) If $\{f_n : n \geqslant 0\}$ in (32.18) is also ergodic, then θ is $P\mathbf{f}^{-1}$-ergodic by (32.14.a). Therefore, $\frac{1}{n}\sum_{i=0}^{n-1}\pi_i \to \int_{R^\infty}\pi_0 d(P\mathbf{f}^{-1}) = Ef_0\; P\mathbf{f}^{-1}$-a.s. by (31.21.a), and so $\frac{1}{n}\sum_{i=0}^{n-1} f_i \to Ef_0\; P$-a.s.

(b) Theorem (32.18) yields Kolmogorov's SLLN. Indeed, if $f_n, n \geqslant 0$, are i.i.d. random variables, then $\{f_n : n \geqslant 0\}$ is stationary by (32.3.a). Moreover, $\{f_n : n \geqslant 0\}$ is ergodic, since $\mathbf{f}^{-1}(\mathcal{I}_\theta) \subset \mathbf{f}^{-1}(\mathcal{T}) = \cap_{n\geqslant 0}\sigma(f_n, f_{n+1}, \ldots) \subset \mathcal{O}_P$ by (32.10.b) and the Kolmogorov zero-one law (18.20), where $\mathcal{T}$ is as in (32.10.b). (Alternatively, as $P\mathbf{f}^{-1} = \otimes_{n\geqslant 0} P f_n^{-1}$ by (18.29), θ is a one-sided Bernoulli shift, which is $P\mathbf{f}^{-1}$-ergodic by (31.15.b), and so $\{f_n : n \geqslant 0\}$ is ergodic by (32.14.a).) Then $\frac{1}{n}\sum_{i=0}^{n-1} f_i \overset{a.s.}{\to} Ef_0$ by (a).

(c) Even in the special case of ergodicity, (32.18) contains strictly Kolmogorov's SLLN, because there are many stationary and ergodic sequences $\{f_n : n \geqslant 1\}$ for which $f_n, n \geqslant 1$, are not independent. For instance, consider the one-sided Markov shift θ defined in terms of the vector $\mathbf{p} = (p_y)_{y\in Y}$ and the matrix $\Pi = (p_{yy'})_{y,y'\in Y}$ where $Y \subset R$ is finite, $p_y > 0$ for all $y \in Y$, and $p_{y_0 y_1} \neq p_{y_1}$ for some $y_0, y_1 \in Y$. Then $\{\pi_n : n \geqslant 1\}$ is stationary by (32.3.b), but $\pi_n, n \geqslant 1$, are not independent. Moreover, if Π is irreducible, then θ is ergodic by (31.27), and so is $\{\pi_n : n \geqslant 1\}$.

The von Neumann ergodic theorem, spelled out for stationary sequences, is as follows.

Theorem 32.20. *If $\{f_n : n \geqslant 0\}$ is a stationary sequence of random variables with $f_0 \in \mathfrak{L}^p$, where $1 \leqslant p < \infty$, then $\frac{1}{n}\sum_{i=0}^{n-1} f_i \overset{\mathfrak{L}^p}{\to} E[f_0|\mathbf{f}^{-1}(\mathcal{I}_\theta)]$.*

Proof. In view of (29.7), (11.29) and (31.32), we have

$$\int_X \left|\frac{1}{n}\sum_{i=0}^{n-1} f_i - E_P[f_0|\mathbf{f}^{-1}(\mathcal{I}_\theta)]\right|^p dP$$

$$= \int_X \left|\left(\frac{1}{n}\sum_{i=0}^{n-1}\pi_i\right)\circ\mathbf{f} - E_{P\mathbf{f}^{-1}}[\pi_0|\mathcal{I}_\theta]\circ\mathbf{f}\right|^p dP$$

$$= \int_{R^\infty} \left|\frac{1}{n}\sum_{i=0}^{n-1}\pi_i - E_{P\mathbf{f}^{-1}}[\pi_0|\mathcal{I}_\theta]\right|^p d(P\mathbf{f}^{-1}) \to 0. \qquad \square$$

Remarks 32.21. (a) If $\{f_n : n \geqslant 0\}$ in (32.20) is also ergodic, then $\frac{1}{n}\sum_{i=0}^{n-1} f_i \overset{\mathfrak{L}^p}{\to} Ef_0$.

(b) If $f_n, n \geqslant 0$, are i.i.d. random variables with $f_0 \in \mathfrak{L}^p$, where $1 \leqslant p < \infty$, it follows from (a) that $\frac{1}{n}\sum_{i=0}^{n-1} f_i \overset{\mathfrak{L}^p}{\to} Ef_0$.

Exercise 32.22. Let $\mathcal{P} = \{A_i : i \in I\} \subset \mathcal{A}$ be a countable partition of X, and let $T : X \to X$ be a measurable transformation. For $n \geqslant 0$ and $x \in X$, define $f_n(x) = i$ if $T^n(x) \in A_i$. Prove the following.

(a) If T is P-preserving, then $\{f_n : n \geqslant 0\}$ is a stationary sequence.
(b) If T is ergodic, then $\{f_n : n \geqslant 0\}$ is ergodic.

Exercise 32.23. For $i \in I \neq \emptyset$, let $(Y^{(i)}, \mathcal{B}^{(i)})$ be a measurable space, let $f_n^{(i)} : X \to Y^{(i)}, n \geqslant 0$, be random elements, and assume that $\{f_n^{(i)} : n \geqslant 0\}$ is stationary. Prove that $\{(f_n^{(i)})_{i \in I} : n \geqslant 0\}$ is a stationary sequence whenever $\sigma(f_n^{(i)}, n \geqslant 0), i \in I$, are independent.

Exercise 32.24. If $\{f_n : n \geqslant 0\}$ is stationary and f_0 is independent of $\sigma(f_1, f_2, \ldots)$, then $f_n, n \geqslant 0$, are independent.

Exercise 32.25. Let $\{f_n : n \geqslant 0\}$ be a stationary sequence, let $\{B_n : n \geqslant 0\} \subset \mathcal{B}$, and let $B \in \mathcal{B}$. Prove the following.
(a) $P(\cup_{n\geqslant 0}\{f_n \in B_n\}) = P(\cup_{n\geqslant 1}\{f_n \in B_{n-1}\})$.
(b) $\cup_{n\geqslant 0}\{f_n \in B\} \overset{a.s.}{=} \limsup_n\{f_n \in B\}$ and $\cap_{n\geqslant 0}\{f_n \in B\} \overset{a.s.}{=} \liminf_n\{f_n \in B\}$.
(c) There is $A \in \mathcal{A}$ with $P(A) = 1$ such that, for $x \in A$, either $f_n(x) \in B$ for all $n \geqslant 0$ or $f_n(x) \in B^c$ for infinitely many n.

Exercise 32.26. Let $\{f_n : n \geqslant 0\}$ be as in (32.20), and let $k \in N$. Show that $\frac{1}{n}\sum_{i=0}^{n-1} f_i \overset{a.s.}{\to} E[f_k|\mathbf{f}^{-1}(\mathcal{I}_\theta)]$ and $\frac{1}{n}\sum_{i=0}^{n-1} f_i \overset{\mathfrak{L}^p}{\to} E[f_k|\mathbf{f}^{-1}(\mathcal{I}_\theta)]$. [Use (31.62).]

Exercise 32.27. Let $\{f_n : n \geqslant 0\}$ be a stationary sequence. Prove that it is ergodic if and only if, for any $l \geqslant 1$ and $B \in \otimes_{i=0}^{l-1}\mathcal{B}_i$, $\frac{1}{n}\sum_{i=0}^{n-1} 1_B \circ (f_i, \ldots, f_{i+l-1}) \overset{a.s.}{\to} P((f_0, \ldots, f_{l-1}) \in B)$. [Hint. For the sufficiency, write $\frac{1}{n}\sum_{i=0}^{n-1} 1_B \circ (f_i, \ldots, f_{i+l-1}) = \frac{1}{n}\sum_{i=0}^{n-1} 1_{\pi^{-1}_{\{0,\ldots,l-1\}}(B)} \circ \theta^i \circ \mathbf{f}$ and apply (31.23) to verify that θ is $P\mathbf{f}^{-1}$-ergodic.]

List of Symbols

$[a]$ [largest integer $\leqslant a$]
A^- [adherence]
A° [interior]
A^c [complement]
A^d [derived set]
$\mathcal{A}_J$ [σ-algebra in **(7.20)**]
$\mathcal{A}^\mu$ [σ-algebra in **(9.13)**]
$]a, b[, [a, b[,]a, b], [a, b]$ [intervals]
$A - B$ [set difference]
$A \Delta B$ [symmetric set difference]
$A \sqsubset B$, $A \sim B$ [relations in **(9.36)**]
$A \overset{a.s.}{\subset} B$, $A \overset{a.s.}{=} B$ [relations in **(17.2)**]
$A \cap \mathcal{B}$ [induced family]
$\bigvee_{i \in I} \mathcal{A}_i$, $\bigvee_{i=1}^{n} \mathcal{A}_i$, $\mathcal{A}_1 \vee \ldots \vee \mathcal{A}_n$ [σ-algebra in **(18.12)**]
$\bigotimes_{i \in I} \mathcal{A}_i$, $\bigotimes_{i=1}^{n} \mathcal{A}_i$, $\mathcal{A}_1 \otimes \ldots \otimes \mathcal{A}_n$ [product σ-algebra]
$\bigoplus_{i \in I} \mathcal{A}_i$ [sum σ-algebra]
$A_{(x_j)_{j \in J}}$ [set section]

$\mathcal{B}_0$, $\mathcal{B}_\infty$ [σ-algebras above **(30.1)**]
$\mathcal{B}_0(X)$ [Baire σ-algebra]
$(\mathcal{B}_1, \mathcal{B}_2, \mathcal{B}_3)$ [Markov triple]
$b(k; n, p)$ [probability related to the binomial distribution]
$\mathcal{B}_\tau$ [pre-$\tau\sigma$-algebra]
$\mathcal{B}(X)$ [Borel σ-algebra]
$\mathfrak{B}(X)$ [metric space of all bounded functions from X into K]
$\mathfrak{B}_r(X)$ [metric space of all bounded functions from X into R]

C [complex numbers]
$C(F)$ [set of continuity points of F]
$c(\mathcal{M})$ [monotone class generated by $\mathcal{M}$]
$\mathrm{Cov}[f, g]$ [covariance]
$\mathfrak{C}_c(X)$ [metric space of all bounded continuous functions from X into C]

$f = d\nu/d\mu, d\nu = f d\mu, \nu = f\mu$ [Radon-Nikodým derivative]

$\int\limits_X f d\mu, \int\limits_A f d\mu, \int\limits_A f(x)d\mu(x), \int\limits_A f(x)\mu(dx), \int\limits_a^b f(x)dx,$

$\int\limits_A f(x_1, \ldots, x_n)dx_1 \ldots dx_n$ [integral]

$\int\limits_A f dF, \int\limits_A f(x)dF(x), \int\limits_a^b f dF, \int\limits_a^b f(x)dF(x), \int\limits_{-\infty}^{\infty} f dF, \int\limits_{-\infty}^{\infty} f(x)dF(x)$
[Lebesgue-Stieltjes integral]

$F_1 * F_2$ [convolution]

$f_n \xrightarrow{a.e.} f$ [a.e. convergence]

$f_n \xrightarrow{D} f, f_n \xrightarrow{D} F, f_n \xrightarrow{D} P$ [convergence in distribution]

$f_n \xrightarrow{\mathfrak{L}^p} f$ [converge in $\mathfrak{L}^p$]

$f_n \xrightarrow{\mu} f$ [convergence in measure]

$f_n \xrightarrow{\mathbf{P}} f$ [convergence in probability]

$f_n \xrightarrow{u} f$ [uniform convergence]

$F_n \xrightarrow{w} F$ [weak convergence]

$f \sim g$ [equivalent relation on $\mathfrak{P}$]

$(f_i)_{i \in I}, (f_1, \ldots, f_n)$ [Cartesian product of functions]

$\{f_n^* : n \in N\}$ [symmetrized $\{f_n : n \in N\}$]

$F_{nk}, \varphi_{nk}, m_{nk}$ [functions in **(23.22)**]

$\widetilde{f}_{nk}, \widetilde{F}_{nk}, \widetilde{\varphi}_{nk}$ [functions above **(23.25)**]

$\mathbf{f}^{-1}(\mathcal{T})$ [tail σ-algebra]

$f_{(x_j)_{j \in J}}$ [function section]

$f : X \to Y$ [function]

$g^{1/n}$ [distinguished n-th root]

$g \circ f$ [composition]

$G_{m,\sigma}$ [Gaussian distribution function]

$G_n \xrightarrow{c} G$ [complete convergence]

$\mathcal{I}, \mathcal{I}_T$ [invariant σ-algebra]

$\mathcal{I}', \mathcal{I}'_T$ [a.s. invariant σ-algebra]

i_A [inclusion function]

$i_A^{-1}(\mathcal{B})$ [induced family]

i.o. [infinitely often]

$\inf\limits_{i \in I} y_i, \bigwedge_{i=1}^n y_i, y_1 \wedge \ldots \wedge y_n$ [infimum]

K [R or C]

$\mathfrak{L}^p, \mathfrak{L}^p(X, \mathcal{A}, P)$ [function space in **(13.1)**]

$L^p, L^p(X, \mathcal{A}, P)$ [metric space above **(13.14)**]

$\mathfrak{L}^\infty$ [function space in **(13.19)**]
L^∞ [metric space in **(13.20)**]
$\mathfrak{L}^0$ [function space in **(15.25)**]
L^0 [metric space in **(15.25)**]
$\lim_{\mathbf{x}\searrow\mathbf{a}} f(\mathbf{x}), \lim_{\mathbf{x}\nearrow\mathbf{a}} f(\mathbf{x})$ [right-hand limit, left-hand limit]
$\limsup_n A_n, \liminf_n A_n$ [superior limit of $\{A_n : n \geqslant 1\}$, inferior limit of $\{A_n : n \geqslant 1\}$]
$\limsup_n x_n, \liminf_n x_n$ [limit superior of $\{x_n : n \in N\}$, limit inferior of $\{x_n : n \in N\}$]
L_n, b_n [numbers above **(28.11)**]
$\text{Log}\, g$ [distinguished logarithm]
$\log_2 x$ [1 or $\log\log x$, according as $0 < x \leqslant e$ or $x > e$]

$\overline{\mu}$ [completion of μ]
μ^* [outer measure]
$\mu^+, \mu^-, |\mu|$ [positive variation of μ, negative variation of μ, total variation of μ]
μ-a.e. [μ-almost everywhere]
μ_A [restriction of μ to $A \cap \mathcal{A}$]
$\mu_{\mathcal{B}}$ [restriction of μ to $\mathcal{B}$]
$m[f]$ [median]
μf^{-1} [image of μ under f]
μ_F [Lebesgue-Stieltjes measure on $\mathcal{B}(R)$ induced by F]
$\bigoplus_{i\in I} \mu_i$ [sum measure]
$\bigotimes_{i=1}^{n} \mu_i, \mu_1 \otimes \ldots \otimes \mu_n$ [product of σ-finite measures]
$\mu \vee \nu$ [$(\mu - \nu)^+ + \nu$]
$\mu \wedge \nu$ [$-(\mu - \nu)^- + \nu$]

N [positive integers]
$\nu \ll \mu$ [ν is absolutely continuous with respect to μ]
$\nu \sim \mu$ [equivalent measures]
$\nu \perp \mu$ [ν is singular with respect to μ]

$\mathcal{O}_P$ [σ-algebra in **(18.17)**]

P [space of all probabilities on $\mathcal{A}$, $\mathcal{B}(X)$]
P^*, E^*, Var^* [probability, expectation, variance above **(23.20)**]
$\mathbf{p}, \Pi$ [vector, stochastic matrix in **(31.3)**]
$P^{\mathcal{B}}(A), P(A|\mathcal{B}), P(A|g), P(A|g_1, \ldots, g_n)$ [conditional probability]
P-a.s., a.s. [P-almost surely, almost surely]
P_B [conditional probability given B]
$P^{\mathcal{B}}$ [regular conditional probability]

ψ_f [function above **(22.1)**]
$\psi(t;\gamma,G)$, $\varphi(t;\gamma,G)$ [functions above **(23.13)**]
$\bigotimes_{i\in I} P_i$ [product of probabilities]
π^I_J, π^I_i, π_i [projection]
$p(k;\lambda)$ [probability related to the Poisson distribution]
$P_n \Rightarrow P$ [weak convergence]
$P^{\mathcal{P}}(A)$ [conditional probability given $\mathcal{P}$]
$P \otimes Q$ [probability in **(29.39)**]
$\mathcal{P}(X)$ [power set of X]

Q [rational numbers]
Q_+ [$Q \cap]0,\infty[$]

R [real numbers]
$\overline{R}$ [extended real numbers]
R^* [$[-\infty,\infty[$ or $]-\infty,\infty]$]
$\mathbb{R}$ [relation]
$\mathbb{R}^*$ [inverse relation]
$\mathrm{Re}\, f$, $\mathrm{Im}\, f$ [real part, imaginary part]

$\mathfrak{S}$, $\mathfrak{S}'$ [classes of all simple functions]
$\sigma[f]$ [standard deviation]
$\sigma(f_i, i\in I), \sigma(f_1,\ldots,f_n)$ [σ-algebra generated by functions]
$\sigma(\mathcal{M})$ [σ-algebra generated by $\mathcal{M}$]
S^*_n [sum in **(25.18)**]
$S_n(f)$ [sum in **(31.9)**]
$\sup_k$, $\sum_k$, $\prod_k$ [notation in **(23.22)**]
$\sup_{i\in I} y_i$, $\bigvee_{i=1}^n y_i$, $\bigwedge_{i=1}^n y_i$, $y_1 \vee \ldots \vee y_n$ [supremum]
$S(x,r)$, $S_d(x,r)$ [open sphere]
$\overline{S}(x,r)$, $\overline{S}_d(x,r)$ [closed sphere]

T, T^0, T^n [notation above **(31.1)**]
θ [shift transformation]
$\mathcal{T}$ [topology]
$\mathcal{T}_d$ [topology generated by d]
$\tau(\mathcal{S})$ [topology generated by $\mathcal{S}$]

$U_{a,b}$, $U^{(m)}_{a,b}$ [number of upcrossings of $[a,b]$]
$U_\varepsilon(\mathbf{x})$ [open set in **(4.8)**]

$\mathcal{V}$ [topology in **(22.7)**]
$\mathrm{Var}\, f$, $\mathrm{Var}[f]$ [variance]

$\mathcal{W}$ [weak topologyin P]
$W_P(f, \varepsilon)$ [open set in **(22.1)**]

$\widetilde{x}$ [equivalent class]
$X/\sim$ [family of all equivalent classes]
$x + A$, $A + x$, xA, Ax [sets in **(7.60)**]
$(X, \mathcal{A})$ [measurable space]
$(X, \mathcal{A}, \mu)$ [measure space]
$(X, \mathcal{A}, P)$ [probability space]
(X, d) [metric space]
$(X, \mathcal{T})$ [topological space]
$\{x_i : i \in I\}$ [indexed set]
$\prod_{i \in I} X_i$, $X_1 \times \cdots \times X_n$, X^I, X^∞, X^n [Cartesian product of sets]

Z [integers]
$\varnothing$ [void set]
$\square$ [end of proof]
1_A [indicator function]
$\bigcup \mathcal{A}$ [union of all sets of $\mathcal{A}$]
$\bigcap \mathcal{A}$ [intersection of all sets of $\mathcal{A}$]
$\rightarrow$, $\nrightarrow$ [convergence, negated convergence]